MULTISENSORY PERCEPTION

MULTISENSORY PERCEPTION

From Laboratory to Clinic

Edited by

K. Sathian
Department of Neurology, Milton S. Hershey Medical Center & Penn State College of Medicine, 30 Hope Drive, Hershey, PA

V.S. Ramachandran
Gilman Drive, Mandler Hall, UC San Diego La Jolla, CA

ELSEVIER

ACADEMIC PRESS
An imprint of Elsevier

Academic Press is an imprint of Elsevier
125 London Wall, London EC2Y 5AS, United Kingdom
525 B Street, Suite 1650, San Diego, CA 92101, United States
50 Hampshire Street, 5th Floor, Cambridge, MA 02139, United States
The Boulevard, Langford Lane, Kidlington, Oxford OX5 1GB, United Kingdom

Library of Congress Cataloging-in-Publication Data
A catalog record for this book is available from the Library of Congress

British Library Cataloguing-in-Publication Data
A catalogue record for this book is available from the British Library

ISBN: 978-0-12-812492-5

For information on all Academic Press publications visit our website at
https://www.elsevier.com/books-and-journals

Publisher: Nikki Levy
Acquisition Editor: Melanie Tucker
Editorial Project Manager: Kristi Anderson
Production Project Manager: Bharatwaj Varatharajan
Cover Designer: Matthew Limbert

Typeset by TNQ Technologies

Contents

II

Multisensory interactions

7. Visuo-haptic object perception

Simon Lacey and K. Sathian

8. Multisensory processes in body ownership

Henrik H. Ehrsson

9. Visual—vestibular interactions

Aasef G. Shaikh, David S. Zee, Jeffrey Taube, and Amir Kheradmand

10. Multisensory flavor perception: A cognitive neuroscience perspective

Charles Spence

11. Audiovisual crossmodal correspondences: behavioral consequences and neural underpinnings

Charles Spence and K. Sathian

12. How do crossmodal correspondences and multisensory processes relate to synesthesia?

David Brang and Vilayanur S. Ramachandran

13. Synesthesia: the current state of the field

Jamie Ward and Julia Simner

17. Neurodevelopmental and neuropsychiatric disorders affecting multisensory processes

Carissa J. Cascio, David M. Simon, Lauren K. Bryant, Gabriella DiCarlo, and Mark T. Wallace

18. Disorders of body representation

Laura K. Case, Marco Solcà, Olaf Blanke, and Nathan Faivre

Contributors

Eric Altschuler Department of Physical Medicine and Rehabilitation, Metropolitan Hospital, New York, NY, United States

Amir Amedi Department of Medical Neurobiology, Institute for Medical Research Israel-Canada, Faculty of Medicine, Hebrew University of Jerusalem, Hadassah Ein-Kerem, Jerusalem, Israel; Department of Cognitive Science, Faculty of Humanities, Hebrew University of Jerusalem, Hadassah Ein-Kerem, Jerusalem, Israel

Olaf Blanke Center for Neuroprosthetics, Laboratory of Cognitive Neuroscience, Brain-Mind Institute, École Polytechnique Fédérale de Lausanne-EPFL, Lausanne, Switzerland; Department of Neurology, University Hospital Geneva, Geneva, Switzerland

Nadia Bolognini Department of Psychology & NeuroMi, University of Milan – Bicocca, Milano, Italy; Istituto Auxologico Italiano, IRCCS, Laboratory of Neuropsychology, Milano, Italy

David Brang University of Michigan, Ann Arbor, MI, United States

Andrew J. Bremner School of Psychology, University of Birmingham, Birmingham, United Kingdom

Lauren K. Bryant Microsoft, Inc. Seattle, WA, United States

Blake E. Butler Department of Psychology, National Centre for Audiology, Brain and Mind Institute, University of Western Ontario, London, ON, Canada

Carissa J. Cascio Vanderbilt University Medical Center, Nashville, TN, United States

Laura K. Case Pain and Integrative Neuroscience Branch, National Center for Complementary and Integrative Health, Bethesda, MD, United States

Chaipat Chunharas Department of Psychology, Center for Brain and Cognition, University of California, San Diego, CA, United States

Gabriella DiCarlo Vanderbilt University School of Medicine, Nashville, TN, United States

Christian Dohle MEDIAN Klinik Berlin-Kladow, Berlin, Germany; Center for Stroke Research Berlin, Charité – University Medicine Berlin, Berlin, Germany

Henrik H. Ehrsson Department of Neuroscience, Karolinska Institutet, Stockholm, Sweden

Nathan Faivre Center for Neuroprosthetics, Laboratory of Cognitive Neuroscience, Brain-Mind Institute, École Polytechnique Fédérale de Lausanne-EPFL, Lausanne, Switzerland; Laboratoire de Psychologie et Neurocognition, LPNC CNRS 5105, Université Grenoble Alpes, France

Matthew Fulkerson Department of Philosophy, University of California, San Diego, CA, United States

Hannah Glick Department of Speech, Language, & Hearing Science, Institute of Cognitive Science, Center for Neuroscience, University of Colorado at Boulder, Boulder, CO, United States

Radhika S. Gosavi Department of Educational Psychology, University of Wisconsin–Madison, Madison, WI, United States

Benedetta Heimler Department of Medical Neurobiology, Institute for Medical Research Israel-Canada, Faculty of Medicine, Hebrew University of Jerusalem, Hadassah Ein-Kerem, Jerusalem, Israel

Edward M. Hubbard Department of Educational Psychology, University of Wisconsin–Madison, Madison, WI, United States

Amir Kheradmand Department of Neurology, The Johns Hopkins University, Baltimore, MD, United States

Simon Lacey Department of Neurology, Neural & Behavioral Sciences, and Psychology, Pennsylvania State University, Hershey, PA, United States; Departments of Neurology and Psychology, Emory University, Atlanta, GA, United States

David J. Lewkowicz Haskins Laboratories, New Haven, CT, United States

Stephen G. Lomber Departments of Physiology and Pharmacology and Psychology, National Centre for Audiology, Brain and Mind Institute, University of Western Ontario, London, ON, Canada

Zeve Marcus Department of Psychology, Center for Brain and Cognition, University of California, San Diego, CA, United States

Pawel J. Matusz The LINE (Laboratory for Investigative Neurophysiology), Department of Radiology, University Hospital Center and University of Lausanne, Lausanne, Switzerland; Information Systems Institute at the University of Applied Sciences Western Switzerland (HES-SO Valais), Sierre, Switzerland; Department of Hearing and Speech Sciences, Vanderbilt University Medical Center, Nashville, TN, United States

David Meijer Computational Cognitive Neuroimaging Laboratory, Computational Neuroscience and Cognitive Robotics Centre, University of Birmingham, Birmingham, United Kingdom

Micah M. Murray The LINE (Laboratory for Investigative Neurophysiology), Department of Radiology, University Hospital Center and University of Lausanne, Lausanne, Switzerland; Department of Hearing and Speech Sciences, Vanderbilt University Medical Center, Nashville, TN, United States; Department of Ophthalmology, University of Lausanne and Fondation Asile des Aveugles, Lausanne, Switzerland; The EEG Brain Mapping Core, Center for Biomedical Imaging (CIBM), University Hospital Center and University of Lausanne, Lausanne, Switzerland

Uta Noppeney Computational Cognitive Neuroimaging Laboratory, Computational Neuroscience and Cognitive Robotics Centre, University of Birmingham, Birmingham, United Kingdom

Vilayanur S. Ramachandran Department of Psychology and Center for Brain and Cognition, University of California, San Diego, CA, United States

Benjamin A. Rowland Department of Neurobiology and Anatomy, Wake Forest School of Medicine, Winston–Salem, NC, United States

K. Sathian Department of Neurology, Neural & Behavioral Sciences, and Psychology, Pennsylvania State University, Hershey, PA, United States; Departments of Neurology and Psychology, Emory University, Atlanta, GA, United States

Aasef G. Shaikh Department of Neurology, University Hospitals, Cleveland VA Medical Center, Case Western Reserve University, Cleveland, OH, United States

Anu Sharma Department of Speech, Language, & Hearing Science, Institute of Cognitive Science, Center for Neuroscience, University of Colorado at Boulder, Boulder, CO, United States

Julia Simner School of Psychology, University of Sussex, Falmer, Brighton, United Kingdom

David M. Simon Axial Healthcare, Nashville, TN, United States

Marco Solcà Center for Neuroprosthetics, Laboratory of Cognitive Neuroscience, Brain-Mind Institute, École Polytechnique Fédérale de Lausanne-EPFL, Lausanne, Switzerland; Department of Psychiatry, University Hospital Geneva, Geneva, Switzerland

Charles Spence Crossmodal Research Laboratory, Department of Experimental Psychology, Oxford University, Oxford, United Kingdom

Barry E. Stein Department of Neurobiology and Anatomy, Wake Forest School of Medicine, Winston–Salem, NC, United States

Jeffrey Taube Department of Psychological and Brain Sciences, Dartmouth College, Hanover, NH, United States

Giuseppe Vallar Department of Psychology & NeuroMi, University of Milan – Bicocca, Milano, Italy; Istituto Auxologico Italiano, IRCCS, Laboratory of Neuropsychology, Milano, Italy

Mark T. Wallace Department of Hearing and Speech Sciences, Vanderbilt University Medical Center, Nashville, TN, United States; Department of Psychology, Vanderbilt University, Nashville, TN, United States; Kennedy Center for Research on Human Development, Vanderbilt University, Nashville, TN, United States; Department of Psychiatry, Vanderbilt University, Nashville, TN, United States; Vanderbilt Brain Institute, Vanderbilt University, Nashville, TN, United States

Jamie Ward School of Psychology, University of Sussex, Falmer, Brighton, United Kingdom

David S. Zee Department of Neurology, The Johns Hopkins University, Baltimore, MD, United States

Preface

Aristotle postulated the existence of a sixth sense—"a shared sense, responsible for unifying, distinguishing and coordinating the five senses."[1] Nowadays, rather than conceptualizing an Aristotelian "supersense," we think more in terms of integrating information across the various sensory systems to create experiential unity in terms of perception, motor responses, memory, and so on. While unicellular organisms may respond to multiple types of input, e.g., mechanical and chemical, the evolutionary acquisition of layers of complexity led to the emergence of modularity with highly specialized sense organs for each kind of input. It is widely accepted that we have five senses, i.e., five independent sensory channels streaming different types of information about our environment into our nervous system. These are commonly recognized as vision, hearing, touch, taste, and smell. To these *externally* directed senses, we could add a number of channels carrying information about the *internal* environment of the body, such as pain (signaling the risk of impending tissue damage), proprioception (signals about moving body parts), the vestibular sense (conveying information about the orientation and motion of the head), and interoception (providing information about internal organs). With such an array of sensory systems comes the challenge of integrating the distributed information into a unified whole, as was recognized by Aristotle. Such integration depends critically on connections that evolved between the sensory systems to complement their modularity. Traditional teaching in neuroscience emphasized a hierarchy of information processing in the cerebral cortex, with primary sensory areas feeding into unisensory association areas and eventually converging on high-order multisensory regions.[2] Over the last few decades of the preceding century, our understanding of multisensory processes has been revolutionized, and we now know that multisensory processing is ubiquitous in the brain. Thus, some have asked whether the entire neocortex should be thought of as multisensory.[3] Viewed from this perspective, the study of intersensory integration may provide valuable clues to the very nature of the self and of consciousness. The importance of multisensory information is underscored by its relevance in literature: Flannery O'Connor reveals the teaching of Flaubert, transmitted via another writer whose name remains undisclosed, "that it takes at least three activated sensuous strokes to make an object real."[4]

It is worth reflecting on the elements of what is considered to be an important scientific advance, assuming that the findings are replicable. First, the new finding is often surprising, i.e., unexpected based on current understanding. Second, despite being unexpected, it must be explicable by known or at least candidate mechanisms, so that one can connect the new with well-established scientific principles. Third, in biology, the newly discovered structure or process should serve a "purpose" that can be understood, e.g., in functional, evolutionary, or pathophysiological terms. Finally, the discovery typically raises more questions than it answers, stimulating further research. Studies of phantom limbs in the 1990s exemplify these critical elements. Although the existence of phantom limb sensations, and even sensations referred from other body parts to the phantom, had been known for nearly a century, the new finding was that the phantom hand could be topographically mapped on the face in terms of the stimuli evoking sensations in each part of the phantom.[5] This allowed linking of the perceptual phenomena to the recently reported "massive" cortical reorganization by Pons and colleagues,[6] in which deafferentation of an entire upper limb in monkeys led to "invasion" of the representation of the arm by that of the face, in primary somatosensory cortex. In turn, this led to the idea that the genesis of phantom sensations, including pain, might be related to such cortical remapping. Of course, this led to many more questions and investigations, including the idea of using "mirror therapy," an intervention probably involving multisensory as well as motor processes, to ameliorate phantom pain, other kinds of pain, neglect, and hemiparesis, as elaborated in the chapter by Dohle, Altschuler, and Ramachandran in this volume. The discoveries in many laboratories around the world, by a host of investigators with their fellows and students, that multisensory processing is not limited to high-level regions of the cerebral cortex collectively constitute another striking example of an important scientific advance.

In addition to the obvious clinical utility of interventions such as mirror therapy, Ramachandran's work together with that of many others has entailed a radical revision in our thinking about brain function in three separate contexts: intermodular interactions, embodiment, and the malleability of connections. The "old" model of brain function represented in traditional teaching assumes that sensory information is initially processed by highly specialized modules (e.g., for color, form, stereo, and even for high-level information pertaining to faces etc.), each module computing certain aspects of this information and relaying it to the next module higher up for further processing. According to this serial hierarchical bucket-brigade model of brain function, the modules are largely hardwired and autonomous; sensory information is relayed stage by stage and eventually culminates in the climax of object recognition. Ramachandran's work flatly contradicts this.[7] According to him, the modules are highly malleable and in a constant state of dynamic equilibrium with the environment they are immersed in—constantly changing in response to environmental challenges. Furthermore, the partially digested information from each module of the hierarchy is sent back to earlier modules, which in turn biases subsequent processing. In addition to these top-down and lateral cross-modular interactions (such as mirrors influencing phantom pain), there are also interactions with skin and bones (as in the condition known as reflex sympathetic dystrophy or complex regional pain syndrome), and indeed with other brains (mirror neurons). As a result of this new approach to brain function, we have come to regard brain activity as a shifting, fluctuating mosaic, culminating occasionally in stable, yet dynamic, neural coalitions which we call perception. These findings are broadly consistent with the contributions of many researchers reviewed in the chapters of this book, the focus of these writings being how multisensory perception ensues from such dynamic neural coalitions.

In this volume, the current state of knowledge on multisensory processes is surveyed by a number of experts. The explosion of interest in the field explains why we are unable to cover each and every topic, as interesting as it might be. We have therefore chosen to provide a somewhat selective survey that, we hope, illustrates the kind of exciting advances that have been made. The authors of the various chapters also consider some of the open questions that might be amenable to experimental solution. The 20 chapters are organized into three sections: the first covers foundational material; the second reviews a variety of multisensory interactions; and the third deals with clinical applications.

The opening chapter by Ramachandran, Marcus, and Chunharas sets the stage in a wide-ranging account of a number of seemingly unrelated phenomena that, in fact, might share common mechanisms, e.g., associations between word sounds and visual forms, metaphors, and synesthesia. Links are drawn between these phenomena and the mirror neuron system, the evolution of language, and disorders such as dyslexia, and a number of predictions are made that can be tested empirically. A recurring theme in this article is the attempt to link neuroanatomy with phenomenology—as seen through an evolutionary lens (not just naïve hyperadaptation, but the opportunistic coopting of preexisting structures that were selected for other reasons). Next, Fulkerson considers fundamental philosophical questions about how the senses are defined, individuated, and integrated into unified percepts. Stein and Rowland then review animal studies of the development of neurophysiological processes that underlie multisensory integration in the superior colliculus, sketching how these processes are shaped by postnatal experience and descending input from the cerebral cortex. This is complemented by the material of the next chapter, in which Lewkowicz and Bremner describe the role of experience and attention in molding the emergence of audiovisual perception in human infants and the slow development of perception of the self during childhood by progressive integration of visual and tactile inputs. Meijer and Noppeney follow with an analysis of computational models of multisensory perception, reviewing the evidence that the brain computes Bayesian causal inference estimates. In the last chapter of the first section, Matusz, Wallace, and Murray summarize the benefits of multisensory information for learning and memory and explore its neurophysiological basis.

The second section begins with a review by Lacey and Sathian of interactions between vision and touch in object perception and their neuroimaging correlates. Molyneux famously asked Locke whether restoring sight to a congenitally blind person would allow visual recognition of objects known previously by touch alone.[8] This would be expected if there are modality-independent object representations, as suggested by the work reviewed by Lacey and Sathian, and indeed, the answer to Molyneux's question appears to be negative, although after a few days, visual recognition becomes possible, implying rapid learning after sight restoration.[9] The next chapter comprises Ehrsson's outline of the important contribution of multisensory congruency to body ownership and the relevant neural substrates in frontoparietal cortex. Visual-vestibular interactions are the next topic: Shaikh, Zee, Taube. and Kheradmand address the role of parietotemporal cortex in their mediation. Spence then discusses the blending of a number of different sensory inputs to give rise to our perception of flavor; work in this area has major implications for consumer behavior. The next series of four chapters addresses the twin issues of crossmodal correspondences and synesthesia. Spence and Sathian provide a review of crossmodal correspondences and what is understood of their neural basis.

Next, Brang and Ramachandran explore the relationship between synesthesia and cross-modal correspondences, including a curious condition called "calendar synesthesia," in which the subject sees the annual calendar in an idiosyncratic shape (e.g., U-shaped, L-shaped, hula-hoop-shaped). Ward and Simner follow with an update on thinking about synesthesia in terms of the combination of genetic and environmental influences and the impact of synesthesia on perception, imagery, and memory. To round out the series, Gosavi and Hubbard focus on memory and demonstrate that superior long-term memory in synesthetes derives from earlier stages of memory processes, i.e., sensory and working memory.

Heimler and Amedi kick off the last section with a description of task-selective but modality-independent neural organization that has emerged from comparing congenitally blind with sighted people and show the relevance of such organization for sensory substitution. In a parallel chapter, Lomber, Butler, Glick, and Sharma review neurophysiological studies in animal models of deafness and corresponding neuroplasticity in humans associated with aging and with cochlear implantation in deaf children. Cascio, Simon, Bryant, DiCarlo, and Wallace deal with atypical multisensory processing as seen in autism, exemplifying neurodevelopmental disorders, and schizophrenia, as an example of neuropsychiatric disorders. Body image disturbances occur in a variety of neurological and psychiatric conditions; their relationship to dysfunctional multisensory processing is considered by Case, Solcà, Blanke, and Faivre. Multisensory rehabilitative interventions have been shown to improve function in both hemianopia and hemispatial neglect, as reviewed in the chapter by Bolognini and Vallar. Finally, Dohle, Altschuler, and Ramachandran review the progress that has been made in the application of mirror therapy to conditions such as various pain syndromes and hemiparesis and address the mechanisms by which this therapy may work as well as variations in its clinical use.

Our work on this book has been a labor of love. We, the editors, have been long-time colleagues (although we have not previously collaborated formally) and have thoroughly enjoyed putting this book together. We thank the authors of all the chapters for their willing contributions, thoughtful and insightful writings and careful revisions, and their patience with our editorial reviews. We are also grateful to the editorial staff, particularly Kristi Anderson and Melanie Tucker, for keeping us on track.

K. Sathian
V.S. Ramachandran

References

1. Howes D. http://sixthsensereader.org/about-the-book/abcderium-index/common-sens/; 2017.
2. Amaral DG. The functional organization of perception and movement. In: Kandel ER, Schwartz JH, Jessell TM, et al., eds. *Principles of Neural Science*. 5th ed. New York: McGraw Hill Medical; 2013:356–369.
3. Ghazanfar A, Schroeder C. Is neocortex essentially multisensory? *Trends Cogn Sci*. 2006;10:278–285.
4. O'Connor F. The nature and aim of fiction. In: Fitzgerald S, Farrar FR, eds. *Mystery and Manners: Occasional Prose*. New York: Strauss and Giroux; 1970:63–86.
5. Ramachandran VS, Rogers-Ramachandran D, Stewart M. Perceptual correlates of massive cortical reorganization. *Science*. 1992;258:1159–1160.
6. Pons TP, Garraghty PE, Ommaya AK, et al. Massive cortical reorganization after sensory deafferentation in adult macaques. *Science*. 1991;252:1857–1860.
7. Ramachandran VS. Plasticity and functional recovery in neurology. *Clin Med*. 2005;5:368–373.
8. Morgan MJ. *Molyneux's Question: Vision, Touch and the Philosophy of Perception*. Oxford, UK: Cambridge Univ Press; 1977.
9. Held R, Ostrovsky Y, deGelder B, et al. The newly sighted fail to match seen with felt. *Nat Neurosci*. 2011;14:551–553.

SECTION I

Foundations of multisensory perception

CHAPTER

1

Bouba-Kiki: cross-domain resonance and the origins of synesthesia, metaphor, and words in the human mind

Vilayanur S. Ramachandran, Zeve Marcus, Chaipat Chunharas

Department of Psychology and Center for Brain and Cognition, University of California, San Diego, CA, United States

All is name and form. ***Sarvam Naam'am Rupam (Sanskrit text, circa 2000 BC)***

Any monkey can reach for a banana; only a human can reach for the stars. **V.S. Ramachandran**

Introduction

In science, one is often forced to choose between precise answers to boring or trivial questions (e.g., what is the number of cone cells in the eye?) and vague answers to big questions (what is consciousness?). However, now and then, we have a precise answer to a big question and hit the jackpot, as in the discovery of DNA and its relationship to heredity. Psychology is replete with failed attempts at category 2 (precise answers to big questions), but they may have nonetheless paved the way for more sophisticated theories in the future. The ideas in this chapter, dealing with crossmodal interactions as a central feature of the human mind, may serve a similar purpose.

Although our chapter deals with a diverse range of psychological phenomena, the main theme running through it is an evolutionary and neurological approach to understanding metaphor, sound-form symbolism, and creativity—traditionally the domain of cognitive psychology. In particular, we emphasize the importance of pruning genes in enabling or disabling the cross-domain resonance that underlies much of normal human cognition as well as abnormalities such as synesthesia and—more speculatively—autism.

Multisensory Perception
https://doi.org/10.1016/B978-0-12-812492-5.00001-2

A central feature of the nervous system is its ability to integrate information from different sources. "Primitive" brains such as those of lower vertebrates do not display the striking degree of modular architecture seen in primates, but with the further sophisticated development of more complex sensory systems (including subdivisions within vision, touch, hearing, smell, and taste), even more specialized brain structures emerged. These structures were adapted to the unique computational needs of each sensory system—resulting in modularity. Yet, paradoxically further along in our evolution, the barriers between the modules began to partially dissolve, accompanied by the expansion of cortical "association" areas (as long recognized by neurologists and psychologists[1,2]). With even further evolution, modules again started interacting powerfully with each other—thereby not only improving the signal-to-noise ratio—but facilitating more sophisticated styles of computation (consider the analogy of two drunks neither of whom can cross the road independently—but, by leaning on each other—they can stumble across with relative ease). The dissolution of barriers was mediated by connections, which enabled several kinds of crossmodal interactions, of which two are most obvious.

First, such interactions enable the so-called "binding" of different attributes of an object (e.g., the color, shape, and smell of an apple) into a perceptual whole. Second, they allow the abstraction of common attributes across modules—the central focus of this chapter. A striking example of this is sound-form symbolism (the Bouba-Kiki effect): if, when viewing Fig. 1.1, we ask you which shape, A or B, corresponds most obviously to the sound "Bouba" and which one to the sound "Kiki," you would probably, like most people, select A as Bouba and B as Kiki—even though you have never been taught these associations before. In this paper, we use this little-known association as a jumping-off point to speculate on a number of seemingly unrelated, but compelling, aspects of the human mind ranging from word origins, protolanguage, prehension, gesturing, and mirror neurons, to metaphor, esthetics, visual imagery, numbers, and calendars.

Individual components of the idea are to be found scattered in the vast literature of psychology, but the manner in which we have put them together, we believe, may have some novel features. We present our views, not in a strict logical order, but partly as historical narrative—the sequence in which the ideas emerged.

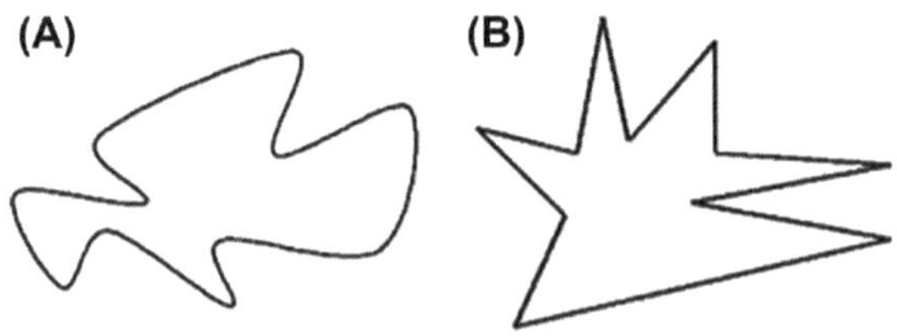

FIGURE 1.1 Examples of kiki and bouba. Subjects tend to map the name kiki onto the figure on the left because of the sharp inflection of the visual shape, and tend to map bouba to the right figure because of the more rounded shape.

Synesthesia

The term synesthesia was coined by Francis Galton, first cousin of Charles Darwin, in the 19th century. He noticed that a certain proportion of the general population had a perceptual quirk despite being completely normal in other respects. When looking at a printed number (e.g., "5"), or letter, they would always see it tinged a specific color such as red, chartreuse, indigo, etc. This is known as grapheme-color synesthesia. The particular color evoked remains constant for any given synesthete[3], but is different across synesthetes. Galton also noticed that the trend runs in families and may therefore have a genetic basis. Since his time, it has been noticed that the incidence of the phenomenon is seven times higher in poets, artists, and novelists[4,5] than in the general population.

It had long been assumed that synesthesia was very rare, occurring in 0.1%—0.01% of the population, but we now accept that it's much more common—as many as 1 in 100 to 1 in 40 people have synesthesia. It also turns out that there are at least two kinds of synesthetes, loosely referred to as projectors and associators. Projectors, constituting about 10%—15%, claim to actually see the color spatially superimposed on the grapheme (letter or number), whereas associators "see" the color only in their mind's eye.

Although long regarded as a curiosity, there has been a tremendous resurgence of interest in synesthesia in the last decade (see also Chapters 12—14) as a result of several groups—including our own (VSR, Edward Hubbard, David Brang, Nicholas Root, Romke Rouw) working in parallel and often converging on the same insights; Julia Simner, Jamie Ward, Anina Rich, Jason Mattingley, Roi Kadosh, Philip M. Merikle, Mike J. Dixon, Daniel Smilek Dov Sagiv, and Bruce Bridgeman. Such a convergence rarely happens in psychology-fueling optimism that we are on the right track ("fools rush in where angels fear to tread"). Before this resurgence, a recurring question in the last 100 years of research in synesthesia had been whether it is an early perceptual-sensory phenomenon or a high-level cognitive association (as when you see "yellow" when looking at a black-and-white picture of a banana). The precise articulation of this question and the answer was first attempted by us[6—8] when we made several observations that support the view, that in at least a subset of synesthetes, the colors evoked are sensory in nature, and are (probably) associated with qualia. Although this possibility had been raised before, no attempt had been made to prove the point using multiple strategies, which was our goal.[6,7,9]

1) In a matrix of black 5s, it is ordinarily difficult to segregate a cluster of 2s (mirror images of 5s), because they are made up of the same features—vertical and horizontal lines (Fig. 1.2). It is difficult for normal subjects to discern the shape conveyed by the 2s, but some projector synesthetes do so more rapidly and accurately, presumably because the evoked color facilitates the segregation. This segregation (popout) suggests that in some synesthetes (projectors) the colors evoked are sensory or perceptual in nature.[6,7,10,11] High-level features do not lead to segregation.[12—14]

2) We have seen at least two individuals who are color anomalous—because of deficient S-cone pigments in their retinas, but could nonetheless see what they charmingly referred to as "Martian colors," when looking at printed black numbers, even though the same colors could not be seen in the real world. Perhaps the graphemes make it

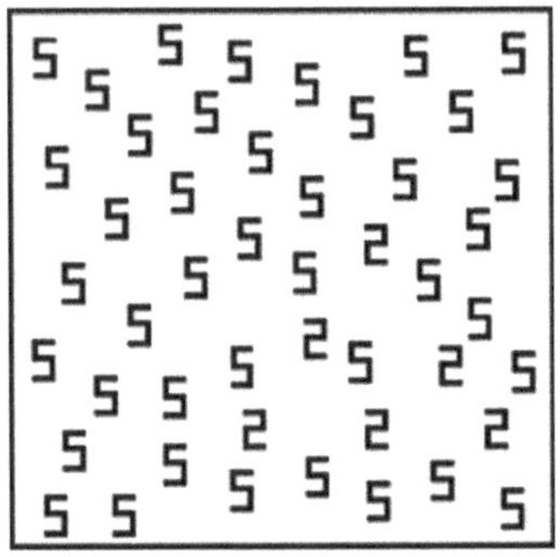

FIGURE 1.2 A matrix of randomly placed, computer-generated '2's with an embedded triangle composed of computer-generated '5's.

possible to bypass the need for differential activation of retinal cones entirely and lead to direct cross activation of color neurons in V4. This supports the view that synesthesia is caused by sensory cross activation rather than childhood memory associations. You cannot form memory associations of percepts that you have never seen in the world.

3) Even for synesthetes with normal color vision, different regions of a single letter can be tinged different colors, an observation that suggests a "hardware" glitch rather than a cognitive effect. In addition to this, they will sometimes report seeing "alien colors," the sort which they do not experience in natural scenes.[6–8]

4) The saturation of synesthetically induced color shows some sensitivity to elementary physical features of the image such as contrast[15] and the precise font of the letter.[a]

5) In a small subset of synesthetes, the saturation of the evoked colors also varies with visual field eccentricity and laterality.[16] This again suggests a sensory phenomenon rather than a high-level memory association; memories do not vary with retinal location.

6) In Fig. 1.3, most normal people do not see the letters hidden between the blocks, but after 30–60 seconds, they can recognize the message conveyed by the letters. Some projector synesthetes (such as EA) see the color before seeing the letters and subsequently infer what the letters must be, which in turn facilitates visual identification of the letters.

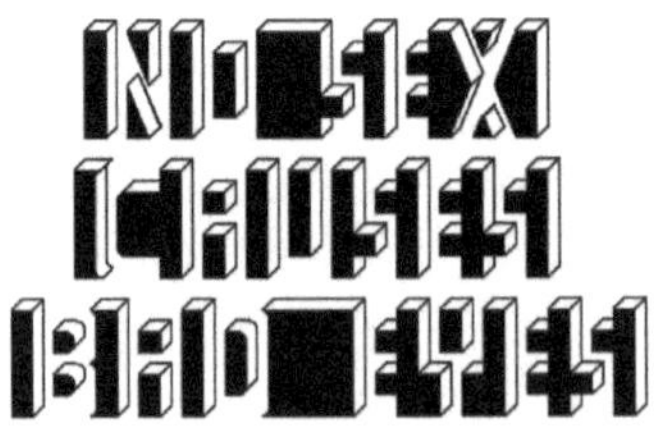

FIGURE 1.3 Try reading the words. If you have trouble, squint your eyes and move 10–15 ft away.

[a] Additionally, in 2011, we showed that the distribution of color–grapheme associations were not random. Within a synesthete, graphemes which have similar features (e.g., curves vs. angles) are more likely to have similar colors associated with them.[6,7,105,106]

The hidden letters evoke the synesthetic colors in the fusiform gyrus as a result of cross activation, before they are consciously recognized.[17] This also argues for an early sensory process.

7) Many synesthetes report that they actually see the colors spatially superposed on the number and in the one case we examined carefully (subject EA) the color spread outside the boundary of the letter—as though it had been spray painted. But if we drew an ameboid irregular outline surrounding the letter, the spreading of the color cloud was blocked by the outline! If the ameba was large, the same quota of color spread over a wider area, appearing less saturated. If a hole was punched in the outline of the amoeba, the color "leaked out." Even an illusory edge was sufficient to block the color spread (Fig. 1.4). The degree of confidence and precision with which these judgments were made by the subjects would be hard to explain in terms of high-level mental associations (there may, of course, be differences with this between projectors and associators). The effect possibly also involves reciprocal connections between the fusiform color area V4, the fusiform grapheme area, and illusory contours processed in V2. EA also reported that the spreading color had a dusty texture, which is consistent with the fact that some V4 neurons are also tuned to visual texture.
8) Early in our research, we noted[6,7] that only Arabic numbers evoke colors, not Roman ones, implying that the colors are evoked by the shape of the number rather than the concept.
9) Our evidence from brain imaging, using functional magnetic resonance imaging (fMRI)[18] and magnetoencephalography (MEG),[19] shows activation of color area V4 (see below) when the synesthete is shown black-and-white graphemes. An early hint of cross activation was first observed by Geoffrey Gray, Michael Morgan, and colleagues in phoneme-color synesthesia.[20]

Even more striking is the demonstration that one can actually observe increased white matter within V4 (as predicted by us[6,7]), using diffusion tensor imaging (DTI),[21] providing a neural substrate for grapheme-color synesthesia.

It is also known that synesthetic colors can lead to Stroop interference, e.g., if the letter G normally evokes red, the subject is slower to report the letter if presented in a different color (such as green).[22] This shows synesthesia is automatic, but does not necessarily show that it is sensory. Indeed, Stroop interference can occur at any stage in information processing,[23] even at the level of motor output! (Try humming a high pitch while moving your head down.)

FIGURE 1.4 Illusory edges enclosing the grapheme 'A'.

Physiology of synesthesia

What causes synesthesia? To account for the phenomenon of synesthesia (grapheme color), we proposed the sensory cross-activation hypothesis[6,7] which simply postulates the following: when neurons mediating one sense modality (e.g., form) are activated by a real stimulus, there is spontaneous cross activation of neurons that would ordinarily be mediating another modality (e.g., color) or attribute. Cross activation can occur due to abnormal anatomical connections or disinhibition of preexisting connections—including feedback projections—between brain modules.[24]

Why would such connections for cross activation exist? We postulated that in a fetus or early infancy, there are redundant connections between brain modules, perhaps even far-flung ones—which are then subsequently pruned under the control of specific, "pruning genes"—resulting in the characteristic modular organization in the adult brain. One example of this is the segregation of the grapheme and color areas of the fusiform gyrus. If there is a mutation that causes defective pruning, and the mutation happens to be expressed (due to transcription factors) in a specific brain location, for example, only in the fusiform gyrus, the result is grapheme-color synesthesia. The same model can account partly for more rare forms of synesthesia such as the man who tasted shapes,[25] or "texture–emotion synesthesia,"[26] both of which could involve cross activation between the insula (taste/emotion) and secondary somatosensory cortex (S2; texture)—or the intraparietal sulcus (IPS) and the anterior parts of the supramarginal gyrus (SMG) (shapes).

As noted earlier, brain-imaging studies (fMRI), which allow precise localization of V4 activity, support our contention that "projector" synesthetes have cross activation between color and form in the fusiform gyrus.[18] Even more compelling evidence for our hypothesis is the recent finding—using DTI—showing that there is an actual increase of axons in the fusiform gyri of projector synesthetes,[21,27] whereas, in associators, the increase is seen in the inferior parietal cortex, precisely as predicted by our model.[6–8]Indeed, there is an overall increase in long-range resting-state potentials throughout the brains of synesthetes,[26] and consistent with this, we have found that grapheme-color synesthetes show enhanced crossmodal interactions between modalities other than color and form (e.g., auditory and visual modalities[28]; see also Lacey et al., who showed enhanced Bouba-Kiki effects in synesthetes[29]). Thus, unlike neurotypical individuals, synesthetes as a whole appear to have a widespread enhancement of sensory—and perhaps conceptual—cross activation.[6,7] In our scheme, this would also explain the increase, not only in creativity and memory but also in the persistent tendency of synesthetes to form strange associations between seemingly unrelated attributes and ideas. For example, "4 is a warm, even-tempered, puppy-dog kind of guy. A Labrador, if you will. He likes being outdoors and gives really good hugs. He recycles and rides bikes, but isn't all up in your face about it." This is a far cry from the anatomical precision that we postulated for certain types of synesthesias (e.g., grapheme color or number lines—see last section of this chapter). Such strange introspections as quoted above that lie at the boundary between sensation and cognition add a dimension of complexity to the synesthesia problem, which makes it at once both frustratingly challenging, yet providing a novel foothold to approaching some of the most enigmatic and treasured aspects of our minds. As Holmes told Watson, "It is this, that lifts this case from the regions of the commonplace."

Indeed, listening to some of the more outlandish remarks of synesthetes, one realizes that even though synesthesia has already begun to provide us with valuable insights into the mechanisms underlying brain function, we have barely scratched the surface of the problem.

Fulfilling at least a subset of the nine criteria (loosely defined) described earlier and the associated (presumed) activation of V4 appears to be true mainly for projectors (constituting 10%–15% of synesthetes). In associators, the color is evoked not by graphemes (visual shapes of numbers), but by their MEANING—such as the abstract idea of numerical sequence. This would encompass the ordinary sequence of natural integers as well as days of the week or months of the year (which many associators experience as being colored). We suggested that perhaps in many of these associators, the cross activation occurs higher up in the sensory processing stream. It is also worth noting that in some of them the color evoked by the first letter of a word tints the entire word (see below[6–8]).

But where is "higher up"? A reasonable candidate is the IPS—a cortical area that receives input from V4[30] and is adjacent to the angular gyrus (AG) (Fig. 1.5). The AG, as we shall see, represents numbers, fingers (as well as sequences they embody), and—possibly, in conjunction with the hippocampus—calendars (see last section). It also receives input from the area processing "form" (including graphemes) in the fusiform gyrus. Thus, there is plenty of opportunity for direct cross activation.

A second possibility is that unidentified back projections from the AG to V4 activate color-sensitive neurons. This indirect activation would explain the "weak" qualia experienced by associators in contrast to projectors ("I see the colors in my mind's eye" vs. "I see it painted on the letter"). Finally, in addition to the visual grapheme area in the anterior fusiform, there appears

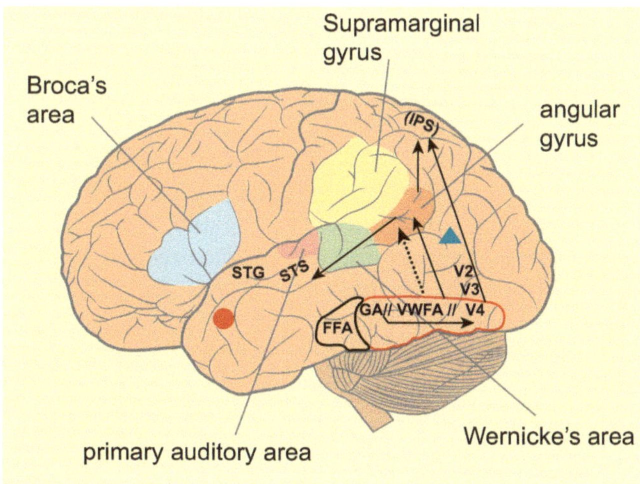

FIGURE 1.5 Schematic diagram of areas involved in grapheme-color synesthesias: *FFA*, fusiform face area; Dot, •, on temporal lobe, amygdala. *Blue triangle* marks the insula. *IPS*, intraparietal sulcus. *STG*, superior temporal gyrus. *STS*, superior temporal sulcus. *GA*, grapheme area (representing visual shapes of the alphabet and numbers). *VWFA*, visual word form area. *V4*, cortical color area. FFA, GA, VWFA, and V4 are all tucked away in the fusiform gyrus. The *arrows* depict cross activation between form areas and color areas in the brain, both in projectors and associators (see text). Back projections from the angular gyrus down to the fusiform gyrus also play a role but are not depicted.

to be a visual word form area (VWFA) in the middle fusiform that encodes whole words including their pronunciation, and it's possible some associators have cross activation arising there.

The cascade of activation in different types of synesthesia can be further teased apart by MEG (which has exquisite temporal resolution). Experiments spearheaded in our laboratory by David Brang showed that V4 is activated 5 ms after the graphee area, rather than the much longer delay of the kind that would be required if back projections from the AG were involved.[19]

We must be cautious, however, not to overemphasize the hardwired genetic approach to synesthesia. We are sometimes misquoted as having shown that "synesthesia is genetic"—a sentence that is biologically meaningless. None of us is born with graphemes. We have always maintained that what the synesthesia gene(s) confers is a propensity to link arbitrary shapes and colors.

Creativity

As noted above, as the result of transcription factors, the "cross-wiring" gene might be selectively expressed in a single brain region, leading to a specific form of synesthesia (e.g., grapheme color). But what would happen if the gene were more diffusely expressed? The result would be an excess of cross wiring throughout the brain. If abstract concepts are also represented in specific brain regions, such diffuse cross wiring would confer a propensity to link seemingly unrelated concepts represented in far-flung brain areas; the basis of metaphor (as in describing "sharp cheese" or that "Juliet is the sun"). Hence, the higher incidence of synesthesia in artists, poets, and novelists, as shown by us,[5] who all have the ability to link unrelated ideas. This "hidden agenda" might explain the high prevalence of the otherwise useless synesthesia gene.[6,7] This situation is analogous to the manner in which the sickle-cell anemia gene survived in the Mediterranean—despite it being lethal in the double-recessive form—because the single recessive gene confers immunity from malaria (making it evolutionarily advantageous). We also noted[6–8] that the "selective gene-expression" hypothesis would also explain the common observation, recently confirmed,[31] that if you have one type of synesthesia you are also more likely to have another. The gene in question would confer a propensity for defective pruning throughout the brain. As a result of transcription factors, the gene might be expressed in a patchy manner in far-flung brain regions, predicting a higher than chance incidence of multiple synesthesias in a single individual. Finally, as predicted by our model,[6,7] the hyperconnectivity throughout the brain results in increased resting-state potentials across large regions of the cortex.[26] Additionally, we have found that grapheme-color synesthetes show enhanced crossmodal interactions even between auditory and visual modalities.[28] Thus, there appear to be widespread differences between the brains of synesthetes and nonsynesthetes, consistent with our cross-activation model.

Paradoxes of synesthesia

As the Bard said,

> Hang up philosophy! Unless philosophy can make a Juliet

The intersensory cross-activation model was what a physicist would call a "working hypothesis," but even in our very earliest studies we noted that this "phrenological" account could not be the whole story.[6–8] For example, in Fig. 1.6, the central letter can be seen as

FIGURE 1.6 The perception of the middle character is influenced by the letters that surround it. When presented with the ambiguous H/A form in THE CAT, report that they experienced different colors for the H and the A, even though the physical stimulus was identical in both cases.

H or A depending on whether you group it vertically (THE) or horizontally (CAT). The CATEGORIZATION of the letter can alter the color evoked, presumably because the context shifts the activation of form-extracting networks in the fusiform gyrus.

Another challenge to the simple version of the cross-activation model comes from a more exotic form of synesthesia known as grapheme personification, in which each number is seen as tinged with its own personality.[6–8,32] For example, one subject, EA, describes "An 8 is very noble and kind of held together, almost like a parent figure to 5. 9 is a brown-haired guy, and he's pretty calm—but he's really into seven." Are we to assume then that the grapheme map cross activates a "personality map" somewhere in the brain? This notion is not as outlandish as it seems. It is not inconceivable that there are "primary" personality traits—maybe a few dozen—from which we construct the spectrum of many hundreds, a large number—to be sure—but not so large that it cannot be represented in brain maps, perhaps in the vicinity of the superior temporal sulcus (STS). If so, number personification becomes less mysterious.

Third, there is the gender assignment effect. Some synesthetes claim that each grapheme has its own stable gender—male or female. Along with their colleagues, Maina Amin[33] and Julia Simner[34] have shown, elegantly, that the effect is genuine, using Stroop interference as a probe. Perhaps this represents an amplification of a binarizing propensity that exists even in normal brains (e.g., good-evil; black-white; big-small; yin-yang, etc., supported by the binaries in Daoism), a strategy for efficient coding seen even in the center-surround organization of orientation-specific cells in V1. It strains the imagination to think of maps for binarizing becoming cross wired with graphemes. On the other hand, even normal people can be persuaded to assign genders to graphemes and it's not outlandish to propose that such tendencies—perhaps based on curved versus sharp features—are enhanced in synesthetes. It would be interesting to see if synesthetes follow the same patterns of assignment—which would lend support to the idea that there are synesthetic propensities in us all.

Fourth, there is the curious phenomenon we first drew attention to in Ref. 6,7, dubbed the first letter precedence effect. Many synesthetes report that when looking at a word, the entire word is imbued with the first color, and in the one synesthete whom we tested, JC, this was true even if the first letter was silent, as in "psalm." Thus, if "p" was red, the entire word, "pat," was red and so was "psalm." It's hard enough to think of grapheme nodes cross-activating colors, but even harder to think of patterns of activity representing whole words ("wordemes") in small neuronal clusters (as in the VWFA) being able to cross activate color neurons.

Finally, there is the riddle of qualia. One of our "star" projector synesthetes (EA) had, in addition to calendar line (described at the end of this chapter) and grapheme-color associations, a strong propensity to see colors in response to taste. Intriguingly, she claimed to experience the color quale as being LOCALIZED in her mouth instead of in her mind's eye (e.g., "salt tastes blue—my mouth feels blue"). It is tempting to regard her experiencing vision on her tongue as an atavism. In early evolution, "primitive senses," like touch and taste, were indeed localized on the body surface. And, remote sensing through vision was a relatively late invention. Thus, her seeing colors "in" her mouth may be a result of reverting to the default mode of sense localization in evolution.

Such phenomena must surely have relevance to the philosophical problem of quale that was also implicit in our discussion of Martian colors. Indeed, seeing EA reminded us that we must remain open to the possibility that some of the axiomatic foundations of sensory physiology such as Müller's law and place coding (that quale depend on which neuron fires where—not on patterns of nerve impulses) may need to be challenged and revised.

There are no simple answers to these questions, but whatever direction the inquiry takes us, synesthesia is bound to illuminate the basic coding strategies used by the brain in representing objects, concepts, and events in the external world.

Bouba-Kiki

The key characteristic of synesthesia—even among associators—is its inevitable quality—it cannot be suppressed (unlike one's tendency to think of a chariot when one merely hears the word Cinderella). We wondered, "Are there other 'compulsory' associations seen in most normal people and not just in a quirky minority?". The obvious candidate is "sound-form symbolism"—now sometimes called the Bouba-Kiki effect.

Inspired by the work of Köhler[35] and Werner and Caplan,[36] we presented the two shapes shown in Fig. 1.7A and B, to subjects who were unaware of the purpose of the experiment. They were told that, "Just as each English alphabet has a certain shape and associated sound (e.g., "A"—is the sound Aaaaaa and "B" has an associated sound "Bbbbbb"), let's pretend for fun, that these two abstract shapes are from a Martian alphabet each with its own sound. There is no correct answer—we would just like you to tell us what you think. Which one is Bouba? And which one is Kiki?" (95% of subjects report the right being Kiki, and the left being Bouba, even though none of them is Martian, and none has ever seen these shapes before). We are now in the process of creating a UCSD Sound-Form Symbolism Battery, of which some examples can be seen in Fig. 1.8.

To explain the Bouba-Kiki finding,[7] we suggested that the jaggedness of the contours of the Kiki shape mimic the sudden inflection of the sound, and indeed, of the tongue on the palate—whereas the ameboid Bouba shape resonates with the gentle undulation of the lips.

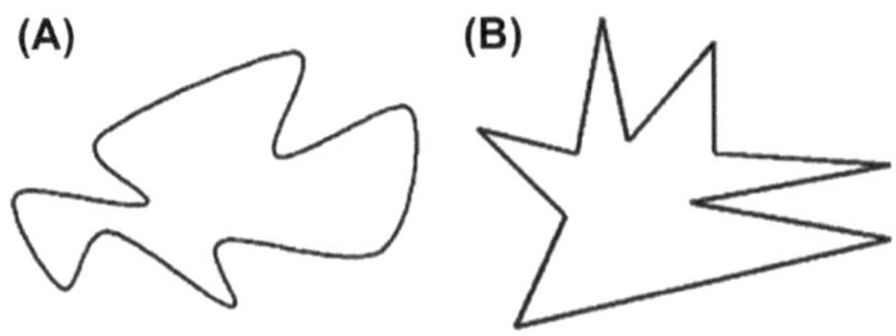

FIGURE 1.7 Examples of kiki and bouba.

FIGURE 1.8 Examples of figures used in the UCSD Sound-Form symbolism battery. Sound correspondences: (A) /ʃ/; (B) ⟨r⟩ (C) /s/; (D).

For over 50 years, the phenomenon was regarded as little more than a curiosity, and some dissenters even pointed to similarities between "Bouba" and the English alphabet B (jagged edges of "Kiki" to K) as an explanation. This objection was refuted by two additional experiments. First, we used new stimuli (Fig. 1.8A–C), which are, respectively, Shhhhhhh, RRRrrrrrr, and SSSssssssss, and Fig. 1.8D, which shows O (as in Oh), or Oo (as in "Oops"), with Ah (as in "army") obtaining similar results even though they bear no resemblance to English. Secondly, we presented the Bouba-Kiki stimuli to Tamilians of Southern India and obtained identical results. From these results, it appears that, perhaps the letters b, k, etc., are chosen to represent rounded versus pointed sounds because of crossmodal similarities.

We recognized the potential importance of this demonstration for understanding the emergence of words in evolution. We also pointed out the possible operational analogy between intersensory abstraction and high-level conceptual abstraction or metaphor, as well as sensory-to-motor abstraction as exemplified by the mirror neuron system (MNS) (see following sections). The Bouba-Kiki effect seemed to us to have much broader implications than hitherto believed—and subsequent research seems to have borne this out.

In concluding this section, we note the Bouba-Kiki effect is not an example of synesthesia, although we are often misquoted as having shown that. We have pointed out that in synesthesia the link between shape and color is arbitrary—there is nothing red about "3"—and the associations are idiosyncratic, whereas the shape-sound links of the Bouba-Kiki effect are precisely the opposite. They are nonarbitrary, in that they share the features of roundness and jaggedness across different domains (see Chapter 11). Nevertheless, it is true, that—given the enhanced cross connections in synesthetes' brains—one might predict that they would have an enhanced Bouba-Kiki effect, which does indeed appear to be true.[29]

There is evidence from Daphne Maurer's elegant experiments that the Bouba-Kiki phenomenon is innate.[37] Her research team tested children as young as $2^1/_2$ years (and as old as 5) on the task, many of them still unable to read. Using adults as controls, the outcome was that there were no significant differences between the shape-to-sound correspondences made by the young children, implying an inherent bias to make the choices we adults do with Bouba and Kiki.

Bouba-Kiki and word origins

> That which we call a rose, by any other name, would smell as sweet. ***William Shakespeare***

Would it really?

We noted the link between Bouba-Kiki and conceptual metaphors, but a more important connection is between Bouba-Kiki and the evolutionary/quasicultural origin of words. According to Saussure and orthodox linguists, as there is no physical resemblance between a word and what it denotes, the choice of certain sounds (words) for objects and verbs is completely arbitrary. An example of this can be found with the word "dog": Dog (English), Chien (French), Kutta (Hindi), Maa (Thai), Nai (Tamil); the word is different in different languages.

If so, how did the first word come into existence?[b] Did a group of early *Homo sapiens*, 50K years ago, sit under a tree and say in chorus, "Let's call this an axe; everybody repeat axe." That seems absurd, and fortunately Bouba-Kiki comes to our rescue. If the mapping of sound to shape were not arbitrary, that would enormously reduce the "problem space" by putting constraints on the consensus of word choice. This propensity would have biased word choice and resulted in an early vocabulary of say 40 or 50 words—that would have helped "jump start" protolanguage which—once the whole language network was in place—set the stage for further development of Saussurean-type words and etymological derivatives. One prediction would be that if "modern" words are removed (e.g., scientific jargon) leaving behind more universal words (e.g., body parts, water, fire, fruit tree, sun, beat, pull, push) what remains (even if it is as little as 10% of the lexicon) should show more of a Bouba-Kiki trend and be more similar across cultures.

Consider the following examples: trudge, sludge, and fudge. Saying "trudge" (as in to trudge through the sticky mud) conjures an image of a heavy boot or wooden leg dipping into viscous mud and pulling out gradually, mimicked by the sudden release of the tongue on the palate, a slow pushing and sudden exit, "trudgGGGGGee sluDDDgggee." The same idea holds true when comparing with "tap" or "tip," which are articulated in the mouth by sudden contact and equally sudden withdrawal of the tongue. Or, consider the word "nuDGe," which implies that an object is initially encountering an obstacle, but later overcoming it. Far more intriguing are the higher order examples like juDGe and gruDGe. The judge's verdict is sloshing in the sticky mud of evidence and there is hesitation until he decides and hands out the verdict suddenly ... "Judge." Or, we say "he gruDGingly gave it," another case of initial hesitation. We see here the transition from the sensory Bouba-Kiki to slightly more abstract (nudge) to pure metaphorical (judge). Before we can be sure about this, however, we would need to rule out more standard etymological explanations. It is possible that the initial kick-start came from Bouba-Kiki-like word generation, but subsequent evolution and divergence of words was a result of cultural etymology. The reader is encouraged to try sounding these things aloud, to understand what we are talking about.

Some other examples are worth looking at, such as the repeated emergence of /r/ (alveolar trill) in words which convey some aspect of roughness (e.g., Harsh, Brisk, rough, rogue) versus /s/ words (e.g., smooth, silky, soft). This harks back to "RRRRRRRRRRR" for sawtooth and "SHhhhhhhhhhhh" for a blurred line (Fig. 1.8). Looking for examples such as these can be instructive, if one is lucky, and at the very least, it can be an amusing pastime. The word stinGGGGGy evokes cash adhering metaphorically to the wallet; in Thai, the word for stingy is "khee niaoww," which literally means sticky adhering stools—a reluctance to evacuate the bowel, meaning metaphorically—a disgusting refusal to part with money. In Tamil, the word is "pishi nari"—once again meaning sticky stinking stools.

What remains to be seen is whether the Bouba-Kiki effect played a role in protolanguage taking off, or whether it is a general biasing effect in brain function that incidentally spilled

[b] In this paper, when we refer to "language," we mean protolanguage, especially word meaning and metaphor—not recursion or hierarchical embedding—so there is no contradiction with classical psycholinguistics.

over into language—and merely acted as a catalyst in the origin of our lexicon and understanding of metaphor. Obviously, in tackling these questions, one would first like to know how pervasive a given phenomenon is and to what extent it is cross cultural.

Similar questions arise when considering the evolutionary origin of gesturing. Evolutionary arguments are always hard to tackle, but one way of doing so would be to exploit the link between singing ability and gesture. We have often noticed that south Indian Carnatic musicians (vocal) when improvising scale-sequence variations ("swaras" in Indian music) engage in elaborate gesturing often mimicking the string of notes with hand movements that are ascending or descending but is this an example of true embodied cognition in the sense of the body being a REQUIREMENT for vocalizations of ragas and scales? Or is it merely an accompaniment? If you restrict the head gesticulations and body movements of eminent vocalists such as Semmangudi or Bhimsen Joshi (or any singer), would the quality of music deteriorate? (And if it does, we would have to rule out that it's not just the discomfort from the strait-jacket control by making the singer uncomfortable using some other means.)

Bouba-Kiki and memory consolidation

Orthodox linguists accept as axiomatic the view that the word used to designate an object is purely arbitrary - with no resemblance of any sort to the object. Nouns (for example) can have different names and sounds across languages (Saussure; the same animal is called "dog" in English, Chien in French, kuttah in Hindi, ma in Thai). Exceptions to the Sassurian claim exist but are thought to be rare the (e.g., sounds similar to "mama" for mother and "papa" for father are spoken in a wide range of languages).

Yet, it turns out that the bouba-kiki effect is far more common than anyone realized. Could there be a hidden evolutionary agenda? For example: Does sound-form symbolism enhance memory acquisition and consolidation? If so, it could facilitate a labeling efficiency (congruence between sound and form) which, in turn, would have been conducive with the emergence of proto language.

To answer this - we gave naïve subjects lists composed of a pair of either congruent (e.g., bouba paired with an amoeba like shape) or incongruent (e.g., bouba paired with jagged shape) shapes and words (16 pairs total). They were asked to remember the pairings an hour after training to criterion. During this testing phase, we gave them two choices for every memorized target. For congruent targets, one was identical to what learned, and the other was analogously congruent but not identical. In this condition they mostly chose the identical (what was shown in the learning phase) which rules out the possibility that they were defaulting to the standard sensory sound-form symbolism (the same logic holds for incongruent pairs). As expected, the converse is true for the incongruent pairings.

We found similar results for the consolidation of colored words. Subjects are shown a list of 8 color-word pairs that were either congruent (e.g., 'RED' printed in red ink) or incongruent (e.g., 'YELLOW' printed in blue ink), and asked to pay close attention as they will tested on them later (but not told how they will be tested). In the testing phase, they are shown 16 items and their task was to indicate whether an item is old or new, while rating their confidence on a scale from 1-7. Congruent items were correctly identified as old or new more readily than incongruent items, and congruence is a significant predictor of subject responses.

We conclude that the process that psychologists call consolidation of memories, far from merely being a non-selective strengthening and stabilization of synapses involves editing, classifying etc. of the acquired information across different sensory modalities, unconsciously for hours after exposure. It's a Bartlettian "constructive" process, but unlike Bartlett's stories, which are contaminated by high level semantics we have stripped the stimuli down to bare essentials. This might allow us to more effectively probe the, "laws" of memory. Crucially we since we trained the subjects to criterion, any difference that might be observed between congruent and incongruent pairings cannot be a result of one of the pairs being more strongly acquired to begin with.

Bouba-Kiki and conceptual metaphors

Thanks to the pioneering work of Lakoff and Johnson,[38] Turner,[39] Annaz et al.,[40] Thibodeau and Boroditsky[41] Nunez,[42] and Coulson and Matlock[43]—in the last two decades there has been a growing realization that metaphors are no mere decoration used by poets. They are fundamental to our understanding the human mind. At first sight, Bouba-Kiki would seem to have no relevance to the Bard's, "Juliet is the sun," but they are in fact, both examples of abstraction—extracting the common denominator from seemingly dissimilar conceptual realms. The visual "Kiki," is a bunch of photons hitting particular sets of retinal photoreceptors in parallel and evoking activity in the fusiform gyrus. The sound of "Kiki" is a set of hair cells activated sequentially and evoking a neural pattern in the auditory cortex. The two patterns of neural activity have absolutely nothing obvious in common but still must have something in common. Somehow, "Kiki-ness" is abstracted by some brain network—maybe a "master" modality-free shape map. A metaphor too involves taking two utterly different patterns of neural activity evoked by a glowing ball of fire (sun) and a young woman, which are floridly dissimilar; yet extracting what's in common is practically effortless. Each word evokes a penumbra or halo of associations: for Juliet, female, love, hair, youth, romance etc., which have nothing to do with the sun. But what she has in common with the sun is that she is radiant, warm, nurturing, the center of the cosmos, etc. The overlapping zone between the two halos of associations evoked by the sun and Juliet is the metaphor.

In creative people, in general, and synesthetes in particular (because of possible defective pruning)—the existence of more long-range connections would lead to bigger halos. Hence, there would be a greater opportunity for overlap between those halos, leading to a greater propensity for metaphorical thinking.

Intersensory metaphors

We regard intersensory metaphors as lying halfway between Bouba-Kiki and conceptual metaphors. Examples would include "loud tie" or "sharp cheese." Again, these two examples illustrate two subtypes of intersensory metaphors, abstraction of the quality of magnitude (for loud sounds and bright colors[44] vs. the noncommensurable qualia of sharpness (in taste) that cheddar cheese shares with a sharp knife's tactile qualities). It remains to be seen if this belongs to the Bouba-Kiki category (if cheese is gustatorily sharp in some abstract map that it shares with tactile sharpness). The difference is, with Bouba-Kiki, it is easy to explain, after the fact, why the jagged shape resonates with the sudden sound inflection in Kiki—"it makes sense." The sharpness of cheese, and tactile sharpness, on the other hand, seems to be inherently similar without requiring any explanation outside of the subjective quality.

The same is more strikingly true for "bitter" smells, like that of a crushed beetle, or the sweet smell of nail polish, which are experienced distinctly as smell quale, but nonetheless evoke the distinctly gustatory quale of bitterness or sweetness (see Chapter 10). This intermodal blending of qualia for smell and taste may reflect their evolutionary origins and the underlying anatomical proximity and connectivity. If there is actual anatomical convergence, we predict that after smelling acetone, the threshold for taste of sweetness would be elevated, but not for the taste of (say) sourness. We are here, not merely talking about the standard conjunction of smell and taste in daily life—but the deeper resonance between smell and taste in some abstract code in the brain.

Directionality of metaphors

Intersensory metaphors are common,[45] both in literature and in daily conversation: e.g., "loud shirt," "hot babe," and "sharp cheese." And it is known that there is a weak overall directional trend.

Examples include

note is sour—but taste is not loud
color is loud—but sound is not red
voice is soft—but taste is not loud
note is sharp but a knife is not loud
taste is flat but knife is not sour
color is warm but warmth is not colorful
the smell of flowers is soft, a soft thing is not floral

Notice the hierarchy: touch to taste to smell to hearing to vision. Can it be a coincidence that this follows phylogeny crudely? Tactile receptors occur in jellyfish; taste evolved later—an intermediate step toward remote sensing by smell. In parallel, touch evolved into hearing from lateral lines of fish, while vision arguably coevolved. The picture is clouded by the vicissitudes of convergent and parallel evolution; but the arrows of synesthesia seem to go backward, from phylogenetically recent to older. The culmination of the trend is the use of sensory metaphors for cognitive concepts—e.g., we say "dull thinker," but not "stupid knife"; "blurred meaning" but not "meaningless line," etc.

We have noticed that "regular" synesthesia shows the same directional trends (though color is a counterexample)—if put in a forced-choice situation. Additional research is needed.

Inferior parietal lobule and Bouba-Kiki

What part of the brain is involved in mediating Bouba-Kiki? An obvious candidate is the dominant inferior parietal lobule (IPL), especially its posterior segment, the AG. It is strategically located at the crossroads between the parietal (touch and proprioception), occipital (vision), and temporal lobes (hearing), i.e., optimally for engaging in crossmodal tasks. One might expect it, for instance, to be involved in generalizing a discrimination (e.g., pick square—not circle), learned visually, to the tactile realm. This ability may have originally evolved to its present sophisticated form by the selection pressure of tree climbing in our arboreal ancestors. The creature sees a vertical or oblique tree branch and matches the visually perceived orientation of the branch to the proprioceptively felt position of its own grasping arm.

Visual-kinesthetic abstraction might have sufficed for our ape-like ancestors (and our modern cousins). But as they evolved into species of *Homo*, emphasis shifted toward vocal communication—which until then had served only limbic, guttural, and utterances. The net result may have been the strengthening of auditory input to IPL; leading up to the Bouba-Kiki effect. And that, in turn, set the stage for protowords and metaphors. This is well in line with Norman Geschwind's speculations that vocalizations in primates evolved originally for emotions—evoked via limbic connections—and not for symbolic protolanguage per se. The latter—we suggest—required the intermediate Bouba-Kiki step based on the neocortex. If this is correct, we would predict that even those primates (e.g., chimps) that can crossmodally match visually presented shapes with shapes that are felt out-of-sight, might fail at the Bouba-Kiki task. On the other hand, emotional vocalizations are probably still based on right hemisphere limbic (including anterior cingulate) structures, so that one could regard the preservation of swear words in aphasias as atavisms that reveal the "indelible stamp of our lowly origins" (to quote Darwin).

Not coincidentally, perhaps, the left IPL underwent an accelerated evolution as early primates evolved into hominids—achieving its present day fivefold increase in size. The IPL then split into the SMG and the AG, uniquely in humans. The SMG is specialized for producing and recognizing skilled actions, both of which are compromised in ideational apraxia. The AG, on the other hand, took the specialization in a more abstract direction of Bouba-Kiki, metaphor, protowords, and perhaps higher forms of abstraction in general. It is no wonder then that damage to it produces loss of those quintessentially human attributes—words, spelling, reading, writing, arithmetic, and metaphor. This would explain why any ape can reach for a banana, but only a human can "reach for the stars" or understand what that means.

Evidence for this speculation comes from our preliminary experiments on three patients with lesions in the left IPL. All of them were deficient on the Bouba-Kiki task as well as the interpretation of metaphors—which we have dubbed "metaphor blindness" (despite their having normal intelligence and being able to correctly interpret the classic story of Solomon discovering who the real mother of the baby was). The result needs to be replicated on a larger number of subjects; an N of 3 is hardly adequate.

There may be a further division of labor between left and right IPL: the former being specialized for conceptual metaphors like "Juliet is the sun," and action metaphors like "He reached great heights" or he "grasped the concept," while the latter is for spatial metaphors (e.g., looking "forward" to a brighter tomorrow or a theory being "ahead" of its time).

Even our very sense of self requires cross-domain interaction. The left hemisphere linked to propositional language might be involved in conceptual aspects of self—including one's autobiography—and sense of being anchored in time—here and now—in one's mental calendar. The right IPL on the other hand is concerned with the multimodal integration required for construction of body image. Indeed, the self is not the monolithic entity it conceives itself to be but consists of many components. This is especially evident in somatoparaphrenia—caused by lesions in the right parietal lobe (sometimes including the insula).

For instance, we saw a patient whose left arm was paralyzed, and lying on the table. She insisted repeatedly that it was her brother's arm. When asked where her brother was, she insisted that he was 'hiding under the table' She also affirmed that her own left arm was OK—not paralyzed. Yet, when asked to touch her nose with her left hand she unhesitatingly used her right hand to reach out and grab her lifeless left arm, then used it as a "tool" raising it to touch her nose! Implying, that "somebody" in there, knew that the paralyzed arm was in fact her own, and, additionally, that it was paralyzed. Such observations raise disquieting questions about what it means to be a "single" conscious human being who can simultaneously entertain contradictory beliefs. Like the famous physicist Neils Bohr said in response to his colleague laughing at his horseshoe dangling from his office door: "Well, I don't believe in it but it works anyway." Or, my decision to start praying when I learned of the experiment, given that placebos work, even if you know it is a placebo.

The neural substrate of body image and its disturbances has been studied cleverly by Ehrsson, Blanke, and their colleagues (see Chapters 8 and 18) and may well involve interactions between those circuits—and the limbic system must also play a role, which would also explain why, after induction of the rubber hand illusion, any threat to the rubber hand elicits a galvanic skin response.[46] Extreme examples of disturbances in these very same circuits may underlie disorders like xenomelia—the desire for amputation[47,48] as well as anorexia.[49]

Our emphasis on the importance of the IPL/AG in human cognition is also consistent with the ingenious speculations of Kurt Goldstein,[1] Norman Geschwind,[50] and Daniel Schacter.[51] Geschwind, especially, predicted that crossmodal transfers occur only between vision and proprioception in apes[52]—and hearing was added on later. And as we noted earlier, this cascade of events would have, in turn, led to Bouba-Kiki, which propelled the emergence of protowords and metaphors. And perhaps, many other types of abstraction as well.

As noted earlier, damage to the left AG results in the four components of the Gerstmann's syndrome; dyscalculia (difficulty with arithmetic), finger agnosia, left-right confusion, and agraphia (inability to write), together with anomia, alexia, and we can now add metaphor.[53] It is noteworthy that this relatively small region of cortex has multiple subdivisions and seems to be involved in a disproportionately large number of functions.

Other sequelae of IPL damage

Over the years, we have observed a number of other signs and symptoms that deserve mention even though they elude clear definition:

1) We tested one of our subjects who had anomia; an inability to name even common objects and a tendency to misname them; as commonly seen following left AG lesions. Yet in his case, we noticed that the anomia permeated his very perception of the object that he misnamed. One example of this is that when shown glasses in an empty testing room, he called it "Eye medicine," and proceeded to mime instilling eye drops in his eyes. It's as if a word is not merely a label but a golden key to a whole treasury of associations—and using the wrong word can taint the perception of a misnamed object. It's also known that those with anomia will reluctantly say things like "I know what it is, but its name eludes me," when shown common objects. Astonishingly, our patient, when shown an abstract, misshapen, sculpture, a nonsense shape (Fig. 1.9), said "I know what it is but don't remember its name," and on testing him the second day "I knew its name yesterday but have now forgotten" (reminds us of the Sapir—Whorf hypothesis). It is worth noting that he was from a region in India, where he is unlikely to have been exposed to the idea of abstract art.

2) We have noticed that some patients with left AG lesions, when presented spatially alternating sequence of triangles and circles, proceeded to complete the series by drawing a bulbous intermediate shape. Failing to grasp the alternation rule, they chose an average

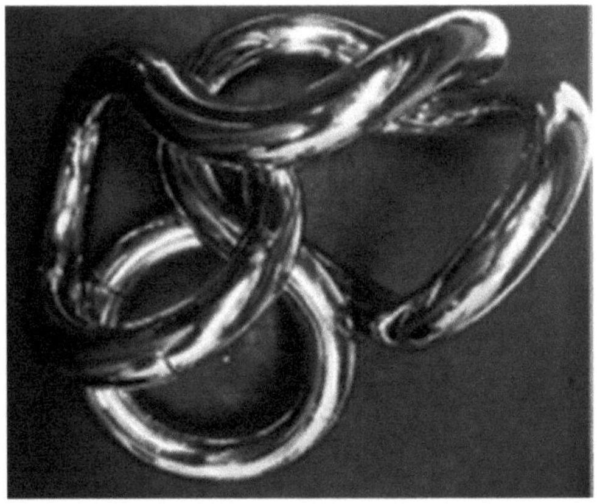

FIGURE 1.9 Example of an abstract 3D "nonsense" shape shown to subjects with anomia.

FIGURE 1.10 An example of a subjects' drawing of an intermediate shape when asked to complete the sequence of alternating shapes.

of the two shapes (Fig. 1.10). Or, sometimes the patient drew a triangle overlapping with a circle.

3) The degree of dyscalculia varies from patient to patient and we recently saw a patient who hardly had any. We then noticed him covertly moving his fingers when asked to do arithmetic. When we held his fingers tight so he could not move them, he lost the ability to perform even simple subtraction (e.g., 23−17) or division. This is a striking example of embodied cognition.

It may not be a coincidence that numbers and fingers are both represented in the AG. We ordinarily learn to count with fingers, but as we grow up, the ability is transferred to "virtual" fingers in the left AG and if the representation of virtual finger wriggling is compromised, one needs to revert to wriggling real fingers again. This idea can be tested, by seeing if patients with complex regional pain syndrome in their right hand (which might lead to retrograde "trophic" changes in the left parietal cortex) have difficulty with arithmetic.

4) The existence of specialized brain modules, e.g., the AG and Wernicke's area, can hardly be denied, but a recurring theme in this article is the surprisingly high extent to which they interact, especially in humans.

When given easy math problems, most patients with dyscalculia confabulate, they confidently produce an incorrect answer. Yet we have seen at least two patients who when asked what is 23 minus 17 started producing incomprehensible Wernicke's speech! This was not simply because of stress because more difficult questions, e.g., about their paralysis, yielded perfectly coherent speech (as you would expect given that their lesion was confined to the IPL-not extending into Wernicke's area). It's as if the mere activation of the malfunctioning arithmetic-mediating circuits in the AG introduces a temporary, "virus" in the Wernicke's area to which it projects; a tendency we have dubbed "lesion contagion."[54]

Our notion is that the selection pressure for the evolution of crossmodal abstraction (or abstraction in general) may initially have come from the need for prehension when our ancestors were grasping branches in the treetops. This is consistent with the observation that in many unrelated phylogenetic lines, there seems to be a correlation between intelligence and prehensile skills (e.g., parrots, crows, elephants, and even invertebrates like the octopus).

Predictions about dyslexia

Dyslexia is a surprisingly common learning disability characterized by a disproportionate difficulty in reading despite having a normal IQ. Intriguingly, histological anomalies have been noted in the left AG.[55] If our speculations are right, dyslexics should perform poorly on crossmodal object retrieval, metaphors, and Bouba-Kiki (they may also have finger agnosia). We recently had the opportunity to test 20 dyslexic children of mean age 9 years and near normal IQ and found that many of them had severe problems SEQUENCING (which is critical for spelling, representing calendars, numbers—even clocks). They had difficulty playing SIMON or using internally imagined calendars to punctuate the week's events in the proper sequence—(see below) a sign we have dubbed "sequence agnosia."

Surprisingly, sequence agnosia also seems to impair their ability to interpret family trees. For example: "John is Joes Father and Janes Son. What is the relationship between Jane and Joe".

Metaphor blindness

If the architecture of the brain is at least partially modular, not just for perceptual modalities (e.g., area MT for visual motion; area V4 for color) but also for more cognitive capacities like humor or metaphor, then it is conceivable that gene mutations could selectively affect these functions.[53] Consistent with this view, we have found that about 5% of the population—despite their near normal IQ—tend to interpret metaphors literally and suggested they may have a mild dysfunction in the IPL. This result needs replication using a larger number of subjects including controls matched for verbal IQ. One would like to know also whether the deficiency simply represents one end of a normal distribution curve. Similarly, there may be specialized hardwired circuits for humor, given the wide prevalence of "humor blindness."

Taxonomy of intersensory effects

Research and speculation on intersensory effects have a long and venerable tradition in the history of psychology and more recently in neuroscience,[56–59] but until about 10 years ago, there was a tendency to confound different types of interdomain effects. We will therefore attempt a rough taxonomy, without any pretentions of theoretical insight.

1) Simple interactions such as probability summation at threshold or the weighted average of two or more suprathreshold inputs.
2) Vetoing:
   If five parallel inputs signal conflicting information and two signals are exactly the same, the brain accepts the latter as correct and the first five are vetoed; one does not take the arithmetic average. When only two cues are present, the statistically more reliable; therefore, less ambiguous cue vetoes the other, e.g., the McGurk effect.[60]
3) Arbitrary memory associations:
   Those acquired solely by virtue of their cooccurrence in the past. For example, when you hear "Cinderella," you think of chariot, pumpkin, or glass slippers. Even though it seems automatic, it is optional rather than compulsory or obligatory; unlike synesthesia in which 5 is red; and there is no choice, and no imagining it as a different color. Similarly, if I showed you a black-and-white photo of an apple, you could imagine it being tinged red, but it lacks qualia and you can just as easily imagine it being tinged green, if you choose to.
4) Synesthesia (both projectors and associators):
   In grapheme-color synesthetes, individual graphemes are seen as tinged with highly specific colors (e.g., six is chartreuse and you cannot will it to be otherwise). This is unlike the tendency we all have of imagining yellow when shown a black-and-white photo of a banana.
   We hasten to add that the boundary between the experiences of associator synesthetes and ordinary memories gets blurred at times. We have observed that chains of associations, which would normally evoke only memories in normal individuals, would sometimes seem to actually evoke qualia-laden sense impressions in some higher synesthetes. Therefore, the merely metaphorical can become quite literal. For example, in one of our subjects, "R is red and red is hot—so R is hot," etc. One wonders whether "hyperconnectivity" (either through sprouting or disinhibition) has affected back projections between different areas in the neural hierarchy. We, and others, have previously suggested that such back projections may play an important role in the genesis of qualia[61,62].
5) Bouba-Kiki effect:
   The abstraction of the common denominator between two otherwise utterly dissimilar entities, e.g., the Kiki visual shape—photons hitting the eye in parallel and evoking a pattern of activity in the fusiform gyrus that is completely different from the pattern evoked in the auditory cortex by the hair cells activated by the sound Kiki. What they share is the sudden jagged inflection or high frequencies in an abstract mathematical Fourier space. Whether there is another master brain area where this abstract quality is represented remains to be seen.
6) Cross-sensory metaphors in normal subjects (of which there are two subtypes):
   a) Loud shirt: where it is obvious what attribute is being abstracted—magnitude estimation.[44]

   b) Cheddar cheese is sharp: why the use of a tactile adjective to describe an otherwise ineffable taste sensation? Could it be Bouba-Kiki in disguise, i.e., the quality of sharpness is abstracted in some higher-dimensional space?

7) Onomatopoeia:

The tendency to use the sound produced by an entity to name it, as in "bow-wow" for dog or "swish" for swirling water. This bears a superficial resemblance to Bouba-Kiki but is in fact the opposite. There is no resemblance whatsoever between the visual appearance of a dog and "bow-wow" or between a falling tree and its "thud" sound, whereas the "Kiki" sound and shape do share an abstract similarity. (Admittedly there are borderline cases, e.g., "suck"—in which the same lip and tongue movement occur whether you ACTUALLY suck or say suck; slurp may be another one.). The evolutionary sequence may have started with literal mimicry of a sound (e.g., a child or hunter barking like a dog to alert neighbors of its presence). In the next stage, it may be conventionalized into something like "bow wow" or "woof woof" which bears some resemblance to the object without pure mimicry. In stage three, etymological and linguistic arbitrariness takes over.

8) Conceptual metaphors:

What is the evolutionary advantage of metaphors? They could not have evolved merely as a literary device. Why not say that Juliet is warm, radiant, or nurturing? One answer might be the economy of words. The "sun" is one word, while warm, nurturing, and radiant are three words (this hardly justifies its widespread appeal). One cannot help thinking that Juliet has many attributes irrelevant to the context … she has hair, a skeleton, is a woman, etc., but the poet is highlighting her unique attributes by using the sun as a metaphor. We have to ask why such highlighting of common attributes should be more powerfully evocative than merely listing them separately. It's as though the very act of abstraction is a general brain propensity that is inherently rewarding (because of its role in thought) and its manifestation as metaphor is a bonus. Another possibility is that words like nurturing, radiant, etc., are more abstract (phylogenetically recent) and therefore less effective in evoking visceral sense impressions and emotions than a simple word like "the sun." (It would also explain why we say "His mind is sharp," but not "This knife is intelligent!") Likewise, we speak of a bright child, but not of a smart candle.

9) Crossmodal tactile to visual shape:

If I show you an abstract, non-nameable complex shape visually and ask you to palpate four objects hidden from view, you would effortlessly choose the correct match. As noted previously, it is known that monkeys fail at this task, whereas chimpanzees succeed, which would be consistent with our evolutionary speculations. We predict that children with autism spectrum disorder (ASD) (see below) and dyslexia who have IPL damage[55] should perform poorly on this task. The matter requires further exploration. There is still some debate over this issue because it is also known that, in monkeys engaged in a crossmodal orientation matching task, orientation-selective V4 neurons coded orientation of tactile gratings when they cued the orientation of visual gratings.[63].

Esthetic Blending

It is obvious that certain tastes blend well with certain smells, whereas others do not. We can conceive of a garlic smell blending with tastes of umami (e.g., cheese), but not with ice cream or even sugar. Ginger, on the other hand, blends well with sweetness. Vanilla works with sugar but not with salt. Some of this is undoubtedly culturally acquired, but some may be hardwired. This awaits cross-cultural testing. Sour seems to work with sweet, whether the sour is conveyed through smell ("citrus smell") or taste. This possibly evolved because of the close association of sweet and sour in a frugivorous primate diet. Whether nail polish remover (acetone—which has a sweetish smell) and sucrose will blend remains to be seen. Salt and sweet typically do not blend although this tendency can be overcome by aggressive marketing (e.g., salted caramel ice cream).

Esthetic blending can be seen within the visual modality (Fig. 1.11) as when the word "tilt" is portrayed in tilted letters of the alphabet. There is a resonance between semantics and the perceptual form of the word (it feels right—opposite of Stroop effects). Similarly, the words "tremble," "fear," or "cold" are often conveyed by wiggly contours that schematically mimic the shaking, forming a spatial graph of rhythmic movement in time. This is not unlike Bouba-Kiki, save the esthetic component. Indeed, the spoken word "tremble" itself involves rhythmic vibration of the tongue /r/ (trilled), and this may not be a coincidence.

The link between Bouba-Kiki and the tilt illusion is not obvious at first, but both are examples of cross-domain resonance. The additional esthetic jolt in the tilt illusion may be because abstraction between hierarchical levels (semantics and visual letter form) is inherently rewarding. Compare the tilt illusion with Fig. 1.12—a large X made of tiny ones, which is interesting, but lacks the esthetic harmony achieved by abstraction between different levels—conceptual to perceptual; it's just a large X made of small ones (obviously, this informal observation needs confirmation from a large number of subjects).

The same "levels of abstraction" principle hold for ordinary metaphors. Compare the example cited earlier: "Juliet is the sun"—a top-down metaphor, to "cheese is sharp." Juliet's beauty (high-level abstraction) is made more vivid by highlighting her radiance and warmth, exploiting the sensory associations evoked by the sun. On the other hand, when saying "cheese is sharp," there is no hierarchical transition between levels of abstraction—it's simply a matter of linking an enigmatic similarity between two sensations—touch and taste.

FIGURE 1.11 The word TILT, with font-style reflecting meaning of the word. A demonstration of resonance between semantics and form (inconicity).

FIGURE 1.12 A large X composed of smaller X's. An example of resonance within a single domain (vision).

As a third example, consider another well-known metaphor, the proverbial "bull in a china shop." A proper use of the metaphor would be "when Joe the atheist walked into a Bible study, he was like a bull in a china shop." Knowing he is an atheist, this is richly evocative. Whereas saying "Johnny had a stroke and stumbled around his house like a bull in a china shop" confounds the literal with the metaphorical. Using a metaphor literally to illustrate the clumsiness of a person is not only ineffective but also comically so.

And, what about the word **MIЯЯOЯ**?

Sex and violence

A tangential question: why are so many swear words threatening violence linked to sex, even across cultures?[64]

Examples include "FU%ˆYOU!" This is especially odd given that sex is a pleasurable activity, why not say "KICK YOU," "BITE YOU," or "claw you"? Or even "beat you" or "hurt you"?

And why the rude "flicking off" gesture of an extended middle finger for aggression? We suggest this link arose in early hominid tribes because of the expressed aggression between rival males competing for females. The alpha male guards his female carefully, especially in the presence of a rival junior male who is offering a "challenge," signaled by his erect penis targeted at or in the presence of the female. In both the alpha male and rival, the erection becomes ritualized into a symbol for aggression just as baring fangs which was originally always a prelude to actual biting—and later evolved into a threat symbol as either a warning or bluff.

Perhaps the loincloth was invented to conceal one's attraction from the alpha male, thereby reducing intermale hostility or allowing for more covert attempts at approaching the female, hence the need for an iconic substitute for the erection conveying aggression. What better choice than the rude finger sign, which visually resembles (iconic, in the Pierce sense) an erect phallus and is easy to produce quickly. The sign is invariant across most cultures.[65]

The mirror neuron system

Neurons in area F5 in the premotor cortex fire when you make volitional movements—with different motor "command neurons" for different movements (e.g., reaching for a peanut, pulling a lever, pushing or putting something in your mouth, etc...).[65a] Intriguingly, a subset of them, 10% fire even when you merely watch another person performing the action. These neurons—dubbed mirror neurons—allow higher centers to say in effect "the same cells are firing as would fire if I were about to reach out for the peanut—so that's what the other person is INTENDING to do." Whether reading the higher-order intentions that are required for constructing a theory of mind (ToM) is based on circuits performing analogous computations is a matter of some debate.

We pointed out[7] that the computation performed by the MNS, crossmodal abstraction from vision maps (appearance of hand reaching out) and maps representing motor command neurons, orchestrating a sequence of muscle twitches in F5 (Broca's area and areas that project to it) is fundamentally the same as the kind of vision to auditory–motor abstraction seen in Bouba-Kiki effects. (Perhaps not coincidentally, neurons with mirroring properties have also been observed in monkey IPL.)

We ordinarily think of maps as topographically organized—with one-to-one correspondence between adjacent points in space or a sensory surface activating adjacent points in brain maps, but one can have more computational maps (e.g., points are adjacent in motion space). In such a map, similar directions and locations of motion are mapped on to a single location in the cortical map, independent of which location in the visual field in which the object is moving. Many maps of abstract dimensions of this kind may exist along with constrained cross-domain mappings between them. A strengthening or distortion of such mappings would explain many types of synesthesia.

We believe (and there is mounting evidence) that the MNS plays a crucial role in many aspects of human behavior and social cognition. However, the extent to which it is important for understanding human cognition has been contested. One obvious function of the MNS is to anticipate impending goal-directed movements which may, in turn, have been an evolutionary prelude to assuming that the person you are interacting with has a mind similar to yours, that you can deceive them, and that they can equally deceive you.

The MNS may also be involved in adopting an allocentric view of novel skilled actions performed by a parent; a necessary prerequisite for IMITATION learning and—therefore—for the rapid Lamarckian dissemination of one-of-a-kind innovations that would have otherwise died with the inventor. The same ability must surely be involved in (A) passing down sophistication of tool use and (B) hafting which sets the stage for multicomponent tool construction (requiring visualizing unusual combinations of parts—with anticipation of its function before the tool is even finished). And notice the similarities between these very abilities, which originally evolved for tool use and for what we usually call "thinking," which suggests that the latter may have evolved from the former.

Sensory mirror neuron system

One component of ToM is empathy. Cells in S2 respond in a topographic manner to stimuli delivered to the contralateral body surface. Again, about 10% fire when I merely watch another person being touched in the same location.[66] The higher brain centers say in effect "the same neuron in S2 is firing as would fire if I were to be caressed"—so I not only know, but also feel what she's going through—and this allows me to empathize emotionally (via output to the insula, amygdala, and other limbic structures). The result is a much more vivid and authentic VR simulation of her caress than what could be achieved through intellectual deduction. Analogously, pain-sensing neurons in your anterior cingulate respond not only to, say, your thumb being poked but to watching another person's thumb being poked. As a result, you can almost, but not quite, feel her pain with its ensuing autonomic arousal, yet you do not shout "ouch" and withdraw your hand. There are two reasons for this:

1) The receptors in your thumb send a null signal to your MNS network saying you are NOT being poked and the net result is empathy that lacks the sensory quale of pain.
2) Inhibition from prefrontal areas (damage to which leads to miming other's actions due to disinhibition of the motor MNS). A counterintuitive prediction would be that if an arm amputee was to look at another person's intact thumb being caressed or poked—he would

literally experience the caress or poke in his phantom thumb. We tested three amputees and found this to be true, a curious form of INTERSUBJECT synesthesia! One of them even shouted "ouch" and withdrew his hand when watching us poke a third person.[67]

Recently, the concept of mirror neurons has found clinical application in patients with certain types of phantom pain. For example, when we asked a patient to watch his wife massage her own hand, the result was a phantom massage in his phantom arm, which relieved his pain. The observation shows that the MNS hypothesis can actually lead to predictions for clinical application. This circumvents arcane philosophical debates as to whether the discovery of the MNS goes any further than the original ToM postulate of psychologists. No one could have predicted this clinical application starting from the abstract idea of a ToM module (we hasten to add, that the observation needs to be replicated in blinded placebo-controlled trials).

Lastly, there appears to be a congenital variant of the same phenomenon, which was first discovered by Sarah Blakemore and Jamie Ward,[68] a condition in which an otherwise normal (neurotypical) person actually experiences touch when watching another person being touched. We found recently that even subtle qualities like wetness were referred—if patient TC watched water dribbled on a volunteer's hand she shook her own hand as if to dry it off and grabbed a towel in the vicinity. And if the volunteer's eye was threatened by a looming object, TC displayed a blink reflex and avoidant movements, which suggests that there is an MNS for involuntary reflexive movements, not just volitional ones. Perhaps most surprisingly, she could not stop laughing when she watched the experimenter being tickled ... interpersonal gargalesthesia[69]!

Out-of-body experience

Our model might also explain some out-of-body experience (OBE). The simulations of the MNS partly involve seeing the world from another person's perspective (optically speaking, taking an allocentric, vantage point). In monkeys and apes, this is triggered by seeing another creature performing an action, but as a human, you can volitionally imagine yourself looking at the world—or indeed your own body—from another visual vantage point—a prelude to an OBE. However, you do not feel you are literally leaving your body because of partial inhibition from the prefrontal cortex, as well as somatic afferents anchoring your body. A stroke that affects this prefrontal inhibition circuit may well explain OBEs seen in a clinical context.[70]

OBEs are also experienced by 30% of the population during REM sleep, which has never previously been explained. We suggest that it results from a massive functional deafferentation that characterizes REM.[71] The widespread prevalence of this phenomenon in the normal population might, in addition to the intellectual desire for immortality, explain the tenacity of belief in a noncorporeal soul among (otherwise) intelligent people.

Gestures, imitation, synkinesia, and the MNS

Both the Bouba-Kiki effect and the MNS have relevance to the gestural theory of language origins, which has seen a resurgence in recent years[76–80] after having been ignored for decades.

MNS-style computation of cross-domain resonance is surely involved in the imitation of words or phonemes—something humans excel at. It's possible this is initially done by trial and error, you hear "pa," try various combinations of lip and tongue movements while vocalizing, compare the sound you produce with the heard sound, shift the lip and tongue to reduce the difference, passing through successive iterations until the error is zero. This type of learning may be required for initially setting up the imitation system, but once set up, new sounds can be imitated without error-based training.[c] For example, if I say "pafacagula" is the Eskimo word for horses, you can reproduce it instantly, as can an Eskimo child if his father points to a horse and says the name.

In addition, there is the phenomenon we have dubbed "synkinesia" first observed by Darwin. He noticed that when we merely watch someone use a pair of scissors, we unconsciously clench and unclench our jaws! This provides a crucial link we are looking for; it is entirely possible that a preexisting cross-domain resonance exists between the map in motor and premotor cortex that represents twitching muscles of the tongue, lips, vocal cords, etc... (for articulation and phonation)—and the map just below it that represents the hands. This would have been an important prelude for translating every gesture into words, especially if accompanied by air blown through the larynx, the whole process culminating in spoken words. The first early attempts would be an impoverished version, but it would have set the stage for emergence of speech.[d]

Consider a concrete example. The gesture for "COME" which involves making a curve of your fingers and flexion toward your own body with an upward facing palm may be a ritualized version of the movement required for actually pulling a person toward you and the converse holds for "GO." In Step 2, through synkinesia, the "come hither" gesture is mimed by the tongue's motion on the palate and the pushing away action becomes ritualized into the palm thrusting out which in turn is mimed by the outward protrusion or pouting of lips. Yet another step would have been a transition to pronouns (e.g., "YOU," "THOU," "NEE" in Tamil, or "THU" in Thai). Another example is using the thumb and index finger approaching each other to mime a small object; the finger movements being echoed by a small mouth orifice for producing words like "teeny weeny," "tiny," "un peu," "diminutive," "chinna." In all cases, the tongue tip makes a small contact behind the incisors. Compare this with "large," "enormous," or "grande," where the mouth opens more and the tongue contact is no longer small; and you move your hands and arms apart.

It is also noteworthy that sometimes a gesture itself can be metaphorical,[e] as when you make a precision grip with thumb and index finger when indicating a precise point during

[c] Similar interactions have been shown for vocalization and lip movements in babies. Andy Meltzoff has shown that newborn infants can mimic their mother sticking out her tongue. While this could be a reflex, more sophisticated facial mirroring would almost certainly require hardwired MNS-like computations as you cannot use error feedback; you cannot see your own face. We have previously suggested that one might be able to use the Meltzoff test for the early screening for ASD in high-risk babies.[1]

[d] There is evidence from clinical neurology of a phylogenetic/anatomical link between hand and mouth and the proximity of these in the Penfield map may not be coincidental. What is not clear, however, is whether this is merely an exaggerated manifestation of a central coherence mechanism for all motor output (e.g., you cannot move your leg or head clockwise, while moving your hand counterclockwise).[107]

[e] We have previously shown that suppression of the mu wave in your EEG occurs not only when you make a volitional movement but also when watching someone make a movement. Surprisingly, mu wave modulation occurs even when listening to the word "grasp" whether spoken literally, e.g., "the balloon was too big to grasp" or, metaphorically, "the equation was too complex to grasp."[108,109] The anatomical basis of metaphors has also been elegantly explored by Krish Sathian and coworkers[110,111] using an ingenious experimental approach.

an argument or clench your fist to indicate determination. A similar argument could be made for the wrinkling of facial muscles above the nasal bridge—including eyebrows—which you do when reading tiny print (to change corneal curvature) which evolved to cope with OPTICAL lack of clarity, but then evolved in into a signal for intellectual lack of clarity—puzzlement.

One final point: when listening to someone's spoken words, e.g., "teeny weeny," you would unconsciously and silently tend to echo it (echolalia) and—through synkinesia—this would activate neurons which generate a pincer-like gesture. This in turn evolved from an actual precision grip between thumb and index finger. The whole cascade culminates in an embodied comprehension of the shared meaning of the word. This type of sensory/motor/intersubject resonance would also lead to shared emotions and empathy. It would help explain our finding[81] that it's hard to judge if someone else is smiling or not if you have a pencil gripped horizontally in your mouth. When one says the words, comprehension is beyond semantics, it is embodied.

Autism

ASD is a developmental disorder characterized by highly specific cognitive and motor symptoms.[82,83] The former includes a poverty of social interactions, diminished empathy, reduction in language, poor eye contact, absence of pretend play, and (sometimes) poor imitation. When we first heard of Rizzolatti's work over 25 years ago, we were struck by the fact that these are precisely the functions that the MNS would be mediating or at least contributing to.[65a]

We proposed that the cause of the major symptoms of ASD is a dysfunctional MNS. The hypothesis can be defended on grounds of theoretical plausibility; the fact that the deficits seen are exactly what you would expect from the known list of properties of the MNS. No other theory comes even close. But is there any empirical evidence for the theory? In some preliminary experiments, we obtained EEG recordings and saw strong hints of MNS dysfunction.[84,85] Subsequent EEG recordings and fMRI by more than a dozen groups have yielded mixed results with no more than half reporting MNS deficits. So, one could say the results are suggestive, but not compelling. It may turn out that in many ASD cases, the dysfunction is in the distal projection zones of the MNS rather than the MNS itself. Alternately, ASD may be a heterogeneous disorder with only some subtypes showing MNS abnormalities.

Taking the link between Bouba-Kiki and MNS a step further, we now propose that many aspects of ASD might result from impoverished cross-domain interactions. One result of such failure would be MNS dysfunction with ensuing problems with imitation, empathy, pretend play, etc… The theory also predicts a deficiency in abstraction in general, e.g., those with ASD should have difficulty with metaphors and the Bouba-Kiki task.[86,87] The idea seems plausible to us, but we must bear in mind that a theory that is too general and which explains too many phenomena provides as much grounds for skepticism as one that explains too little.

We tested this in 10 high-functioning ASD subjects whose IQs were above 80 and verbal IQs were close to 100. Remarkably, their performance on Bouba-Kiki was at chance level.[88]

Obviously, these experiments need to be replicated with a control comparison group, but it hints strongly at the possibility that a reduced cross-domain abstraction—and resulting "literal—mindedness"—might be the core deficit in ASD. If so, one could regard ASD as being, in a sense, the "mirror image" of synesthesia. However, the number of subjects was small and the experiment needs replication. Our speculations in this regard are reminiscent of the ingenious theory of autism proposed by Uta Frith.[89]

We have studied a synesthete (TK) who sees colored halos around people—with different facial expressions evoking different colors, which we measured using detection thresholds for actual colored spots presented inside or outside the halos.[90] He also had Asperger's syndrome and during his boyhood had great difficulty not only in recognizing peoples' expressions but also in introspecting on his own emotions. But starting from when he was 10, his parents taught him to use the evoked colors to help him label each emotion correctly, thereby paving the way to actually experiencing the emotion—albeit to a limited extent. His original synesthetic links may have been exclusively between the "facial expression area" in the STS[91] and color areas in V4, but that enabled him to adopt the strategy of mapping his emotions onto his hardwired color space to create a taxonomy of emotions. Significantly, even now, he sees the color first before the emotion is perceived or experienced (e.g., "I feel green today, so I must be happy" or "His halo is purple—he's arrogant," etc.). He even blended emotions in a predictable manner; red was anger, blue was pride, and purple was arrogant, suggesting a systematic—rather than random—mapping of color space on to emotions. Connections between STS, V4, and the insula may be involved in mediating these effects (with some nuances provided via frontal cortex). One wonders whether one could "educate" other ASD children by exploiting color labels as was done with TK (in a manner analogous to our use of color to label altitude or temperature in geological maps).

Number lines

"What is a number, that Man might know it. And a man, that he may know a number -*Warren S. McCullough.*

Another example of variant cross-domain mapping might occur in what are usually called number lines. Numbers can denote both quantity and sequence in space (e.g., ruler) and time (clock) in daily life. If you ask a nonsynesthete to visualize (say) the numbers 1 to 20, he will conjure up a visual image of the numbers laid out vaguely left to right—this is what is called a number line. If asked to report which of two numbers, e.g., seven or eight is bigger, as quickly as possible, he takes longer to reply correctly than if he is asked if three or eight is bigger—the "number distance" effect. The RTs vary linearly with the difference between the numbers, implying that instead of using an algorithm or lookup table, the brain uses an analog spatial representation of the numbers and consequently finds it confusing to deal with adjacent numbers.[92]

In 1880, Galton noticed that about 1% of the population had a highly convoluted number line[93] (Fig. 1.13) instead of the usual straight line (left to right). Sometimes, the line would

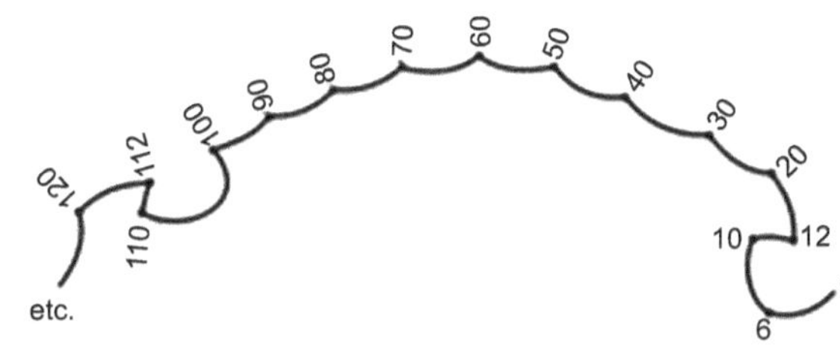

FIGURE 1.13 A graphic representation of a synesthete's number line.

actually double back on itself so, for example, 34 could be nearer in Cartesian space to 5 than to 36.[f]

During the last century or so, there was the usual skepticism about the very existence of the phenomenon. But, perhaps for the first time since Galton, we were able to establish the "reality" of convoluted number lines using the number–distance effect.[94,95] We obtained RTs (reaction times) for reporting which of two numbers is bigger. The question was would the RT depend on numerical distance along the convoluted line or on SPATIAL cartesian distance? The answer was a compromise between the two distances, implying at the very least that number lines are indeed authentic. These preliminary studies were followed up more meticulously by other groups, with similar results, although the dust has yet to settle completely.

What causes the convoluted number line effect? A number represents many things—11 apples, 11 minutes, the 11th day of Christmas—but what they have in common are the separate notions of order and quantity. These are very abstract qualities, and our apish brains surely were not under selective pressure to handle arithmetic per se. Studies of hunter-gatherer societies suggest that our prehistoric ancestors probably had names for a few small numbers—perhaps up to 10, the number of our fingers—but more advanced and flexible counting systems are cultural inventions of historical times; there simply would not have been enough time for the brain to evolve a "lookup table" or number module starting from scratch. On the other hand (no pun intended), the brain's representation of space is almost as ancient as that of our piscine Devonian ancestors from 300 million years ago. Given the opportunistic nature of evolution, it is possible that the most convenient way to represent abstract numerical ideas, including sequence, is to map them onto a preexisting map of visual space. Given that the parietal lobe originally evolved to represent space, is it a surprise that calculations are also computed there, especially in the AG? This is a prime example of what might have been a unique step in human evolution.

We would like to argue that further specialization might have occurred in our space-mapping parietal lobes. The left AG might be involved in representing sequence ordinality. The right AG might be specialized for quantity and magnitude. The simplest way to spatially map out a numerical sequence in the brain would be a straight line from left to right. This, in turn, might be mapped onto notions of quantity represented in the right hemisphere. But

[f] As with calendar lines, the number line is usually not head-centered. When subjects tilted their heads 90° (the ear resting on the shoulder), the number line continued to remain vertical with respect to gravity. Curiously, in those individuals where the number line in convoluted with portions tilted in 3D space, the graphemes also tilted in 3D space to conform to twists in their lines!.

now, let us assume that the gene that allows such remapping of sequence—and quantity—on visual space is mutated. The result might be a convoluted number line of the kind you see in number-space synesthetes.[6–8] Perhaps other types of sequences—such as months or weeks—are also mediated by the left AG. If this is correct, we should expect that a patient with a stroke in this area might have difficulty in quickly telling you whether, for example, October comes after or before August ("sequence agnosia").

Calendar lines

That brings us to what is called the calendar line.[1,6,96,97] We have observed that a small percentage of people, when asked to visualize the months of the year or days of the week in front of them, see them arranged as a circle in front of their body, usually in the horizontal plane, and in 3D space.[98] This suggests that, just as with numerical sequences, the abstract idea of time is also spatially represented in the brain.[g] Whether the months progress clockwise or anticlockwise seems to depend on handedness.[99]

We then did four simple clinching experiments that proved, for the first time since Galton observed the effect, that calendar lines are not a vague cognitive construct. In fact, they are perfectly "authentic" and probably activate neurons in the very earliest stages of sensory processing (Fig. 1.14).

1) We asked subjects to recite alternate months of the year, backwards—prompted with an arbitrary starting point by the experimenter (e.g., prompt: February; Answer: December, October, August, June etc....). Most of us do this algorithmically and (therefore, slowly) sometimes taking twice the time backward as forward. This difference in speed is reduced significantly in calendar synesthetes (about 50%) as was their overall reading time, presumably because they are just reading months' names off the calendar conjured up in their minds and "displayed" in front of them.
2) We had a normal person look at "noise" or static on a TV screen (dynamic twinkling black-and-white dots). We then drew a square or circle or triangle using a thin black felt-pen on a sheet of plexiglass and superposed the plexiglass on the screen. Most people see the twinkle near the thin outline as being made of coarser darker dots—especially inside the figure, and if you jiggle the plexiglass, the dots move along with it.[100] Surprisingly, if the subject projected his/her calendar on the screen, the same effect was seen (but not in the control condition—in which he/she merely imagined the outline of a square or a calendar different from his own). It's as if the mental calendar behaves like a real physical calendar—which contradicts "hard" AI approaches that emphasize symbolic descriptions.
3) Astonishingly, if a subject with a rectangular calendar projected his calendar onto MacKay's rays or bicycle spokes (Fig. 1.15A), the lines defining the boundaries of the calendar were seen bulging outward, in the same way that real lines do (Fig. 1.15B depicts

[g] Intriguingly, although we have seen subjects with calendar and number lines, we have never encountered an individual with an alphabet line. This may be because such phenomena can only arise when there is some abstract dimension like quantity, ordinality, or cardinality, being mapped onto space and the mapping process goes awry. On the other hand, alphabets have no logical sequence and are arbitrary (the relative position of letters have no significance and convey no information).

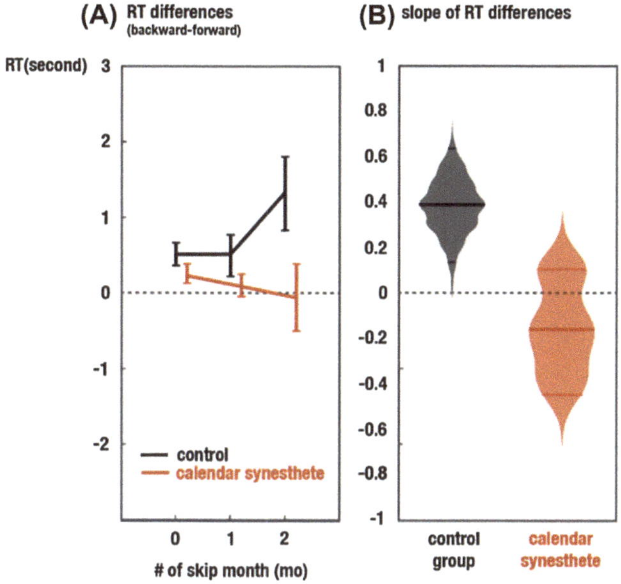

FIGURE 1.14 Visualized comparison of synesthete and control participants' performance on the backward recitation task.

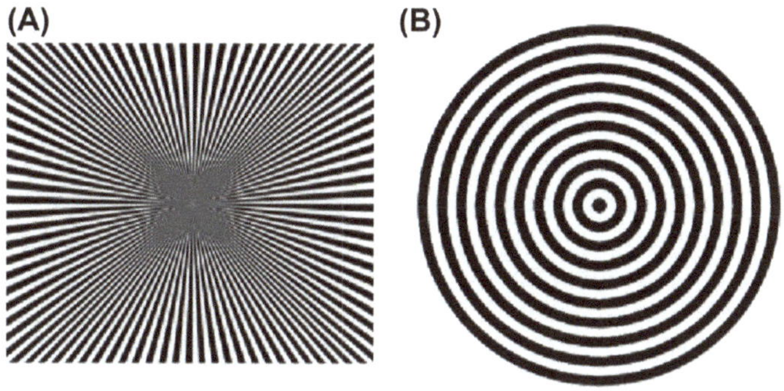

FIGURE 1.15 (A) Depicts pattern of radiating black lines (MacKay's rays) (B) Concentric circles.

concentric circles of the kind used in OpArt). Conversely, if it was superposed on concentric circles, the rectangle looked distorted into an hour-glass shape. In other words, the lines defining the calendar were behaving like actual physical lines exciting the retina.

We hasten to add that, as is the case in grapheme-color synesthesia, there appear to be two groups—projectors and associators (although the distribution may not be bimodal). Our experiments were performed, for the most part, on the projectors. Despite the variability, the evidence suggests that the mental calendar, instead of being represented in abstract propositional form, is projected back perhaps as far down as area 17 almost in pictorial form. This might seem like a silly thing to do because there is no homunculus watching, but in fact, it makes sense if you consider the following analogy.

Consider the following story: a monk gets up at dawn at the bottom of a hill, walks uphill along a tortuous, irregular path with pauses, varying speeds, fits and starts, reaching the top at exactly dusk. He sleeps overnight, gets up at dawn, and reaches the bottom once again at dusk, with a new sequence of randomly varying speeds. Can you prove that even though the variations of speed are uncorrelated in the two directions, there is an exact point in space and time where he passes in both journeys? People often struggle with this problem mathematically, to no avail, but you can explain it to a smart 10-year-old using graphs. If you simply plot the uphill and downhill trajectories of the monk, the two lines have to intersect at one particular point no matter how irregular the graphs are. The riddle and its solution illustrate the efficacy of pictorial representations in solving abstract problems. The brain may take advantage of this principle in dealing with concepts like time, including calendars.

For an even simpler solution, imagine another monk was coming up at the same time in the opposite direction; it is instantly obvious that our monk will cross the other monk's path at some point, and this would be equally true if our monk walked back along the same path he had walked the previous day! Now is this a visual solution (whether or not graphical) or a propositional one? To answer this, we need to understand what thinking is, and we do not!

As noted above, when SS projects her calendar on a ray pattern, the tips of the U diverge. And when projected atop concentric circles, they converge. On the other hand, when the rays and circles alternate, the tips of the calendar appear to actually *move* inward and outward—suggesting, again, that the mental calendar must be processed early, certainly earlier than the motion-processing area of visual cortex (MT).

We also asked the subject to close her left eye and adjust her distance from the monitor and gaze direction so that the blind spot would coincide with a portion of her calendar. We wondered: would it get chopped down the middle? Would the corresponding month (say, March), when falling on the blind spot disappear? Or would lines representing the month fill in the blind region? Sure enough, it fills in, but only partially. Its color and luminance differed from the rest of the calendar.

More recently, we investigated a number-line synesthete, EA, who had a hula hoop like calendar line wrapped around her body, with her own viewpoint fixed on 31 January, midnight, no matter what the current month is. This spatial calendar, she claimed, helped her organize her memories and "place" events in her life to provide coherence. When we had her tilt her head 90° coronally, so that her right ear touched her right shoulder, her calendar remained horizontal, implying vestibular correction. When she rotated her head sideways, 90°, the calendar did not move with her head, it remained facing forward stuck to her chest. This in itself is surprising but even more extraordinarily, she said when she turned her head right she could no longer "see" the left portion of her calendar and memory for salient events became fuzzy, whereas when she turned left those same memories became vivid again. This is a striking example of embodied cognition; of access to memories being gated by signals—both afferent and referent—from neck muscles (and vestibular input, see below). It remains to be seen whether the same effect might occur in more rudimentary form, even in nonsynesthetes.

Neural basis of mental calendars

It is unlikely the calendar "resides" in any one brain region, but we suggest that the AG plays an important role. We could not help but notice that it has connections via the inferior longitudinal fasciculus with regions rich in place cells[101] (i.e., the hippocampus) and grid cells[102] (i.e., entorhinal cortex). If so, patients with left AG lesions and amnesiacs with hippocampal damage would be expected to have some degree of "calendar agnosia." We recently had the opportunity to examine 20 school-aged children with dyslexia, which revealed that a substantial number of them had this syndrome. This is consistent with the observation that dyslexia involves disruption of cortical layers in the AG. Even the observation reported above, of access to memories fading with head direction, can be explained in terms of head direction cells—whose activity is strikingly modulated by vestibular input and input from neck muscles. This allows the subject to "get their bearings," before navigating the environment utilizing their calendar (we have, in fact, shown that position of the calendar can be modulated by stimulation of the vestibular nerve using galvanic vestibular stimulation).

We should add that, although these findings are intriguing, we have barely scratched the surface of how the brain actually constructs calendars. Consider our subject RR, who experienced a vivid, ribbon-shaped calendar line in the form of a U, floating in front of her in the coronal plane, the lower convexity of the U a foot away from her manubriosternal junction. She could see the lettering of names of the months (including size and fonts). We then asked her to adopt a different, imaginary, vantage point of her calendar (e.g., as if to inspect the calendar from behind). She said she would "try," then added "yes, I can see all the months but—my god—why are they all mirror-reversed," but then did a double take and stated "OF course! this must be true" (we first reported this mirror-reversal effect in 2006).[103] This is puzzling enough, but then she noted that the upper half of the left side of the U was slightly twisted in such a manner that from her particular angle of inspection (standing behind the calendar) she saw the words on that segment alone as nonreversed. One would expect this from simple laws of optics—but why this internal "coded" representation would be constrained by such laws is wholly bewildering. If the brain creates abstract representations, why the ribbon? And why the twist? And why the reversal of letters? As Haldane said "the brain is not only queerer than we think—it's queerer than we can think." Seeing a twisted calendar with appropriately realigned or reversed letters is surprising because it is a logical equivalent of changing the soldering in an Apple laptop and seeing a meaningful distortion of an image (say seeing an orange rather than an Apple) on the screen as oppose to uninterpretable garbage. Of course, this assumes the distinction between hardware and software in our descriptions of the brain is a valid one. **This leads to the question: Could she spell those words backwards more easily you or I? (e.g., "R E B O T C O" when asked to spell October backwards) when she is behind the reversed limb.**

Conclusion

We conclude by acknowledging that many ideas in this essay were foreshadowed by the great Victorian biologist Alfred Russell Wallace[104] who, independently of Darwin, put forth the idea that natural selection was the driving force of evolution. Wallace pointed out that words depicting continuous motion—run, fly, swim, move, flow—often have continuous first consonants—ffffffffff iiiiiiiiiii mmmmmmmmmm nnnnnnnnn—and conversely, stopped consonants like b, d, g, k, p, and t characterize abrupt words like kick, stab, bat, pat, put etc… Had he only pursued his line of reasoning a few months more, he might have saved us the trouble of writing this article.

Better understanding intersensory interactions is also important for addressing the so-called binding problem, which is relevant to studying consciousness. We have various brain modules extracting information from the external world and computing it at different speeds, resulting in a constantly fluctuating mosaic of activity, and yet we feel like a single person with unitary experiences. This coherencing of initially asynchronous inputs is obviously relevant to the unity of conscious experience.

We have used our poetic license (endorsed by editorial privilege) as an opportunity to express some highly speculative ideas that may, perhaps, have important implications. Furthermore, our goal has been to provide an impressionistic survey of our ideas on cross-sensory interactions rather than a comprehensive review, of the kind elegantly dealt with by other authors in this book. We suspect that many of our speculations will be disproved in the near future, but even if that were the case, they would hopefully have served as a catalyst to new inquiry.

As Darwin said:

> False facts are highly injurious to the progress of science, for they often long endure; but false views, if supported by some evidence, do little harm, as everyone takes a salutary pleasure in proving their falseness; and when this is done, one path towards error is closed and the road to truth is often at the same time opened.

We thank former members of our team at UCSD who spearheaded much of the research reported here—Edward Hubbard, Lindsay Oberman, David Brang, Paul McGeoch, and Eric Altschuler. We also thank colleagues and students for many stimulating discussions—Nick Root, Laura Case, and Herb Lurie (Herb had independently conceived of the MNS theory of ASD); Gerald Arcilla, Lance Stone, Jamie Pineda, Seanna Coulson, Krish Sathian, Roland Lee, Geoff Boynton, Ming Xuang, John Smythies, and Jack Pettigrew.

References

1. Ramachandran VS. *The Tell-Tale Brain: A Neuroscientist's Quest for What Makes Us Human*. New York: W.W. Norton; 2011.
2. Goldstein K. *The Organism: A Holistic Approach to Biology Derived from Pathological Data in Man*. New York: American Book Company; 1939.
3. Baron-Cohen S, Harrison J, Goldstein LH, Wyke M. Coloured speech perception: is synaesthesia what happens when modularity breaks down? *Perception*. 1993;22(4):419–426.
4. Domino G. Synaesthesia and creativity in fine arts students: an empirical look. *Creativ Res J*. 1989;2(1–2):17–29.

5. Mulvenna C, Hubbard EM, Ramachandran VS, Pollick F. The relationship between synaesthesia and creativity. In: *Eleventh Annual Meeting of the Cognitive Neuroscience Society. San Francisco, CA*. 2004.
6. Ramachandran VS, Hubbard EM. Psychophysical investigations into the neural basis of synaesthesia. *Proc R Soc Lond*. 2001;268:979–983.
7. Ramachandran VS, Hubbard EM. Synaesthesia – a window into perception, thought and language. *J Conscious Stud*. 2001;8(12):3–34.
8. Ramachandran VS, Hubbard EM. The phenomenology of synaesthesia. *J Conscious Stud*. 2003;10(8):49–57.
9. Ramachandran VS, Brang D. *Scholarpedia*. 2008;3(6):3981.
10. Palmeri TJ, Blake R, Marois R, Flanery MA, Whetsell W. The perceptual reality of synesthetic colors. *Proc Natl Acad Sci USA*. 2002;99:4127–4131.
11. Ward J, Jonas C, Dienes Z, Seth A. Graphene-colour synaesthesia improves detection of embedded shapes, but without pre-attentive 'pop-out' of synaesthetic colour. *Proc Biol Sci*. 2010;277:1021–1026.
12. Treisman A. Perceptual grouping and attention in visual search for features and for objects. *J Exp Psychol Hum Percept Perform*. 1982;8(2):194–214.
13. Julesz B. *Foundations of Cyclopean Perception*. IL: University of Chicago Press; 1971.
14. Beck J. Effect of orientation and of shape similarity on perceptual grouping. *Percept Psychophys*. 1966;1:300–302.
15. Hubbard EM, Manohar S, Ramachandran VS. Contrast affects the strength of synesthetic colors. *Cortex*. 2006;42(2).
16. Brang D, Ramachandran VS. Visual field heterogeneity, laterality, and eidetic imagery in synesthesia. *Neurocase*. 2010;16(2):169–174. First published on: November 6, 2009.
17. Ramachandran VS, Seckel E. Graphemes evoke synesthetic colors even before they are consciously recognized. *Perception*. 01/2011;40(4):490–492. https://doi.org/10.1068/p6861.
18. Hubbard EM, Arman AC, Ramachandran VS, Boyton GM. Individual differences among graphene-colour synaesthetes: brain behavior correlations. *Neuron*. 2005;45:975–985. https://doi.org/10.1016/j.neuron.2005.02.008.
19. Brang D, Hubbard EM, Coulson S, Huang MX, Song T, Ramachandran VS. Magnetoencepalography reveals early activation of V4 in graphene color synesthesia. *Neuroimage*. 2010;53(1):268–274.
20. Nunn JA, Gregory LJ, Brammer M, et al. Functional magnetic resonance imaging of synesthesia: activation of V4/V8 by spoken words. *Nat Neurosci*. 2002;5:371–375.
21. Rouw R, Scholte HS, Colizoli O. Brain areas involved in synaesthesia: a review. *J Neuropsychol*. 2011;5(2):214–242.
22. Mattingley JB, Rich AN, Yelland G, Bradshaw JL. Unconscious priming eliminates automatic binding of colour and alphanumeric form in synesthesia. *Nature*. 2001;410(6828):580–582.
23. MacLeod CM. Half a century of research on the Stroop effect: an integrative review. *Psychol Bull*. 1991;109(2):163–203.
24. Armel KC, Ramachandran VS. Acquired synesthesia in retinitis pigmentosa. *Neurocase*. 1999;5:293–296. https://doi.org/10.1080/13554799908411982.
25. Cytowic RE. *The Man Who Tasted Shapes*. 1st ed. Cambridge, Mass: MIT Press.; 1998.
26. Rothen N, Devin BT. Increased resting state network connectivity in synesthesia: evidence for a neural basis of synesthetic consistency. *J Neurosci*. 2012;32(40):13641–13643. https://doi.org/10.1523/JNEUROSCI.3577-12.2012.
27. Weiss PH, Fink GR. Graphene-colour synaesthetes show increased grey matter volumes of parietal and fusiform cortex. *Brain*. 2009;132:65–70. https://doi.org/10.1093/brain/awn304.
28. Brang D, Williams LE, Ramachandran VS. Graphene-color synesthetes show enhanced crossmodal processing between auditory and visual modalities. *Cortex*. 2012;48:630–637.
29. Lacey S, Martinez M, McCormick K, Sathian K. Synesthesia strengthens sound-symbolic cross-modal correspondences. *Eur J Neurosci*. 2016;44:2716–2721.
30. Cheadle SW, Zeki S. The role of parietal cortex in the formation of color and motion based concepts. *Front Hum Neurosci*. 2014;8:535. https://doi.org/10.3389/fnhum.2014.00535.
31. Tomson SN, Avidan N, Lee K, et al. The genetics of colored sequence synaesthesia: suggestive evidence of linkage to 16q and genetic heterogeneity for the condition. *Behav Brain Res*. 2011;223:48–52.
32. Simner J, Holenstein E. Ordinal linguistic personification as a variant of synesthesia. *J Cogn Neurosci*. 2007;19:694–703.

33. Amin M, Olu-Lafe O, Claessen LE, et al. Understanding graphene personification: a social synaesthesia? *J Neuropsychol*. 2011;5(2):255–282.
34. Simner J, Gartner O, Taylor MD. Cross-modal personality attributions in synaesthetes and non-synaesthetes. *J Neuropsychol*. 2011;5(2):283–301.
35. Köhler W. *Gestalt Psychology*. New York: Liveright; 1929.
36. Werner H, Kaplan B. *Symbol Formation: An Organismic-Developmental Approach to Language and the Expression of Thought*. New York, NY: John Wiley & Sons, Inc.; 1963.
37. Maurer D, Pathman T, Mondloch CJ. The shape of boubas: sound–shape correspondences in toddlers and adults. *Dev Sci*. 2006;9:316–322. https://doi.org/10.1111/j.1467-7687.2006.00495.x.
38. Lakoff G, Johnson MH. *Metaphors We Live by*. Chicago: University of Chicago Press; 1980.
39. Lakoff G, Turner M. *More Than Cool Reason: A Field Guide to Poetic Metaphor*. Chicago: University of Chicago Press; 1989.
40. Annaz D, Van Herwegen J, Thomas M, Fishman R, Karmiloff-Smith A, Rundblad G. Comprehension of metaphor and metonymy in children with Williams syndrome. *Int J Lang Commun Disord*. 2009;44(6):962–978.
41. Thibodeau PH, Boroditsky L. Metaphors we think with: the role of metaphor in reasoning. *PLoS One*. 2011;6(2):e16782. https://doi.org/10.1371/journal.pone.0016782.
42. Lakoff G, Núñez RE. *Where Mathematics Comes from. How the Embodied Mind Brings Mathematics into Being*. New York: Basic Books; 2000.
43. Coulson S, Matlock T. Metaphor and the space structuring model. *Metaphor Symbol*. 2001;16:295–316.
44. Marks LE. Magnitude estimation and sensory matching. *Percept Psychophys*. 1988;43:511–525.
45. Day S. Synaesthesia and synaesthetic metaphors. *Psyche*. 1996;2(32). http://psyche.cs.monash.edu.au/v2/psyche-2-32-day.html.
46. Armel KC, Ramachandran VS. Projecting sensations to external objects: evidence from skin conductance response. *Proc R Soc Lond B Biol Sci*. 2003;270:1499–1506.
47. McGeoch PD, Brang D, Song T, Lee RR, Huang M, Ramachandran VS. Xenomelia: a new right parietal lobe syndrome. *J Neurol Neurosurg Psychiatry*. 2011;82(12):1314–1319.
48. Brang D, McGeoch PD, Ramachandran VS. Apotemnophilia: a neurological disorder. *Neuroreport*. 2008;19(13):1305–1306.
49. Case L, Wilson RC, Ramachandran VS. Diminished size–weight illusion in anorexia nervosa: evidence for visuo-proprioceptive integration deficit. *Exp Brain Res*. 2012;217(1):79–87.
50. Geschwind N. Disconnexion syndromes in animals and man. *Brain*. 1965;88(237–294):585–644.
51. Schacter DL, Addis DR, Buckner RL. Remembering the past to imagine the future: the prospective brain. *Nat Rev Neurosci*. 2007;8:657–661.
52. Davenport RK, Rogers CM. Intermodal equivalence of stimuli in apes. *Science*. 1970;168:279–280.
53. Jalal B, Ramachandran VS. A pilot investigation of "metaphor blindness" in a college student population. *Med Hypotheses*. 2014;82(6):648–651.
54. Ramachandran VS. *A Brief Tour of Human Consciousness*. New York: Pi Press; 2004.
55. Kaufmann WE, Galaburda AM. Cerebrocortical microdysgenesis in neurologically normal subjects: a histopathologic study. *Neurology*. 1989;39:238–244.
56. Calvert GA, Spence C, Stein BE. *The Handbook of Multisensory Processes*. Cambridge, MA: The MIT Press; 2004.
57. Holmes NP, Spence C. The body schema and multisensory representation (s) of peripersonal space. *Cogn Process*. 2004;5(2):94–105.
58. Rock I, Victor J. Vision and touch: an experimentally created conflict between the two senses. *Science*. 1964;143(3606):594–596.
59. Zangaladze A, Epstein CM, Grafton ST, Sathian K. Involvement of visual cortex in tactile discrimination of orientation. *Nature*. 1999;401:587–590.
60. McGurk H, MacDonald J. Hearing lips and seeing voices. *Nature*. 1976;264(5588):746–748. https://doi.org/10.1038/264746a0. PMID 1012311.
61. Ramachandran VS, Blakeslee S. *Phantoms in the Brain: Probing the Mysteries of the Human Mind*. New York: William Morrow; 1998.
62. Humphrey N. *Consciousness Regained*. Oxford: Oxford University Press; 1982.
63. Haenny PR, Maunsell JH, Schiller PH. State-dependent activity in monkey visual cortex. II. Retinal and extraretinal factors in V4. *Exp Brain Res*. 1988;69:245–259.

64. Rozin P, Berman L, Royzman E. Biases in use of positive and negative words across twenty natural languages. *Cognit Emot*. 2010;24(3):536–548.
65. Morris D. *Bodytalk: The Meaning of Human Gestures*. New York: Crown Trade Paperbacks; 1995.
65a. Rizzolatti G, Craighero L. "The mirror-neuron system" (PDF). *Annual Review of Neuroscience*. 2004;27(1):169–192.
66. Keysers C, Gazzola V. Towards a unifying neural theory of social cognition. *Prog Brain Res*. 2006;156:379–401.
67. Ramachandran VS, Brang D. Sensations evoked in patients with amputation from watching an individual whose corresponding intact limb is being touched. *Arch Neurol*. 2009;66:1281–1284.
68. Blackmore SJ, Bristow D, Bird G, Frith C, Ward J. Somatosensory activations during the observation of touch and a case of vision-touch synaesthesia. *Brain*. 2005;128:1571–1583. https://doi.org/10.1093/brain/awh500.
69. Ramachandran VS, Sellers CA. Interpersonal gargalesthesia. *Neurocase*. 2018:1–7. PMID: 29693492.
70. Ramachandran VS, Hirstein W. The perception of phantom limbs. The D. O. Hebb lecture. *Brain*. 1998;121:1603–1630. https://doi.org/10.1093/brain/121.9.1603.
71. Jalal B, Ramachandran VS. Sleep paralysis and "the bedroom intruder": the role of the right superior parietal, phantom pain and body image projection. *Med Hypotheses*. 2014;83(6):755–757.
72. Altschuler EL, Ramachandran VS. A simple method to stand outside oneself. *Perception*. 2007;36(4):632–634.
73. Ehrsson HH. The experimental induction of out-of-body experiences. *Science*. 2007;317:1048. https://doi.org/10.1126/science.1142175.
74. Blanke O, Metzinger T. Full-body illusions and minimal phenomenal selfhood. *Trends Cognit Sci*. 2009;13:7–13.
75. Libet B, Gleason CA, Wright EW, Pearl DK. Time of conscious intention to act in relation to onset of cerebral activity (Readiness-Potential) – the unconscious initiation of a freely voluntary act. *Brain*. 1983;106:623–642.
76. Corballis MC. *From Hand to Mouth: The Origins of Language*. Princeton, NJ: Princeton University Press; 2002.
77. Corballis MC. The origins of modernity: was autonomous speech the critical factor? *Psychol Rev*. 2004;111:543–552.
78. Rizzolatti G, Arbib MA. Language within our grasp. *Trends Neurosci*. 1998;21:188–194.
79. Donald M. *Origins of the Modern Mind: Three Stages in the Evolution of Culture and Cognition*. Harvard; 1991.
80. Goldin-Meadows S. The role of gesture in communication and thinking. *Trends Cognit Sci*. 2001;3(11):419–429.
81. Oberman L, Winkielman P, Ramachandran VS. Face to face: blocking facial mimicry can selectively impair recognition of emotional expressions. *Soc Neurosci*. 2007;2:167–178.
82. Kanner L. Autistic disturbances of affective contact. *Nerv Child*. 1943;2:217–250.
83. Frith U, Happé F. Theory of mind and self consciousness: what is it like to be autistic? *Mind Lang*. 1999;14:1–22.
84. Altschuler EL, Vankov A, Hubbard EM, Roberts E, Ramachandran VS, Pineda JA. Mu wave blocking by observation of movement and its possible use as a tool to study theory of other minds. *Soc Neurosci Abstr*. 2000;26 67:23.
85. Oberman L, Hubbard EM, McCleery JP, Altschuler EL, Ramachandran VS, Pineda JA. EEG evidence for mirror neuron dysfunction in autism spectrum disorders. *Cogn Brain Res*. 2005;24:190–198.
86. Ramachandran VS, Oberman LM. Broken mirrors: a theory of autism. *Sci Am*. 2006;295(5):62–69.
87. Ramachandran VS, Hubbard EM. Hearing colors, tasting shapes. *Sci Am*. 2003;288:43–49.
88. Oberman L, Ramachandran VS. Preliminary evidence for deficits in multisensory integration in autism spectrum disorders: the mirror neuron hypothesis. *Soc Neurosci*. 2008;3:348–355.
89. Frith U. *Autism: Explaining the Enigma*. Oxford: Blackwell; 1989.
90. Ramachandran VS, Miller LE, Livingstone M, Brang D. Colored halos around faces and emotion-evoked colors. A new form of synesthesia. *Neurocase*. 2012;18(4):352–358.
91. Perrett DI, Smith PA, Potter DD, et al. Neurones responsive to faces in the temporal cortex: studies of functional organization, sensitivity to identity and relation to perception. *Hum Neurobiol*. 1984;3:197–208.
92. Dehaene S, Dupoux E, Mehler J. Is numerical comparison digital? Analogical and symbolic effects in two-digit number comparison. *J Exp Psychol Hum Percept Perform*. 1990;16(3):626–641.
93. Galton F. Visualized numerals. *Nature*. 1880;2:252–256.
94. Azoulai S, Hubbard EM, Ramachandran VS. Does synesthesia contribute to mathematical savant skills. *J Cogn Neurosci*. 2005;69.
95. Hubbard EM, Escudero MR, Ramachandran VS. The numerical distance effect demonstrates the reality of synesthetic number lines. In: *Eleventh Annual Meeting of the Cognitive Neuroscience Society*. 2004. San Francisco, CA.

96. Sagiv N, Simner J, Collins J, Butterworth B, Ward J. What is the relationship between synaesthesia and visuo-spatial number forms? *Cognition*. 2006;101:114–128.
97. Smilek D, Callejas A, Dixon MJ, Merikle PM. Ovals of time: time-space associations in synaesthesia. *Conscious Cognit*. 2007;16:507–519.
98. Brang D, Teuscher U, Ramachandran VS, Coulson S. Temporal sequences, synesthetic mappings, and cultural biases: the geography of time. *Conscious Cognit*. 2010;19:311–320.
99. Brang D, Teuscher U, Miller L, Ramachandran VS, Coulson S. Handedness and calendar orientations in time-space synaesthesia. *J Neuropsychol*. 2011;5:323–332.
100. MacKay DM. *Nature*. 1957;180:849–850.
101. O'Keefe J, Nadel L. *The Hippocampus Is a Cognitive Map*. Oxford: Clarendon Press; 1978.
102. Hafting T, Fyhn M, Molden S, Moser M-B, Moser EI. Microstructure of a spatial map in the entorhinal cortex. *Nature*. 2005;436(7052):801–806. https://doi.org/10.1038/nature03721. PMID 15965463. Bibcode:2005Natur.436..801H.
103. Ramachandran VS. V.S Ramachandran, Computational neuroscience. In: van Hemmen JL, Sejnowski TJ, eds. *23 Problems in Systems Neuroscience*. New York, NY, US: Oxford University Press; 2006.
104. Wallace AR. On the law which has regulated the introduction of new species. *Ann Mag Nat Hist*. 1855;16:184–196.
105. Brang D, Rouw R, Ramachandran VS, Coulson S. Similarly shaped letters evoke similar colors in graphene–color synesthesia. *Neuropsychologia*. 2011;49:1355–1358. https://doi.org/10.1016/j.neuropsychologia.2011.01.002.
106. Hubbard EM, Azoulai S, Ambrosio P, Ramachandran VS. *Phoneme and grapheme based patterns in synesthetic colors. Twelfth Annual meeting of the Cognitive Neuroscience Society.*. NY: New York; 2005, April.
107. Altschuler EL, Ramachandran VS. Can you do this?: extremely difficult interbodypart coordination tasks and implications for central limitations on control of coordination. *Perception*. 2006;35:281–283.
108. Troyer M, Curley L, Miller LE, Saygin AP, Bergen B. Action verbs are processed differently in metaphorical and literal sentences depending on the semantic match of visual primes. *Front Hum Neurosci*. 2014;8:982.
109. Oberman L, Ramachandran VS. The simulating social mind: the role of the mirror neuron system and simulation in the social and communicative deficits of autism spectrum disorders. *Psychol Bull*. 2007;133:310–327.
110. Lacey S, Stilla R, Sathian K. Metaphorically feeling: comprehending textural metaphors activates somatosensory cortex. *Brain Lang*. 2012;120:416–421.
111. Lacey S, Stilla R, Deshpande G, et al. Engagement of the left extrastriate body area during body-part metaphor comprehension. *Brain Lang*. 2017;166:1–18.

Further reading

1. Ramachandran VS, Brang D. Tactile-emotion synesthesia. *Neurocase*. 2008;14(5):390–399.

CHAPTER

2

Philosophical insights

Matthew Fulkerson

Department of Philosophy, University of California, San Diego, CA, United States

Starting point: the problem of individuating the senses

Multisensory interactions are interactions between senses and sensory systems. Understanding the nature of these interactions would seem to require that we have some understanding of just what these systems are that are doing the interacting. The very terms used to describe these interactions ("crossmodal," "multimodal," "multisensory," etc.) suggest that there is a relatively clear grasp of the individual senses and that multisensory perception is nothing more than the combination or interaction between these senses. Unfortunately, things are a lot more complicated. What, after all, are the senses? It seems we need some answer to this question to properly investigate the nature of sensory interactions (which, after all, on initial reflection seem to be interactions *between* the senses). While it would be nice if we could set aside this question and just investigate multisensory interactions themselves, it turns out that sensory interactions put a lot of pressure on any plausible account of individual modalities. In this first section, we examine some of these theoretical issues.

In ordinary circumstances, it seems obvious that we have distinct senses, and what they are. We seem to have a set of distinct sensory capacities—sight, smell, hearing, touch, taste, and maybe several others—and these capacities seem unique and self-contained. There is an intuitive grasp of what it means to *see* something, and the various ways in which that differs from *hearing* something. For example, we see with our eyes, and we hear with our ears. In addition, visual experiences feel different from auditory ones. Seeing brings awareness of colors; hearing awareness of sounds. Finally, the senses can be negatively defined and understood by their absence. We can lose our sight or our hearing or smell while maintaining our other modalities (more or less) intact. For all of these reasons, we seem to have an intuitive understanding of ourselves as possessing a fixed number of sensory capacities (the traditional five senses, and some additional ones often including vestibular awareness and kinesthesis). This intuitive notion has also contributed in part to a longstanding visuocentrism in discussions of sensory experience. For example, in fostering the idea that vision might plausibly stand in for other senses, that investigations in one modality have implications for all of the other senses, etc.[1,2]

Multisensory Perception
https://doi.org/10.1016/B978-0-12-812492-5.00002-4

As soon as we start trying to formulate a more precise characterization of the senses, however, we run into problems.[a] Giving anything like a robust account of what counts as a sensory modality has proven to be a far more difficult project than it would seem from the longstanding assumptions about separate sensory modalities. For each proposed criterion for distinguishing the senses from one another, there are a host of worries and counterexamples. For instance, the most obvious attempt might be to distinguish the senses by appeal to their distinct organs. But then it can be difficult to define "the organ" for any sensory modality, because each of the traditional senses is actually a complex network of interconnected systems involving many different physiological components (Macpherson[5] covers these problems in detail). There is much more to vision than the eyes, and once we look at the details of these additional processing streams, we see that there are interconnections between each of the sensory modalities and many other systems. For instance, there are well-known reorganizations of these areas. Cheung et al. 2009[9] found that tactile information processing activates early visual areas in a visually impaired subject reading Braille. Are these activations in areas that are usually visual still *visual* or are they now part of touch because they are serving awareness through the hands? What about examples discussed by others showing that even in normal subjects there are activations in visual areas caused by tactile inputs (many examples of such interactions can be found in[10]; see chapter by Lacey and Sathian, this volume)?

A simple appeal to organs does not easily work to define the senses. The same issue arises if we try to appeal only to phenomenology. There are deep interactions between sensory modalities that make a difference in the phenomenology of the experience, yet ordinarily these are changes that we are not aware of. It can thus be difficult to tease apart what we think of as unisensory experiences from experiences that are the result of sensory[11] interaction.[b] While we can plausibly point to distinct organs (however "organ" is defined) in the case of vision or hearing, there is no single organ we can point to in other cases like in touch or flavor awareness.[12,13] We might like to point to the skin as the organ of touch, but the skin contains many receptors that do not seem like part of touch (nociceptors that code for pain, and receptors that code for itch, tingle, chemoreceptors, etc.), and many receptors crucial for touch are not found in the skin (stretch receptors in the muscles and joints, for instance). Flavor is a combination of activation of retronasal olfactory, taste, tactile (both pressure and thermal), and trigeminal chemosensory receptors.[12,14] The philosophical landscape is littered with attempts to define and differentiate the senses, and all face serious difficulties.

To fully understand the nature of multisensory interactions, it would seem that we need some answers here. We need to understand what the senses are before we start looking at how they interact. In addition, doing so has many other important advantages. First, especially for philosophers, perception is investigated as the key point of contact connecting our conscious experiences with the external world. It is thus one of our primary sources of knowledge, and so getting a sense of our perceptual capacities and how they are individuated and related to one another is crucially important. In addition, there is the issue of how to understand consciousness, the felt quality of our experiences. Much of our consciousness seems grounded in our perceptual awareness, and getting an account of how the senses

[a] There is a large literature on these problems. For a good start, see the discussions in.[3–8]

[b] The recent volume by Bennett and Hill[11] contains several excellent papers on this issue.

are related to each other and connected in experience is an important part of understanding the nature of conscious experience. And, finally, many philosophers have seen the problem of sensory individuation as a pressing concern in the general philosophy of psychology. This is part of a general concern about defining and categorizing mental states. There have been many discussions in recent years rethinking what counts as a mental kind, and how we individuate them. For instance, philosophers in recent years have been trying to define and offer empirically informed accounts of such central mental kinds as attention, memory, motivation, deliberation, theory of mind, and so on. Getting a clear sense of perceptual kinds seems like an important first step in getting clear on the nature of perception itself and all of the more general questions that involves (like those mentioned above).

While there were already problems with many attempts at providing necessary criteria for individuating the senses, these problems are magnified by pervasive sensory[15] interactions.[c] For instance, the senses interact even at very early stages of sensory processing, and there are pervasive interactions at all levels of processing.[16] These show up most clearly in multisensory illusions.[17] The significance of these multisensory interactions for philosophy and cognitive science was noted in especially nice detail by Casey O'Callaghan in a series of important papers.[2,18,19] O'Callaghan noted that because cases like the McGurk effect[20,21] and the motion-bounce illusion[22] involve one sensory modality making a difference to the content and character of another, there must be some shared content or common code between the senses at the earliest levels of sensory processing.[d] Advances in functional imaging techniques have also revealed that large swaths of sensory cortex, as well as many subcortical attentional and orienting areas, are multisensory.[16] The traditional division of brain areas into discrete areas for each sense is not correct, and recent studies have found large projections from typically tactual and auditory regions into visual regions and vice versa. Experiences which we once thought to be entirely unimodal in fact are influenced and altered by processing that occurs in the other sensory modalities.

Consider the sense of touch. One of the constituents of touch is cutaneous stimulations, or *tactile* sensations. These constituents themselves are not modular or the product of discrete modules. We cannot treat the class of pure cutaneous stimulations as purely unimodal because such experiences involve several functionally distinct sensory processes and are therefore at least weakly multimodal. There are a number of very different sensory transducers in the skin, and these include the classes of mechanoreceptors, thermoreceptors, and chemoreceptors.[23] These different classes of sensory transducers have widely different receptive field properties and can be selectively (thus, functionally) dissociated. For instance, one can lose the ability to sense cutaneous thermal stimulation, yet retain the ability to sense cutaneous pressure.[24] But these disparate classes of receptors do not perform their processing in isolation from the rest. There are deep and important interactions between thermal receptors, pressure receptors, and muscle receptors, for instance, leading to such effects as cold objects appearing heavier than warm ones. There are two functions being served in these cases

[c] I discuss some of these issues in my recent paper.[15]

[d] This point is connected to the idea of "common sensibles," the view that some features are experienced by more than one sense (number, shape, size, location). These common sensibles were contrasted with the "proper sensibles" that defined each modality (color, sound, smell, hot and cold, flavor). This notion was first expounded by Aristotle in his *De Anima*.

by the distinct classes of transducers, but there is little chance that these systems are wholly isolated or would count as Fodorian modules. It's not clear if we could ever reach a level where there were discrete modules which allow for a plausible definition of multimodality as being simply an experience generated by the processing of at least two modules: all sensory experiences would satisfy that constraint. But cutaneous perceptions have one feature that not just any weak multimodal experience has, they are typically *unified* in experience.[25] When we touch a cold surface, we experience the coldness and the texture and the solidity as aspects of one perceptual object. The object is a unified perception which is associated with a range of different sensory features.

The nature of these interactions makes it difficult to imagine that we could offer any robust account of the individual sensory modalities. Where does this leave us? As I see it, there are three main options. First, there is *sensory revisionism* (the senses are not all what we thought they were). Second, there is *sensory eliminativism* (there are, strictly speaking, no senses). And, finally, there is *sensory pluralism* (there are many distinct notions of the senses). Let us consider them in turn.

The orthodox view of the senses tracks pretty closely to Jerry Fodor's view of the senses as modular "input systems."[26] On this conception, the senses are informationally encapsulated, stimulus-driven, hardwired, fast, and domain-specific, turning out to match pretty closely with what we thought they were all along: the five traditional senses and a few others added to the mix. There have been many challenges to this view over the years, especially from the problem of top-down influences on perception from cognition, so-called *cognitive penetration*. Still, the dominant view for a long time was that the senses were distinct, informationally isolated systems that only start interacting at the level of their outputs. Multisensory interactions, for obvious reasons, undermine this model and force us to find alternative views.

Sensory revisionism is the view that the senses are not quite what we thought they were. Instead of unique channels of sensory information, the senses are something else. What could this be? Various options have been put forward here. J.J. Gibson,[27] for instance, thought of the senses as complex, integrated forms of experience that were constituted by complex motor interactions with the environment. On this view, and the many sensorimotor and ecological views inspired by it, the senses are action-oriented ways of interacting with the world, not separate channels of perceptual awareness. There is not, on this view, a clear separation between the senses, but instead different modes or activities associated with certain kinds of awareness. It is on such a view that someone like Alva Noë[28] can think of a tactual-visual substitution system as a legitimate form of *seeing*. These views, however, generally come with a host of other commitments, some of which may not be palatable to everyone (they tend, for example, to have very austere and restricted conceptions of sensory representation). Other options here might include giving a more robust functional criterion for sensory individuation, or a set of more complex criteria that must be met. At the moment, none of these alternatives seem to have caught on.

The second possibility is sensory elimination. The idea is that the category of the senses is a loose or informal way of talking, a pattern entrenched in our folk ways of doing psychology. The claim then is that advances in our understanding of perceptual systems have revealed that, strictly speaking, there are no such things as the senses. The common or folk understanding of separate sensory systems for each of the traditional modalities has turned out on this view to be mistaken. Instead of separate senses, for instance, we might have more entangled collections of sensory interactions all the way down, at every level of sensory processing. Such an eliminationist move has been common in theorizing about the mind, especially for the so-called "propositional attitudes" like belief and hope. According to eliminativists, talk of such states should be replaced by more accurate physiological or functional descriptions of the underlying cognitive states. When talking about the senses, the idea is that, strictly speaking, there is no single, separate, obvious entity that is picked out by the term "vision." Instead, we might have motion processing centers and vision-for-action pathways, texture and color processors, and various systems for binding such details and other systems for helping integrate these systems with other perceptual and nonperceptual systems. Instead of individual senses then, on such a view we would have much more complicated interacting systems and that is all. This view seems to be held by some.[29] The downside is that it gives up on the notion of the senses (makes it theoretically inert).

There is another option, and it happens to be my preferred view: when we talk about "senses" we mean different things. This is a result of a looseness in our ordinary language, and a failure to be precise when turning to more exact scientific inquiry. The idea is that our ordinary notion of the senses has several components or meanings, and when we switch to more precise scientific language, it turns out that there is no single thing we mean by the word "senses." In addition, there is nothing there, either. The difference between my view and eliminationist views is that on my view there are robust entities picked out by our sensory terminology. It's just that we sometimes have different things in mind when we talk about the senses. In context, these different things will end up referring to different features of the world. For instance, sometimes we mean to refer only to a particular phenomenology, independently of how it was processed. On this view, we might think of any experience that involves only awareness of colors and visible shapes and textures as "visual." In other contexts, we might be especially interested in informational interactions and so a visual experience that was generated in part by influences from another modality (the motion-bounce illusion, for instance) might count as "multisensory" or "audio-visual." As we will see, a view like this also works well for understanding the many different forms of multisensory interactions. The key is not in counterexamples for any one view, but in realizing that often people are just talking past each other and not really disagreeing with each other at all.

A taxonomy of sensory interactions

Philosophers have spent some time recently trying to work out a more precise language for thinking about and categorizing multisensory interactions. Fiona MacPherson,[30] for instance, has offered a detailed and rigorous account of crossmodal experiences. As she writes, "Distinguishing these different sorts of experience is a useful exercise, for then we can begin to consider which kind of experience is occurring in various cases when two or more sensory modalities are in operation or when multisensory integration takes place" (431). She starts by characterizing pure unimodal experiences. These are experiences that are unimodal according to four criteria. First, they have phenomenal character (the "what-it's-like" to have that experience; the way it feels) associated with only a single modality. Second, they have representational content associated with only a single modality. Third, they are caused by proximal stimuli. Finally, they are produced by the sensory system associated with that modality. In addition to these unimodal experiences, we can characterize experiences as unimodal when they satisfy any one of these criteria (and these are thus criterion experiences). For instance, we can have unimodal$_{\text{representational}}$ or unimodal$_{\text{phenomenal}}$ experiences. MacPherson describes two classes of crossmodal experiences: crossmodal$_{\text{across}}$ and crossmodal$_{\text{within}}$ experiences. The former refers to experiences that satisfy at least two of the unimodal criteria above with respect to two different modalities. An example she mentions is the experience of phosphenes caused by tactual pressure. The other category she defines like this: "Crossmodal$_{\text{within}}$ experiences are ones that are not unimodal$_{\text{criterion}}$ experiences with respect to at least one of the criteria for individuating the senses" (445). The resulting categories allow for a robust series of very specific descriptions of crossmodal experiences. For instance, there can be crossmodal$_{\text{within-sensory-system}}$ which are experiences caused by the interaction of more than one sensory system (for example, the McGurk Effect, which is a change in our auditory experience (a "within-sensory-system" experience) but that has been generated by inputs from vision ("crossmodal"). These categories are useful and capture something interesting about the complexity of different forms of sensory interaction. The only respect in which these categories fall short (if they do) is that they do not obviously track the forms of interaction that seem most clear in the empirical literature. For instance, consider cases of sensory facilitation or inverse effectiveness, especially those that are highly task-sensitive. It's not obvious that such individual cases fit naturally into these categories in ways that might be explanatorily useful. (Again, this is not to deny the usefulness of these categories in general.)

Another set of taxonomic criteria for multimodal experiences is provided by Robert Briscoe.[31] In particular, he describes perceptual experiences as potentially multisensory when they are sometimes simultaneously "co-conscious." This way of being multisensory contrasts with more robust ways in which our experience may not be specific to any single modality. For instance, when hearing a sound source there may be a more integrated form of awareness beyond the mere concatenation of two distinct experiences.

Casey O'Callaghan has been instrumental in addressing the issues raised by multisensory perception. In a recent paper,[19] he outlines six grades of multisensory awareness, showing how we can start from more basic interactions to more robust forms of interaction. He starts with minimally multisensory awareness, which involves awareness in different modalities in a subject at the same time. Next he considers coordinated multisensory awareness, which involves some connection or interaction between the distinct forms of sensory awareness. Then follows intermodal binding awareness, multisensory awareness of novel feature instances, multisensory awareness of novel feature types, and finally novel awareness in a modality. These grades of interaction suggest many of the different ways in which sensory systems can interact with one another, leading to ever closer forms of interaction and unity in experience.

In previous work, I too have tried to provide something like an initial taxonomy of multisensory interactions.[25,30] Let us start with what I will call the "intuitive account" of multisensory experience (which, as you will see, builds on the discussion above about sensory modalities). The intuitive account simply says that a perceptual experience that involves more than one sense is multisensory. This account is intuitive inasmuch as it relies on a commonsense understanding of what counts as a sense; that is, it essentially assumes that the senses are well-understood, separate, and that we have a good grasp of how to individuate them. Unfortunately, this intuitive view is inadequate as a general account of multimodal experience. For one, the project of individuating the individual senses is fraught with difficulties, as we have seen, and cannot be completely separated from our account of multimodal experience. It may turn out that there are many more or fewer senses than we thought. For instance, if it turns out that there are two separate visual streams (for which there is evidence), then what we now call "vision" might be better understood as multimodal. The question of which experiences are the multimodal ones thus cannot appeal to our current intuitive conception of the senses. We need an independent criterion for deciding cases, not just for senses but also for experiences. Second, we now know that the senses are far more deeply intertwined and interdependent than the intuitive account allows. The intuitive account leaves it up in the air how to individuate experiences as well as senses. The intuitive account thus threatens circularity or implausibility.

One might have thought that it would be easy to tell whether you were experiencing something with one sense or many. But such awareness is not possible; it is simply not possible to type multimodal experiences by the quality and character of the experience. Now, there are cases where we might notice, for instance, that there are two or more phenomenal elements to our overall experience and, on that basis, describe it as multimodal. Or there might be an influence on attention, such that we could shift our focus from one element or object to another. But in many other cases, we will have no obvious access to the interactions between the senses. In addition, appeal to introspection seems to appeal to the intuitive notion of individual senses. How we divide up the constituent elements might just fall into our folk notion of the senses. We could try to strengthen the claim by appeal to particular phenomenal properties. One might think, for instance, that multimodal experiences could be characterized by some difference in their phenomenal properties, that a unisensory experience just *feels* different from a multimodal one. After all, a visual experience feels different from an auditory one, and perhaps their interactions similarly have a unique feel. We could then appeal to this difference in felt quality to characterize a multimodal experience as one involving associations between unisensory-feeling experiences.

This approach quickly runs into difficulties. Setting aside worries about sensory substitution or nonqualitative perception, in standard cases, no phenomenal property reveals that an experience is multimodal. For instance, it is possible that some of our phenomenally unitary senses—maybe touch and flavor—are themselves multimodal in important respects. So too, a seemingly unisensory experience can be influenced by other perceptual experiences without the subject realizing it, leading to a range of crossmodal influences. For example, a visual experience can be influenced by processing in nonvisual areas,[22] and an auditory experience can be altered by visual input.[32] In these cases, the subject has no introspective access to the multimodal nature of these experiences.

There seems to be little chance for a phenomenal property that could differentiate unisensory from multimodal experiences. The senses are involved in a range of interactions, and they can influence each other at many levels of processing. While such interactions can sometimes lead to differences in the qualitative character of the experiences, this is not always the case. An experience may very well have a unisensory qualitative character but still be multimodal in nature in some other respect.

We can start to better understand multimodal interactions with a simplified[33] example.[e] Consider two perceptual experiences that are unrelated to each other: looking at a tomato while hearing a trumpet from the other side of the room. Intuitively, these seem to be two different sensory experiences. The two experiences have a kind of independence; if you stop looking at the tomato, the sound of the trumpet does not change, and vice versa. This is a cooccurrence of two otherwise unrelated experiences. They possess what I will call *apperceptive unity*—they occur in the same subject at the same time, and nothing more. In a minimal sense, there is a single overall experience that is multisensory—the subject is able to conjoin features of one with features of the other. Perhaps sometimes this is what we mean when we talk about our experience of the world being multimodal. At the same time, in the same subject, we are able to experience different things going on in our immediate environment. I can hear a conversation across the room while looking at the book in front of me, etc. There is a sense, however, in which this minimal notion of multimodal experience—which seems to motivate the intuitive account—fails to capture what is most important about sensory combinations.

Paradigm multimodal experiences—the kind studied in experimental settings—seem to involve more than cooccurrence. They almost all involve some kind of association or interaction. Suppose, for example, that the tomato was emitting a foul odor, easily picked up from across the room. In this case, there is a strong association between the visual and olfactory experiences. For one, the smell is originating from the tomato. If the tomato were moved to a new location, the smell would move there as well (likely because the smells are in or from places or things). If we change our perspective on the tomato, by moving to a new location relative to it, then both the olfactory and visual experiences change accordingly. For instance, if we move closer to the tomato (perhaps to see if it has gone bad), the awful smell would become stronger and more distinct. Unlike apperceptive cases, these two sensory experiences are strongly related to one another. Changes to one experience are closely aligned with changes in the other, and both are related to our exploratory actions. These connections provide robust, overlapping information about the world. We might, for instance, come to believe that the tomato was taken out of the trash if we once again detect its stench. Such convergence occurs frequently in perceptual experience, and such interactions provide a stronger form of multimodal interaction that I will call an *associative relation*. These relations involve something more than apperceptive unity; the smelly tomato involves temporal and spatial coherence, causal interactions (the tomato causes the smell), and influences on our perceptual knowledge and exploratory actions. Mere cooccurrence is one thing; the truly interesting forms of multimodal perception, at minimum, seem to involve a *convergence* or coordination of information between different sources.

Now, I should make clear what I mean by "interesting" here. Certainly, an interaction can be interesting in itself, simply because it involves a novel kind of interaction or influence, or it helps explain some other more robust kind of interaction. I do not deny that, in this sense, all of our sensory experiences can be interesting. What I mean is that among our sensory interactions, some involve strong influences and interactions that influence, enhance, and alter our experience of the world. In understanding *these* cases, some experiential interactions are more

[e] The following discussion closely follows some of my previous discussions of these distinctions, especially in the early parts of.[33]

interesting than others. While it is certainly useful to know how two otherwise unrelated experiences can be subjectively connected (certainly important for understanding many pathologies), the focus here is on those interaction that underlie our rich experience of the world around us.

Phenomenal typing (at the level of experience) does not get us very far and, like the intuitive view of the senses, fails to individuate any interesting notion of multimodal interaction. We can, however, motivate a very interesting conception of multimodal interaction by appeal to the functional characteristics of sensory experiences (or more accurately, of the underlying systems that generate them). We can characterize an experience as multimodal if it involves (that is, if it is caused or generated by) functionally distinct sensory systems. We can define two systems to be functionally distinct if the proper functioning of one system is independent of the other.[26] That is, if we can *double-dissociate* two systems, by showing that one can be maintained in the absence of the other, and vice versa, then the two systems can be shown to be functionally distinct. In cognitive science, there is a long tradition, highlighted by Fodor,[26] of treating the senses as functionally distinct modules.

The involvement of functionally distinct systems seems to be a good criterion of multimodal experience:

> **Functional Dissociation Criterion (FDC)**: An experience *E* is multimodal if it is produced by two or more functionally distinct sensory systems. Two sensory systems are functionally distinct if they can be double-dissociated from one another.

It seems that many claims about multimodal experience appeal to some form of FDC. The senses involve distinct systems with dedicated functions and outputs. Consider again our previous example, of seeing a tomato while hearing a trumpet blare from across the room. This experience involves separate functional systems. We can still see the tomato just fine, even if we plug our ears. If we instead close our eyes, we can still hear the trumpet. The overall experience seems to be a kind of conjunction, combining outputs or elements contributed by two independent systems. In general, this model seems to capture an important way of understanding multimodal experience. For many cases, the FDC is all that we mean in saying that an experience is multimodal. Unfortunately, the FDC fails to capture some important distinctions and cases.

First, the FDC makes no distinction between apperceptive unities and experiences that have some coordination or influence on each other. This means that the central distinction between multimodal and crossmodal experiences is very difficult to capture through functional properties alone (both equally involve functionally distinct systems, but the nature of the interactions is very different). This requires that some additional constraints be placed on the criterion, but it is not clear what those constraints ought to be. The big worry is that unisensory experiences are not immune from multimodal proliferation: even within a single modality there are functionally distinct, doubly dissociable processing streams. Visual motion and color, for instance, are largely dissociable: one can lose the ability to experience motion but retain color experience, and one can lose color experience but retain the experience of motion. Similar dissociations can be demonstrated in all of the perceptual modalities, across a wide range of features. Unless strong additional constraints are added to the FDC, then very nearly every sensory experience, even within a single modality, would be classified as multimodal.

Let us return to the earlier example of the tomato and its smell. Instead of looking at low-level sensory processing, consider instead the representational contents (or information) involved in the experience. When we both see and smell the tomato, we seem to be getting very different kinds of information about the world. From our visual system, we learn of colors, shapes, sizes, and locations. From our olfactory system, we learn of the many characteristics of the tomato odor. The two sensory systems seem to be picking up distinct information about the world and carrying that information along distinct sensory channels. Sometimes, it seems, we consider multimodal experiences to involve the operation of distinct informational channels.

Dretske[34] argued that the sensory systems could be individuated as distinct channels of information. His suggestion offers a means of characterizing multimodal perception that can be simply stated:

Shared Content Criterion (SCC): A perceptual experience E is multimodal if it has content c_1 (or information i_1) from source m via channel x, and content c_2 (or information i_2) from source n via channel y, where x and y are distinct channels.

As an example, consider the McGurk effect.[20,21] This effect occurs when we perceive a phoneme that has been altered by being associated with a mismatched lip movement. For instance, if the sound /ba/ is produced along with the lip movements that make the sound /ga/, it usually results in an auditory experience of the sound /da/. The visual information about the source of the sound alters the auditory character of the sound. The visual experience thus dramatically influences the auditory perception. For this to occur, there must be some content shared between the two modalities. As Casey O'Callaghan[2] writes:

> [I]t requires recognizing both a component of experiential content and an aspect of perceptual phenomenology that are shared by distinct perceptual modalities. Perceptual experience thus cannot be understood exclusively in modality-specific terms (317).

One might suppose that, just as the SCC says, multimodal experiences are those that involve information carried from distinct sources.

As with the FDC, there are many times when our primary concern in understanding a perceptual experience is to distinguish the informational channels involved or to distinguish the representational characteristics involved. In other words, we might be most interested in the content and how it was generated. This might be the primary concern, for instance, if we come across a novel species and we want to figure out how many senses it has, and how they interact with one another. We might start with a gross functional decomposition (in the spirit of the FDC), but eventually we want to tease apart the various channels, figure out what they represent, and then understand how they influence or interact with one another.

One immediate concern is that if informational channels are defined functionally, then the SCC will just collapse into the FDC. Supposing that we can offer an independent construal of an information channel (perhaps by appeal to Dretskean or Fodorian considerations), it would still seem that the SCC fails to capture the structure of many interesting multimodal interactions. Again, it seems that too many perceptual experiences would be classed as multimodal. As we noted, the tomato and trumpet case would count as multimodal, as would the tomato and smell cases. In addition, once again supposedly unisensory experiences would be classed as multimodal. A typical visual experience involves separate processing channels, involving distinct bits of information for visual features like motion, shape, texture, and color.

The McGurk effect, like most paradigm crossmodal illusions, involves an associative relation between experiences: an auditory experience (hearing a phoneme) is directly related to a visual experience (seeing a lip movement). One of the constituent experiences—the visual one—seems to make a difference in the nature of the auditory experience. Call such a case one of *experiential dependence*: the precise content and character of one experience depends on its association with another. One question immediately raised by experiential dependence is how we should understand the nature of the dependence. Is it the *experiences* as such that alter one another or is it lower-level processing that leads to the influences? This is a difficult question, but it is highly probable that no clear answer will present itself, because both low-level connections between sensory processing areas along with aspects of the experience itself play a role in the associative relations. In the McGurk effect, the auditory experience of a /da/ experientially depends on the associated visual experience and/or its underlying processing; without it, we would have heard a /ba/. In other words, the visual experience changes the sound of the phoneme. More generally, we could test for experiential dependence by appeal to a weak counterfactual association grounded in the relations between the experiences: an experience E_1 experientially depends on an experience E_2, just in case, if E_2 were to be altered or removed, it would make a difference in the content and character of E_1. Such dependence is not found, for instance, with the smelly tomato. Closing one's eyes does not alter or influence the smell, and plugging our nose does not alter the visual appearance of the tomato. Association does not imply dependence.

Experiential dependence is simple enough in philosophical terms; there is a counterfactual dependence between two experiences (and/or their underlying processing). The dependence is in most instances only in one direction, but two experiences could influence each other. Even overlooking the messy question of what, exactly, is doing the influencing, just typing the relevant experiences presents a difficulty. Suppose we type experiences by their contents or characters. This is common enough, but it means we will have some difficulty working out which experiences are connected to which others. We might not recognize that an experience depends on another until we do some careful studies to determine, say, that auditory processing is playing a role in our visual experience (in certain cases). The actual relations, in other words, are going to be quite messy and the formulation above is something of an idealization. It assumes that we can easily tease the constituents apart, and determine which influence which, when in fact it will often be difficult to do. The main possibility is that instead of a crossmodal relation, two or more sets of processing lead to a single, blended experience. This seems to be the case in the generation of flavors, where processing in taste and smell areas lead to what certainly seems to be a single, robust experience (albeit with distinct aspects). It is no easy task to determine whether there is a relation of experiential dependence between two processing streams or whether these two streams in some sense combine and blend their processing, leading to a certain kind of unified experience. Let us consider in a little more detail the possibility of these kinds of multisensory blends.

Multisensory blends

Sometimes an experience will influence another, making a difference to its content and character. Experiences can enter into such relations in a variety of ways, from being temporally or spatially aligned to having appropriate contents to having very low-level, seemingly hardwired interactions. These relations are characterized by a broad counterfactual relation: if one experience was not present, then the other would have been different. But sometimes, there is processing between multiple systems that leads to what seems to be a single experience. If one of the constituent processing streams is removed, the resulting experience is completely altered. For instance, suppose that while eating you were to suddenly lose all sense of smell. How would we describe the resulting effect? It does not seem correct to say that the *taste* of the food was altered by the smell. This is because the smell does not merely alter a standard taste experience but leads to an entirely novel experience that can only be produced by the joint contribution of both smell and taste.[12] People who lose their sense of smell do not merely have different taste experiences, they have completely impoverished and bland flavor experiences. The existence of the novelty, and the unique relationship between the associated processing streams, means that these sorts of multisensory influences are quite different in kind from the sorts of crossmodal illusions we considered above.

The difficult element in considering these sorts of blends is that we need a way, distinct from our way of typing experiences generally, to say when and how multiple processing streams lead to what should be considered a single overall experience. Not only this, but once we admit multisensory blends, we face an even more troubling difficulty: how to distinguish the individual sensory modalities, which also involve a blending of multiple lower-level sensory systems, from multisensory blends. That is, flavor intuitively seems like a multimodal experience resulting from the blending of the sense of smell and the sense of taste. But if taste and smell themselves are a kind of blend resulting from the combination of distinct lower-level processing, then it would seem that all of our sensory experiences are, in this sense, multimodal. This is a real possibility, of course, but it would be nice if we could keep the various multimodal interactions distinct from unisensory experiences. Thus, the main challenge facing a typology of multimodal perception is figuring out how to determine the constituent experiences involved in associative relations and cases of experiential dependence and keep them distinct from multisensory blends and from unisensory experiences. In the next several sections, I will say something about how we might individuate these experiential constituents.

Binding and unisensory perception

The individual senses seem, at first, to be stable, coherent forms of experience. The intuitive view of multimodal experience suggests that these stable forms of experience enter into relations with one another, and that these relations fully characterize multimodal interactions. But there are many problems with such a view. For one, the notion of *experiences* interacting is not a very clear one. What does it mean for an experience to blend or influence another? Second, given the nature of multimodal interactions (along the lines of FDC and SCC), the individual senses themselves seem to be characterized as multimodal forms of experience. Touch, for example, involves multiple sensory channels, with functionally distinct parts, and distinct stimuli. If we adopt a more multimodal perspective on perceptual experience, how do we preserve something interesting about unisensory experience?

My thought is that the importance of the individual senses rests in the *structure* of their representations—on relations of sensory individuation, binding, grouping.[8] Let me briefly explain how such a characterization might work. At some level of sensory processing, it seems that the outputs of certain subsystems are "packaged up" into relatively coherent and stable representations, and that these packaged representations are what eventually become associated with one another. One clear case of such packaging is in the strong association of perceptual features in experience, sometimes called "feature binding." The individual senses all seem to involve a strong form of such binding. When we look at the world, the objects of our visual experience are seen to possess a range of distinct features. These assignments of features to objects seem to be much like predication, we see that *o is F*, where F specifies a range of visual features. Multimodal experiences, while they may involve coordinating many sensory features, do not seem to package or bind these features together in quite the same way.

If this turns out to be right, then it is the structure of perceptual contents (specifically their representational structure) that provides a means of distinguishing unisensory from multimodal experiences. Such a feature cuts across many of the individual elements we have already discussed, for feature binding involves precise interactions between low-level subsystems, alignment of contents (it is individual objects that seem to be the bearers of sensory features), and a kind of unity of our actual experience. We can measure and test for sensory binding, and we can use these bound representations in higher-level operations such as object perception, grouping, and recognition. While there are a range of distinct processes that can be described as multimodal, there are also a set of distinct forms of interaction that capture what is most interesting and important about our unisensory perceptual experiences.

Are all experiences multimodal?

We seem justified in thinking that all perceptual experiences are in some sense multimodal, in that even so-called unisensory experiences involve, albeit at lower levels of processing, the very same kind of interaction and blending of information from different processing areas that occurs in paradigm multimodal cases. In addition, we know that even the individual modalities function along with the other senses; they are actively seeking and coordinating their activities with the other modalities. Despite this, there is some utility in thinking of the individual modalities as a kind of limit or maximally coherent set of sensory experiences, one that for purposes of investigation or explanation can be isolated from the activities of the other modalities. On my view, the level of feature binding most closely aligns with the level that we usually associate with the traditional sensory modalities. At this level, the modalities are highly bundled, highly coherent units of experience. Dissociations are rare and the contents of such experiences are very tightly constrained. The real lesson, however, is that even at this level these systems are not wholly separate from the other senses and not themselves complexes built up from very tangled constituents. The philosophical and psychological data go a long way toward explaining exactly how these combinations happen.

Some of the evidence includes the similarity in the structure of the combinations between intra- and intermodal experience and the inherent (increased) modularity of the perceptual processing. I think that this view shows that all of our experiences are multisensory or rather that the idea of a unisensory experience is in fact either an abstraction or a shorthand for a cluster of maximally integrated experiences. The point is simply that such clusters are malleable. The senses are fluid and there are no set boundaries between them, only sets of experiences which are more typically integrated with one another than with sets of others.

My claim here then is that, taken in totality, all of our perceptual experiences seem to be multimodal, in that there is no such thing as an essentially unitary sensory experience, one completely isolated from other sensory systems (and other systems which might contribute to them). The claim is that our multisensory experiences, like our unisensory ones, are unified and bound, but are also the result of slightly more complex combinations of different sensory systems. The crucial point is that such combinations are not a wholly separate mechanism than those which operate over binding within the individual senses. The important and critical question to investigate then is not whether or not some experience is multimodal, but in what *sense* it is multimodal. What really matters is not the distinction between unisensory and multisensory experiences (an issue that has concerned many philosophers, for too long I think). Instead, what is important concerns the variety and nature of the many forms of sensory interaction that together make up our experience of the world.

This chapter has only touched on the most general and most abstract issues concerning considerations of multisensory experience. The rest of this volume will of course concern many more specific issues and concerns raised by multisensory awareness, and these too will I am sure prove to be of deep theoretical and philosophical interest.

References

1. O'Callaghan C. *Sounds*. OUP Oxford; 2010.
2. O'Callaghan C. Seeing what you hear: crossmodal illusions and perception. *Phil Issues*. 2008;18(1):316–338.
3. Stokes D, Matthen M, Biggs S. *Perception and its Modalities*. Oxford University Press; 2014.
4. Macpherson F. *The Senses*. OUP; 2011.
5. Macpherson F. Taxonomising the senses. *Philos Stud*. 2010;153(1):123–142.
6. Keeley B. Making sense of the senses: individuating modalities in humans and other animals. *J Philos*. 2002;99(1):5–28.
7. Grice HP. *Some Remarks About the Senses. Vision and Mind: Selected Readings in the Philosophy of Perception*. February 1962, 35-35.
8. Nudds M. The significance of the senses. *Proc Aristot Soc*. 2003;104(1):31–51.
9. Cheung S-H, Fang F, He S, Legge GE. Retinotopically specific reorganization of visual cortex for tactile pattern recognition. *Curr Biol*. 2009;19(7):1–6.
10. Calvert G, Thesen T. Multisensory integration: methodological approaches and emerging principles in the human brain. *J Physiol Paris*. 2004;98(1–3):191–205.
11. Bennett D, Hill C. *Sensory Integration and the Unity of Consciousness*. MIT Press; 2014.
12. Auvray M, Spence C. The multisensory perception of flavor. *Conscious Cognit*. 2008;17(3):1016–1031.
13. Lederman SJ, Klatzky RL. Haptic perception: a tutorial. *Atten Percept Psychophys*. 2009;71(7):1439–1459.
14. Smith B. Taste, philosophical perspectives. In: Pashler H, ed. *Encyclopedia of the Mind*. Thousand Oaks: Sage Publications, Inc.; 2009:731–734.
15. Fulkerson M. *Rethinking the Senses and Their Interactions: The Case for Sensory Pluralism*. December 2014:1–14.
16. Ghazanfar AA, Schroeder CE. Is neocortex essentially multisensory? *Trends Cognit Sci*. 2006;10(6):278–285.

17. Cinel C, Humphreys GW, Poli R. Crossmodal illusory conjunctions between vision and touch. *J Exp Psychol Hum Percept Perform*. 2002;28(5):1243–1266.
18. O'Callaghan C. Perception and multimodality. In: Margolis E, Samuels R, Stich S, eds. *Oxford Handbook of Philosophy of Cognitive Science*. Oxford University Press; 2012.
19. O'Callaghan C. Grades of multisensory awareness. *Mind Lang*. 2017;32(2):155–181.
20. McGurk H, MacDonald J. Hearing lips and seeing voices. *Nature*. 1976;264:746–748.
21. Munhall KG, Hove ten MW, Brammer M, ParE M. Audiovisual integration of speech in a bistable illusion. *Curr Biol*. 2009;19(9):1–5.
22. Sekuler R, Sekuler AB, Lau R. Sound alters visual motion perception. *Nature*. 1997;385(6614):308.
23. Lumpkin EA, Caterina MJ. Mechanisms of sensory transduction in the skin. *Nature*. 2007;445(7130):858–865.
24. Jones LA, Lederman SJ. *Human Hand Function*. New York: Oxford University Press; 2006.
25. Fulkerson M. The unity of haptic touch. *Phil Psychol*. 2011;24(4):493–516.
26. Fodor JA. *The Modularity of Mind: An Essay on Faculty Psychology*. MIT press; 1983.
27. Gibson JJ. *The Senses Considered as Perceptual Systems*. Westport, Connecticut: Greenwood Press Reprint; 1966.
28. Noë A. *Action in Perception*. Cambridge, MA: MIT; 2004.
29. Shimojo S, Shams L. Sensory modalities are not separate modalities: plasticity and interactions. *Curr Opin Neurobiol*. 2001;11(4):505–509.
30. Macpherson F. Crossmodal experiences. *Proc Aristot Soc*. 2011;111(3pt.3):429–468.
31. Briscoe RE. Multisensory processing and perceptual consciousness: part I. *Philos Compass*. 2016;11(2):121–133. https://doi.org/10.1111/phc3.12227.
32. Driver J, Spence C. Multisensory perception: beyond modularity and convergence. *Curr Biol*. 2000;10(20):731–735.
33. Fulkerson M. *The First Sense*. MIT Press; 2014.
34. Dretske FI. *Knowledge & the Flow of Information*. MA): MIT Press; 1983.

CHAPTER

3

Neural development of multisensory integration

Barry E. Stein, Benjamin A. Rowland

Department of Neurobiology and Anatomy, Wake Forest School of Medicine, Winston–Salem, NC, United States

Overview

All animals have multiple senses, each tuned to a very different form of environmental energy. Having an array of such sensors enhances the brain's ability to detect, locate, and identify external events in a variety of circumstances[1–6] (see also Ref. 7 for reviews). Each sense has a primary projection pathway in the central nervous system composed of hierarchically organized structures that are largely devoted to processing its specific information. A great deal of attention has been devoted to studying this common organizational feature of sensory pathways, an effort that has provided us with a rich body of information about their commonalities and idiosyncrasies, as well as the neural features that define their information processing capabilities.

Because of their different sensitivities, the multiple senses can substitute or compensate for one another when they are individually compromised; for example, when we use somatosensation to feel our way through a dark room that we might otherwise visually navigate. Moreover, environmental events of interest are often simultaneously registered by more than one sense. Multisensory neurons pool this information across sensory modalities to enhance operational effectiveness (i.e., detection, localization, or identification accuracy) beyond that achievable by one sense in isolation. This pooling occurs in all chordates, at multiple levels of the neuraxis, and within circuits that have very different functional roles; even within structures traditionally defined as "unisensory"[8–34]. Interest in multisensory processing has been growing exponentially and this book, which targets both basic scientists and clinicians, is a testament to the breadth of interest that this subject has engendered and the possibility that its principles might be used to ameliorate sensory deficits (see Section III of this volume). This is a theme to which we will return toward the end of this chapter.

It is important to note that, as far as we know, multisensory processing is always in operation (see Chapter 2). The brain is continuously aggregating and integrating signals that are

Multisensory Perception
https://doi.org/10.1016/B978-0-12-812492-5.00003-6

likely derived from the same event, as well as segregating those that are likely derived from different events. Although we are generally not conscious of these operations, and often think that we make our decisions about what is "out there" based on input from a single sense, those seemingly "unisensory" judgments are always being modified by cues from other senses—either by dramatically altering perception (e.g., Ref. 35) or more subtly by increasing or decreasing the access of modality-specific signals to a particular neural circuit.

The products of multisensory integration

The key operation of multisensory integration is the transformation of unisensory inputs into a multisensory product that is unique from its component (i.e., unisensory) parts. This applies to both the physiological process that takes place in individual multisensory neurons and the behavioral consequences of this neural process. From an operational perspective, the neural product of a neuron's multisensory transform is calculated by comparing the number of impulses evoked by each modality-specific stimulus in a bisensory pair (e.g., visual or auditory) to the number evoked by their combination (see Ref. 36). To characterize the nature of the underlying neural computation, the magnitude of an enhanced multisensory response can also be compared to the sum of the unisensory responses (e.g., Refs. 37,38). These comparisons can reveal the product of multisensory integration to be less than their sum (subadditive), equal to their sum (additive), or greater than their sum (superadditive). Behavioral measures usually quantify changes in detection, localization accuracy, and/or identification (e.g., Refs. 39–42) (Fig. 3.1). Multisensory integration also yields changes in the speed of responses (neuronal response latency) and changes in response reliability[43,44]; (e.g., Refs. 13, 45).

It is useful to distinguish "multisensory integration" as functionally distinct from other forms of multisensory processing; for example, those in which comparisons are made across modalities with regard to specific stimulus features such as shape or rhythm.[46] In these cases of "crossmodal matching," the unisensory comparators must retain their individual identity so that their crossmodal equivalencies or differences can be assessed. The mechanisms underlying these forms of multisensory processing, their developmental ontogeny, and their neural substrates are outside the scope of the present discussion.

Multisensory integration in individual SC neurons

One of the best-known sites of primary multisensory convergence is the intermediate and deep layer superior colliculus (SC) neuron (the overlying superficial layers are purely visual).[48] As shown in the cat, the majority of these neurons send descending projections to the brainstem and spinal cord to facilitate the role of this midbrain structure in detecting, localizing, and orienting to contralateral targets.[39,41,42,49–56] The basic design of the SC and its functional organization has largely been conserved across mammalian species, reflecting a variety of modifications of its nonmammalian progenitor, or homologue, the optic tectum.[48,57] Its functional role is well served by the maplike representation of its visual, auditory, and somatosensory space. These individual topographic maps are in spatial register with one another, and in spatial register with a corresponding motor map.[161] As a result, activation of any SC locus, by any individual cue or any crossmodal combination of concordant cues, represents a potential sensory target and can initiate the movements needed to turn

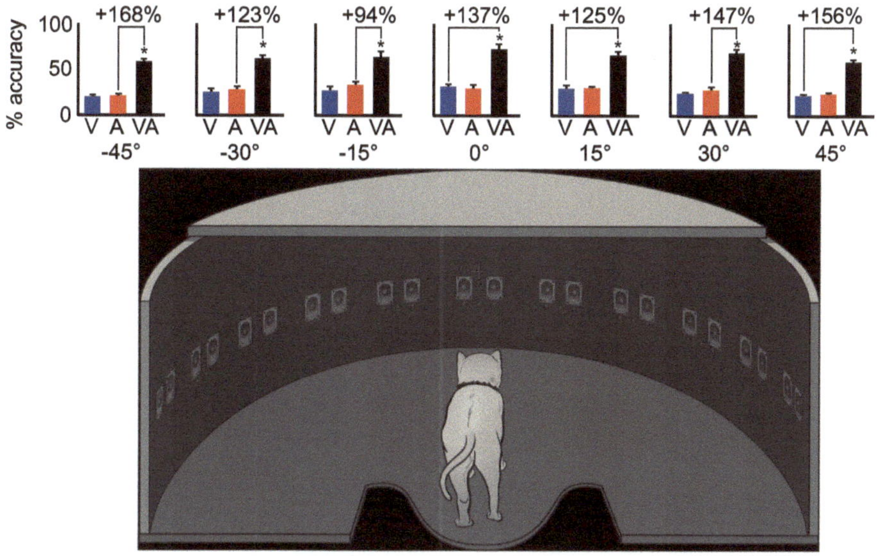

FIGURE 3.1 **An example of the behavioral products of multisensory integration.** Animals are trained to localize weakly effective and brief visual (flashes) and auditory (broadband noise bursts) stimuli at multiple locations within a perimetry apparatus. Bar graphs at each location indicate localization accuracy (% correct responses) for the visual (V) and auditory (A) cues presented alone or in spatiotemporal concordance (VA). The crossmodal stimulus complex (VA) is much more accurately detected and localized at each location.

toward it (Fig. 3.2). In this obvious way, physiological multisensory integration can amplify (or degrade) the likelihood of this behavioral response and the accuracy with which external cues are detected and localized[39,40]; e.g., Refs. 41,42. This elegant system has been used as a model to understand the principles that govern multisensory integration at the level of the single neuron, as well as the multisensory transform itself (cf.[47,58]).

Because only one orientation response can be initiated at any given moment, there are many cues that compete for access to the circuitry of the SC. Multisensory combinations that are spatiotemporally concordant (i.e., cues at the same target location) enhance SC responses (multisensory enhancement), and those that are spatially disparate degrade SC responses (multisensory depression).[36] Temporally disparate cues yield no response enhancement and can produce depression.[10] These observations are frequently referenced as the spatial and temporal principles of multisensory integration. Because the magnitude of multisensory enhancement (as a proportion of the largest unisensory response) increases as the effectiveness of the component stimuli decreases (i.e., "inverse effectiveness," see Ref. 34,36), the biggest benefits are associated with the integration of weakly effective cues.[59]

The native state of multisensory integration is competitive

From a straightforward mechanical perspective, one might expect that a neuron receiving two sensory inputs would have an inherent capability to integrate them and yield an enhanced multisensory product. This is not the case. Neonatal multisensory SC neurons

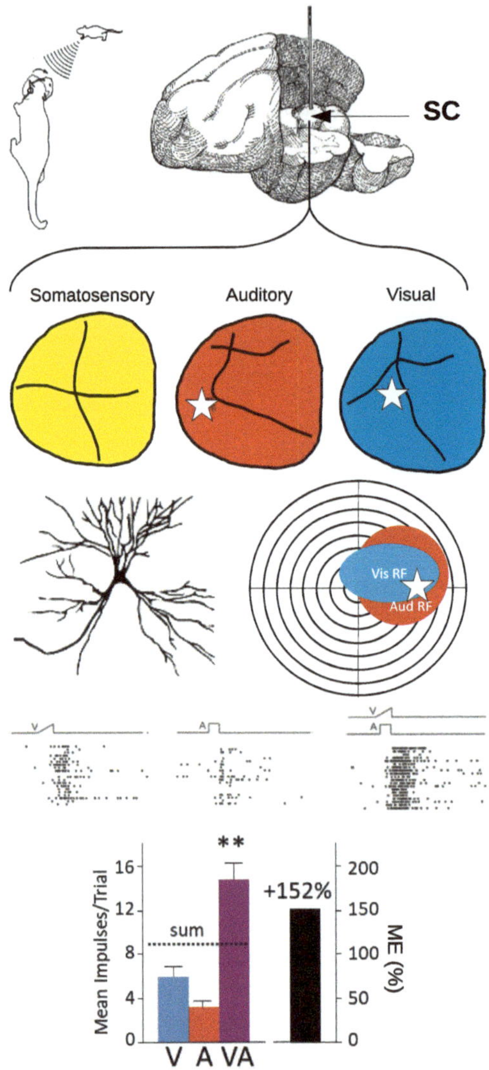

FIGURE 3.2 **Multisensory integration within the superior colliculus (SC) enhances the behavioral and physiological responses to external events.** Top: the detection and localization of salient environmental cues is facilitated by the SC, shown here in a cutaway of the cat brain. Middle: within the SC are topographic maps of sensory space for the somatosensory, auditory, and visual modalities that are overlapping, so that signals (*stars*) derived from the same event registered by more than one modality will be routed onto common multisensory neurons. The illustrated visual–auditory neuron has overlapping receptive fields for each of its responsive modalities (drawn on polar plot of space). Bottom: impulse rasters and summary histograms of response magnitude illustrate the magnitude of this neuron's response to visual (V) and auditory (A) stimuli presented alone or together in spatiotemporal concordance (VA). The proportionate difference between the multisensory and largest unisensory response (here, V) magnitudes is summarized by the metric of multisensory enhancement (ME, black bar). Here it is 152%. *Figure components adapted from Stein BE, Meredith MA.* The Merging of the Senses. *Cambridge, MA: MIT Press; 1993 and Stein BE, Stanford TR, Rowland BA. Development of multisensory integration from the perspective of the individual neuron.* Nat Rev Neurosci. *2014;15:520–535.*

show no evidence of this enhancement capability despite responding reliably to multiple senses. In addition, there is a minority of multisensory neurons in the adult SC that appear to be incapable of generating an enhanced multisensory response.[60–63] In these cases, the response to a bisensory stimulus pair is no greater than the largest response to its component stimuli when presented individually, and in fact, can be lower. This suggests a suppressive or inhibitory multisensory effect, as would be expected if the cues are competing with one another for the circuitry of the SC. This appears to be the default multisensory condition, one that is normally overridden during development.

Developing multisensory integration and the principles that govern it

At birth, the cat SC, the primary model for studies of the development of multisensory integration, has no multisensory neurons. Its sensory-responsive neurons are driven only by tactile cues.[64] SC responses to tactile cues are already seen during late fetal stages and appear to help prepare the neonate for locating the nipple soon after birth.[65] The concentration of their receptive fields around the mouth, nose, and whisker pad appear to be appropriate for this task. The cat's visual and auditory systems are nonfunctional at birth, and the eyelids and ear canals are still closed. It takes several days before the ear canals open and SC neurons become responsive to auditory stimuli. It takes even longer for the visual representation to develop. The eyes open 7–11 days after birth, but even at this time the ocular media are not clear and light entering the eye is substantially degraded before reaching the retina. The ocular media clear rapidly and weak visual responses can be evoked from a minority of neurons in the superficial SC at about this time. However, neurons in the deeper aspects of the structure where the multisensory neurons are located do not respond to visual cues until approximately 3 weeks postnatal (Fig. 3.3).[66] On initial appearance, the visual response properties are immature; for example, their visual receptive fields are very large, and their response latencies are very long.[67,68]

Achieving the full representation of all possible multisensory SC convergence patterns among the visual, auditory, and somatosensory inputs takes many weeks. However, as noted above, while these initial multisensory neurons can respond well to stimuli from multiple sensory modalities, they lack the ability to integrate these inputs: spatiotemporally concordant multisensory cues do not elicit enhanced responses. Achieving their adult-like multisensory integration capabilities takes months.[60] The maturational disconnect between the appearance of multisensory neurons and *integrating* multisensory neurons is especially evident in the newborn rhesus monkey.[69] This is a precocial species and, like us, its eyes and ear canals are already open at birth. It can see and hear reasonably well and already has multisensory neurons in its SC. But, like the neonatal cat, multisensory responses in the newborn rhesus monkey are no greater than their responses to one of the stimuli individually. Instead, it is as if the neonatal multisensory neuron acts only as a common conduit for different sensory inputs to access the same SC motor plan.

During early development, there are rapid changes taking place in the unisensory properties of both unisensory and multisensory SC neurons (when identifiable). These changes have been documented most extensively in the cat. Among these, sensory changes are the shortening of

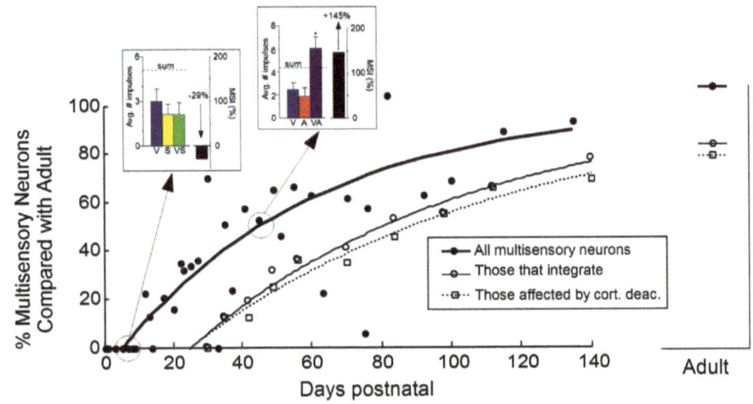

FIGURE 3.3 **Developmental profile of multisensory integration in the cat superior colliculus (SC).** Multisensory neurons first appear in the second postnatal week and increase steadily thereafter, nearing adult levels by postnatal day 140 (*thick solid line, filled circles*). However, multisensory integration capabilities are not seen until the fourth week of life (*thin solid line, open circles*). Thereafter, their incidence increases roughly in parallel with the total incidence of multisensory neurons. The multisensory responses of neurons with multisensory integration capabilities can almost always be depressed by deactivating neurons in association cortex from which descending cortico-collicular afferents originate (*dashed line, unfilled squares*). Although the timing of these three developmental trajectories is parallel, the delay in the development of multisensory integration reinforces the idea that the ability to respond to multiple modalities and the ability to integrate the information they provide are different phenomena, mediated by related but not identical developmental processes. *From Stein BE, Stanford TR, Rowland BA. Development of multisensory integration from the perspective of the individual neuron.* Nat Rev Neurosci. *2014;15:520–535.*

response latencies and the contraction of receptive fields.[60,64,67,68] As will be discussed later, these changes are mostly attributable to sensory-specific (i.e., rather than multisensory) experience. In typical development, the contraction of receptive fields in multisensory neurons is associated with a concomitant increase in their crossmodal overlap; that is, it is as if the multiple receptive fields of a multisensory neuron contract into alignment with one another. As discussed below, this emergent feature can be very sensitive to multisensory experience.

How experience leads to the development of SC multisensory integration capabilities

Given the functional sensorimotor behaviors of the kitten, it may seem surprising at first that the period required to develop multisensory integration capabilities lasts for months. However, the finding that the performance benefits of multisensory integration are slow to develop is consistent with the idea that the critical circuit must be crafted based on multisensory experience, experience that is adequate to distinguish among stimulus configurations that are derived from the same event and those that are derived from different events. It is also consistent with similarly delayed multisensory development observed in humans[70,71] (see Chapter 4). We assumed that this process reflects an evolving calculation of the likelihood that inputs having a specific relationship (e.g., relaying spatiotemporally concordant cues) are derived from common, rather than different, events. If so, development can be thought of as a progression in which evidence accrues favoring one calculation over the other for the various multisensory configurations encountered. This reasoning led to a series of experiments to examine this assumption. In each experimental series, the frequency of an animal's early experience with different multisensory configurations was manipulated experimentally.

The most dramatic approach to altering the frequency with which, for example, visual and auditory cues are combined is to restrict vision or audition and thereby eliminate all visual–auditory experience. The simplest way to do this during early development is to rear animals in darkness, a technique that has been used previously to examine the impact of visual experience on the development of neurons within the primary visual projection system.[72–86]

Dark-rearing: To ensure the absence of visual experience, animals were born and raised in the dark. When tested as adults, they possessed maps for each of the three sensory modalities, and all of the normal multisensory convergence patterns, albeit there were some shifts in the percentage of each (e.g.,Ref. 84). Visual–auditory neurons were common, and most neurons had very large receptive fields, suggesting that their maturation was incomplete. Most important in the current context was that, when presented with a variety of spatiotemporally concordant visual–auditory stimuli, their responses were no better than those elicited by the most effective of those modality-specific component stimuli presented alone.[63,84,86] In other words, their multisensory neurons lacked multisensory enhancement capabilities, as did those in the neonate. This was consistent with the hypothesis that without multisensory experience, SC neurons never acquire the ability to associate cues from different senses and, therefore, cannot integrate them. However, the SC is a visually dominant structure and will presumably lean heavily on visual input for development. It was quite possible that the lack of multisensory experience was not at the root of this multisensory deficit. Rather, the absence of visual experience might have precluded the normal maturation of the underlying architecture necessary for these neurons to develop normal adult information processing capabilities that would be necessary for a variety of functions, including multisensory integration. If so, then minimizing visual–auditory experience without compromising vision might reveal this. To this end, animals were reared with masking sound.

Noise-rearing: In this case, animals were born in a "noise" room in which audio speakers provided white noise from all directions (i.e., omnidirectional).[87,88] This masked all but the loudest transient auditory cues. The illumination in the room was kept at normal levels, so there was no interference with vision. Just as visual receptive fields failed to contract to normal size during dark-rearing, auditory receptive fields failed to contract to normal size during omnidirectional noise-rearing—and, once again, visual–auditory neurons failed to develop their normal ability to integrate visual–auditory cues to produce enhanced multisensory responses. They too responded no more robustly to the bisensory cue combination than they did to the more effective of the two component cues, often approximating an average of the two. The effect on multisensory integration was the same as seen after dark-rearing. These data strongly support the contention that it is experience with multisensory cues that is critical for the development of multisensory integration. That these effects were due to the absence of concomitant audio–visual experiences was confirmed in a random-rearing paradigm (see below).

Random-rearing: In this case, animals were reared in the dark, but each weekday they were individually provided with several hours of controlled experience in which visual cues and auditory cues appeared frequently, but independently.[90] The cues were randomized in timing and location so that while the animal received both visual and auditory cues, they did not appear in combination. A control group was housed in the same dark room and each animal was exposed to the same visual and auditory cues for the same durations, but the cues were always in spatiotemporal synchrony. The former group failed to develop SC visual–auditory multisensory integration capabilities, but the latter did develop these capabilities, responding to these bisensory stimuli very much like neurons in normal animals do (Fig. 3.4).

Collectively, these observations strongly support the contention that multisensory experience is essential for the maturation of multisensory integration capabilities and suggest that the brain uses its multisensory experience to craft the mechanisms necessary to use the different senses synergistically. A natural question prompted by these results is whether this development truly reflects (as hypothesized above) the calculation of a likelihood of association from experience or whether it reflects the construction or emergence of a general capability for multisensory integration within these circuits. To address this issue, the multisensory response properties of trisensory neurons in animals reared with sensory restriction were examined.

Trisensory neurons: Neurotypic trisensory neurons can integrate all combinations of crossmodal cues. However, trisensory neurons in dark- and noise-reared animals would

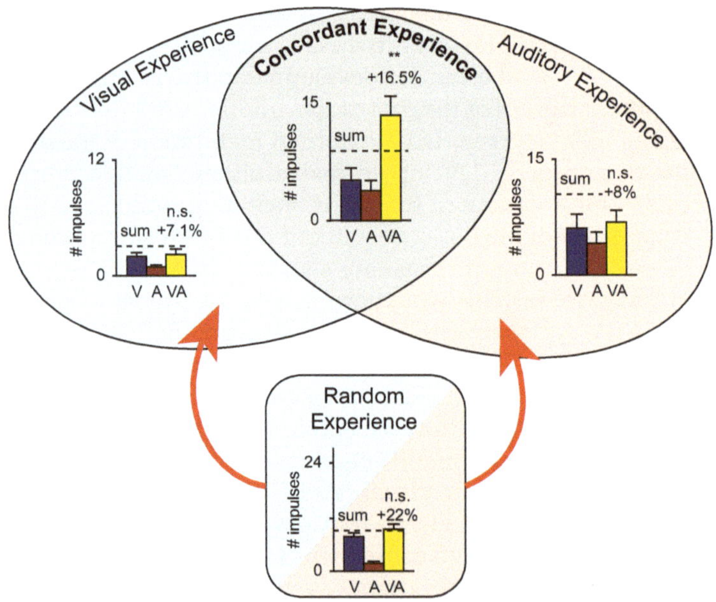

FIGURE 3.4 **Development of superior colliculus (SC) multisensory integration depends on experience with concordant crossmodal cues.** Depicted are responses of exemplar neurons illustrating common outcomes of four different rearing conditions. Shown for each exemplar are summary histograms of visual (V), auditory (A), and multisensory (VA) responses. Rearing with visual experience but with degraded auditory experience (i.e., noise-rearing, left), or with auditory experience but without visual experience (i.e., dark-rearing, right), or with random visual and auditory experience (bottom) yielded multisensory SC neurons lacking multisensory integration capabilities. However, rearing with concordant visual–auditory experience (top middle) allowed SC neurons to develop their multisensory integration capabilities.

have had experience only with specific combinations of modalities (e.g., auditory-somatosensory in the dark, visual-somatosensory in the omnidirectional sound room) and never with visual–auditory combinations. If the trisensory neurons were found to integrate cue combinations that were excluded by their rearing environment, this would be consistent with the idea that a general mechanism for multisensory integration is developed by any crossmodal experience, but that it is not specifically adapted to environmental features. If only specific pairs (i.e., those experienced) were integrated, this would indicate that development is specifically adapted to experience and that multiple (and conflicting) rules for multisensory integration could be simultaneously expressed by the same multisensory neurons. In a metastudy of dark-reared and noise-reared animals, support was found for the latter hypothesis. When neurons had experience with a particular cue combination, they could integrate that combination with roughly the same level of enhancement as a neurotypic animal. However, combinations of cues that were not present in the rearing environment did not elicit enhanced responses.[89]

The evidence reviewed above indicates that the functional properties of multisensory neurons are adapted by experience (Fig. 3.5). But does this mean that they can adapt their responses to the particular features of that experience or does multisensory experience simply "unlock" a basic functional capability in the circuit? To address this issue, dark-reared animals were provided with visual–auditory experience in which the cues were temporally aligned, but at different locations (i.e., an "anomalous" rearing condition).

Anomalous-rearing: Animals reared and housed in the dark received a periodic visual cue (flash of light) from an LED on one wall of the enclosure and a synchronous auditory cue (broadband noise burst) from a speaker on another wall.[91] Such a consistently synchronized, yet spatially disparate, combination of auditory and visual cues is unlikely to be derived from the same natural event, and both cues could not be used simultaneously to initiate the same goal-directed SC-mediated orientation response. In this context, the multisensory experience was operationally defined as "anomalous." When the animals were examined as adults, two results were apparent. The first is that the majority of neurons looked very much like neurons in dark-reared animals that had not received any visual–auditory experience. The second observation was that a minority, but substantial, number of neurons had anomalous receptive fields and multisensory enhancement capabilities.

In these latter neurons, the receptive fields were smaller than in the dark-reared animal but their centers were displaced laterally, sometimes with little overlap, as if reflecting the disparity of the cues in the rearing condition. Particularly interesting in this context was that the neurons that appeared affected by the rearing condition could integrate visual and auditory cues but showed response enhancement with spatially disparate cues, because these cues fell within their spatially displaced receptive fields. Thus, the bisensory stimulus configuration that would elicit response depression in neurotypic neurons produced response enhancement. This showed that the neurons can encode the specific features of their multisensory experience.

Because this capability was expressed in only a minority of neurons, it suggests that there is some inherent bias toward multisensory cues that are in spatial alignment. This makes intuitive sense as it not only is consistent with the inherent, albeit rough, topographic arrangement of afferents from the different senses even in the neonate but with the functional role

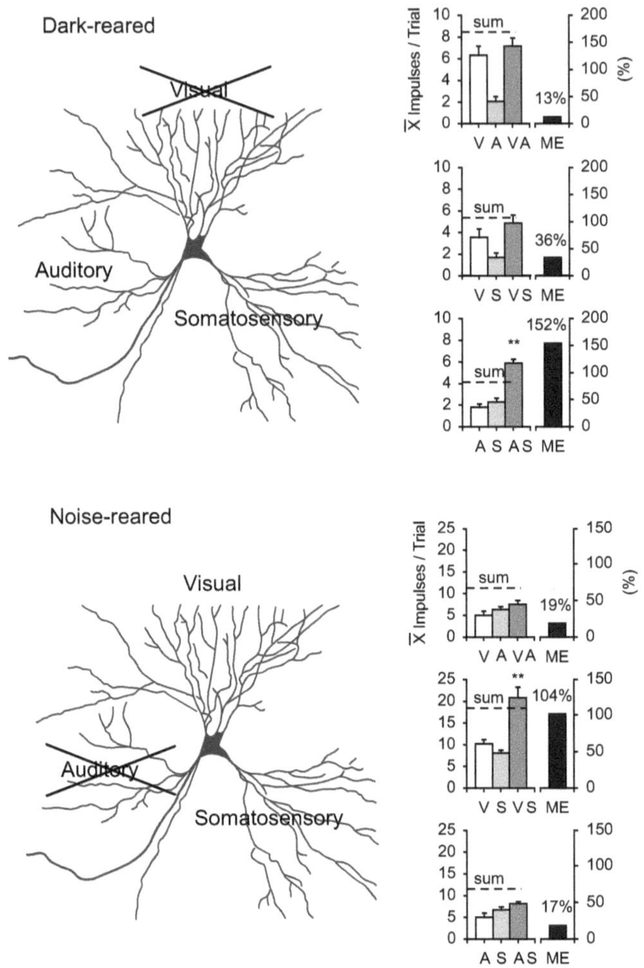

FIGURE 3.5 **Exemplar adult multisensory superior colliculus (SC) neurons recorded from animals reared in complete darkness (top, "dark-reared") or in omnidirectional masking sound (bottom, "noise-reared").** Both rearing environments preclude patterned experience with a sensory modality (icons, left) and, as a consequence, preclude experience with bisensory pairs involving the compromised modality. Despite being sensitive to multiple sensory modalities, these neurons failed to generate enhanced responses to bisensory pairs that were excluded from early life experience. Thus, the dark-reared neuron did not integrate visual–auditory or visual–somatosensory pairs, and the noise-reared neuron did not integrate visual–auditory or auditory–somatosensory pairs. Note that bisensory pairs that were not restricted elicited typically enhanced multisensory responses. *Adapted from Xu J, Yu L, Stanford TR, Rowland BA, Stein BE. What does a neuron learn from multisensory experience?* J Neurophysiol. *2015;113:883–889.*

of the SC in detecting a singular event as the target of a gaze shifts—and singular events are most likely to produce spatially aligned stimuli.

Whether this inherent bias extends to stimulus features other than space is not yet known. Similarly, it is not known whether other forms of "register" exist between the senses in this context. Studies in both nonhuman primates and in human subjects have suggested that the semantic register, or matching, of multisensory stimuli is also an important factor determining the multisensory product in cortex.[21,92–96] That such higher-order features as semantics, or identity, play a substantial role in cortical multisensory integration is not surprising. Whether such features also play a role in a structure like the SC, which is involved in more rudimentary tasks (detecting and locating and targeting an event for an orientation movement) is not yet known.

Incorporating multisensory experience into the circuit

Although there are an impressive number of subcortical and cortical tectopetal regions that provide the SC with sensory information (see review in Ref. 97) and play roles in determining its response properties and topographic organization, the inputs from association cortex have a special role in SC multisensory processing. In the cat, the areas of relevance include the anterior ectosylvian sulcus (AES) and an adjacent area, the rostral aspect of the lateral suprasylvian sulcus (rLS).[98–101] Although the functional organization of the AES has been studied more extensively than that of the rLS because more SC neurons depend on it,[40,101–105] they are assumed to have similar roles in mediating the development of SC multisensory integration.

The AES is organized into three regions that are largely unisensory, one visual: the anterior ectosylvian visual area (AEV), one auditory: the auditory field of the AES (FAES), and one somatosensory: the fourth somatosensory area (SIV).[98–101,106–111] At the borders between these regions are many multisensory neurons that represent combinations of the bordering region sensitivities. Thus, the border of AEV and FAES contains many visual–auditory neurons. However, neither the multisensory neurons scattered in the unisensory regions nor the multisensory neurons in the border regions project to the SC. Only unisensory AES neurons project to the SC.[112] This primarily ipsilateral projection provides convergent inputs from modalities that match the convergent patterns their target neurons receive from other sources. For example, SC neurons that receive visual and auditory inputs from non-AES sources also receive AES inputs from its AEV and FAES subregions.

In the normal adult animal, these cortical influences are essential for SC multisensory enhancement. Lesioning or reversibly deactivating these association cortical regions eliminates the enhanced responses of SC neurons and multisensory enhancements in SC-mediated localization behaviors.[100–102,105,113–115] In many cases, the resulting multisensory response approximates an average of the responses to those two component stimuli—a decrease from the most robust of them individually.[116] Interestingly, deactivating each of the matching subregions within the AES (e.g., AEV or FAES for a visual–auditory SC neuron) has essentially the same result as deactivating the entire AES[105]; see Fig. 3.6). This has been interpreted as indicating that AES-SC afferents interact synergistically in influencing the SC multisensory process. While the biophysical basis for this synergy remains unknown, some computational models have supported speculation about how this may result from unique synaptic configurations,[117–119] many of which can be accommodated by the complex nature of the AES-SC projection[111]; e.g., Ref. 120.

The AES and rLS play an important role not only in mediating SC multisensory integration in the adult but also in crafting the needed architecture during development. During early life, there appears to be some functional redundancy between these regions: if either the

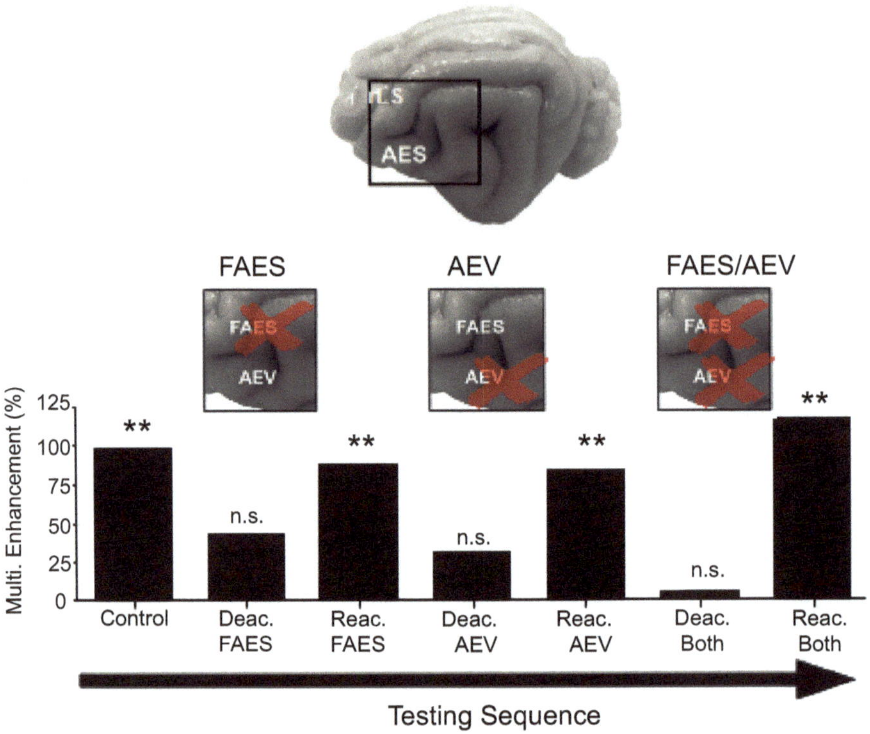

FIGURE 3.6 **A synergy between unisensory subregions of association cortex drives multisensory integration capabilities in the superior colliculus (SC).** Depicted is the degree of multisensory enhancement in a typical visual–auditory SC neuron when the auditory (FAES) and/or visual (AEV) subregion of AES was reversibly deactivated. Deactivation of either subregion alone eliminated multisensory enhancement, rendering the multisensory response no better than the largest unisensory response. Subsequent reactivation after each deactivation series restored the neuron's functional capabilities. *Adapted from Alvarado JC, Stanford TR, Rowland BA, Vaughan JW, Stein BE. Multisensory integration in the superior colliculus requires synergy among corticocollicular inputs.* J Neurosci. *2009;29:6580–6592.*

AES or the rLS is lost during early life, the role of the other can expand to compensate, and the acquisition of multisensory enhancement capabilities in SC neurons and SC-mediated behavior can appear to be normal.[103,104] When both are lost, this capability never develops. But as noted above, the apparent redundancy between the AES and rLS does not extend to the adult, in which the loss of influences from the AES is not compensated by the rLS, or any other cortical area.[40,105] Many of the corticotectal projections from SIV, and presumably those from AEV and FAES, are already present at birth[121] but their synapses are not yet formed or are immature at this time. These synapses appear to be configured and reconfigured gradually as multisensory experience is acquired.[120,122]

Multisensory experience and influences from cortex: two interrelated developmental events

Thus, the two primary factors involved in the development of SC multisensory integration capabilities appear to be experience with multisensory events and functional inputs from association cortex. This is unlikely to be coincidental. One possibility is that the tectopetal projections from association cortex (i.e., AES, rLS) serve as the portal through which multisensory experience affects the multisensory circuit. Thus, projection patterns of these afferents form a crucial substrate for the expression of multisensory integration capabilities. There are several additional points of evidence consistent with this idea.

As noted above, neonatal lesions of both the AES and rLS preclude the development of SC multisensory integration capabilities and multisensory enhancement in associated orientation behaviors.[103,104] However, such lesions allow a multitude of opportunistic anatomical changes to take place in the circuit. To explore the simple absence of the functional influences of association cortex on this developmental process, association cortex was deactivated during development. In these experiments, a pledget infused with muscimol (an inhibitor of neuronal activity) was implanted over association cortex during postnatal weeks 4–12. This eliminated the ability of its neurons to respond to sensory (e.g., bisensory) cues just at the time at which multisensory integration capabilities are normally being developed in SC neurons. Behavioral tests were later conducted when the animals were mature, long after the pledgets were depleted of muscimol or were removed and cortex was once again active. Even when reaching adulthood, the animals failed to exhibit multisensory integration capabilities in their behavioral or SC physiological responses.[123] These observations are consistent with the idea that association cortex is coding crossmodal experiences and then sharing them with the SC so that its multisensory integration capabilities can develop.

Interestingly, in the absence of visual–auditory experience, this pathway does not fail to develop, as might be expected from the absence of multisensory integration capabilities in SC neurons. Indeed, in dark-reared animals, the projection develops an even stronger influence on SC multisensory responses than in normally reared animals. But its influence is much less specific: it now enhances responses to all visual and auditory stimuli. There is no preferential enhancement of multisensory responses. Presumably, crossmodal experience exerts its effect on multisensory integration by shaping the AES-SC projection.[86]

How might the AES-SC projection be shaped by experience to promote the instantiation of SC multisensory integration capabilities? Although the specific mechanisms are unknown, the above evidence suggests that some variant of an associative or Hebbian-based learning principle is likely to be responsible. In such a model, convergent afferents relaying unisensory information from AES target a common SC neuron and become strengthened when their presynaptic activation reliably precedes the activation of the SC target.[124,125] There is some empirical evidence for the operation of these algorithms in the SC[126] and, as reviewed below, some direct evidence for its action on the afferents of SC multisensory neurons in the adult.

Multisensory plasticity in adulthood

The ability of the nervous system to accommodate sensory experience does not end at the onset of adulthood. Although the extent to which this multisensory circuit is sensitive to adult experience has received little attention, several studies have noted that there are such adaptive accommodations.[7] For example, temporally offsetting visual and auditory stimuli so that they are just outside the window for integration elicits two seemingly independent responses from multisensory neurons. But, after repeatedly presenting this sequence of bisensory stimuli, neurons begin to alter their responses, as if they "learned" that they were related. The first response increases in magnitude and duration, while the latency of the second response shortens until the two responses become more like a single, longer-duration, and more robust response[127]; see also Fig. 3.8. This apparent fusion has been explained by the same Hebbian learning rule that is thought to lead to the acquisition of multisensory integration capabilities during normal development. In short, the presynaptic activity induced by the first of two sequential stimuli increases in effect when it induces postsynaptic activity, and it does so with greater potentiation in those synaptic weights when the second stimulus is present.

A second observation of multisensory plasticity was encountered when spatiotemporally concordant bisensory cues were repeatedly presented. This mimics the situation in which an organism continues interacting with the source of that bisensory input. The effect of this repeated multisensory stimulation was to increase the effectiveness of the individual sensory inputs. These plastic changes were most strikingly observed in "covert" multisensory neurons, which were individually responsive to only a single modality but generated enhanced responses to spatiotemporally concordant bisensory stimuli. Repeated bisensory stimulation had the effect of rendering the neuron overtly responsive to the previously

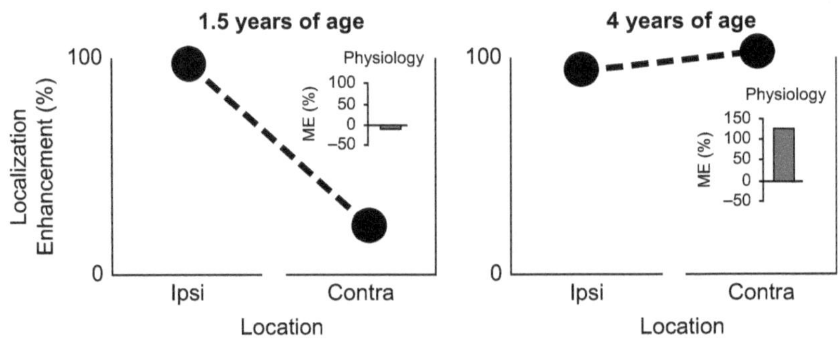

FIGURE 3.7 **Temporary deactivation of association cortex during early life delayed the acquisition of superior colliculus (SC) multisensory integration capabilities.** Unilateral muscimol–infused implants rendered AES and rLS inactive between postnatal weeks 4–8. When animals were tested on a localization task approximately 1.5 yrs later (left), they failed to show multisensory performance enhancements to stimuli in contralateral space but responded normally to stimuli on the opposite side. This lack of enhancement in behavior was paralleled by a lack of enhancement in the physiological responses of ipsilateral SC neurons to multisensory stimuli (inset). However, when animals were retested on the same task at 4 years of age (right), performance on both sides of space was equivalent, and the physiological defects in multisensory integration appeared to have resolved (inset). *ME*, multisensory enhancement. *Adapted from Rowland BA, Jiang W, Stein BE. Brief cortical deactivation early in life has long-lasting effects on multisensory behavior.* J Neurosci. *2014;34:7198–7202.*

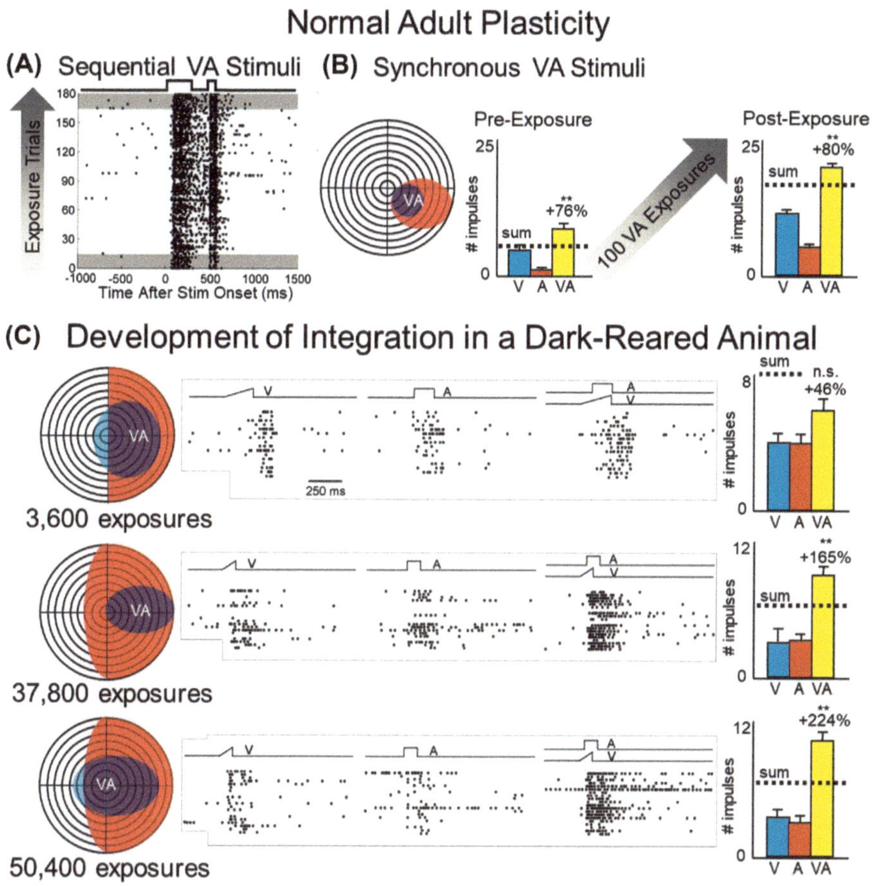

FIGURE 3.8 **Examples of adult plasticity in multisensory integration generated by different stimulus configurations.** (A) Repeated sequential visual–auditory stimuli increased the magnitude and duration of responses to the first stimulus and decreased the latency of the second. This minimized the gap between their discharge trains. (B) Repeated presentation of spatiotemporally concordant visual–auditory stimuli increased multisensory and unisensory response magnitudes, effectively "activating" unisensory channels that did not generate suprathreshold responses before exposure (in the example, the auditory channel is activated). (C) Neurons in dark-reared animals normally do not express visual–auditory integration capabilities. However, these capabilities can be rapidly entrained by intense exposure to concordant visual–auditory stimuli. Note, that receptive fields retain immature (i.e., large) sizes with this impoverished sensory experience despite developing their integrative capability. *Adapted from Yu L, Rowland BA, Stein BE. Initiating the development of multisensory integration by manipulating sensory experience.* J Neurosci. *2010;30:4904–4913 and Yu L, Stein BE, Rowland BA. Adult plasticity in multisensory neurons: short-term experience-dependent changes in the superior colliculus.* J Neurosci. *2009;29:15910–15922; Yu et al. (2009).*

"silent" sensory modality. Now each modality-specific stimulus elicited responses when presented alone.[86]

Observations of adult plasticity in multisensory processing raised the interesting possibility that animals lacking multisensory integration capabilities (e.g., as a consequence of early sensory deprivation or cortical deactivation, as described above) might possibly be "rehabilitated" as adults by multisensory experience. This possibility would provide hope that similar rehabilitation might be induced in human patients with deficits in multisensory integration. Such deficits are associated with a number of developmental anomalies such as autism, dyslexia, and sensory processing disorder[128–132] and are also apparent after corrective surgery for congenital cataracts and/or hearing disorders.[133–135] Many of these multisensory deficits are described in detail in Chapters 15–17.

Clinical implications

As described above, if early multisensory experience is impaired by natural causes (e.g., if congenital cataracts or deafness compromise visual–auditory experience), or through experimental manipulation (e.g., rearing in darkness or with masking noise), multisensory integration capabilities will be compromised. In those experimental circumstances, SC neurons fail to develop their characteristic visual–auditory integration capabilities regardless of how visual–auditory experience is restricted. But these defects are not irreversible: controlled, consistent visual–auditory experience later in life can be used to entrain these capabilities in compromised adults (Fig. 3.8). Interestingly, the acquisition was found to be very rapid and was induced even in anesthetized animals.[63,88]

It is possible that this recovery was because sensory restriction extended the sensitive period characteristic of early life and/or because of an inherent plasticity in the system that is present even in adulthood, or some combination of the two. Yet regardless of which factors were at play, the system retains enough plasticity to use later visual–auditory experience to rapidly acquire the ability to integrate those multisensory cues despite the lack of any external reward associated with the stimulus combination or any required response, and even while anesthetized. The efficiency of the process even under these circumstances was great enough to achieve nearly the same magnitudes of response enhancement observed in animals reared in normal conditions and to generalize to stimuli with different physical characteristics. It was also resilient, being maintained in the absence of continued experience with these multisensory cues. Based on these observations, it appears safe to say that multisensory integration capabilities will be instantiated if the multisensory cues to which the animal has access maintain a consistent spatiotemporal relationship. In short, experience with the same cue combination is a key factor.

These observations raise a perplexing question in light of the other results described above. Why is it that the animals that had their association cortex deactivated for a comparatively brief period during development did not also acquire these SC capabilities rapidly once deactivation was ended? Their deficits persisted for more than a year in a normal environment and were not resolved for a long period thereafter (when tested again 4 years after the deactivation period they appeared to be normal in this regard, Fig. 3.7). Similarly persistent defects have been observed in noise-reared animals placed in a normal environment for up to a year.[88] The answer to this question is not obvious. The normal environment is rich in multisensory experiences of every variety and would seem to be just the condition to facilitate forming new multisensory associations. However, it may have been that it was the very richness of the multisensory experiences in the natural environment that was problematic. Even when caused by the same event, bisensory (e.g., visual–auditory) stimuli often vary in their spatiotemporal relationships to one another. Different distances from the receiver will exaggerate differences between the arrival times of these stimuli; different angles will alter their apparent spatial alignment; different times of day, background conditions, etc., will alter the context in which the event is experienced. Furthermore, the individual visual and auditory cues can appear independent of one another and/or be coupled with other sensory cues. The list of possible complications is long. But, in the training environment, the visual and auditory stimuli were invariant. There was but one stimulus configuration, with no unisensory intervening trials, and any possible competing visual and/or auditory background stimuli were minimized.

Of course, the normal environment is the condition in which a normally developing neonate builds its multisensory integration capabilities. While the complications listed above undoubtedly increase the time its brain needs to learn the multisensory associations, it does ensure that these variations are part of the associations formed and, thus, are well tolerated. On the other hand, the adult brain may not tolerate such variations very well, which might help explain why humans with early visual or auditory deficits that preclude the multisensory experience needed to develop their multisensory integration capabilities exhibit multisensory deficits that could persist for many years after the initial sensory deficit has been corrected.[133–135]

The suggestion that the neonatal brain "tolerates" variations in the exact multisensory relationship needs more discussion because it is not necessarily insensitive to those variations. Rather, it may even encode the statistical nature of those variations in its tuning functions. For example, experience with spatial variation may determine the nature of a neuron's response to different crossmodal disparities: it may have a higher probability of responding to stimuli with disparities that are most frequently encountered or it may respond to each disparity (within its RFs) with magnitudes of enhancement that reflect the frequency with which it was encountered. That this is a reasonable possibility is indicated by the results of experiments in sensory-restricted animals that were later exposed to an invariant, spatiotemporally concordant bisensory stimulus combination. First, this training paradigm proved to be effective only in cases in which the exposure site fell within both receptive fields of a visual–auditory neuron (Fig. 3.9). Second, these neurons developed a "proximity effect" that was not observed in normally reared animals: there was a preference for stimuli that were in exact spatial concordance.[90] Crossmodal stimuli that progressively deviated from spatial concordance produced progressively lower enhancement magnitudes. Yet in normally reared animals, as long as the stimuli fall within the neuron's overlapping receptive fields, there is no systematic relationship between their separation and the magnitude of enhancement they generate.

Using the principles of multisensory integration to ameliorate hemianopia

The promise of multisensory training programs for rehabilitating sensory deficits, even those manifested in only one sensory modality, is underscored by their effectiveness in ameliorating another condition: hemianopia (see Chapter 19).

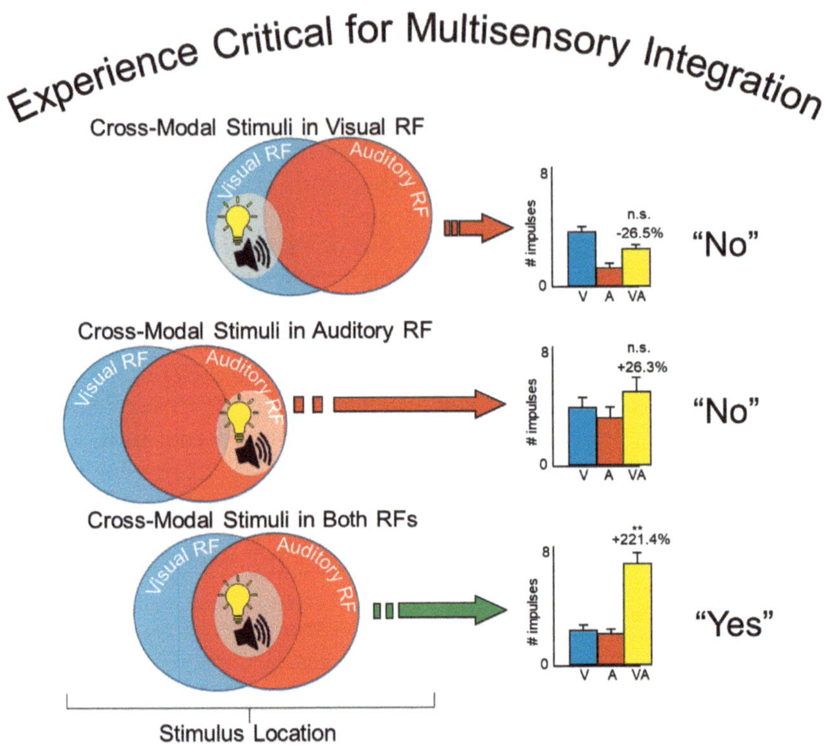

FIGURE 3.9 **Spatial specificity for adult development of multisensory integration capabilities.** Sampling from multiple neurons in dark-reared cats given exposure to spatially and temporally concordant bisensory stimuli revealed that individual neurons must experience both components of the bisensory events for them to develop multisensory integration capabilities. If, for example, the auditory (top) or visual (middle) cue falls outside a neuron's receptive field, repeated presentation of both cues will not drive the neuron to develop multisensory integration capabilities: the neuron's experience is unisensory. Only when the two cues fall within their respective receptive fields (bottom) does their repeated presentation lead to the development of the capacity for multisensory integration. *Adapted from Stein BE, Stanford TR, Rowland BA. Development of multisensory integration from the perspective of the individual neuron.* Nat Rev Neurosci. *2014;15:520–535.*

A common effect of visual cortex damage is a failure to attend to visual cues in the contralesional hemifield. The individual is hemianopic and acts as if blind in that hemifield. Although many of the visual inputs to the ipsilesional SC are lost following the cortical lesion,[136] visual afferents from other visual structures, including the retina and thalamus, are retained.[137–141] Thus, the inability of the SC to foster the detection of visual stimuli in the opposite hemifield and induce visually guided orientation responses to them seems perplexing. Presumably, the excitatory–inhibitory balance between inputs from cortex and from other regions (e.g., the opposite hemisphere) is disrupted, thereby compromising the functional integrity of visually responsive SC neurons and rendering them incapable of supporting SC-mediated visuomotor responses (e.g., Refs. 142–146).

As described in detail above, many of the visually responsive neurons in the deep SC are also responsive to auditory and/or somatosensory cues and their sensory responses are quite plastic. Some that appear unresponsive to visual stimuli can become responsive to those stimuli after experiencing repeated presentations of concordant visual–auditory cues. Based on this observation, and previous studies in which nonvisual cues could facilitate visuomotor behavior in hemianopic patients,[147–150] we hypothesized that the cortical lesion degraded the visual responsiveness of ipsilesional SC neurons, that "training" with concordant auditory–visual cues could reverse this impairment in multisensory neurons that retained visual afferents from regions other than the lesioned cortex, and that visual cues in the compromised visual field would thereafter activate these neurons and elicit SC-mediated orientation behavior. If so, this multisensory training paradigm would provide a simple, noninvasive way to ameliorate the hemianopic consequences of the cortical lesion.

To test this possibility, cats were trained on the standard visual orientation task to approach a visual stimulus at any of 15 degrees spaced locations between 105 degrees left and right of visual fixation (0°). Then visual cortex was ablated in one hemisphere, a lesion that included all contiguous regions of visual cortex. The degeneration of the ipsilesional lateral geniculate nucleus revealed the effectiveness of the lesion, as did the profound neglect of all visual stimuli in the contralesional hemifield. In contrast to this contralesional blindness, the animals responded well to visual stimuli in the control (i.e., ipsilesional) hemifield and to auditory and tactile cues in both hemifields. This was consistent with physiological findings. The SC on the side opposite to the ablated cortex appeared to be normal, but the ipsilesional SC was not.[151]

Physiological abnormalities in the ipsilesional SC were not immediately apparent. When an electrode was lowered into its superficial layers which, as noted earlier, are purely visual, robust visual responses were evident. Although these neurons are not directly involved in the visuomotor role of the SC,[152] it was still surprising that they were not linked to any overt visual behaviors. However, their responsiveness showed that visual afferents from regions other than the damaged cortex were still capable of activating SC neurons. In striking contrast, however, the underlying multisensory layers, which are directly involved in the visuomotor roles of the SC, were largely devoid of visually responsive neurons. What must have previously been visually responsive multisensory neurons appeared to have become either unisensory auditory or somatosensory or nonvisual multisensory (auditory–somatosensory). The few visually responsive neurons that were encountered in these layers were clustered toward the rostral pole of the structure (normally representing central space) and most had abnormal receptive fields. Their centers and much of their excitatory regions were located in the "wrong" (i.e., opposite, or normal) hemifield. Indeed, there was a dramatic and abrupt shift in the location of the visual receptive field centers as soon as the electrode passed through the superficial layers. This substantial shift in receptive fields produced a clear misalignment of superficial-deep layer receptive fields in the same vertical electrode penetration. This is not seen in normal animals. Because much of their receptive fields were primarily in the opposite or "seeing"' hemifield they were of little help in dealing with visual cues in the compromised hemifield. In short, visual stimuli in nearly all of contralesional space (except in the most central region) were ineffective in activating neurons in the ipsilesional SC or in supporting the behaviors they normally initiate. It was now evident why animals failed to react to visual stimuli in the blind hemifield even though the SC was structurally intact: the cortex normally subserving it had been removed and the critical midbrain element of the visuomotor circuit had been shut down.

To test the possibility that bisensory (auditory–visual) training could restore the lost visual responsiveness in the SC, animals were given training sessions five times/week. The animal's task in these sessions was to orient to and approach an auditory–visual stimulus presented 45 degrees into the hemianopic field as shown in Fig. 3.10. Because their hearing had not been disrupted, and the many auditory responsive neurons in the SC could still engage its sensorimotor circuitry, this was readily accomplished. The daily sessions included both the bisensory cue combination (n = 45–60), interleaved with trials in which the visual stimulus was presented alone in the homologous location in the normal field (n = 10), and catch trials in which no stimulus was presented (n = 10). The animals had no trouble responding appropriately to the different cues. At the end of each day of training a set of visual "probe" trials was run in the hemianopic field. After approximately 4 weeks of training, each of the animals now showed orientation to a visual stimulus when presented alone in the previously blind hemifield and did so at all locations. The hemianopia had been ameliorated.

Restoring the ability to respond to visual events in the previously blind hemifield was accompanied by the restoration of the visual responsiveness of neurons in the previously unresponsive SC. Presumably they were supporting the restored visual behavior. The vast majority of them were demonstrably multisensory, an observation that is consistent with the assumption that a training paradigm that normally engages the process of SC multisensory integration can induce long-term changes in the sensitivity of these neurons to their different (in this case, visual) inputs.

If this reasoning is sound, one should be able to disrupt this new functional circuit in predictable ways. For example, as discussed earlier, SC multisensory enhancement depends on influences from AES. When these influences are disrupted, ipsilateral SC neurons lose their enhancement responses to bimodal stimuli and the animal loses its overt multisensory performance benefits. Presumably then, removing the ipsilesional AES would eliminate the functional integrity of the rehabilitated circuit and reinstate the hemianopia even though AES normally plays no obvious role in visual orientation. This is just what happened in the three animals that were tested. The AES lesion had an effect it does not normally have, rendering these animals blind in contralesional space yet again (Figs. 3.11 and 3.12).

It is assumed that one effect of the training paradigm was to enhance the effect of the AES visual inputs to the ipsilesional SC; even though the visual activity of the AES would have been compromised by the cortical lesion, AES receives much of its visual input from extrastriate visual cortex.[106,108,110] In addition, it is also likely that the AES enhanced the effectiveness of visual inputs from other regions such as those coming directly from the retina[138] and/or those relayed from the overlying superficial visual layer SC neurons[140,153] which retained their visual responsiveness after the cortical lesion. These superficial layer visual neurons also likely project to the AES via the thalamus[154] and, given their sophisticated visual capabilities, may have an outsized role in this rehabilitative effect (e.g., Refs. 155,156). However, this remains to be determined. The AES may also accomplish its role indirectly, by modulating the inhibitory effect of the basal ganglia on SC neurons.[17,157–159] The basal ganglia play critical roles in action selection and learning[159] and previous studies have shown that lesions interrupting the basal ganglia influence on the deep SC can also reverse hemianopia.[144]

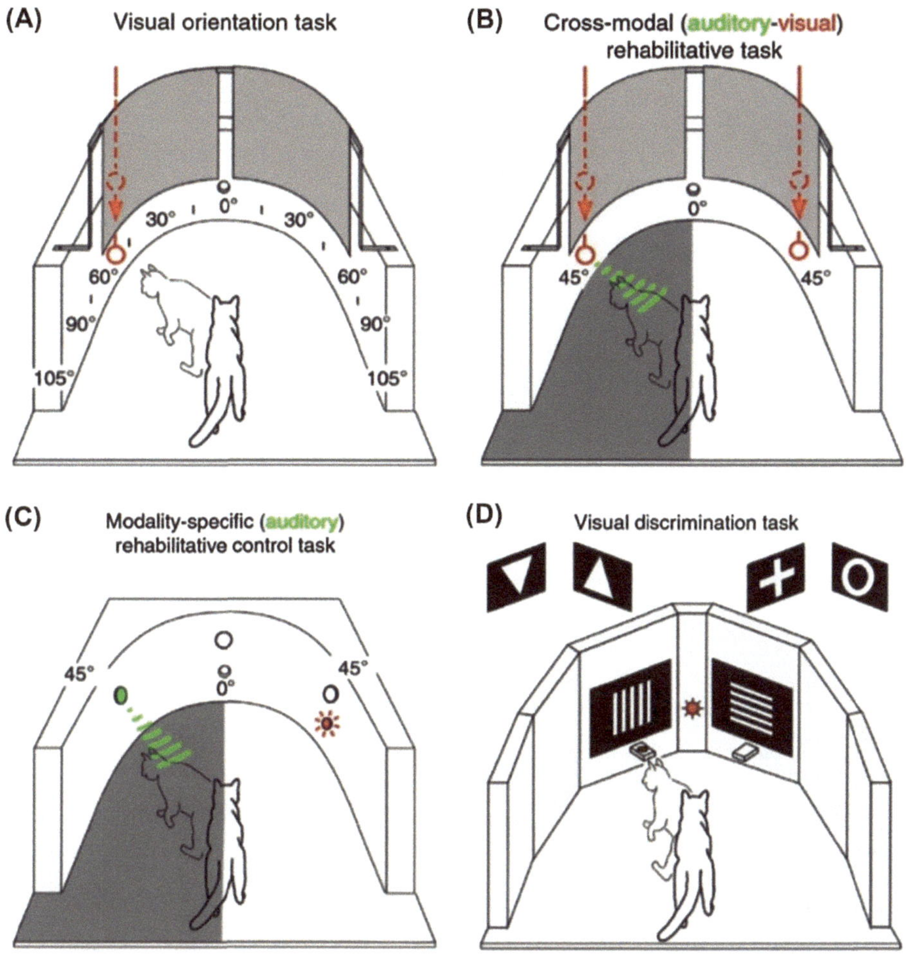

FIGURE 3.10 **Four different tasks were utilized in the present series of behavioral experiments.** (A) Animals were trained to fixate ahead (0°) and then orient to and approach a visual cue (a white sphere emerging from behind a black curtain) in the visual field quasirandomly at each 15 degrees of visual eccentricity. Food rewards for correct orientation were delivered at the matching eccentricity. During catch trials (no cue) and for trials in which the cue was not attended, the animal moved straight ahead to receive a food reward at the 0 degree position. When criterion performance was achieved (>90% averaged across tested eccentricities), all contiguous regions of visual cortex were removed unilaterally, and animals were later retested. (B) For bimodal rehabilitative training, animals made repeated, food rewarded orienting and approach movements to a spatially and temporally coincident auditory–visual cue presented at 45 degrees into the hemianopic field (gray shaded area). To preclude conditioning, on some trials, a visual cue was presented alone at 45 degrees in the intact hemifield. If no cue was delivered, the animal moved toward the fixation point for the food reward. Food rewards were delivered at the site of cue presentation. (C) As a control, animals made repeated food rewarded orientation and approach movements to an auditory cue presented alone at 45 degrees into the hemianopic field (gray shaded area). Auditory cues were delivered by a speaker and approximated the intensity and duration of auditory cues used for bisensory training. To preclude conditioning, on some trials, a visual cue was presented alone at 45 degrees in the intact hemifield. If no cue was delivered, the animal moved toward the fixation point for the food reward. Rewards were delivered at the site of cue presentation. (D) To evaluate visual capabilities beyond basic orientation, select animals also were trained preoperatively in a forced-choice paradigm to discriminate vertical versus horizontal gratings of different spatial frequencies and simple form discrimination (triangle orientation; cross vs. annulus). Computer-generated white-on-black background discriminanda were presented on two LCD screens. The animal fixated on the central fixation point and was rewarded for correct movements toward the target discriminanda. Food rewards were delivered at the site of cue presentation. *From Jiang H, Stein BE, McHaffie JG. Multisensory training reverses midbrain lesion-induced changes and ameliorates haemianopia.* Nat Commun. *2015;6:7263.*

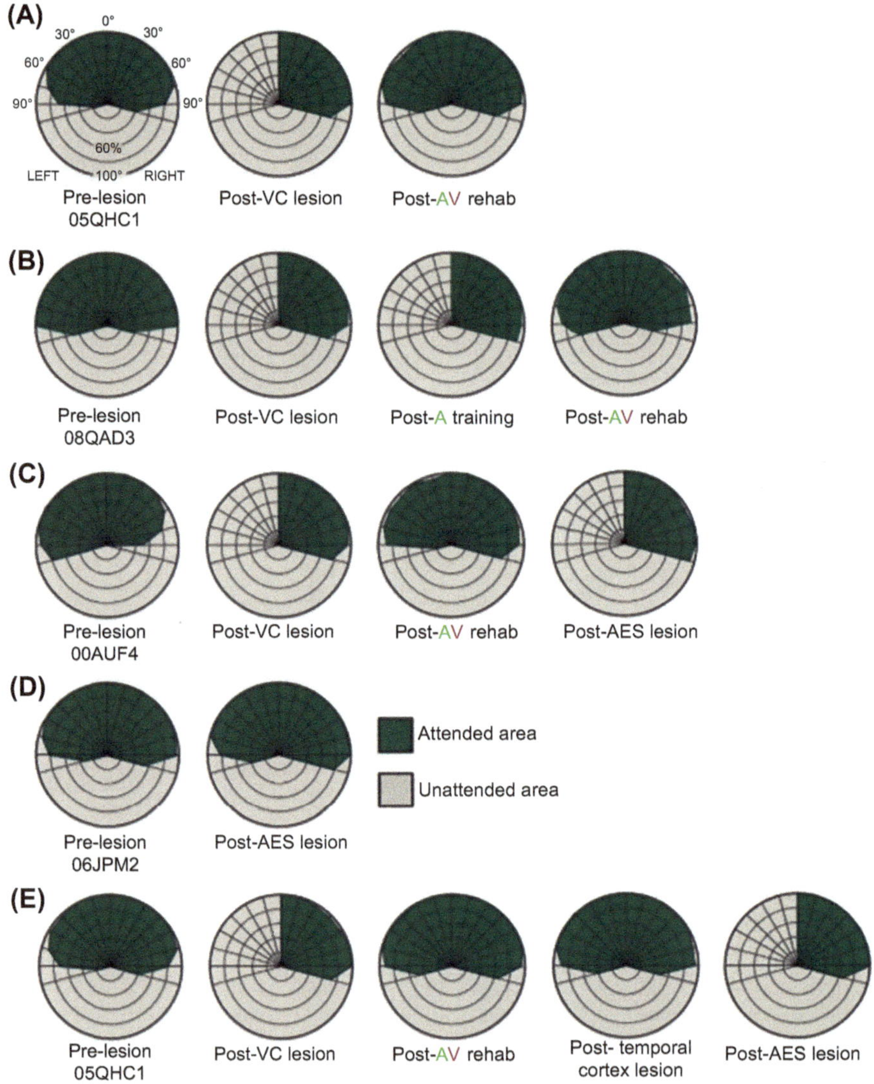

FIGURE 3.11 **Polar plots of visual orientation capabilities for five representative animals following visual cortex lesions, bisensory training, and other experimental manipulations.** (A) Postlesion training with bisensory cues ameliorated visual cortex lesion-induced hemianopia. Each circle in the polar plot represents 10% correct responses and visual performance is shown in green. Note that the normal visual field (left schematic) was compromised by a right cortical lesion and vision was lost in the left hemifield (middle), but recovered after AV training (right). (B) Training with auditory cues alone failed to produce this result, but subsequent AV training reinstated vision. (C) The reinstated vision was lost following removal of ipsilesional AES cortex. (D) Despite the importance of AES in rehabilitated animals, its loss in normal animals has no effect on visual orientation. (E) The reinstated vision is unaffected by lesions to other (i.e., temporal cortex) cortical regions. *From Jiang H, Stein BE, McHaffie JG. Multisensory training reverses midbrain lesion-induced changes and ameliorates haemianopia.* Nat Commun. *2015;6:7263.*

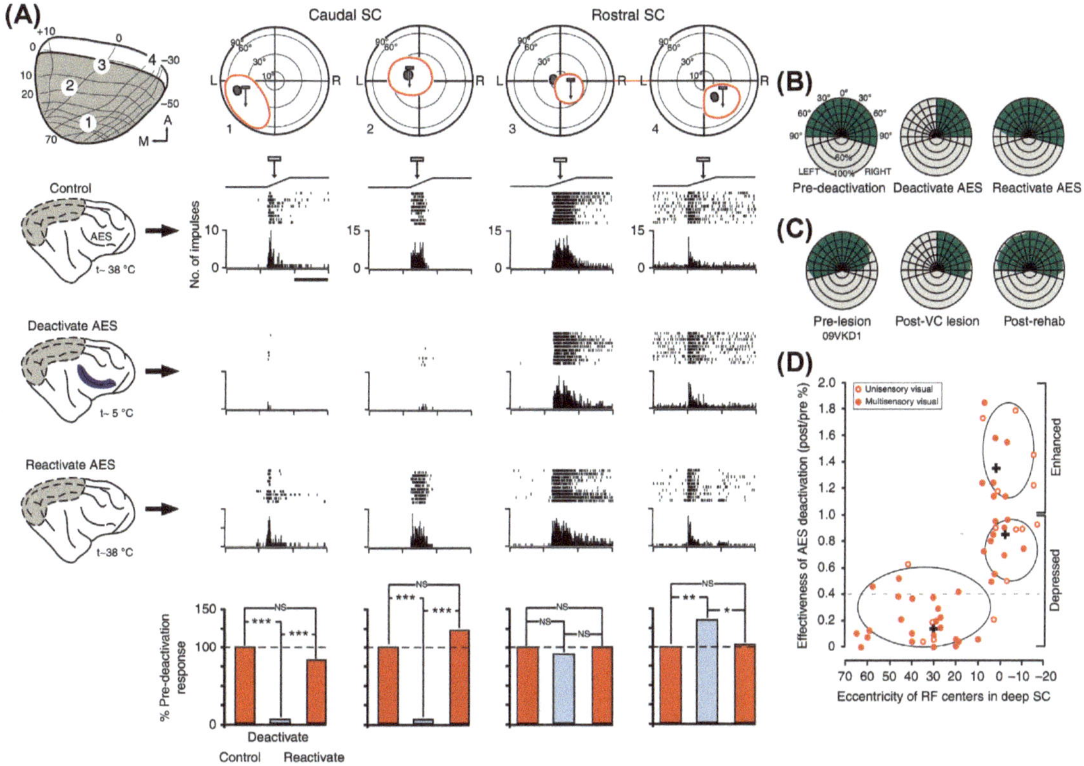

FIGURE 3.12 **Deactivation of AES cortex differentially modulates the visual responsiveness of deep superior colliculus (SC) neurons in rehabilitated animals.** (A) Shown at the top are the visual receptive fields (RF) (*red ovoids*) of four neurons recorded at different topographical loci in the ipsilesional deep SC of rehabilitated animals (left). The movement of a bar of light through the RF (*downward arrow*) is represented by the ramp above the response rasters and histograms. Each neuron's response to three different experimental conditions (control, right AES deactivation, and right AES reactivation) is shown in vertical columns, with the percent of response modulation shown in the bar graphs. Note that deactivation of AES cortex produced a variety of effects, ranging from enhancement (i.e., neuron 4) to a minimal change (i.e., neuron 3) to profound attenuation (i.e., neurons 1 and 2). (B) Following completion of electrophysiological experiments, the animal was evaluated for visual orientation capabilities before, during, and following AES deactivation. Deactivation of the right AES during the behavioral task produced a profound contralesional hemianopia, consistent with the profound attenuation of visually evoked activity in the ipsilesional SC observed previously. Similarly, AES reactivation reinstated visual orientation in the transiently hemianopic field, mirroring the reemergence of visually evoked activity in deep SC neurons after AES reactivation. (C) The loss and reemergence of visual orientation competencies observed during AES deactivation/reactivation recapitulated the behavior observed in this same animal during behavioral testing before, and following multisensory rehabilitation. (D) Illustration of the effects of AES deactivation on evoked visual responsiveness as a function of receptive field eccentricity observed in deactivation experiments (n = 2). K-means cluster analysis revealed that three distinct populations were differentially distributed in the SC. The ellipses delimit 95% confidence levels for each population, with the individual population mean denoted by black +. The greatest response attenuation was observed for neurons representing regions of visual space where orientation behaviors had been lost following visual cortex lesions. By contrast, neurons representing central visual space (and largely unaffected by visual cortex lesions) were either modestly affected or had their visual responsiveness enhanced by AES deactivation. Although both visual unisensory (open symbols) and visual multisensory (closed symbols) neurons were affected by AES deactivation, the majority were multisensory. The horizontal dashed line indicates maximum attenuation of visual responsiveness seen previously in intact animals (see Ref. 29), indicating that AES has a more profound influence on visual activity in rehabilitated animals. *From Jiang H, Stein BE, McHaffie JG. Multisensory training reverses midbrain lesion-induced changes and ameliorates haemianopia.* Nat Commun. *2015;6:7263.*

Postrehabilitation visual capabilities

One intriguing observation made in these rehabilitated animals was their capacity for sophisticated visual behavior. Such behavior is not generally considered the province of the SC. This became evident when two of the rehabilitated animals were trained in a forced-choice discrimination task as shown in Fig. 3.10D.

In one task, they were required to discriminate between horizontal and vertical stripes. They did so just as well as they did before the lesions when the striations were at spatial frequencies of 0.28–2.0 Hz. They began to perform poorly only when the spatial frequencies were substantially higher (3.0–4.0 Hz, Fig. 3.13). Similarly, one of these animals was trained to discriminate an upright triangle from an inverted one and an annulus from a cross.

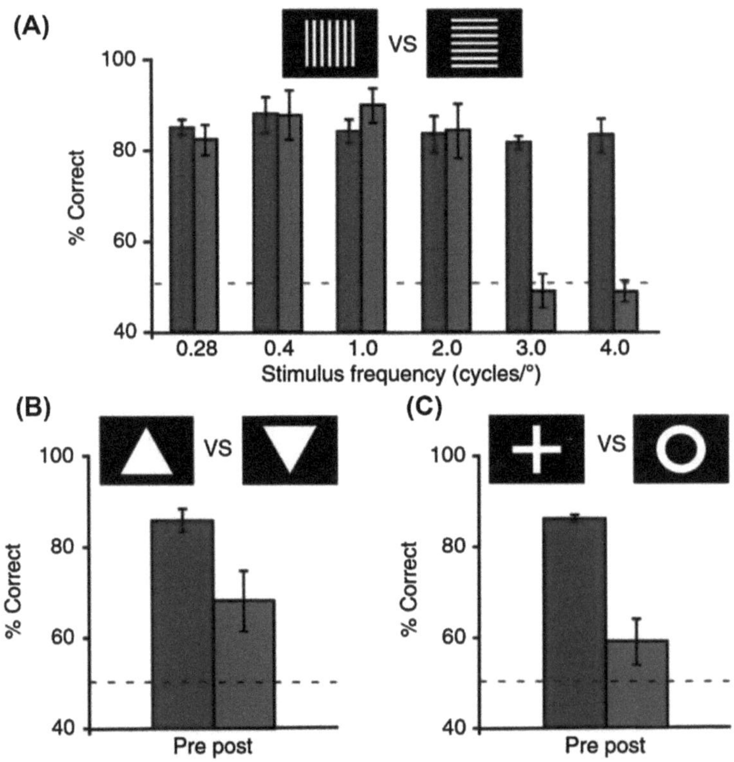

FIGURE 3.13 **Visual discriminative capabilities of rehabilitated animals.** This animal was trained in the orientation task as well as three forced-choice visual discrimination tasks (vertical vs. horizontal gratings; upright vs. inverted triangle; cross vs. annulus) before visual cortical extirpation. Prelesion capabilities are shown in green; post-crossmodal training performance in gray. (A) Before visual cortex removal, the animal performed the spatial frequency task with a high level of proficiency, performing well above chance levels discriminating vertical versus horizontal gratings of spatial frequencies up to the limit tested (4.0 cycles/°). By contrast, after bisensoryl training and the reinstatement of visual orientation behaviors, the animal reacquired the ability to discriminate such stimuli but never achieved the previous level of proficiency, with discriminations of highest levels tested (3.0, 4.0 c/d) falling to chance levels. (B) and (C) Similarly, while performance with more complex stimuli (triangle orientation; cross vs. annulus) did not achieve prelesion levels, these, too, were above chance level. These data indicate that bisensory training facilitates the reacquisition of previously learned discriminations. *From Jiang H, Stein BE, McHaffie JG. Multisensory training reverses midbrain lesion-induced changes and ameliorates haemianopia.* Nat Commun. *2015;6:7263.*

Although its performance was a bit lower than before the lesion, it performed far better than expected. Current concepts of brain organization tend to emphasize the idea that a progressive encephalization of sensory function accompanied the mammalian invention of a neocortex, an evolutionary change that shifted higher-order visual functions out of the midbrain. Whether these sophisticated visual capabilities were actually retained by the mammalian midbrain and simply revealed in this circumstance, or were a consequence of reorganizing its remaining visual circuitry, is not clear. But given that concepts of functional segregation in the brain often lean on evidence from studies using ablation techniques, and cortical ablation can produce excitotoxic damage in its target structures thereby extending the functional consequences of the lesion,[160] it is quite likely that the capabilities of these phylogenetically older circuits have been underestimated.

Acknowledgments

Portions of the work described here were supported by NIH grants EY016716 and NS036916, the Tab Williams Foundation, and the Wallace Foundation.

References

1. Stein BE, London N, Wilkinson LK, Price DD. Enhancement of perceived visual intensity by auditory stimuli: a psychophysical analysis. *J Cogn Neurosci*. 1996;8:497–506.
2. Ernst MO, Banks MS. Humans integrate visual and haptic information in a statistically optimal fashion. *Nature*. 2002;415:429–433.
3. Lovelace CT, Stein BE, Wallace MT. An irrelevant light enhances auditory detection in humans: a psychophysical analysis of multisensory integration in stimulus detection. *Brain Res Cogn Brain Res*. 2003;17:447–453.
4. Alais D, Burr D. The ventriloquist effect results from near-optimal bimodal integration. *Curr Biol*. 2004;14:257–262.
5. Shams L, Ma WJ, Beierholm U. Sound-induced flash illusion as an optimal percept. *Neuroreport*. 2005;16:1923–1927.
6. Jaekl PM, Harris LR. Sounds can affect visual perception mediated primarily by the parvocellular pathway. *Vis Neurosci*. 2009;26:477–486.
7. Stein BE, ed. *The New Handbook of Multisensory Processing*. Cambridge, Mass: MIT Press; 2012.
8. Meredith MA, Stein BE. Interactions among converging sensory inputs in the superior colliculus. *Science*. 1983;221:389–391.
9. King AJ, Palmer AR. Integration of visual and auditory information in bimodal neurones in the Guinea-pig superior colliculus. *Exp Brain Res*. 1985;60:492–500.
10. Meredith MA, Nemitz JW, Stein BE. Determinants of multisensory integration in superior colliculus neurons. I. Temporal factors. *J Neurosci*. 1987;7:3215–3229.
11. Binns KE, Salt TE. Importance of NMDA receptors for multimodal integration in the deep layers of the cat superior colliculus. *J Neurophysiol*. 1996;75:920–930.
12. Wallace MT, Stein BE. Sensory organization of the superior colliculus in cat and monkey. *Prog Brain Res*. 1996;112:301–311.
13. Bell AH, Corneil BD, Meredith MA, Munoz DP. The influence of stimulus properties on multisensory processing in the awake primate superior colliculus. *Can J Exp Psychol*. 2001;55:123–132.
14. Wallace M, Ramachandran R, Stein B. A revised view of sensory cortical parcellation. *Proc Natl Acad Sci USA*. 2004;101:2167–2172.
15. Ghazanfar AA, Schroeder CE. Is neocortex essentially multisensory? *Trends Cognit Sci*. 2006;10:278–285.
16. Jain R, Shore S. External inferior colliculus integrates trigeminal and acoustic information: unit responses to trigeminal nucleus and acoustic stimulation in the Guinea pig. *Neurosci Lett*. 2006;395:71–75.

17. Nagy A, Eördegh G, Paróczy Z, Márkus Z, Benedek G. Multisensory integration in the basal ganglia. *Eur J Neurosci*. 2006;24:917–924.
18. Avillac M, Ben Hamed S, Duhamel J-R. Multisensory integration in the ventral intraparietal area of the macaque monkey. *J Neurosci*. 2007;27:1922–1932.
19. Bizley JK, Nodal FR, Bajo VM, Nelken I, King AJ. Physiological and anatomical evidence for multisensory interactions in auditory cortex. *Cerebr Cortex*. 2007;17:2172–2189.
20. Lakatos P, Chen C-M, O'Connell MN, Mills A, Schroeder CE. Neuronal oscillations and multisensory interaction in primary auditory cortex. *Neuron*. 2007;53:279–292.
21. Romanski LM. Representation and integration of auditory and visual stimuli in the primate ventral lateral prefrontal cortex. *Cerebr Cortex*. 2007;17(Suppl 1):i61–i69.
22. Winkowski DE, Knudsen EI. Top-down control of multimodal sensitivity in the barn owl optic tectum. *J Neurosci*. 2007;27:13279–13291.
23. Bizley JK, King AJ. Visual-auditory spatial processing in auditory cortical neurons. *Brain Res*. 2008;1242:24–36.
24. Reches A, Gutfreund Y. Auditory and multisensory responses in the tectofugal pathway of the barn owl. *J Neurosci*. 2009;29:9602–9613.
25. Zahar Y, Reches A, Gutfreund Y. Multisensory enhancement in the optic tectum of the barn owl: spike count and spike timing. *J Neurophysiol*. 2009;101:2380–2394.
26. Fetsch CR, Pouget A, DeAngelis GC, Angelaki DE. Neural correlates of reliability-based cue weighting during multisensory integration. *Nat Neurosci*. 2012;15:146–154.
27. Lippert MT, Takagaki K, Kayser C, Ohl FW. Asymmetric multisensory interactions of visual and somatosensory responses in a region of the rat parietal cortex. *PLoS One*. 2013;8:e63631.
28. Reig R, Silberberg G. Multisensory integration in the mouse striatum. *Neuron*. 2014;83:1200–1212.
29. Ishikawa T, Shimuta M, Häusser M. Multimodal sensory integration in single cerebellar granule cells in vivo. *Elife*. 2015;4.
30. Costa M, Piché M, Lepore F, Guillemot J-P. Age-related audiovisual interactions in the superior colliculus of the rat. *Neuroscience*. 2016;320:19–29.
31. Felch DL, Khakhalin AS, Aizenman CD. Multisensory integration in the developing tectum is constrained by the balance of excitation and inhibition. *Elife*. 2016;5.
32. Kardamakis AA, Pérez-Fernández J, Grillner S. Spatiotemporal interplay between multisensory excitation and recruited inhibition in the lamprey optic tectum. *Elife*. 2016;5.
33. Bieler M, Sieben K, Cichon N, Schildt S, Röder B, Hanganu-Opatz IL. Rate and temporal coding convey multisensory information in primary sensory cortices. *eNeuro*. 2017;4.
34. Truszkowski TL, Carrillo OA, Bleier J, et al. A cellular mechanism for inverse effectiveness in multisensory integration. *Elife*. 2017;6.
35. McGurk H, MacDonald J. Hearing lips and seeing voices. *Nature*. 1976;264:746–748.
36. Meredith MA, Stein BE. Spatial factors determine the activity of multisensory neurons in cat superior colliculus. *Brain Res*. 1986;365:350–354.
37. Perrault TJ, Vaughan JW, Stein BE, Wallace MT. Superior colliculus neurons use distinct operational modes in the integration of multisensory stimuli. *J Neurophysiol*. 2005;93:2575–2586.
38. Stanford TR, Quessy S, Stein BE. Evaluating the operations underlying multisensory integration in the cat superior colliculus. *J Neurosci*. 2005;25:6499–6508.
39. Stein BE, Meredith MA, Huneycutt WS, McDade L. Behavioral indices of multisensory integration: orientation to visual cues is affected by auditory stimuli. *J Cogn Neurosci*. 1989;1:12–24.
40. Wilkinson LK, Meredith MA, Stein BE. The role of anterior ectosylvian cortex in cross-modality orientation and approach behavior. *Exp Brain Res*. 1996;112:1–10.
41. Burnett LR, Stein BE, Chaponis D, Wallace MT. Superior colliculus lesions preferentially disrupt multisensory orientation. *Neuroscience*. 2004;124:535–547.
42. Gingras G, Rowland BA, Stein BE. The differing impact of multisensory and unisensory integration on behavior. *J Neurosci*. 2009;29:4897–4902.
43. Corneil BD, Munoz DP. The influence of auditory and visual distractors on human orienting gaze shifts. *J Neurosci*. 1996;16:8193–8207.
44. Goldring JE, Dorris MC, Corneil BD, Ballantyne PA, Munoz DP. Combined eye-head gaze shifts to visual and auditory targets in humans. *Exp Brain Res*. 1996;111:68–78.

45. Rowland BA, Quessy S, Stanford TR, Stein BE. Multisensory integration shortens physiological response latencies. *J Neurosci*. 2007;27:5879–5884.
46. Stein BE, Burr D, Constantinidis C, et al. Semantic confusion regarding the development of multisensory integration: a practical solution. *Eur J Neurosci*. 2010;31:1713–1720.
47. Stein BE, Stanford TR, Rowland BA. Development of multisensory integration from the perspective of the individual neuron. *Nat Rev Neurosci*. 2014;15:520–535.
48. Stein BE, Meredith MA. *The Merging of the Senses*. Cambridge, MA: MIT Press; 1993.
49. Sprague JM, Meikle TH. The role of the superior colliculus in visually guided behavior. *Exp Neurol*. 1965;11:115–146.
50. Stein BE, Goldberg SJ, Clamann HP. The control of eye movements by the superior colliculus in the alert cat. *Brain Res*. 1976;118:469–474.
51. Stein BE, Clamann HP. Control of pinna movements and sensorimotor register in cat superior colliculus. *Brain Behav Evol*. 1981;19:180–192.
52. Munoz DP, Pélisson D, Guitton D. Movement of neural activity on the superior colliculus motor map during gaze shifts. *Science*. 1991;251:1358–1360.
53. Paré M, Crommelinck M, Guitton D. Gaze shifts evoked by stimulation of the superior colliculus in the head-free cat conform to the motor map but also depend on stimulus strength and fixation activity. *Exp Brain Res*. 1994;101:123–139.
54. Lomber SG, Payne BR, Cornwell P. Role of the superior colliculus in analyses of space: superficial and intermediate layer contributions to visual orienting, auditory orienting, and visuospatial discriminations during unilateral and bilateral deactivations. *J Comp Neurol*. 2001;441:44–57.
55. Guillaume A, Pélisson D. Kinematics and eye-head coordination of gaze shifts evoked from different sites in the superior colliculus of the cat. *J Physiol*. 2006;577:779–794.
56. Rowland B, Stanford T, Stein B. A Bayesian model unifies multisensory spatial localization with the physiological properties of the superior colliculus. *Exp Brain Res*. 2007;180:153–161.
57. Knudsen EI. Auditory and visual maps of space in the optic tectum of the owl. *J Neurosci*. 1982;2:1177–1194.
58. Miller RL, Stein BE, Rowland BA. Multisensory integration uses a real-time unisensory-multisensory transform. *J Neurosci*. 2017;37:5183–5194.
59. Stein BE, Stanford TR, Ramachandran R, Perrault Jr TJ, Rowland BA. Challenges in quantifying multisensory integration: alternative criteria, models, and inverse effectiveness. *Exp Brain Res*. 2009;198:113–126.
60. Wallace MT, Stein BE. Development of multisensory neurons and multisensory integration in cat superior colliculus. *J Neurosci*. 1997;17:2429–2444.
61. Wallace MT, Stein BE. Onset of cross-modal synthesis in the neonatal superior colliculus is gated by the development of cortical influences. *J Neurophysiol*. 2000;83:3578–3582.
62. Stein BE. The development of a dialogue between cortex and midbrain to integrate multisensory information. *Exp Brain Res*. 2005;166:305–315.
63. Yu L, Rowland BA, Stein BE. Initiating the development of multisensory integration by manipulating sensory experience. *J Neurosci*. 2010;30:4904–4913.
64. Stein BE, Labos E, Kruger L. Sequence of changes in properties of neurons of superior colliculus of the kitten during maturation. *J Neurophysiol*. 1973;36:667–679.
65. Larson MA, Stein BE. The use of tactile and olfactory cues in neonatal orientation and localization of the nipple. *Dev Psychobiol*. 1984;17:423–436.
66. Kao CQ, McHaffie JG, Meredith MA, Stein BE. Functional development of a central visual map in cat. *J Neurophysiol*. 1994;72:266–272.
67. Stein BE, Lábos E, Kruger L. Letter: long-lasting discharge properties of neurons in the kitten midbrain. *Vis Res*. 1973;13:2615–2619.
68. Stein BE, Labos E, Kruger L. Determinants of response latency in neurons of superior colliculus in kittens. *J Neurophysiol*. 1973;36:680–689.
69. Wallace MT, Stein BE. Sensory and multisensory responses in the newborn monkey superior colliculus. *J Neurosci*. 2001;21:8886–8894.
70. Gori M, Giuliana L, Sandini G, Burr D. Visual size perception and haptic calibration during development. *Dev Sci*. 2012;15:854–862.

71. Gori M, Sandini G, Burr D. Development of visuo-auditory integration in space and time. *Front Integr Neurosci.* 2012;6:77.
72. Buisseret P, Gary-Bobo E. Development of visual cortical orientation specificity after dark-rearing: role of extra-ocular proprioception. *Neurosci Lett.* 1979;13:259–263.
73. Crabtree JW, Riesen AH. Effects of the duration of dark rearing on visually guided behavior in the kitten. *Dev Psychobiol.* 1979;12:291–303.
74. Lund RD, Mitchell DE. The effects of dark-rearing on visual callosal connections of cats. *Brain Res.* 1979;167:172–175.
75. Rauschecker JP, Singer W. The effects of early visual experience on the cat's visual cortex and their possible explanation by Hebb synapses. *J Physiol.* 1981;310:215–239.
76. Shinkman PG, Isley MR, Rogers DC. Prolonged dark rearing and development of interocular orientation disparity in visual cortex. *J Neurophysiol.* 1983;49:717–729.
77. Wilkinson M, Shaw C, Khan I, Cynader M. Ontogenesis of beta-adrenergic binding sites in kitten visual cortex and the effects of visual deprivation. *Brain Res.* 1983;283:349–352.
78. Lehmkuhle S, Sherman SM, Kratz KE. Spatial contrast sensitivity of dark-reared cats with striate cortex lesions. *J Neurosci.* 1984;4:2419–2424.
79. Blakemore C, Price DJ. Effects of dark-rearing on the development of area 18 of the cat's visual cortex. *J Physiol.* 1987;384:293–309.
80. Withington-Wray DJ, Binns KE, Keating MJ. The maturation of the superior collicular map of auditory space in the Guinea pig is disrupted by developmental visual deprivation. *Eur J Neurosci.* 1990;2:682–692.
81. Fox K, Daw N, Sato H, Czepita D. Dark-rearing delays the loss of NMDA-receptor function in kitten visual cortex. *Nature.* 1991;350:342–344.
82. Withington DJ. The effect of binocular lid suture on auditory responses in the Guinea-pig superior colliculus. *Neurosci Lett.* 1992;136:153–156.
83. Rauschecker JP, Kniepert U. Auditory localization behaviour in visually deprived cats. *Eur J Neurosci.* 1994;6:149–160.
84. Wallace MT, Perrault Jr TJ, Hairston WD, Stein BE. Visual experience is necessary for the development of multisensory integration. *J Neurosci.* 2004;24:9580–9584.
85. Carriere BN, Royal DW, Perrault TJ, et al. Visual deprivation alters the development of cortical multisensory integration. *J Neurophysiol.* 2007;98:2858–2867.
86. Yu L, Xu J, Rowland BA, Stein BE. Development of cortical influences on superior colliculus multisensory neurons: effects of dark-rearing. *Eur J Neurosci.* 2013.
87. Xu J, Yu L, Rowland BA, Stanford TR, Stein BE. Noise-rearing disrupts the maturation of multisensory integration. *Eur J Neurosci.* 2014;39:602–613.
88. Xu J, Yu L, Rowland BA, Stein BE. The normal environment delays the development of multisensory integration. *Sci Rep.* 2017;7:4772.
89. Xu J, Yu L, Stanford TR, Rowland BA, Stein BE. What does a neuron learn from multisensory experience? *J Neurophysiol.* 2015;113:883–889.
90. Xu J, Yu L, Rowland BA, Stanford TR, Stein BE. Incorporating cross-modal statistics in the development and maintenance of multisensory integration. *J Neurosci.* 2012;32:2287–2298.
91. Wallace MT, Stein BE. Early experience determines how the senses will interact. *J Neurophysiol.* 2007;97:921–926.
92. De Gelder B, Bertelson P. Multisensory integration, perception and ecological validity. *Trends Cognit Sci.* 2003;7:460–467.
93. Beauchamp MS, Lee KE, Argall BD, Martin A. Integration of auditory and visual information about objects in superior temporal sulcus. *Neuron.* 2004;41:809–823.
94. Barraclough NE, Xiao D, Baker CI, Oram MW, Perrett DI. Integration of visual and auditory information by superior temporal sulcus neurons responsive to the sight of actions. *J Cogn Neurosci.* 2005;17:377–391.
95. Baier B, Kleinschmidt A, Müller NG. Cross-modal processing in early visual and auditory cortices depends on expected statistical relationship of multisensory information. *J Neurosci.* 2006;26:12260–12265.
96. Fiebelkorn IC, Foxe JJ, Molholm S. Dual mechanisms for the cross-sensory spread of attention: how much do learned associations matter? *Cerebr Cortex.* 2010;20:109–120.

97. Butler BE, Chabot N, Lomber SG. A quantitative comparison of the hemispheric, areal, and laminar origins of sensory and motor cortical projections to the superior colliculus of the cat: corticotectal inputs to cat superior colliculus. *J Comp Neurol*. 2016;524:2623–2642.
98. Clemo HR, Stein BE. Topographic organization of somatosensory corticotectal influences in cat. *J Neurophysiol*. 1984;51:843–858.
99. Clemo HR, Stein BE. Effects of cooling somatosensory cortex on response properties of tactile cells in the superior colliculus. *J Neurophysiol*. 1986;55:1352–1368.
100. Wallace MT, Stein BE. Cross-modal synthesis in the midbrain depends on input from cortex. *J Neurophysiol*. 1994;71:429–432.
101. Jiang W, Wallace MT, Jiang H, Vaughan JW, Stein BE. Two cortical areas mediate multisensory integration in superior colliculus neurons. *J Neurophysiol*. 2001;85:506–522.
102. Jiang W, Jiang H, Stein BE. Two corticotectal areas facilitate multisensory orientation behavior. *J Cogn Neurosci*. 2002;14:1240–1255.
103. Jiang W, Jiang H, Stein BE. Neonatal cortical ablation disrupts multisensory development in superior colliculus. *J Neurophysiol*. 2006;95:1380–1396.
104. Jiang W, Jiang H, Rowland BA, Stein BE. Multisensory orientation behavior is disrupted by neonatal cortical ablation. *J Neurophysiol*. 2007;97:557–562.
105. Alvarado JC, Stanford TR, Rowland BA, Vaughan JW, Stein BE. Multisensory integration in the superior colliculus requires synergy among corticocollicular inputs. *J Neurosci*. 2009;29:6580–6592.
106. Mucke L, Norita M, Benedek G, Creutzfeldt O. Physiologic and anatomic investigation of a visual cortical area situated in the ventral bank of the anterior ectosylvian sulcus of the cat. *Exp Brain Res*. 1982;46:1–11.
107. Stein BE, Spencer RF, Edwards SB. Corticotectal and corticothalamic efferent projections of SIV somatosensory cortex in cat. *J Neurophysiol*. 1983;50:896–909.
108. Norita M, Mucke L, Benedek G, Albowitz B, Katoh Y, Creutzfeldt OD. Connections of the anterior ectosylvian visual area (AEV). *Exp Brain Res*. 1986;62:225–240.
109. Meredith MA, Clemo HR. Auditory cortical projection from the anterior ectosylvian sulcus (Field AES) to the superior colliculus in the cat: an anatomical and electrophysiological study. *J Comp Neurol*. 1989;289:687–707.
110. Scannell JW, Sengpiel F, Tovée MJ, Benson PJ, Blakemore C, Young MP. Visual motion processing in the anterior ectosylvian sulcus of the cat. *J Neurophysiol*. 1996;76:895–907.
111. Harting JK, Feig S, Van Lieshout DP. Cortical somatosensory and trigeminal inputs to the cat superior colliculus: light and electron microscopic analyses. *J Comp Neurol*. 1997;388:313–326.
112. Wallace MT, Meredith MA, Stein BE. Converging influences from visual, auditory, and somatosensory cortices onto output neurons of the superior colliculus. *J Neurophysiol*. 1993;69:1797–1809.
113. Stein BE, Wallace MW, Stanford TR, Jiang W. Cortex governs multisensory integration in the midbrain. *Neuroscientist*. 2002;8:306–314.
114. Alvarado JC, Vaughan JW, Stanford TR, Stein BE. Multisensory versus unisensory integration: contrasting modes in the superior colliculus. *J Neurophysiol*. 2007;97:3193–3205.
115. Alvarado JC, Stanford TR, Vaughan JW, Stein BE. Cortex mediates multisensory but not unisensory integration in superior colliculus. *J Neurosci*. 2007;27:12775–12786.
116. Alvarado JC, Rowland BA, Stanford TR, Stein BE. A neural network model of multisensory integration also accounts for unisensory integration in superior colliculus. *Brain Res*. 2008;1242:13–23.
117. Anastasio TJ, Patton PE, Belkacem-Boussaid K. Using Bayes' rule to model multisensory enhancement in the superior colliculus. *Neural Comput*. 2000;12:1165–1187.
118. Rowland BA, Stanford TR, Stein BE. A model of the neural mechanisms underlying multisensory integration in the superior colliculus. *Perception*. 2007;36:1431–1443.
119. Cuppini C, Ursino M, Magosso E, Rowland BA, Stein BE. An emergent model of multisensory integration in superior colliculus neurons. *Front Integr Neurosci*. 2010;4:6.
120. Fuentes-Santamaria V, Alvarado JC, Stein BE, McHaffie JG. Cortex contacts both output neurons and nitrergic interneurons in the superior colliculus: direct and indirect routes for multisensory integration. *Cerebr Cortex*. 2008;18:1640–1652.
121. McHaffie JG, Kruger L, Clemo HR, Stein BE. Corticothalamic and corticotectal somatosensory projections from the anterior ectosylvian sulcus (SIV cortex) in neonatal cats: an anatomical demonstration with HRP and ^{3}H-leucine. *J Comp Neurol*. 1988;274:115–126.

122. Fuentes-Santamaria V, McHaffie JG, Stein BE. Maturation of multisensory integration in the superior colliculus: expression of nitric oxide synthase and neurofilament SMI-32. *Brain Res*. 2008;1242:45–53.
123. Rowland BA, Jiang W, Stein BE. Brief cortical deactivation early in life has long-lasting effects on multisensory behavior. *J Neurosci*. 2014;34:7198–7202.
124. Cuppini C, Stein BE, Rowland BA, Magosso E, Ursino M. A computational study of multisensory maturation in the superior colliculus (SC). *Exp Brain Res*. 2011;213:341–349.
125. Cuppini C, Magosso E, Rowland B, Stein B, Ursino M. Hebbian mechanisms help explain development of multisensory integration in the superior colliculus: a neural network model. *Biol Cybern*. 2012;106:691–713.
126. Bi G, Poo M. Synaptic modification by correlated activity: Hebb's postulate revisited. *Annu Rev Neurosci*. 2001;24:139–166.
127. Yu L, Stein BE, Rowland BA. Adult plasticity in multisensory neurons: short-term experience-dependent changes in the superior colliculus. *J Neurosci*. 2009;29:15910–15922.
128. Brett-Green BA, Miller LJ, Schoen SA, Nielsen DM. An exploratory event-related potential study of multisensory integration in sensory over-responsive children. *Brain Res*. 2010;1321:67–77.
129. Brandwein AB, Foxe JJ, Butler JS, et al. The development of multisensory integration in high-functioning autism: high-density electrical mapping and psychophysical measures reveal impairments in the processing of audiovisual inputs. *Cerebr Cortex*. 2013;23:1329–1341.
130. Fiebelkorn IC, Foxe JJ, McCourt ME, Dumas KN, Molholm S. Atypical category processing and hemispheric asymmetries in high-functioning children with autism: revealed through high-density EEG mapping. *Cortex*. 2013;49:1259–1267.
131. Stevenson RA, Siemann JK, Schneider BC, et al. Multisensory temporal integration in autism spectrum disorders. *J Neurosci*. 2014;34:691–697.
132. van Laarhoven T, Keetels M, Schakel L, Vroomen J. Audio-visual speech in noise perception in dyslexia. *Dev Sci*; 2016. Available at: http://doi.wiley.com/10.1111/desc.12504.
133. Schorr EA, Fox NA, van Wassenhove V, Knudsen EI. Auditory-visual fusion in speech perception in children with cochlear implants. *Proc Natl Acad Sci USA*. 2005;102:18748–18750.
134. Putzar L, Hötting K, Röder B. Early visual deprivation affects the development of face recognition and of audio-visual speech perception. *Restor Neurol Neurosci*. 2010;28:251–257.
135. Putzar L, Gondan M, Röder B. Basic multisensory functions can be acquired after congenital visual pattern deprivation in humans. *Dev Neuropsychol*. 2012;37:697–711.
136. Ogasawara K, McHaffie JG, Stein BE. Two visual corticotectal systems in cat. *J Neurophysiol*. 1984;52:1226–1245.
137. Edwards SB, Ginsburgh CL, Henkel CK, Stein BE. Sources of subcortical projections to the superior colliculus in the cat. *J Comp Neurol*. 1979;184:309–329.
138. Berson DM, McIlwain JT. Retinal Y-cell activation of deep-layer cells in superior colliculus of the cat. *J Neurophysiol*. 1982;47:700–714.
139. Huerta MF, Harting JK. The projection from the nucleus of the posterior commissure to the superior colliculus of the cat: patch-like endings within the intermediate and deep grey layers. *Brain Res*. 1982;238:426–432.
140. Rhoades RW, Mooney RD, Rohrer WH, Nikoletseas MM, Fish SE. Organization of the projection from the superficial to the deep layers of the hamster's superior colliculus as demonstrated by the anterograde transport of Phaseolus vulgaris leucoagglutinin. *J Comp Neurol*. 1989;283:54–70.
141. May PJ. The mammalian superior colliculus: laminar structure and connections. *Prog Brain Res*. 2006;151:321–378.
142. Sprague JM. Interaction of cortex and superior colliculus in mediation of visually guided behavior in the cat. *Science*. 1966;153:1544–1547.
143. Sherman SM. Monocularly deprived cats: improvement of the deprived eye's vision by visual decortication. *Science*. 1974;186:267–269.
144. Wallace SF, Rosenquist AC, Sprague JM. Recovery from cortical blindness mediated by destruction of nontectotectal fibers in the commissure of the superior colliculus in the cat. *J Comp Neurol*. 1989;284:429–450.
145. Wallace SF, Rosenquist AC, Sprague JM. Ibotenic acid lesions of the lateral substantia nigra restore visual orientation behavior in the hemianopic cat. *J Comp Neurol*. 1990;296:222–252.
146. Durmer JS, Rosenquist AC. Ibotenic acid lesions in the pedunculopontine region result in recovery of visual orienting in the hemianopic cat. *Neuroscience*. 2001;106:765–781.

147. Bolognini N, Rasi F, Coccia M, Làdavas E. Visual search improvement in hemianopic patients after audio-visual stimulation. *Brain*. 2005;128:2830–2842.
148. Leo F, Bolognini N, Passamonti C, Stein BE, Làdavas E. Cross-modal localization in hemianopia: new insights on multisensory integration. *Brain*. 2008;131:855–865.
149. Passamonti C, Bertini C, Làdavas E. Audio-visual stimulation improves oculomotor patterns in patients with hemianopia. *Neuropsychologia*. 2009;47:546–555.
150. Dundon NM, Bertini C, Làdavas E, Sabel BA, Gall C. Visual rehabilitation: visual scanning, multisensory stimulation and vision restoration trainings. *Front Behav Neurosci*. 2015;9:192.
151. Jiang H, Stein BE, McHaffie JG. Multisensory training reverses midbrain lesion-induced changes and ameliorates haemianopia. *Nat Commun*. 2015;6:7263.
152. Casagrande VA, Harting JK, Hall WC, Diamond IT, Martin GF. Superior colliculus of the tree shrew: a structural and functional subdivision into superficial and deep layers. *Science*. 1972;177:444–447.
153. Behan M, Appell PP. Intrinsic circuitry in the cat superior colliculus: projections from the superficial layers. *J Comp Neurol*. 1992;315:230–243.
154. Sparks DL, Hartwich-Young R. The deep layers of the superior colliculus. *Rev Oculomot Res*. 1989;3:213–255.
155. Girman SV, Lund RD. Most superficial sublamina of rat superior colliculus: neuronal response properties and correlates with perceptual figure-ground segregation. *J Neurophysiol*. 2007;98:161–177.
156. Tamietto M, Cauda F, Corazzini LL, et al. Collicular vision guides nonconscious behavior. *J Cogn Neurosci*. 2010;22:888–902.
157. McHaffie JG, Norita M, Dunning DD, Stein BE. Corticotectal relationships: direct and "indirect" corticotectal pathways. *Prog Brain Res*. 1993;95:139–150.
158. Redgrave P, Coizet V, Comoli E, et al. Interactions between the midbrain superior colliculus and the basal ganglia. *Front Neuroanat*. 2010;4.
159. Redgrave P, Vautrelle N, Reynolds JNJ. Functional properties of the basal ganglia's re-entrant loop architecture: selection and reinforcement. *Neuroscience*. 2011;198:138–151.
160. Jiang H, Stein BE, McHaffie JG. Cortical lesion-induced visual hemineglect is prevented by NMDA antagonist pretreatment. *J Neurosci*. 2009;29:6917–6925.
161. Stein BE, Stanford TR. Multisensory integration: current issues from the perspective of the single neuron. *Nature Reviews Neuroscience*. 2008;9:255–266.

CHAPTER

4

The development of multisensory processes for perceiving the environment and the self

David J. Lewkowicz[1], *Andrew J. Bremner*[2]

[1]Haskins Laboratories, New Haven, CT, United States; [2]School of Psychology, University of Birmingham, Birmingham, United Kingdom

Our external and internal worlds are specified by concurrent sources of electromagnetic, mechanical, and chemosensory energy. When transduced by our sensory systems, these various sources of physical energy result in visual, auditory, somatosensory, vestibular, olfactory, and gustatory sensations. This mélange of different combinations of multisensory inputs provides us with concurrent and often equivalent information that, when integrated, enables us to experience unitary and psychologically meaningful objects, events, and self-generated actions.[1,2]

When and how does the ability to extract multisensory unity develop in humans? Two classic views have provided the theoretical framework for thinking about this question. The developmental integration view holds that, at birth, infants do not perceive the unity of their multisensory world and that they only acquire it gradually through experience.[3,4] In contrast, the developmental differentiation view holds that, at birth, infants can perceive some forms of unity by perceiving some of the amodal invariants that are available in their perceptual array and that, with growth and experience, they gradually become capable of perceiving increasingly more complex forms of multisensory unity.[5] Adjudicating which theoretical framework best represents the development of multisensory processing is critical because most domains of human functioning, including perception, cognition, emotion, social interaction, and sensorimotor integration, depend critically on our ability to perceive multisensory unity.[4–7]

This chapter provides an overview of what is currently known about the development of multisensory processing across two broad domains of perception. In the first part of this chapter, we will examine how infants and children process audiovisual information specifying the movements, locations, and utterances of objects and people in their external environment. The

Multisensory Perception
https://doi.org/10.1016/B978-0-12-812492-5.00004-8

emphasis in this section will be on the effects of early experience on the development of these abilities and on the role of selective attention (SA) in infant responsiveness to audiovisual information. In the second part of this chapter, we describe what is currently known about the development of infants' and young children's ability to perceive their own bodies and selves and how they fit into their physical and social surroundings: multisensory perceptual abilities which typically involve tactile inputs. Overall, we show that multisensory abilities unfold over an extended period of development and that experience plays an important role in these processes.

Infant perception of the audiovisual attributes specifying inanimate and animate objects

The perception of unitary and psychologically meaningful inanimate and animate objects and events depends on our ability to extract multisensory unity. The challenge for infants is that they must perceive such unity despite their neural and behavioral immaturity and their general lack of experience. Fortunately, the world provides a variety of physical cues that can readily provide the basis for the perception of multisensory unity. In the audiovisual domain, objects and events are usually specified by concurrent modality-specific perceptual attributes as well as by modality-independent attributes (these are also referred to as amodal attributes in Gibson's ecological theory of perception). For example, people are usually specified by various modality-specific attributes such as the color, shape, and texture of their face and the pitch and timbre of their voice. In addition, people are usually specified by various modality-independent qualitative attributes such as affect, gender, and identity and by various modality-independent temporal attributes such as the duration, tempo, and rhythm of their vocalizations and concurrent facial articulations. In general, constellations of modality-specific attributes must be learned through association to be perceived as belonging to unitary objects and events because they are unique to their specific modalities. In contrast, modality-independent attributes are often considered to be perceived directly because they can be perceived regardless of whether we are looking or listening.

The modality-specific and modality-independent multisensory attributes that usually specify people, objects, and events provide a great deal of redundant multisensory information. This is advantageous because multisensory redundancy is known to increase perceptual salience and this, in turn, is known to enhance perception, attention, learning, and memory across the life span in humans[8–12] and in nonhuman species.[2,10–15] To illustrate, at the neural level, studies have found that some superior colliculus neurons exhibit superadditive responses to multisensory as opposed to unisensory inputs, meaning that they respond much more vigorously to multisensory than unisensory inputs.[16] At the behavioral level, studies have found that detection of speech is better when it can be simultaneously heard and seen,[17,18] that comprehension of noisy auditory speech is better when it can be lip-read,[17,19–21] and that processing of ambiguous auditory or silent speech is facilitated by conscious[22,23] and unconscious lip-reading.[24]

In addition to inducing redundancy effects, multisensory inputs can induce a variety of unique illusions. These include (a) the McGurk effect in which the perception of an audible speech articulation is altered by a conflicting visual speech articulation, (b) the ventriloquism illusion in which the perception of the location of a sound is mislocalized to an ostensible but not actual visual sound source,[25–27] (c) the double-flash illusion[28,29] where a single flash is seen as two flashes when it is accompanied by two sounds presented in rapid succession, (d) the stream-bounce illusion[30,31] where two objects moving in opposite directions along the same path are perceived as bouncing against one another if a sound occurs when they intersect as they pass through one another, and (e) the rubber hand illusion,[32–34] in which the sight of a fake hand being stroked in synchrony with tactile strokes felt on the participant's real hand result in the experience of ownership over the fake hand and mislocalization of the participant's real hand toward the fake hand (this last illusion is discussed in detail in the Chapter 8; see also Chapter 18).

Overall, multisensory redundancy and illusion effects provide clear evidence that multisensory inputs induce unique perceptual experiences. This means that it is imperative to study the development of multisensory functions in their own right if we are to gain a more complete understanding of the development of perception, action, cognition, and social behavior.

Development of audiovisual processing in infancy

It is somewhat difficult to organize the extant findings on infant processing of audiovisual information because the studies in this area have investigated processing in multiple domains and because many of them are demonstration studies. Of course, this reflects the natural progression of a field where initial studies aim to test the big theoretical questions. Here, the big question has been whether infants can perceive the coherence of their multisensory world or, as famously claimed by William James, do they just perceive "blooming, buzzing confusion." Overall, studies have found that infants begin life with some relatively primitive audiovisual processing abilities, that these abilities improve gradually during infancy and beyond, and that experience plays a key role in their development.[35–37]

Studies with newborns have found that their response to visual stimulation is influenced by auditory stimulation,[38] that they are sensitive to intensity-based audiovisual equivalence,[39] that they detect the temporal synchrony of auditory and visual inputs,[40] and that they can associate arbitrary auditory and visual stimulus attributes based on their temporal synchrony.[41] Together, these findings illustrate that newborn infants enter the postnatal world with relatively low-level, rudimentary multisensory processing abilities. The existence of these early abilities challenges to some extent the idea of the blooming, buzzing confusion advocated by William James. Nonetheless, these basic abilities limit young infants' perception of multisensory unity. This is evident in the fact that even though newborns perceive the multisensory unity of facial and vocal articulations based on their temporal coincidence, they do not accomplish this task by detecting the unity of higher-level multisensory identity cues.[40] That is, they do not detect the link between a particular face defined by its specific features and the specific spectral characteristics of the concurrent vocalization.

Despite the fact that newborns only possess a rudimentary multisensory-unity detection system, this system provides them with a powerful initial gateway for the discovery of the unity of their everyday audiovisual world. Its power is illustrated by the fact that infants continue to rely on it to detect a variety of multisensory relations. For example, 3- to

4-month-old infants continue to rely on their ability to detect temporal audiovisual synchrony to perceive relatively simple spatiotemporal audiovisual relations, including those between bouncing objects and their impact sounds[42–47] and those between specific human faces and voices.[48,49] Interestingly, even though infants continue to rely on low-level synchrony cues for the detection of multisensory unity, with growth and increasing perceptual differentiation, they begin to perceive higher-level forms of unity. A developmentally early example of this is the fact that as early as 2 months of age, infants exhibit the ability to perceive the equivalence of isolated auditory and visual vowels based solely on their identity.[50–52] Another, but developmentally later example, is the finding that whereas young infants (e.g., 3- and 5-month olds) depend on audiovisual temporal synchrony cues to learn arbitrary multisensory associations (e.g., color-pitch or shape-sound), older infants (e.g., 7-month olds) no longer depend on them for successful learning and memory.[53,54] These two examples illustrate the general theoretical principle that, across the life span, multisensory processing is the result of the complex interaction of stimulus-, task-, and experience-dependent factors.[35,36,55] This means that cataloging the emergence of the ability to perceive multisensory unity based on a single factor (e.g., age) is inappropriate. Instead, multiple factors must be taken into account when considering the detection of a specific multisensory unity cue.

Keeping the above theoretical caveats in mind, it is interesting to note that, as infants continue to grow and acquire perceptual experience, their ability to perceive audiovisual unity continues to improve. For instance, whereas at 4 months of age infants do not perceive the spatiotemporal unity of moving objects and sounds inherent in illusion-inducing audiovisual events, they do perceive it by 6 and 8 months of age.[30] Similarly, by 8 months of age, infants begin to exhibit the ability to perceive more complex multisensory unity cues such as distance,[56] gender,[57–60] and affect.[61] In addition, even though infants as young as 2 months of age exhibit faster spatial localization to combined (and synchronized) audiovisual spatial cues relative to auditory-only or visual-only cues, it is not until 8 months of age that infants begin to exhibit the ability to integrate auditory and visual localization cues in an adult-like, nonlinear manner.[62] Crucially, this newly emerged ability at 8 months reflects a growing reliance on simultaneously coordinated spatial and temporal cues. This conclusion is supported by findings that 2- and 4-month olds associate sounds and objects based on their temporal synchrony even if they are not colocated in space but that 6- and 8-month olds only associate objects and sounds if they are temporally synchronized and colocated.[63]

Finally, by the end of the first year of life, infants begin to exhibit relatively sophisticated multisensory processing abilities. For example, by 12 months of age, infants begin to perceive the equivalence of auditory and visual cues specifying linguistic identity.[64] Likewise, by 12–14 months of age, infants begin to perceive the equivalence of native and nonnative auditory and visual fluent speech and they can even perceive it when the native auditory and visual speech streams are temporally desynchronized but not when the nonnative streams are desynchronized.[65] These findings are interesting in light of the findings cited earlier indicating that infants as young as 2 months of age can perceive the unity of isolated auditory and visual speech syllables. Together, these two sets of findings illustrate the theoretical principle that multisensory processing is stimulus-, task-, and experience-dependent. Specifically, they show that stimulus-processing/task difficulty partly determines when in development infants can first detect the coherence of audiovisual speech and that early linguistic experience determines perception of multisensory unity.

Effects of early experience on audiovisual processing

In the previous section, we have shown that multisensory processing begins as a relatively primitive capacity, that it improves gradually across early development, and that early experience plays a key role in its developmental emergence. In this section, we review evidence from various species showing that early experience with coordinated auditory and visual inputs plays a critical role in setting up multisensory processing mechanisms.

One example that illustrates the effects of early experience on multisensory responsiveness in human infants comes from a study of infant response to audiovisual affect. In this study, it was found that infants as young as 3.5 months of age could match the auditory and visual attributes of affect produced by their caregiver but that only 7-month-old infants could match such attributes when they were produced by a stranger.[61,66] Deprivation studies also provide evidence of the importance of early experience. For example, studies have found that individuals who are deprived of visual input for several months after birth due to congenital cataracts—these are removed in infancy—exhibit audiovisual integration deficits in adulthood.[67,68] Similarly, studies have found that individuals who are born deaf and who are deprived of auditory input for several months until they receive a cochlear implant exhibit difficulties in identification of fluent audiovisual speech and deficits in audiovisual speech integration[69,70] (see Chapter 16). Finally, studies have found that children with autism, who typically attend less to their social partners' faces, exhibit poor audiovisual speech integration[71] and impaired perception of audiovisual temporal relations[72,73] (see Chapter 17).

Together, these behavioral findings suggest that typical experience with spatiotemporally coordinated auditory and visual inputs is essential for setting up appropriate multisensory processing mechanisms. This, of course, implies that the nervous system is initially plastic and open to the effects of early experience. Indeed, animal studies that have manipulated the specific type of spatiotemporal experience early in life either through deprivation or enrichment, or studies that have rerouted neural connections early in life, have found a great deal of plasticity and an openness to experience. For example, it has been found that the multisensory neurons in the superior colliculus that normally integrate auditory, visual, and somatosensory localization cues are not present in newborn cats; these animals only develop typical integrative capacities if they are exposed to spatiotemporally coordinated auditory and visual inputs early in development[74,75] (see Chapter 3). Similarly, studies show that the spatial tuning and calibration of the neural map of auditory space in newborn ferrets and barn owls requires that the animals be exposed to concurrent, spatiotemporally aligned, audiovisual inputs early in life.[76,77]

The plasticity found at lower levels of the nervous system has also been found at the cortical level across different species, including humans.[78] For instance, when normal visual input in neonatal ferrets is rerouted to the auditory cortex, neurons in the primary auditory cortex become responsive to visual input, develop organized orientation-selective modules normally found in visual cortex,[79] and support visually appropriate behavioral responsiveness.[80] This sort of crossmodal reorganization early in development also occurs when a cortical area is deprived of its expected sensory input. For example, studies in young cats show that low-level neural circuits in the auditory cortex exhibit facilitated responsiveness to visual and somatosensory inputs when the auditory cortex is deprived of its expected auditory input[81] (see Chapter 16). Finally, crossmodal cortical reorganization effects have been found in humans. For example, adults who are deprived of patterned visual input in infancy due to congenital cataracts exhibit responsiveness to auditory inputs in occipital cortex[82] (see Chapter 15).

Early experience and audiovisual processing: multisensory perceptual narrowing

The classic theoretical views on the development of multisensory processing hold that early experience plays a role in the improvement and broadening of multisensory perceptual capacity but do not admit to the possibility that early experience may simultaneously have a regressive effect on such capacity. Such regressive effects can be found in altricial species who usually have a poorly tuned perceptual system at birth and who depend on early experience with species-specific sensory inputs to tune their perceptual system to match the exigencies of their ecological setting. Gottlieb's[83,84] studies of canalization in birds provide a classic example of such tuning. Gottlieb showed that mallard hatchlings' preference for their species-specific maternal call depends on prenatal exposure to their mother's calls, other siblings' embryonic vocalizations, and self-produced embryonic vocalizations.

Humans are an altricial species and, thus, regression is also likely to play an important role in human perceptual development. Indeed, studies of human infants have found evidence of regression in the form of perceptual narrowing. This has been found across multiple unisensory domains, including phonetic, face, and music processing. Importantly, in humans, the experience-dependent narrowing occurs after birth. In general, the narrowing leads to a decline in sensitivity to nonnative perceptual inputs—due to the lack of experience with such inputs[85,86]—and to a concurrent emergence of specialization and expertise for native inputs—due to the acquisition of increasingly greater experience with native inputs. Perceptual narrowing is illustrated by findings that (a) younger infants (between birth and around 6–8 months of age) can discriminate many of the consonant distinctions found across the world's languages but that older infants only discriminate native ones, (b) that younger infants can discriminate same- and other-species faces and own- and other-race faces but that older infants can only discriminate same-species and own-race faces (see Minar and Lewkowicz[87] for evidence of no narrowing to dynamic/vocalizing other-race faces), and (c) that younger infants can discriminate the metrical structure of non-Western and Western musical rhythms but that older Western infants can only discriminate Western rhythms (for a review of this work, see Lewkowicz[35]).

Given the domain-general nature of unisensory perceptual narrowing, Lewkowicz and Ghazanfar[88] investigated the possibility that perceptual narrowing might also be a pansensory process. They exposed 4-, 6-, 8-, and 10-month-old infants to side-by-side faces of the same rhesus monkey producing two different calls (i.e., a coo and a grunt) and compared visual preferences to the visible calls in silence versus preferences in the presence of one of the matching audible calls. Results showed that the 4- and 6-month-old infants matched the visible and audible calls by looking at the visible call more in the presence than in the absence of the matching audible call. By contrast, the 8- and 10-month-old infants did not match, indicating that multisensory perception has narrowed by this time in development. In a follow-up study, Lewkowicz et al.[89] showed that the successful matching found in the younger infants by Lewkowicz and Ghazanfar[88] was not due to the detection of audible and visible identity cues but to the detection of the temporal synchrony between the visible and audible calls. This was evident in the fact that younger infants no longer matched when the calls were desynchronized. Importantly, Lewkowicz et al.[89] also showed that the decline in cross-species matching persists into the second year of life in that neither 12- nor 18-month-old infants performed cross-species multisensory matching. In a second follow-up study, Lewkowicz et al.[40] examined matching of monkey faces and vocalizations in newborn infants and found that newborns also match visible and audible monkey calls and that they, like 4- and 6-month-olds, accomplish this task by relying on the temporal synchrony of the visible and audible calls. Finally, Grossmann et al.[90] used event-related potentials to study the neural signature of multisensory perceptual narrowing in infancy and, consistent with the behavioral data, found evidence of narrowing in the brain of 8-month-old but not of 4-month-old infants.

The cross-species multisensory narrowing results prompted Lewkowicz and colleagues[91] to ask whether multisensory perceptual narrowing extends to audiovisual speech processing. To do so, they examined infant ability to match the visible and audible syllables /ba/ and /va/ in English-learning and Spanish-learning infants at 6 and 11 months of age. They expected that the Spanish-learning infants would not match at the older age because /ba/ and /va/ are not perceptually distinct for Spanish speakers and because these Spanish-learning infants become native-language perceptual experts by 11 months of age. To examine matching, infants were shown side-by-side faces of the same person articulating a /ba/ on one side and a /va/ on the other in silence, then they were familiarized with either one or the other audible syllable in the absence of the visible syllables, and then they again saw the two visible-only syllables in silence. As predicted, results indicated that English-learning infants preferred the matching visible syllable following familiarization with the audible syllable at both ages but that the Spanish-learning infants only matched the syllables at 6 months of age. These findings, together with those from the cross-species multisensory matching studies, show that multisensory perceptual narrowing is a pan-sensory process.

Development of selective attention to multisensory inputs

As is apparent from the above review, evidence indicates that multisensory processing abilities emerge gradually in infancy and that experience plays a key role in their developmental emergence. Of course, such abilities are not useful unless infants are able to select relevant multisensory information for processing. In principle, doing so is challenging because our everyday environment is usually cluttered with multiple objects. Some of these objects are stationary and silent, while others are dynamic and produce sounds. To detect the latter types of objects, to learn about them, and to generate appropriate actions with respect to them, one needs a SA system that can not only rapidly and efficiently select such objects for processing but that can detect the multisensory redundancy inherent in them.

Until recently, it was not known when such a system begins operating in infancy. This is because most studies of the development of SA in infancy to date have focused on the development of visual attention. These studies have found that between birth and around 6 months of age, infants respond reflexively to visual stimulation and that after 6 months of age they begin to respond to it in a goal-directed, voluntary fashion.[92–99] In other words, infant's visual attention is largely under the control of exogenous factors initially but by around 6 months of age endogenous factors begin to control attention as well.

Recent studies by Lewkowicz and colleagues have investigated infant SA to multisensory inputs and have found that the exogenous to endogenous shift occurs in response to audiovisual inputs as well. In addition, these studies have identified unique patterns of shifting attention across early development that are specifically the result of multisensory processing. In these studies, Lewkowicz and colleagues examined infant SA to talking faces to determine which parts of a talking face attract infant SA and whether and why the part of greatest interest changes with development. In the first of these studies, Lewkowicz and Hansen-Tift[100] examined SA to a talker's eyes and mouth, in 4-, 6-, 8-, 10-, and 12-month-old infants and adults, while a female talker spoke in either the participants' native language or in a nonnative language. As Fig. 4.1 shows, there were dramatic age-related shifts in SA: at 4 months of age infants attended more to the talker's eyes regardless of whether the talker spoke in the native or in a nonnative language, whereas by 8 and 10 months of age infants shifted their attention to the talker's mouth. Furthermore, by 12 months of age, infants no longer attended more to the talker's mouth when she spoke in the infants' native language, but they still attended more to the talker's mouth when she spoke in a nonnative language. In contrast, adults attended more to the talker's eyes regardless of language spoken. Importantly, the task here involved passive exposure to talking faces (see below for adults' response when instructed to process specific information).

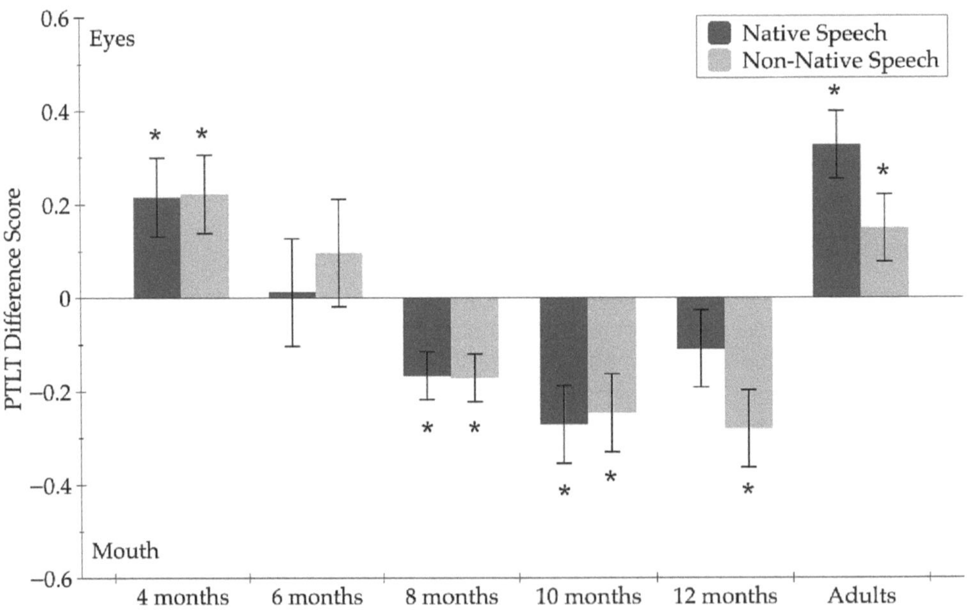

FIGURE 4.1 Mean PTLT (proportion of total looking time) difference scores as a function of age and type of language. The scores reflect the difference between the amount of time infants attended to the eyes out of the total amount of time they attended to the face minus the amount of time they attended to the mouth out of the total amount of time they attended to the face. A negative score indicates greater attention to the mouth. Error bars represent SEMs and asterisks indicate statistical significance.

The shift in SA to the talker's mouth by 8–10 months of age is of special developmental interest for several reasons. First, the shift occurs at the same time that endogenous visual attention emerges.[93] This suggests that the shift to the talker's mouth reflects the emergence of endogenous SA to audiovisual information. Second, the shift in SA to the mouth occurs at the same time that canonical babbling emerges. Given that the emergence of babbling signals the onset of infant interest in speech production, this suggests that the new focus on the talker's mouth enables infants to lip-read for the first time and, in the process, gain direct access to highly salient redundant audiovisual speech cues. Third, the fact that infant lip-reading at 8–10 months of age is not affected by the specific language spoken indicates that SA to audiovisual speech is initially nonspecific. Finally, the fact that 12-month-old infants no longer lip-read when exposed to native audiovisual speech but that they lip-read when exposed to nonnative speech indicates that SA becomes more specialized by 12 months of age. This increasing specialization is the result of the emergence of native-language expertise and a concurrent narrowing of the ability to process nonnative audiovisual speech.[101] The net effect of this newly emerged specialization for native audiovisual speech is twofold. On the one hand, it enables infants of this age to dispense with direct access to redundant audiovisual speech cues because they now have the expertise to process native audiovisual speech. On the other hand, they are now compelled to continue relying on the greater salience of redundantly specified nonnative audiovisual speech (i.e., lip-read) to disambiguate what has now become unfamiliar speech to them.

Subsequent studies by Lewkowicz and his colleagues have replicated and extended their initial findings. Specifically, Pons et al.[102] tested Spanish- and/or Catalan-learning infants in the same task and replicated the developmental pattern of shifting SA found in English-learning infants by Lewkowicz and Hansen-Tift.[100] Pons et al.[102] also replicated the finding that 12-month-old monolingual infants (in this case Spanish- or Catalan-learning infants) lip-read nonnative audiovisual speech but not native speech. Finally, Pons et al.[102] found that bilingual Spanish/Catalan-learning infants rely even more on lip-reading than do monolingual infants, suggesting that bilinguals rely more on the redundancy of audiovisual speech to overcome the cognitive challenge of learning two languages. In a second follow-up study, Lewkowicz and colleagues[103] tested the role of audiovisual synchrony in infant SA to talking faces and found that infants tune rapidly to the normally synchronous (i.e., redundant) audible and visible attributes of everyday speech. This was evident in the fact that their preference for the talker's mouth was not affected by a desynchronization of the auditory and visual speech streams at 8 months of age but that it was affected by it by 10 months of age. Finally, in a third study extending the infant findings, Barenholtz, Mavica, and Lewkowicz[22] asked whether language familiarity continues to modulate attention to the redundant audiovisual speech cues inherent in a talker's mouth into adulthood. They tracked eye gaze in monolingual and bilingual participants while they performed an explicit audiovisual speech-encoding task. Results indicated that the monolinguals exhibited more lip-reading when the talker spoke in an unfamiliar than in a familiar language but that the bilinguals who were familiar with both languages exhibited equal amounts of lip-reading.

Together, the results from the SA studies demonstrate that a nonspecific, endogenous SA system emerges after 6 months of age, that it rapidly tunes to multisensory redundancy, that it specializes as infants acquire increasing perceptual and cognitive expertise, and that it becomes an essential part of our cognitive toolkit into adulthood. In addition, the infant SA findings provide key insights into the role of early experience in speech and language acquisition. The fact that monolingual infants cease lip-reading native audiovisual speech at 12 months but that they continue lip-reading nonnative audiovisual speech indicates that lack of experience with nonnative speech compels infants to continue relying on the greater salience of redundantly specified audiovisual speech. Similarly, the fact that bilingual infants rely more on lip-reading than do monolingual infants shows that early linguistic experiences modulate the degree to which infants rely on the redundancy inherent in audiovisual speech. Interestingly, these two sets of findings on the effects of early experience on SA to audiovisual fluent speech are consistent with findings from studies of matching of native or nonnative auditory and visual speech.[65] The latter show that 12- to 14-month-old English-learning infants can match temporally synchronized and desynchronized native-language auditory and visual speech streams but that they only match synchronized nonnative language speech streams.

Multisensory processes for perceiving the body, self, and the environment at hand in infancy and childhood

Our sense of our own bodies and selves is rooted in tactile experience,[104,105] and touch is the first of our senses to develop. Behavioral responses to tactile stimulation are seen from the first prenatal trimester and before other sensory responses.[106] Consequently, touch provides the sensory scaffold on which we build the perceptual map of our own bodies and sense of self. Touch also gives us access to the environment at hand through haptics,[107] and in addition has social-affective functions.[108] Despite this, surveys of the literature indicate that the numbers of articles reporting on tactile perceptual development are dwarfed by those reporting on the development of the other senses.[109] In this section, we will focus on the development of the multisensory processes involving touch which underlie body representations and haptic experience of the environment. Consistent with previous arguments,[110] we will argue that these skills emerge gradually because the sensorimotor abilities which provide the necessary experiences have an extended developmental trajectory. For the purpose of this chapter, we will adopt a broad definition of touch including not just cutaneous sensation but also interoceptive cues (which we will not spend much time on) and proprioception (which we will spend some time on).[a]

[a] Interoceptors provide information (typically unconscious) that is used to maintain organ function, homeostasis, digestion, and respiration. Proprioception provides information about how our body and limbs are arrayed, or moving, in space, and are mediated by receptors in the muscles, tendons, and joints.

Visual–haptic object perception in infancy

Can newborn infants perceive objects as the same regardless of whether they see or touch them? In other words, can they perceive the crossmodal equivalence between an object's visual and tactile attributes? This question goes to the heart of whether early experience is necessary for the perception of crossmodal unity and, consequently, has fascinated philosophers since the 17th century (e.g., Molyneux's famous question to Locke[b]). Despite its long history, the first empirical attempts to answer this question and discern the development of crossmodal transfer in infancy did not occur until the 1970s.[111] It is now clear that infants are capable of tactile-visual crossmodal matching rather early in postnatal development. For instance, Rose and colleagues confirmed in a series of studies that bidirectional transfer between touch and vision can be observed at 6 months of age,[112] but they also demonstrated that infants become progressively more efficient at this kind of crossmodal encoding and recognition between 6 and 12 months.[113] Even more remarkably, some studies have demonstrated an ability to perceive commonalities between touch and vision from soon after birth in both manual[114,115] and oral[116] haptic exploratory contexts. There has been some difficulty replicating oral-visual transfer at 1 month,[117] but there is also some evidence for manual-visual transfer in newborns.[115]

The fact that infants can make visual–haptic matches only days after birth suggests that minimal multisensory experience is required to perceive a unitary multisensory world. There is evidence, however, that is inconsistent with this conclusion. Held et al.'s[118] recent study of crossmodal transfer following the removal of congenital cataracts demonstrates that the ability to make visual-tactile crossmodal matches is not evident right after surgery (despite the fact that these patients can make visual–visual and tactile–tactile matches). Rather, visual–tactile crossmodal transfer emerges over a period of several days following the onset of visual experience. On the face of it, this finding contradicts evidence that newborn infants are able to make visual–haptic transfers at birth.[115] Of course, it is possible that infants learn visual–haptic transfer quickly enough to be able to demonstrate it a few days after birth. It is also possible that the kinds of visual–haptic transfer that newborns demonstrate are experience independent, but simpler than those which adults have to learn following the onset of sight.[118] It is interesting to speculate about the potential differences between the acquisition of visual–haptic transfer of shape in infancy and adulthood (following the onset of sight after congenital blindness). Congenitally blind adults will have mastered a range of "haptic exploratory procedures" giving them access to both simple and complex shapes in haptics.[107] By contrast, young infants have a much more limited repertoire of haptic exploratory

[b] In 1688, John Locke (1632–1704) published a question put to him by William Molyneux (1656–98) which henceforth became known as Molyneux's question. This question asked whether an adult born blind, if restored to sight might be able to identify and distinguish, with vision alone, objects that he/she had previously only experienced through touch. Molyneux, Locke, and a little later George Berkeley (1685–1753) were in agreement that the answer was "no." Berkeley, in particular, was of the view that the senses were quite distinct in nature providing completely different forms of information that were irreconcilable without the construction of crossmodal associations through experience. The no answer quite clearly implies an account of development in which infants and children have to learn to associate previously separate sensations across multiple sense modalities. On the other hand, a positive answer implies that infants and adults have perceptual access to a coherent multisensory environment independent of experience. Since its posing, Molyneux's question has received considerable attention in philosophy and, later, psychology. It remains of seminal importance.

procedures,[119] and so we might expect that there will be a slower development of haptic–visual transfer of more complex shape properties across infancy and childhood (for some evidence of this, see Refs. 113,114,120).

Visual–haptic coordination can develop in more subtle ways than straightforward improvement in crossmodal matching and memory. Gori et al.[121] have worked with young children to investigate the development of the ability to integrate visual and haptic object attributes that can be simultaneously seen and felt. They asked children and adults to identify which of two successively presented objects (the children simultaneously saw and felt each object, respectively) was the larger. When adults are presented with multiple cues to the same stimulus dimension, their estimates tend to combine estimates from the component modalities which are weighted according to their relative reliabilities.[122,123] For instance, in one particular experimental scenario which Gori et al.[121] used, visual size cues are the most reliable with respect to object size, and so adults tend to weight them more highly than haptic cues. Given that this optimal weighting of touch and vision determines the combined estimate, adults perform significantly better when *both* visual and haptic cues are presented. However, before 8 years of age, Gori et al.[121] found that children did not show this multisensory advantage—they were not able to optimally integrate haptics and vision according to their relative reliabilities. Furthermore, when presented with conflicting haptic and visual cues, young children's estimates appeared to be dominated by haptics, even though that was not the most reliable cue for them.

Gori et al.[121] propose an interesting explanation for why the development of optimal multisensory integration may take so long; they suggest that the optimal weighting of the senses is delayed so as to provide a point of reference against which to calibrate the senses in multisensory tasks. Because the body and limbs continue to grow over the course of childhood and adolescence, this recalibration is constantly required, and, so, according to Gori et al.,[121] optimal weighting does not take place until the body has stopped growing substantially. Although there is no direct support as yet for the idea of multisensory integration being delayed for this reason, a number of studies have backed up the idea that crossmodal calibration is an important aspect of multisensory development during childhood. Gori et al.[124] show that congenitally blind children show poor haptic orientation discrimination relative to their sighted peers. What might explain this counterintuitive finding? Gori et al.[124] argue that congenitally blind children are poorer at haptic orientation discrimination because they have not had access to visual orientation cues during development which would have otherwise calibrated their perception of haptic orientation.

As well as being calibrated by vision in development, more recent findings from Gori et al.[125] suggest that in some contexts haptics help calibrate our visual discriminations. Gori et al.[125] presented visual size discrimination tasks to adults and children at different distances. They found that inside the reaching space, visual size discriminations were fairly accurate, whereas outside of reaching space, children were biased to underestimate size and adults were biased to overestimate size. This suggests that haptic cues to size are able to recalibrate visual cues to the same when visual stimuli are close enough to the body to be able to receive both haptic and visual estimates. Gori et al.[121] point out that haptic cues provide "direct" size information, rather than the "indirect" cues provided by vision, which have to be interpreted and although reliable in nature can be subject to biases. This may be why haptic cues play this calibrating role.

In summary, the ability to make links between what we sense about objects with our hands and what we see appears to be something which develops early for some simple shapes/textures with fairly minimal experience.[115,118] Later in development, as children's sensorimotor haptic exploratory procedures, and as their ability to resolve fine detail in all sense modalities develops, coordination of haptics with vision seems likely to improve. In addition, crossmodal calibration processes in which vision helps improve the estimates we make with haptics (and vice versa) appear to be an important aspect of early to mid-childhood. However, after 8 years or so, children start to integrate separate sensory cues to achieve an optimal multisensory estimate concerning objects.[121]

Multisensory body perception in infancy and childhood

Adults form representations of limb and body position resulting from the combination and integration of information from somatosensory receptors with visual and auditory information in multisensory cortical and subcortical regions.[105] Recent research shows that there is some degree of competence in early infancy at perceiving the colocation of tactile/proprioceptive and visual[126–129] and tactile and auditory events.[130]

It is remarkable that newborns show some ability to colocate touch to vision.[126] However, it is also clear that the interactions between touch and vision underlying body perception develop substantially in the first year of life and beyond. Strikingly, young infants' responses to tactile stimuli presented in isolation (i.e., without a concurrent visual/auditory cue) indicate that, until 6 months of age, infants do not refer touches on their own body to a location in external space[131] (see Fig. 4.2). Research conducted with individuals with visual impairments indicates that the processes whereby we refer cutaneous touches on our body surface to locations in external space result from visual experience: whereas sighted and late blind adults automatically refer touch to external spatial coordinates, congenitally blind adults do not.[132] Evidence from adults who have had their sight restored by the removal of congenital cataracts in early development indicates that there is a sensitive period in which this automatic external coding of touch between 5 and 24 months is established.[133,134] This is consistent with the trajectory of development between 4 and 6 months of age shown by Begum Ali et al.[131] Perhaps the emergence of successful reaching to visual objects from 5 months drives infants' learning about the relationship of tactile stimulation to visual events in external space.

It is not surprising that spatiotemporal links between touch and vision should develop gradually. Irrespective of age, the spatial relationships between touch and vision/audition can vary dramatically from moment to moment (e.g., each time our arm changes position, tactile coordinates on the hands shift with respect to auditory and visual space). These difficulties are amplified during development, as the spatial layout of the limbs and the body changes substantially in early prenatal and postnatal life (as indeed do the number and variety of postural changes made in the service of skilled movement).

One particular hurdle for young infants involves developing an ability to update the location of a touch (and a limb) when the body moves into unfamiliar postures. At 6 months of age, we know that when infants are locating touches presented in isolation, they do not take account of current limb position and seemingly fall back on canonical body posture information (as evidenced by their crossed-hands and crossed-feet deficits when responding to

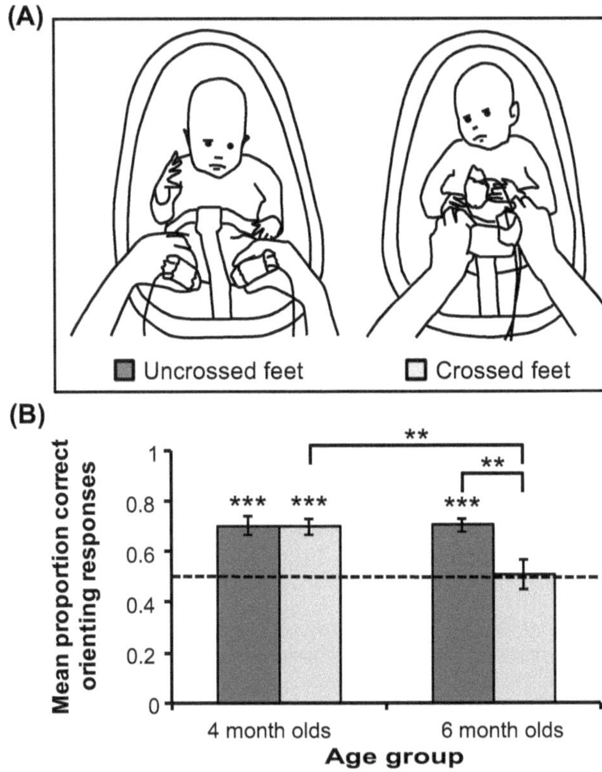

FIGURE 4.2 Probing infants' spatial representations of touch by crossing the legs.[131] (A) An infant participant in the uncrossed and crossed feet postures. The tactors, attached to the infants' feet using a cohesive bandage, were controlled remotely. The experimenter held the infant's feet in the assigned posture during tactile stimulation. (B) Mean proportion of correct first unilateral foot movements to vibrotactile stimuli (error bars indicate the standard error of the mean). The 6-month olds showed a crossed-feet deficit, whereas the 4-month olds performed equivalently across conditions, matching the best performance of the 6-month olds. Significant comparisons are indicated (*$P < .05$, **$P < .01$, ***$P < .001$). The crossed feet deficit in tactile localization is considered to result from referral of the touch on the foot to the location in external space where that foot would normally rest. *Figure reprinted with permission from Begum Ali J, Spence C, Bremner AJ. Human infants' ability to perceive touch in external space develops postnatally.* Curr Biol. *2015;25:R978–R979.*

touches on their hands and feet[131,135] (see Fig. 4.2 of this chapter regarding Begum Ali et al.'s findings)). However, by 10 months of age, Bremner et al.[135] found that infants maintain a consistent level of accuracy in their visual and manual orienting responses across both familiar and unfamiliar postures of the arms. So between 6 and 10 months of age, it seems that an ability to incorporate sensory information about current limb posture into representations of the external location of tactile events (referred to here as "postural remapping") develops significantly. This work raises questions about the processes that underlie the improvements in behavioral performance just described. One question is whether developmental improvements in postural remapping reflect improvements in the spatial *perception* of bodily events across different postures or whether they reflect improvements in the infants' coordination of their orienting *responses*. The second question is what sensory information infants are using to update their representations of limb position and remap their behavioral responses. A study by Rigato et al.[136] helps address both of these questions.

Rigato and colleagues[136] investigated modulatory effects of arm posture on somatosensory evoked potentials (SEPs) recorded from the scalp in 6.5- and 10-month-old infants (see Fig. 4.3). When presented with tactile stimuli, the 6.5-month olds showed no reliable effect of posture on their SEPs; it was as if these younger infants processed tactile events in the same way irrespective of the posture of their limbs, mirroring the findings from behavioral studies that this age group tended to respond to the same external location irrespective of limb posture. However, the older 10-month olds, like adults,[137] showed significant postural modulations of somatosensory processing. Thus, improvements in tactile localization across limb postures may be underpinned by the increased role of postural information in somatosensory processing seen in the ERPs. Importantly, the modulatory effects of posture seen at 10 months occur (as they do in adults[137]) early in somatosensory processing (~60–120 ms). In answer to the first question raised earlier, this suggests that improvements in somatosensory processing and postural remapping at 10 months of age occur largely at the perceptual end of neural processing in somatosensory cortex.

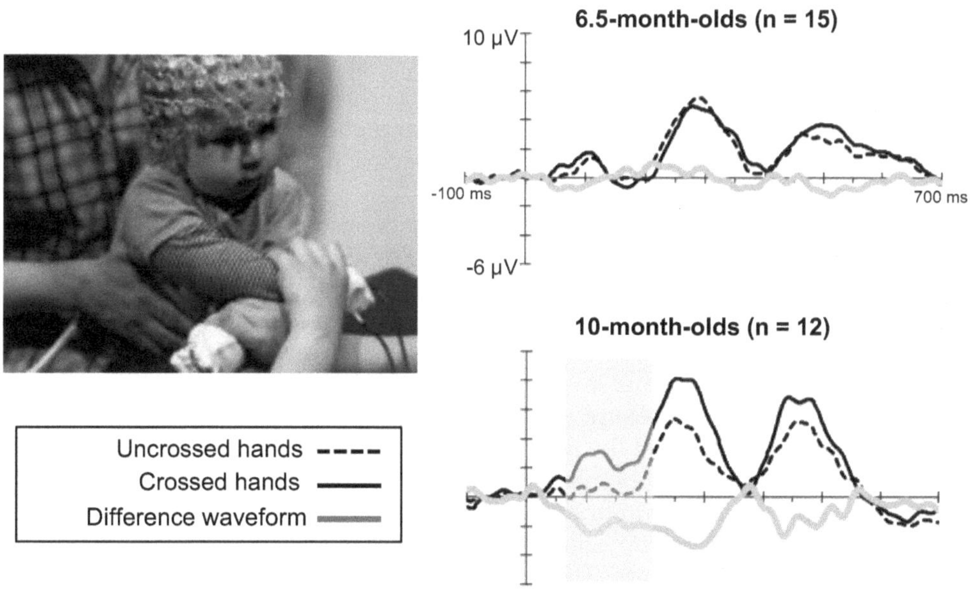

FIGURE 4.3 Researching the early development of body representations via somatosensory evoked potentials (SEPs). Rigato and colleagues[136] examined the effect of arm posture on SEPs. The infants (as pictured on the left) received 200 ms vibrotactile stimuli to the hand in either an uncrossed- or a crossed-hands arm posture (the crossed-hands posture is shown). As in adults,[137] 10-month olds' SEPs were modulated by posture from early in somatosensory processing, indicating a somatosensory postural remapping. Younger infants (6.5-month olds), however, demonstrated SEPs which were unaffected by arm posture. *Figure reprinted with permission from Bremner AJ. Developing body representations in early life: combining somatosensation and vision to perceive the interface between the body and the world.* Dev Med Child Neurol. *2016;58 (suppl 4):12–16.*

In addressing the second question raised earlier, a further experiment investigated what sensory information concerning limb position drives the postural modulation of somatosensory processing seen at 10 months. Tactile stimuli were presented again, but this time with the infants' arms obscured from view by a cloak. In this case, the 10-month olds showed no effect of posture on their SEPs. It seems that at this point in development, *visual* cues to hand posture are required to remap tactile space.

Thus, by 10 months of age, infants appear to be able not just to represent the location of touches in the external environment but also to dynamically update tactile localization across changes in limb posture. Physical and behavioral changes, like becoming able to cross the hands over the body midline, necessitate continuous adaptation of multisensory body representations in early life.[121] Correspondingly, there is now growing evidence that visual–tactile integration in body representations undergoes a gradual honing across infancy and childhood.[138–141]

Touch and its multisensory interactions are intrinsically tied-up with the developing self. The relationship between multisensory body perception and experience of the bodily self is a subject of recent scrutiny by researchers studying mature adults.[105,142] Pertinent for developmental psychology, this literature shows that the ability to detect multisensory correspondences between touch, proprioception, and vision (which as we have seen is present very early in life[126,128,130]) does not necessarily imply the experience of limb or body ownership. Nonetheless, such cues appear to be the clearest driver of a sense of body ownership from 4 years to adulthood.[140] Given that young infants, and even newborns, are sensitive to tactile–visual and tactile–auditory spatiotemporal correspondences, it is tempting to speculate that feelings of ownership over the body are an early acquisition. However, changes in the way in which the body is perceived in infancy and childhood[131,140] may mean that the content of a sense of ownership (i.e., what kind of body we feel ownership over) changes substantially in early development.

One line of inquiry into the development of self-perception is ripe for exploitation, namely the role of interoceptive touch. Awareness of interoceptive signals (e.g., one's heartbeat or respiratory activity) has been shown to be important in grounding a stable sense of self on the body.[105] Maister and colleagues[143] have recently developed a method to measure infants' perception of their own heartbeat, demonstrating some ability at this from 5 months of age. It will be interesting to see how this line of inquiry develops, as interoceptive awareness is perhaps the least ambiguous measure we have of self-perception in early life.

Although we have emphasized the role which touch plays in self-perception, it is increasingly clear across a range of studies from developmental psychology and cognitive neuroscience that touch also plays a role in bridging the perceptual gap between self and others. As adults, we have a transparent appreciation of the meaning of tactile experiences when we observe those happening to others. This "vicarious mapping" is a part of our capacity to share the experiences of others (i.e., empathy), a critical aspect of human behavior.[144] One characteristic of the mature human brain that lends itself to this vicarious sensory empathy is the recruitment of similar brain regions when a state is experienced and when that state is observed in others.[145] For instance, seeing other people being touched, or touching objects, activates similar brain areas as when we experience touch ourselves.[146,147] A study by Rigato and colleagues[148] demonstrates that 4-month-old infants' SEPs measured in scalp EEG are modulated by visual observations of touches to another person's hand (see Fig. 4.4).

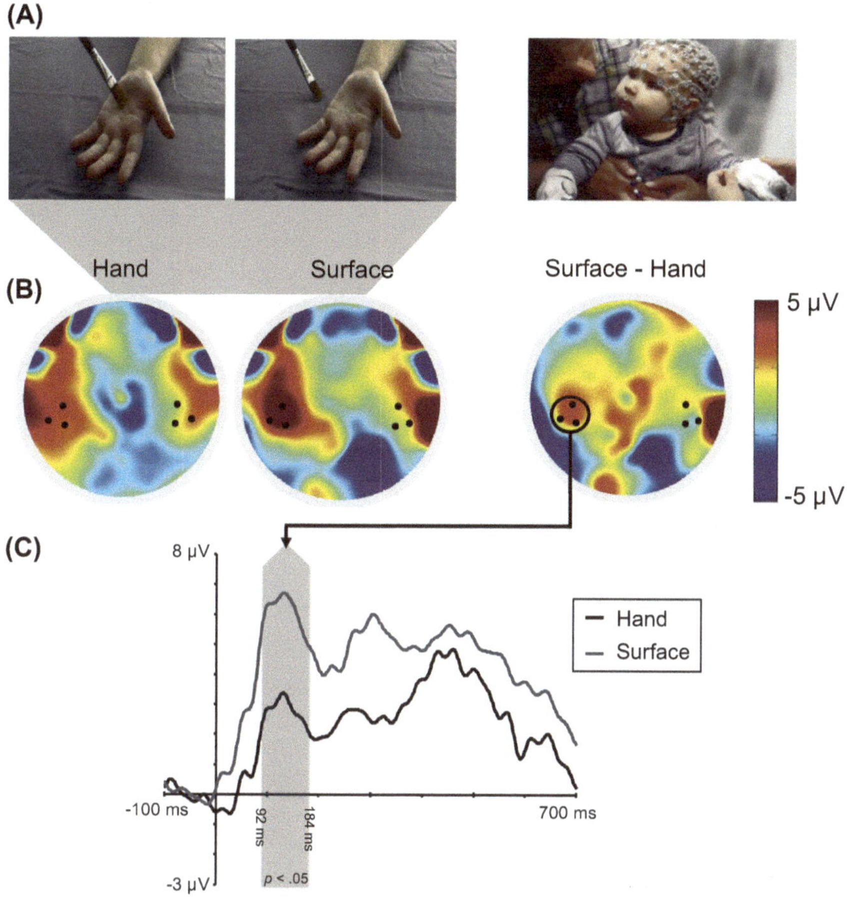

FIGURE 4.4 Visually presented touches on another person's hand modulated somatosensory evoked potentials (SEPs) in 4-month-old infants. (A) Rigato et al.[148] showed 4-month-old infants visual touches to either a hand or a surface simultaneously with vibrotactile stimuli to one of the infants' hands. (B) Grand average topographical representations of the voltage distribution over the scalp in the hand and surface conditions between 92 and 184 ms after the stimulus. With a surface–hand difference map to the right. (C) Grand average SEPs on contralateral sites (CP3/CP4), showing a statistically reliable effect of condition on the SEP amplitude between 92 and 184 ms. *Figure reprinted with permission from Rigato S, Banissy M, Romanska A, Thomas R, van Velzen J, Bremner AJ. Cortical signatures of vicarious tactile experience in four-month-old infants.* Dev Cogn Neurosci. *2017. https://doi.org/10.1016/j.dcn.2017.09.003.*

That seen touches to a body part can influence cortical signatures of the perceptual stages of somatosensory processing in 4-month-old infants is consistent with the assertion that vicarious mapping plays a foundational role in early social perceptual development.[149,150] These findings also underscore the importance of tactile and multisensory tactile–visual processes in early social perception[109,126,151] where multisensory interactions have typically been studied in the audiovisual domain. But what is the precise basis of the visual–tactile mapping that we have observed? One possibility is that by 4 months of age infants are sensitive to the highly socially specific properties of the visual event specifying a touch occurring to an observed hand. However, given that we have only drawn the relatively crude comparison of a hand being touched versus a surface being touched, there is the potential that more low-level aspects of the visual event served to modulate somatosensory processing in the fashion which we observed.

One question concerns whether infants will show the same visual influences on somatosensory processing across a range of different orientations of a hand (or other body part). A further question concerns whether the effect we have observed is general across social and nonsocial stimuli. Further experiments in which effects of touches to a hand versus touches to an inanimate object (e.g., a toy) are compared would help shed light on the extent to which the visual modulations of somatosensory processing we have observed are a specifically social crossmodal mapping process or might also underlie the development of the crossmodal basis of representing or experiencing ownership over one's own body.[152] It is important to note however, that even if the effect we have observed is, in fact, a crossmodal mapping which generalizes across social and nonsocial domains, it might nonetheless play an important role in driving the development of the vicarious mappings of touch which serve as social perceptual adaptations in the adult brain.[105]

Summary

In this chapter, we have described the current state of the art in research on the development of multisensory processes underlying the perception of the environment and of the self from infancy to adulthood. In the first part of the chapter, we showed that multisensory processing emerges gradually and that SA to multisensory inputs changes dramatically during infancy. Specifically, we showed that infants begin life with relatively rudimentary multisensory processing abilities and that these abilities improve and become increasingly more specialized with growth and the acquisition of experience. We also showed that the developmental changes in infant multisensory processing are accompanied by major changes in infant SA to multisensory information. These latter changes consist of shifting developmental patterns of SA that reflect the emergence of the ability to voluntarily deploy SA to sources of multisensory redundancy and that this new attentional strategy becomes incorporated into a new perceptual/cognitive toolkit that helps infants and adults solve challenging processing tasks. In the second part of this chapter, we discussed the development of an ability to perceive the multisensory body, perceptual representations which are typically underpinned by interactions between tactile inputs and sensory inputs to other senses (including vision and audition). The key messages from this area of research are that mature interactions between touch and vision, in particular, develop gradually over postnatal development, from infancy right up to late childhood and beyond. This gradual developmental trajectory may be at least in part because mature interactions between touch and vision depend to a large extent on experiences which are contingent on physical development and the development of sensorimotor abilities, both of which continue well beyond infancy.

A common theme across both parts of this chapter is the role of experience in the typical development of multisensory processes. The labile nature of multisensory development as a result of experience suggests that the role of multisensory processes should be carefully considered across a wide range of fields in which development is important. For instance, recent findings in children with neurodevelopmental disorders are starting to reveal striking and significant atypicalities in multisensory processing in disorders including autism spectrum disorder, developmental coordination disorder, and developmental dyslexia. Tracing the developmental processes underlying typical and atypical multisensory abilities will be crucial for gaining a better understanding of the phenotypes of these disorders, not to mention the developmental origins of mature typical perceptual abilities.

References

1. Marks L. *The Unity of the Senses*. New York: Academic Press; 1978.
2. Stein BE, Meredith MA. *The Merging of the Senses*. Cambridge, MA: The MIT Press; 1993.
3. Birch HG, Lefford A. Intersensory development in children. *Monogr Soc Res Child Dev*. 1963;25(5):1–48.
4. Piaget J. *The Origins of Intelligence in Children*. New York: International Universities Press; 1952.
5. Gibson EJ. *Principles of Perceptual Learning and Development*. 1969. New York: Appleton.
6. Lewkowicz DJ. The development of intersensory temporal perception: an epigenetic systems/limitations view. *Psychol Bull*. 2000;126(2):281–308.
7. Thelen E, Smith LB. *A Dynamic Systems Approach to the Development of Cognition and Action*. Cambridge, MA: MIT Press; 1994.
8. Bahrick LE, Lickliter R, Flom R. Intersensory redundancy guides the development of selective attention, perception, and cognition in infancy. *Curr Dir Psychol Sci*. 2004;13(3):99–102.
9. Lewkowicz DJ, Kraebel K. The value of multimodal redundancy in the development of intersensory perception. In: Calvert G, Spence C, Stein B, eds. *Handbook of Multisensory Processing*. Cambridge: MIT Press; 2004:655–678.
10. Partan S, Marler P. Communication goes multimodal. *Science*. 1999;283(5406):1272–1273.
11. Rowe C. Receiver psychology and the evolution of multicomponent signals. *Anim Behav*. 1999;58:921–931.
12. Stein BE, Stanford TR. Multisensory integration: current issues from the perspective of the single neuron. *Nat Rev Neurosci*. 2008;9(4):255–266.
13. Ghazanfar AA, Schroeder CE. Is neocortex essentially multisensory? *Trends Cognit Sci*. 2006;10(6):278–285.
14. Stein BE. *The New Handbook of Multisensory Processing*. MIT Press; 2012.
15. Wallace MT, Ramachandran R, Stein BE. A revised view of sensory cortical parcellation. *Proc Natl Acad Sci USA*. 2004;101(7):2167–2172. https://doi.org/10.1073/pnas.0305697101.
16. Stein BE, Stanford TR, Rowland BA. Development of multisensory integration from the perspective of the individual neuron. *Nat Rev Neurosci*. 2014;15(8):520.
17. Sumby WH, Pollack I. Visual contribution to speech intelligibility in noise. *J Acoust Soc Am*. 1954;26:212–215.
18. Summerfield AQ. Use of visual information in phonetic perception. *Phonetica*. 1979;36:314–331.
19. Driver J. Enhancement of selective listening of illusory mislocation of speech sounds due to lip-reading. *Nature*. 1996;381(6577):66–68.
20. Summerfield Q. Lipreading and audio-visual speech perception. *Philos Trans R Soc Lond Ser B Biol Sci*. 1992;335(1273):71–78.
21. Vatikiotis-Bateson E, Eigsti I-M, Yano S, Munhall KG. Eye movement of perceivers during audiovisual speech perception. *Percept Psychophys*. 1998;60(6):926–940.
22. Barenholtz E, Mavica L, Lewkowicz DJ. Language familiarity modulates relative attention to the eyes and mouth of a talker. *Cognition*. 2016;147:100–105.
23. Lansing CR, McConkie GW. Word identification and eye fixation locations in visual and visual-plus-auditory presentations of spoken sentences. *Percept Psychophys*. 2003;65(4):536–552.
24. Plass J, Guzman-Martinez E, Ortega L, Grabowecky M, Suzuki S. Lip reading without awareness. *Psychol Sci*. 2014;25(9):1835–1837. https://doi.org/10.1177/0956797614542132.
25. Jack CE, Thurlow WR. Effects of degree of visual association and angle of displacement on the "ventriloquism" effect. *Percept Mot Skills*. 1973;37:967–979.
26. Radeau M. Ventriloquism against audio-visual speech: or, where Japanese-speaking barn owls might help. *Cah Psychol*. 1994;13(1):124–140.
27. Recanzone GH. Interactions of auditory and visual stimuli in space and time. *Hear Res*. 2009;258(1–2):89–99. https://doi.org/10.1016/j.heares.2009.04.009. S0378-5955(09)00096-3 [pii].

28. Foss-Feig JH, Kwakye LD, Cascio CJ, et al. An extended multisensory temporal binding window in autism spectrum disorders. *Exp Brain Res*. 2010;203(2):381–389.
29. Shams L, Kamitani Y, Shimojo S. What you see is what you hear. *Nature*. 2000;408(6814):788.
30. Scheier C, Lewkowicz DJ, Shimojo S. Sound induces perceptual reorganization of an ambiguous motion display in human infants. *Dev Sci*. 2003;6:233–244.
31. Sekuler R, Sekuler AB, Lau R. Sound alters visual motion perception. *Nature*. 1997;385:308.
32. Botvinick M, Cohen J. Rubber hands 'feel' touch that eyes see. *Nature*. 1998;391(6669):756. https://doi.org/10.1038/35784.
33. Cowie D, Sterling S, Bremner AJ. The development of multisensory body representation and awareness continues to 10years of age: evidence from the rubber hand illusion. *J Exp Child Psychol*. 2016;142:230–238. https://doi.org/10.1016/j.jecp.2015.10.003.
34. Tsakiris M, Haggard P. The rubber hand illusion revisited: visuotactile integration and self-attribution. *J Exp Psychol Hum Percept Perform*. 2005;31:80–91.
35. Lewkowicz DJ. Early experience and multisensory perceptual narrowing. *Dev Psychobiol*. 2014;56(2):292–315.
36. Lewkowicz DJ, Ghazanfar AA. The emergence of multisensory systems through perceptual narrowing. *Trends Cognit Sci*. 2009;13(11):470–478.
37. Soto-Faraco S, Calabresi M, Navarra J, Werker J, Lewkowicz DJ. *Development of Audiovisual Speech Perception Multisensory Development*. Oxford: Oxford University Press; 2012:207–228.
38. Lewkowicz DJ, Turkewitz G. Intersensory interaction in newborns: modification of visual preferences following exposure to sound. *Child Dev*. 1981;52(3):827–832.
39. Lewkowicz DJ, Turkewitz G. Cross-modal equivalence in early infancy: auditory-visual intensity matching. *Dev Psychol*. 1980;16:597–607.
40. Lewkowicz DJ, Leo I, Simion F. Intersensory perception at birth: newborns match non-human primate faces & voices. *Infancy*. 2010;15(1):46–60.
41. Slater A, Quinn PC, Brown E, Hayes R. Intermodal perception at birth: intersensory redundancy guides newborn infants' learning of arbitrary auditory-visual pairings. *Dev Sci*. 1999;2(3):333–338.
42. Bahrick LE. Infants' perception of substance and temporal synchrony in multimodal events. *Infant Behav Dev*. 1983;6(4):429–451.
43. Bahrick LE. Intermodal learning in infancy: learning on the basis of two kinds of invariant relations in audible and visible events. *Child Dev*. 1988;59:197–209.
44. Lewkowicz DJ. Infants' response to temporally based intersensory equivalence: the effect of synchronous sounds on visual preferences for moving stimuli. *Infant Behav Dev*. 1992;15(3):297–324.
45. Lewkowicz DJ. Infants' responsiveness to the auditory and visual attributes of a sounding/moving stimulus. *Percept Psychophys*. 1992;52(5):519–528.
46. Lewkowicz DJ. Perception of auditory-visual temporal synchrony in human infants. *J Exp Psychol Hum Percept Perform*. 1996;22(5):1094–1106.
47. Spelke ES, Born WS, Chu F. Perception of moving, sounding objects by four-month-old infants. *Perception*. 1983;12(6):719–732.
48. Bahrick LE, Hernandez-Reif M, Flom R. The development of infant learning about specific face-voice relations. *Dev Psychol*. 2005;41(3):541–552.
49. Brookes H, Slater A, Quinn PC, Lewkowicz DJ, Hayes R, Brown E. Three-month-old infants learn arbitrary auditory-visual pairings between voices and faces. *Infant Child Dev*. 2001;10(1–2):75–82.
50. Kuhl PK, Meltzoff AN. The bimodal perception of speech in infancy. *Science*. 1982;218(4577):1138–1141.
51. Patterson M,L, Werker JF. Matching phonetic information in lips and voice is robust in 4.5-month-old infants. *Infant Behav Dev*. 1999;22(2):237–247.
52. Patterson M,L, Werker JF. Two-month-old infants match phonetic information in lips and voice. *Dev Sci*. 2003;6(2):191–196.
53. Bahrick LE. The development of infants' sensitivity to arbitrary intermodal relations. *Ecol Psychol*. 1994;6(2):111–123.
54. Morrongiello BA, Lasenby J, Lee N. Infants' learning, memory, and generalization of learning for bimodal events. *J Exp Child Psychol*. 2003;84(1):1–19.
55. Murray MM, Lewkowicz DJ, Amedi A, Wallace MT. Multisensory processes: a balancing act across the lifespan. *Trends Neurosci*. 2016. https://doi.org/10.1016/j.tins.2016.05.003.

56. Walker-Andrews AS, Lennon EM. Auditory-visual perception of changing distance by human infants. *Child Dev*. 1985;56(3):544–548.
57. Hillairet de Boisferon A, Dupierrix E, Quinn PC, et al. Perception of multisensory gender coherence in 6 and 9-month-old infants. *Infancy*. 2015;20(6):661–674.
58. Patterson M,L, Werker JF. Infants' ability to match dynamic phonetic and gender information in the face and voice. *J Exp Child Psychol*. 2002;81(1):93–115.
59. Poulin-Dubois D, Serbin LA, Kenyon B, Derbyshire A. Infants' intermodal knowledge about gender. *Dev Psychol*. 1994;30(3):436–442.
60. Walker-Andrews AS, Bahrick LE, Raglioni SS, Diaz I. Infants' bimodal perception of gender. *Ecol Psychol*. 1991;3(2):55–75.
61. Walker-Andrews AS. Intermodal perception of expressive behaviors: relation of eye and voice? *Dev Psychol*. 1986;22:373–377.
62. Neil PA, Chee-Ruiter C, Scheier C, Lewkowicz DJ, Shimojo S. Development of multisensory spatial integration and perception in humans. *Dev Sci*. 2006;9(5):454–464.
63. Morrongiello BA, Fenwick KD, Nutley T. Developmental changes in associations between auditory-visual events. *Infant Behav Dev*. 1998;21(4):613–626.
64. Lewkowicz DJ, Pons F. Recognition of amodal language identity emerges in infancy. *Int J Behav Dev*. 2013;37(2):90–94.
65. Lewkowicz DJ, Minar NJ, Tift AH, Brandon M. Perception of the multisensory coherence of fluent audiovisual speech in infancy: its emergence and the role of experience. *J Exp Child Psychol*. 2015;130:147–162. https://doi.org/10.1016/j.jecp.2014.10.006.
66. Kahana-Kalman R, Walker-Andrews AS. The role of person familiarity in young infants' perception of emotional expressions. *Child Dev*. 2001;72(2):352–369.
67. Lewkowicz DJ, Röder B. The effects of experience on the development of multisensory processing. In: Stein B, ed. *The New Handbook of Multisensory Processing*. Cambridge, MA: MIT Press; 2012.
68. Putzar L, Goerendt I, Lange K, Rösler F, Röder B. Early visual deprivation impairs multisensory interactions in humans. *Nat Neurosci*. 2007;10(10):1243–1245.
69. Bergeson TR, Pisoni DB, Davis RAO. Development of audiovisual comprehension skills in prelingually deaf children with cochlear implants. *Ear Hear*. 2005;26(2):149.
70. Schorr EA, Fox NA, van Wassenhove V, Knudsen EI. Auditory-visual fusion in speech perception in children with cochlear implants. *Proc Natl Acad Sci USA*. 2005;102(51):18748–18750.
71. Irwin JR, Tornatore LA, Brancazio L, Whalen DH. Can children with autism spectrum disorders "hear" a speaking face? *Child Dev*. 2011;82(5):1397–1403. https://doi.org/10.1111/j.1467-8624.2011.01619.x.
72. Bebko JM, Weiss JA, Demark JL, Gomez P. Discrimination of temporal synchrony in intermodal events by children with autism and children with developmental disabilities without autism. *J Child Psychol Psychiatry*. 2006;47(1):88–98.
73. Kwakye LD, Foss-Feig JH, Cascio CJ, Stone WL, Wallace MT. Altered auditory and multisensory temporal processing in autism spectrum disorders. *Front Integr Neurosci*. 2010;4:129.
74. Wallace MT, Stein BE. Early experience determines how the senses will interact. *J Neurophysiol*. 2007;97(1):921.
75. Xu J, Yu L, Rowland BA, Stanford TR, Stein BE. Incorporating cross-modal statistics in the development and maintenance of multisensory integration. *J Neurosci*. 2012;32(7):2287–2298.
76. King AJ, Hutchings ME, Moore DR, Blakemore C. Developmental plasticity in the visual and auditory representations in the mammalian superior colliculus. *Nature*. 1988;332(6159):73–76.
77. Knudsen EI, Brainard MS. Visual instruction of the neural map of auditory space in the developing optic tectum. *Science*. 1991;253(5015):85–87.
78. Lee H-K, Whitt JL. Cross-modal synaptic plasticity in adult primary sensory cortices. *Curr Opin Neurobiol*. 2015;35:119–126.
79. Sharma J, Angelucci A, Sur M. Induction of visual orientation modules in auditory cortex. *Nature*. 2000;404:841–847.
80. von Melchner L, Pallas SL, Sur M. Visual behaviour mediated by retinal projections directed to the auditory pathway. *Nature*. 2000;404(6780):871–876.
81. Meredith MA, Lomber SG. Species-dependent role of crossmodal connectivity among the primary sensory cortices. *Hear Res*. 2017;343:83–91.

82. Collignon O, Dormal G, de Heering A, Lepore F, Lewis TL, Maurer D. Long-lasting crossmodal cortical reorganization triggered by brief postnatal visual deprivation. *Curr Biol*. 2015;25(18):2379–2383.
83. Gottlieb G. Experiential canalization of behavioral development: results. *Dev Psychol*. 1991;27(1):35–39.
84. Gottlieb G. Experiential canalization of behavioral development: theory. *Dev Psychol*. 1991;27(1):4–13.
85. Pascalis O, Scott LS, Kelly DJ, et al. Plasticity of face processing in infancy. *Proc Natl Acad Sci USA*. 2005;102(14):5297–5300.
86. Scott LS, Monesson A. The origin of biases in face perception. *Psychol Sci*. 2009;20(6):676–680.
87. Minar NJ, Lewkowicz DJ. Overcoming the other-race effect in infancy with multisensory redundancy: 10–12-month-olds discriminate dynamic other-race faces producing speech. *Dev Sci*. 2017. https://doi.org/10.1111/desc.12604.
88. Lewkowicz DJ, Ghazanfar AA. The decline of cross-species intersensory perception in human infants. *Proc Natl Acad Sci USA*. 2006;103(17):6771–6774.
89. Lewkowicz DJ, Sowinski R, Place S. The decline of cross-species intersensory perception in human infants: underlying mechanisms and its developmental persistence. *Brain Res*. 2008;1242:291–302.
90. Grossmann T, Missana M, Friederici AD, Ghazanfar AA. Neural correlates of perceptual narrowing in cross-species face-voice matching. *Dev Sci*. 2012;15(6):830–839.
91. Pons F, Lewkowicz DJ, Soto-Faraco S, Sebastian-Galles N. Narrowing of intersensory speech perception in infancy. *Proc Natl Acad Sci USA*. 2009;106(26):10598–10602. https://doi.org/10.1073/pnas.0904134106.
92. Amso D, Scerif G. The attentive brain: insights from developmental cognitive neuroscience. *Nat Rev Neurosci*. 2015;16(10):606–619.
93. Colombo J. The development of visual attention in infancy. *Annu Rev Psychol*. 2001;52:337–367.
94. Colombo J. Infant attention grows up: the emergence of a developmental cognitive neuroscience perspective. *Curr Dir Psychol Sci*. 2002;11(6):196–200. https://doi.org/10.1111/1467-8721.00199.
95. Richards JE, Reynolds GD, Courage ML. The neural bases of infant attention. *Curr Dir Psychol Sci*. 2010;19(1):41.
96. Rothbart MK, Posner MI. The developing brain in a multitasking world. *Dev Rev*. 2015;35:42–63.
97. Ruff HA, Rothbart MK. *Attention in Early Development: Themes and Variations*. New York, NY: Oxford University Press; 1996.
98. Yu C, Smith LB. Joint attention without gaze following: human infants and their parents coordinate visual attention to objects through eye-hand coordination. *PLoS One*. 2013;8(11):e79659.
99. Yu C, Smith LB. The social origins of sustained attention in one-year-old human infants. *Curr Biol*. 2016;26(9):1235–1240.
100. Lewkowicz DJ, Hansen-Tift AM. Infants deploy selective attention to the mouth of a talking face when learning speech. *Proc Natl Acad Sci USA*. 2012;109(5):1431–1436.
101. Maurer D, Werker JF. Perceptual narrowing during infancy: a comparison of language and faces. *Dev Psychobiol*. 2014;56(2):154–178.
102. Pons F, Bosch L, Lewkowicz DJ. Bilingualism modulates infants' selective attention to the mouth of a talking face. *Psychol Sci*. 2015;26(4):490–498.
103. Hillairet de Boisferon A, Tift AH, Minar NJ, Lewkowicz DJ. Selective attention to a talker's mouth in infancy: role of audiovisual temporal synchrony and linguistic experience. *Dev Sci*. 2017;20(3). https://doi.org/10.1111/desc.12381.
104. Longo MR. Distorted body representations in healthy cognition. *Q J Exp Psychol*. 2017;70:378–388.
105. Tsakiris M. The multisensory basis of the self: from body to identity to others. *Q J Exp Psychol*. 2017;70:597–609.
106. Humphrey T. Some correlations between the appearance of human fetal reflexes and the development of the nervous system. *Prog Brain Res*. 1964;4:93–135.
107. Lederman SJ, Klatzky RL. Haptic perception: a tutorial. *Atten Percept Psychophys*. 2009;71:1439–1459.
108. Morrison I, Löken LS, Olausson H. The skin as a social organ. *Exp Brain Res*. 2009;204:305–314.
109. Bremner AJ, Spence C. The development of tactile perception. *Adv Child Dev Behav*. 2017;52:227–268.
110. Bremner AJ. The origins of body representations in early life. In: Vignemont Fd, Alsmith A, eds. *The Subject Matter: Self Consciousness and the Body*. Cambridge, M. A.: MIT Press; 2017.
111. Bryant PE, Jones P, Claxton V, Perkins GH. Recognition of shapes across modalities by infants. *Nature*. 1972;240:303–304.
112. Rose SA, Gottfried AW, Bridger WH. Cross-modal transfer in 6-month-old infants. *Dev Psychol*. 1981;17(5):661–669.

113. Rose SA. From hand to eye: findings and issues in infant cross-modal transfer. In: Lewkowicz DJ, Lickliter R, eds. *The Development of Intersensory Perception: Comparative Perspectives*. Hillsdale, NJ: Lawrence Erlbaum Associates; 1994:265–284.
114. Sann C, Streri A. Perception of object shape and texture in human newborns: evidence from cross-modal transfer tasks. *Dev Sci*. 2007;10:399–410.
115. Streri A. Crossmodal interactions in the human newborn: new answers to Molyneux's question. In: Bremner AJ, Lewkowicz DJ, Spence C, eds. *Multisensory Development*. Oxford, UK: Oxford University Press; 2012:88–112.
116. Meltzoff AN, Borton RW. Intermodal matching by human neonates. *Nature*. 1979;282(5737):403–404.
117. Maurer D, Stager C, Mondloch C. Cross-modal transfer of shape is difficult to demonstrate in one-month-olds. *Child Dev*. 1999;70(5):1047–1057.
118. Held R, Ostrovsky Y, de Gelder B, et al. The newly sighted fail to match seen with felt. *Nat Neurosci*. 2011;14:551–553.
119. Bushnell EW, Boudreau JP. Motor development and the mind: the potential role of motor abilities as a determinant of aspects of perceptual development. *Child Dev*. 1993;64:1005–1021.
120. Kalagher H, Jones SS. Developmental change in young children's use of haptic information in a visual task: the role of hand movements. *J Exp Child Psychol*. 2011;108:293–307.
121. Gori M, Viva MM, Sandini G, Burr DC. Young children do not integrate visual and haptic form information. *Curr Biol*. 2008;18:694–698.
122. Alais D, Burr D. The ventriloquist effect results from near-optimal bimodal integration. *Curr Biol*. 2004;14(3):257–262.
123. Ernst MO, Banks MS. Humans integrate visual and haptic information in a statistically optimal fashion. *Nature*. 2002;415:429–433.
124. Gori M, Sandini G, Martinoli C, Burr D. Poor haptic orientation discrimination in nonsighted children may reflect disruption of cross-sensory calibration. *Curr Biol*. 2010;20:223–225.
125. Gori M, Giuliana L, Sandini G, Burr D. Visual size perception and haptic calibration during development. *Dev Sci*. 2012;15:854–862.
126. Filippetti ML, Johnson MH, Lloyd-Fox S, Dragovic D, Farroni T. Body perception in newborns. *Curr Biol*. 2013;23:2413–2416.
127. Freier L, Mason L, Bremner AJ. Perception of visual-tactile co-location in the first year of life. *Dev Psychol*. 2016;52:2184–2190.
128. Rochat P. Self-perception and action in infancy. *Exp Brain Res*. 1998;123:102–109.
129. Zmyj N, Jank J, Schütz-Bosbach S, Daum MM. Detection of visual-tactile contingency in the first year after birth. *Cognition*. 2011;120:82–89.
130. Thomas R, Misra R, Akkunt E, Ho C, Spence C, Bremner AJ. Sensitivity to auditory-tactile colocation in early infancy. *Dev Sci*. 2017. https://doi.org/10.1111/desc.12597.
131. Begum Ali J, Spence C, Bremner AJ. Human infants' ability to perceive touch in external space develops postnatally. *Curr Biol*. 2015;25:R978–R979.
132. Röder B, Rösler F, Spence C. Early vision impairs tactile perception in the blind. *Curr Biol*. 2004;14(2):121–124. S0960982203009849 [pii].
133. Azañón E, Camacho K, Morales M, Longo MR. The sensitive period for tactile remapping does not include early infancy. *Child dev*. 2017. https://doi.org/10.1111/cdev.12813.
134. Ley P, Bottari D, Shenoy BH, Kekunnaya R, Röder B. Partial recovery of visual-spatial remapping of touch after restoring vision in a congenitally blind man. *Neuropsychologia*. 2013;51:1119–1123.
135. Bremner AJ, Mareschal D, Lloyd-Fox S, Spence C. Spatial localization of touch in the first year of life: early influence of a visual code, and the development of remapping across changes in limb position. *J Exp Psychol Gen*. 2008;137:149–162.
136. Rigato S, Begum Ali J, van Velzen J, Bremner AJ. The neural basis of somatosensory remapping develops in human infancy. *Curr Biol*. 2014;24:1222–1226.
137. Rigato S, Bremner AJ, Mason L, Pickering A, Davis R, van Velzen J. The electrophysiological time course of somatosensory spatial remapping: vision of the hands modulates effects of posture on somatosensory evoked potentials. *Eur J Neurosci*. 2013;38:2884–2892.
138. Cowie D, Makin T, Bremner AJ. Children's responses to the rubber hand illusion reveal dissociable pathways in body representations. *Psychol Sci*. 2013;24:762–769.

139. Cowie D, McKenna A, Bremner AJ, Aspell JE. The development of bodily self-consciousness: changing responses to the Full Body Illusion in childhood. *Dev Sci*. 2017:e12557. https://doi.org/10.1111/desc.12557.
140. Cowie D, McKenna A, Bremner AJ, Aspell JE, Sterling S. The development of bodily self-consciousness: changing responses to the full body illusion in childhood. *Dev Sci*. 2016;142:230–238.
141. Nardini M, Begus K, Mareschal D. Multisensory uncertainty reduction for hand localization in children and adults. *J Exp Psychol Hum Percept Perform*. 2013;39:773–787.
142. Moseley GL, Gallace A, Spence C. Bodily illusions in health and disease: physiological and clinical perspectives and the concept of a cortical 'body matrix'. *Neurosci Biobehav Rev*. 2012;36:34–46.
143. Maister L, Tang T, Tsakiris M. Neurobehavioral evidence of interoceptive sensitivity in early infancy. *eLife*. 2017;6.
144. Bird G, Viding E. The self to other model of empathy: providing a new framework for understanding empathy impairments in psychopathy, autism, and alexithymia. *Neurosci Biobehav Rev*. 2014;47:520–532. https://doi.org/10.1016/j.neubiorev.2014.09.021.
145. Keysers C, Gazzola V. Expanding the mirror: vicarious activity for actions, emotions, and sensations. *Curr Opin Neurobiol*. 2009;19(6):666–671. https://doi.org/10.1016/j.conb.2009.10.006.
146. Blakemore SJ, Bristow D, Bird G, Frith C, Ward J. Somatosensory activations during the observation of touch and a case of vision–touch synaesthesia. *Brain*. 2005;128(7):1571–1583. https://doi.org/10.1093/brain/awh500.
147. Bufalari I, Aprile T, Avenanti A, Russo F, Aglioti SM. Empathy for pain and touch in the human somatosensory cortex. *Cerebr Cortex*. 2007;17:2553–2561.
148. Rigato S, Banissy M, Romanska A, Thomas R, van Velzen J, Bremner AJ. Cortical signatures of vicarious tactile experience in four-month-old infants. *Dev Cogn Neurosci*. 2017. https://doi.org/10.1016/j.dcn.2017.09.003.
149. Marshall PJ, Meltzoff AN. Body maps in the infant brain. *Trends Cognit Sci*. 2015;19:499–505.
150. Meltzoff AN, Ramírez RR, Saby JN, Larson E, Taulu S, Marshall PJ. Infant brain responses to felt and observed touch of hands and feet: an MEG study. *Dev Sci*. 2018. https://doi.org/10.1111/desc.12651.
151. Fairhurst MT, Grossmann T. Physiological and behavioral responses reveal 9-month-old infants' sensitivity to pleasant touch. *Psychol Sci*. 2014;25:1124–1131.
152. Zeller D, Litvak V, Friston KJ, Classen J. Sensory processing and the rubber hand illusion-an evoked potentials study. *J Cogn Neurosci*. 2015;27:573–582.
153. Bremner AJ. Developing body representations in early life: combining somatosensation and vision to perceive the interface between the body and the world. *Dev Med Child Neurol*. 2016;58(Suppl 4):12–16.

CHAPTER

5

Computational models of multisensory integration

David Meijer, Uta Noppeney

Computational Cognitive Neuroimaging Laboratory, Computational Neuroscience and Cognitive Robotics Centre, University of Birmingham, Birmingham, United Kingdom

Introduction

Various sensory organs continuously provide our brains with uncertain information about our environment. Critically, every sensor has its specific limitations. For example, the sensitivity of our eyes' photoreceptors is optimized for use during daylight (e.g., photoreceptor sensitivity of nocturnal insects is much higher[1]). Our ears are specialized for detecting differences in sound pitch, but they provide only imprecise estimates for the location of a sound's source.

Imagine you are in a dimly lit bedroom at night and you hear the sound of a mosquito. To obtain the most precise estimate of the mosquito's location, the brain should combine uncertain spatial information furnished by the auditory and visual senses. Critically, the brain should integrate sensory signals only when they pertain to the same event, but process them independently when they come from different events. For example, we are all familiar with those vague black spots on the wall that look annoyingly like mosquitos in the dark. These immobile black spots should not be integrated with the mosquito's buzzing sound around the head. In short, to generate a coherent percept of the environment, the brain needs to infer whether or not sensory signals are caused by common or independent sources. This process has been termed multisensory causal inference.[2]

In this chapter, we will explore the computational operations that the brain may use to solve these two challenges involved in multisensory perception, i.e., (1) how to weight and integrate signals that come from a common source into a unified percept and (2) how to infer whether signals come from common or independent sources.

In the first section, we will introduce the normative Bayesian framework focusing on perception based on input from a single sensory channel and prior expectations. In the second section, we will describe how the brain integrates signals from multiple sensory channels pertaining to the same event into a unified percept (i.e., so-called forced fusion model). In the

Multisensory Perception
https://doi.org/10.1016/B978-0-12-812492-5.00005-X

third section, we will explore the more general case of multisensory perception in the face of uncertainty about the world's causal structure, i.e., uncertainty about whether signals are caused by common or independent sources. Hence, this final case combines the two challenges facing the brain in a multisensory world: causal inference and weighted sensory integration. Each section first describes the normative Bayesian model and then briefly reviews the empirical evidence that shows the extent to which data from human or nonhuman primates are in accordance with those computational principles.

Combining information from a single sensory channel with prior knowledge

Any sensory signal that reaches the cerebral cortex is inevitably contaminated with various sources of noise. Let us consider how an observer can estimate the location of an event for spatial orienting from visual inputs. An observer's eyes are bombarded with photons, and each eye's lens refracts the photons such that a ray of focused light hits the retina. There, photoreceptors and ganglion cells transform the electromagnetic radiation into action potentials. This eventually, via several synapses, results in an activity pattern in the visual cortex. Importantly, noise may be introduced at each of those processing stages. The eye's view can be partially obscured by a dirty window, and its lens is unlikely to be perfectly in focus; the transformation from photons to action potentials functions in bulk[3]; and synaptic transmission is a probabilistic process.[4] In short, the sensory organs and systems provide the brain only with an uncertain or noisy estimate of a particular property (e.g., spatial location) of events and objects in the outside world.

To constrain perceptual inference, the observer can combine the noisy sensory evidence with *prior* knowledge or expectations. For example, in our natural environment, it is very unlikely to observe a concave human face, where the tip of the nose faces away from the observer. When an observer is shown the inside of a mask, the brain often falsely interprets the image such that the nose is perceived to be facing the observer. The visual hollow-face illusion, as this effect was dubbed, is only one of many examples where prior knowledge affects our perception.[5]

The normative Bayesian framework in neuroscience posits that the brain forms a probabilistic generative model of the sensory inputs that is inverted during perceptual inference (= recognition model). Bayesian probability theory offers a precise formulation of how observers should combine uncertain information such as different sorts of noisy sensory evidence and prior knowledge to form the most reliable representation of the world. It thus sets a benchmark of a so-called "ideal observer" or optimal performance given a particular loss function against which an organism's neural and behavioral responses can be compared.

Fig. 5.1A shows the graphical model that illustrates the generative process for the spatial localization example above based on a single sensory channel and prior knowledge. A hidden source at the true location S generates a noisy sensory signal representation X. The true location S is sampled from a prior distribution, which is often assumed to be a Gaussian with mean μ: $S \sim N(\mu_{prior}, \sigma^2_{prior})$. The sensory signal is corrupted by noise, i.e., sampled from a Gaussian centered on the true source location: $x \sim N(S, \sigma^2_{sensory})$. The generative model defines the probability of each sensory input given a particular source location $P(x|S)$. During perception, the observer needs to invert this generative model to compute the posterior probability $P(S|x)$, i.e., the probability of a spatial location given the sensory input x, by combining sensory evidence and prior knowledge. According to Bayes' rule, the posterior probability of

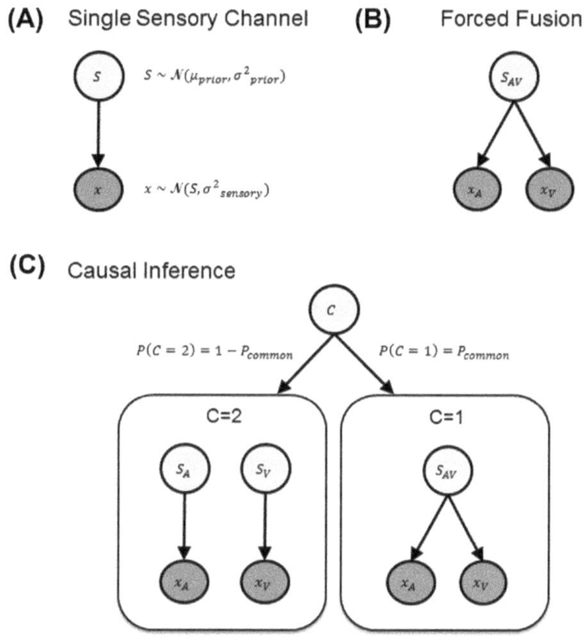

FIGURE 5.1 **Generative models corresponding to the three different cases.** (A) Single sensory signal: a hidden source generates a sensory signal that is corrupted by noise. (B) Forced fusion: a hidden source generates two sensory signals (e.g., auditory and visual) that are independently corrupted by noise. (C) Causal inference model explicitly models the potential causal structures that could have generated the two sensory signals (e.g., auditory and visual). In the full segregation model component (left), two independent hidden sources generate the auditory and visual signals. In the forced fusion model component, a common source generates two sensory signals (e.g., auditory and visual). A Bayesian causal inference estimate combines the estimates obtained from those two model components using a specific decision function (e.g., model averaging). *Adapted from Kording KP, Beierholm U, Ma WJ, Quartz S, Tenenbaum JB, Shams L. Causal inference in multisensory perception.* PLoS One. *2007;2(9):e943.*

a spatial location given a particular sensory input, $P(S|x)$, is proportional to the product of the likelihood $P(x|S)$ and the prior $P(S)$:

$$P(S|x) = \frac{P(x|S) * P(S)}{P(x)} \propto P(x|S) * P(S) \tag{5.1}$$

The normalization constant $P(x)$ can be obtained from the product of the likelihood function and the prior by marginalizing (i.e., integrating) over all possible locations S:

$$P(x) = \int P(x|S) * P(S) * dS \tag{5.2}$$

The observer then needs to minimize a particular loss function that specifies the cost of selecting the estimate $\hat{S}$ given the true location S to report a final point estimate. For instance, using the squared error loss function, the observer would report the mean of the posterior distribution as the final spatial estimate. By contrast, using a zero-one loss function, the observer would report the maximum a posteriori estimate (MAP), i.e., the mode of the posterior distribution. Critically, under Gaussian assumptions of both prior and likelihood, the posterior mean and mode are identical, i.e., both loss functions yield the same final estimate. However, asymmetric posterior distributions lead to different estimates for the posterior mean and MAP.[6,7]

Priors can emerge at multiple timescales potentially ranging from seconds to evolutionary times. For instance, during evolution, certain hardwired neural priors may have emerged as a result of selection pressures.[8] Likewise, other hardwired priors may be fine-tuned during neurodevelopment when the immature brain is exposed to the statistics of the sensory inputs.[9] Finally, the brain is thought to rapidly adjust priors to changes in the input statistics across and perhaps even within trials where the posterior of the current trial or time point forms the prior for the next trial or time point.[10,11] Priors are critical to constrain perceptual inference in the face of uncertainty resulting from noise, occlusion, etc. As we will derive in greater detail in the next "forced fusion" section, the influence of the prior on the final posterior estimate should be greatest if the sensory input is noisy and uncertain. This is because different sorts of evidence (e.g., prior vs. sensory evidence or different sensory evidences) should be combined in a manner weighted by their relative reliabilities (see Forced fusion: integrating sensory signals that come from a common source section for details).

Priors can be formed about all sorts of properties such as spatial location, shape, speed, etc. Indeed, numerous studies have demonstrated how prior knowledge or expectations shape and *bias* perceptual inference in our natural environment or designed experimental settings: the light-from-above prior (objects with ambiguous depth seem to face forward if the shadow is below them[12]), the circularity assumption (we tend to think that an object's depth is equal to its width[13]), the foveal bias (relevant objects are more likely to appear in the center of our field of view[14,15]), the slow speed preference (most objects do not move or tend to move slowly[16,17]), and the cardinal orientation prior (vertical and horizontal orientations can be more frequently found[18]). In the latter example, the experimentally determined probabilities of the human prior distribution for orientations were shown to match the environmental statistics for orientations that were found in a large set of photographs.[18] In addition to the long-term priors, the brain can also rapidly adapt priors to the dynamics of statistical regularities. In laboratory experiments, participants may learn the distribution from which the stimuli are sampled (e.g., the range of stimulus durations in a time-interval estimation task[19]). In the real world, they can adopt prior distributions that apply to a particular situation (e.g., the typical velocities for a ball in a game of tennis[20]). Multiple studies have also shown that the biasing influence of the prior is—as expected (see above)—inversely related to the reliability of the sensory stimuli.[16–20]

At the neural level, a recent functional magnetic resonance imaging (fMRI) study has shown that the brain estimates the reliability or precision of sensory representations in primary visual cortex (V1) on a trial-by-trial basis.[21] Participants were presented with visual gratings that varied in their orientation across trials. On each trial, they indicated the perceived orientation using a rotating bar. Critically, even though no external noise was added to the stimuli, the precision of sensory representations in V1 may vary across trials because of internal neural noise. Indeed, the uncertainty estimated from the activity patterns in the visual cortex varied across trials. Moreover, it correlated positively with the variance of participants' responses and negatively with their orientation errors. The results of this study[21] suggest that sensory cortices represent stimulus uncertainty on a trial-by-trial basis and that this uncertainty affects behavioral performance, as predicted by probabilistic models of Bayesian inference.

Forced fusion: integrating sensory signals that come from a common source

Many events and objects in the natural environment can be perceived concurrently by multiple senses that are each specialized for specific features of the outside world. Signals from different senses can provide complementary information. For instance, honey can be perceived as yellow by vision, but tastes sweet. Alternatively, multiple senses can provide redundant information about the same physical property such as spatial location. Thus, we can locate a puncture in a bicycle's inner tube by vision, audition, or touch (i.e., seeing, hearing, or feeling where the air flows out of the tube). In the case of redundant information across the senses, multisensory perception enables the observer to form a more precise or reliable (reliability being the inverse of variance) estimate of the environmental property in question by integrating evidence across the senses.

Fig. 5.1B shows the generative model for spatial localization based on redundant auditory and visual information. The generative model assumes one single source at the true location S_{AV} that emits two internal sensory signals; in this case, a visual and an auditory signal: x_A and x_V. As we do not allow for the two signals to be generated by two independent sources, we refer to this generative model as the forced fusion scenario, where optimal performance can be obtained by mandatory sensory integration. Again, as in the unisensory case, we assume that the auditory and visual signals, x_A and x_V, are corrupted by independent Gaussian noise. Hence, we sample x_A and x_V independently according to $x_A \sim N(S_{AV}, \sigma^2{}_A)$ and $x_V \sim N(S_{AV}, \sigma^2{}_V)$.

During perceptual inference, the observer needs to compute the posterior probability of the spatial location given auditory and visual inputs according to Bayes' theorem:

$$P(S_{AV}|x_A, x_V) = \frac{P(x_A, x_V|S_{AV}) * P(S_{AV})}{P(x_A, x_V)} \propto P(x_A, x_V|S_{AV}) * P(S_{AV}) \tag{5.3}$$

Furthermore, as auditory and visual inputs are assumed to be conditionally independent (i.e., independent noise assumption across sensory channels), we can factorize the likelihood:[22]

$$P(S_{AV}|x_A, x_V) \propto P(x_A|S_{AV}) * P(x_V|S_{AV}) * P(S_{AV}) \tag{5.4}$$

Furthermore, most studies in multisensory integration assume an uninformative or flat prior $P(S_{AV})$, where we can ignore the influence of the prior. As a result, the maximum a posteriori estimate turns into a maximum likelihood estimate:

$$P(S_{AV}|x_A, x_V) \propto P(x_A|S_{AV}) * P(x_V|S_{AV}) \tag{5.5}$$

Assuming independent Gaussian noise and uninformative priors, the optimal, most precise (i.e., most reliable or with minimum variance) audiovisual estimate $\widehat{S}_{AV}$ can be computed as a reliability-weighted linear average of the two unisensory estimates[22,23]:

$$\widehat{S}_{AV} = w_A \widehat{S}_A + w_V \widehat{S}_V \quad \text{with } w_A = \frac{r_A}{r_A + r_V} \quad \text{and} \quad w_V = \frac{r_V}{r_A + r_V} = 1 - w_A \tag{5.6}$$

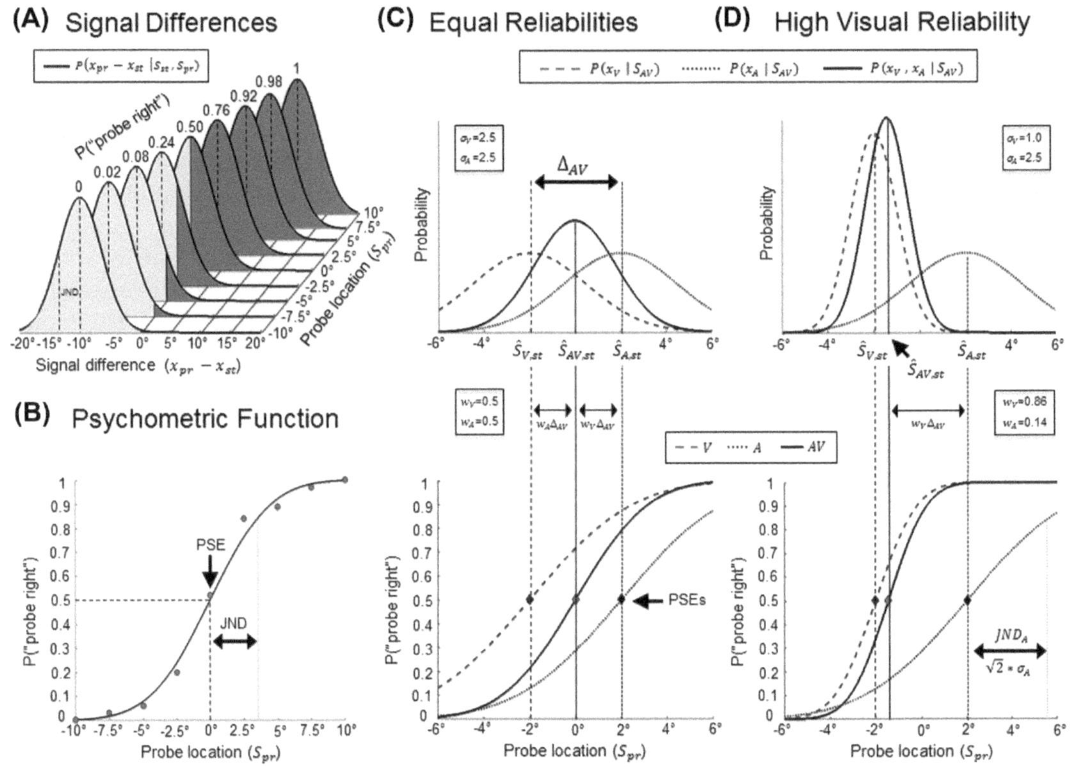

FIGURE 5.2 **Forced fusion model, maximum likelihood estimation, and psychometric perturbation analysis.** (A) Signal detection theoretic analysis of a 2IFC spatial discrimination task. For each true probe stimulus location S_{pr} (and standard stimulus location S_{st} at 0 degree), the observer computes a spatial estimate of the probe signal (x_{pr}) relative to the standard signal (x_{st}): i.e., the spatial signal difference $x_{pr} - x_{st}$. Because of trial-specific external and internal noise affecting both standard and probe stimuli, the signal difference is assumed to vary from trial to trial for identical true stimuli locations, S_{pr} and S_{st}, according to a Gaussian probability distribution with a standard deviation of $\sqrt{2} * \sigma_{sensory}$ that defines the summed sensory noise of the standard and probe stimuli. The observer provides a "probe right" discrimination response when the spatial signal difference is greater than zero degrees visual angle (i.e., $x_{pr} - x_{st} > 0°$). (B) Psychometric function. For the data of panel A, a cumulative Gaussian shows the probability (or fraction of trials; *gray circles*, including measurement noise) of "probe right" responses as a function of the true probe location S_{pr}. The probability "probe right" (in B) corresponds directly to the integral (i.e., dark shaded area in (A) of the Gaussian probability distribution (in A) where $x_{pr} - x_{st} > 0°$). The point of subjective equality (PSE) refers to the probe location associated with *P("probe right")* = 0.5. The just noticeable difference (JND) refers to the difference in probe stimulus locations at the two thresholds: *P("probe right")* = 0.5 and *P("probe right")* ≈ 0.84. In a 2IFC task, the JND (in B) is equal to the standard deviation of the Gaussian probability distribution of signal differences (in A): *i.e.* $JND = \sqrt{2} * \sigma_{sensory}$. C and D. Maximum likelihood estimation (MLE) under forced fusion assumptions: the observer is presented with an audiovisual conflict stimulus (Δ_{AV}), i.e., the visual signal is presented at $-\frac{1}{2}\Delta_{AV}$ and the auditory signal is presented at $+\frac{1}{2}\Delta_{AV}$, as the standard in the first interval and an audiovisual congruent stimulus as the probe in the second interval. The Gaussians (top) show the likelihood functions and unbiased spatial estimates (i.e., maximum likelihood estimates; *vertical lines*) from the standard stimulus separately for the visual signal ($x_V = S_{V,st} = -\frac{1}{2}\Delta_{AV}$, *dashed*), the auditory signal ($x_A = S_{A,st} = +\frac{1}{2}\Delta_{AV}$, *dotted*), and the combined audiovisual signal as obtained from MLE-based integration (Eqs. 5.6 and 5.7, *solid*). The means of the Gaussian likelihood functions for the audiovisual conflict stimuli (top) can be estimated as the PSEs of the cumulative Gaussians (bottom) obtained from auditory, visual, and audiovisual 2IFC trials where the audiovisual spatial conflict stimulus is presented as the standard stimulus (i.e., see above $S_{st} = \pm\frac{1}{2}\Delta_{AV}$) and the probe stimulus is presented at variable degrees

where the reliability is defined as the inverse of the Gaussian's variance: $r = \frac{1}{\sigma^2}$. Moreover, the reliability of this audiovisual estimate can be expressed as the sum of the two unisensory reliabilities:

$$r_{AV} = r_A + r_V \text{ which is equivalent } \sigma^2_{AV} = \frac{\sigma^2_A * \sigma^2_V}{\sigma^2_A + \sigma^2_V} \quad (5.7)$$

Hence, the reliability of the audiovisual estimate is greater than (or equal to) the maximal reliabilities of the unisensory estimates. Eq. (5.7) shows formally that multisensory integration increases the precision of the percept. The maximal multisensory variance reduction by a factor of 2 can be obtained when the variances of the two sensory signals are equal.

In summary, the maximum likelihood estimation (MLE) model under forced-fusion assumptions makes two critical predictions for human multisensory perception performance. First, the variance associated with the multisensory percept is smaller than (or equal to) the minimal variance of the unisensory percepts (Eq. 5.7). Second, the multisensory percept is obtained by integrating sensory inputs weighted by their relative reliabilities (Eq. 5.6).

In the following, we will describe the standard psychophysical approach[23,24] that allows us to test whether human behavior is in accordance with these two MLE predictions. The main steps for testing each of the two MLE predictions involve (1) estimating the unisensory variances from perceptual performance on unisensory trials, (2) using Eqs. (5.6) and (5.7) to make parameter-free MLE predictions about the multisensory variance and the sensory weights applied during multisensory integration, and (3) comparing these predictions with the multisensory variances and weights empirically measured during multisensory perceptual performance. We will use an audiovisual spatial discrimination task as an example.[25]

To investigate whether audiovisual integration of spatial inputs leads to the MLE-predicted variance reduction, we need to measure the variances associated with auditory, visual, and audiovisual percepts. The empirical variances for these percepts (e.g., spatial estimates) can be estimated from participants' responses in a two-interval forced choice (2IFC) paradigm. On each trial, the observer is presented with a standard stimulus in the first interval at zero degrees ($S_{st} = 0°$) and a probe stimulus in the second interval at variable degrees of visual angle along the azimuth (S_{pr}). Standard and probe stimuli are both presented in the visual, auditory, or audiovisual modalities. The observer discriminates whether the probe stimulus is on the left or right side of the standard. Next, we fit psychometric functions, i.e., a cumulative Gaussian (ψ), to the percentage "perceived right" responses as a function of the visual angle of the presented probe separately for the visual, auditory, and audiovisual conditions (e.g., using MLE for fitting[26]; see Fig. 5.2A and B).

of visual angle. (C) For equal visual and auditory reliabilities, the means of the Gaussian likelihood functions and the PSEs of the corresponding cumulative Gaussian psychometric functions are equal to the average of the auditory and visual means or PSEs. (D) If the visual reliability is greater (i.e., visual variance is smaller) than the auditory one, the visual signal is assigned a greater weight. As a result, the mean of the audiovisual estimate is closer to the visual than the auditory estimate. As shown in the figure, we can estimate the sensory weights from the PSEs of the psychometric functions of the unisensory visual, unisensory auditory, and audiovisual conflict conditions in a 2IFC task. *Adapted from Ernst MO, Banks MS. Humans integrate visual and haptic information in a statistically optimal fashion.* Nature. *2002;415(6870):429–433.*

$$\psi(S_{pr}) = \frac{\beta}{\sqrt{2\pi}} \int_{-\infty}^{S_{pr}} \exp\left(-\frac{\beta^2 (S_{pr} - \alpha)^2}{2}\right) \tag{5.8}$$

where α is the point of subjective equality (PSE), i.e., the probe location where the psychometric function equals 0.5, and it is equally likely for the observer to perceive the probe left or right of the standard. Furthermore, the just noticeable difference (JND), i.e., the difference in probe locations between the PSE and the point where the psychometric function equals ~0.84, is given by $\frac{1}{\beta}$. Importantly, as shown in Fig. 5.2A and B, the PSE and JND obtained from the psychometric function as a cumulative Gaussian correspond directly to the mean (μ) and standard deviation (σ) of the Gaussian distribution that describes the perceptual noise for the auditory, visual, or audiovisual spatial estimates.[27] More specifically, as we used a 2IFC paradigm in which sensory noise of both standard and probe contribute equally to the signal differences ($x_{pr} - x_{st}$), we can compute the perceptual variance for the auditory, visual, and audiovisual conditions from the JNDs of their psychometric functions according to $JND^2 = 2\sigma^2$. Using Eq. (5.7), we can then assess whether the empirically measured AV variance is in accordance with the MLE-predicted AV variance computed from the unisensory auditory and visual variances.

To investigate whether observers integrate sensory signals weighted by their relative reliabilities as predicted by MLE, we use a so-called perturbation analysis.[28] For the perturbation analysis, we need to introduce a small nonnoticeable conflict between the auditory and visual signals of the audiovisual standard stimulus (n.b. no audiovisual conflict is introduced for the probe stimulus). For instance, we can shift the auditory signal by $+\frac{1}{2}\Delta_{AV}$ and the visual signal by $-\frac{1}{2}\Delta_{AV}$ relative to $S_{AV,st}$ congruent (=0°). If the auditory and visual signals are equally reliable and hence equally weighted in the AV spatial estimate, the perceived AV location of the conflict AV stimulus is equal to the perceived location of the corresponding congruent AV stimulus (see Fig. 5.2C, top panel). Yet, if the visual reliability is greater than the auditory reliability, the perceived location (i.e., spatial estimate) for the AV conflict stimulus should be biased toward the true location of the visual signal (i.e., in the above case shifted toward the left; see Fig. 5.2D, top panel) and vice versa for greater auditory reliability. The more frequently reported visual bias on the perceived sound location has been coined the ventriloquist effect, a perceptual illusion known since ancient times. Yet, the opposite bias from audition to vision can also emerge if the visual signal is rendered less reliable.[25] To summarize, the crossmodal bias operating from vision to audition and vice versa provides us with information about the relative sensory weights applied during multisensory integration. Formally, we can quantify the weights applied to the auditory and visual signals from the PSEs of the psychometric functions obtained from the AV conflict conditions by rewriting Eq. (5.6) (see Fig. 5.2C and D, lower panels):[23]

$$w_{A,emp} = \frac{PSE_{\Delta AV} - S_{V,st}}{S_{A,st} - S_{V,st}} \quad \text{with } w_V = 1 - w_A \tag{5.9}$$

Note that this equation implicitly assumes that unisensory auditory and visual perception are unbiased (i.e., the PSEs of the unisensory psychometric functions are equal to zero). These empirical sensory weights can then be statistically compared with the MLE-predicted weights computed from the JNDs of the unisensory psychometric functions according to Eq. (5.6).

Critically, measuring the sensory weight requires a difference in the location of unisensory component signals, i.e., the presentation of incongruent audiovisual signals. While a greater intersensory conflict may enable a more reliable estimation of sensory weights, it progressively violates the forced fusion assumption and makes it less likely that observers assume a common source for the sensory signals. As a rule of thumb, a Δ_{AV} equal to the JND of the more reliable sensory signal has been proposed to be adequate.[24]

Numerous psychophysical studies have suggested that human observers integrate two sensory signals near-optimally, i.e., as predicted by the forced fusion model outlined above. For instance, near-optimal integration has been shown for visual-tactile size estimates in a seminal study by Ernst and Banks.[23] Four participants judged, by looking and/or feeling, whether the height of a raised ridge stimulus was taller than a standard comparison height. The true height of the ridge varied with small deviations from the standard height on a trial-by-trial basis. The used apparatus allowed the researchers to independently decrease the visual reliability by addition of visual noise at four different levels. Psychometric functions were fit to the unisensory and bisensory responses such that MLE-predicted and empirical weights and variances could be compared (as described above). Results indicated that the visual variance increased and visual weights decreased with increasing visual noise levels (as predicted by Eq. 5.6). Importantly, the empirical visual weights and visual-haptic variances were similar to the MLE-predicted weights and variances for all four noise levels (with a notably clear bisensory variance reduction when the visual and haptic perceptual reliability were similar; Eq. 5.7); thus suggesting that visual and haptic sensory signals were integrated in (near-) optimal fashion.[23] A follow-up experiment by the same group, using similar stimuli and apparatus, replicated the finding of an optimal variance reduction for visual-tactile size estimates (in conditions with negligible spatial disparity between the two sensory-specific cues).[29] Other examples of multisensory integration for which human behavior was shown to be in line with MLE include audiovisual location estimates,[25] audiovisual frequency discrimination,[30,31] visual-tactile object-shape judgments,[32] audiovisual duration estimates,[33] and audiovisual motion-speed discrimination.[34]

At the neural level, neurophysiological studies in nonhuman primates have shown that neural populations[35] and single neurons[36,37] integrate sensory signals weighted by their reliabilities in line with MLE predictions in visual-vestibular motion discrimination tasks. Furthermore, Fetsch et al.[35] showed that the variances and sensory weights obtained from decoding spiking rates in a population of multisensory neurons were qualitatively comparable with the variances and weights observed at the behavioral level. At a more implementational level, these authors have proposed the divisive normalization model.[38,39] This normalization model mediates reliability-weighted sensory integration, because the activity of each neuron is normalized by the activity of the entire pool of neurons.

Additional evidence in support of reliability-weighted multisensory integration at the neural level comes from several human fMRI studies showing that the connectivity between unisensory regions and association regions such as the superior temporal sulcus depends on the relative audiovisual reliabilities in speech recognition tasks.[40,41] Likewise, the blood oxygenation level-dependent response induced by somatosensory inputs in parietal areas was modulated by the reliability of concurrent visual input during a visuohaptic size discrimination task.[42]

Despite considerable evidence in support of MLE-optimal integration in human and nonhuman primates, accumulating research has also revealed situations where the sensory weights and reduction in multisensory variance are not fully consistent with the predictions of MLE. These findings highlight assumptions and limitations of the standard MLE forced fusion model for multisensory perception.

Focusing on the sensory weights, numerous studies have shown that human observers overweight a particular sensory modality in a range of tasks. Most prominently, in contrast to the classical study by Alais and Burr[25] showing MLE-optimal auditory and visual weights in spatial localization, Battaglia et al.[43] reported that observers rely more strongly on visual than auditory signals for spatial localization. Likewise, a series of studies have shown auditory overweighting in audiovisual temporal judgment tasks,[44,45] vestibular overweighting in visual-vestibular self-motion tasks,[46,47] visual overweighting in a visual-vestibular self-rotation task,[48] and haptic overweighting in a visual-haptic slant discrimination task.[49] In all of those studies, the sensory modality that is overweighted was the modality that is usually more reliable for this particular task in everyday experiences. One may therefore argue that the brain adjusts the weights of the sensory inputs not only based on the input's current reliability but also imposes a modality-specific reliability prior that reflects the modality's reliability for a particular property or task in everyday life.[43,45]

With respect to the second MLE prediction of multisensory variance reduction, numerous studies, covering a variety of sensory modalities and tasks, have also shown a decrease in multisensory variance that is smaller than predicted by the forced fusion model (Eq. 5.7). For example, this was shown for audiovisual interval duration judgments,[44] audiovisual speed discrimination,[50] visual-haptic slant discrimination,[49] and visual-haptic size and depth estimation tasks.[51,52] This "suboptimal" integration performance can be explained by several key assumptions of the forced fusion model that may not hold in our natural environment. First, the forced fusion model assumes that two signals are necessarily generated by one single source. However, in the real world, sensory signals can be generated by common or independent sources, leading to uncertainty about the world's causal structure (see next section). Likewise, in some experimental settings, the observer may take into account this causal uncertainty, in particular if conflict trials are included or artificial stimuli are used that do not enhance the observer's forced fusion or common source assumptions.[50,51] Second, the MLE model assumes that the sensory noise is independent between sensory modalities.[22] This assumption may be violated in some multisensory estimation tasks where dependencies exist between sensory modalities as a result of crossmodal adaptive calibration (e.g., auditory spatial estimates can be recalibrated by synchronous visual signals through a process that is different from multisensory integration).[51,53–56] Third, the MLE model does not include additional sources of noise that may be added after integration, e.g., during decision-making and response selection.[44,52]

Causal inference: accounting for observer's uncertainty about the world's causal structure

The forced fusion model presented in the previous section accommodates only the special case of where two signals come from a common source. As a result, it can only model that two signals are integrated in a mandatory fashion. Yet, in our natural environment, our senses are bombarded with many different signals. In this more naturalistic scenario, an observer should bind signals into one coherent and unified percept only when they come from a common source, but he needs to treat them separately when they come from independent sources. Critically, the observer does not know the causal structure underlying the sensory signals. Instead, he needs to infer whether signals come from common or independent sources from the signals themselves. A range of correspondence cues such as temporal coincidence and correlations, spatial colocation, and higher-order cues such as semantic, phonological, metaphoric, etc., correspondences (see Chapter 11)[57–69] are critical cues informing observers about whether signals come from a common source and should thus be integrated. Hence, multisensory perception in our natural environment relies on solving the so-called causal inference problem.[2] It requires observers not only to deal with uncertainty about perceptual estimates but also with causal uncertainty, i.e., their uncertainty about the world's causal structure.

Spatial ventriloquism is a prominent audiovisual perceptual illusion that illustrates not only reliability-weighted integration (see Forced fusion: integrating sensory signals that come from a common source section) but also how the brain arbitrates between integration and segregation in the face of uncertainty about the causal structure of the world. At small spatial disparities, the perceived location of an auditory event (e.g., the voice of a puppeteer) shifts toward the location of a temporally correlated but spatially displaced visual event (e.g., the facial movements of the puppet) and vice versa depending on the relative auditory and visual reliabilities as described in the forced fusion section.[25] This spatial biasing (i.e., the ventriloquist effect) breaks down or is at least attenuated at large spatial disparities and audiovisual asynchronies when it is unlikely that auditory and visual signals are caused by a common source.[70–74]

Initial modeling approaches introduced coupling priors to allow signals from different senses to bias each other without being integrated into one single unified percept.[75,76] More recently, Körding et al.[7] (and simultaneously Sato et al.[77]) proposed a Bayesian causal inference model that explicitly models the potential causal structures (i.e., common source or independent sources) that could have generated the sensory signals. Fig. 5.1C shows the generative model for Bayesian causal inference in an audiovisual spatial ventriloquist paradigm and localization task.

The generative model of Bayesian causal inference assumes that common ($C = 1$) or independent ($C = 2$) sources are determined by sampling from a binomial distribution with $P(C = 1)$ equal to the common-source prior P_{common}. The common source prior thus quantifies the observers' "unity assumption"[78] or prior tendency to integrate signals from different sensory modalities into one unified percept.

For a common source, the "true" location S_{AV} is drawn from the spatial prior distribution $N(\mu_{prior}, \sigma^2_{prior})$. For two independent causes, the "true" auditory (S_A) and visual (S_V) locations are drawn independently from this spatial prior distribution. The spatial prior distribution models an observer's prior expectations of where events may happen (see Combining information from a single sensory channel with prior knowledge section). For instance, we can model a central bias or expectation that events happen in the center of the visual field[14,15] by setting $\mu_{prior} = 0°$ and adjusting its strength in terms of the variance σ^2_{prior}.

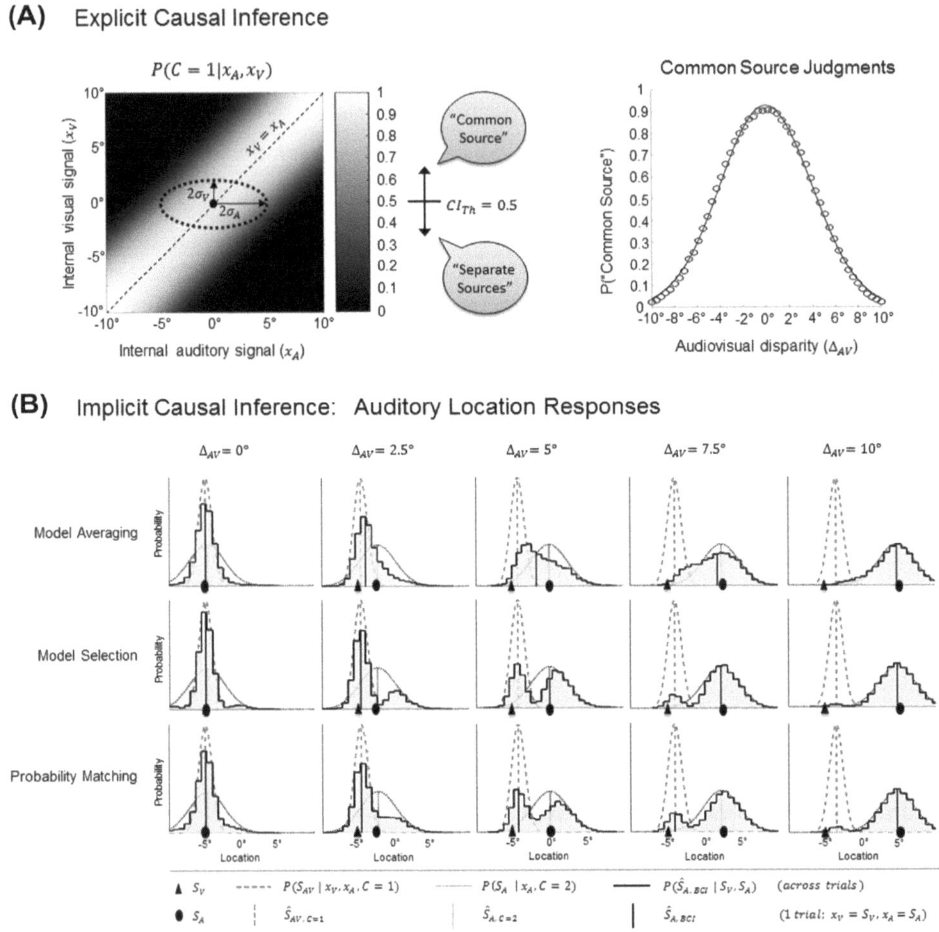

FIGURE 5.3 **Explicit and Implicit Bayesian Causal Inference.** (A) Explicit causal inference. The posterior probability of a common source $P(C=1|x_A,x_V)$ is shown as a function of the internal auditory and visual signals (x_A and x_V). It decreases for increasing spatial disparities between the internal audiovisual signals. The observer is assumed to report a common source if the posterior probability for a common source is greater than a threshold CI_{th} (e.g., if $P(C=1|x_A,x_V)>0.5$). Critically, even if the true auditory and visual source locations are identical (i.e., $S_A=S_V$), the internal visual and auditory signals can differ because of internal and external noise (e.g., the area circumscribed by the *dashed black circle* covers 95% of the bivariate Gaussian probability distribution $P(x_A,x_V|S_A=0°,S_V=0°)$). Right panel: probability of a common source judgment (across trials) as a function of spatial disparity Δ_{AV} between the auditory and visual sources (S_A and S_V) as predicted by the Bayesian causal inference model (see text). (B) Implicit causal inference. Auditory location responses: simulated auditory location responses as a function of audiovisual spatial disparity (Δ_{AV}, columns 1 to 5) according to Bayesian causal inference for the three decision functions: model averaging (top row), model selection (middle row), and probability matching (bottom row). The black triangles indicate the true visual source location S_V and the black disks the true auditory source location S_A. For one trial per panel with $x_A=S_A$ and $x_V=S_V$: the dashed lines show the audiovisual posterior probability distributions $P(S_{AV}|x_A,x_V,C=1)$ and audiovisual spatial estimates $\widehat{S}_{AV,C=1}$ (i.e., maximum a posteriori estimates; vertical lines) for the forced fusion model component. The dotted lines show the auditory posterior probability distributions $P(S_A|x_A,C=2)$ and auditory spatial estimates $\widehat{S}_{A,C=2}$ for the full segregation model component. Finally, the *vertical solid lines* indicate the Bayesian causal inference estimate $\widehat{S}_{A,BCI}$. The solid lines

Finally, exactly as in the unisensory and the forced fusion cases, noise is introduced independently for each sensory modality by drawing the sensory inputs x_A and x_V independently from normal distributions centered on the true auditory (or visual) locations with parameter σ_A (or σ_V). Thus, σ_A and σ_V define the noise (i.e., reliability) of the inputs in each sensory modality.

In total, the generative model includes the following free parameters: the common-source prior P_{common}, the spatial prior standard deviation σ_{prior}, the auditory standard deviation σ_A, and the visual standard deviation σ_V.

Given this probabilistic generative model, the observer needs to infer the causal structure that has generated the sensory inputs (i.e., common source or causal judgment) and the location of the auditory and/or visual inputs (i.e., spatial localization task). Critically, as we will see below, an observer's spatial estimates inherently depend on his strategy of how to deal with his uncertainty about the underlying causal structure. In other words, the observer's implicit causal inference codetermines his spatial estimate during a localization task.

The posterior probability of the underlying causal structure can be inferred by combining the common-source prior with the sensory evidence according to Bayes' rule:[7]

$$P(C = 1|x_A, x_V) = \frac{P(x_A, x_V|C = 1) * P_{common}}{P(x_A, x_V)} \tag{5.10}$$

In explicit causal inference tasks (e.g., common source or congruency judgments), observers may thus report common or independent sources by applying a fixed threshold (e.g., $CI_{Th} = 0.5$) to the posterior probability of a common source:

$$\widehat{C} = \begin{cases} 1 & if \quad P(C = 1|x_A, x_V) \geq CI_{Th} \\ 2 & if \quad P(C = 1|x_A, x_V) < CI_{Th} \end{cases} \tag{5.11}$$

As expected and shown in Fig. 5.3A, the posterior probability for a common source decreases with increasing spatial disparity between the auditory and visual signals. Indeed, numerous studies have demonstrated that participants are less likely to perceive signals as coming from a common source for large intersensory conflicts such as audiovisual spatial disparity or temporal asynchrony.[62–64,70–73,79,80]

Critically, the estimate of the auditory and visual source location needs to be formed depending on the underlying causal structure: in the case of a known common source ($C = 1$), the optimal estimate of the audiovisual location is a reliability-weighted average of the auditory and visual percepts and the spatial prior (i.e., this is the forced fusion estimate

delineating the gray shaded area define the probability distributions (i.e., normalized histograms) of the Bayesian causal inference estimates across many trials $P\left(\widehat{S}_{A,BCI}|S_A, S_V\right)$. The distributions were generated from 10,000 randomly sampled x_A, x_V for each combination of S_A, S_V, with the parameters for visual noise: $\sigma_V = 1°$, auditory noise: $\sigma_A = 2.5°$, central spatial prior distribution: $\mu_{prior} = 0°$ and $\sigma_{prior} = 10°$, and common source prior: $P_{common} = 0.5$ (n.b. the same parameter values were used in panel A). *Adapted from Wozny DR, Beierholm UR, Shams L. Probability matching as a computational strategy used in perception.* PLoS Comput Biol. *2010;6(8).*

of Forced fusion: integrating sensory signals that come from a common source section, Eq. (5.6), with addition of the spatial prior):

$$\widehat{S}_{AV,C=1} = \frac{\frac{x_A}{\sigma_A^2} + \frac{x_V}{\sigma_V^2} + \frac{\mu_{prior}}{\sigma_{prior}^2}}{\frac{1}{\sigma_A^2} + \frac{1}{\sigma_V^2} + \frac{1}{\sigma_{prior}^2}} \tag{5.12}$$

In the case of known independent sources ($C = 2$), the optimal estimates of the auditory and visual signal locations (for the auditory and visual location report, respectively) are independent from each other (i.e., the so-called full segregation estimates).

$$\widehat{S}_{A,C=2} = \frac{\frac{x_A}{\sigma_A^2} + \frac{\mu_{prior}}{\sigma_{prior}^2}}{\frac{1}{\sigma_A^2} + \frac{1}{\sigma_{prior}^2}} \text{ and } \widehat{S}_{V,C=2} = \frac{\frac{x_V}{\sigma_V^2} + \frac{\mu_{prior}}{\sigma_{prior}^2}}{\frac{1}{\sigma_V^2} + \frac{1}{\sigma_{prior}^2}} \tag{5.13}$$

Critically, the observer does not know the underlying causal structure and hence needs to provide a final estimate of the auditory and visual locations that account for this causal uncertainty. More specifically, the observer can combine the estimates under the two causal structures using various decision functions such as "model averaging," "model selection," or "probability matching,"[81] as described below.

According to the "model averaging" strategy, the observer accounts for his causal uncertainty by combining the integrated, forced fusion spatial estimate with the segregated, task-relevant unisensory spatial estimate (i.e., either auditory or visual; whichever needs to be reported) weighted in proportion to the posterior probability of the underlying causal structures. This strategy minimizes the error about the spatial estimates under the assumption of a squared loss function.[7]

$$\widehat{S}_A = P(C = 1|x_A, x_V) * \widehat{S}_{AV,\,C=1} + (1 - P(C = 1|x_A, x_V)) * \widehat{S}_{A,\,C=2} \tag{5.14}$$

$$\widehat{S}_V = P(C = 1|x_A, x_V) * \widehat{S}_{AV,\,C=1} + (1 - P(C = 1|x_A, x_V)) * \widehat{S}_{V,\,C=2} \tag{5.15}$$

According to the "model selection" strategy, the observer reports the auditory ($\widehat{S}_A$) or visual ($\widehat{S}_V$) spatial estimate selectively from the more likely causal structure. This strategy minimizes the error about the inferred causal structures, as well as the error about the spatial estimates given the inferred causal structures.

$$\widehat{S}_A = \begin{cases} \widehat{S}_{AV,C=1} & \text{if } P(C = 1|x_A, x_V) \geq 0.5 \\ \widehat{S}_{A,C=2} & \text{if } P(C = 1|x_A, x_V) < 0.5 \end{cases} \tag{5.16}$$

$$\widehat{S}_V = \begin{cases} \widehat{S}_{AV,C=1} & if \quad P(C = 1|x_A, x_V) \geq 0.5 \\ \widehat{S}_{V,C=2} & if \quad P(C = 1|x_A, x_V) < 0.5 \end{cases} \tag{5.17}$$

According to "probability matching," the observer reports the spatial estimate of one causal structure stochastically selected in proportion to its posterior probability.

$$\widehat{S}_A = \begin{cases} \widehat{S}_{AV,C=1} & if \quad P(C = 1|x_A, x_V) \geq \alpha \\ \widehat{S}_{A,C=2} & if \quad P(C = 1|x_A, x_V) < \alpha \end{cases} \quad \text{with } \alpha \sim Uniform(0, 1) \tag{5.18}$$

$$\widehat{S}_V = \begin{cases} \widehat{S}_{AV,C=1} & if \quad P(C = 1|x_A, x_V) \geq \alpha \\ \widehat{S}_{V,C=2} & if \quad P(C = 1|x_A, x_V) < \alpha \end{cases} \quad \text{with } \alpha \sim Uniform(0, 1) \tag{5.19}$$

As illustrated in Fig. 5.3B, Bayesian causal inference transitions gracefully between sensory integration and segregation as a function of intersensory conflict irrespective of the specific decision function. In other words, while the forced fusion model allows only for a linear combination of the sensory signals ($\widehat{S}_{AV,C=1}$ in Fig. 5.3B), Bayesian causal inference models ($\widehat{S}_{A,\ BCI}$) combine sensory signals nonlinearly as a function of intersensory conflict. They predominantly integrate sensory signals approximately in line with forced fusion models, when the conflict is small, but attenuate integration for large conflicts. Numerous studies since the inception of multisensory integration as a research field in its own right have provided qualitative evidence for the computational principles governing Bayesian causal inference. For instance, several studies have demonstrated an inverted U-shape function for % perceived synchronous or the McGurk effect as a function of audiovisual synchrony of speech signals.[60,62,63,67]

Over the past decade, accumulating research has also quantitatively compared human behavior with the predictions of Bayesian causal inference in a range of tasks including audiovisual spatial localization,[7,15,74,77,79–86] audiovisual temporal discrimination,[86–88] visual-vestibular heading estimation,[89] audiovisual speech recognition,[90] audiovisual distance perception,[91] and audiovisuo–tactile numerosity judgments.[92] In the following, we discuss the role of (1) reliability of the sensory inputs, (2) the common source prior, and (3) the decision function in Bayesian causal inference.

To investigate the influence of sensory reliability on how human observers arbitrate between sensory integration and segregation, Rohe and Noppeney[79] presented participants with auditory and visual spatial signals at multiple spatial disparities and visual reliabilities. In a dual task, observers performed Bayesian causal inference implicitly for auditory spatial localization and explicitly for common source judgment. The study showed that visual reliability shapes multisensory integration not only by determining the relative sensory weights but also by defining the spatial integration window. As expected by Bayesian causal inference, highly reliable visual signals sensitized observers to audiovisual disparity thereby sharpening the spatial integration window.[79]

In addition to bottom-up sensory signals, Bayesian causal inference depends on the so-called "common source prior," embodying an observer's prior expectations that two signals are caused by a common source. This raises the question whether these common source priors are hardwired in an individual, specifically for a particular task and stimulus characteristics. For instance, in a conversational setting with a single speaker, we should be more inclined to integrate his/her facial movements with the syllables he/she is uttering for improved speech comprehension. By contrast, in a busy pub where we are bombarded with many conflicting auditory and visual speech signals, unconstrained information integration would be detrimental. In a first study, Odegaard and Shams[86] showed that common source priors are relatively stable across time (also see Beierholm et al.[85]), yet task-specific. More specifically, they did not generalize from a spatial ventriloquism task to a double flash illusion task. Yet, in a follow-up study where they dynamically manipulated the probability of audiovisual signals being synchronous and colocated, in a ventriloquist paradigm, they demonstrated that observers dynamically adapt their common source priors to the environmental statistics.[74] Indeed, dynamic adjustment of common source priors had also previously been shown during audiovisual speech perception.[93–95]

Finally, Wozny et al.[81] investigated in a large cohort of more than 100 observers, whether observers are more likely to use model averaging, model selection, or probability matching as decisional functions in Bayesian causal inference. Surprisingly, they demonstrated that human observers predominantly use probability matching in audiovisual spatial localization. While probability matching may be thought of as being suboptimal for static environments, humans have been shown to use this strategy in a variety of cognitive tasks (e.g., reward learning[96,97]). The authors proposed that probability matching may be a useful strategy to explore potential causal structures in a dynamic environment. In summary, accumulating psychophysical research has shown that human perception is governed qualitatively and to some extent quantitatively by the principles of Bayesian causal inference, raising the question of how the brain may compute Bayesian causal inference.

At the neural level, extensive neurophysiological and neuroimaging evidence has demonstrated that multisensory integration, as indexed by multisensory response enhancement or suppression relative to the unisensory responses, depends on a temporal and spatial window of integration.[98,99] Spatial windows of integration may be related to neuronal receptive field properties. By contrast, temporal windows of integration may rely on computation of temporal correlations (e.g., see recent model using the Hassenstein-Reichardt detector[100]) and have recently been associated with brain oscillations.[101–103] Models for the neural implementations of Bayesian causal inference have been proposed, but their biological plausibility still needs to be shown.[104–107]

At the neural systems level, two recent neuroimaging studies by Rohe and Noppeney[82,83] investigated how the brain accomplishes Bayesian causal inference by combining psychophysics, fMRI, Bayesian modeling, and multivariate decoding. On each trial, participants localized audiovisual signals that varied in spatial discrepancy and visual reliability. The studies demonstrated that the brain computes Bayesian causal inference by encoding multiple spatial estimates across the cortical hierarchy. At the bottom of the hierarchy, in auditory and visual cortical areas, location is represented on the basis that the two signals are generated by independent sources (= segregation). At the next stage, in the posterior intraparietal sulcus, location is estimated under the assumption that the two signals are from a common source (= forced fusion). It is only at the top of the hierarchy, in the anterior intraparietal sulcus, that the uncertainty about whether signals are generated by common or independent sources is taken into account. As predicted by Bayesian causal inference, the final location is computed by combining the segregation and the forced fusion estimates, weighted by the posterior probabilities of common and independent sources.

Conclusions

Bayesian models of perceptual inference define how an observer should integrate uncertain sensory signals to provide an accurate and reliable percept of our environment. They thus set a benchmark of an ideal observer against which human perceptual performance can be compared. Forced fusion models and psychophysical studies have highlighted that human observers integrate sensory signals that come from a common source weighted approximately in proportion to their relative reliabilities. More recent models of Bayesian causal inference account for an observer's uncertainty about the world's causal structure by explicitly modeling whether sensory signals come from common or independent sources. A final Bayesian causal inference estimate is then obtained by combining the estimates under the assumptions of common or independent sources according to various decision functions. Accumulating psychophysical and neuroimaging evidence has recently suggested that human observers perform spatial localization and speech recognition tasks in line with the principles of Bayesian causal inference.

Acknowledgments

This research was funded by ERC-2012-StG_20111109 multsens.

References

1. Honkanen A, Immonen EV, Salmela I, Heimonen K, Weckstrom M. Insect photoreceptor adaptations to night vision. *Philos Trans R Soc Lond B Biol Sci*. 2017;372(1717).
2. Shams L, Beierholm UR. Causal inference in perception. *Trends Cognit Sci*. 2010;14(9):425–432.
3. Barlow HB. Retinal noise and absolute threshold. *J Opt Soc Am*. 1956;46(8):634–639.
4. Stevens CF. Neurotransmitter release at central synapses. *Neuron*. 2003;40(2):381–388.
5. Gregory RL. Knowledge in perception and illusion. *Philos Trans R Soc Lond B Biol Sci*. 1997;352(1358):1121–1127.
6. Yuille AL, Bulthoff HH. Bayesian decision theory and psychophysics. In: David CK, Whitman R, eds. *Perception as Bayesian Inference*. Cambridge University Press; 1996:123–161.
7. Kording KP, Beierholm U, Ma WJ, Quartz S, Tenenbaum JB, Shams L. Causal inference in multisensory perception. *PLoS One*. 2007;2(9):e943.
8. Geisler WS, Diehl RL. Bayesian natural selection and the evolution of perceptual systems. *Philos Trans R Soc Lond B Biol Sci*. 2002;357(1420):419–448.
9. Gopnik A, Tenenbaum JB. Bayesian networks, Bayesian learning and cognitive development. *Dev Sci*. 2007;10(3):281–287.
10. Roach NW, McGraw PV, Whitaker DJ, Heron J. Generalization of prior information for rapid Bayesian time estimation. *Proc Natl Acad Sci U S A*. 2017;114(2):412–417.
11. Di Luca M, Rhodes D. Optimal perceived timing: integrating sensory information with dynamically updated expectations. *Sci Rep*. 2016;6:28563.
12. Mamassian P, Landy MS. Interaction of visual prior constraints. *Vis Res*. 2001;41(20):2653–2668.
13. Jacobs RA. Optimal integration of texture and motion cues to depth. *Vis Res*. 1999;39(21):3621–3629.
14. Kerzel D. Memory for the position of stationary objects: disentangling foveal bias and memory averaging. *Vis Res*. 2002;42(2):159–167.

15. Odegaard B, Wozny DR, Shams L. Biases in visual, auditory, and audiovisual perception of space. *PLoS Comput Biol*. 2015;11(12):e1004649.
16. Weiss Y, Simoncelli EP, Adelson EH. Motion illusions as optimal percepts. *Nat Neurosci*. 2002;5(6):598–604.
17. Stocker AA, Simoncelli EP. Noise characteristics and prior expectations in human visual speed perception. *Nat Neurosci*. 2006;9(4):578–585.
18. Girshick AR, Landy MS, Simoncelli EP. Cardinal rules: visual orientation perception reflects knowledge of environmental statistics. *Nat Neurosci*. 2011;14(7):926–932.
19. Jazayeri M, Shadlen MN. Temporal context calibrates interval timing. *Nat Neurosci*. 2010;13(8):1020–1026.
20. Kording KP, Wolpert DM. Bayesian integration in sensorimotor learning. *Nature*. 2004;427(6971):244–247.
21. van Bergen RS, Ma WJ, Pratte MS, Jehee JF. Sensory uncertainty decoded from visual cortex predicts behavior. *Nat Neurosci*. 2015;18(12):1728–1730.
22. Oruc I, Maloney LT, Landy MS. Weighted linear cue combination with possibly correlated error. *Vis Res*. 2003;43(23):2451–2468.
23. Ernst MO, Banks MS. Humans integrate visual and haptic information in a statistically optimal fashion. *Nature*. 2002;415(6870):429–433.
24. Rohde M, van Dam LCJ, Ernst MO. Statistically optimal multisensory cue integration: a practical tutorial. *Multisensory Res*. 2016;29(4–5):279–317.
25. Alais D, Burr D. The ventriloquist effect results from near-optimal bimodal integration. *Curr Biol*. 2004;14(3):257–262.
26. Kingdom FAA, Prins N. Chapter 4 - Psychometric Functions*. In: *Psychophysics*. 2nd ed. San Diego: Academic Press; 2016:55–117.
27. Acuna DE, Berniker M, Fernandes HL, Kording KP. Using psychophysics to ask if the brain samples or maximizes. *J Vis*. 2015;15(3).
28. Young MJ, Landy MS, Maloney LT. A perturbation analysis of depth perception from combinations of texture and motion cues. *Vis Res*. 1993;33(18):2685–2696.
29. Gepshtein S, Burge J, Ernst MO, Banks MS. The combination of vision and touch depends on spatial proximity. *J Vis*. 2005;5(11):1013–1023.
30. Raposo D, Sheppard JP, Schrater PR, Churchland AK. Multisensory decision-making in rats and humans. *J Neurosci*. 2012;32(11):3726–3735.
31. Sheppard JP, Raposo D, Churchland AK. Dynamic weighting of multisensory stimuli shapes decision-making in rats and humans. *J Vis*. 2013;13(6).
32. Helbig HB, Ernst MO. Optimal integration of shape information from vision and touch. *Exp Brain Res*. 2007;179(4):595–606.
33. Hartcher-O'Brien J, Di Luca M, Ernst MO. The duration of uncertain times: audiovisual information about intervals is integrated in a statistically optimal fashion. *PLoS One*. 2014;9(3):e89339.
34. Mendonca C, Santos JA, Lopez-Moliner J. The benefit of multisensory integration with biological motion signals. *Exp Brain Res*. 2011;213(2–3):185–192.
35. Fetsch CR, Pouget A, DeAngelis GC, Angelaki DE. Neural correlates of reliability-based cue weighting during multisensory integration. *Nat Neurosci*. 2011;15(1):146–154.
36. Gu Y, Angelaki DE, Deangelis GC. Neural correlates of multisensory cue integration in macaque MSTd. *Nat Neurosci*. 2008;11(10):1201–1210.
37. Morgan ML, Deangelis GC, Angelaki DE. Multisensory integration in macaque visual cortex depends on cue reliability. *Neuron*. 2008;59(4):662–673.
38. Ohshiro T, Angelaki DE, DeAngelis GC. A normalization model of multisensory integration. *Nat Neurosci*. 2011;14(6):775–782.
39. Ohshiro T, Angelaki DE, DeAngelis GC. A neural signature of divisive normalization at the level of multisensory integration in primate cortex. *Neuron*. 2017;95(2):399–411 e398.
40. Beauchamp MS, Pasalar S, Ro T. Neural substrates of reliability-weighted visual-tactile multisensory integration. *Front Syst Neurosci*. 2010;4:25.
41. Nath AR, Beauchamp MS. Dynamic changes in superior temporal sulcus connectivity during perception of noisy audiovisual speech. *J Neurosci*. 2011;31(5):1704–1714.
42. Helbig HB, Ernst MO, Ricciardi E, et al. The neural mechanisms of reliability weighted integration of shape information from vision and touch. *Neuroimage*. 2012;60(2):1063–1072.

43. Battaglia PW, Jacobs RA, Aslin RN. Bayesian integration of visual and auditory signals for spatial localization. *J Opt Soc Am A*. 2003;20(7):1391–1397.
44. Burr D, Banks MS, Morrone MC. Auditory dominance over vision in the perception of interval duration. *Exp Brain Res*. 2009;198(1):49–57.
45. Maiworm M, Röder B. Suboptimal auditory dominance in audiovisual integration of temporalcues. *Tsinghua Sci Technol*. 2011;16(2):121–132.
46. Fetsch CR, Turner AH, DeAngelis GC, Angelaki DE. Dynamic reweighting of visual and vestibular cues during self-motion perception. *J Neurosci*. 2009;29(49):15601–15612.
47. Butler JS, Smith ST, Campos JL, Bulthoff HH. Bayesian integration of visual and vestibular signals for heading. *J Vis*. 2010;10(11):23.
48. Prsa M, Gale S, Blanke O. Self-motion leads to mandatory cue fusion across sensory modalities. *J Neurophysiol*. 2012;108(8):2282–2291.
49. Rosas P, Wagemans J, Ernst MO, Wichmann FA. Texture and haptic cues in slant discrimination: reliability-based cue weighting without statistically optimal cue combination. *J Opt Soc Am A*. 2005;22(5):801–809.
50. Bentvelzen A, Leung J, Alais D. Discriminating audiovisual speed: optimal integration of speed defaults to probability summation when component reliabilities diverge. *Perception*. 2009;38(7):966–987.
51. Gepshtein S, Banks MS. Viewing geometry determines how vision and haptics combine in size perception. *Curr Biol*. 2003;13(6):483–488.
52. Battaglia PW, Kersten D, Schrater PR. How haptic size sensations improve distance perception. *PLoS Comput Biol*. 2011;7(6):e1002080.
53. Jacobs RA. What determines visual cue reliability? *Trends Cognit Sci*. 2002;6(8):345–350.
54. Gori M, Sciutti A, Burr D, Sandini G. Direct and indirect haptic calibration of visual size judgments. *PLoS One*. 2011;6(10):e25599.
55. Wozny DR, Shams L. Recalibration of auditory space following milliseconds of cross-modal discrepancy. *J Neurosci*. 2011;31(12):4607–4612.
56. Ernst MO. Optimal multisensory integration: assumptions and limits. In: Stein BE, ed. *The New Handbook of Multisensory Processes*. Cambridge, Massachusetts: MIT Press; 2012:527–544.
57. Warren DH, Welch RB, McCarthy TJ. The role of visual-auditory "compellingness" in the ventriloquism effect: implications for transitivity among the spatial senses. *Percept Psychophys*. 1981;30(6):557–564.
58. Bishop CW, Miller LM. Speech cues contribute to audiovisual spatial integration. *PLoS One*. 2011;6(8):e24016.
59. Kanaya S, Yokosawa K. Perceptual congruency of audio-visual speech affects ventriloquism with bilateral visual stimuli. *Psychon Bull Rev*. 2011;18(1):123–128.
60. Lee H, Noppeney U. Long-term music training tunes how the brain temporally binds signals from multiple senses. *Proc Natl Acad Sci U S A*. 2011;108(51). E1441-1450.
61. Lee H, Noppeney U. Temporal prediction errors in visual and auditory cortices. *Curr Biol*. 2014;24(8):R309–R310.
62. van Wassenhove V, Grant KW, Poeppel D. Temporal window of integration in auditory-visual speech perception. *Neuropsychologia*. 2007;45(3):598–607.
63. Soto-Faraco S, Alsius A. Deconstructing the McGurk-MacDonald illusion. *J Exp Psychol Hum Percept Perform*. 2009;35(2):580–587.
64. Stevenson RA, Fister JK, Barnett ZP, Nidiffer AR, Wallace MT. Interactions between the spatial and temporal stimulus factors that influence multisensory integration in human performance. *Exp Brain Res*. 2012;219(1):121–137.
65. Noppeney U, Josephs O, Hocking J, Price CJ, Friston KJ. The effect of prior visual information on recognition of speech and sounds. *Cerebr Cortex*. 2008;18(3):598–609.
66. Adam R, Noppeney U. Prior auditory information shapes visual category-selectivity in ventral occipito-temporal cortex. *Neuroimage*. 2010;52(4):1592–1602.
67. Maier JX, Di Luca M, Noppeney U. Audiovisual asynchrony detection in human speech. *J Exp Psychol Hum Percept Perform*. 2011;37(1):245–256.
68. Parise CV, Spence C. 'When birds of a feather flock together': synesthetic correspondences modulate audiovisual integration in non-synesthetes. *PLoS One*. 2009;4(5):e5664.
69. Parise CV, Spence C, Ernst MO. When correlation implies causation in multisensory integration. *Curr Biol*. 2012;22(1):46–49.

70. Slutsky DA, Recanzone GH. Temporal and spatial dependency of the ventriloquism effect. *Neuroreport*. 2001;12(1):7–10.
71. Lewald J, Guski R. Cross-modal perceptual integration of spatially and temporally disparate auditory and visual stimuli. *Cogn Brain Res*. 2003;16(3):468–478.
72. Hairston WD, Wallace MT, Vaughan JW, Stein BE, Norris JL, Schirillo JA. Visual localization ability influences cross-modal bias. *J Cogn Neurosci*. 2003;15(1):20–29.
73. Wallace MT, Roberson GE, Hairston WD, Stein BE, Vaughan JW, Schirillo JA. Unifying multisensory signals across time and space. *Exp Brain Res*. 2004;158(2):252–258.
74. Odegaard B, Wozny DR, Shams L. A simple and efficient method to enhance audiovisual binding tendencies. *PeerJ*. 2017;5:e3143.
75. Bresciani JP, Dammeier F, Ernst MO. Vision and touch are automatically integrated for the perception of sequences of events. *J Vis*. 2006;6(5):554–564.
76. Roach NW, Heron J, McGraw PV. Resolving multisensory conflict: a strategy for balancing the costs and benefits of audio-visual integration. *Proc Biol Sci*. 2006;273(1598):2159–2168.
77. Sato Y, Toyoizumi T, Aihara K. Bayesian inference explains perception of unity and ventriloquism aftereffect: identification of common sources of audiovisual stimuli. *Neural Comput*. 2007;19(12):3335–3355.
78. Chen YC, Spence C. Assessing the role of the 'unity assumption' on multisensory integration: a review. *Front Psychol*. 2017;8:445.
79. Rohe T, Noppeney U. Sensory reliability shapes perceptual inference via two mechanisms. *J Vis*. 2015;15(5):22.
80. Bosen AK, Fleming JT, Brown SE, Allen PD, O'Neill WE, Paige GD. Comparison of congruence judgment and auditory localization tasks for assessing the spatial limits of visual capture. *Biol Cybern*. 2016;110(6):455–471.
81. Wozny DR, Beierholm UR, Shams L. Probability matching as a computational strategy used in perception. *PLoS Comput Biol*. 2010;6(8).
82. Rohe T, Noppeney U. Cortical hierarchies perform Bayesian causal inference in multisensory perception. *PLoS Biol*. 2015;13(2):e1002073.
83. Rohe T, Noppeney U. Distinct computational principles govern multisensory integration in primary sensory and association cortices. *Curr Biol*. 2016;26(4):509–514.
84. Natarajan R, Murray I, Shams L, Zemel RS. Characterizing response behavior in multisensory perception with conflicting cues. In: Koller D, Schuurmans D, Bengio Y, Bottou L, eds. *Advances in Neural Information Processing Systems*. Vol. 21. MIT Press; 2009:1153–1160.
85. Beierholm UR, Quartz SR, Shams L. Bayesian priors are encoded independently from likelihoods in human multisensory perception. *J Vis*. 2009;9(5), 23 21-29.
86. Odegaard B, Shams L. The brain's tendency to bind audiovisual signals is stable but not general. *Psychol Sci*. 2016;27(4):583–591.
87. Magnotti JF, Ma WJ, Beauchamp MS. Causal inference of asynchronous audiovisual speech. *Front Psychol*. 2013;4:798.
88. McGovern DP, Roudaia E, Newell FN, Roach NW. Perceptual learning shapes multisensory causal inference via two distinct mechanisms. *Sci Rep*. 2016;6:24673.
89. de Winkel KN, Katliar M, Bulthoff HH. Causal inference in multisensory heading estimation. *PLoS One*. 2017;12(1):e0169676.
90. Magnotti JF, Beauchamp MS. A causal inference model explains perception of the McGurk effect and other incongruent audiovisual speech. *PLoS Comput Biol*. 2017;13(2):e1005229.
91. Mendonca C, Mandelli P, Pulkki V. Modeling the perception of audiovisual distance: Bayesian causal inference and other models. *PLoS One*. 2016;11(12):e0165391.
92. Wozny DR, Beierholm UR, Shams L. Human trimodal perception follows optimal statistical inference. *J Vis*. 2008;8(3), 24 21-11.
93. Nahorna O, Berthommier F, Schwartz JL. Binding and unbinding the auditory and visual streams in the McGurk effect. *J Acoust Soc Am*. 2012;132(2):1061–1077.
94. Nahorna O, Berthommier F, Schwartz JL. Audio-visual speech scene analysis: characterization of the dynamics of unbinding and rebinding the McGurk effect. *J Acoust Soc Am*. 2015;137(1):362–377.
95. Gau R, Noppeney U. How prior expectations shape multisensory perception. *Neuroimage*. 2016;124(Pt A):876–886.

96. Erev I, Roth AE. Maximization, learning, and economic behavior. *Proc Natl Acad Sci U S A*. 2014;111(Suppl. 3):10818–10825.
97. Vul E, Goodman N, Griffiths TL, Tenenbaum JB. One and done? Optimal decisions from very few samples. *Cogn Sci*. 2014;38(4):599–637.
98. Meredith MA, Nemitz JW, Stein BE. Determinants of multisensory integration in superior colliculus neurons. I. Temporal factors. *J Neurosci*. 1987;7(10):3215–3229.
99. Meredith MA, Stein BE. Spatial determinants of multisensory integration in cat superior colliculus neurons. *J Neurophysiol*. 1996;75(5):1843–1857.
100. Parise CV, Ernst MO. Correlation detection as a general mechanism for multisensory integration. *Nat Commun*. 2016;7:11543.
101. Cecere R, Rees G, Romei V. Individual differences in alpha frequency drive crossmodal illusory perception. *Curr Biol*. 2015;25(2):231–235.
102. Samaha J, Postle BR. The speed of alpha-band oscillations predicts the temporal resolution of visual perception. *Curr Biol*. 2015;25(22):2985–2990.
103. Thakur B, Mukherjee A, Sen A, Banerjee A. A dynamical framework to relate perceptual variability with multisensory information processing. *Sci Rep*. 2016;6:31280.
104. Ma WJ, Rahmati M. Towards a neural implementation of causal inference in cue combination. *Multisensory Res*. 2013;26(1–2):159–176.
105. Spratling MW. A neural implementation of Bayesian inference based on predictive coding. *Connect Sci*. 2016;28(4):346–383.
106. Yu Z, Chen F, Dong J, Dai Q. Sampling-based causal inference in cue combination and its neural implementation. *Neurocomputing*. 2016;175:155–165.
107. Cuppini C, Shams L, Magosso E, Ursino M. A biologically inspired neurocomputational model for audiovisual integration and causal inference. *Eur J Neurosci*. 2017;46(9):2481–2498.

CHAPTER

6

Multisensory contributions to object recognition and memory across the life span

Pawel J. Matusz[1,2,3], *Mark T. Wallace*[3,4,5,6,7], *Micah M. Murray*[1,3,8,9]

[1]The LINE (Laboratory for Investigative Neurophysiology), Department of Radiology, University Hospital Center and University of Lausanne, Lausanne, Switzerland; [2]Information Systems Institute at the University of Applied Sciences Western Switzerland (HES-SO Valais), Sierre, Switzerland; [3]Department of Hearing and Speech Sciences, Vanderbilt University Medical Center, Nashville, TN, United States; [4]Department of Psychology, Vanderbilt University, Nashville, TN, United States; [5]Kennedy Center for Research on Human Development, Vanderbilt University, Nashville, TN, United States; [6]Department of Psychiatry, Vanderbilt University, Nashville, TN, United States; [7]Vanderbilt Brain Institute, Vanderbilt University, Nashville, TN, United States; [8]Department of Ophthalmology, University of Lausanne and Fondation Asile des Aveugles, Lausanne, Switzerland; [9]The EEG Brain Mapping Core, Center for Biomedical Imaging (CIBM), University Hospital Center and University of Lausanne, Lausanne, Switzerland

Introduction

We are all familiar with this situation, typical for conferences and other networking-oriented professional meetings: you are a novice (e.g., a first-year graduate student in neuroscience, psychology, etc.) and you decide to go to the preconference reception. Naturally, as you do not know anyone there, you arrive to the event with another young colleague from your lab, Casey. You nervously hold onto your drink and hover around one end of the table with snacks. You comment on how busy the place is, list the presentations you are looking forward to hearing, and discuss the places you would like to visit while in the conference

Multisensory Perception
https://doi.org/10.1016/B978-0-12-812492-5.00006-1

city. Suddenly someone comments "apparently they have a fantastic museum of modern art here." You and Casey turn around and you both instantly recognize an internationally renowned neuroscientist, Professor Alex Broderick. You are star-struck, but your friend Casey quickly recovers and thanks the Professor for advice as the museum was not on your list. Casey then introduces themselves, gesticulates at you to do the same—which you do, hesitatingly and quietly—and goes on saying which laboratory you two work in and the project you two are working on. Professor Broderick congratulates you two on an interesting research program, stating that they quickly pursued it around 20 years ago, during their graduate studies, but then stopped. A typical silence ensues, awkward smiles are exchanged, and Professor Broderick excuses themselves saying they have to join their colleagues. Two days later, you and Casey bump into Professor Broderick again during the conference breakfast. Whose name do you think Professor Broderick remembers more easily: yours or Casey's?

This is just one of numerous examples that one can provide to demonstrate a fundamental principle regarding information processing and learning in the real world: they occur in contexts where information stimulating multiple senses at once is commonplace. Over the last decades, many fundamental mechanisms and principles have been revealed with respect to how memory-related processes support our abilities to perceive and interact with objects and individuals in the outside world.[1–5] However, these discoveries were based on unisensory, typically visual or auditory, research. This leaves open the question of the extent to which this knowledge generalizes to everyday environments, which, among possessing other important attributes, such as unpredictability, or noisiness of stimulation, are inherently multisensory. More recent research in this area demonstrated that processing stimulus attributes, at least in the case of naturalistic objects (e.g., alphanumeric symbols, identities), activates the same regions in the brain regardless of the modality of stimulation (visual, auditory, tactile, etc.), consistent with the brain inherently representing objects in a multisensory fashion.[6–8] However, even these studies are limited to the extent that they did not directly measure the extent of improvements (or impairments, depending on the task context) elicited in object recognition by multisensory relative to unisensory information.

In this chapter, we first review the interactions between multisensory processes and the traditional processes involved in object recognition as well as learning. We then focus on one such line of systematic investigation, that is, the processes governing the efficacy of single multisensory experiences in influencing memory for both visual and auditory objects. We summarize the main findings emerging from this research and situate them within the broader context of literature on learning in multisensory contexts. We then identify the underlying mechanisms and conditions sufficient for multisensory-induced memory improvements (at least in some contexts) and the implications of these processes for information processing in healthy individuals across the life span as well as in atypical and clinical populations.

Multisensory contributions to object recognition

What do we know about the interplay between multisensory and memory functions from the point of view how they can interact to influence brain and behavioral responses in everyday situations? Objects in the real world are typically complex and familiar, at least in terms of their semantic categories: from voices and faces through animate and inanimate objects (animals and tools, respectively) to alphanumeric, symbolic objects such as digits, letters, and words (see an equally broad definition of object in other reviews[9]). As such, the extent to which our brains can detect, extract, and benefit from redundancies in object-defining attributes within multisensory stimuli provides us with an important advantage during perceiving and interacting with objects in everyday situations. Early studies in the area have confirmed that long-term memory-dependent processes based on semantic congruence do improve perception.[10] Since then, multisensory processes associated with different categories of naturalistic objects have been demonstrated to influence object recognition, selective attention, memory, and other cognitive functions (for detailed reviews see Refs. 9,11–14). As such, semantic congruence seems to be an important principle governing multisensory integrative processes,[10] complementing—in real-world environments—the "classic" principles, focused on the spatial and temporal coincidence of signals across the senses.[15] Behavioral benefits of such memory-based multisensory processes are typically accompanied by and/or are directly related to the activity of a network centered around the superior temporal cortex (STC) and inferior parietal cortex (IPC) (e.g.,[16–22]). The dominating viewpoint is that these brain areas are themselves involved in and are the locus of the integration of object features into unified representations (see also Refs. 23,24). Others posit that brain regions such as the superior temporal sulcus (STS) serve as a relay of unisensory information to other brain areas where these are finally integrated into a consolidated object representation.[25,26] For example, some views proposed that multisensory representations of object-related information are typically located in the visual cortices, which is taken to indicate the predominance of vision in object processes (e.g.,[27,28]; but see Ref. 29). Notably, other brain areas are also implicated, e.g., the planum temporale (speech/script, e.g.,19) and lateral occipital cortex (LOC) (object recognition involving touch; reviewed in, Ref. 9 see chapter by Lacey and Sathian, this volume). Frontal cortices (inferior and dorsolateral prefrontal areas) are typically engaged only by incongruent and/or unfamiliar audiovisual associations (for a comprehensive review, see Ref. 30). We would emphasize that these propositions are not mutually exclusive, and multiple circuits and varieties of representations are likely to coexist.

However, in the large majority of these studies, multisensory information was central to the task, i.e., participants were advised to use information across multiple senses to perform a given task. This leaves open a crucial question as to whether multisensory processes can influence perception and behavior with objects when the multisensory information is not central to the performed task. In one study on this topic, we have demonstrated that peripheral audiovisual distractors interfere with an attention-demanding task such as visual search and do so to the same extent when the search task is easy or difficult. We demonstrated that these findings generalize across both simpler (color-defined objects) and more complex (letters) stimuli.[31] Notably, in these studies, targets and distractors always shared their features, suggesting the potential dependence of these effects on the goals of the observer. Furthermore, in real-world settings, where stimuli are dynamic, the detection of semantic multisensory congruence and consequent behavioral facilitation might be more dependent on the available attentional resources. In a setup with multiple visual speakers and a single voice, Alsius and Soto-Faraco[32] showed that detection of audiovisual face/voice congruence is dependent on the number of simultaneously presented faces, indicating the importance of available attentional resources (see also Ref. 33). Similarly, the McGurk illusion (i.e., perceiving a novel auditory syllable from mismatching auditory and visual syllables[34]) elicited by a task-irrelevant stimulus is reduced when attention is directed away, toward a concurrent attention-demanding task[35,36] (see Ref. 37 for absence of event-related potential (ERP) indices of the McGurk illusion in such contexts). To summarize, these findings provide evidence that multisensory processes can influence object recognition, even in naturalistic, cluttered settings. At the same time, these influences seem to be at least partly contingent on available attentional resources and/or goals of the individual. This idea is supported by the relative late latency with which semantic memory-dependent multisensory processes engage the brain (>100 ms poststimulus[18,27,38–42]) and their strong dependence on the task.[43–45] These effects have important clinical implications in terms of utility of such processes in supporting recovery of neurocognitive functions during rehabilitation. An area that has been relatively less researched is the extent to which memory-dependent multisensory processes can influence learning.

Multisensory learning as the norm rather than an exception

The circumstances under which multisensory memory traces impact subsequent unisensory retrieval is directly related to the extent to which multisensory processes can be utilized to support learning in real-world settings, as well as in rehabilitation. However, the precise nature of these circumstances remains largely unresolved. This question falls into the broader research framework focusing on the general differences in learning across unisensory (visual, auditory) and multisensory (audiovisual) settings (e.g.,[46–48]) (Fig. 6.1A). Research involving a wide variety of stimuli has consistently demonstrated that learning in multisensory settings is more effective and efficient than in comparable unisensory settings (reviewed in Ref. 47). For example, during coherent motion detection and discrimination, perceptual learning involving auditory–visual stimuli is more effective than visual training.[49] Individuals undergoing audiovisual training, compared with those undergoing purely visual training, learned faster not only overall, across the whole training involving 10 sessions, but this advantage was already visible within the first training session. These and other studies clearly demonstrate that the brain's perceptual skills and cognitive functions are particularly attuned to multisensory processes. To the extent that multisensory attuning is a general property of brain functions, fundamental memory processes, such as encoding, storage, and retrieval of information, would all be facilitated in the context of multisensory information, whereas unisensory information is typically suboptimal, to the extent that the computational brain architecture in place is not utilized to its full extent under unisensory conditions. In another study involving visual motion discrimination,[50] benefits of audio–visual training over purely visual training were found exclusively in a group that trained with congruent multisensory information (auditory and visual stimuli were moving in the same direction) but not in the incongruent condition group (two types of stimuli moving in opposite directions). While research on perceptual learning has provided important insights into the efficacy and potential circumstances promoting the benefits elicited by multisensory processes in learning, simplified and artificial stimuli were typically employed, thus leaving unclear whether these findings generalize to settings involving more naturalistic objects.

Predominantly two types of paradigms have been utilized to study the efficacy of multisensory learning on the ability to recognize unisensory (typically visual) objects. In the first paradigm, effortful and extended multisensory training preceded unisensory object

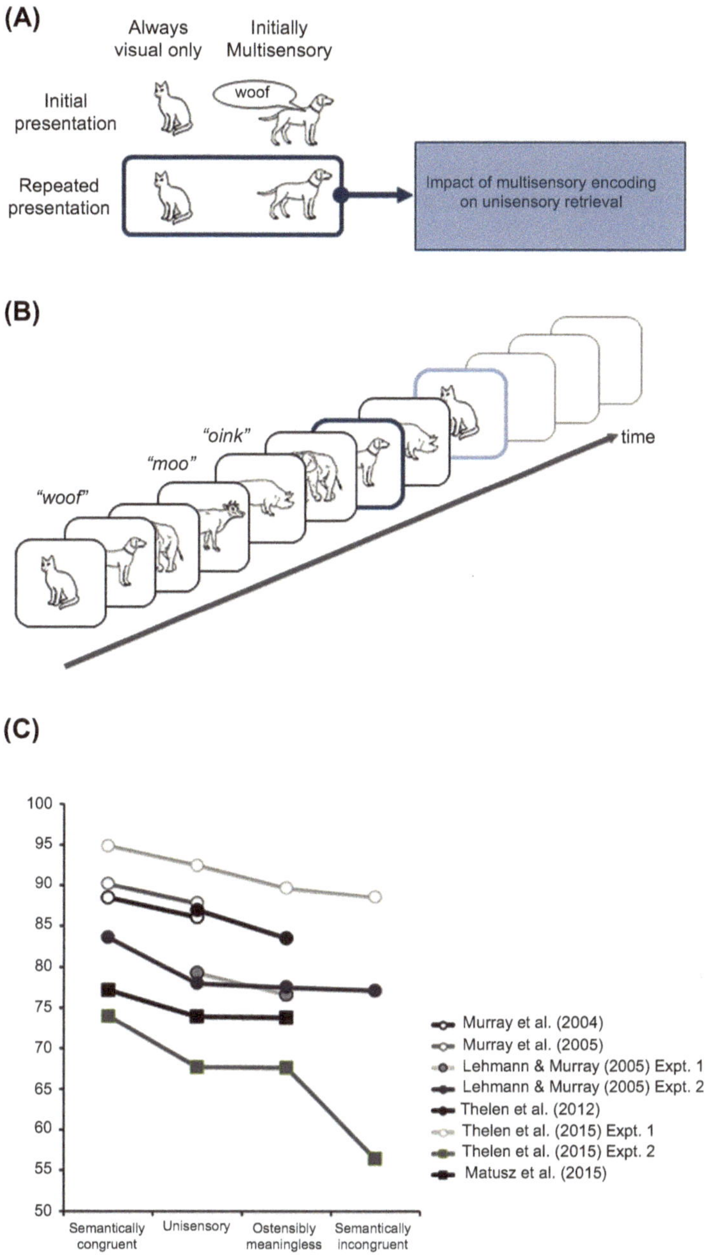

FIGURE 6.1 (A) A general schematic of how the impact of multisensory encoding on later unisensory recognition may be investigated. (B) A schematic of a continuous recognition task requiring participants to indicate whether the image was presented for the first or repeated time. Whether or not the image was presented with a sound was task-irrelevant. (C) Summary of behavioral findings. Accuracy for the various repeated presentations are displayed. Lines with circular markers refer to studies where the task was performed in the visual modality, whereas lines with square markers refer to studies where the task was performed in the auditory modality. Across studies, it can be seen that stimuli that had been initially presented in a semantically congruent multisensory context result in higher accuracy than stimuli that had only been experienced in a unisensory context. Other had-been multisensory contexts generally result in no difference or even performance impairment relative to the unisensory context.

recognition, with the two clearly separated into two sessions.[51,52] Participants were required to explicitly remember the unisensory versus multisensory context in which a stimulus appeared during training. At the brain level, visual words presented previously with matching sounds activated auditory cortices[51] (see Ref. 52 for similar findings involving images of naturalistic objects). These results were taken as supporting the so-called theory of reintegration.[53] This theory postulates that networks active during encoding are reactivated during retrieval. That is, presentation of a single element of a consolidated memory suffices to (re) activate the representation of the whole experience. However, it is noteworthy that in these studies, stimuli learned in a multisensory context were remembered less well than those learned in unisensory, visual contexts. Other findings in this area are more in line with behavioral benefits of multisensory learning on memory. For example, in a study by von Kriegstein and Giraud,[54] participants learned to associate semantically congruent multisensory pairings (faces and voices) as well as arbitrary multisensory pairings (voices and written names, and ringtones paired with cell phones or names). Subsequently, purely auditory voices were better recognized when they were initially paired with faces rather than written names, and the face–voice associations elicited enhanced functional coupling between the anterior temporal and fusiform brain regions involved in processing voices and faces, respectively. Sounds (ringtones) from arbitrary pairings showed no similar results.

The other type of paradigm provides a more consistent picture on the circumstances facilitating multisensory memory improvements (Fig. 6.1B). The task requires the participants to indicate whether the presented image (or sound, in the case of the auditory version of the task), such as that of a cow, is being presented for the first time or is a repeated stimulus ("old/new" task). On its initial presentation, the image is either presented alone (visual-only trials) or together with a matching sound, such a "moo" for an image of a cow. This paradigm, used extensively by our group over the past ~15 years,[55] has a number of distinctive features that distinguish it from the first type of paradigm, while at the same time, in our opinion, being closer to the settings in which multisensory processes exert their effect on learning and memory in everyday situations. First and perhaps most crucially, encoding and retrieval trials are randomly intermixed with each other within every block of trials. Second, the encoding and retrieval are separated only by a short interval of time (up to 1 minute, see below). Third, due to the focus on the episode (i.e., seeing an image or sound for the first time or subsequently) rather than on the image, the memory processes engaged by objects can be studied, without the potential confounds from focusing attention in a top-down fashion on the object identity/category. Fourth and relatedly, the multisensory information, similarly to the experiment of von Kriegsten and Giraud,[54] is irrelevant to the task itself, which allows for more rigorous investigation, unconfounded by top-down attentional processes, of the effects of distinct multisensory processes.

To more closely emulate information processing in naturalistic environments, we manipulated the type of senses engaged, their task relevance, and the relationship between the two crossmodal stimuli, as well as a variety of other factors that could determine the efficacy of multisensory memories.[56] The mounting evidence from our group and, more recently, other laboratories has provided novel insights into the behavioral and brain mechanisms guiding memory and information processing in everyday situations. One take-home message from this research, which we summarize in the following section, is that a single exposure to multisensory pairings suffices to improve subsequent recognition (i.e., memory) of a unisensory stimulus and that these benefits generalize across vision and hearing.[24,29,55,57–61]

From multisensory learning to unisensory object memory

When is multisensory memory better than memory based on unisensory experiences?

In our "old/new" continuous recognition task, the improvements from multisensory contexts on object memory are visible in discrimination accuracy, with no comparable benefits found on response speed. For example, when an image is presented initially (and thus encoded) in a multisensory context, its discrimination as "old" versus "new" is more accurate, with these benefits found across all participants when the stimuli across two senses are semantically congruent. Across experiments, the magnitude of multisensory benefits imparted on memory by semantic congruence ranged between 2.5% and 9% compared with performance for purely visual or auditory trials (Fig. 6.1C). In experiments where the task design permitted the calculation of more direct measures of sensory processes as distinct from decision bias (i.e., d′[62]), the benefits rose to 12% improvement.[61] Statistically, these multisensory benefits, reported until now in six published studies in >100 participants, include effects ranging from small to large ($\eta^2_p = 0.14–0.63$; for similar size of effects in studies involving setups with separate exposure and recall, see Refs. 63–67). The semantic congruence-induced benefits seem robust against a variety of factors that can be considered typical for everyday situations. For example, the multisensory benefits were reported in an fMRI study[57] where the usual lag between the initial and repeated presentations (5 seconds) increased 10-fold to allow for intertrial intervals long enough to accommodate the requirements for the acquisition of fMRI data. Of particular note is that the multisensory, audiovisual benefits were found despite the ubiquitous scanner noise.

We urge the reader to recognize the importance of these findings to applications in everyday situations: benefits for episodic memory from long-term multisensory associations of semantic features of naturalistic objects transpire without the explicit will of the individual. That is, it is not necessary for the individual to focus explicitly on the multisensory nature of the object for the benefits to be present. It seems that the crucial factor here is the established and preserved multisensory representation of these features. In the case of pairings of attributes that do not match the same object, in our paradigm, multisensory contexts are detrimental or at most highly variable in terms of their effect on object memory. Perhaps least surprisingly, semantically *incongruent* initial multisensory contexts (e.g., image of a cow with a barking sound) impair object recognition, relative to unisensory (both visual and auditory) contexts, with these impairments being of similar magnitudes to the respective multisensory benefits (4%–16.5% accuracy decrease). Multisensory contexts that involve arbitrary, nonsemantically related pairings, such as an image of a cow with a simple "pip," overall also impair recognition, but these decrements are less strong (3%–4% accuracy decrease). This can be explained by the fact that roughly half the participants benefit from such meaningless multisensory contexts, while the other half are impaired by them, and this proportion is similar irrespective of whether the task is visual or auditory.[24] In the case of multisensory contexts that involve arbitrary, nonsemantically related pairings, the benefits (when observed) ranged between 0.5%–7% and 2.5%–10% performance improvement for visual and auditory conditions, respectively. We discuss these differences and the potential mechanisms underlying the differential multisensory-induced memory effects in Section 3.2.

The demonstration of benefits from multisensory contexts involving semantically congruent pairings for object memory provides important advances to our understanding of how brain functional organization as well as how such crucial cognitive functions as memory and semantic knowledge likely operate in naturalistic, multisensory settings. These novel insights are treated in detail elsewhere (see Ref. 56 for a comprehensive discussion). We succinctly summarize them here in Section 4, where we discuss potential applications of our findings to developmental and clinical domains.

What do multisensory benefits for memory depend on and how/why do they vary across individuals?

To summarize the main characteristics of the multisensory benefits in object memory emerging from the work of our group and others, these benefits occur even following a single exposure to a multisensory context. These memory effects remain present for approximately 1 minute, are robust to the intervening presentation of several other test items, and are most uniformly present for multisensory pairings that are semantically congruent. There are several factors that influence both the presence and strength of multisensory benefits.

Semantic congruence elicits benefits for both visual and auditory object memory. There is a continued interest in how sensory and more higher-level cognitive processes differ between hearing and vision, the two senses so critically determining our abilities to interact with the environment.[68–73] Consequently, research is increasingly focusing on how the task-relevant sense influences multisensory processing in a bottom-up manner (e.g.,[74–76]). In one of our studies,[60] we have compared the strength of multisensory benefits in visual and auditory tasks within the same individuals, as evidence generally points to memory for auditory objects being weaker than that for visual objects[42,77,78]: Consistent with previous research, memory for sounds was generally weaker than for images (67% vs. 92% accuracy, respectively). However, as expected, auditory memory benefits from semantically congruent pairings were fourfold larger than the visual benefits (8.8% vs. 2.2% accuracy improvement). As the same individuals took part in the two tasks, which were quite similar to each other, these results are in line with the principle of "inverse effectiveness," i.e., multisensory benefits are often stronger in contexts where the inputs are weakly effective[79–82] (see chapter by Stein and Rowland, this volume), thus extending this principle from the originally studied context of instantaneous perception to the context of memory function.

While semantic congruence and the specific task- or goal-relevant sense are factors determining the efficacy of multisensory benefits in object memory, there seem to exist relatively important interindividual differences in the benefits exhibited in our paradigm within healthy adults and across populations. In a study from our laboratory, Thelen et al.[24] systematically analyzed the bimodal distribution of benefits versus impairments from initial, multisensory, meaningless contexts. We found this bimodal distribution in two separate samples that each performed either the visual or the auditory version of the old/new task. As we recorded EEG activity and analyzed ERPs within the electrical neuroimaging framework in these as well as other studies (see Section 3.3), we were able to shed some light on the potential differences in the two subpopulations (i.e., those benefitting or not from multisensory contexts). When we analyzed responses (separately for participants in the visual and those in the auditory task) to the initial multisensory meaningless stimuli, we found that multisensory benefits versus impairments were associated with enhanced versus reduced strength of brain response to the multisensory stimuli, respectively. Additionally, despite differences in latency, the same brain area, that is the IPC, differentiated between those improving and impaired across both visual and auditory memory tasks. Notably, the two groups did not show any differences in processing unisensory visual or auditory information. The two groups also did not differ in terms of their performance metrics (accuracy and response speed), suggesting that the differences did not emerge from general distractibility. At the same time, the relatively long latency of the differences at the brain level (>150 ms poststimulus[25]) suggests the involvement of selective attention processes (e.g., see Refs. 83,84 for evidence of audiovisual modulation of unisensory responses at these latencies). Also the localization of the origin of these differences to IPC, an area to be involved in multisensory processes as well as top-down attentional control (e.g.,[21]), suggests the differences may lie in the way multisensory processes influence selective attention processes linked to encoding of the object information. One line of current efforts in the lab is to determine whether interindividual differences, such as those demonstrated here, originate because some individuals have a higher propensity to integrate multisensory information, irrespective of the stimulus type, stimulus combination, or even task. As will be discussed in Section 4.1, such differences do seem to be present in the population and also emerge relatively early in life.

What are the cognitive and brain mechanisms governing multisensory benefits in memory?

Our paradigm provides an access point for a particular example of memory processes. Our paradigm focuses on episodic memory (are you seeing this object for the first time or was it shown earlier?). As the task uses naturalistic objects (tools, animals), the involvement of semantic (multisensory) object memory is expected. Finally, the multisensory processes that we are investigating are those activated outside of the individual's attentional focus and goals, as the task is always unisensory (e.g.,[41,85] for studies of similar effects of task-irrelevant multisensory processes on selective visual attention). In this section, we will first discuss results related to the brain responses and mechanisms accompanying the discussed behavioral benefits, and will then contrast these with the mechanisms for multisensory benefits in memory proposed by studies involving paradigms that facilitate effortful and explicit, rather than incidental and implicit, encoding of multisensory stimuli.

Brain correlates of implicit multisensory benefits in memory

Most of our research on the brain mechanisms governing the observed multisensory benefits pertains to visual memory. Besides the study on interindividual differences in adults,[86] all of our analyses focused on the brain responses to *repeated* stimuli. Across studies involving ERPs and fMRI, we found consistently that the LOC responds more strongly to repeated presentations of images of naturalistic objects that initially appeared with semantically congruent sounds, relative to images always presented alone. Using ERPs,[55] we demonstrated two different topographies, indicating that statistically distinct brain networks are activated in response to the two types of repeated images, as early as in the first 60–135 ms poststimulus (with later differences found as well; at ~210–260 ms and 318–390 ms). Using source localization techniques, we revealed these effects to be driven by enhanced responses within the LOC for images previously seen in a multisensory versus visual-only context.

We have also determined the brain loci of differences in impaired recognition for images that were previously presented in meaningless multisensory contexts. Images were paired with one and the same tone[58] or, in later studies, with a distinct tone (with tones modulated in their spectral composition, amplitude envelope, and waveform type[29]). ERP differences underlying the behavioral impairments were observed as early as 100 ms poststimulus and, as reviewed above, were driven by changes in ERP topography and changes in the underlying configurations of brain sources. Importantly, these effects were yet again localized to a small cluster within the LOC (right), as well as a larger cluster in the posterior STS. Notably, the LOC activity was now *weak*er for multisensory contexts, while STS activations were stronger, for previously multisensory compared with visual conditions. There were also topographic differences in responses at 270–310 ms, with these differences now localized to the right middle temporal cortex; the strength of the response of this area was directly related to the magnitude of memory impairment. Thus, the exact brain areas activated during a visual memory task are determined not by the mere presence of multisensory contexts but rather the "sign" of their effect on visual object memory.

Consistent with the marked differences in the extent of behavioral benefits for visual and auditory memory,[60] quite different networks of brain areas as well as mechanisms seem to orchestrate the multisensory benefits in auditory versus visual memory. In a study where EEG was recorded from participants performing the old/new task on sounds,[61] the ERP differences associated with previous multisensory semantically congruent contexts on auditory memory were found at 35–85 ms poststimulus. Notably, right IPC, right STC, and the right inferior occipital cortex and left frontal cortex supported multisensory-induced benefits in auditory memory. Crucially, right IPC and right STC—the two areas whose activity modulated in a manner consistent with the pattern of observed behavioral benefits—showed *suppressed* responses to previously multisensory semantically congruent sounds compared with sounds just presented alone, despite the former eliciting behavioral benefits. This direction of brain responses suggests potential involvement of a response suppression mechanism, proposed to govern short-term learning within auditory cortices.[87,88]

Multisensory representations of objects in the brain

Collectively, the results discussed in Section 3.3.1 bear important implications for our understanding of the way in which naturalistic objects are represented in the brain and how these representations are accessed. A consistent finding emerging across our fMRI and EEG studies is that the representations of task-relevant objects were affected early during brain processing by whether previous object presentations involved multisensory contexts or not. That is, networks responsible for the processing of unisensory stimuli have access to multisensory memory representations early on in sensory–cognitive processing. Notably, using source estimation techniques, we demonstrated that this access is reflected by brain activity within nominally unisensory object-recognition brain areas (accompanied by IPC activity in a task involving memory for sounds).

Our proposal is that these early ERP modulations reflect rapid reactivation of distinct multisensory (audio–visual) and visual or auditory object representations affected in the course of encoding during initial stimulus presentation. Several lines of evidence support this idea. First, it is now well established that unisensory objection-recognition regions—LOC in the case of vision and STC in the case of hearing—exhibit auditory–visual convergence (e.g., reviewed in Refs. 13,14,30). Second, multisensory object representations are present in these areas and are distinguishable from their unisensory counterparts. Studies recording from microelectrodes in monkey posterior inferotemporal (IT) cortex (LOC is believed to be the human homologue of IT), as well as visual area V4, show selective delay-period responses on a delayed match-to-sample task for specific multisensory and unisensory pairings (e.g.,[89–92]; see also Refs. 93–95). The IT and V4 neurons were selectively responsive to unisensory stimuli as a function of the learned association, i.e., whether a given visual stimulus appeared with another visual stimulus or rather an auditory stimulus. Notably, these neurons were selectively responsive to a given learned association.[89] While we recognize that our findings can be influenced by the initial multisensory experiences impacting unisensory representations (which may not be mutually exclusive with our proposal above), these single-cell recordings support the idea that there are distinct representations of unisensory and multisensory associations within patches of the IT cortex (see Ref. 96 for findings of "patchy," uni/multisensory organization of areas bordering between multisensory and unisensory areas of STC).

Our work extends this body of knowledge in several important ways. First, the multisensory representations can be accessed in a fashion largely independent of the goals of the observer, to the extent that only one sensory modality was ever important to perform the task in our paradigm, while the sensory signals from the task-irrelevant modality would not be expected consciously to provide an advantage in the task. Second, these multisensory representations are established or accessed within the cortices of the sense relevant to the task. Third, these representations and/or their activation (i.e., memory traces) can be accessed after a time delay. In other words, object categorization based on past experiences, at least at early brain processing stages, is supported by processes within the task-relevant cortices that likely operate on multisensory representations.

These processes are unlikely to be similar to those engaged by the effortful encoding paradigms utilized by early studies, which provided discrepant findings on the benefits of semantically congruent multisensory processes in unisensory memory (e.g.,[51,52,97]). Across these studies, the areas activated during memory encoding and retrieval closely overlapped. The findings from these studies were regarded as evidence for the "reintegration" account[53] proposing that consolidated memory leads to the reactivation of both the task-relevant cortices (here, visual) as well as the task-irrelevant cortices (here, auditory) despite the presence of only task-relevant stimuli. However, these findings need to be qualified. For one, in the study of Nyberg et al.,[51] the absence of activations in other than in the auditory cortices can be due the fact that the brain areas activated during the encoding stage in this study served as regions of interest for analyses of brain responses at retrieval. Second, given the tasks explicitly required the participants to recall if a given word was learned with a sound, the activation of auditory cortices is consistent with the participants utilizing mental imagery to aid their memory recall. Notwithstanding, the paradigms involving effortful encoding and recall, and those utilized by us, focusing on the implicit activation and influence of multisensory processes on continuous encoding/retrieval, are likely to rely on different types of object representations. These paradigmatic differences could help to reconcile our findings with those proposing the critical role of the medial temporal cortices (especially perirhinal cortex) in governing the binding of semantic multisensory features into coherent object representations.[25,94] This notion is based predominantly on lesion studies, showing that disconnection of the perirhinal cortex results in impaired performance in the delayed match-to-sample task, in line with both encoding and retrieval relying on the integrity of this particular area. What characterizes all of these discussed studies is the likely (yet uncontrolled) role in individuals' abilities to attend to and encode into memory the crossmodal feature pairings. In contrast, in our task, encoding was focused solely on the one, task-relevant sense. As such, these other studies might be building or engaging much richer representations than those targeted by the continuous recognition task. Repetition priming may be another mechanism at play within a continuous recognition task (see, in vision, e.g.,[98]; in hearing. e.g.,[70,88]). However, we would contend that repetition priming alone could not account fully for our effects (cf.[55]). Instead, we reiterate, our findings are likely to be driven by multisensory representations of naturalistic objects, residing in the early cortices of the task-relevant sense that can be activated by task-irrelevant but semantically congruent stimuli, with this activation improving memory for the unisensory task-relevant counterparts of these objects during repeated presentations.

There are several lines of evidence against the multisensory benefits we have observed being driven predominantly by the initial multisensory experiences impacting unisensory representations. First, there were no accuracy differences when initial trials were unisensory or multisensory, when all multisensory conditions (including semantically incongruent and congruent) were considered, that would indicate the presence of multisensory perception benefits. In fact, reaction times were consistently and significantly *slower* for multisensory than unisensory initial trials, suggesting some performance *costs* of initial multisensory presentations, despite the later accuracy improvements during unisensory recognition. The same pattern was observed in both visual and auditory tasks (cf. Figure 2 in Ref. 59). Second, while the initial-presentation responses did not modulate reliably according to the manipulated conditions, these manipulations were highly effective in influencing both behavioral and brain responses to repeated presentations. Third, the topographic ERP differences and the differential modulation across visual[29,55] and auditory cortices[61] in response to repeated-presentation stimuli as a function of the initial multisensory contexts goes against a generic, increased top-down (memory-/goal-driven) attention and/or salience mechanism influencing the processing of the initial-presentation stimuli. Lastly, the study of Thelen et al.[99] clearly demonstrated that the extent to which the initial meaningless multisensory contexts benefitted versus impaired participants was predicted only by brain responses to the multisensory, not unisensory, initial stimuli. If perceptual processing was the driver of the memory enhancements, one would have predicted an overall stronger response to both multisensory and unisensory stimuli in the individuals exhibiting multisensory memory benefits versus impairments, yet no such general group differences were found.

Broader implications: multisensory processes scaffold cognition across the life span

We now first succinctly summarize the theoretical implications of our findings for models of multisensory processing as well as those of memory. We then focus on their potential practical applications for supporting development, education, and well-being across the life span within the healthy population as well as their rehabilitative potential in atypical and clinical populations.

Theoretical implications of the interplay between multisensory processes and memory functions

First, we demonstrate that the products of multisensory processes persist over time. This research fits with and extends the larger body of research focused on learning in multisensory contexts, based on congruencies across features from simple object categories.[22,47,100] This work points also to the importance of an individual's sensory experience, both long- and short-term, in influencing responses to unisensory and multisensory objects, a topic that we have treated in detail elsewhere.[14,101] This framework is consistent with other research aiming to clarify the interplay of stimulus-driven and top-down attentional control processes that jointly shape memory performance.[102]

Second, our findings challenge some of the most fundamental principles proposed to govern memory functions that have been derived from research based on purely visual stimulation. Traditional research suggests that memory performance is maximal when we retrieve information in similar contexts to those in which we have encoded it.[2,103] These principles may not generalize beyond visual contexts to naturalistic contexts, where notions such as conceptual novelty versus physical familiarity come into play.[104] When considered together with the implicit nature of the multisensory benefits that we have observed, multisensory processes based on the detection of semantic congruence and thus based on the activation of long-term memory associations might be particularly ubiquitous in their influences in everyday environments. Furthermore, the observed benefits are likely specific to multisensory processes, rather than any particular object-related feature (e.g., visual) redundancy. Effects of multisensory versus unisensory redundancy are confirmed by research, across both humans and nonhuman animals, focused on perception[105,106] as well as memory.[63,89]

Lastly, our findings bear important implications for models of functional brain organization, by providing independent evidence for the inherently multisensory nature of object representations.[8,14,101] Moreover, our findings would suggest that simultaneity may be a sufficient condition for reaping multisensory benefits for learning and memory with objects in the real world (at least in the case of semantically congruent information). The majority of models of multisensory processing is based on simple stimuli and their spatiotemporal cooccurrence (see chapter by Stein andand Rowland, this volume; cf.,[13,14,107,108] for reviews on the role of audiovisual simultaneity detection in modulating instantaneous perception and selective attention). Our findings suggest that in everyday life the efficacy of these processes to benefit behavior might be limited (but see Refs. 15,85,109,110). To better understand the importance of multisensory processes in supporting cognitive functions in everyday environments, research in our laboratories for some time has been focused on understanding how multisensory processes influence cognitive functions in populations other than healthy typical adults.

Outlook: the importance of multisensory processes in public health

One significant line of active research by our group focuses on the idea that a person's capacity to integrate multisensory information, such as during a simple detection task, may scale up directly to the extent to which this person utilizes multisensory experiences to facilitate object recognition and memory. That is, does one's ability to benefit from multisensory contexts in a memory task rely on a more general capacity to integrate multisensory signals, including even simple beeps and flashes presented simultaneously at the same location? One shortcoming of our extant work is that all responses were related to different components within the same paradigm (i.e., initial vs. repeated exposures during a continuous recognition task). It thus remains unclear whether links between multisensory processes are still found when measured using two or multiple tasks (each with their own stimulus set, goals, and attentional demands).

The extent to which multisensory integrative capacity maps onto specific behavioral metrics and brain mechanisms is equally unclear. Thus far, in the continuous recognition paradigm, we have reported a link between brain activity at one point in time and behavior at a subsequent time point on the same task. We are now enriching those findings by studying relationship between multisensory processes and other metrics of behavior. We have recently explored directly the scaling of multisensory benefits across separate laboratory tasks of detection and recognition memory as well as the links of such benefits with clinical metrics of working memory and fluid intelligence (Denervaud et al., under review). In schoolchildren, like the adults described above, we observed that the magnitude of multisensory benefits on a simple detection task positively predicted the magnitude of benefits of multisensory encoding on the continuous recognition task we have described throughout this chapter. In addition, such multisensory benefits also predicted working memory scores on the ascending digits' tasks and fluid intelligence scores as measured using Raven's Progressive Matrices. Our findings show that the scaffolding that low-level multisensory processes provide for higher-order memory and cognition is already established during childhood. One consequence is that typical models of cognitive development will surely need to better incorporate the role of multisensory processes; with a likely impact on education practices. They might also open exciting opportunities to facilitate early learning through multisensory programs. More generally, these data suggest that simpler and more resource-effective sensory-based methods can provide direct insights into the integrity of cognitive functions in schoolchildren.

We have likewise applied a similar approach in aging. Behavior on a simple multisensory detection task can predict memory performance measured with a standardized questionnaire indexing memory function (the Mini-Mental State Examination).[111] Specifically, we have demonstrated that an index combining the extent of an older person's sensory preference for auditory or visual stimuli (i.e., sensory dominance) and the extent of their multisensory benefits, both of which are derived from the same audio–visual detection task, can accurately diagnose a person as belonging to the healthy elderly versus mild cognitive impairment group. Crucially, our task requires no specialist or trained personnel, is fast (<5 minutes), and requires no calibration of stimuli (sensory dominance and multisensory gain were linearly correlated suggesting that alternations in stimulus effectiveness would have comparable effects on both metrics). Nonetheless, it provided diagnostic values similar to more traditional neuropsychological tests that were designed specifically for such assessments and which require both financial and human infrastructure to administer and interpret.

Conclusion

We hope the reader has gained a sense for how the common understanding of learning and memory must be expanded into a multisensory framework.[101] Multisensory processes are not limited to instantaneous perceptions, though certainly important for such perceptions. Rather, they also extend over time and can impact what is typically viewed as a strictly unisensory perception. Similarly, the manner in which we have traditionally conceived of mental representations of objects, and how these representations are established and maintained, has been expanded to incorporate multisensory processes and crossmodal plasticity. In these regards, a multisensory framework can facilitate attempts to link laboratory-based research with more realistic and ecological settings and applications (for a broader discussion see Ref. 15). This emerging research includes, but is not limited to, two crucial domains of everyday life, i.e., education and rehabilitation of sensory and cognitive functions.

Acknowledgments

PJM receives support from the Swiss National Science Foundation (PZ00P1_174150) and Pierre Mercier Foundation. MTW receives support from the National Institutes of Health (grants CA183492, DC014114, HD083211, MH109225), from the Simons Foundation Autism Research Initiative, and from the Wallace Foundation. MMM receives support from the Swiss National Science Foundation (grants 320030-149982 and 320030-169206 as well as National Centre of Competence in Research project "SYNAPSY, The Synaptic Bases of Mental Disease" [project 51AU40-125759]) and a grantor advised by Carigest S.A. PJM and MMM are both supported by the Fondation Asile des Aveugles.

References

1. Fuster JM. Unit activity in prefrontal cortex during delayed-response performance: neuronal correlates of transient memory. *J Neurophysiol*. 1973;36(1):61–78. https://doi.org/10.1152/jn.1973.36.1.61.
2. Baddeley A, Eysenchm MW, Anderson MC. *Memory*. Psychology Press; 2009.
3. Gazzaley A, Nobre AC. Top-down modulation: bridging selective attention and working memory. *Trends Cognit Sci*. 2012;16(2):129–135. https://doi.org/10.1016/j.tics.2011.11.014.
4. Rugg MD, Thompson-Schill SL. Moving forward with fMRI data. *Perspect Psychol Sci*. 2013;8(1):84–87. https://doi.org/10.1177/1745691612469030.
5. Summerfield C, de Lange FP. Expectation in perceptual decision making: neural and computational mechanisms. *Nat Rev Neurosci*. 2014;15(11):745–756. https://doi.org/10.1038/nrn3838.
6. Mahon BZ, Anzellotti S, Schwarzbach J, Zampini M, Caramazza A. Category-specific organization in the human brain does not require visual experience. *Neuron*. 2009;63(3):397–405. https://doi.org/10.1016/j.neuron.2009.07.012.
7. Striem-Amit E, Cohen L, Dehaene S, Amedi A. Reading with sounds: sensory substitution selectively activates the visual word form area in the blind. *Neuron*. 2012;76(3):640–652. https://doi.org/10.1016/j.neuron.2012.08.026.
8. Reich L, Maidenbaum S, Amedi A. The brain as a flexible task machine: implications for visual rehabilitation using noninvasive vs. invasive approaches. *Curr Opin Neurol*. 2012;25(1):86–95. https://doi.org/10.1097/WCO.0b013e32834ed723.
9. Amedi A, Von Kriegstein K, Van Atteveldt NM, Beauchamp MS, Naumer MJ. Functional imaging of human crossmodal identification and object recognition. *Exp Brain Res*. 2005;166(3–4):559–571. https://doi.org/10.1007/s00221-005-2396-5.
10. Laurienti PJ, Kraft R a, Maldjian J a, Burdette JH, Wallace MT. Semantic congruence is a critical factor in multisensory behavioral performance. *Exp Brain Res*. 2004;158(4):405–414. https://doi.org/10.1007/s00221-004-1913-2.
11. Ghazanfar AA, Schroeder CE. Is neocortex essentially multisensory? *Trends Cognit Sci*. 2006;10(6):278–285. http://www.ncbi.nlm.nih.gov/entrez/query.fcgi?cmd=Retrieve&db=PubMed&dopt=Citation&list_uids=16713325.
12. Naumer MJ, Doehrmann O, Müller NG, Muckli L, Kaiser J, Hein G. Cortical plasticity of audio-visual object representations. *Cerebr Cortex*. 2009;19(7):1641–1653. https://doi.org/10.1093/cercor/bhn200.

13. Murray MM, Thelen A, Thut G, Romei V, Martuzzi R, Matusz PJ. The multisensory function of the human primary visual cortex. *Neuropsychologia*. 2016;83(1):161–169. https://doi.org/10.1016/j.neuropsychologia.2015.08.011.
14. ten Oever S, Romei V, van Atteveldt N, Soto-Faraco S, Murray MM, Matusz PJ. The COGs (context, object, and goals) in multisensory processing. *Exp Brain Res*. 2016;234(5):1307–1323. https://doi.org/10.1007/s00221-016-4590-z.
15. Matusz PJ, Dikker S, Huth AG, Perrodin C. Are we ready for real-world neuroscience? *J Cogn Neurosci*. 2018:1–13. https://doi.org/10.1162/jocn_e_01276.
16. Calvert G a, Campbell R, Brammer MJ. Evidence from functional magnetic resonance imaging of crossmodal binding in the human heteromodal cortex. *Curr Biol*. 2000;10(11):649–657. http://www.ncbi.nlm.nih.gov/pubmed/10837246.
17. Calvert G a. Crossmodal processing in the human brain: insights from functional neuroimaging studies. *Cerebr Cortex*. 2001;11(12):1110–1123. http://www.ncbi.nlm.nih.gov/pubmed/11709482.
18. Raij T, Uutela K, Hari R. Audiovisual integration of letters in the human brain. *Neuron*. 2000;28(2):617–625. http://www.ncbi.nlm.nih.gov/pubmed/11144369.
19. Van Atteveldt N, Formisano E, Goebel R, Blomert L. Integration of letters and speech sounds in the human brain. *Neuron*. 2004;43(2):271–282. https://doi.org/10.1016/j.neuron.2004.06.025.
20. Beauchamp MS, Lee KE, Argall BD, Martin A. Integration of auditory and visual information about objects in superior temporal sulcus. *Neuron*. 2004;41(5):809–823. http://www.ncbi.nlm.nih.gov/pubmed/15813999.
21. Werner S, Noppeney U. Distinct functional contributions of primary sensory and association areas to audiovisual integration in object categorization. *J Neurosci*. 2010;30(7):2662–2675. https://doi.org/10.1523/JNEUROSCI.5091-09.2010.
22. Powers AR, Hevey M a, Wallace MT. Neural correlates of multisensory perceptual learning. *J Neurosci*. 2012;32(18):6263–6274. https://doi.org/10.1523/JNEUROSCI.6138-11.2012.
23. Tanabe HC, Honda M, Sadato N. Functionally segregated neural substrates for arbitrary audiovisual paired-association learning. *J Neurosci*. 2005;25(27):6409–6418. https://doi.org/10.1523/JNEUROSCI.0636-05.2005.
24. Thelen A, Matusz PJ, Murray MM. Multisensory context portends object memory. *Curr Biol*. 2014;24(16):320030. https://doi.org/10.1016/j.cub.2014.06.040.
25. Taylor KI, Moss HE, Stamatakis E a, Tyler LK. Binding crossmodal object features in perirhinal cortex. *Proc Natl Acad Sci U S A*. 2006;103(21):8239–8244. https://doi.org/10.1073/pnas.0509704103.
26. Hocking J, Price CJ. The role of the posterior superior temporal sulcus in audiovisual processing. *Cerebr Cortex*. 2008;18(10):2439–2449. https://doi.org/10.1093/cercor/bhn007.
27. Molholm S. Multisensory visual-auditory object recognition in humans: a high-density electrical mapping study. *Cerebr Cortex*. 2004;14(4):452–465. https://doi.org/10.1093/cercor/bhh007.
28. Diaconescu AO, Hasher L, McIntosh AR. Visual dominance and multisensory integration changes with age. *Neuroimage*. 2013;65:152–166. https://doi.org/10.1016/j.neuroimage.2012.09.057.
29. Thelen A, Cappe C, Murray MM. Electrical neuroimaging of memory discrimination based on single-trial multisensory learning. *Neuroimage*. 2012;62(3):1478–1488. https://doi.org/10.1016/j.neuroimage.2012.05.027.
30. Doehrmann O, Naumer MJ. Semantics and the multisensory brain: how meaning modulates processes of audio-visual integration. *Brain Res*. 2008;1242:136–150. https://doi.org/10.1016/j.brainres.2008.03.071.
31. Matusz PJ, Broadbent H, Ferrari J, Forrest B, Merkley R, Scerif G. Multi-modal distraction: insights from children's limited attention. *Cognition*. 2015;136:156–165. https://doi.org/10.1016/j.cognition.2014.11.031.
32. Alsius A, Soto-Faraco S. Searching for audiovisual correspondence in multiple speaker scenarios. *Exp Brain Res*. 2011;213(2–3):175–183. https://doi.org/10.1007/s00221-011-2624-0.
33. Morís Fernández L, Visser M, Ventura-Campos N, Ávila C, Soto-Faraco S. Top-down attention regulates the neural expression of audiovisual integration. *Neuroimage*. 2015;119:272–285. https://doi.org/10.1016/j.neuroimage.2015.06.052.
34. McGurk H, MacDonald J. Hearing lips and seeing voices. *Nature*. 1976;264(5588):746–748. http://www.ncbi.nlm.nih.gov/pubmed/1012311.
35. Alsius A, Navarra J, Campbell R, Soto-Faraco S. Audiovisual integration of speech falters under high attention demands. *Curr Biol*. 2005;15(9):839–843. https://doi.org/10.1016/j.cub.2005.03.046.
36. Alsius A, Navarra J, Soto-Faraco S. Attention to touch weakens audiovisual speech integration. *Exp Brain Res*. 2007;183(3):399–404. https://doi.org/10.1007/s00221-007-1110-1.

37. Alsius A, Möttönen R, Sams ME, Soto-Faraco S, Tiippana K. Effect of attentional load on audiovisual speech perception: evidence from ERPs. *Front Psychol*. 2014;5:727. https://doi.org/10.3389/fpsyg.2014.00727.
38. Senkowski D, Saint-Amour D, Kelly SP, Foxe JJ. Multisensory processing of naturalistic objects in motion: a high-density electrical mapping and source estimation study. *Neuroimage*. 2007;36(3):877–888. https://doi.org/10.1016/j.neuroimage.2007.01.053.
39. van Wassenhove V, Grant KW, Poeppel D. Visual speech speeds up the neural processing of auditory speech neurophysiological basis of multisensory integration. 2004.
40. Stekelenburg JJ, Vroomen J. Neural correlates of multisensory integration of ecologically valid audiovisual events. *J Cogn Neurosci*. 2007;19(12):1964–1973. https://doi.org/10.1162/jocn.2007.19.12.1964.
41. Diaconescu AO, Alain C, McIntosh AR. The co-occurrence of multisensory facilitation and cross-modal conflict in the human brain. *J Neurophysiol*. 2011;106(6):2896–2909. https://doi.org/10.1152/jn.00303.2011.
42. Yuval-Greenberg S, Deouell LY. What you see is not (always) what you hear: induced gamma band responses reflect cross-modal interactions in familiar object recognition. *J Neurosci*. 2007;27(5):1090–1096. https://doi.org/10.1523/JNEUROSCI.4828-06.2007.
43. van Atteveldt NM, Formisano E, Goebel R, Blomert L. Top-down task effects overrule automatic multisensory responses to letter-sound pairs in auditory association cortex. *Neuroimage*. 2007;36(4):1345–1360. https://doi.org/10.1016/j.neuroimage.2007.03.065.
44. van Atteveldt NM, Peterson BS, Schroeder CE. Contextual control of audiovisual integration in low-level sensory cortices. *Hum Brain Mapp*. 2014;35(5):2394–2411. https://doi.org/10.1002/hbm.22336.
45. Stevenson R a, Wallace MT. Multisensory temporal integration: task and stimulus dependencies. *Exp Brain Res*. 2013;227(2):249–261. https://doi.org/10.1007/s00221-013-3507-3.
46. Murray MM, Sperdin HF. Single-trial multisensory learning and memory retrieval:1-30.
47. Shams L, Seitz AR. Benefits of multisensory learning. *Trends Cognit Sci*. 2008;12(11):411–417. https://doi.org/10.1016/j.tics.2008.07.006.
48. Shams L, Kim R. *Cross-Modal Facilitation of Unisensory Learning*. 2011:475–482.
49. Seitz AR, Kim R, Shams L. Sound facilitates visual learning. *Curr Biol*. 2006;16(14):1422–1427. https://doi.org/10.1016/j.cub.2006.05.048.
50. Kim RS, Seitz AR, Shams L. Benefits of stimulus congruency for multisensory facilitation of visual learning. *PLoS One*. 2008;3(1):e1532. https://doi.org/10.1371/journal.pone.0001532.
51. Nyberg L, Habib R, McIntosh a R, Tulving E. Reactivation of encoding-related brain activity during memory retrieval. *Proc Natl Acad Sci U S A*. 2000;97(20):11120–11124. http://www.pubmedcentral.nih.gov/articlerender.fcgi?artid=27158&tool=pmcentrez&rendertype=abstract.
52. Wheeler ME, Petersen SE, Buckner RL. Memory's echo: vivid remembering reactivates sensory-specific cortex. *Proc Natl Acad Sci U S A*. 2000;97(20):11125–11129. https://doi.org/10.1073/pnas.97.20.11125.
53. Hamilton W. *Lectures on Metaphysics and Logic*. Boston: Gould and Lincoln; 1859.
54. von Kriegstein K, Giraud A-L. Implicit multisensory associations influence voice recognition. *PLoS Biol*. 2006;4(10):e326. https://doi.org/10.1371/journal.pbio.0040326.
55. Murray MM, Michel CM, Grave de Peralta R, et al. Rapid discrimination of visual and multisensory memories revealed by electrical neuroimaging. *Neuroimage*. 2004;21(1):125–135. https://doi.org/10.1016/j.neuroimage.2003.09.035.
56. Matusz PJ, Wallace MT, Murray MM. A multisensory perspective on object memory. *Neuropsychologia*. 2017;105:243–252. https://doi.org/10.1016/j.neuropsychologia.2017.04.008.
57. Murray MM, Foxe JJ, Wylie GR. The brain uses single-trial multisensory memories to discriminate without awareness. *Neuroimage*. 2005;27(2):473–478. https://doi.org/10.1016/j.neuroimage.2005.04.016.
58. Lehmann S, Murray MM. The role of multisensory memories in unisensory object discrimination. *Cogn Brain Res*. 2005;24(2):326–334. https://doi.org/10.1016/j.cogbrainres.2005.02.005.
59. Murray MM, Thelen A. The efficacy of single-trial multisensory memories. *Multisensory Res*. 2013;26(5):483–502. https://doi.org/10.1163/22134808-00002426.
60. Thelen A, Talsma D, Murray MM. Single-trial multisensory memories affect later auditory and visual object discrimination. *Cognition*. 2015;138:148–160. https://doi.org/10.1016/j.cognition.2015.02.003.
61. PJPJ M, Thelen A, Amrein S, Geiser E, Anken J, Murray MMMM. The role of auditory cortices in the retrieval of single-trial auditory-visual object memories. *Eur J Neurosci*. 2015;41(5):699–708. https://doi.org/10.1111/ejn.12804.

62. Macmillan NA, Creelman CD. *Detection Theory: A User's Guide*. 2nd ed. Lawrence Erlbaum Associates Publishers; 2005.
63. Heikkilä J, Alho K, Hyvönen H, Tiippana K. Audiovisual semantic congruency during encoding enhances memory performance. *Exp Psychol*. 2015;62(2):123–130. https://doi.org/10.1027/1618-3169/a000279.
64. Heikkilä J, Tiippana K. School-aged children can benefit from audiovisual semantic congruency during memory encoding. *Exp Brain Res*. 2015. https://doi.org/10.1007/s00221-015-4341-6.
65. Naghavi HR, Eriksson J, Larsson A, Nyberg L. Cortical regions underlying successful encoding of semantically congruent and incongruent associations between common auditory and visual objects. *Neurosci Lett*. 2011;505(2):191–195. https://doi.org/10.1016/j.neulet.2011.10.022.
66. Ueno D, Masumoto K, Sutani K, Iwaki S. Latency of modality-specific reactivation of auditory and visual information during episodic memory retrieval. *Neuroreport*. 2015;26(6):303–308. https://doi.org/10.1097/WNR.0000000000000325.
67. Moran ZD, Bachman P, Pham P, Cho SH, Cannon TD, Shams L. Multisensory encoding improves auditory recognition. *Multisensory Res*. 2013;26(6):581–592. https://doi.org/10.1027/1618-3169/a000279.
68. Murray MM, Camen C, Gonzalez Andino SL, Bovet P, Clarke S. Rapid brain discrimination of sounds of objects. *J Neurosci*. 2006;26(4):1293–1302. https://doi.org/10.1523/JNEUROSCI.4511-05.2006.
69. De Lucia M, Camen C, Clarke S, Murray MM. The role of actions in auditory object discrimination. *Neuroimage*. 2009;48(2):475–485. https://doi.org/10.1016/j.neuroimage.2009.06.041.
70. De Lucia M, Cocchi L, Martuzzi R, Meuli RA, Clarke S, Murray MM. Perceptual and semantic contributions to repetition priming of environmental sounds. *Cerebr Cortex*. 2010;20(7):1676–1684. https://doi.org/10.1093/cercor/bhp230.
71. Meo R De, Matusz PJ, Knebel J-F, Murray MM, Thompson WR, Clarke S. What makes medical students better listeners? *Curr Biol*. 2016;26(13):R519–R520. https://doi.org/10.1016/j.cub.2016.05.024.
72. Matusz PJ, Retsa C, Murray MM. The context-contingent nature of cross-modal activations of the visual cortex. *Neuroimage*. 2016;125. https://doi.org/10.1016/j.neuroimage.2015.11.016.
73. Retsa C, Matusz PJ, Schnupp JWH, Murray MM. What's what in auditory cortices? *Neuroimage*. 2018;176. https://doi.org/10.1016/j.neuroimage.2018.04.028.
74. Romei V, Murray MM, Cappe C, Thut G. Preperceptual and stimulus-selective enhancement of low-level human visual cortex excitability by sounds. *Curr Biol*. 2009;19(21):1799–1805. https://doi.org/10.1016/j.cub.2009.09.027.
75. Romei V, Murray MMMM, Cappe C, Thut G. The contributions of sensory dominance and attentional bias to cross-modal enhancement of visual cortex excitability. *J Cogn Neurosci*. 2013;25(7):1122–1135. https://doi.org/10.1162/jocn_a_00367.
76. Schmid C, Büchel C, Rose M. The neural basis of visual dominance in the context of audio-visual object processing. *Neuroimage*. 2011;55(1):304–311. https://doi.org/10.1016/j.neuroimage.2010.11.051.
77. Cohen M a, Horowitz TS, Wolfe JM. Auditory recognition memory is inferior to visual recognition memory. *Proc Natl Acad Sci U S A*. 2009;106(14):6008–6010. https://doi.org/10.1073/pnas.0811884106.
78. Yuval-Greenberg S, Deouell LY. The dog's meow: asymmetrical interaction in cross-modal object recognition. *Exp Brain Res*. 2009;193(4):603–614. https://doi.org/10.1007/s00221-008-1664-6.
79. Stein BE, Meredith MA. *The Merging of the Senses*. MIT Press; 1993.
80. Wallace MT, Ramachandran R, Stein BE. A revised view of sensory cortical parcellation. *Proc Natl Acad Sci U S A*. 2004;101(7):2167–2172. https://doi.org/10.1073/pnas.0305697101.
81. James TW, Stevenson RA, Kim S. *Inverse Effectiveness and BOLD fMRI*. 2011:207–222.
82. Holmes NP. The law of inverse effectiveness in neurons and behaviour: multisensory integration versus normal variability. *Neuropsychologia*. 2007;45(14):3340–3345. https://doi.org/10.1016/j.neuropsychologia.2007.05.025.
83. Matusz PJ, Eimer M. Top-down control of audiovisual search by bimodal search templates. *Psychophysiology*. 2013;50:996–1009. https://doi.org/10.1111/psyp.12086.
84. Matusz PJ, Turoman N, Tivadar RI, Retsa C, Murray MM. Brain and cognitive mechanisms of top-down attentional control in a multisensory world: benefits of electrical neuroimaging. *J Cogn Neurosci*. 2018.
85. Matusz PJ, Eimer M. Multisensory enhancement of attentional capture in visual search. *Psychon Bull Rev*. 2011;18(5):904–909. https://doi.org/10.3758/s13423-011-0131-8.
86. Thelen A, Matusz PJ, Murray MM. Multisensory context portends object memory. *Curr Biol*. 2014;24(16):1–10. https://doi.org/10.1016/j.cub.2014.06.040.

87. Bergerbest D, Ghahremani DG, Gabrieli JDE. Neural correlates of auditory repetition priming: reduced fMRI activation in the auditory cortex. *J Cogn Neurosci.* 2004;16(6):966–977. https://doi.org/10.1162/0898929041502760.
88. Murray MM, Camen C, Spierer L, Clarke S. Plasticity in representations of environmental sounds revealed by electrical neuroimaging. *Neuroimage.* 2008;39(2):847–856. https://doi.org/10.1016/j.neuroimage.2007.09.002.
89. Gibson JR, Maunsell JH. Sensory modality specificity of neural activity related to memory in visual cortex. *J Neurophysiol.* 1997;78(3):1263–1275. http://www.ncbi.nlm.nih.gov/pubmed/9310418.
90. Colombo M, Gross CG. Responses of inferior temporal cortex and hippocampal neurons during delayed matching to sample in monkeys (*Macaca fascicularis*). *Behav Neurosci.* 1994;108(3):443–455. http://www.ncbi.nlm.nih.gov/pubmed/7917038.
91. Haenny PE, Maunsell JH, Schiller PH. State dependent activity in monkey visual cortex. II. Retinal and extraretinal factors in V4. *Exp Brain Res.* 1988;69(2):245–259. http://www.ncbi.nlm.nih.gov/pubmed/3345806.
92. Maunsell JH, Sclar G, Nealey TA, DePriest DD. Extraretinal representations in area V4 in the macaque monkey. *Vis Neurosci.* 1991;7(6):561–573. http://www.ncbi.nlm.nih.gov/pubmed/1772806.
93. Goulet S, Murray EA. Neural substrates of crossmodal association memory in monkeys: the amygdala versus the anterior rhinal cortex. *Behav Neurosci.* 2001;115(2):271–284. http://www.ncbi.nlm.nih.gov/pubmed/11345954.
94. Murray EA, Bussey TJ. Perceptual – mnemonic functions of the perirhinal cortex. *Trends Cogn Sci.* 1999;3(4):142–151.
95. Murray EA, Bussey TJ, Saksida LM. Visual perception and memory: a new view of medial temporal lobe function in primates and rodents. *Annu Rev Neurosci.* 2007;30:99–122. https://doi.org/10.1146/annurev.neuro.29.051605.113046.
96. Beauchamp MS, Argall BD, Bodurka J, Duyn JH, Martin A. Unraveling multisensory integration: patchy organization within human STS multisensory cortex. *Nat Neurosci.* 2004;7(11):1190–1192. https://doi.org/10.1038/nn1333.
97. Von Kriegstein K, Giraud AL. Implicit multisensory associations influence voice recognition. *PLoS Biol.* 2006;4(10):1809–1820. https://doi.org/10.1371/journal.pbio.0040326.
98. Doniger GM, Foxe JJ, Schroeder CE, Murray MM, Higgins B a, Javitt DC. Visual perceptual learning in human object recognition areas: a repetition priming study using high-density electrical mapping. *Neuroimage.* 2001;13(2):305–313. https://doi.org/10.1006/nimg.2000.0684.
99. Thelen A, Matusz PJ, Murray MM. Multisensory context portends object memory. *Curr Biol.* 2014;24(16):R734–R735. https://doi.org/10.1016/j.cub.2014.06.040.
100. Powers AR, Hillock AR, Wallace MT. Perceptual training narrows the temporal window of multisensory binding. *J Neurosci.* 2009;29(39):12265–12274. https://doi.org/10.1523/JNEUROSCI.3501-09.2009.
101. Murray MM, Lewkowicz DJ, Amedi A, Wallace MT. Multisensory processes: a balancing act across the lifespan. *Trends Neurosci.* 2016;39(8):567–579. https://doi.org/10.1016/j.tins.2016.05.003.
102. Matusz PJ, Traczyk J, Sobkow A, Strelau J. Individual differences in emotional reactivity moderate the strength of the relationship between attentional and implicit-memory biases towards threat-related stimuli. *J Cogn Psychol.* 2015;27(6):715–724. https://doi.org/10.1080/20445911.2015.1027210.
103. Smith SM, Vela E. Environmental context-dependent memory: a review and meta-analysis. *Psychon Bull Rev.* 2001;8(2):203–220. http://www.ncbi.nlm.nih.gov/pubmed/11495110.
104. Reggev N, Bein O, Maril A. Distinct neural suppression and encoding effects for conceptual novelty and familiarity. *J Cogn Neurosci.* 2016;28(10):1455–1470. https://doi.org/10.1162/jocn_a_00994.
105. Gingras G, Rowland BA, Stein BE. The differing impact of multisensory and unisensory integration on behavior. *J Neurosci.* 2009;29(15):4897–4902. https://doi.org/10.1523/JNEUROSCI.4120-08.2009.
106. Chen YC, Spence C. The crossmodal facilitation of visual object representations by sound: evidence from the backward masking paradigm. *J Exp Psychol Hum Percept Perform.* 2011;37(6):1784–1802. https://doi.org/10.1037/a0025638.
107. De Meo R, Murray MM, Clarke S, Matusz PJ. Top-down control and early multisensory processes: chicken vs. egg. *Front Integr Neurosci.* 2015;9. https://doi.org/10.3389/fnint.2015.00017.
108. Tang X, Wu J, Shen Y. The interactions of multisensory integration with endogenous and exogenous attention. *Neurosci Biobehav Rev.* 2016;61:208–224. https://doi.org/10.1016/j.neubiorev.2015.11.002.

109. Van der Burg E, Olivers CNL, Bronkhorst AW, Theeuwes J. Pip and pop: nonspatial auditory signals improve spatial visual search. *J Exp Psychol Hum Percept Perform*. 2008;34(5):1053–1065. https://doi.org/10.1037/0096-1523.34.5.1053.
110. Van der Burg E, Talsma D, Olivers CNL, Hickey C, Theeuwes J. Early multisensory interactions affect the competition among multiple visual objects. *Neuroimage*. 2011;55(3):1208–1218. https://doi.org/10.1016/j.neuroimage.2010.12.068.
111. Murray MM, Eardley AF, Edginton T, Oyekan R, Smyth E, Matusz PJ. Sensory dominance and multisensory integration as screening tools in aging. *Sci Rep*. 2018;8(1):8901. https://doi.org/10.1038/s41598-018-27288-2.

SECTION II

Multisensory interactions

CHAPTER

7

Visuo-haptic object perception

Simon Lacey[1,2], *K. Sathian*[1,2]

[1]Department of Neurology, Neural & Behavioral Sciences, and Psychology, Pennsylvania State University, Hershey, PA, United States; [2]Departments of Neurology and Psychology, Emory University, Atlanta, GA, United States

Introduction

The field of object perception and recognition has been dominated by vision science even though in real life these are invariably multisensory processes. For a long time, haptics was the "Cinderella" modality,[1] with the other senses left even further behind.[2,3] Now, times have changed: following seminal work by Klatzky and Lederman in the 1980s[4–6] ([7,8] for reviews), haptics has developed as a field in its own right, alongside renewed interest in the psychophysics and neurophysiology of tactile perception.[9–12] The 1990s ushered in an explosion of interest in multisensory interactions at both behavioral and neural levels. The old orthodoxy of a brain organized around discrete unisensory inputs has yielded to a growing consensus around the concept of a "metamodal" brain whose regions are organized around specific tasks and receive multisensory inputs.[13–15] In each subsequent section of this chapter, we review the interactions and commonalities between vision and touch for a particular aspect of multisensory object processing, firstly from a behavioral standpoint and then moving on to consider the neural basis. A review of basic within-modal haptic and crossmodal visuo-haptic object recognition sets the stage. Because grouping like objects together is one way to facilitate recognition, we then move to a consideration of recent work on the similarities between visual and haptic categorization and crossmodal transfer of category knowledge. Changes in object orientation pose a significant computational challenge to object recognition; we show that this challenge is solved by a multisensory representation that underlies both crossmodal recognition and view-independence. We next address individual differences in visuo-haptic representations and, finally, review the evidence for our working conceptual model of multisensory visuo-haptic object recognition.

Multisensory Perception
https://doi.org/10.1016/B978-0-12-812492-5.00007-3

Before proceeding to the main focus of this chapter on visuo-haptic object perception, we lay some groundwork with a few examples of what from a classical perspective is considered "crossmodal" recruitment of "visual" cortex during tactile tasks. Because this topic has been extensively reviewed recently,[11] only a brief summary is given here. Our laboratory provided the first demonstration of activity of a visual cortical area during normal tactile perception, in a positron emission tomographic study in humans.[16] In this study, we found that tactile discrimination of the orientation of gratings on the fingerpad, relative to a control task requiring tactile discrimination of grating groove width, recruited a focus in extrastriate visual cortex, close to the parieto-occipital fissure. This focus, corresponding to the location of human V6,[17] had previously been shown to be active during visual discrimination of grating orientation.[18] Using transcranial magnetic stimulation (TMS) to transiently interfere with the function of this parieto-occipital region, we were able to disrupt tactile performance on grating orientation discrimination,[19] signifying that this area is truly involved functionally, rather than the activation merely being an epiphenomenon. Analogously, other neocortical regions originally described as exhibiting selectivity for specific visual attributes are also active when these attributes are sensed nonvisually. For instance, the human MT complex (hMT+), a visual cortical region well known to encode visual motion, is also active during tactile[20–23] and auditory motion perception.[24] Together, these findings suggest that hMT+ functions as a modality-independent motion processor. Similar findings have been reported for convergent processing of various other visuo-haptic object properties, as will be reviewed later in this chapter.

Haptic and visuo-haptic object recognition

Behavioral studies

The speed and accuracy of visual object recognition is well established. Haptic recognition is also highly accurate, at least for familiar objects, with 96% correctly named overall: 68% in less than 3 seconds and 94% within 5 seconds.[4] Although slower than visual recognition, longer haptic response times are likely due to the longer time required to explore larger items compared with essentially parallel visual processing across the retinal image[4]; in fact, in some contexts, a "haptic glance" of less than 1 second is sufficient.[25] Interestingly, haptic object recognition is possible with the feet as well as the hands, albeit more slowly and less accurately, with hand and foot performance being highly correlated across individuals.[26] Haptic identification begins with a "grasp and lift" stage that provides low-level information about an object and proceeds to a series of hand movements, known as "exploratory procedures" (EPs), which extract more precise information.[27] EPs are property-specific; for example, texture information is extracted by lateral motion while precise shape information is provided by contour following.[6] These object properties differ in their salience to haptic processing depending on the context. Under instructions that emphasize haptic processing, salience progressively decreases in this order: texture > shape > hardness; but under neutral instructions, the order changes to hardness > texture > shape.[5] Note that in simultaneous visual and haptic perception, and in haptic perception under instructions to use concurrent visual imagery, the salience order under neutral instructions is reversed to shape > texture > hardness/size.[5]

Crossmodal visuo-haptic object recognition, while fairly accurate, generally comes at a cost compared with within-modal recognition.[28–30] Broadly speaking, crossmodal recognition is also asymmetric, being better when visual encoding is followed by haptic recognition rather than the reverse,[31–37] but the reasons for this are unclear. Interestingly, implicit memory does not appear to be subject to this asymmetry as crossmodal priming is symmetric,[33–35] although this may have been facilitated by verbal encoding strategies in these studies. Some have proposed that constraints imposed by different stages of motor development underlie the crossmodal asymmetry observed in infants,[36] but this cannot be a complete explanation because the asymmetry persists into adulthood.[31,34,35] One possibility is that the visual and haptic systems differ in how well they encode shape information because of competition from other, more salient, modality-specific object properties. Thus, in the haptic-visual crossmodal condition, it might be more difficult to encode shape because, as noted above, hardness and texture information are more salient to the haptic modality than shape. This effect could be reduced by the use of concurrent visual imagery in which shape information might be more prominent as it is common to vision and touch. We should also note that when vision and touch are employed simultaneously, properties that have different weights in vision and touch could be optimally combined on the basis of maximum likelihood estimates.[38–41]

Alternatively, differences in visual and haptic memory capacity might explain crossmodal asymmetry. Compared with visual working memory, haptic working memory capacity appears to be limited and variable and may be more error prone as a result.[42] Another difference is that haptic representations simply decay faster and more abruptly than visual representations. Instead of a gradual decline over time, the haptic decay function appears to occur entirely in a band of 15–30 seconds poststimulus[43](see also Craddock & Lawson,[44] who found no decline in performance at 15 seconds although longer intervals were not tested). If haptically encoded representations have substantially decayed by the time visual recognition is tested, this might explain poor performance in the haptic-visual condition. On the other hand, robust long-term haptic representations can be formed both with and without overt instructions to memorize objects[45] and other studies have shown little difference between haptic-visual and visual-haptic recognition after delays of up to 30 seconds[46,47] or even a week.[48] Thus, an explanation of the asymmetry in terms of a simple function of haptic memory properties is likely insufficient.

Finally, the underlying neural activity may be asymmetric between the two crossmodal conditions. In a match-to-sample task, there was a selectively greater response in bilateral lateral occipital complex (LOC), fusiform gyrus (FG), and anterior intraparietal sulcus (aIPS) for crossmodal, compared with unimodal, object matching when haptic targets followed visual samples, and more strongly still when the haptic target and visual sample were congruent rather than incongruent.[49] However, these regions showed no such increase for visual targets in either crossmodal or unimodal conditions.[49] This asymmetric increase in activation in the visual-haptic condition suggests that, in terms of multisensory binding of shape information, the haptic modality integrates previously presented visual information more than vision integrates previous haptic information.[49]

The neural basis of visuo-haptic object processing

Segregated ventral "what" and dorsal "where" pathways

Visual object processing broadly divides along a ventral "what" pathway that deals with object identity and perception for recognition and a dorsal "where/how" pathway dealing with object location and perception for action, e.g., reaching and grasping[50,51] (but see Ref. 52 for a proposal of more distributed shape processing in these pathways). This functional division is apparently not unique to vision, because similar ventral and dorsal pathways mediating perception of object identity and location, respectively, have been demonstrated for the auditory system.[53–55] In the somatosensory system, the action-perception dichotomy was suggested to underlie segregation of processing pathways into a dorsally directed stream targeting posterior parietal cortex and a ventrally directed pathway ending in insular cortex[56]—the latter pathway was originally proposed by Mishkin[57] as analogous to the ventral visual pathway.

An early functional magnetic resonance imaging (fMRI) study found that haptic object recognition activated frontal cortical areas as well as inferior parietal cortex, while a haptic object location task activated superior parietal regions.[58] However, given the salience of texture to touch,[5] we reasoned that texture would be a better marker of haptic object identity. Accordingly, a later fMRI study from our laboratory[59] found that, while both visual and haptic location judgments involved a similar dorsal pathway comprising large sectors of the intraparietal sulcus (IPS) and frontal eye fields (FEFs) bilaterally, haptic texture perception engaged extensive areas of the more ventrally located parietal operculum, which contains higher-order (i.e., nonprimary) regions of somatosensory cortex. In addition, we found that parts of extrastriate (i.e., nonprimary) visual cortex processed texture in both visual and haptic modalities.[59,60] Moreover, several of these bisensory areas showed correlations of activation magnitude between the visual and haptic perception of either texture or location, further indicating some shared, task-specific cortical processing across modalities.[59] These findings were extended by another group who showed that activation magnitudes in early visual cortex not only scaled with the interdot spacing of dot patterns but were also modulated by the presence of matching haptic input[61]; multivariate decoding of roughness using multivoxel pattern analysis (MVPA), however, was restricted to higher-order somatosensory (parietal opercular) and visual (inferotemporally in the collateral sulcus) cortex.[62] Conversely, a separate group showed that visual presentation of textures cuing tactile associations, e.g., gloss, roughness, and matte, could be decoded in higher-order secondary somatosensory cortex.[63]

Multisensory processing of object shape

Cortical areas in both the ventral and dorsal visual pathways are functionally involved during haptic tasks analogous to the visual tasks for which they are known to be specialized (for reviews see Ref. 64–67). The second paragraph of this chapter outlined evidence for such multisensory processing in parieto-occipital cortex (human V6) for orientation discrimination and the human MT complex for motion perception. Even early areas in the human visual pathway, which project to both dorsal and ventral streams, respond to changes in haptic shape, suggesting that haptic shape perception might involve the entire ventral stream.[68] However, in this and many other studies, visual imagery cannot be excluded as an explanation for visual cortical recruitment (see below). Nonetheless, even if visual imagery mediates visual cortical activation during haptic perception, this implies the existence of polysynaptic cortical pathways between somatosensory and visual cortices in humans, as shown in the macaque using network analysis.[69] Most studies of visuo-haptic object processing have concentrated on higher-level visual areas and in particular the LOC, an object-selective region in the ventral visual pathway.[70] A subregion of LOC, termed LOtv, also responds selectively to objects in touch as well as vision,[60,71,72] processing both haptic 3-D[60,71,73] and tactile 2-D stimuli.[74,75]

The role of the LOC in auditory object processing is less clear. On the one hand, the LOC does not respond during object recognition cued by object-specific sounds which might be expected to evoke the shape of that object.[72] On the other hand, James et al.[15] showed that the physical shape of objects, either rods or balls, could be reliably inferred from the impact sounds made by dropping these objects onto a hard surface, and that the LOC was more activated during these inferences than when participants had to categorize the sounds by material (either wooden or metallic). A more recent study showed that object sounds, but not voices, evoked activity in LOC in both early blind and sighted participants.[76] In both these studies, though, LOC activity in sighted participants could have reflected visual imagery—we return to the potential role of visual imagery in a later section.

From a neurorehabilitation perspective, however, the LOC does respond to auditory shape information generated by a visual–auditory sensory substitution device (SSD)[77] (see Chapter 15), which converts visual information into an auditory stream or "soundscape" via a specific algorithm in which the visual horizontal axis is represented by auditory duration and stereo panning, the visual vertical axis by variations in tone frequency, and pixel brightness by variations in tone loudness. Both sighted and blind humans can learn to recognize objects by extracting shape information from SSD soundscapes although extensive training is required.[77] It is important to note that the LOC only responds to soundscapes created according to the algorithm—and which therefore represent shape in a principled way—and not when participants learn soundscapes that are arbitrarily associated with particular objects.[77] The LOC can therefore be regarded as processing geometric shape information independently of the sensory modality used to acquire it.

Visuo-haptic object processing has also been demonstrated in several parietal regions: in particular, the aIPS is active during perception of both object shape and location, with coactivation of the LOC for shape and the FEF for location.[59,60,78] We suggested that the role of the IPS in shape perception is to compute spatial relations between object parts to assemble a global object representation.[14,59] The postcentral sulcus, corresponding to Brodmann's area 2 of primary somatosensory cortex (S1)[79] and traditionally considered exclusively somatosensory, also shows visuo-haptic shape selectivity[60] consistent with earlier neurophysiological studies that suggested visual responsiveness in parts of S1.[80,81]

Multisensory responses found with fMRI might result from visuo-haptic integration by multisensory neurons, or, alternatively, they might reflect distinct unisensory neuronal populations that are intermingled. To distinguish these possibilities in the case of visuo-haptic responses, Tal and Amedi[82] employed the technique of fMRI adaptation (fMR-A), which utilizes the repetition suppression effect, i.e., attenuation of the blood oxygenation level-dependent (BOLD) signal by repeating the same stimulus and thus revealing neuronal selectivity profiles.[83,84] For stimuli that were presented visually and then haptically, a robust crossmodal adaptation effect was found in the LOC and the aIPS, as well as ventral premotor cortex and the right anterior insula, suggesting that these areas were integrating multisensory inputs at the neuronal level. However, other regions active during both visual and haptic presentations failed to show crossmodal adaptation, implying that multisensory responses in these regions were likely derived from intermingled unisensory neurons. However, it has been argued that fMR-A effects may not reliably index neuronal selectivity,[85] and thus it will be necessary to verify the findings of Tal and Amedi[81] with converging evidence using other methods.

When visual cortical areas are active during haptic or tactile tasks, it is critical to determine whether this activation is functionally relevant, i.e., whether it is *necessary* for task performance or merely a by-product of other processes. Although much work is still required, two lines of evidence support the functional relevance of visual cortical activation during touch. Firstly, case studies indicate that the LOC is necessary for both haptic and visual shape perception. A lesion to the left occipito-temporal cortex, which likely included the LOC, resulted in both tactile and visual agnosia (inability to recognize objects) even though somatosensory cortex and basic somatosensory function were intact.[86] Another patient with bilateral LOC lesions was unable to learn new objects either visually or haptically.[87] These case studies are consistent with the existence of a shared multisensory representation in the LOC. By contrast, a patient with bilateral occipitotemporal lesions that included LOC had visual, but not haptic, agnosia[88] although this might reflect preserved haptic shape processing in other regions, e.g., IPS, and/or some degree of cortical reorganization because the patient was tested 12 years post-stroke.

As described earlier in the case of V6 in relation to tactile orientation discrimination,[19] "virtual lesions" can be created using TMS to temporarily deactivate specific, functionally defined, cortical areas.[89] Repetitive TMS (rTMS) over the LOC has been used to infer a critical role in visual shape processing,[90,91] but haptic shape perception has only been investigated with rTMS over regions of the IPS. rTMS over the left aIPS impaired visual-haptic, but not haptic-visual, shape matching using the right hand, but shape matching with the left hand during rTMS over the right aIPS was unaffected in either crossmodal condition.[92] Thus, the precise roles of the IPS and LOC in multisensory shape processing have yet to be fully worked out.

Object categorization

Behavioral studies

Categorization facilitates object recognition and is critical for much of higher-order cognition.[93] As in vision, haptics also exhibits categorical perception, i.e., discriminability increases sharply when objects belong to different categories and decreases when they belong to the same category.[94] However, visual and haptic categorization are not entirely alike because object properties are differentially weighted depending on the modality,[95–97] broadly consistent with the differential perceptual salience of object properties laid out by

Klatzky et al.[5] For example, shape is more important than texture for visual categorization, whereas in haptic and bisensory categorization, shape and texture are approximately equally weighted.[95] These studies suggest a close connection between vision and haptics in terms of similarity mechanisms for categorization, supported by the finding of symmetric crossmodal transfer of category information following either visual or haptic category learning, and which generalizes to new objects from these categories.[98] Category structure, i.e., ordinal relationships and category boundaries,[95] also transfers crossmodally[99] which is interesting because the ordering of each item within the category is (at least in the studies reviewed here) perceptually driven. Thus, it may be that a shared multisensory representation underlies crossmodal categorization, as has been suggested for crossmodal recognition[14,66] (see next section).

A series of recent studies using multidimensional scaling analysis has shown that visual and haptic similarity ratings and categorization result in visual and haptic perceptual spaces (i.e., topological representations of the *perceived* (dis)similarity along a given dimension) that are highly congruent for both novel[95,96,100,101] and familiar objects.[97] This was so for unisensory as well as bisensory presentations,[95] for both 2D visual objects and 3D haptic objects,[96,100] and under a variety of visual and haptic exploratory conditions.[96] These highly congruent visual and haptic perceptual spaces also showed high fidelity to the physical object space (i.e., isomorphic topological representations of the *actual* (dis)similarity along a given dimension).[96,100] The isomorphism between perceptual and physical spaces was, furthermore, independent of the type of categorization task.[100,101]

Neural correlates of visuo-haptic object categorization

There has been rather limited neural study of crossmodal category—selective representations. An early study, using MVPA of fMRI data, demonstrated that selectivity for particular categories of man-made objects was correlated across vision and touch in a region of inferotemporal cortex.[102] More recent MVPA studies have shown that category information for visually presented objects can be reliably decoded in primary somatosensory cortex (S1) bilaterally, and in the right secondary somatosensory cortex (S2).[103] Interestingly, such decoding was only possible for familiar, and not unfamiliar, objects; this suggests prior visuo-haptic experience of objects was necessary for category information to be encoded in somatosensory cortex and that visual presentation alone was insufficient. When objects (mostly familiar but including one category for unfamiliar shapes) were presented haptically, category information could be decoded not only from somatosensory cortex but also from the LOC,[104] further confirming the LOC as a multisensory center for shape processing. Crossmodal categorization performance—at least for visually learned and haptically tested categories—reflects individual differences in the microstructure of the inferior longitudinal fasciculus and the frontotemporal part of the superior longitudinal fasciculus (SLF), white matter tracts previously associated with visual object processing and visuospatial and featural attention, respectively[105] (note that in a separate study of within-modal haptic categorization, performance was correlated with individual differences in the frontoparietal SLF[106]).

View-dependence

Behavioral studies

Changes in the orientation of an object present a significant computational challenge to visual object recognition because they change the related sensory input, i.e., the retinal pattern. Therefore, an important goal of sensory systems is to achieve perceptual constancy so that objects can be recognized independently of such changes. Visual object recognition is considered view-dependent if rotating an object away from its original orientation impairs subsequent recognition and view-independent if not.[107] Intuitively, one might expect that haptic recognition would be view-independent because the hands can contact an object from different sides simultaneously, acquiring information about several different "views" at the same time. But numerous studies have now shown that haptic object recognition is also view-dependent,[30,44,108–114] although the reasons for this remain unclear. Interestingly, different types of change in orientation affect vision and touch in different ways. When objects are rotated about the x- and y-axes, i.e., in depth (Fig. 7.1), visual object recognition is slower and less accurate than when objects are rotated about the z-axis, i.e., in the picture plane, and this is true for both 2D[112] and 3D stimuli.[30] By contrast, haptic recognition is

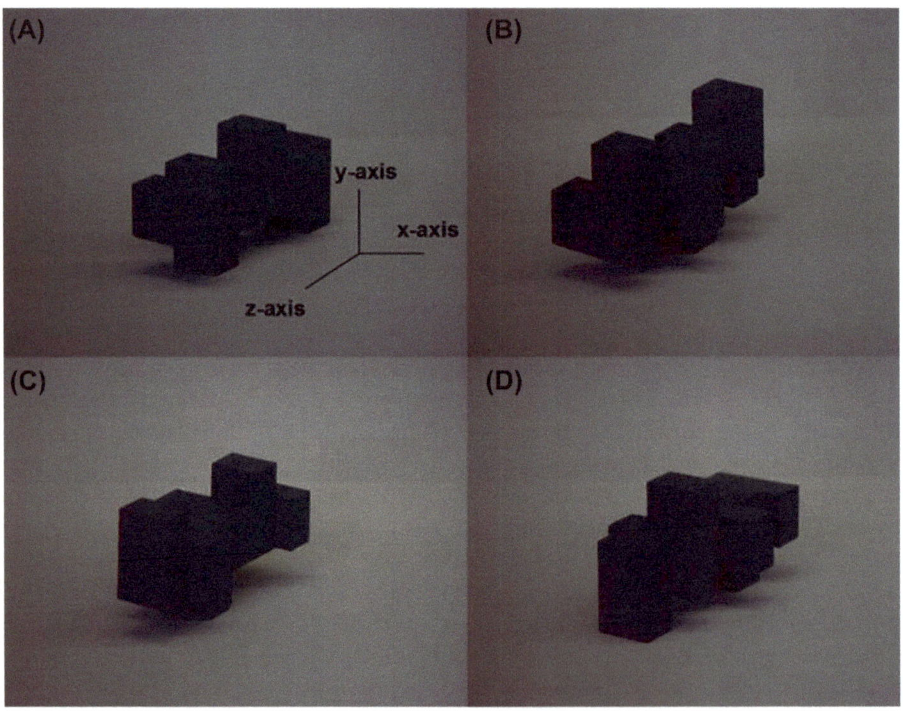

FIGURE 7.1 Example 3D unfamiliar object shown (A) in its original orientation and rotated 180 degrees about the (B) z-axis, (C) x-axis (both are rotations in depth), and (D) the y-axis (a rotation in the picture-plane). *From reference Lacey S, Peters A, Sathian K. Cross-modal object representation is viewpoint-independent.* PLoS ONE. *2007;2:e890. https://doi.org/10.1371/journal.pone0000890.*

equally impaired by rotation about any axis,[30] suggesting that the basis for visual and haptic view-dependence is different in each modality. One possible explanation is that vision and haptics differ in whether or not a surface is occluded by rotation. For vision, rotating an object can involve not only a transformation in perceptual shape but also occlusion of one or more surfaces—unless the observer physically changes position relative to the object.[115,116] Compare, for example, Fig. 7.1A and C—rotation about the x-axis means that the object is now upside down and the former top surface is no longer visible, i.e., it becomes occluded. In haptic exploration, provided the object is small enough, no surface is necessarily occluded because the hands are free to move over all surfaces of an object and to manipulate it into different orientations relative to the hand. Thus, no single axis of rotation should be more or less disruptive than another due to surface occlusion, so that haptic recognition only has to deal with a shape transformation. Further work is required to examine whether this explanation is, in fact, correct.

View-dependence is typically observed for unfamiliar objects. Visual recognition tends to become view-independent as object familiarity increases,[117,118] as does haptic recognition, even where there is a change in the hand used to explore the object.[119] An exception to this is when a familiar object is typically seen in one specific orientation known as a canonical view, for example, the front view of a house.[120] View-independence may still occur for a limited range of orientations around the canonical view, but visual recognition is impaired for radically noncanonical views, for example, a teapot seen from directly above.[117,118,120] Haptic recognition also reverts to view-dependence for noncanonical orientations.[112] However, vision and haptics differ in what constitutes a canonical view. For vision, the preferred view is one in which the object is aligned at 45 degrees to the observer,[120] while for the haptic canonical view, objects are generally aligned either parallel or orthogonal to the body midline.[121] Canonical views may facilitate view-independent recognition either because they provide the most structural information about an object or because they most closely match a stored representation, but the end result is the same for both vision and haptics.[112,121]

Orientation changes in visuo-haptic crossmodal recognition might be expected to be doubly disruptive because there is not only a change in orientation but also a change in modality between study and test. Again, this intuition is incorrect, because crossmodal recognition is view-independent even for unfamiliar objects that are highly similar (Fig. 7.1), whether visual study is followed by haptic test or vice versa, and whatever the axis of rotation.[30,110,111,122] For familiar objects, crossmodal view-independence has been shown in the haptic-visual, but not visual-haptic, condition.[113] However, the familiar objects used in this particular study were a mixture of scale models (e.g., bed, shark) and actual-size objects (e.g., jug, pencil); crossmodal asymmetry may have arisen because the scale models were of objects that would typically be more familiar visually than haptically, resulting in increased error rates when visually familiar objects had to be recognized by touch. Additional research on the potentially disruptive effects of differential familiarity is merited.

A curious finding is that knowledge of the test modality does not appear to help achieve within-modal view-independence. Both visual and haptic within-modal recognitions were view-dependent when participants were told the test modality, whereas crossmodal recognition was view-independent.[110,111] By contrast, when participants were not told the test modality, both within- and crossmodal recognition were view-independent.[110] At first glance, this is puzzling: one would expect that knowledge of the test modality would confer an advantage. However, when participants knew that they would be tested crossmodally (visual-haptic only), their eye movements showed longer and more diffuse fixations during encoding compared with when they knew the test would be within modal.[110] It is possible that, on the "principle of least commitment,"[123] the same pattern of eye movements occurs when the test modality is *not* known (i.e., it is not possible to commit to an outcome), extracting as much information as possible and resulting in both within- and crossmodal view-independence. Further examination of eye movements during both crossmodal conditions would be valuable, because these differences in eye movement patterns could serve as behavioral markers for the multisensory view-independent representation discussed next.

There are several ways in which crossmodal view-independence could arise, the most straightforward being that the unisensory visual and haptic view-dependent representations are directly integrated into a multisensory view-independent representation (Fig. 7.2A). Alternatively, crossmodal view-independence might be gated by unisensory view-independent representations (Fig. 7.2B). In a perceptual learning study designed to distinguish between these competing explanations, we found that view-independence acquired by learning in one modality transferred completely and symmetrically to the other without

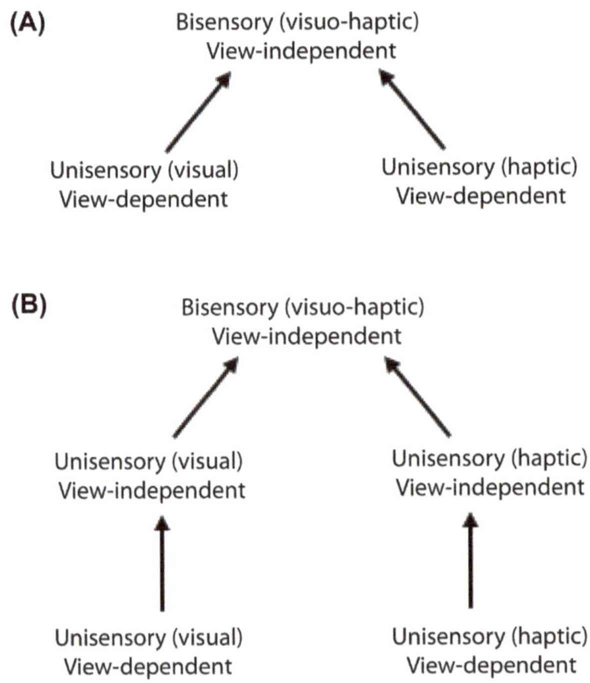

FIGURE 7.2 Alternative models of visuo-haptic view-independence: (A) direct integration of the unisensory view-dependent representations into a multisensory view-independent representation; (B) bisensory view-independence gated by separate, unisensory view-independent representations. Evidence supports (A) the direct integration model. *From reference Lacey S, Pappas M, Kreps A, et al. Perceptual learning of view-independence in visuo-haptic object representations. Exp Brain Res. 2009;198:329-337.*

additional training; we inferred, therefore, that both visual and haptic within-modal view-independence rely on the same view-independent representation.[109] Furthermore, both visual and haptic within-modal view-independence resulted from crossmodal training (whether haptic-visual or visual-haptic). We therefore concluded that visuo-haptic view-independence is supported by a single multisensory representation that directly integrates the unisensory view-dependent representations[109] as depicted in Fig. 7.2A, consistent with models that have been proposed for vision.[124] Thus, the same representation appears to support both crossmodal recognition and view-independence (whether within- or crossmodal).

View-(in)dependent cortical regions

The cortical locus of the multisensory view-independent representation is currently uncertain. Evidence for visual view-independence in the LOC is mixed: as might be expected, unfamiliar objects produce view-dependent LOC responses[125] and familiar objects produce view-independent responses.[126–128] By contrast, one study found view-dependence in the LOC even for familiar objects, although in this study there was position-independence,[129] whereas another found view-independence for both familiar and unfamiliar objects.[130] A developmental fMRI study found view-independent LOC responses for unfamiliar objects but only in adults and not in adolescents or children.[131] A recent TMS study suggests that the LOC is functionally involved in view-independent recognition of 2D shape,[132] but TMS effects were only seen for the smaller of two rotations tested (20 degrees and 70 degrees) and further work is required to investigate this finding more fully. Responses in the FG are also variable with the left FG being less sensitive to orientation changes than the right FG.[133,134]

Various parietal regions show visual view-dependent responses, e.g., the IPS[130] and a parieto-occipital area.[126] Superior parietal cortex is view-dependent during mental rotation but not visual object recognition.[125,135] As these regions are in the dorsal pathway for object location and processing for action (see above), view-dependent responses in these regions are not surprising.[50,51] Actions such as reaching and grasping will be affected by changes in object orientation, for example, by requiring a different or unusual grip, and consistent with this, lateral parieto-occipital cortex shows view-dependent responses for graspable, but not for nongraspable objects.[136]

To our knowledge, there are no neuroimaging studies of haptic or crossmodal processing of stimuli across changes in orientation to date. Although James et al.[137] varied object orientation, this study concentrated on haptic-to-visual priming rather than the crossmodal response to same versus different orientations per se. Additionally, there is much work to be done on the effect of orientation changes when shape information is derived from the auditory soundscapes produced by SSDs and also when the options for haptically interacting with an object are altered by a change in orientation.

Individual differences in visuo-haptic representations

A critical question for object recognition is what information is contained in the neural representations that support it. Neuroimaging studies show that visual shape, color, and texture are processed in different cerebral cortical areas,[138,139] but behavioral evidence indicates that these structural (shape) and surface (color, texture, etc.) properties are integrated in visual object representations.[140] Shape recognition was found to be impaired if the color of an object or its part-color combinations changed between study and test, but not if the background color against which objects were presented was changed.[140] This effect could therefore be isolated to the object representation, indicating that this representation contains both shape and color information.[140] These findings were extended by a later study from our laboratory which showed that both visual and haptic within-modal object discrimination are impaired by changes in surface texture.[122] Thus, haptic representations also integrate structural and surface properties and, furthermore, information about surface properties in visual representations is not limited to modality-specific properties like color. To investigate whether surface properties are integrated into the multisensory representation underlying crossmodal object recognition, we tested object discrimination across changes in texture, orientation (thus requiring access to the view-independent multisensory representation discussed above), or both. Consistent with earlier studies,[30,110,111] crossmodal object discrimination was view-independent provided there was no change in texture; but performance was reduced to chance levels if there was a texture change, whether orientation also changed or not.[122] Interestingly, there was wide individual variation in the effect of texture changes, and we surmised that this might have arisen from individual differences in object representations that can be conveniently indexed by preferences for different kinds of imagery.

Visual imagery can be divided into two subtypes: "object imagery" (involving vivid and detailed pictorial images that deal with the literal appearance of objects in terms of shape, color, brightness, etc.) and "spatial imagery" (involving schematic images that concentrate on the spatial relations of objects, their component parts, and spatial transformations).[141–143] An experimentally important difference is that while both subtypes encode the structural property of shape, object imagery also includes surface property information about color, texture, etc., while spatial imagery does not. To establish whether object and spatial imagery differences occur in touch as well as vision, we required participants to discriminate shape across changes in texture, and texture across changes in shape (Fig. 7.3), in both visual and haptic within-modal conditions. We found that spatial imagers could discriminate shape despite changes in texture but not vice versa,[144] presumably because their images tend not to encode surface properties. By contrast, object imagers could discriminate texture despite changes in shape, but not the reverse,[144] indicating that texture, a surface property, is integrated into their shape representations. Importantly, visual and haptic performance did not differ significantly on either task and performance largely reflected both self-reports of visual imagery preference and scores on the Object and Spatial Imagery Questionnaire (OSIQ).[143] Thus, the object-spatial imagery continuum characterizes haptics as well as vision, and individual differences in imagery preference along this continuum affect the extent to which surface properties are integrated into object representations.[144] A reanalysis of the texture-change condition in our earlier crossmodal study[122] showed that performance was indeed related to imagery preference: both object and spatial imagers showed crossmodal view-independence but object imagers were impaired by texture changes, whereas spatial imagers were not.[144] In addition, the extent of the impairment was correlated with OSIQ scores such that greater preference for object imagery was associated with greater impairment by texture changes; surface properties are therefore likely only integrated into the multisensory representation by object imagers.[144] Moreover, spatial imagery preference correlated

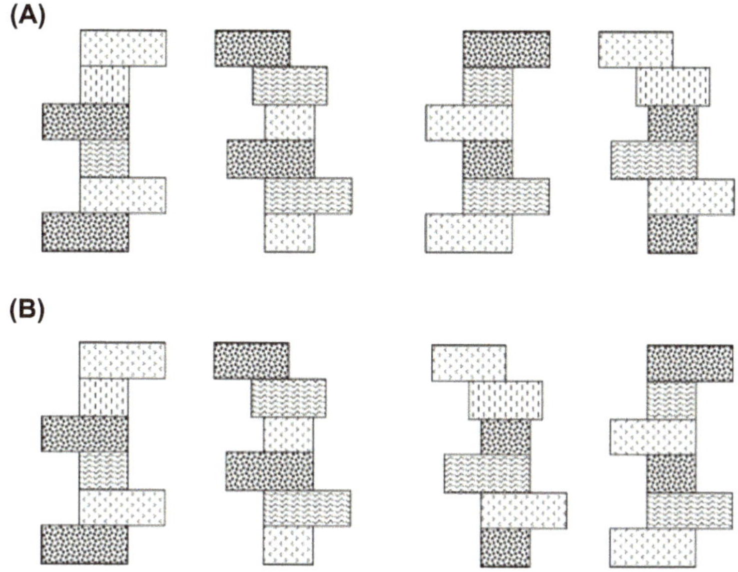

FIGURE 7.3 (A) Schematic example of Shapes 1 and 2 with (left pair) original texture schemes and (right pair) the texture schemes exchanged. (B) Example of Textures 1 and 2 with (left pair) original shapes and (right pair) the shapes exchanged. *From reference Lacey S, Lin JB, Sathian K. Object and spatial imagery dimensions in visuo-haptic representations. Exp Brain Res. 2011;213:267-273.*

with the accuracy of crossmodal object recognition.[30] It appears, then, that the multisensory representation has some features that are stable across individuals, like view-independence, and some that vary across individuals, such as integration of surface property information, correlating with individual differences in imagery preference. Finally, it appears that object and spatial imagery preferences generalize across vision and touch within individuals because further analysis of the data in Lacey et al.[144] showed that visual and haptic performances were highly correlated in both the shape and texture change conditions.[145]

Neural differences between object and spatial imagers

To date, work on the neural correlates of object and spatial imagery is extremely sparse. An early fMRI study employing an object property verification task showed that activation magnitudes in bilateral LOC and right dorsolateral prefrontal cortex were lower for object imagers than spatial imagers, suggesting that greater object imagery ability was accompanied by more efficient use of neural resources.[146] However, the object properties to be verified were spatial, e.g., horizontal and vertical symmetry, parallelism, etc., so it is not clear that this was a task that particularly favored object imagery. Furthermore, this study lacked complementarity as it did not test object and spatial imagers on a task requiring spatial imagery. Future work should test object and spatial imagers using tasks that reflect each imagery preference's strengths and weaknesses in a more principled way and go beyond the visual modality to test these in both haptic and crossmodal conditions.

A model of visuo-haptic multisensory object representation

During tactile discrimination of simple shape stimuli, activity in S1 propagates to the LOC as early as 150 ms after stimulus onset, which is about the same timeframe in which S2 activity appears and is consistent with feedforward projections to LOC.[147,148] In a tactile microspatial discrimination task, LOC activity was associated with feedforward propagation from S1 in a beta-band oscillatory network.[148] These findings suggest that somatosensory input can activate the LOC, although this could certainly be along polysynaptic pathways. Similarly, a patient with bilateral ventral occipito-temporal lesions, but with sparing of the dorsal part of the LOC that likely included the multisensory subregion, showed visual agnosia but intact haptic object recognition with associated activation of the intact dorsal part of the LOC, suggesting that somatosensory input could directly activate this region.[149]

Alternatively, the effect of haptic perception on the LOC might be more indirect via the evocation of visual imagery of the felt object, thus resulting in "top-down" activation of the LOC[16]; consistent with this hypothesis, LOC activity during visual imagery has been shown in many studies: For example, auditorily cued mental imagery of familiar objects resulted in left LOC activity in both blind and sighted participants, where shape information would arise mainly from haptic experience for the blind and mainly from visual experience for the sighted.[150] The left LOC is also active during retrieval of geometric and material object properties from memory.[151] Furthermore, haptic shape-selective activation magnitudes in the right LOC were highly correlated with ratings of visual imagery vividness.[73]

In several studies from our lab, we have explicitly tested the visual imagery hypothesis discussed above[152–154]; these studies have provided evidence for a model of visuo-haptic multisensory object representation first proposed by Lacey et al.[14] In this model, object representations in the LOC can be flexibly accessed either bottom-up or top-down, depending on object familiarity, and independently of the input modality. Because there is no stored representation for unfamiliar objects, these have to be fully explored during haptic shape processing to compute global shape and to relate component parts to one another. The model proposes that this occurs in a bottom-up pathway from somatosensory cortex to the LOC, facilitated by spatial imagery processes occurring in the IPS that compute part relationships and thence global shape. By contrast, there are indeed stored representations of familiar objects, for which global shape can be inferred more easily, perhaps from distinctive features or one diagnostic part; haptic exploration rapidly acquires enough information to trigger a stored visual image and generate a hypothesis about its identity, as has been proposed for vision.[155] The model proposes that this occurs in a top-down pathway from prefrontal cortex to LOC, involving primarily object imagery processes (though spatial imagery may still have a role in processing familiar objects, for example, in view-independent recognition).

We tested this model using analyses of intertask correlations of activation magnitude between haptic shape perception and both visual object[153] and spatial imagery[154] as well as analyses of effective connectivity.[152,154] We hypothesized that task-specific activation magnitudes should be correlated across participants if the tasks relied on similar processes, and that this should also be reflected in similar patterns of effective connectivity across tasks. In contrast to previous studies, we used object and spatial imagery tasks that required an explicitly recorded response to ensure that participants engaged in imagery throughout each scan. In each study, participants performed haptic shape discrimination tasks using familiar and unfamiliar objects. We found that object familiarity modulated intertask correlations between haptic shape perception and visual object imagery, as predicted by our model.[14] However, relatively few regions showed intertask correlations between spatial imagery and haptic perception of either familiar or unfamiliar shape and, contrary to the model predictions, parietal foci appeared in both sets of correlations.[154] This finding suggests that spatial imagery is relevant to haptic shape perception independently of object familiarity, whereas our earlier study of object imagery suggested that this was more strongly associated with haptic perception of familiar, than unfamiliar, shape.[153]

We also conducted effective connectivity analyses, based on the inferred neuronal activity derived from deconvolving the hemodynamic response out of the observed BOLD signals.[156] These analyses supported the broad architecture of the model: the spatial imagery network shared much more commonality with the network associated with unfamiliar, compared with familiar, shape perception, while the object imagery network shared much more commonality with familiar, than unfamiliar, shape perception.[154] More specific aspects of the model were also supported. For example, the model proposes that the component parts of an unfamiliar object are assembled into a representation of global shape via a "bottom-up" pathway facilitated by spatial imagery processes.[14] In the network common to spatial imagery and unfamiliar haptic shape perception, the LOC is driven by multiple parietal foci with cross talk between posterior parietal and somatosensory foci, consistent with the predicted role for cortex in and around the IPS in spatial imagery and via interactions with somatosensory cortex, although these could not be characterized as entirely bottom-up.[153] There was a much smaller network common to spatial imagery and haptic perception of familiar shape, from which the IPS and somatosensory interactions were absent. By contrast, the model predicts that the network common to object imagery and haptic perception of familiar shape perception is characterized by top-down pathways from prefrontal areas.[14] This prediction was supported by the finding that, in the network shared by object imagery and haptic perception of familiar shape, a focus in the left inferior frontal gyrus drove bilateral LOC activity, whereas these pathways were absent from the extremely sparse network common to object imagery and unfamiliar haptic shape perception.[154]

In the current version of our model for haptic shape perception (Fig. 7.4)[67,154], the LOC is driven bottom-up from primary somatosensory cortex as well as top-down via object imagery processes from prefrontal cortex, with additional input in both cases from the IPS that likely reflects spatial imagery processes. The bottom-up route is more important for haptic perception of unfamiliar objects, whereas the top-down route is more important for haptic perception of familiar objects. While this set of studies examined the contributions of object and spatial imagery processes to familiar and unfamiliar object perception, respectively, it remains for future work to assess the impact of individual differences in object and spatial imagery on these processes and paths.

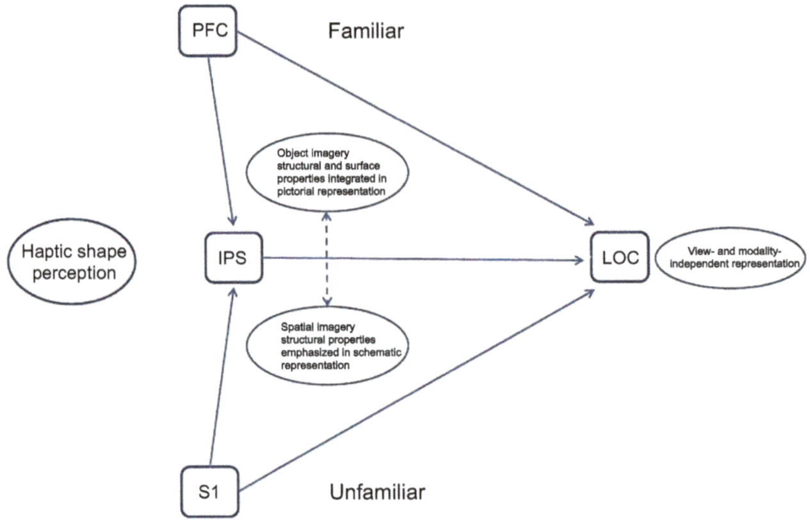

FIGURE 7.4 Schematic model of haptic object representation in lateral occipital complex (LOC) modulated by object familiarity and imagery type. For unfamiliar objects, the LOC is driven bottom-up from somatosensory cortex (S1) supported by spatial imagery processes in the IPS. For familiar objects, the LOC is driven top-down from prefrontal cortex (PFC) via object imagery processes. Thus, the LOC houses an object representation that is flexibly accessible, via both vision and touch, both bottom-up and top-down, and which is additionally view-independent. *From reference Lacey S, Sathian K. Visuo-haptic multisensory object recognition, categorization, and representation. Front Psychol. 2014;5:730, doi:10.3389/fpsyg.2014.00730.*

Conclusion

The evidence reviewed in this chapter shows that the visual and haptic modalities are deeply intertwined at almost every level of object processing. They share highly similar and transferable perceptual spaces for object categorization, common object representations in crossmodal and view-independent recognition, and common dimensions in imagery preferences. These behavioral similarities are underpinned by multisensory neural substrates and complex interactions between bottom-up and top-down processes as well as between object and spatial imagery. Nonetheless, there is still much to be done to provide a detailed account of visuo-haptic multisensory behavior and its underlying mechanisms and how this understanding can be put to use, for example, in the service of neurorehabilitation, particularly for those with sensory deprivation of various sorts.

Acknowledgments

Support to KS from the National Eye Institute at the NIH, the National Science Foundation, and the Veterans Administration is gratefully acknowledged.

References

1. Wagman JB. Taking the other Cinderella to the ball: a review of Psychology of Touch and Blindness by Morton A. Heller & Edouard Gentaz. *IEEE Trans Haptics*. 2015;8:339–340.

2. Gallace A, Spence C. The cognitive and neural correlates of tactile memory. *Psychol Bull.* 2009;135:380–406.
3. Gallace A. Somesthetic mental imagery. In: Lacey S, Lawson R, eds. *Multisensory Imagery*. New York: Springer: Springer; 2013:29–50.
4. Klatzky RL, Lederman SJ, Metzger VA. Identifying objects by touch: an 'expert system'. *Percept Psychophys.* 1985;37:299–302.
5. Klatzky RL, Lederman SJ, Reed CL. There's more to touch than meets the eye: the salience of object attributes for haptics with and without vision. *J Exp Psychol: Gen.* 1987;116:356–369.
6. Lederman SJ, Klatzky RL. Hand movements: a window into haptic object recognition. *Cogn Psychol.* 1987;19:342–368.
7. Klatzky RL, Lederman SJ. Haptic object perception: spatial dimensionality and relation to vision. *Phil Trans R Soc B.* 2011;366:3097–3105.
8. Lederman SJ, Klatzky RL. Haptic perception: a tutorial. *Atten Percept Psychophys.* 2009;71:1439–1459.
9. Johnson KO, Hsaio SS. Neural mechanisms of tactual form and texture perception. *Annu Rev Neurosci.* 1992;15:227–250.
10. Sathian K. Tactile sensing of surface features. *Trends Neurosci.* 1989;12:513–519.
11. Sathian K. Analysis of haptic information in the cerebral cortex. *J Neurophysiol.* 2016;116:1795–1806.
12. Saal HP, Bensmaia SJ. Touch is a team effort: interplay of submodalities in cutaneous sensibility. *Trends Neurosci.* 2014;37:689–697.
13. Pascual-Leone A, Hamilton RH. The metamodal organization of the brain. *Prog Brain Res.* 2001;134:427–445.
14. Lacey S, Tal N, Amedi A, et al. A putative model of multisensory object representation. *Brain Topogr.* 2009;21:269–274.
15. James TW, Stevenson RW, Kim S, et al. Shape from sound: evidence for a shape operator in the lateral occipital cortex. *Neuropsychologia.* 2011;49:1807–1815.
16. Sathian K, Zangaladze A, Hoffman JM, et al. Feeling with the mind's eye. *Neuroreport.* 1997;8:3877–3881.
17. Pitzalis S, Galletti C, Huang RS, et al. Wide-field retinotopy defines human cortical visual area V6. *J Neurosci.* 2006;26:7962–7973.
18. Sergent J, Ohta S, MacDonald B. Functional neuroanatomy of face and object processing. A positron emission tomography study. *Brain.* 1992;115:15–36.
19. Zangaladze A, Epstein CM, Grafton ST, et al. Involvement of visual cortex in tactile discrimination of orientation. *Nature.* 1999;401:587–590.
20. Hagen MC, Franzén O, McGlone F, et al. Tactile motion activates the human middle temporal/V5 (MT/V5) complex. *Eur J Neurosci.* 2002;16:957–964.
21. Blake R, Sobel KV, James TW. Neural synergy between kinetic vision and touch. *Psychol Sci.* 2004;15:397–402.
22. Summers IR, Francis ST, Bowtell RW, et al. A functional-magnetic-resonance-imaging investigation of cortical activation from moving vibrotactile stimuli on the fingertip. *J Acoust Soc Am.* 2009;125:1033–1039.
23. van Kamenade BM, Seymour K, Wacker E, et al. Tactile and visual motion processing in hMT+/V5. *NeuroImage.* 2014;84:420–427.
24. Poirier C, Collignon O, De Volder AG, et al. Specific activation of the V5 brain area by auditory motion processing: an fMRI study. *Cogn Brain Res.* 2005;25:650–658.
25. Klatzky RL, Lederman SJ. Identifying objects from a haptic glance. *Percept Psychophys.* 1995;57:1111–1123.
26. Lawson R. Recognizing familiar objects by hand and foot: haptic shape perception generalizes to inputs from unusual locations and untrained body parts. *Atten Percept Psychophys.* 2014;76:541–558.
27. Klatzky RJ, Lederman SJ. Stages of manual exploration in haptic object identification. *Percept Psychophys.* 1992;52:661–670.
28. Bushnell EW, Baxt C. Children's haptic and cross-modal recognition with familiar and unfamiliar objects. *J Exp Psychol: Human.* 1999;25:1867–1881.
29. Casey SJ, Newell FN. Are representations of unfamiliar faces independent of encoding modality? *Neuropsychologia.* 2007;45:506–513.
30. Lacey S, Peters A, Sathian K. Cross-modal object representation is viewpoint-independent. *PLoS ONE.* 2007;2:e890. https://doi.org/10.1371/journal.pone0000890.
31. Lacey S, Campbell C. Mental representation in visual/haptic crossmodal memory: evidence from interference effects. *Q J Exp Psychol.* 2006;59:361–376.

32. Nabeta T, Kawahara J. Congruency effect of presentation modality on false recognition of haptic and visual objects. *Memory*. 2006;14:307–315.
33. Reales JM, Ballesteros S. Implicit and explicit memory for visual and haptic objects: cross-modal priming depends on structural descriptions. *J Exp Psychol: Learn*. 1999;25:644–663.
34. Easton RD, Greene AJ, Srinivas K. Transfer between vision and haptics: memory for 2-D patterns and 3-D objects. *Psychon Bull Rev*. 1997;4:403–410.
35. Easton RD, Srinivas K, Greene AJ. Do vision and haptics share common representations? Implicit and explicit memory within and between modalities. *J Exp Psychol*. 1997;23:153–163.
36. Streri A, Molina M. Constraints on intermodal transfer between touch and vision in infancy. In: Lewkowicz DJ, Lickliter R, eds. *The Development of Intersensory Perception: Comparative Perspectives*. Hove, UK: Lawrence Erlbaum Associates; 1994:285–307.
37. Jones B. The developmental significance of cross-modal matching. In: Walk RD, Pick Jr HL, eds. *Intersensory Perception and Sensory Integration*. New York: Plenum Press; 1981:108–136.
38. Ernst MO, Banks MS. Humans integrate visual and haptic information in a statistically optimal fashion. *Nature*. 2002;415:429–433.
39. Helbig HB, Ernst MO. Optimal integration of shape information from vision and touch. *Exp Brain Res*. 2007;179:595–606.
40. Helbig HB, Ernst MO, Ricciardi E, et al. The neural mechanisms of reliability weighted integration of shape information from vision and touch. *NeuroImage*. 2012;60:1063–1072.
41. Takahashi C, Watt SJ. Visual-haptic integration with pliers and tongs: signal 'weights' take account of changes in haptic sensitivity caused by different tools. *Front Psychol*. 2014;5:109. https://doi.org/10.3389/fpsyg.2014.00109.
42. Bliss I, Hämäläinen H. Different working memory capacity in normal young adults for visual and tactile letter recognition task. *Scand J Psychol*. 2005;46:247–251.
43. Kiphart MJ, Hughes JL, Simmons J, et al. Short-term haptic memory for complex objects. *Bull Psychon Soc*. 1992;30:212–214.
44. Craddock M, Lawson R. The effects of temporal delay and orientation on haptic object recognition. *Atten Percept Psychophys*. 2010;72:1975–1980.
45. Hutmacher F, Kuhbandner C. Long-term memory for haptically explored objects: fidelity, durability, incidental encoding, and cross-modal transfer. *Psychol Sci*. 2018;29:2031–2038.
46. Woods AT, O'Modhrain S, Newell FN. The effect of temporal delay and spatial differences on crossmodal object recognition. *Cognit Affect Behav Neurosci*. 2004;4:260–269.
47. Garvill J, Molander B. Effects of standard modality, comparison modality and retention interval on matching of form. *Scand J Psychol*. 1973;14:203–206.
48. Pensky AEC, Johnson KA, Haag S, et al. Delayed memory for visual-haptic exploration of objects. *Psychon Bull Rev*. 2008;15:574–580.
49. Kassuba T, Klinge C, Hölig C, et al. Vision holds a greater share in visuo-haptic object recognition than touch. *NeuroImage*. 2013;65:59–68.
50. Ungerleider LG, Mishkin M. Two cortical visual systems. In: Ingle DJ, Goodale MA, Mansfield RJW, eds. *Analysis of Visual Behavior*. Cambridge, MA: MIT Press; 1982:549–586.
51. Goodale MA, Milner AD. Separate visual pathways for perception and action. *Trends Neurosci*. 1992;15:20–25.
52. Freud E, Culham JC, Plaut DC, et al. The large-scale organization of shape processing in the ventral and dorsal pathways. *eLife*. 2017;6:e27576. https://doi.org/10.7554/eLife.27576.
53. Alain C, Arnott SR, Hevenor S, et al. "What" and "where" in the human auditory system. *Proc Natl Acad Sci USA*. 2001;98:12301–12306.
54. Arnott SR, Binns MA, Grady CL, et al. Assessing the auditory dual pathway model in humans. *NeuroImage*. 2004;22:401–408.
55. Rauscheker JP, Tian B. Mechanisms and streams for processing of "what" and "where" in auditory cortex. *Proc Natl Acad Sci USA*. 2000;97:11800–11806.
56. Dijkerman HC, de Haan EHF. Somatosensory processes subserving perception and action. *Behav Brain Sci*. 2007;30:189–239.
57. Mishkin M. Analogous neural models for tactual and visual learning. *Neuropsychologia*. 1979;47:139–151.
58. Reed CL, Klatzky RL, Halgren E. What vs. where in touch: an fMRI study. *NeuroImage*. 2005;25:718–726.

59. Sathian K, Lacey S, Stilla R, et al. Dual pathways for somatosensory information processing. *NeuroImage*. 2011;57:462–475.
60. Stilla R, Sathian K. Selective visuo-haptic processing of shape and texture. *Hum Brain Mapp*. 2008;29:1123–1138.
61. Eck J, Kaas AL, Goebel R. Crossmodal interactions of haptic and visual texture information in early sensory cortex. *NeuroImage*. 2013;75:123–135.
62. Eck J, Kaas AL, Mulders JL, et al. The effect of task instruction on haptic texture processing: the neural underpinning of roughness and spatial density perception. *Cerebr Cortex*. 2016;26:384–401.
63. Sun H-C, Welchman AE, Change DHF, Di Luca M. Look but don't touch: visual cues to surface structure drive somatosensory cortex. *NeuroImage*. 2016;128:353–361.
64. Amedi A, von Kriegstein K, van Atteveldt NM, et al. Functional imaging of human crossmodal identification and object recognition. *Exp Brain Res*. 2005;166:559–571.
65. Sathian K, Lacey S. Journeying beyond classical somatosensory cortex. *Can J Exp Psychol*. 2007;61:254–264.
66. Lacey S, Sathian K. Multisensory object representation: insights from studies of vision and touch. *Prog Brain Res*. 2011;191:165–176.
67. Lacey S, Sathian K. Visuo-haptic multisensory object recognition, categorization and representation. *Front Psychol*. 2014;5:730. https://doi.org/10.3389/fpsyg.2014.00730.
68. Snow JC, Strother L, Humphreys GW. Haptic shape processing in visual cortex. *J Cogn Neurosci*. 2014;26:1154–1167.
69. Négyessy L, Nepusz T, Kocsis L, et al. Prediction of the main cortical areas and connections involved in the tactile function of the visual cortex by network analysis. *Eur J Neurosci*. 2006;23:1919–1930.
70. Malach R, Reppas JB, Benson RR, et al. Object-related activity revealed by functional magnetic resonance imaging in human occipital cortex. *Proc Natl Acad Sci USA*. 1995;92:8135–8139.
71. Amedi A, Malach R, Hendler T, et al. Visuo-haptic object-related activation in the ventral visual pathway. *Nat Neurosci*. 2001;4:324–330.
72. Amedi A, Jacobson G, Hendler T, et al. Convergence of visual and tactile shape processing in the human lateral occipital complex. *Cerebr Cortex*. 2002;12:1202–1212.
73. Zhang M, Weisser VD, Stilla R, et al. Multisensory cortical processing of object shape and its relation to mental imagery. *Cognit Affect Behav Neurosci*. 2004;4:251–259.
74. Stoesz M, Zhang M, Weisser VD, et al. Neural networks active during tactile form perception: common and differential activity during macrospatial and microspatial tasks. *Int J Psychophysiol*. 2003;50:41–49.
75. Prather SC, Votaw JR, Sathian K. Task-specific recruitment of dorsal and ventral visual areas during tactile perception. *Neuropsychologia*. 2004;42:1079–1087.
76. Dormal G, Pelland M, Rezk M, et al. Functional preference for objects sounds and voices in the brain of early blind and sighted individuals. *J Cogn Neurosci*. 2018;30:86–106. https://doi.org/10.1162/jocn_a_01186.
77. Amedi A, Stern WM, Camprodon JA, et al. Shape conveyed by visual-to-auditory sensory substitution activates the lateral occipital complex. *Nat Neurosci*. 2007;10:687–689.
78. Saito DN, Okada T, Morita Y, et al. Tactile-visual cross-modal shape matching: a functional MRI study. *Cogn Brain Res*. 2003;17:14–25.
79. Grefkes C, Geyer S, Schormann T, et al. Human somatosensory area 2: observer-independent cytoarchitectonic mapping, interindividual variability, and population map. *NeuroImage*. 2001;14:617–631.
80. Iriki A, Tanaka M, Iwamura Y. Attention-induced neuronal activity in the monkey somatosensory cortex revealed by pupillometrics. *Neurosci Res*. 1996;25:173–181.
81. Zhou Y-D, Fuster JM. Neuronal activity of somatosensory cortex in a cross-modal (visuo-haptic) memory task. *Exp Brain Res*. 1997;116:551–555.
82. Tal N, Amedi A. Multisensory visual-tactile object related network in humans: insights gained using a novel crossmodal adaptation approach. *Exp Brain Res*. 2009;198:165–182.
83. Grill-Spector K, Henson R, Martin A. Repetition and the brain: neural models of stimulus-specific effects. *Trends Cognit Sci*. 2006;10:14–23.
84. Krekelberg B, Boynton GM, van Wezel RJ. Adaptation: from single cells to BOLD signals. *Trends Neurosci*. 2006;29:250–256.
85. Mur M, Ruff DA, Bodurka J, et al. Face-identity change activation outside the face system: "release from adaptation" may not always indicate neuronal selectivity. *Cerebr Cortex*. 2010;20:2027–2042.

86. Feinberg TE, Rothi LJ, Heilman KM. Multimodal agnosia after unilateral left hemisphere lesion. *Neurology*. 1986;36:864–867.
87. James TW, James KH, Humphrey GK, et al. Do visual and tactile object representations share the same neural substrate? In: Heller MA, Ballesteros S, eds. *Touch and Blindness: Psychology and Neuroscience*. Mahwah, NJ: Lawrence Erlbaum Associates; 2006:139–155.
88. Snow JC, Goodale MA, Culham JC. Preserved haptic shape processing after bilateral LOC lesions. *J Neurosci*. 2015;35:13745–13760.
89. Sack AT. Transcranial magnetic stimulation, causal structure-function mapping and networks of functional relevance. *Curr Opin Neurobiol*. 2006;16:593–599.
90. Mullin CR, Steeves JKE. TMS to the lateral occipital cortex disrupts object processing but facilitates scene processing. *J Cogn Neurosci*. 2011;23:4174–4184.
91. Chouinard PA, Meena DK, Whitwel RL, et al. A TMS investigation on the role of lateral occipital complex and caudal intraparietal sulcus in the perception of object form and orientation. *J Cogn Neurosci*. 2017;29:881–895.
92. Buelte D, Meister IG, Staedtgen M, et al. The role of the anterior intraparietal sulcus in crossmodal processing of object features in humans: an rTMS study. *Brain Res*. 2008;1217:110–118.
93. Graf M. Categorization and object shape. In: Glatzeder BM, Goel V, von Müller A, eds. *Towards a Theory of Thinking: Building Blocks for a Conceptual Framework*. Berlin: Springer-Verlag; 2010:73–101.
94. Gaiβert N, Waterkamp S, Fleming RW, et al. Haptic categorical perception of shape. *PLoS ONE*. 2012;7:e43062.
95. Cooke T, Jäkel F, Wallraven C, et al. Multimodal similarity and categorization of novel, three-dimensional objects. *Neuropsychologia*. 2007;45:484–495.
96. Gaiβert N, Wallraven C, Bülthoff HH. Visual and haptic perceptual spaces show high similarity in humans. *J Vis*. 2010;10:2. https://doi.org/10.1167/10.11.2.
97. Gaiβert N, Wallraven C. Categorizing natural objects: a comparison of the visual and haptic modalities. *Exp Brain Res*. 2012;216:123–134.
98. Yildirim I, Jacobs RA. Transfer of object category knowledge across visual and haptic modalities: experimental and computational studies. *Cognition*. 2013;126:135–148.
99. Wallraven C, Bülthoff HH, Waterkamp S, et al. The eyes grasp, the hands see: metric category knowledge transfers between vision and touch. *Psychon Bull Rev*. 2014;21:976–985.
100. Gaiβert N, Wallraven C, Bülthoff HH. Analyzing perceptual representations of complex, parametrically-defined shapes using MDS. *Lect Notes Comput Sci*. 2008;5024:265–274.
101. Gaiβert N, Bülthoff HH, Wallraven C. Similarity and categorization: from vision to touch. *Acta Psychol*. 2011;138:219–230.
102. Pietrini P, Furey ML, Ricciardi E, et al. Beyond sensory images: object-based representation in the human ventral pathway. *Proc Natl Acad Sci USA*. 2004;101:5658–5663.
103. Smith FW, Goodale MA. Decoding visual object representations in early somatosensory cortex. *Cerebr Cortex*. 2015;25:1020–1031.
104. Hernández-Pérez R, Cuaya LV, Rojas-Hortelano E, et al. Tactile object categories can be decoded from the paireal and lateral-occipital cortices. *Neuroscience*. 2017;352:226–235.
105. Masson HL, Wallraven C, Petit L. "Can touch this": cross-modal shape categorization performance is associated with microstructural characteristics of white matter association pathways. *Hum Brain Mapp*. 2017;38:842–854.
106. Masson HL, Kang H-M, Petit L, Wallraven C. Neuroanatomical correlates of haptic object processing: combines evidence from tractography and functional neuroimaging. *Brain Struct Funct*. 2018;223:619–633.
107. Peissig JJ, Tarr MJ. Visual object recognition: do we know more now than we did 20 years ago? *Annu Rev Psychol*. 2007;58:75–96.
108. Newell FN, Ernst MO, Tjan BS, et al. View dependence in visual and haptic object recognition. *Psychol Sci*. 2001;12:37–42.
109. Lacey S, Pappas M, Kreps A, et al. Perceptual learning of view-independence in visuo-haptic object representations. *Exp Brain Res*. 2009;198:329–337.
110. Ueda Y, Saiki J. Viewpoint independence in visual and haptic object recognition. *Jpn J Psychon Sci*. 2007;26:11–19.
111. Ueda Y, Saiki J. Characteristics of eye movements in 3-D object learning: comparison between within-modal and cross-modal object recognition. *Perception*. 2012;41:1289–1298.

112. Craddock M, Lawson R. Repetition priming and the haptic recognition of familiar and unfamiliar objects. *Percept Psychophys*. 2008;70:1350–1365.
113. Lawson R. A comparison of the effects of depth rotation on visual and haptic three-dimensional object recognition. *J Exp Psychol: Human*. 2009;35:911–930.
114. Lawson R. An investigation into the cause of orientation-sensitivity in haptic object recognition. *Seeing Perceiving*. 2011;24:293–314.
115. Pasqualotto A, Finucane C, Newell FN. Visual and haptic representations of scenes are updated with observer movement. *Exp Brain Res*. 2005;166:481–488.
116. Pasqualotto A, Newell FN. The role of visual experience on the representation and updating of novel haptic scenes. *Brain Cogn*. 2007;65:184–194.
117. Tarr MJ, Pinker S. Mental rotation and orientation dependence in shape recognition. *Cogn Psychol*. 1989;21:233–282.
118. Bülthoff I, Newell FN. The role of familiarity in the recognition of static and dynamic objects. *Prog Brain Res*. 2006;154:315–325.
119. Craddock M, Lawson R. Do left and right matter for haptic recognition of familiar objects? *Perception*. 2009;38:1355–1376.
120. Palmer SE, Rosch E, Chase P. Canonical perspective and the perception of objects. In: Long J, Baddeley A, eds. *Attention and Performance IX*. Hillsdale, NJ: Lawrence Erlbaum Associates; 1981:135–151.
121. Woods AT, Moore A, Newell FN. Canonical views in haptic object perception. *Perception*. 2008;37:1867–1878.
122. Lacey S, Hall J, Sathian K. Are surface properties integrated into visuo-haptic object representations? *Eur J Neurosci*. 2010;31:1882–1888.
123. Marr D. Early processing of visual information. *Philos T Roy Soc*. 1976;275:483–524.
124. Riesenhuber M, Poggio T. Hierarchical models of object recognition in cortex. *Nat Neurosci*. 1999;2:1019–1025.
125. Gauthier I, Hayward WG, Tarr MJ, et al. BOLD activity during mental rotation and view-dependent object recognition. *Neuron*. 2002;34:161–171.
126. Valyear KF, Culham JC, Sharif N, et al. A double dissociation between sensitivity to changes in object identity and object orientation in the ventral and dorsal streams: a human fMRI study. *Neuropsychologia*. 2006;44:218–228.
127. Eger E, Ashburner J, Haynes J-D, et al. fMRI activity patterns in human LOC carry information about object exemplars within category. *J Cogn Neurosci*. 2008;20:356–370.
128. Pourtois G, Schwarz S, Spiridon M, et al. Object representations for multiple visual categories overlap in lateral occipital and medial fusiform cortex. *Cerebr Cortex*. 2009;19:1806–1819.
129. Grill-Spector K, Kushnir T, Edelman S, et al. Differential processing of objects under various viewing conditions in the human lateral occipital complex. *Neuron*. 1999;24:187–203.
130. James TW, Humphrey GK, Gati JS, et al. Differential effects of view on object-driven activation in dorsal and ventral streams. *Neuron*. 2002;35:793–801.
131. Nishimura M, Scherf KS, Zachariou V, et al. Size precedes view: developmental emergence of invariant object representations in lateral occipital complex. *J Cogn Neurosci*. 2015;27:474–491.
132. Silvanto J, Schwarzkopf DS, Gilaie-Dotan S, et al. Differing causal roles for lateral occipital complex and occipital face area in invariant shape recognition. *Eur J Neurosci*. 2010;32:165–171.
133. Andresen DR, Vinberg J, Grill-Spector K. The representation of object viewpoint in human visual cortex. *NeuroImage*. 2009;45:522–536.
134. Harvey DY, Burgund ED. Neural adaptation across viewpoint and exemplar in fusiform cortex. *Brain Cogn*. 2012;80:33–44.
135. Wilson K, Farah MJ. Distinct patterns of viewpoint-dependent BOLD activity during common-object recognition and mental rotation. *Perception*. 2006;35:1351–1366.
136. Rice NJ, Valyear KF, Goodale MA, et al. Orientation sensitivity to graspable objects: an fMRI adaptation study. *NeuroImage*. 2007;36:T87–T93.
137. James TW, Humphrey GK, Gati JS, et al. Haptic study of three-dimensional objects activates extrastriate visual areas. *Neuropsychologia*. 2002;40:1706–1714.
138. Cant JS, Goodale MA. Attention to form or surface properties modulates different regions of human occipitotemporal cortex. *Cerebr Cortex*. 2007;17:713–731.

139. Cant JS, Arnott SR, Goodale MA. fMR-adaptation reveals separate processing regions for the perception of form and texture in the human ventral stream. *Exp Brain Res*. 2009;192:391–405.
140. Nicholson KG, Humphrey GK. The effect of colour congruency on shape discriminations of novel objects. *Perception*. 2003;32:339–353.
141. Kozhevnikov M, Hegarty M, Mayer RE. Revising the visualiser-verbaliser dimension: evidence for two types of visualisers. *Cogn Instr*. 2002;20:47–77.
142. Kozhevnikov M, Kosslyn SM, Shephard J. Spatial versus object visualisers: a new characterisation of cognitive style. *Mem Cogn*. 2005;33:710–726.
143. Blajenkova O, Kozhevnikov M, Motes MA. Object-spatial imagery: a new self-report imagery questionnaire. *Appl Cognit Psychol*. 2006;20:239–263.
144. Lacey S, Lin JB, Sathian K. Object and spatial imagery dimensions in visuo-haptic representations. *Exp Brain Res*. 2011;213:267–273.
145. Lacey S, Feng H, Caesar E, et al. Preferences for integrative versus schematic sensory imagery across modalities. *Soc Neurosci Abstr*. 2013, 571.10.
146. Motes MA, Malach R, Kozhevnikov M. Object-processing neural efficiency differentiates object from spatial visualizers. *Neuroreport*. 2008;19:1727–1731.
147. Lucan JN, Foxe JJ, Gomez-Ramirez M, et al. Tactile shape discrimination recruits human lateral occipital complex during early perceptual processing. *Hum Brain Mapp*. 2010;31:1813–1821.
148. Adhikari BM, Sathian K, Epstein CM, et al. Oscillatory activity in neocortical networks during tactile discrimination near the limit of spatial acuity. *NeuroImage*. 2014;91:300–310.
149. Allen HA, Humphreys GW. Direct tactile stimulation of dorsal occipito-temporal cortex in a visual agnosic. *Curr Biol*. 2009;19:1044–1049.
150. De Volder AG, Toyama H, Kimura Y, et al. Auditory triggered mental imagery of shape involves visual association areas in early blind humans. *NeuroImage*. 2001;14:129–139.
151. Newman SD, Klatzky RL, Lederman SJ, et al. Imagining material versus geometric properties of objects: an fMRI study. *Cogn Brain Res*. 2005;23:235–246.
152. Deshpande G, Hu X, Lacey S, et al. Object familiarity modulates effective connectivity during haptic shape perception. *NeuroImage*. 2010;49:1991–2000.
153. Lacey S, Flueckiger P, Stilla R, et al. Object familiarity modulates the relationship between visual object imagery and haptic shape perception. *NeuroImage*. 2010;49:1977–1990.
154. Lacey S, Stilla R, Sreenivasan K, et al. Spatial imagery in haptic shape perception. *Neuropsychologia*. 2014;60:144–158.
155. Bar M. The proactive brain: using analogies and associations to generate predictions. *Trends Cognit Sci*. 2007;11:280–289.
156. Sathian K, Deshpande G, Stilla R. Neural changes with tactile learning reflect decision-level reweighting of perceptual readout. *J Neurosci*. 2013;33:5387–5398.

CHAPTER

8

Multisensory processes in body ownership

Henrik H. Ehrsson

Department of Neuroscience, Karolinska Institutet, Stockholm, Sweden

Introduction

On looking at one's own hands, one immediately recognizes to whom they belong. One automatically senses that these extremities are part of one's own body without the need to actively think about it. This percept of a body part or entire body as belonging to oneself is called the sense (or feeling) of *body ownership*.[1–5] The perceptual distinction between what is part of one's body and what is not is critical for survival and thus fundamentally important for human perception, action, and cognition. Moreover, the sense of one's own body is the centerpiece of human conscious experience because it is from the perspective of this bodily self that a person becomes aware of his or her surroundings. Importantly, the sense of body ownership is multisensory in nature and cannot be easily reduced to a single sensory modality. The crucial aspect seems to be how all the impressions from the different sensory modalities come together into a coherent percept of a single owned body part (or body). This includes the feeling of the skin stretching around the digits and joints, the feeling of coolness or warmth at the surface of the skin, the pressure and tension on the muscles and tendons, and perhaps an ache in the wrist from yesterday's tennis match. Thus, the various sensations originating from the body and reaching the brain through different peripheral and central pathways are effortlessly blended together into coherent percepts of one's own body parts. The way the sense of body ownership emerges from the combination of individual sensory inputs is thus somewhat similar to the way one recognizes the taste of one's favorite red wine as a distinct yet rich composite of various tastes, smells, and visual impressions. When viewed in this way, the term "*sense* of body ownership" is a misnomer, strictly speaking, because it is not referring to a single basic sense analogous to touch or olfaction, but rather a *perception* of one's own body that arises from interpretation of different kinds of afferent sensory signals. In this chapter, we will consider the problem of body ownership from the perspective of multisensory perception and integration, focusing

Multisensory Perception
https://doi.org/10.1016/B978-0-12-812492-5.00008-5

on recent behavioral and neuroimaging studies in humans that have addressed this issue experimentally.

A good starting point for our discussions is a set of interesting observations published in the neurological literature. We know that people with damage to their frontal and parietal lobes can sometimes fail to recognize their paralyzed limbs as their own (asomatognosia: loss of ownership of a limb; somatoparaphrenia: denial of ownership of a limb and confabulatory attribution of this limb to another person).[6–11] This inability cannot be explained by basic sensory impairments of vision or touch, as these individuals do see that they are looking at a hand and can sometimes perceive somatic stimuli applied to the affected limb. Instead, these neurological observations suggest that nonprimary areas in the frontal and parietal association cortices that have the capacity to integrate visual and somatosensory impressions are responsible for generating the feeling of owning limbs. However, it is difficult to make inferences about the specific perceptual processes and neuronal mechanisms involved in body ownership from neuropsychological studies alone because the lesions are typically large, involving multiple areas and damaging the underlying white matter that connects different parts of the brain.

Twenty years ago, a paper was published that sparked the modern interest in experimental studies of body ownership[12] (for an earlier anecdotal observation see Ref. 13). In this article, now cited over 1500 times according to Web of Science, the authors described a fascinating perceptual illusion called the rubber hand illusion.[12] To elicit this illusion, the experimenter keeps the participant's real hand out of the field of vision (behind a screen, under the table, or under a box) while a realistic life-sized rubber hand is placed in front of the participant (Fig. 8.1A). The experimenter—or an apparatus—uses two small paintbrushes to stroke the rubber hand and the participant's hidden hand, synchronizing the timing of the brushing as accurately as possible. After a short period of repeated stroking, approximately 10 seconds on average,[14,15] the majority of people perceive that the rubber hand is their own hand and that it senses the touches of the paintbrush. The illusion can be quantified subjectively with questionnaires and visual analogue rating scales,[12,16] behaviorally as changes in reported hand location toward the location of the rubber hand ("proprioceptive drift"),[12,17]

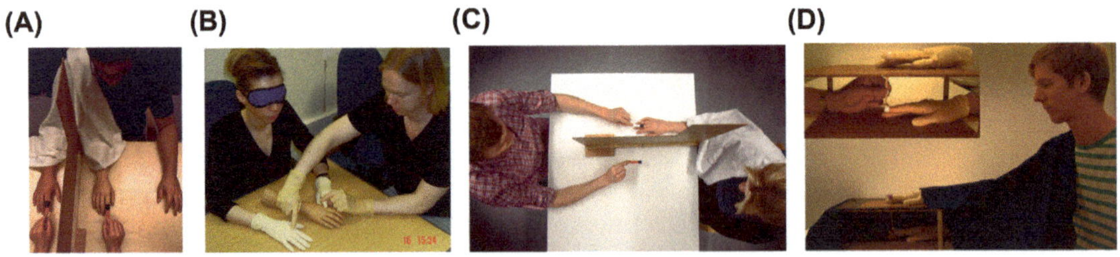

FIGURE 8.1 Induction of the classical version of the rubber hand illusion[12] with synchronous brushstrokes applied to the rubber hand, which is in view, and the participant's real hand, which is hidden (A). The somatic rubber hand illusion[70] with a blindfolded participant is elicited by having the participant touch the rubber hand with her left index finger at the same time as she receives touches on her real right hand (B). The invisible hand illusion[42] was induced by stroking "empty space" and the participant's hidden real hand with very well-matched spatiotemporal stroking patterns (C). The moving rubber hand illusion[58] was produced by congruent movements of the rubber hand's index finger and the participant's real index finger (D).

and physiologically by recording skin conductance responses when the rubber hand is physically threatened or "injured"[18] (for further discussions of these and other proposed methods, see Ref. 2,19–23). What made this illusion so popular was that it provided scientists with a model system in which to start to experiment with the sense of body ownership: one moment the participant is looking at a lifeless piece of rubber, and the next moment the rubber hand "comes alive" and becomes part of the participant's perceived body, with the subjective feelings associated with a real limb. This explicit change in the sense of ownership of the rubber hand is what sets this illusion apart from other well-known body illusions that involve stretching, moving, or dislocating limbs and body parts.[24–27] By systematically investigating the perceptual rules that govern the rubber hand illusion and similar body ownership illusions (see below) and by clarifying the associated neuronal substrates of such illusory percepts with human brain imaging experiments, a great deal can be learned about how the sense of ownership is generated under more natural conditions. In this chapter, we will consider body ownership from a multisensory perspective, focusing on the multisensory processes that underpin this perceptual phenomenon. In particular, we will review studies that show that body ownership shares many perceptual and neural similarities with fundamental principles of multisensory integration, and we will discuss the parsimonious hypothesis that body ownership can be explained as coherent multisensory perception of one's own body. We will not have time to discuss the fascinating philosophical implications of experimental body ownership research.[28] Alternative models of body ownership that are not based on multisensory perception,[29] models in which multisensory interactions are considered only at the initial stage in a more complex cognitive architecture,[3] and self-models based on predictive coding principles[30–32] are also beyond the scope of the present discussion. The overall structure of this chapter is that we will start by reviewing behavioral studies of the limb ownership and rubber hand illusions, move on to the topic of functional magnetic resonance imaging (fMRI) studies that have sought the neural correlates of limb ownership, and, finally, turn to the issue of how a person comes to experience ownership of an entire body.

Perceptual rules of body ownership

One important goal for behavioral studies of body ownership has been to work out the basic perceptual rules that determine the elicitation of ownership sensations. By clarifying the fundamental constraints on the factors that elicit body ownership illusions, we can ascertain what kinds of processes are likely to mediate this perceptual phenomenon. As will be discussed in detail below, the basic rules of body ownership bear striking similarities to the temporal and spatial congruency principles that determine multisensory integration in general,[33–35] see chapter by Stein and Rowland, this volume. These principles state that when two (or more) signals from two (or more) different sensory modalities occur at the same time (temporal principle) and in the same place (spatial principle), they will be integrated, and multisensory perceptual unity will be experienced. The congruency of other stimulus features that influence multisensory binding, such as texture, shape, and orientation,[35,36] also influences body ownership. The observation that the sense of body ownership is governed by the same principles as multisensory perception suggests that multisensory integration mechanisms play a critical role in this self-perceptual phenomenon.

Temporal rule

The feeling of ownership of a limb depends on the temporal synchrony of multisensory cues from that limb. Rubber hand illusion studies have repeatedly shown that the illusion is abolished under asynchronous control conditions in which visual and tactile stimulation have a temporal mismatch on the order of 500–1000 ms.[14,17,18,37] When the temporal delay between the visual and tactile stimulations was systematically varied, delays longer than 300 ms were found to significantly reduce the illusion[38–40] (Fig. 8.2B). Moreover, the degree of asynchrony necessary to eliminate the rubber hand illusion is related to the individual's general perceptive sensitivity to asynchrony during visuotactile stimulation.[41] Thus, there exists a systematic relationship between the temporal constraint of the rubber hand illusion and the temporal binding window of vision and touch. The importance of the temporal congruency of the somatosensory and visual events in the rubber hand illusion fits well with the temporal congruency principle in multisensory integration.[33,34]

It should be made clear that the elicitation of the rubber hand illusion requires a series of correlated synchronous (or near synchronous) visuotactile stimuli applied to the hands, usually on the order of at least five to six seen and felt strokes in participants susceptible to the illusion.[14,42] Anecdotally, an irregular pattern of simultaneous strokes gives a stronger illusion than a strictly regular pattern gives.[43] Thus, the temporal structure of the correlated signals seems to be a factor that influences the illusion over and above temporal coincidence.

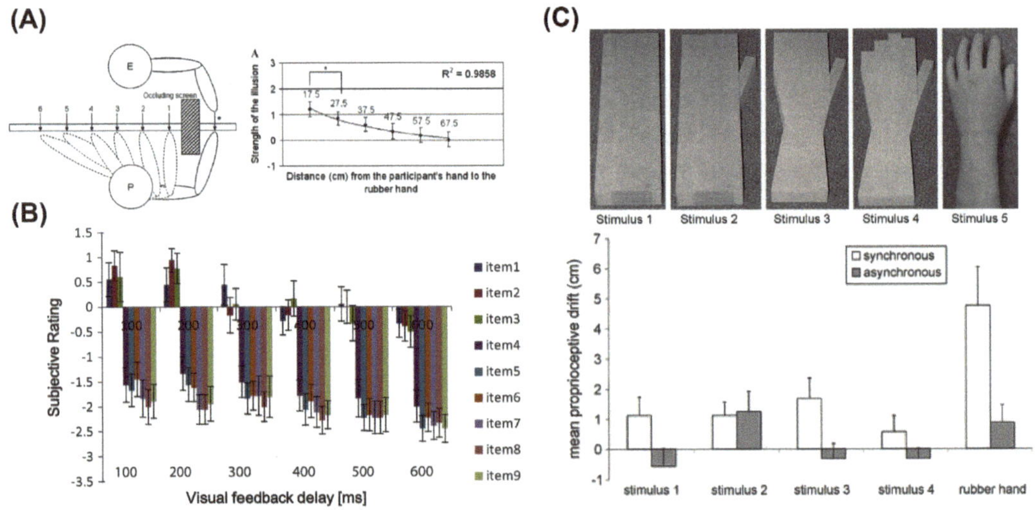

FIGURE 8.2 The distance rule of body ownership (A). The rubber hand illusion is strongest for distances less than or equal to 27.5 cm between the rubber hand and the real hand.[15] The temporal congruency rule of body ownership (B). The rubber hand illusion is reduced as the asynchrony of seen and felt strokes increases, with asynchronies of more than 300 ms breaking the illusion.[39] Items one to nine relate to the statements in a questionnaire, where statements one to three refer to the rubber hand illusion, and the others are controls for suggestibility and task compliance. The humanoid shape rule of body ownership (C). Only the realistic-looking prosthetic hand elicits the rubber hand illusion (top panel; stimulus 5, the rubber hand), and the other wooden objects with varying degrees of resemblance to a hand do not[66]; the lower panel shows a behavioral index of the illusion (proprioceptive drift) for each of the five objects tested.

This is in line with studies on audiovisual integration[44–46] that have shown that temporal correlation influences multisensory integration by strengthening the inference that the signals have a common cause. Future experiments should characterize exactly how the fine-grained patterns of temporally correlated multisensory signals influence body ownership.

Spatial rule(s)

The rubber hand illusion also depends on the spatial congruence of the various sensory signals. There are several spatial factors that influence the rubber hand illusion, such as the relative directions and locations of the tactile and visual stimulation on the seen rubber hand and the unseen real hand, the relative orientations of the two limbs, and the distance between the real hand and the rubber hand (*distance rule*), which will be discussed first. Lloyd[15] parametrically varied the distance between the rubber hand and the hidden real hand, and she found a significant decrease in the strength of the illusion for distances greater than 27.5 cm in the horizontal plane (Fig. 8.2A). Kalckert and Ehrsson[47] obtained a very similar result when examining the effect of vertical distance between the hands. These are interesting observations because the dimensions approximately match the extent of *peripersonal space* (the space immediately surrounding our bodies[48]) around the upper limb as estimated in electrophysiological[49,50] and neuropsychological[51] studies. In addition, Preston found that the strength of the illusion is affected not only by the distance between the rubber hand and the participant's real hand but also by the distance between the model hand and the participant's trunk and by whether the model hand was placed laterally or medially with respect to the participant's hand[52]; all these observations are compatible with multisensory processing in various body part–centered spatial reference frames (a combination of the hand and the trunk, for example). The fact that the limb ownership illusion can be maintained with a slowly elongating virtual arm[53] does not falsify the spatial distance rule because, in this case, peripersonal space probably gradually extends outward along with the flexible representation of the stretched arm.[25] Collectively, the above observations fit with the notion that peripersonal space is a basic constraint of body ownership illusions, which, in turn, suggests that the integration of multisensory signals in body ownership occurs in spatial reference frames centered on the body and its parts.

Another important aspect of the spatial principle is that the seen and felt brushstrokes must be in the same *direction* on the rubber hand and the real hand.[42,54,55] Importantly, "same direction" is defined with respect to a spatial reference frame centered on the hands, such that if, say, the right rubber hand is rotated 20 degrees counterclockwise (toward the body midline) while the hidden real hand is oriented straight ahead, then the strokes applied to the rubber hand must also be rotated 20 degrees counterclockwise to maintain spatial congruence with the straight strokes applied to the real hand.[54] This corresponds well with the notion that the illusion requires multisensory processing performed in body part–centered spatial coordinate systems.[50,56]

However, the rubber hand illusion is not constrained only by spatial factors relating to vision and touch. The spatial congruency between visual and proprioceptive information about the orientation of the hand and arm is also important (*orientation rule*). When the rubber hand is rotated by 90 degrees[17,57] or by 180 degrees[14,58] with respect to the real hand, the rubber hand illusion is abolished. Similarly, when the participant's real hand is rotated

90 degrees medially while the participant looks at a video image of his or her hand pointing straight ahead, i.e., 90 degrees mismatch between the seen and felt hand positions, the feeling of ownership of the seen hand is eliminated.[55] In addition to orientation congruency, the "anatomical plausibility" of the rubber hand is also important. Ide presented the rubber hand rotated by 0 degrees, 45 degrees, 90 degrees, 180 degrees, 225 degrees, 270 degrees, and 315 degrees and found significantly stronger illusions for angles that were anatomically plausible in that they were easy to mimic with the real hand (0 degrees, 45 degrees, 90 degrees, 315 degrees) than for implausible angles (180 degrees, 225 degrees, 270 degrees).[59] As expected, the strongest illusion was seen for 0 degrees, when the rubber hand was presented in the same orientation as the hidden real hand. In a similar vein, the illusion is extinguished when a left rather than right rubber hand[17] or a right rubber foot rather than a right rubber hand[60] is used in experiments involving the participant's right hand. One recent study even suggests that, with a certain set of experimental procedures, the rubber hand illusion can be elicited solely by the congruent orientation of the seen and felt hands, without any application of synchronous brush stroking[61] (however, no such effect was seen in Ref. 62). Intriguingly, one recent paper argued that a weak rubber hand illusion could be elicited when the rubber hand and the real hand were placed in different postures (palm up vs. palm down).[63] However, the questionnaire ratings of ownership were relatively low for incongruent postures, and no significant behavioral effect was observed on the proprioceptive drift test. Moreover, this study did not directly compare congruent and incongruent hand postures, meaning that the effect of postural congruency was not tested. Thus, collectively, the available data on the matter strongly suggest that the match between the seen and felt orientations of the hands is an important spatial factor in the rubber hand illusion.

Tactile congruence rule

After discussing the effects of spatiotemporal congruence, we turn to congruencies among other stimulus properties, beginning with tactile congruence between the seen and felt touches. A recent study has found that the rubber hand illusion depends on the congruency of the tools used to stroke the rubber hand and the real hand.[64] Tactile incongruence was created by touching the rubber hand with a pencil and the real hand with a paintbrush, and this led to significant reductions in the strength of the illusion. Interestingly, when more similar tools were used that differed only in terms of roughness or smoothness—a paintbrush versus a mascara brush,[64] or a piece of cotton versus a rough sponge[65]—no significant effect on the illusion was observed. It is not entirely clear why no effect was observed for these subtler incongruencies. This observation might reflect the limited sensitivity of the questionnaires and proprioceptive drift tests used to quantify limb ownership, or, more interestingly, it could suggest that there exists a "window of integration" in the dimension of tactile congruency, such that less pronounced incongruencies are tolerated but more substantial incongruencies break the illusion. The tactile congruence rule is based on the multisensory congruency between the seen and felt objects touching the hands in terms of texture and macrogeometric features; thus, similar to multisensory perception in general,[35,36] body ownership is influenced not only by spatiotemporal coincidence but also by congruencies among other stimulus properties.

Humanoid shape rule

The rubber hand illusion also depends on the congruency of the shape and spatial configuration of the observed and felt limbs. When a rigid object is shown, it is clear that the object must visually resemble a human hand for the participants to be able to experience the rubber hand illusion. Objects that do not look like a human hand, such as blocks of wood or sticks, do not elicit significant limb ownership illusions.[17,42,66] One important study tested the rubber hand illusion with a series of five different objects, ranging from a plain wooden block that did not resemble a human limb to handlike wooden objects and a prosthetic hand, and found that the participants experienced a sense of ownership only for the realistic-looking prosthetic hand[66] (Fig. 8.2C). Thus, the object to be experienced as part of one's physical self must resemble a human hand in terms of its shape and structure (at least when it is a rigid physical object,[42] see further below). We know that it is the humanoid shape of the object that matters, rather than the material or color, because the illusion works well with wooden hands,[37] metallic robotic hands,[67] digital images of real hands,[55] and hands of different skin colors.[68] From a multisensory perspective, the humanoid shape rule is another example of multisensory congruency in which the shapes of the seen and felt hands are matching or nonmatching. Alternatively, this constraint could be seen as a top-down effect related to the semantic incongruency[69] between the noncorporeal object in view and the participant's real limb. A third explanation that has been put forward is that the viewed object must fit with a reference model of the body that contains structural information about body parts.[66] However, shape congruency might be the most parsimonious explanation because the rubber hand illusion can be induced in a portion of empty space. In this rather counterintuitive "invisible hand illusion,"[42] the scientist outlines the contours of an invisible hand by moving the brush in empty space straight in front of the participants while corresponding strokes are being applied to the hidden real hand (Fig. 8.1C). If these strokes are very well timed and carefully matched in terms of their spatial trajectories, then the participants will perceive that they have an invisible hand that is sensing the touches of the paintbrush. Importantly, in this case, the illusion relies entirely on the spatiotemporal correlations between vision and touch, even in the face of gross semantic incongruence between the portion of empty space and the participant's real hand. Importantly, however, the shape of the "contours" of the invisible hand and the shape of the real hand are congruent. The illusion presumably works because the brain is used to situations in which the hand cannot be seen, for example, in the dark or when it is occluded behind other objects. In summary, the basic temporal, spatial, and other stimulus property constraints of the rubber hand illusion fit well with the notion that it is a genuine multisensory illusion and that body ownership can be viewed as multisensory perception.

Multisensory congruency matters, not the particular modality

It appears that no single modality plays the all-decisive role in the elicitation of the rubber hand illusion. Rather, it is the congruent patterns of available multisensory signals that drive this perceptual phenomenon. Thus far, we have mainly considered the classic rubber hand illusion, in which the rubber hand is observed visually while tactile stimulation is applied to the model hand and the participant's real hand. Importantly, however, various versions of the rubber hand illusion have been described that depend on different combinations of sensory modalities. For example, the illusion can be elicited without vision, as in the case of the "somatic rubber hand illusion."[70–73] In this nonvisual version, the participant is blindfolded, and the researcher moves the participant's left index finger so that it is touching a right rubber hand placed in front of him or her (Fig. 8.1B). Simultaneously, the researcher is touching the corresponding part of the participant's right hand, which is placed 15 cm to the left of the rubber hand; after a short period of repeated synchronized tapping, this stimulation triggers the illusion that the participant is directly touching his or her own right hand.[70,71] Thus, in this case, it is the correlated tactile and proprioceptive signals from the two hands that are driving the illusion, without any contribution by vision. Similarly, the illusion can be elicited without touches being applied to the skin, as is the case with the "moving rubber hand illusion,"[58] in which a wooden model hand moves its index finger in synchrony with movements made actively or passively by the participant's own hidden index finger (Fig. 8.1D) (see also Ref. 37,74–76). In this situation, it is the congruency of the seen and felt finger movements that elicits the illusion, without dynamic tactile stimulation by an external object in peripersonal space. The illusion seems to be similarly strong for active and passive movements,[37] which suggests that the ownership effect is driven by visuokinesthetic integration. Moreover, this illusion can be elicited when the skin of the moving finger is anesthetized to eliminate tactile feedback from the stretching skin, which suggests that congruent visual and proprioceptive information is sufficient.[77] Importantly, the moving rubber hand illusion and the somatic rubber hand illusion both obey the basic temporal and spatial rules of the classical illusion.[47,58,70,73] Other sensory modalities can also contribute to the sense of limb ownership, as visuo-interoceptive stimulation[78–80] and auditory feedback[81,82] can modulate the strength of illusory ownership, and congruent visual and thermal stimulation can elicit the rubber hand illusion.[83] Thus, body ownership seems to be shaped by the meaningful combination of all available sensory information from the different sensory modalities rather than being predominantly determined by a particular sensory domain, be it vision, touch, or proprioception.

Multisensory integration of body signals in the cortex: nonhuman primates

Before turning to the recent human neuroimaging studies that have investigated the neural substrate of body ownership, we will first consider the neurophysiological literature on nonhuman primates and discuss some key regions that contain cells with receptive field (RF) properties that makes them particularly good candidates to implement the underlying neuronal mechanisms. Specifically, we will focus on multisensory areas in the frontal and parietal association cortices, particularly the premotor cortex, the cortices lining the intraparietal sulcus (the medial, ventral, and lateral intraparietal areas, also known as the MIP, VIP, and LIP), and the inferior parietal cortex (area 7), where electrophysiological studies in macaque monkeys have described single neurons that respond to visual, tactile, and proprioceptive stimulation.[84] Interestingly, many neurons in the ventral premotor cortex respond to a visual stimulus only when it is presented close to the monkey, i.e., within peripersonal space, which extends approximately 30 cm from the body,[85] and these cells typically have overlapping visual and tactile RFs. Further studies showed that these multisensory cells in the premotor cortex encode peripersonal space in body part–centered coordinate systems[50] because these neurons' RFs were anchored to the arm, such that when the arm moved, the visual RFs of the multisensory neurons moved along with it,[49,50,86] and because the RFs were independent of the position of the monkey's gaze.[50,87] Further studies have revealed a number of frontal and parietal areas with multisensory neurons that show visual and sometimes also auditory RFs with extension limited to the space surrounding the monkey's body; these areas include the VIP,[88–93] parietal area 7b,[88,94] the putamen,[95] the secondary somatosensory cortex,[96] and parietal areas 2 and 5,[97,98] in addition to the abovementioned ventral premotor cortex. Importantly, we know that these kinds of cells can perform multisensory integration[89,97] and that they seem to do so according to the temporal and spatial congruency rules.[89] Moreover, neurons in the premotor cortex and the medial intraparietal area not only respond to dynamic tactile stimuli applied to the body but also show sensitivity to the orientation congruency of seen and felt postures of the upper limb.[97,99] Thus, areas in the frontal and parietal association cortices contain multisensory neurons that obey the temporal and spatial congruency principles of multisensory integration, which makes them good candidates to implement the perceptual rules of body ownership.

Multisensory integration of body signals in the cortex: human neuroimaging

Accumulating neuroimaging studies on healthy volunteers suggests that circuits for multisensory integration of body-related signals in peripersonal space also exist in the human brain. fMRI studies have identified areas in the premotor cortex and intraparietal cortex that respond to both visual and tactile stimulation in relation to specific body parts[14,55,100–104] and to visual stimulation near the hand in peripersonal space.[56,100,105] In two fMRI experiments,[55,104] Gentile and colleagues consistently observed strengthened activation in the ventral premotor cortex, intraparietal cortex, inferior parietal cortex (supramarginal gyrus), and cerebellum in a condition in which the real right index finger was stroked with a small soft ball attached to a stick in the participant's sight (congruent visuotactile condition), compared with (i) when the participants closed their eyes during the application of the strokes (unimodal tactile stimulation) (Fig. 8.3A), (ii) when the small ball was moved 2 cm above the index finger without touching it (unimodal visual stimulation) (Fig. 8.3A), or (iii) when temporally or spatially incongruent visuotactile stimulation was applied (Fig. 8.3A and B). Thus, these regions integrate visual and tactile signals from the upper limb. Moreover, multisensory neuronal populations in several of these areas show selectivity in their response profiles for visual stimuli presented near the hand, consistent with the notion that these groups of cells have RFs restricted to peripersonal space near the hand ("perihand space"). Brozzoli and colleagues used the blood oxygenation level-dependent (BOLD) adaptation technique (i.e., the suppression of BOLD signal related to the repetition of the same stimulus, which reveals neuronal stimulus selectivity at the population level[106]) to show that the active areas in the ventral and dorsal premotor cortex, intraparietal cortex, and inferior parietal cortex (supramarginal cortex) show neural selectivity for visual stimuli presented near the hand[105] and that the premotor and intraparietal responses appeared to be anchored to the hand such that when the hand was placed in a new location, the selective neural responses shifted along with the hand[56] (Fig. 8.3C). Thus, the human premotor and

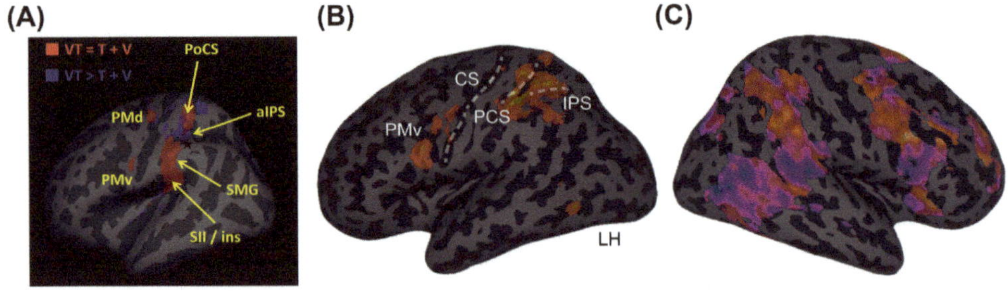

FIGURE 8.3 Areas of the premotor cortex and posterior parietal cortex that show greater fMRI activation during congruent visuotactile stimulation applied to the right hand than during unimodal visual or unimodal tactile stimulation[104] (A). The ventral premotor cortex (PMv) and the cortices lining the intraparietal sulcus (IPS) show greater activation for congruent visuotactile stimulation of the right hand than for temporally and spatially incongruent visuotactile stimulation[55] (B). Areas that show (fMRI BOLD adaptation) selectivity for visual stimulation in space near the hand (yellow) compared with areas that display a lack of such a hand-centered response profile (blue)[56] (C).

intraparietal cortices (and other regions) respond to spatially and temporally congruent multisensory stimulation from the body and the space near the body, in line with the single-cell data from nonhuman primates. In the next section, we will consider how activity in these areas relates to body ownership.

Neuroimaging studies of limb ownership

Neuroimaging experiments conducted in different laboratories over the last 5 years have provided accumulating evidence that the sense of ownership of a limb is associated with activation of multisensory areas in the frontal and parietal lobes.[40,42,55,107–109] Ehrsson and colleagues conducted the first fMRI study of the rubber hand illusion, and these authors found increased BOLD signal in the premotor cortex and intraparietal cortex during the rubber hand illusion compared with relevant control conditions with visuotactile asynchrony and/or spatial incongruency in terms of seen and felt hand orientation[14] (Fig. 8.4A). Moreover, the subjectively rated strength of the illusion across individuals predicted the degree of activity in the bilateral ventral premotor cortex. The activation of the ventral premotor cortex[40,42,55,56,107,108,110] and intraparietal cortex[42,55,56,107,108] has been reproduced in fMRI studies involving various versions of the rubber hand illusion (Fig. 8.4C and D) including the invisible rubber hand illusion[42] and the somatic rubber hand illusion.[70] Interestingly, activity in the bilateral ventral premotor cortex and the bilateral intraparietal cortex, as well as the bilateral inferior parietal cortex (supramarginal gyrus) and the right cerebellum, also reflects the default sense of ownership when participants look at their real hand being stroked, and this activity is reduced when spatial and temporal incongruencies between the visual and tactile stimuli are introduced, which also causes the participants to lose the sense of ownership of their hand[55] (Fig. 8.4B). The stronger this "hand disownership" effect, the greater the reduction in the fMRI signal in the left ventral premotor cortex and right intraparietal cortex. Thus, the activity in the premotor cortex and the intraparietal cortex is not restricted to the cases of illusory ownership of rubber hands but also reflects ownership of real hands.

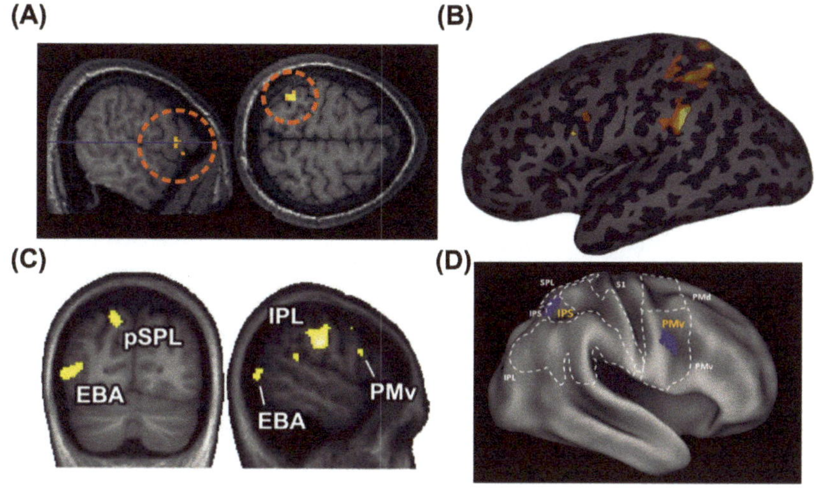

FIGURE 8.4 fMRI activity in the ventral premotor cortex (left) and intraparietal cortex (right) that reflects the sense of ownership of a rubber hand[14] (A). Activation in the ventral premotor cortex, intraparietal cortex, and inferior parietal cortex that reflects the default sense of ownership of one's real hand[55] (B). Ownership of the upper limb driven by orientation congruency of the observed and felt hand is associated with activation in the ventral premotor cortex (PMv), the inferior parietal lobule (IPL), the extrastriate body area (EBA), and the medial aspect of the posterior part of the intraparietal sulcus (here termed the posterior superior parietal lobe, pSPL)[107] (C). Metaanalysis of neuroimaging studies that associate activation (blue) in the ventral premotor cortex (PMv) and intraparietal cortex (IPS) with body ownership in various illusion-based paradigms[109] (D).

fMRI has also been used to investigate whether the rubber hand illusion is associated with a dynamic shift in peripersonal space—from the hidden real hand toward the model hand in view, which is indicated by changes in the visual RFs of a small number of neurons recorded from the posterior parietal cortex when monkeys observed a stuffed monkey arm being stroked in synchrony with strokes applied to the monkey's hidden real arm.[97] In a study by Brozzoli and colleagues,[56] a small ball was repeatedly presented close to the illusorily owned rubber hand, and the BOLD adaptation responses selective to this visual stimulus in perihand space around the upper limb were quantified and compared with a control condition without illusory ownership. The responses indicative of a selective near-hand response in the bilateral ventral premotor cortex, bilateral putamen, and right intraparietal cortex were greater during the rubber hand illusion than without the illusion,[56] an observation that is well in line with the notion that, when the rubber hand is owned, the multisensory neurons encode peripersonal space around the model hand just as they do around the real hand under natural conditions.[56,97,105] Furthermore, the greater the subjective strength of the illusion, the stronger this neuronal shift in perihand space in the ventral premotor cortex,[56] linking the subjective illusion phenomenon to dynamic shifts in the RF properties of multisensory neurons in this area.

In addition to the premotor cortex and the intraparietal cortex as discussed above, recent neuroimaging studies of limb ownership have also consistently found activation in an area tentatively identified as the extrastriate body area (EBA), as well as in the cerebellum. The

EBA is a region located in the lateral occipital cortices (LOCs) that is specialized for the visual processing of body parts.[111] Earlier studies suggested that the visual processing in this region might be influenced by somatosensory information because modulations of LOC activity have been seen during tactile[112–114] and proprioceptive[115,116] stimulation procedures. These crossmodal influences likely stem from sources in the posterior parietal cortex,[117] which is functionally connected to the LOC.[118] In terms of body ownership, several studies have reported significant increases in activation in the LOC during versions of the rubber hand illusion paradigm[42,55,107,108,119] (Fig. 8.4C), and the stronger the subjective ownership illusion, the stronger the LOC activation.[55,107] The 2014 study by Limanowski and Blankenburg[119] is important in this respect because they used a well-established functional localizer procedure to identify the EBA in their particular group of participants, which enabled them to demonstrate that the activity associated with the rubber hand illusion was indeed located in the EBA. One interpretation of this EBA activity is that it reflects crossmodal interplay,[120] whereby congruent tactile and proprioceptive signals influence the visual processing of hand signals via top-down modulations from posterior parietal regions in a way that potentially aids visual self-recognition of the hand. This notion is consistent with the increased effective connectivity found between the intraparietal cortex and the LOC when participants experience ownership of a limb in view.[55,108]

There are also compelling indications that the cerebellum is involved in the rubber hand illusion (not shown in the figures). Several experiments have found activation of the lateral cerebellar hemispheres in the integration of spatially and temporally congruent multisensory signals from the hand in various versions of the rubber hand illusion paradigm.[14,42,55,70,108] This part of the cerebellum receives proprioceptive,[121] visual,[122] and tactile[123] inputs and is responsive to congruent multisensory stimulation of the right upper limb.[104,124] Interestingly, there is increased effective connectivity between this cerebellar region and the intraparietal cortex when people experience ownership of a limb,[55] an observation that is consistent with known anatomical connections between these areas.[122,125] Thus, given the important role of the cerebellum in generating sensory predictions,[126] it can be speculated that the cerebellum is involved in the detection of synchronous multisensory signals and the formation of crossmodal temporal predictions that support the multisensory processing in frontoparietal cortical areas.[55]

Full-body ownership

Thus far, we have considered the sense of ownership of a single limb only. However, the rubber hand illusion can be extended to the entire body, and such studies provide additional information about the relationship between body ownership and multisensory processing. Petkova and Ehrsson described how a full-body ownership illusion could be elicited with the body of a mannequin.[127] To observe the mannequin's body from the natural first-person perspective, the participants wore head-mounted displays (HMDs) connected to two cameras placed side by side to provide a 3D video image feed to the HMDs (Fig. 8.5A). The two cameras were positioned on the head of a mannequin so that the participants were looking down on the mannequin's body with stereoscopic vision.[127] Thus, when the participants wore the HMDs connected to these cameras and looked down, they saw the mannequin's body in the location to that where they would expect to see their own real body

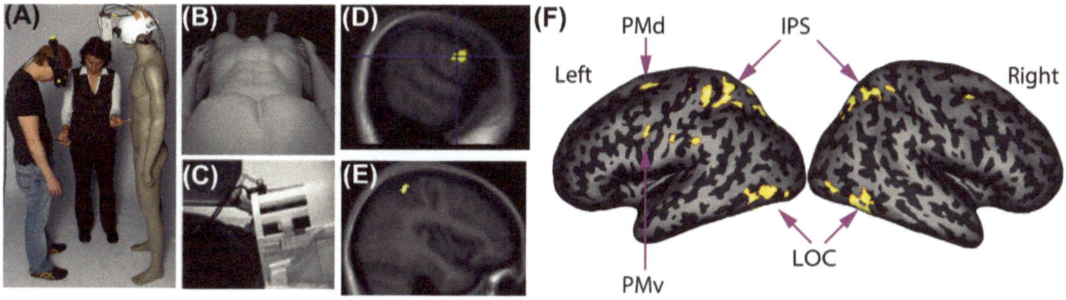

FIGURE 8.5 Elicitation of the full-body ownership illusion with a mannequin, head-mounted displays (HMDs), and synchronized touches[127] (A). In the HMDs, the participants observed the mannequin from the first-person perspective (B), as shown here from an fMRI experiment that associated activation in the ventral premotor cortex (D) and intraparietal cortex (E) with the full-body ownership illusion.[136] The arrangement of the HMDs, the tilted head coil, and the participant's head that is tilted inside the head coil to simulate a natural view of the body during the fMRI experiments (C). Activation associated with illusory full-body ownership of a stranger's body is shown projected onto a "flattened" 3D image of a standard brain[109] (F). *IPS*, Intraparietal sulcus; *PMv*: Ventral premotor cortex; *PMd*, Dorsal premotor cortex; *LOC*, Lateral occipital cortex.

(Fig. 8.5B). When the experimenter used a pair of rods to touch the mannequin's abdomen and the person's real abdomen simultaneously at corresponding sites for 1 minute, this elicited in participants the illusion that the mannequin's body was their own.[127] This effect was quantified with questionnaires and by registering the skin conductance responses when the participants observed a knife cutting the belly of the mannequin.[127] The illusion is robust and works just as well with real human bodies[128–130] and computer-generated avatars[131,132] as with mannequins.

The full-body ownership illusion seems to depend on the same perceptual rules as the rubber hand illusion. Asynchronous stroking of the mannequin body and the participant's real body significantly reduces the illusion, and anecdotal observations suggest that this reducing effect of asynchrony is strengthened when combined with stroking noncorresponding body parts.[133] Furthermore, the mannequin viewed from the first-person point of view needs to be presented in the same orientation in peripersonal space as the real body, so that the body one sees matches the body one senses through proprioception. Presenting the mannequin's body several meters away from the person's real body viewed from the third-person perspective (as one sees another individual[134,135]; see also Ref. 132) or presenting the mannequin in peripersonal space but rotating it so that the head is placed near participants' unseen feet[136] significantly reduces the illusion. This orientation congruency effect is so strong that the illusion can be elicited even without the applications of synchronous touches, especially if the spatial match is so close that the artificial body is "substituted" for the real body.[132] Moreover, the illusion is eliminated when a large block of wood is presented instead of a mannequin,[127,136] in line with the humanoid shape rule; however, as with the invisible hand illusion described above, it is possible to elicit an "invisible-body illusion" if the experimenter very carefully outlines the "contours" of an invisible body by systematically stroking different body parts.[137] Furthermore, as with the rubber hand illusion, the full-body ownership illusion can be elicited by congruent seen and felt movements instead of touches applied with probes or brushes.[127,132,138,139] Thus, the full-body ownership illusion depends on the same temporal, spatial, and humanoid shape congruency principles as the rubber hand illusion described earlier, suggesting the involvement of similar multisensory processes.

This conclusion is further supported by fMRI studies describing how the full-body ownership illusion is associated with activity increases in the ventral premotor cortex[128,136,140] (Fig. 8.5D and F), the intraparietal cortex[128,136,140] (Fig. 8.5E and F), the LOC[128,140] (Fig. 8.5F), the lateral cerebellar hemispheres,[128,136] and the putamen[128,136] compared with various control conditions with temporally and spatially incongruent visuotactile stimulation or with the mannequin replaced by a block of wood or with the mannequin presented from a third-person perspective. Moreover, the degree of activation in the ventral premotor cortex correlates with the strength of the full-body ownership illusion as rated in questionnaires.[128,136,140] Thus, both the perceptual rules and the patterns of brain activation support the notion that similar multisensory processes are involved in limb and full-body ownership.

Still, how are the parts integrated into the whole in own-body perception? One experiences a single coherent body, not a set of disconnected parts, as one's own physical self. The full-body ownership paradigm has been used to address this question. A striking feature of the full-body illusion is that one experiences the *entire* body in view as one's own, not merely the particular part that receives the dynamic visuotactile stimulation.[127,141] This effect can be studied by stimulating one body part (e.g., the right hand) and then probing ownership of specific other body parts (e.g., the abdomen) by using questionnaire ratings or threat-evoked skin conductance responses. The results show how ownership "spreads" from the body part that is stroked to also encompass the others, in line with the entire mannequin's body being experienced as one's own.[127,141] This unitary experience probably requires multisensory perceptual binding of the different limbs and body segments into a single perceptual whole.

In terms of a possible brain mechanism for such full-body perceptual binding, neuronal populations with the capacity to integrate visual, tactile, and proprioceptive signals across body segments might play a key role. Multisensory neurons with RFs that cover two or more body segments have been described in the ventral premotor cortex,[49,85,142] the cortices lining the intraparietal sulcus,[92,143] and the inferior parietal cortex[144]; even neurons that respond to the entire body have been reported.[92,142,144,145] fMRI has been used to search for neural correlates of whole-body perceptual binding effects.[136,141] One approach is to compare the synchronous versus asynchronous stroking of the mannequin's hand in two contexts: either the mannequin's hand was visibly attached to the rest of the mannequin, which triggers the full-body ownership illusion when synchronous stroking is added, or the artificial hand was detached from the mannequin and simply displayed in isolation without any mannequin next to it, which is a condition that does not elicit a full-body ownership illusion even with synchronous visuotactile stimulation.[136] The former condition, with the artificial hand attached to the mannequin's body and the experience of illusory full-body ownership, was associated with increased activation in sections of the dorsal and ventral premotor cortices, the intraparietal cortex, the secondary somatosensory cortex, and the lateral cerebellum. This finding suggests that whole-body multisensory binding requires additional neural processing in these regions over and above the multisensory processing related to the "detached" hand. A multivariate pattern recognition technique called "multivoxel pattern analysis" has been used to further examine how neuronal population responses relate to full-body perceptual binding.[136,141] In these experiments, classifiers were trained to identify fine-grained patterns of activity that were similar regardless of whether the full-body ownership illusion with a mannequin was elicited by synchronously stimulating the abdomen or the right hand,[136] or by stimulating the abdomen, the right hand, or the right foot[141] (compared with control conditions with asynchronous stimulation of the corresponding body parts). Patterns of active voxels that contained information about full-body ownership irrespective of which body part was used to elicit the illusion were identified in the ventral premotor cortex,[136,141] which supports the idea that whole-body ownership percepts are associated with the integration of multisensory signals originating from multiple segments of the body.

Self-identification, mirrors, and the third-person visual perspective

In this chapter, we have focused on the rubber hand illusion and the full-body ownership illusion when the mannequin body is viewed from the first-person perspective. In these cases, vivid and explicit body ownership is experienced, and the initial conflict between the senses is resolved by the formation of a coherent illusory percept of the artificial hand or plastic body in view as one's own. However, multisensory interactions can also influence self-recognition of faces and bodies observed at a distance. When participants are stroked on their face while they are looking at a morphed face on a computer screen[146] or another person's face[147] being touched in synchrony, this visuotactile stimulation promotes recognition of the face as their own compared with asynchronous conditions ("enfacement illusion"). Similarly, synchronous stimulation of the participant's real back and the back of a virtual body presented several meters in front of the participant through an HMD makes the participant self-identify with the virtual body,[148] and this is accompanied by behavioral changes in indices of body representation and peripersonal space.[149–152] These "enfacement illusions" and "full-body illusions" for bodies viewed at a distance from a third-person perspective are very interesting in their own right and have been used to address a wide range of issues related to self-identification (or self-recognition), bodily self-consciousness, and sense of self.[30,153] However, these findings do not contradict the spatial constraints of body ownership that we have discussed in this chapter because even if illusory self-recognition is boosted by visuotactile correlations, the participants still sense the spatial discrepancy between their own body as sensed from the first-person perspective and the artificial body or other person's face they observe at a distance, that is, the perceived spatial conflict is not eliminated[154] and a full-blown body ownership illusion is not experienced as in the rubber hand illusion and full-body ownership illusions from the first-person perspective.[132,134] Interestingly, the rubber hand illusion and the full-body ownership illusion work when the participant sees the rubber hand,[155] the mannequin,[135] or a virtual body[139,156,157] in a mirror placed straight in front of him or her. However, these observations do not falsify the spatial constraints of body ownership, as, in these cases, the visual information from the mirror reflection is automatically projected back to the participant's own body standing in front of the mirror. Importantly, if the mirror is removed and replaced by a rubber hand or mannequin facing the participant, the ownership illusion is significantly reduced, which demonstrates that the mirror transformation plays a critical role in perceived ownership of the body reflected in the mirror.[135,155]

Summary

This chapter has discussed multisensory processes in body ownership. A large body of behavioral and neuroimaging data supports a close relationship between multisensory perception and the sense of ownership of limbs and entire bodies. From behavioral experiments, we have learned that the rubber hand illusion and other limb ownership illusions obey temporal, spatial, and other congruency rules that are related to the properties of the stimuli, which fits well with the congruency principles of multisensory integration. Moreover, illusory changes in body ownership do not seem to depend on, or to be dominated by, a single particular modality; rather, such illusions are the outcome of a flexible integration process in which all available information from the different sensory modalities is used. This is illustrated, for example, by the many different versions of the rubber hand illusion that are based on different combinations of multisensory stimuli. In terms of the human brain, fMRI studies associate changes in the sense of body ownership with increases in activity in multisensory cortical areas such as the premotor cortex and posterior parietal cortex. Importantly, the degree of activity in these areas mirrors the perceptual rules of the rubber hand illusion and shows a systematic relationship with the subjective strength of experienced ownership. The behavioral and neural principles of body ownership have also been extended to the case of the entire body with full-body ownership illusions in which people experience the bodies of mannequins, strangers, and simulated avatars as their own. Collectively, the reviewed studies not only show that multisensory processes play an important role in how we come to experience our body as our own but also suggest that body ownership can be explained as the formation of a coherent multisensory percept of one's body by multisensory integration mechanisms.

Acknowledgments

Henrik Ehrsson is supported by the Swedish Research Council, Hjärnfonden, and Torsten Söderbergs Stiftelse.

References

1. Botvinick M. Neuroscience. Probing the neural basis of body ownership. *Science*. 2004;305(5685):782–783.
2. Ehrsson HH. The concept of body ownership and its relation to multisensory integration. In: Stein BE, ed. *The Handbook of Multisensory Processes*. Cambridge, Massachusetts: MIT Press; 2012:775–792.
3. Tsakiris M. My body in the brain: a neurocognitive model of body-ownership. *Neuropsychologia*. 2010;48(3):703–712.
4. Jeannerod M. The mechanism of self-recognition in humans. *Behav Brain Res*. 2003;142(1–2):1–15.
5. Gallagher II . Philosophical conceptions of the self: implications for cognitive science. *Trends Cogn Sci*. 2000;4(1):14–21.
6. Jenkinson PM, Moro V, Fotopoulou A. Definition: asomatognosia. *Cortex*. 2018;101:300–301.
7. Critchley M. *The Parietal Lobes*. 1953. London.
8. Arzy S, Overney LS, Landis T, Blanke O. Neural mechanisms of embodiment: asomatognosia due to premotor cortex damage. *Arch Neurol*. 2006;63(7):1022–1025.
9. Bottini G, Bisiach E, Sterzi R, Vallar G. Feeling touches in someone else's hand. *Neuroreport*. 2002;13(2):249–252.
10. Feinberg TE, Venneri A, Simone AM, Fan Y, Northoff G. The neuroanatomy of asomatognosia and somatoparaphrenia. *J Neurol Neurosurg Psychiatry*. 2010;81(3):276–281.
11. Feinberg TE, Venneri A. Somatoparaphrenia: evolving theories and concepts. *Cortex*. 2014;61:74–80.

12. Botvinick M, Cohen J. Rubber hands 'feel' touch that eyes see. *Nature*. 1998;391(6669):756.
13. Tastevin J. En partant de l'experience d'Aristotle [Starting from Aristotle's experience]. *L'Encephale*. 1937;1(57–84):140–158.
14. Ehrsson HH, Spence C, Passingham RE. That's my hand! Activity in premotor cortex reflects feeling of ownership of a limb. *Science*. 2004;305(5685):875–877.
15. Lloyd DM. Spatial limits on referred touch to an alien limb may reflect boundaries of visuo-tactile peripersonal space surrounding the hand. *Brain Cogn*. 2007;64(1):104–109.
16. Longo MR, Schuur F, Kammers MP, Tsakiris M, Haggard P. What is embodiment? A psychometric approach. *Cognition*. 2008;107(3):978–998.
17. Tsakiris M, Haggard P. The rubber hand illusion revisited: visuotactile integration and self-attribution. *J Exp Psychol Hum Percept Perform*. 2005;31(1):80–91.
18. Armel KC, Ramachandran VS. Projecting sensations to external objects: evidence from skin conductance response. *Proc Biol Sci*. 2003;270(1523):1499–1506.
19. Abdulkarim Z, Ehrsson HH. No causal link between changes in hand position sense and feeling of limb ownership in the rubber hand illusion. *Atten Percept Psychophys*. 2016;78(2):707–720.
20. Rohde M, Wold A, Karnath HO, Ernst MO. The human touch: skin temperature during the rubber hand illusion in manual and automated stroking procedures. *PLoS One*. 2013;8(11):e80688.
21. Rohde M, Di Luca M, Ernst MO. The rubber hand illusion: feeling of ownership and proprioceptive drift do not go hand in hand. *PLoS One*. 2011;6(6):e21659.
22. Zopf R, Savage G, Williams MA. The crossmodal congruency task as a means to obtain an objective behavioral measure in the rubber hand illusion paradigm. *J Vis Exp*. 2013;77.
23. de Haan AM, Van Stralen HE, Smit M, Keizer A, Van der Stigchel S, Dijkerman HC. No consistent cooling of the real hand in the rubber hand illusion. *Acta Psychol (Amst)*. 2017;179:68–77.
24. Craske B. Perception of impossible limb positions induced by tendon vibration. *Science*. 1977;196(4285):71–73.
25. Lackner JR. Some proprioceptive influences on the perceptual representation of body shape and orientation. *Brain*. 1988;111(Pt 2):281–297.
26. Ramachandran VS, Hirstein W. Three laws of qualia: clues from neurology about the biological functions of consciousness and qualia. *J Conscious Stud*. 1997;4:28.
27. Naito E, Roland PE, Ehrsson HH. I feel my hand moving: a new role of the primary motor cortex in somatic perception of limb movement. *Neuron*. 2002;36(5):979–988.
28. de Vignemont F. Embodiment, ownership and disownership. *Conscious Cognit*. 2011;20(1):82–93.
29. de Vignemont F. Agency and bodily ownership: the bodyguard hypothesis. In: de Vignemont F, Alsmith A, eds. *The Subject's Matter: The Body and Self-Consciousness*. Cambridge, Massachusetts, USA: MIT Press; 2017.
30. Tsakiris M. The multisensory basis of the self: from body to identity to others. *Q J Exp Psychol (Hove)*. 2017;70(4):597–609.
31. Limanowski J, Blankenburg F. Minimal self-models and the free energy principle. *Front Hum Neurosci*. 2013;7:547.
32. Apps MA, Tsakiris M. The free-energy self: a predictive coding account of self-recognition. *Neurosci Biobehav Rev*. 2014;41:85–97.
33. Holmes NP, Spence C. Multisensory integration: space, time and superadditivity. *Curr Biol*. 2005;15(18):R762–R764.
34. Stein BE, Stanford TR. Multisensory integration: current issues from the perspective of the single neuron. *Nat Rev Neurosci*. 2008;9(4):255–266.
35. *The New Handbook of Multisensory Processing*. Cambridge, Massachusetts: The MIT Press; 2012.
36. Chen YC, Spence C. Assessing the role of the 'unity assumption' on multisensory integration: a review. *Front Psychol*. 2017;8:445.
37. Kalkert A, Ehrsson HH. The moving rubber hand illusion revisited: comparing movements and visuotactile stimulation to induce illusory ownership. *Conscious Cognit*. 2017;26:117–132.
38. Shimada S, Fukuda K, Hiraki K. Rubber hand illusion under delayed visual feedback. *PLoS One*. 2009;4(7):e6185.
39. Shimada S, Suzuki T, Yoda N, Hayashi T. Relationship between sensitivity to visuotactile temporal discrepancy and the rubber hand illusion. *Neurosci Res*. 2014;85:33–38.
40. Bekrater-Bodmann R, Foell J, Diers M, et al. The importance of synchrony and temporal order of visual and tactile input for illusory limb ownership experiences - an FMRI study applying virtual reality. *PLoS One*. 2014;9(1):e87013.

41. Costantini M, Robinson J, Migliorati D, Donno B, Ferri F, Northoff G. Temporal limits on rubber hand illusion reflect individuals' temporal resolution in multisensory perception. *Cognition*. 2016;157:39–48.
42. Guterstam A, Gentile G, Ehrsson HH. The invisible hand illusion: multisensory integration leads to the embodiment of a discrete volume of empty space. *J Cogn Neurosci*. 2013;25(7):1078–1099.
43. Petkova VI, Ehrsson HH. When right feels left: referral of touch and ownership between the hands. *PLoS One*. 2009;4(9):e6933.
44. Parise CV, Ernst MO. Correlation detection as a general mechanism for multisensory integration. *Nat Commun*. 2016;7.
45. Parise CV, Spence C, Ernst MO. When correlation implies causation in multisensory integration. *Curr Biol*. 2012;22(1):46–49.
46. Burr D, Silva O, Cicchini GM, Banks MS, Morrone MC. Temporal mechanisms of multimodal binding. *Proc R Soc B Biol Sci*. 2009;276(1663):1761–1769.
47. Kalckert A, Ehrsson HH. The spatial distance rule in the moving and classical rubber hand illusions. *Conscious Cognit*. 2014;30:118–132.
48. Rizzolatti G, Fadiga L, Fogassi L, Gallese V. The space around us. *Science*. 1997;277(5323):190–191.
49. Fogassi L, Gallese V, Fadiga L, Luppino G, Matelli M, Rizzolatti G. Coding of peripersonal space in inferior premotor cortex (area F4). *J Neurophysiol*. 1996;76(1):141–157.
50. Graziano MS, Hu XT, Gross CG. Visuospatial properties of ventral premotor cortex. *J Neurophysiol*. 1997;77(5):2268–2292.
51. Ladavas E, di Pellegrino G, Farne A, Zeloni G. Neuropsychological evidence of an integrated visuotactile representation of peripersonal space in humans. *J Cogn Neurosci*. 1998;10(5):581–589.
52. Preston C. The role of distance from the body and distance from the real hand in ownership and disownership during the rubber hand illusion. *Acta Psychol (Amst)*. 2013;142(2):177–183.
53. Kilteni K, Normand JM, Sanchez-Vives MV, Slater M. Extending body space in immersive virtual reality: a very long arm illusion. *PLoS One*. 2012;7(7):e40867.
54. Costantini M, Haggard P. The rubber hand illusion: sensitivity and reference frame for body ownership. *Conscious Cognit*. 2007;16(2):229–240.
55. Gentile G, Guterstam A, Brozzoli C, Ehrsson HH. Disintegration of multisensory signals from the real hand reduces default limb self-attribution: an fMRI study. *J Neurosci*. 2013;33(33):13350–13366.
56. Brozzoli C, Gentile G, Ehrsson HH. That's near my hand! Parietal and premotor coding of hand-centered space contributes to localization and self-attribution of the hand. *J Neurosci*. 2012;32(42):14573–14582.
57. Pavani F, Spence C, Driver J. Visual capture of touch: out-of-the-body experiences with rubber gloves. *Psychol Sci*. 2000;11(5):353–359.
58. Kalckert A, Ehrsson HH. Moving a rubber hand that feels like your own: a dissociation of ownership and agency. *Front Hum Neurosci*. 2012;6:40.
59. Ide M. The effect of "anatomical plausibility" of hand angle on the rubber-hand illusion. *Perception*. 2013;42(1):103–111.
60. Guterstam A, Petkova VI, Ehrsson HH. The illusion of owning a third arm. *PLoS One*. 2011;6(2):e17208.
61. Samad M, Chung AJ, Shams L. Perception of body ownership is driven by Bayesian sensory inference. *PLoS One*. 2015;10(2):e0117178.
62. Guterstam A, Zeberg H, Ozciftci VM, Ehrsson HH. The magnetic touch illusion: a perceptual correlate of visuotactile integration in peripersonal space. *Cognition*. 2016;155:44–56.
63. Ionta S, Sforza A, Funato M, Blanke O. Anatomically plausible illusory posture affects mental rotation of body parts. *Cognit Affect Behav Neurosci*. 2013;13(1):197–209.
64. Ward J, Mensah A, Junemann K. The rubber hand illusion depends on the tactile congruency of the observed and felt touch. *J Exp Psychol Hum Percept Perform*. 2015;41(5):1203–1208.
65. Schutz-Bosbach S, Tausche P, Weiss C. Roughness perception during the rubber hand illusion. *Brain Cogn*. 2009;70(1):136–144.
66. Tsakiris M, Carpenter L, James D, Fotopoulou A. Hands only illusion: multisensory integration elicits sense of ownership for body parts but not for non-corporeal objects. *Exp Brain Res*. 2010;204(3):343–352.
67. Rosen B, Ehrsson HH, Antfolk C, Cipriani C, Sebelius F, Lundborg G. Referral of sensation to an advanced humanoid robotic hand prosthesis. *Scand J Plast Reconstr Surg Hand Surg*. 2009;43(5):260–266.

68. Farmer H, Tajadura-Jimenez A, Tsakiris M. Beyond the colour of my skin: how skin colour affects the sense of body-ownership. *Conscious Cognit.* 2012;21(3):1242–1256.
69. Hein G, Doehrmann O, Muller NG, Kaiser J, Muckli L, Naumer MJ. Object familiarity and semantic congruency modulate responses in cortical audiovisual integration areas. *J Neurosci.* 2007;27(30):7881–7887.
70. Ehrsson HH, Holmes NP, Passingham RE. Touching a rubber hand: feeling of body ownership is associated with activity in multisensory brain areas. *J Neurosci.* 2005;25(45):10564–10573.
71. Petkova VI, Zetterberg H, Ehrsson HH. Rubber hands feel touch, but not in blind individuals. *PLoS One.* 2012;7(4):e35912.
72. White RC, Aimola Davies AM, Halleen TJ, Davies M. Tactile expectations and the perception of self-touch: an investigation using the rubber hand paradigm. *Conscious Cognit.* 2010;19(2):505–519.
73. Aimola Davies AM, White RC, Davies M. Spatial limits on the nonvisual self-touch illusion and the visual rubber hand illusion: subjective experience of the illusion and proprioceptive drift. *Conscious Cognit.* 2013;22(2):613–636.
74. Dummer T, Picot-Annand A, Neal T, Moore C. Movement and the rubber hand illusion. *Perception.* 2009;38(2):271–280.
75. Tsakiris M, Prabhu G, Haggard P. Having a body versus moving your body: how agency structures body-ownership. *Conscious Cognit.* 2006;15(2):423–432.
76. Sanchez-Vives MV, Spanlang B, Frisoli A, Bergamasco M, Slater M. Virtual hand illusion induced by visuomotor correlations. *PLoS One.* 2010;5(4):e10381.
77. Walsh LD, Moseley GL, Taylor JL, Gandevia SC. Proprioceptive signals contribute to the sense of body ownership. *J Physiol.* 2011;589(Pt 12):3009–3021.
78. Suzuki K, Garfinkel SN, Critchley HD, Seth AK. Multisensory integration across exteroceptive and interoceptive domains modulates self-experience in the rubber-hand illusion. *Neuropsychologia.* 2013;51(13):2909–2917.
79. van Stralen HE, van Zandvoort MJ, Hoppenbrouwers SS, Vissers LM, Kappelle LJ, Dijkerman HC. Affective touch modulates the rubber hand illusion. *Cognition.* 2014;131(1):147–158.
80. Crucianelli L, Metcalf NK, Fotopoulou AK, Jenkinson PM. Bodily pleasure matters: velocity of touch modulates body ownership during the rubber hand illusion. *Front Psychol.* 2013;4:703.
81. Radziun D, Ehrsson HH. Auditory cues influence the rubber hand illusion. *J Exp Psychol Hum Percept Perform.* 2017 (in press).
82. Darnai G, Szolcsanyi T, Hegedus G, et al. Hearing visuo-tactile synchrony - sound-induced proprioceptive drift in the invisible hand illusion. *Br J Psychol.* 2017;108(1):91–106.
83. Trojan J, Fuchs X, Speth SL, Diers M. The rubber hand illusion induced by visual-thermal stimulation. *Sci Rep.* 2018;8(1):12417.
84. Graziano MS, Gross CG, Taylor CSR, Moore T. A system of multimodal areas in the primate brain. In: Spence C, Driver J, eds. *Crossmodal Space and Crossmodal Attention.* Oxford, UK: Oxford University Press; 2004:51–67.
85. Rizzolatti G, Scandolara C, Matelli M, Gentilucci M. Afferent properties of periarcuate neurons in macaque monkeys. II. Visual responses. *Behav Brain Res.* 1981;2(2):147–163.
86. Graziano MS, Yap GS, Gross CG. Coding of visual space by premotor neurons. *Science.* 1994;266(5187):1054–1057.
87. Gentilucci M, Scandolara C, Pigarev IN, Rizzolatti G. Visual responses in the postarcuate cortex (area 6) of the monkey that are independent of eye position. *Exp Brain Res.* 1983;50(2–3):464–468.
88. Ishida H, Nakajima K, Inase M, Murata A. Shared mapping of own and others' bodies in visuotactile bimodal area of monkey parietal cortex. *J Cogn Neurosci.* 2010;22(1):83–96.
89. Avillac M, Ben Hamed S, Duhamel JR. Multisensory integration in the ventral intraparietal area of the macaque monkey. *J Neurosci.* 2007;27(8):1922–1932.
90. Avillac M, Deneve S, Olivier E, Pouget A, Duhamel JR. Reference frames for representing visual and tactile locations in parietal cortex. *Nat Neurosci.* 2005;8(7):941–949.
91. Colby CL, Duhamel JR, Goldberg ME. Ventral intraparietal area of the macaque: anatomic location and visual response properties. *J Neurophysiol.* 1993;69(3):902–914.
92. Duhamel JR, Colby CL, Goldberg ME. Ventral intraparietal area of the macaque: congruent visual and somatic response properties. *J Neurophysiol.* 1998;79(1):126–136.
93. Schlack A, Sterbing-D'Angelo SJ, Hartung K, Hoffmann KP, Bremmer F. Multisensory space representations in the macaque ventral intraparietal area. *J Neurosci.* 2005;25(18):4616–4625.

94. Duhamel JR, Bremmer F, BenHamed S, Graf W. Spatial invariance of visual receptive fields in parietal cortex neurons. *Nature*. 1997;389(6653):845–848.
95. Graziano MS, Gross CG. A bimodal map of space: somatosensory receptive fields in the macaque putamen with corresponding visual receptive fields. *Exp Brain Res*. 1993;97(1):96–109.
96. Hihara S, Taoka M, Tanaka M, Iriki A. Visual responsiveness of neurons in the secondary somatosensory area and its surrounding parietal operculum regions in awake macaque monkeys. *Cerebr Cortex*. 2015;25(11):4535–4550.
97. Graziano MS, Cooke DF, Taylor CS. Coding the location of the arm by sight. *Science*. 2000;290(5497):1782–1786.
98. Iriki A, Tanaka M, Iwamura Y. Coding of modified body schema during tool use by macaque postcentral neurones. *Neuroreport*. 1996;7(14):2325–2330.
99. Graziano MS. Where is my arm? The relative role of vision and proprioception in the neuronal representation of limb position. *Proc Natl Acad Sci U S A*. 1999;96(18):10418–10421.
100. Makin TR, Holmes NP, Zohary E. Is that near my hand? Multisensory representation of peripersonal space in human intraparietal sulcus. *J Neurosci*. 2007;27:731–740.
101. Sereno MI, Huang RS. A human parietal face area contains aligned head-centered visual and tactile maps. *Nat Neurosci*. 2006;9(10):1337–1343.
102. Nakashita S, Saito DN, Kochiyama T, Honda M, Tanabe HC, Sadato N. Tactile-visual integration in the posterior parietal cortex: a functional magnetic resonance imaging study. *Brain Res Bull*. 2008;75(5):513–525.
103. Lloyd DM, Shore DI, Spence C, Calvert GA. Multisensory representation of limb position in human premotor cortex. *Nat Neurosci*. 2003;6:17–18.
104. Gentile G, Petkova VI, Ehrsson HH. Integration of visual and tactile signals from the hand in the human brain: an FMRI study. *J Neurophysiol*. 2011;105(2):910–922.
105. Brozzoli C, Gentile G, Petkova VI, Ehrsson HH. FMRI adaptation reveals a cortical mechanism for the coding of space near the hand. *J Neurosci*. 2011;31(24):9023–9031.
106. Grill-Spector K, Henson R, Martin A. Repetition and the brain: neural models of stimulus-specific effects. *Trends Cogn Sci*. 2006;10(1):14–23.
107. Limanowski J, Blankenburg F. Integration of visual and proprioceptive limb position information in human posterior parietal, premotor, and extrastriate cortex. *J Neurosci*. 2016;36(9):2582–2589.
108. Limanowski J, Blankenburg F. Network activity underlying the illusory self-attribution of a dummy arm. *Hum Brain Mapp*. 2015;36(6):2284–2304.
109. Grivaz P, Blanke O, Serino A. Common and distinct brain regions processing multisensory bodily signals for peripersonal space and body ownership. *Neuroimage*. 2016;147:602–618.
110. Bekrater-Bodmann R, Foell J, Diers M, Flor H. The perceptual and neuronal stability of the rubber hand illusion across contexts and over time. *Brain Res*. 2012;1452:130–139.
111. Downing PE, Jiang Y, Shuman M, Kanwisher N. A cortical area selective for visual processing of the human body. *Science*. 2001;293(5539):2470–2473.
112. Amedi A, Jacobson G, Hendler T, Malach R, Zohary E. Convergence of visual and tactile shape processing in the human lateral occipital complex. *Cerebr Cortex*. 2002;12(11):1202–1212.
113. Kitada R, Johnsrude IS, Kochiyama T, Lederman SJ. Functional specialization and convergence in the occipito-temporal cortex supporting haptic and visual identification of human faces and body parts: an fMRI study. *J Cogn Neurosci*. 2009;21(10):2027–2045.
114. Stilla R, Sathian K. Selective visuo-haptic processing of shape and texture. *Hum Brain Mapp*. 2008;29(10):1123–1138.
115. Astafiev SV, Stanley CM, Shulman GL, Corbetta M. Extrastriate body area in human occipital cortex responds to the performance of motor actions. *Nat Neurosci*. 2004;7(5):542–548.
116. Orlov T, Makin TR, Zohary E. Topographic representation of the human body in the occipitotemporal cortex. *Neuron*. 2010;68(3):586–600.
117. Macaluso E, Frith CD, Driver J. Modulation of human visual cortex by crossmodal spatial attention. *Science*. 2000;289(5482):1206–1208.
118. Peltier S, Stilla R, Mariola E, LaConte S, Hu X, Sathian K. Activity and effective connectivity of parietal and occipital cortical regions during haptic shape perception. *Neuropsychologia*. 2007;45(3):476–483.
119. Limanowski J, Lutti A, Blankenburg F. The extrastriate body area is involved in illusory limb ownership. *Neuroimage*. 2014;86:514–524.

120. Driver J, Noesselt T. Multisensory interplay reveals crossmodal influences on 'sensory-specific' brain regions, neural responses, and judgments. *Neuron*. 2008;57(1):11–23.
121. Murphy JT, MacKay WA, Johnson F. Responses of cerebellar cortical neurons to dynamic proprioceptive inputs from forelimb muscles. *J Neurophysiol*. 1973;36(4):711–723.
122. Glickstein M, Gerrits N, Kralj-Hans I, Mercier B, Stein J, Voogd J. Visual pontocerebellar projections in the macaque. *J Comp Neurol*. 1994;349(1):51–72.
123. Bloedel JR. Cerebellar afferent systems: a review. *Prog Neurobiol*. 1973;2(1):3–68.
124. Naumer MJ, Ratz L, Yalachkov Y, et al. Visuohaptic convergence in a corticocerebellar network. *Eur J Neurosci*. 2010;31(10):1730–1736.
125. Ramnani N. The primate cortico-cerebellar system: anatomy and function. *Nat Rev Neurosci*. 2006;7(7):511–522.
126. Blakemore SJ, Wolpert D, Frith C. Why can't you tickle yourself? *Neuroreport*. 2000;11(11):R11–R16.
127. Petkova VI, Ehrsson HH. If I were you: perceptual illusion of body swapping. *PLoS One*. 2008;3(12):e3832.
128. Preston C, Ehrsson HH. Illusory obesity triggers body dissatisfaction responses in the insula and anterior cingulate cortex. *Cerebr Cortex*. 2016;26(12):4450–4460.
129. Preston C, Ehrsson HH. Illusory changes in body size modulate body satisfaction in a way that is related to nonclinical eating disorder psychopathology. *PLoS One*. 2014;9(1):e85773.
130. Guterstam A, Bjornsdotter M, Bergouignan L, Gentile G, Li TQ, Ehrsson HH. Decoding illusory self-location from activity in the human hippocampus. *Front Hum Neurosci*. 2015;9:412.
131. Slater M, Perez-Marcos D, Ehrsson HH, Sanchez-Vives MV. Inducing illusory ownership of a virtual body. *Front Neurosci*. 2009;3(2):214–220.
132. Maselli A, Slater M. The building blocks of the full body ownership illusion. *Front Hum Neurosci*. 2013;7:83.
133. van der Hoort B, Guterstam A, Ehrsson HH. Being Barbie: the size of one's own body determines the perceived size of the world. *PLoS One*. 2011;6(5):e20195.
134. Petkova VI, Khoshnevis M, Ehrsson HH. The perspective matters! Multisensory integration in ego-centric reference frames determines full-body ownership. *Front Psychol*. 2011;2:35.
135. Preston C, Kuper-Smith BJ, Ehrsson HH. Owning the body in the mirror: the effect of visual perspective and mirror view on the full-body illusion. *Sci Rep*. 2015;5:18345.
136. Petkova VI, Bjornsdotter M, Gentile G, Jonsson T, Li TQ, Ehrsson HH. From part- to whole-body ownership in the multisensory brain. *Curr Biol*. 2011;21(13):1118–1122.
137. Guterstam A, Abdulkarim Z, Ehrsson HH. Illusory ownership of an invisible body reduces autonomic and subjective social anxiety responses. *Sci Rep*. 2015;5:9831.
138. Kilteni K, Bergstrom I, Slater M. Drumming in immersive virtual reality: the body shapes the way we play. *IEEE Trans Vis Comput Graph*. 2013;19(4):597–605.
139. Banakou D, Groten R, Slater M. Illusory ownership of a virtual child body causes overestimation of object sizes and implicit attitude changes. *Proc Natl Acad Sci U S A*. 2013;110(31):12846–12851.
140. Guterstam A, Bjornsdotter M, Gentile G, Ehrsson HH. Posterior cingulate cortex integrates the senses of self-location and body ownership. *Curr Biol*. 2015;25(11):1416–1425.
141. Gentile G, Bjornsdotter M, Petkova VI, Abdulkarim Z, Ehrsson HH. Patterns of neural activity in the human ventral premotor cortex reflect a whole-body multisensory percept. *Neuroimage*. 2015;109:328–340.
142. Graziano MS, Gandhi S. Location of the polysensory zone in the precentral gyrus of anesthetized monkeys. *Exp Brain Res*. 2000;135(2):259–266.
143. Iwamura Y. Hierarchical somatosensory processing. *Curr Opin Neurobiol*. 1998;8(4):522–528.
144. Hyvarinen J. Regional distribution of functions in parietal association area 7 of the monkey. *Brain Res*. 1981;206(2):287–303.
145. Leinonen L, Hyvarinen J, Nyman G, Linnankoski II. Functional properties of neurons in lateral part of associative area 7 in awake monkeys. *Exp Brain Res*. 1979;34(2):299–320.
146. Tsakiris M. Looking for myself: current multisensory input alters self-face recognition. *PLoS One*. 2008;3(12):e4040.
147. Sforza A, Bufalari I, Haggard P, Aglioti SM. My face in yours: visuo-tactile facial stimulation influences sense of identity. *Soc Neurosci*. 2010;5(2):148–162.
148. Lenggenhager B, Tadi T, Metzinger T, Blanke O. Video ergo sum: manipulating bodily self-consciousness. *Science*. 2007;317(5841):1096–1099.

149. Aspell JE, Lenggenhager B, Blanke O. Keeping in touch with one's self: multisensory mechanisms of self-consciousness. *PLoS One*. 2009;4(8):e6488.
150. Ionta S, Heydrich L, Lenggenhager B, et al. Multisensory mechanisms in temporo-parietal cortex support self-location and first-person perspective. *Neuron*. 2011;70(2):363–374.
151. Canzoneri E, di Pellegrino G, Herbelin B, Blanke O, Serino A. Conceptual processing is referenced to the experienced location of the self, not to the location of the physical body. *Cognition*. 2016;154:182–192.
152. Noel JP, Pfeiffer C, Blanke O, Serino A. Peripersonal space as the space of the bodily self. *Cognition*. 2015;144:49–57.
153. Blanke O, Slater M, Serino A. Behavioral, neural, and computational principles of bodily self-consciousness. *Neuron*. 2015;88(1):145–166.
154. Pomes A, Slater M. Drift and ownership toward a distant virtual body. *Front Hum Neurosci*. 2013;7:908.
155. Bertamini M, Berselli N, Bode C, Lawson R, Wong LT. The rubber hand illusion in a mirror. *Conscious Cognit*. 2011;20(4):1108–1119.
156. Slater M, Spanlang B, Sanchez-Vives MV, Blanke O. First person experience of body transfer in virtual reality. *PLoS One*. 2010;5(5):e10564.
157. Gonzalez-Franco M, Perez-Marcos D, Spanlang B, Slater M. The contribution of real-time mirror reflections of motor actions on virtual body ownership in an immersive virtual environment. *IEEE Virtual Reality Conference (VR) (Waltham, MA: IEEE)*. 2010:111–114.

CHAPTER

9

Visual–vestibular interactions

Aasef G. Shaikh[1], *David S. Zee*[2], *Jeffrey Taube*[3], *Amir Kheradmand*[2]

[1]**Department of Neurology, University Hospitals, Cleveland VA Medical Center, Case Western Reserve University, Cleveland, OH, United States; [2]Department of Neurology, The Johns Hopkins University, Baltimore, MD, United States; [3]Department of Psychological and Brain Sciences, Dartmouth College, Hanover, NH, United States**

The human brain uses information from various sensory systems to gauge orientation of the body with respect to the external environment. Our perception of space is based on the image of the external world as registered by various senses and continuously updated and stabilized through sensory feedback from motor activities. In this process, multisensory integration can resolve ambiguities associated with the inherent "noise" from discrete sensory modalities. Accordingly, convergence of visual and vestibular inputs plays a significant role in our perceptions of spatial orientation and motion, which are essential for motor planning and interaction with the external environment. Once movements are generated, the visual–vestibular integration is imperative for optimizing vision and stabilizing the line of sight during movements of the head (i.e., gaze stabilization). Such visual–vestibular interactions are vital for maintaining a coherent perception of spatial orientation during static or dynamic changes in positions of the head and body. In this chapter, we will discuss the basic principles of visual–vestibular interaction within the frameworks of heading (e.g., walking or running) and head tilt with relation to gravity (e.g., a lateral tilt of the head on body). We first describe the fundamental aspects of multisensory integration in these processes along with the underlying physiological and anatomical correlates. We then discuss experimental hypotheses and research findings related to visual–vestibular interaction and outline their clinical applications in human diseases.

Multisensory Perception
https://doi.org/10.1016/B978-0-12-812492-5.00009-7

Visual–vestibular interaction in directional heading

The visual signal as a source of heading perception

As we navigate, the images of the surrounding environment move on the retina—a process referred to as optic flow. It has been recognized for over a century that optic flow provides a robust signal for self-motion.[1] Optic flow can also encode heading direction. For example, straight-ahead movement results in optic flow that radially expands from the epicenter of the visual field. Such radial motion alone is sufficient to induce the illusion of translation within the environment—a phenomenon called linear vection.[2,3]

In the natural world, relying on optic flow alone to determine the direction or velocity of self-motion can pose perceptual ambiguity. The optic flow can represent three possible scenarios. First, it can result from self-motion within the stationary environment. Second, it can arise from apparent movement of the environment even if the observer is stationary. The third and the most frequent scenario is when the observer moves in an environment that is also moving simultaneously—for example, walking in a grocery store among other shoppers. Some psychophysical studies show that the source of optic flow can be inferred by visual analysis; one can use the relative retinal motion of the objects to distinguish their movements during self-motion.[4–8] Nevertheless, nonvisual cues such as vestibular and proprioceptive inputs are critical for the interpretation of optic flow in the presence of object motion.[9–13]

Most natural body movements involve more than one degree of freedom—for example, when the head and the eyes are moving independently. In such circumstances, it becomes nearly impossible to use optic flow alone to determine heading direction because the pattern of optic flow is substantially altered by the movements of the head.[14–16] Remarkably, even in the presence of continuous eye and head movements, perception of self-motion remains largely free of error due to multisensory integration of visual and nonvisual cues such as vestibular signals.[17,18] In this process, vestibular signals are particularly critical because they provide an independent source of information about head movement that can help separate the optic flow induced by self-motion from the optic flow induced by movement within the environment.

The vestibular signal as a source of heading perception

The vestibular system is a robust source of self-motion sensation. The otolith organs in the inner ear transduce translational forces during linear accelerations of the head, whereas the semicircular canals sense forces during angular accelerations.[19–25] In the presence of such linear acceleration sensors, why does the brain need to depend on the additional visual decoding of self-motion from the external environment? The answer lies in the inherent ambiguity of the vestibular signals that encode changes in the head position. According to Einstein's equivalence principle, the otolith organs, like any other accelerometer, encode head acceleration indifferently with respect to the force of gravity. Thus, the otolith signals alone cannot distinguish a linear head translation from a head tilt relative to the direction of gravity.[23–25] To resolve this ambiguity, different signals from more than one motion sensor are required.[19–22,26,27] Accordingly, it has been propo;sed that the brain resolves the tilt-translation ambiguity by combining the angular velocity signals from the semicircular canals and the linear acceleration signals from the otolith organs.[19–22,26] Nevertheless, the central neural mechanism involved in resolution of the tilt-translation ambiguity in humans is still not accurate. For example, a linear translation in the horizontal plane not only leads to a perception of lateral translation but it can also induce a perception of tilt to the side. Such ambiguous perceptions are dangerous for aviators when taking off and landing aircrafts. Major aviation accidents have occurred because of erroneous interpretations of the aircraft's orientation relative to earth. It is therefore critical for the brain to use both visual and vestibular signals to reliably compute the heading direction and body orientation to generate appropriate movements within the environment.

Visual–vestibular interaction in heading perception and its neural correlates

Several contemporary studies have investigated the neural correlates of multisensory integration during heading perception. The brain areas involved in heading perception include three classes of neurons. One class of neurons shows "pure" tuning for visual signals, and hence they are responsive to the optic flow. The middle temporal cortical area[28] and visual area V6 in macaque monkeys contain such neurons that are only tuned to the optic flow. A second group of cells encode "pure" vestibular signals. The parieto-insular vestibular cortex (PIVC) contains such neurons that are tuned to vestibular cues only.[29] The cortical area PIVC corresponds to a part of the parietal operculum in humans, which was found to respond to galvanic vestibular stimulation in a functional imaging study.[30] The third group comprises neurons that have mixed visual and vestibular tuning and that are critical for convergence of visual and vestibular signals. Fig. 9.1 illustrates the cortical areas involved in visual–vestibular integration in nonhuman primates (Fig. 9.1A) and humans (Fig. 9.1B) and their connections (Fig. 9.1C). In the monkey, the dorsomedial superior temporal (MSTd),[17,31,32] ventral intraparietal (VIP), and visual posterior sylvian (VPS) areas contain neurons that encode both visual and vestibular signals.[29,33–37] The neurons in these visual–vestibular sensitive areas, and particularly within MSTd and VIP, are classified into two subtypes. One subtype, called "congruent" neurons, has a similar preferred tuning direction for visual and vestibular signals. The other subtype, called "opposite" neurons, has opposing preferred tuning directions for visual and vestibular signals.[36–38] Each of these subtypes presumably serves different functional roles related to visual–vestibular integration.[39] The "congruent" neurons are more sensitive when visual and vestibular cues are combined and thus they can discriminate small variations in heading direction when both cues are present. Contrary to the congruent neurons, the opposite neurons are least reactive when visual and vestibular cues are combined. Therefore, the activity of the opposite neurons may represent how well vestibular and visual signals are matched. Accordingly, when the optic flow matches the vestibular stimulus, the opposite neurons can become silent. In contrast, mismatch between the optic flow and vestibular stimulus can lead to firing of the opposite neurons (for example, heading in the environment where other surrounding objects are also moving).

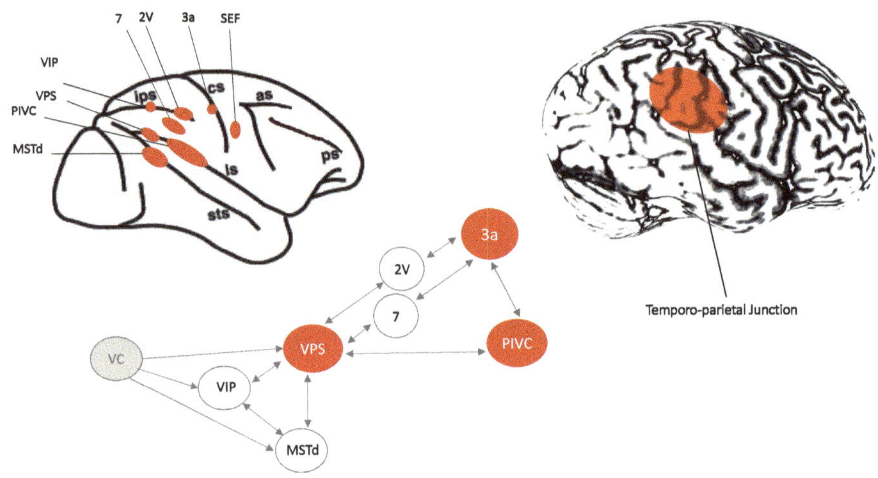

FIGURE 9.1 The main cortical areas that influence vestibular, visual, and multisensory interaction of perceived motion in monkey (A) include area 3a in the somatosensory cortex (neck region), area 7 in the superior parietal lobule, area 2v at the lower tip of the intraparietal sulcus, PIVC, and extrastriate areas including MSTd, VPS, and VIP. Neurons within the PIVC encode the head position in space. Both visual and vestibular signals are recorded from MSTd and VIP, which are implicated in visual and vestibular cue integration for perception of heading direction. VPS, PIVS, and 3a are the main cortical hubs that receive vestibular inputs (shown in red in the bottom diagram). While most of the vestibular cortical areas are interconnected, there is no clear evidence that these areas are organized into a simple hierarchy as are other systems, such as visual and somatosensory cortices. In humans (B), functional imaging, lesion studies, and brain stimulations point to various cortical areas within the posterior insula and the TPJ as the cortical hub for processing vestibular inputs and multisensory integration for various aspects of spatial perception. (C) The cortical areas responsible for visual, vestibular, and visual–vestibular interaction are closely connected into the network. The schematic depicts the connectivity pattern of these regions. *as*, Arcuate sulcus; *cs*, Central sulcus; *ips*, Intraparietal sulcus; *ls*, Lateral sulcus; *MSTd*, Dorsal subdivision of the middle superior temporal cortex; *PIVC*, parieto-insular vestibular cortex; *ps*, Principal sulcus; *SEF*, Supplementary eye field; *sts*, Superior temporal sulcus; *TPJ*, Temporoparietal junction; *VC*, Visual cortex; *VIP*, Ventral intraparietal area; *VPS*, Visual posterior Sylvian area.

Visual–vestibular interaction in tilt perception

We maintain a stable perception of the world in the upright orientation despite frequent changes in the eye, head, and body positions. Such perceptual stability, known as orientation constancy, is a key functional aspect of our spatial orientation. When the neural mechanisms underlying orientation constancy are disrupted, the consequences are often debilitating due to ensuing dizziness, spatial disorientation, and loss of balance. These neural mechanisms are responsible for maintaining a common sensory reference frame by integrating the signals that encode head, eye, and body positions (Fig. 9.2).

Spatial orientation is commonly studied by removing visual cues so that the brain has to rely on other sensory signals that encode body position to determine the orientation of external stimuli. This procedure is the basis for a psychophysical task known as the subjective visual vertical (SVV), in which a visual line is used to report perceived earth–vertical orientation (i.e., upright perception). When the head and body are upright, the sensory reference

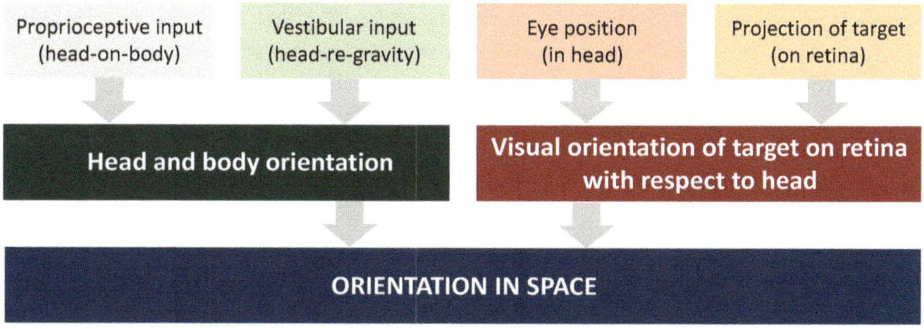

FIGURE 9.2 Sensory inputs from the labyrinths (vestibular system), proprioception (somatosensory system), and vision contribute to perception of spatial orientation. In this process, vestibular inputs encode head position relative to gravity and proprioceptive inputs encode the position of the head relative to the body. Visual orientation of targets is determined by proprioceptive inputs encoding the eye position in the head, the efference copy of the ocular motor signals determining eye position in the orbit, and the orientation of images on the retina. Each sensory input is based on a different reference frame, reflecting a particular perspective from which it is encoded in space.

frames that encode head, eye, and body positions are all aligned along the direction of gravity and SVV errors remain within two degrees of the earth vertical. During lateral whole-body or head-on-body tilts, however, these sensory reference frames become separated and SVV errors increase. Such errors represent challenges for the brain in maintaining a common spatial reference frame in the process of multisensory integration. Accordingly, SVV errors are biased in the tilt direction at angles larger than 60 degrees, reflecting an underestimation of upright orientation with respect to the tilt position, known as the Aubert or A-effect.[40,41] At tilt angles less than 60 degrees, however, SVV errors are often biased in the opposite direction of the tilt, reflecting an overestimation of the upright orientation with respect to the tilt position, known as the Müller or E-effect (E for "Entgegengesetzt").[40–42] These "systematic" errors in upright perception do not correspond with the errors in perception of head tilt position, which remains fairly accurate at all tilt angles.[1–6] Similar SVV errors occur with active head tilts (as opposed to passive tilts), when the brain has access to additional cues such as efference copy signals in the process of encoding sensory information for spatial orientation.[4] Such findings show that spatial orientation is not simply determined by signals that encode head position, and thus further sensory processing and integration must take place within the neural networks that contribute to spatial orientation.

The visual signal for tilt perception

Our environment is rich in visual cues that inform us about earth-vertical orientation. Naturally, visual functions such as orientation discrimination, contrast detection, or visual acuity are more accurate in the vertical and horizontal orientations than in the oblique orientation, a phenomenon known as the oblique effect.[43–45] Visual cues can have strong effects on the accuracy of upright perception in the SVV task. In the rod-and-frame effect the direction of tilt of a rectangular frame biases perception of the SVV as indexed by a rod positioned within the frame.[7–22] Usually, frame orientations close to the baseline SVV error (i.e., without the rod-and-frame effect) result in an "attractor bias" toward the frame orientation, whereas frame tilts of 45 degrees to 90 degrees from the baseline cause a "detractor bias," and frame tilts near 90 degrees from the baseline cause no SVV bias.[46–48] The rod-and-frame effect also depends on the viewing distance and body tilt position. It decreases at larger viewing distances, consistent with reduced reliability of the visual frame as a cue for upright orientation.[48,49] Background visual motion can also induce a strong visual effect on perception of the upright. For example, optokinetic stimulation around the line of sight (i.e., in the roll plane) can induce SVV errors in the direction of the visual rotation.[50–52] When the optokinetic and rod-and-frame effects are combined, the static frame effect is usually more robust, and it can significantly decrease the SVV errors induced by the dynamic effect of optokinetic stimulation.[53,54] Thus, by all accounts, visual inputs are indispensable in the process of multisensory integration for spatial orientation. The significant weight of visual information in this process can improve spatial perception especially during head or body tilts when the noise in vestibular signals increases and the brain is challenged to maintain a common spatial reference frame.

The vestibular signal for tilt perception

The inputs from the otolith organs have a significant weight in spatial orientation by encoding the head tilt position with respect to gravity and driving the compensatory changes in eye position. As a fundamental reference for spatial orientation, gravity is critical in all aspects of our balance, perception, and behavior. Changes in gravity can alter perception of the surrounding environment and result in spatial disorientation. This is often experienced by fighter jet pilots as oculogravic (visual) or somatogravic (nonvisual) tilt illusions due to sudden changes in gravito-inertial forces.[55] Likewise, in microgravity, astronauts often report difficulty distinguishing the spacecraft floors, walls, and ceiling surfaces from one another and have tilt perception errors.[56] In normal gravity, the inherent noise in vestibular signals and low gain of the torsional vestibular–ocular reflex during lateral head tilt are the major sources of error in spatial orientation. Accordingly, SVV errors are primarily modulated by head tilt rather than trunk tilt.[57–61] The brain, however, can maintain spatial orientation either directly by relying on sensory signals that encode head position (e.g., otolith signals) or indirectly through sensory inputs that encode neck and trunk positions (e.g., proprioception).[23] In this context, there is a greater weight of head position signals (e.g., from the otoliths) around the upright position and a greater weight of trunk proprioceptive signals at larger tilt angles.[24,25]

Visual–vestibular interaction and its neural correlates

Visual and vestibular signals are both incorporated into the perception of self-position and spatial orientation. Accordingly, visual–vestibular interaction can be studied using sensory integration models that account for the estimates of head and eye positions. In this approach, spatial orientation can be understood within a Bayesian framework in which the noise in sensory signals and existing sensory information (i.e., the prior in the Bayesian sense) determines the weights of individual modalities in the process of multisensory integration for upright perception (Fig. 9.3; also see chapter by Meijer and Noppeney, this volume). In this Bayesian spatial model, the "*prior*" for head tilt position is the earth-vertical orientation as humans

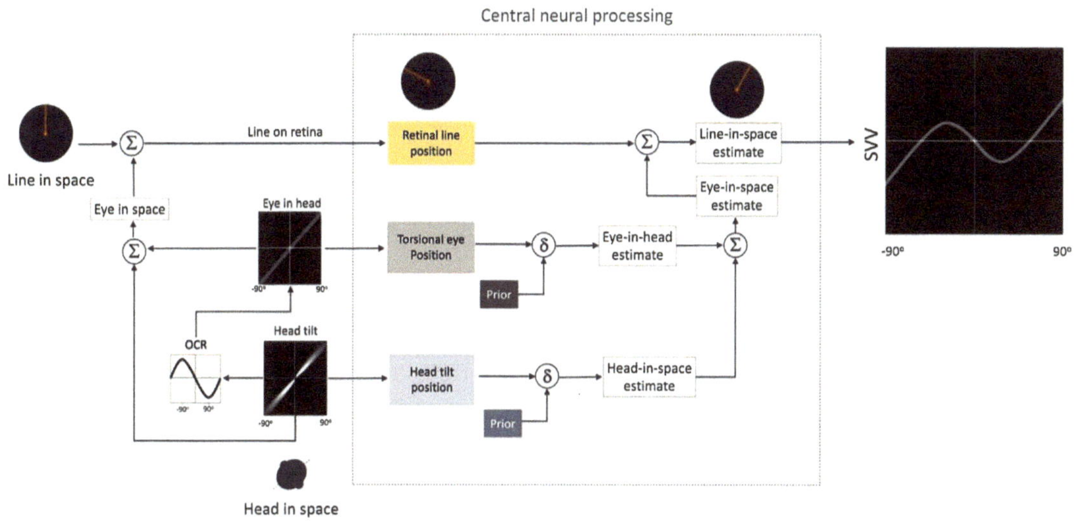

FIGURE 9.3 Schematic presentation of a Bayesian model of spatial perception. A visual stimulus (line in space) is presented in front of a tilted observer (head in space) (a range of ± 90 degrees tilt position is shown in the graph at extreme right). In this model, the internal estimate of the upright orientation (i.e., the SVV) is based on the estimate of a visual line orientation in space (i.e., line-in-space estimate). Head tilt position signals are contaminated by Gaussian noise. Eye-in-head position is driven by the torsional eye position in the opposite direction of the head tilt or the ocular counter-roll, which is contaminated by independent noise. As part of central neural processing, the head-in-space and eye-in-head estimates are derived from the corresponding sensory likelihoods and *priors*. These estimates are integrated into the eye-in-space estimate, which is also combined with the retinal position signals (line on retina) to generate the line-in-space estimate.

spend most of their time in this position. This earth-vertical *prior* reduces the effect of noise in vestibular signals, but in turn it produces a bias in the neural estimate of head tilt position toward the upright position. Thus, the effect of head *prior* can lead to underestimation of the upright orientation with respect to the head tilt position (i.e., the A-effect at large tilt angles). Similarly, at smaller tilt positions, the errors in the brain estimates of torsional eye position can lead to overestimation of the upright orientation with respect to the head tilt position (i.e., the E-effect at smaller tilt angles).[62] The interaction of visual and vestibular signals for spatial orientation has been studied using the rod-and-frame effect during head tilt. The frame effect can increase or decrease the E-effect and A-effect depending on the relative tilt position of the head and the visual frame.[44,47,63] In general, the frame effect is more pronounced at larger body tilts, consistent with increased noise from the otolith inputs and therefore reduced reliability of vestibular signals for spatial orientation.[48,49]

Thus, overall, visual–vestibular interaction for spatial orientation can be studied within the framework of sensory integration for perception of the upright. In this context, the integration of visual and vestibular signals is vital for generating an internal spatial reference frame that anchors "self" with respect to the outside world to maintain orientation constancy. This visual–vestibular integration is affected by the reliability of vestibular signals based on the head and body positions. Accordingly, there are modulatory effects of gravity and visual cues on spatial orientation. In this context, errors of upright perception reflect challenges for the brain in maintaining a common spatial reference frame. The neural substrates for visual—vestibular integration are primarily localized to the cortex of the temporoparietal junction (TPJ). The evidence for involvement of this cortical region comes from transcranial magnetic stimulation (TMS) and anatomical lesion studies, as discussed in subsequent sections.[64–68] The higher-order neural mechanisms within the TPJ cortex must solve the problem of dissociated sensory reference frames in the process of integrating visual and vestibular signals. Currently, little is known about the specific functional roles of these cortical networks and how disruption in one sensory modality may affect processing or integration of other sensory modalities. Future studies should specifically address such sensory contributions with respect to cerebral cortical involvement in spatial orientation.

How is multisensory convergence affected in neurological conditions?

Impaired tilt perception in vestibular and proprioceptive loss

As a multimodal sensory reference, errors of upright perception can be affected by the loss of vestibular and proprioceptive inputs. Patients with vestibular loss often have no E-effect at small tilt angles but have a more pronounced A-effect at larger body tilts, consistent with the reduced weight of head position signals, and consequently a relative underestimation of upright orientation with respect to the tilt position.[69–71] In contrast, patients with proprioceptive loss (affecting the trunk and limbs) have a decreased A-effect at large body tilts, consistent with a reduced weight of body proprioception, and consequently a relative overestimation of upright orientation with respect to the tilt position.[72] These error patterns show how multisensory integration for spatial orientation is affected following the loss of a particular sensory modality. In this process, the weight of visual information for spatial perception is also affected. For example, the rod-and-frame effect can be asymmetric in patients with vestibular loss, with a reduced visual dependence when the frame is tilted toward the intact side, as opposed to a significant frame effect when there is a tilt toward the side of vestibular loss.[73] Similarly, the visual effect of optokinetic stimulation also produces a larger bias toward the side of vestibular loss.[74] These findings indicate that with the loss of vestibular signals, the brain has to rely more on visual information for spatial orientation.

Impaired spatial orientation in stroke

Perception of the upright shares the same neural substrates as those involved in the perception of body orientation, visuospatial attention, heading perception, visual gravitational motion perception, the sense of embodiment, self-localization, and egocentricity.[75–79] These neural networks are involved in multisensory integration and are primarily localized to the TPJ. Notably, a significant vestibular-mediated activation has been found in the nonmotor-dominant human cortex, i.e., the right hemisphere in right-handed individuals and the left hemisphere in left-handed individuals.[80] The involvement of the TPJ in spatial orientation is particularly demonstrated by ischemic lesions in this region that lead to spatial neglect.[64,65] Patients with neglect are not aware of their contralesional hemispace and also show significant deviations in their spatial orientation, reflected by SVV errors away from the side of the lesion.[81,82] In general, TPJ lesions associated with SVV deviations largely involve the inferior parietal lobule and the posterior aspect of the insular cortex. The isolated lesions within the posterior insula, however, are not associated with SVV deviations, suggesting that other cortical areas within the TPJ contribute to multisensory integration for spatial orientation.[83] Consistent with the multisensory role of the TPJ, lesions within this region are also associated with symptoms such as out-of-body experiences or room tilt illusions.[84] Taken together, these findings suggest that the TPJ is involved in multisensory integration that anchors "self" with respect to the surrounding environment and maintains orientation constancy especially with frequent changes in the eye, head, and body positions.

The role of the TPJ in upright perception has also been studied using noninvasive brain stimulation.[66–68] The inhibitory effect of TMS at the posterior aspect of the supramarginal gyrus within the right TPJ results in a shift of SVV errors in the opposite direction of head tilt[66] (Fig. 9.4). The direction of this SVV shift is similar to the "overestimation" errors from the cerebral cortical lesions that involve the TPJ (i.e., a larger E-effect). These SVV deviations are dissociated from changes in eye position during head tilt, showing that the changes in upright perception at the level of cerebral cortex are primarily driven by sensory processes.[85] These findings indicate that, unlike subcortical regions, the TPJ does not directly modulate ocular position. Usually, with caudal brainstem lesions, the torsional eye position deviates toward the side of the lesion, and with rostral brainstem lesions, it deviates away from the side of the lesion.[81,86,87] Thus, in these cases, the errors of upright perception are caused by the altered orientation of the images on the retina, while the errors associated with TPJ

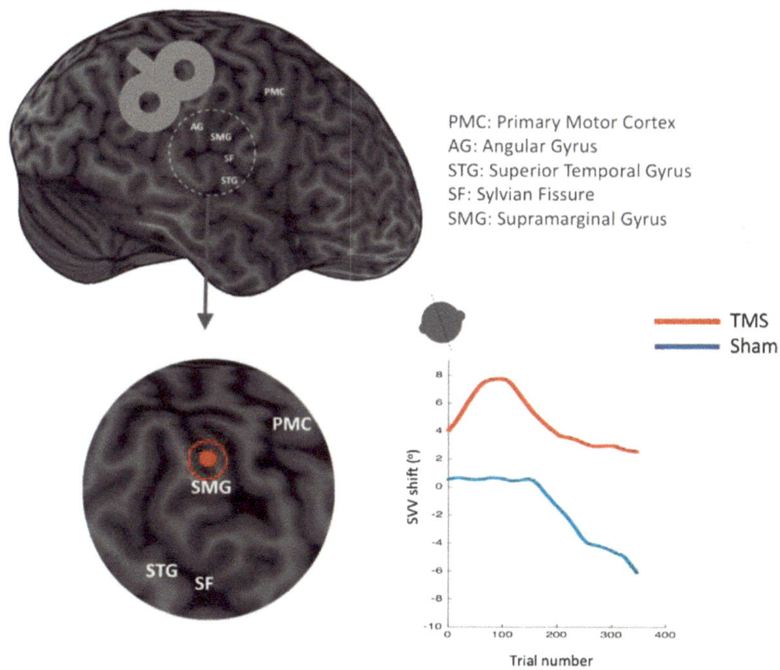

FIGURE 9.4 Transcranial magnetic stimulation (TMS) at the supramarginal gyrus (SMG). Subjective visual vertical (SVV) results are shown in a subject during 20 degrees left head tilt after TMS and sham (i.e., no TMS) stimulations. TMS at SMG, in contrast to the sham stimulation, results in an SVV shift in the opposite direction of the head tilt (i.e., an increase in the E-effect). Positive values on the graph indicate SVV shifts to the right and negative values indicate shifts to the left.

dysfunction are primarily linked to sensory processes involved in perception of spatial orientation.

Persistent SVV errors are often linked to poor balance following stroke, especially in patients with right hemispheric involvement.[26–28,88] A subset of patients with cortical infarctions and large SVV deviations also have robust postural deviations and actively resist corrections of their false postural orientation—a phenomenon known as "pusher syndrome."[89] Patients with pusher behavior have longer-lasting SVV errors and often are unable to learn to walk again even with proper assistance.[90] The pusher behavior is highly correlated with neglect symptoms, suggesting that these patients actively align their body with their erroneous spatial orientation.[91,92]

Future perspective—translating the concepts of visuo–vestibular integration in clinical neuroscience

Elegant contemporary studies in the last two decades have delineated the physiology of visual–vestibular convergence in nonhuman primates. The natural next step is to translate this rich knowledge into understanding the pathophysiology of common human disorders and to put forward novel strategies for treatment or rehabilitation. One example of human disease that is thought to cause impaired interaction between multiple sensory systems is Parkinson's disease. One of the many consequences of the neuronal degeneration in Parkinson's disease is abnormal output from the basal ganglia. Patients with Parkinson's disease frequently experience abnormal abrupt cessation of walking (freezing of gait) while passing through a doorway or in a narrow hallway. One hypothetical explanation for freezing of gait is that the narrowness of the hallway leads to a relative increase in optic flow because the visual image from nearby objects moves faster on the retina compared with the objects that are distant. Therefore, the brain is faced with conflicting visual and vestibular signals. We speculate that normally when such a conflict arises, our brain relies on previous experience, the *priors*, to make a decision as to which signal to "believe." In contrast, patients with Parkinson's disease are unable to combine *prior* information with current information to perform accurately in a perceptual task.[93–95] We speculate that ineffective utilization of the *prior* when visual and vestibular signals are at odds might result in gait slowing or even freezing (Fig. 9.5). Future experimental studies are needed to address this prediction.

Parkinson's disease is increasingly recognized as a disorder altering cognition, perception, and higher-order visuospatial functions.[96–104] Besides optic flow, the perceived location of one's goal may influence one's path of movement.[105] The egocentric reference point divides space into two lateral hemifields with respect to the midline of the trunk.[106] Such organization provides a framework for spatial orientation and goal-directed actions such as walking.[107] A shift in the perception of the egocentric midline is seen in patients with Parkinson's disease[102,108]: such patients often veer to one side while walking or driving a car. This lateralized bias could be the result of impaired heading perception. Although peripheral vestibular function and brainstem vestibulo-ocular reflexes are normal, it is unclear whether the functions of cortical and cerebellar vestibular networks are affected in Parkinson's disease. Two independent pathways can affect central vestibular function in Parkinson's disease. One pathway involves the cerebellum. A major output station of the basal ganglia, particularly in the indirect

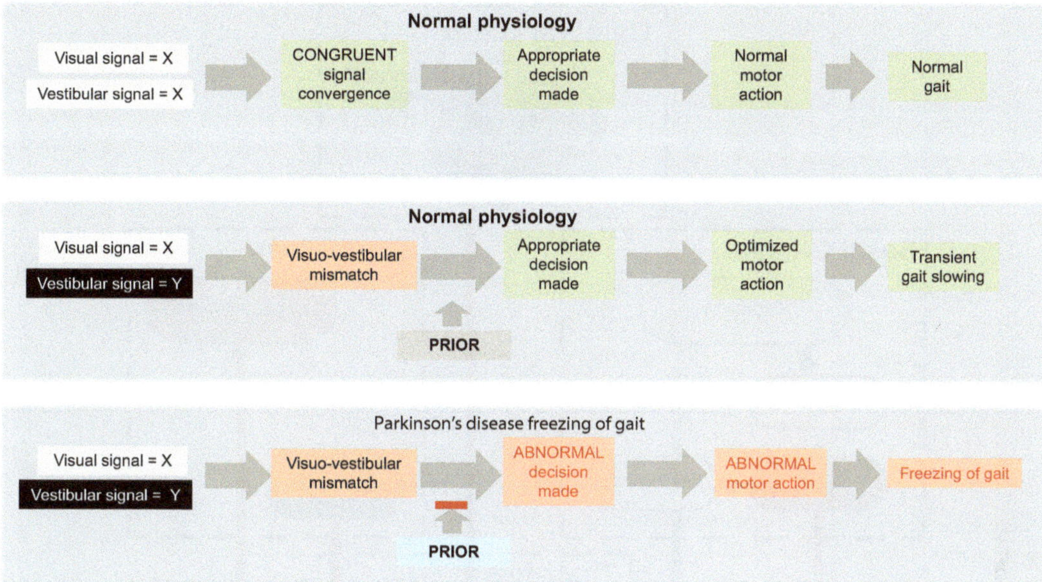

FIGURE 9.5 Hypothetical description of effects of visuo−vestibular mismatch and influence of *priors* in pathogenesis of gait disorder in Parkinson's disease.

pathway, is the subthalamic nucleus, which has projections to the cerebellar cortex by way of the pontine nuclei—the subthalamo-cerebellar tract (orange boxes in Fig. 9.6).[109,110] Abnormal bursting and oscillatory activity in the subthalamic nucleus in patients with Parkinson's disease could affect cerebellar function.[111,112] In turn, impaired cerebellar activity can affect motion perception via the cerebello-thalamic tract, which connects the deep cerebellar nuclei with the thalamus (green boxes in Fig. 9.6), and the onward thalamic connection to the parietotemporal cortex (MST and VIP).[113−126] The second pathway involves the connections of the substantia nigra pars reticulata and thalamus—the nigro-thalamo-cortical pathway.[116,127] The substantia nigra *pars reticulata*, another output station of the basal ganglia, projects to the parietotemporal cortex by way of the thalamus. We speculate that abnormal activity in the basal ganglia due to Parkinson's disease could also be transmitted directly to the thalamus and influence visual−vestibular interaction in parietotemporal cortex, which functionally affects the perception of heading. More studies are needed to investigate the influence of abnormal basal ganglia output on motion perception in Parkinson's disease.

A population of cells in the limbic system also serves perception of heading direction. These cells, referred to as head direction cells, discharge as a function of head direction in the horizontal plane, independent of the animal's location or its ongoing behavior.[128,129] Head direction cells have been identified in many brain areas (mostly within the limbic system) and are found in a number of species including humans.[130] An intact vestibular system is critical for being able to record activity from head direction cells,[131−133] and the head direction signal is hypothesized to be generated subcortically—possibly between the connections of the dorsal tegmental nucleus and lateral mammillary nucleus. From the lateral

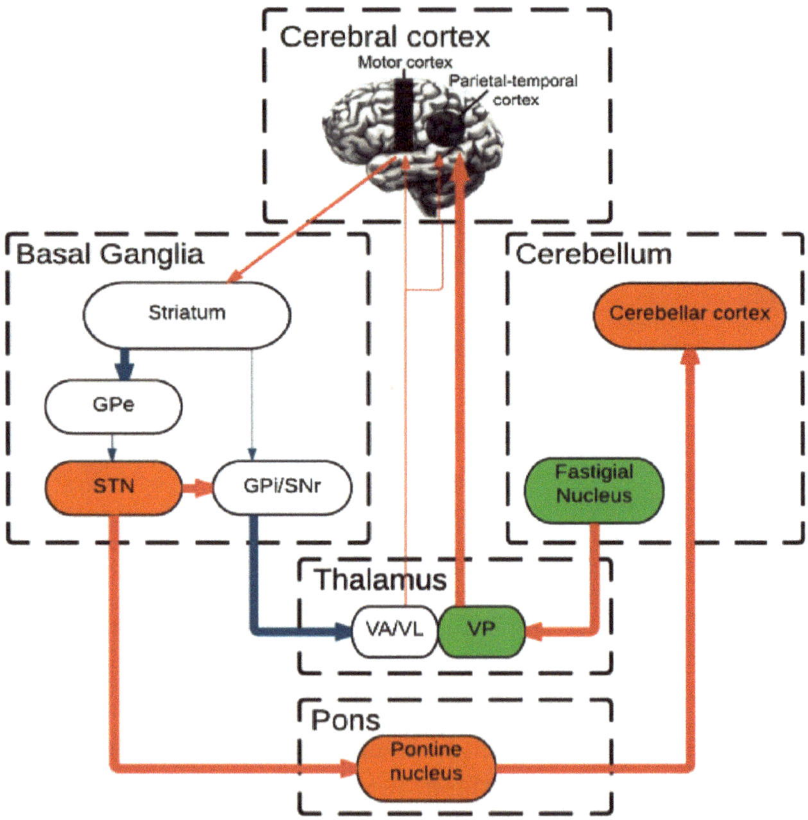

FIGURE 9.6 Circuits interconnecting the cerebral cortex, basal ganglia, and cerebellum. Red arrows represent excitatory connections; blue arrows represent inhibitory connections. Orange boxes represent the subthalamo-cerebellar pathway, green boxes represent the cerebello-thalamic pathway, and white boxes represent traditional direct/indirect basal ganglia pathways in PD. *GPe and GPi*, external and internal segments of globus pallidus; *SNr*, substantia nigra pars reticulata; *STN*, Subthalamic nucleus; *VA, VL, and VP*, ventral anterior, ventral lateral, and ventral posterior nuclei of the thalamus, respectively. *Courtesy of S. Ozinga.*

mammillary nucleus, the head direction signal is projected to the anterodorsal nucleus of the thalamus and then to many structures in the cerebral cortex.[128,134] Recent experiments in rats have found that lesions of the dorsal striatum, an area often disrupted in Parkinson's disease, did not lead to impairment in the responses of head direction cells,[135] suggesting that by extension to humans, Parkinson's disease patients may not have an impaired sense of direction because of a dysfunctional head direction cell network. One important question that remains unanswered is whether generating a bias in a normally functioning heading direction network can influence spatial navigation behavior such as veering. A direct clinical implication for these experiments could lead to utilization of technologies such as deep brain stimulation to compensate for navigational deficits in neurodegenerative disorders affecting the cerebellum or basal ganglia.

Summary

Motion perception relies on signals from multiple sources including the visual, vestibular, and proprioceptive systems. The information from an individual sensory system alone is not reliable; it is frequently ambiguous or in conflict with other signals. Multisensory convergence is critical for the resolution of the sensory ambiguity and conflict. One approach to understanding multisensory interaction between visual and vestibular signals for reliable motion perception is based on implementation of a Bayesian framework in which the brain uses previous experience—the *"prior"*—to determine the weights of individual sensory modalities. The brain must determine which sensory signals to "believe" when they are in conflict. Visual and vestibular signals converge in the parietotemporal cortex and lesions of this region, often seen in stroke victims, can lead to spatial neglect and impaired perception of spatial orientation. Likewise, noninvasive brain stimulation of parietotemporal cortex can affect perceived spatial orientation. Finally, we hypothesize that the navigational abnormalities in patients with Parkinson's disease, such as freezing of gait and veering, may relate to abnormal visual–vestibular interaction in the parietotemporal cortex. The study of patients with various neurological diseases is a fruitful area of translational research to understand how the brain maintains a veridical sense of motion and position of the body in the external environment.

References

1. Gibson JJ. Optical motions and transformations as stimuli for visual perception. *Psychol Rev.* 1957;64(5):288–295.
2. Berthoz A, Pavard B, Young LR. Perception of linear horizontal self-motion induced by peripheral vision (linearvection) basic characteristics and visual-vestibular interactions. *Exp Brain Res.* 1975;23(5):471–489.
3. Young LR, Dichgans J, Murphy R, et al. Interaction of optokinetic and vestibular stimuli in motion perception. *Acta Otolaryngol.* 1973;76(1):24–31.
4. Rushton SK, Warren PA. Moving observers, relative retinal motion and the detection of object movement. *Curr Biol.* 2005;15(14):R542–R543. https://doi.org/10.1016/j.cub.2005.07.020.
5. Warren PA, Rushton SK. Perception of object trajectory: parsing retinal motion into self and object movement components. *J Vis.* 2007;7(11):2 1–11. https://doi.org/10.1167/7.11.2.
6. Warren PA, Rushton SK. Evidence for flow-parsing in radial flow displays. *Vision Res.* 2008;48(5):655–663. https://doi.org/10.1016/j.visres.2007.10.023.
7. Warren PA, Rushton SK. Optic flow processing for the assessment of object movement during ego movement. *Curr Biol.* 2009;19(18):1555–1560. https://doi.org/10.1016/j.cub.2009.07.057.
8. Warren PA, Rushton SK. Perception of scene-relative object movement: optic flow parsing and the contribution of monocular depth cues. *Vision Res.* 2009;49(11):1406–1419.
9. Wexler M, Panerai F, Lamouret I, et al. Self-motion and the perception of stationary objects. *Nature.* 2001;409(6816):85–88. https://doi.org/10.1038/35051081.
10. Morvan C, Wexler M. Reference frames in early motion detection. *J Vis.* 2005;5(2):131–138. https://doi.org/10.1167/5.2.4.
11. Morvan C, Wexler M. The nonlinear structure of motion perception during smooth eye movements. *J Vis.* 2009;9(7):1. https://doi.org/10.1167/9.7.1.
12. Wexler M, Glennerster A, Cavanagh P, et al. Default perception of high-speed motion. *Proc Natl Acad Sci U S A.* 2013;110(17):7080–7085. https://doi.org/10.1073/pnas.1213997110.
13. Dupin L, Wexler M. Motion perception by a moving observer in a three-dimensional environment. *J Vis.* 2013;13(2):15. https://doi.org/10.1167/13.2.15.
14. Royden CS, Banks MS, Crowell JA. The perception of heading during eye movements. *Nature.* 1992;360(6404):583–585. https://doi.org/10.1038/360583a0.

15. Royden CS, Crowell JA, Banks MS. Estimating heading during eye movements. *Vision Res.* 1994;34(23):3197–3214.
16. Crowell JA, Banks MS, Shenoy KV, et al. Visual self-motion perception during head turns. *Nat Neurosci.* 1998;1(8):732–737. https://doi.org/10.1038/3732.
17. Page WK, Duffy CJ. MST neuronal responses to heading direction during pursuit eye movements. *J Neurophysiol.* 1999;81(2):596–610.
18. Shenoy KV, Bradley DC, Andersen RA. Influence of gaze rotation on the visual response of primate MSTd neurons. *J Neurophysiol.* 1999;81(6):2764–2786.
19. Angelaki DE, Shaikh AG, Green AM, et al. Neurons compute internal models of the physical laws of motion. *Nature.* 2004;430(6999):560–564. https://doi.org/10.1038/nature02754.
20. Shaikh AG, Meng H, Angelaki DE. Multiple reference frames for motion in the primate cerebellum. *J Neurosci.* 2004;24(19):4491–4497. https://doi.org/10.1523/JNEUROSCI.0109-04.2004.
21. Shaikh AG, Green AM, Ghasia FF, et al. Sensory convergence solves a motion ambiguity problem. *Curr Biol.* 2005;15(18):1657–1662. https://doi.org/10.1016/j.cub.2005.08.009.
22. Green AM, Shaikh AG, Angelaki DE. Sensory vestibular contributions to constructing internal models of self-motion. *J Neural Eng.* 2005;2(3):S164–S179. https://doi.org/10.1088/1741-2560/2/3/S02.
23. Fernandez C, Goldberg JM. Physiology of peripheral neurons innervating otolith organs of the squirrel monkey. III. Response dynamics. *J Neurophysiol.* 1976;39(5):996–1008.
24. Fernandez C, Goldberg JM. Physiology of peripheral neurons innervating otolith organs of the squirrel monkey. II. Directional selectivity and force-response relations. *J Neurophysiol.* 1976;39(5):985–995.
25. Fernandez C, Goldberg JM. Physiology of peripheral neurons innervating otolith organs of the squirrel monkey. I. Response to static tilts and to long-duration centrifugal force. *J Neurophysiol.* 1976;39(5):970–984.
26. Shaikh AG, Ghasia FF, Dickman JD, et al. Properties of cerebellar fastigial neurons during translation, rotation, and eye movements. *J Neurophysiol.* 2005;93(2):853–863. https://doi.org/10.1152/jn.00879.2004.
27. Yakusheva TA, Shaikh AG, Green AM, et al. Purkinje cells in posterior cerebellar vermis encode motion in an inertial reference frame. *Neuron.* 2007;54(6):973–985. https://doi.org/10.1016/j.neuron.2007.06.003.
28. Chowdhury SA, Takahashi K, DeAngelis GC, et al. Does the middle temporal area carry vestibular signals related to self-motion? *J Neurosci.* 2009;29(38):12020–12030. https://doi.org/10.1523/JNEUROSCI.0004-09.2009.
29. Chen A, DeAngelis GC, Angelaki DE. Macaque parieto-insular vestibular cortex: responses to self-motion and optic flow. *J Neurosci.* 2010;30(8):3022–3042. https://doi.org/10.1523/JNEUROSCI.4029-09.2010.
30. Eickhoff SB, Weiss PH, Amunts K, Fink GR, Zilles K. Identifying human parieto-insular vestibular cortex using fMRI and cytoarchitectonic mapping. *Hum Brain Mapp.* 2006;27:611–621.
31. Page WK, Duffy CJ. Heading representation in MST: sensory interactions and population encoding. *J Neurophysiol.* 2003;89(4):1994–2013. https://doi.org/10.1152/jn.00493.2002.
32. Gu Y, Watkins PV, Angelaki DE, et al. Visual and nonvisual contributions to three-dimensional heading selectivity in the medial superior temporal area. *J Neurosci.* 2006;26(1):73–85. https://doi.org/10.1523/JNEUROSCI.2356-05.2006.
33. Chen A, DeAngelis GC, Angelaki DE. Representation of vestibular and visual cues to self-motion in ventral intraparietal cortex. *J Neurosci.* 2011;31(33):12036–12052. https://doi.org/10.1523/JNEUROSCI.0395-11.2011.
34. Chen A, DeAngelis GC, Angelaki DE. Convergence of vestibular and visual self-motion signals in an area of the posterior sylvian fissure. *J Neurosci.* 2011;31(32):11617–11627. https://doi.org/10.1523/JNEUROSCI.1266-11.2011.
35. Chen X, DeAngelis GC, Angelaki DE. Eye-centered representation of optic flow tuning in the ventral intraparietal area. *J Neurosci.* 2013;33(47):18574–18582. https://doi.org/10.1523/JNEUROSCI.2837-13.2013.
36. Chen X, Deangelis GC, Angelaki DE. Diverse spatial reference frames of vestibular signals in parietal cortex. *Neuron.* 2013;80(5):1310–1321. https://doi.org/10.1016/j.neuron.2013.09.006.
37. Chen A, Deangelis GC, Angelaki DE. Functional specializations of the ventral intraparietal area for multisensory heading discrimination. *J Neurosci.* 2013;33(8):3567–3581. https://doi.org/10.1523/JNEUROSCI.4522-12.2013.
38. Gu Y, Angelaki DE, Deangelis GC. Neural correlates of multisensory cue integration in macaque MSTd. *Nat Neurosci.* 2008;11(10):1201–1210. https://doi.org/10.1038/nn.2191.

39. Gu Y, Angelaki DE, DeAngelis GC. Contribution of correlated noise and selective decoding to choice probability measurements in extrastriate visual cortex. *Elife*. 2014;3. https://doi.org/10.7554/eLife.02670.
40. Van Beuzekom AD, Van Gisbergen JA. Properties of the internal representation of gravity inferred from spatial-direction and body-tilt estimates. *J Neurophysiol*. 2000;84(1):11–27. https://doi.org/10.1152/jn.2000.84.1.11/Fb.
41. Mittelstaedt H. A new solution to the problem of the subjective vertical. *Naturwissenschaften*. 1983;70(6):272–281.
42. Tarnutzer AA, Bockisch C, Straumann D, et al. Gravity dependence of subjective visual vertical variability. *J Neurophysiol*. 2009;102(3):1657–1671. https://doi.org/10.1152/jn.00007.2008.
43. Barnett-Cowan M, Harris LR. Perceived self-orientation in allocentric and egocentric space: effects of visual and physical tilt on saccadic and tactile measures. *Brain Res*. 2008;1242:231–243. https://doi.org/10.1016/j.brainres.2008.07.075.
44. Dyde RT, Jenkin MR, Harris LR. The subjective visual vertical and the perceptual upright. *Exp Brain Res*. 2006;173(4):612–622. https://doi.org/10.1007/s00221-006-0405-y.
45. Haji-Khamneh B, Harris LR. How different types of scenes affect the subjective visual vertical (SVV) and the perceptual upright (PU). *Vision Res*. 2010;50(17):1720–1727. https://doi.org/10.1016/j.visres.2010.05.027.
46. Li W, Matin L. Visually perceived vertical (VPV): induced changes in orientation by 1-line and 2-line roll-tilted and pitched visual fields. *Vision Res*. 2005;45(15):2037–2057. https://doi.org/10.1016/j.visres.2005.01.014.
47. Vingerhoets RA, De Vrijer M, Van Gisbergen JA, et al. Fusion of visual and vestibular tilt cues in the perception of visual vertical. *J Neurophysiol*. 2009;101(3):1321–1333. https://doi.org/10.1152/jn.90725.2008.
48. Alberts BB, Selen LP, Bertolini G, et al. Dissociating vestibular and somatosensory contributions to spatial orientation. *J Neurophysiol*. 2016;116(1):30–40. https://doi.org/10.1152/jn.00056.2016.
49. Zoccolotti P, Antonucci G, Goodenough DR, et al. The role of frame size on vertical and horizontal observers in the rod-and-frame illusion. *Acta Psychol*. 1992;79(2):171–187.
50. Bronstein AM, Yardley L, Moore AP, et al. Visually and posturally mediated tilt illusion in Parkinson's disease and in labyrinthine defective subjects. *Neurology*. 1996;47(3):651–656.
51. Dichgans J, Diener HC, Brandt T. Optokinetic-graviceptive interaction in different head positions. *Acta Otolaryngol*. 1974;78(5–6):391–398.
52. Ward BK, Bockisch CJ, Caramia N, et al. Gravity dependence of the effect of optokinetic stimulation on the subjective visual vertical. *J Neurophysiol*. 2017;117(5):1948–1958. https://doi.org/10.1152/jn.00303.2016.
53. Lubeck AJ, Bos JE, Stins JF. Framing visual roll-motion affects postural sway and the subjective visual vertical. *Atten Percept Psychophys*. 2016;78(8):2612–2620. https://doi.org/10.3758/s13414-016-1150-3.
54. Pavlou M, Quinn C, Murray K, et al. The effect of repeated visual motion stimuli on visual dependence and postural control in normal subjects. *Gait Posture*. 2011;33(1):113–118. https://doi.org/10.1016/j.gaitpost.2010.10.085.
55. Graybiel A. Oculogravic illusion. *AMA Arch Ophthalmol*. 1952;48(5):605–615.
56. Clement G, Moore ST, Raphan T, et al. Perception of tilt (somatogravic illusion) in response to sustained linear acceleration during space flight. *Exp Brain Res*. 2001;138(4):410–418.
57. Alberts BB, de Brouwer AJ, Selen LP, et al. A Bayesian account of visual-vestibular interactions in the rod-and-frame task. *eNeuro*. 2016;3(5). https://doi.org/10.1523/ENEURO.0093-16.2016.
58. Wade NJ, Day RH. Development and dissipation of a visual spatial aftereffect from prolonged head tilt. *J Exp Psychol*. 1968;76(3):439–443.
59. Clement G, Eckardt J. Influence of the gravitational vertical on geometric visual illusions. *Acta Astronaut*. 2005;56(9–12):911–917.
60. Clemens IA, De Vrijer M, Selen LP, et al. Multisensory processing in spatial orientation: an inverse probabilistic approach. *J Neurosci*. 2011;31(14):5365–5377. https://doi.org/10.1523/JNEUROSCI.6472-10.2011.
61. McKenna GJ, Peng GC, Zee DS. Neck muscle vibration alters visually perceived roll in normals. *J Assoc Res Otolaryngol*. 2004;5(1):25–31. https://doi.org/10.1007/s10162-003-4005-2.
62. De Vrijer M, Medendorp WP, Van Gisbergen JA. Accuracy-precision trade-off in visual orientation constancy. *J Vis*. 2009;9(2):9 1–15. https://doi.org/10.1167/9.2.9.
63. Corbett JE, Enns JT. Observer pitch and roll influence: the rod and frame illusion. *Psychon Bull Rev*. 2006;13(1):160–165.
64. Karnath HO, Dieterich M. Spatial neglect–a vestibular disorder? *Brain*. 2006;129(Pt 2):293–305. https://doi.org/10.1093/brain/awh698.

65. De Renzi E, Faglioni P, Scotti G. Judgment of spatial orientation in patients with focal brain damage. *J Neurol Neurosurg Psychiatry*. 1971;34(5):489—495.
66. Kheradmand A, Lasker A, Zee DS. Transcranial magnetic stimulation (TMS) of the supramarginal gyrus: a window to perception of upright. *Cerebr Cortex*. 2015;25(3):765—771. https://doi.org/10.1093/cercor/bht267.
67. Fiori F, Candidi M, Acciarino A, et al. The right temporoparietal junction plays a causal role in maintaining the internal representation of verticality. *J Neurophysiol*. 2015;114(5):2983—2990. https://doi.org/10.1152/jn.00289.2015.
68. Santos-Pontelli TE, Rimoli BP, Favoretto DB, et al. Polarity-dependent misperception of subjective visual vertical during and after transcranial direct current stimulation (tDCS). *PLoS One*. 2016;11(3):e0152331. https://doi.org/10.1371/journal.pone.0152331.
69. Dai MJ, Curthoys IS, Halmagyi GM. Linear acceleration perception in the roll plane before and after unilateral vestibular neurectomy. *Exp Brain Res*. 1989;77(2):315—328.
70. Lopez C, Lacour M, Leonard J, et al. How body position changes visual vertical perception after unilateral vestibular loss. *Neuropsychologia*. 2008;46(9):2435—2440. https://doi.org/10.1016/j.neuropsychologia.2008.03.017.
71. Muller JA, Bockisch CJ, Tarnutzer AA. Spatial orientation in patients with chronic unilateral vestibular hypofunction is ipsilesionally distorted. *Clin Neurophysiol*. 2016;127(10):3243—3251. https://doi.org/10.1016/j.clinph.2016.07.010.
72. Yardley L. Contribution of somatosensory information to perception of the visual vertical with body tilt and rotating visual field. *Percept Psychophys*. 1990;48(2):131—134.
73. Lopez C, Lacour M, Magnan J, et al. Visual field dependence-independence before and after unilateral vestibular loss. *Neuroreport*. 2006;17(8):797—803. https://doi.org/10.1097/01.wnr.0000221843.58373.c8.
74. Goto F, Kobayashi H, Saito A, et al. Compensatory changes in static and dynamic subjective visual vertical in patients following vestibular schwanoma surgery. *Auris Nasus Larynx*. 2003;30(1):29—33.
75. Indovina I, Maffei V, Bosco G, et al. Representation of visual gravitational motion in the human vestibular cortex. *Science*. 2005;308(5720):416—419. https://doi.org/10.1126/science.1107961.
76. Bosco G, Carrozzo M, Lacquaniti F. Contributions of the human temporoparietal junction and MT/V5+ to the timing of interception revealed by transcranial magnetic stimulation. *J Neurosci*. 2008;28(46):12071—12084. https://doi.org/10.1523/JNEUROSCI.2869-08.2008.
77. Donaldson PH, Rinehart NJ, Enticott PG. Noninvasive stimulation of the temporoparietal junction: a systematic review. *Neurosci Biobehav Rev*. 2015;55:547—572. https://doi.org/10.1016/j.neubiorev.2015.05.017.
78. Kaski D, Quadir S, Nigmatullina Y, et al. Temporoparietal encoding of space and time during vestibular-guided orientation. *Brain*. 2016;139(Pt 2):392—403. https://doi.org/10.1093/brain/awv370.
79. Silani G, Lamm C, Ruff CC, et al. Right supramarginal gyrus is crucial to overcome emotional egocentricity bias in social judgments. *J Neurosci*. 2013;33(39):15466—15476. https://doi.org/10.1523/JNEUROSCI.1488-13.2013.
80. Dieterich M, Bense S, Lutz S, et al. Dominance for vestibular cortical function in the non-dominant hemisphere. *Cerebr Cortex*. 2003;13(9):994—1007.
81. Brandt T, Dieterich M, Danek A. Vestibular cortex lesions affect the perception of verticality. *Ann Neurol*. 1994;35(4):403—412. https://doi.org/10.1002/ana.410350406.
82. Piscicelli C, Barra J, Davoine P, et al. Inter- and intra-rater reliability of the visual vertical in Subacute stroke. *Stroke*. 2015;46(7):1979—1983. https://doi.org/10.1161/STROKEAHA.115.009610.
83. Baier B, Conrad J, Zu Eulenburg P, et al. Insular strokes cause no vestibular deficits. *Stroke*. 2013;44(9):2604—2606. https://doi.org/10.1161/STROKEAHA.113.001816.
84. Lopez C, Halje P, Blanke O. Body ownership and embodiment: vestibular and multisensory mechanisms. *Neurophysiol Clin*. 2008;38(3):149—161. https://doi.org/10.1016/j.neucli.2007.12.006.
85. Otero-Millan J, Winnick A, Kheradmand A. *Transcranial Magnetic Stimulation of Supramarginal Gyrus Alters Perception of Upright Without Changing Ocular Torsion*. San Diego, CA: Society for Neuroscience; 2016.
86. Dieterich M, Brandt T. Ocular torsion and tilt of subjective visual vertical are sensitive brainstem signs. *Ann Neurol*. 1993;33(3):292—299. https://doi.org/10.1002/ana.410330311.
87. Brandt T, Dieterich M. Vestibular syndromes in the roll plane: topographic diagnosis from brainstem to cortex. *Ann Neurol*. 1994;36(3):337—347. https://doi.org/10.1002/ana.410360304.
88. Merfeld DM, Zupan L, Peterka RJ. Humans use internal models to estimate gravity and linear acceleration. *Nature*. 1999;398(6728):615—618. https://doi.org/10.1038/19303.

89. Santos-Pontelli TE, Pontes-Neto OM, de Araujo DB, et al. Persistent pusher behavior after a stroke. *Clinics*. 2011;66(12):2169–2171.
90. Mansfield A, Fraser L, Rajachandrakumar R, et al. Is perception of vertical impaired in individuals with chronic stroke with a history of 'pushing'? *Neurosci Lett*. 2015;590:172–177. https://doi.org/10.1016/j.neulet.2015.02.007.
91. Perennou DA, Amblard B, Laassel el M, et al. Understanding the pusher behavior of some stroke patients with spatial deficits: a pilot study. *Arch Phys Med Rehabil*. 2002;83(4):570–575.
92. Perennou DA, Mazibrada G, Chauvineau V, et al. Lateropulsion, pushing and verticality perception in hemisphere stroke: a causal relationship? *Brain*. 2008;131(Pt 9):2401–2413. https://doi.org/10.1093/brain/awn170.
93. Perugini A, Ditterich J, Basso Michele A. Patients with Parkinson's disease show impaired use of priors in conditions of sensory uncertainty. *Curr Biol*. 2016;26(14):1902–1910. https://doi.org/10.1016/j.cub.2016.05.039.
94. Perugini A, Basso MA. Perceptual decisions based on previously learned information are independent of dopaminergic tone. *J Neurophysiol*. 2018 Mar 1;119(3):849–861.
95. Herz Damian M, Bogacz R, Brown P. Neuroscience: impaired decision-making in Parkinson's disease. *Curr Biol*. 2016;26(14):R671–R673. https://doi.org/10.1016/j.cub.2016.05.075.
96. Amick MM, Cronin-Golomb A, Gilmore GC. Visual processing of rapidly presented stimuli is normalized in Parkinson's disease when proximal stimulus strength is enhanced. *Vis Res*. 2003;43(26):2827–2835.
97. Amick MM, Grace J, Chou KL. Body side of motor symptom onset in Parkinson's disease is associated with memory performance. *J Int Neuropsychol Soc*. 2006;12(5):736–740. https://doi.org/10.1017/S1355617706060875.
98. Bodis-Wollner I, Marx MS, Mitra S, et al. Visual dysfunction in Parkinson's disease. Loss in spatiotemporal contrast sensitivity. *Brain*. 1987;110(Pt 6):1675–1698.
99. Cronin-Golomb A, Braun AE. Visuospatial dysfunction and problem solving in Parkinson's disease. *Neuropsychology*. 1997;11(1):44–52.
100. Davidsdottir S, Cronin-Golomb A, Lee A. Visual and spatial symptoms in Parkinson's disease. *Vision Res*. 2005;45(10):1285–1296. https://doi.org/10.1016/j.visres.2004.11.006.
101. Lee AC, Harris JP, Atkinson EA, et al. Disruption of estimation of body-scaled aperture width in Hemiparkinson's disease. *Neuropsychologia*. 2001;39(10):1097–1104.
102. Lee AC, Harris JP, Atkinson EA, et al. Evidence from a line bisection task for visuospatial neglect in left hemiparkinson's disease. *Vision Res*. 2001;41(20):2677–2686.
103. Lee AC, Harris JP, Calvert JE. Impairments of mental rotation in Parkinson's disease. *Neuropsychologia*. 1998;36(1):109–114.
104. Schendan HE, Amick MM, Cronin-Golomb A. Role of a lateralized parietal-basal ganglia circuit in hierarchical pattern perception: evidence from Parkinson's disease. *Behav Neurosci*. 2009;123(1):125–136. https://doi.org/10.1037/a0013734.
105. Rushton SK, Harris JM, Lloyd MR, et al. Guidance of locomotion on foot uses perceived target location rather than optic flow. *Curr Biol*. 1998;8(21):1191–1194.
106. Karnath HO, Schenkel P, Fischer B. Trunk orientation as the determining factor of the 'contralateral' deficit in the neglect syndrome and as the physical anchor of the internal representation of body orientation in space. *Brain*. 1991;114(Pt 4):1997–2014.
107. Hasselbach-Heitzeg MM, Reuter-Lorenz PA. Egocentric body-centered coordinates modulate visuomotor performance. *Neuropsychologia*. 2002;40(11):1822–1833.
108. Karnath HO. Subjective body orientation in neglect and the interactive contribution of neck muscle proprioception and vestibular stimulation. *Brain*. 1994;117(Pt 5):1001–1012.
109. Bostan AC, Dum RP, Strick PL. The basal ganglia communicate with the cerebellum. *Proc Natl Acad Sci U S A*. 2010;107(18):8452–8456. https://doi.org/10.1073/pnas.1000496107.
110. Bostan AC, Strick PL. The cerebellum and basal ganglia are interconnected. *Neuropsychol Rev*. 2010;20(3):261–270. https://doi.org/10.1007/s11065-010-9143-9.
111. Rascol O, Sabatini U, Fabre N, et al. The ipsilateral cerebellar hemisphere is overactive during hand movements in akinetic parkinsonian patients. *Brain*. 1997;120(Pt 1):103–110.
112. Yu H, Sternad D, Corcos DM, et al. Role of hyperactive cerebellum and motor cortex in Parkinson's disease. *Neuroimage*. 2007;35(1):222–233. https://doi.org/10.1016/j.neuroimage.2006.11.047.
113. Hoshi E, Tremblay L, Feger J, et al. The cerebellum communicates with the basal ganglia. *Nat Neurosci*. 2005;8(11):1491–1493. https://doi.org/10.1038/nn1544.

114. Ni Z, Pinto AD, Lang AE, et al. Involvement of the cerebellothalamocortical pathway in Parkinson disease. *Ann Neurol*. 2010;68(6):816–824. https://doi.org/10.1002/ana.22221.
115. Liedgren SR, Milne AC, Schwarz DW, et al. Representation of vestibular afferents in somatosensory thalamic nuclei of the squirrel monkey (*Saimiri sciureus*). *J Neurophysiol*. 1976;39(3):601–612.
116. Middleton FA, Strick PL. Basal ganglia output and cognition: evidence from anatomical, behavioral, and clinical studies. *Brain Cogn*. 2000;42(2):183–200. https://doi.org/10.1006/brcg.1999.1099.
117. Akbarian S, Grusser OJ, Guldin WO. Thalamic connections of the vestibular cortical fields in the squirrel monkey (*Saimiri sciureus*). *J Comp Neurol*. 1992;326(3):423–441. https://doi.org/10.1002/cne.903260308.
118. Bremmer F, Klam F, Duhamel JR, et al. Visual-vestibular interactive responses in the macaque ventral intraparietal area (VIP). *Eur J Neurosci*. 2002;16(8):1569–1586.
119. Gu Y, DeAngelis GC, Angelaki DE. A functional link between area MSTd and heading perception based on vestibular signals. *Nat Neurosci*. 2007;10(8):1038–1047. https://doi.org/10.1038/nn1935.
120. Lenz FA, Tasker RR, Kwan HC, et al. Single unit analysis of the human ventral thalamic nuclear group: correlation of thalamic "tremor cells" with the 3–6 Hz component of parkinsonian tremor. *J Neurosci*. 1988;8(3):754–764.
121. Shaikh AG, Straumann D, Palla A. Motion illusion-evidence towards human vestibulo-thalamic projections. *Cerebellum*. 2017;16(3):656–663. https://doi.org/10.1007/s12311-017-0844-y.
122. Dahlem K, Valko Y, Schmahmann JD, et al. Cerebellar contributions to self-motion perception: evidence from patients with congenital cerebellar agenesis. *J Neurophysiol*. 2016;115(5):2280–2285. https://doi.org/10.1152/jn.00763.2015.
123. Bertolini G, Ramat S, Bockisch CJ, et al. Is vestibular self-motion perception controlled by the velocity storage? Insights from patients with chronic degeneration of the vestibulo-cerebellum. *PLoS One*. 2012;7(6):e36763. https://doi.org/10.1371/journal.pone.0036763.
124. Bertolini G, Wicki A, Baumann CR, et al. Impaired tilt perception in Parkinson's disease: a central vestibular integration failure. *PLoS One*. 2015;10(4):e0124253. https://doi.org/10.1371/journal.pone.0124253.
125. Shaikh AG, Palla A, Marti S, et al. Role of cerebellum in motion perception and vestibulo-ocular reflex-similarities and disparities. *Cerebellum*. 2013;12(1):97–107. https://doi.org/10.1007/s12311-012-0401-7.
126. Tarnutzer AA, Palla A, Marti S, et al. Hypertrophy of the inferior olivary nucleus impacts perception of gravity. *Front Neurol*. 2012;3:79. https://doi.org/10.3389/fneur.2012.00079.
127. Stiles L, Smith PF. The vestibular-basal ganglia connection: balancing motor control. *Brain Res*. 2015;1597:180–188. https://doi.org/10.1016/j.brainres.2014.11.063.
128. Taube JS. The head direction signal: origins and sensory-motor integration. *Annu Rev Neurosci*. 2007;30:181–207. https://doi.org/10.1146/annurev.neuro.29.051605.112854.
129. Taube JS, Muller RU, Ranck Jr JB. Head-direction cells recorded from the postsubiculum in freely moving rats. I. Description and quantitative analysis. *J Neurosci*. 1990;10(2):420–435.
130. Baumann O, Mattingley JB. Medial parietal cortex encodes perceived heading direction in humans. *J Neurosci*. 2010;30(39):12897–12901. https://doi.org/10.1523/JNEUROSCI.3077-10.2010.
131. Stackman RW, Taube JS. Firing properties of head direction cells in the rat anterior thalamic nucleus: dependence on vestibular input. *J Neurosci*. 1997;17(11):4349–4358.
132. Muir GM, Brown JE, Carey JP, et al. Disruption of the head direction cell signal after occlusion of the semicircular canals in the freely moving chinchilla. *J Neurosci*. 2009;29(46):14521–14533. https://doi.org/10.1523/JNEUROSCI.3450-09.2009.
133. Yoder RM, Taube JS. Head direction cell activity in mice: robust directional signal depends on intact otolith organs. *J Neurosci*. 2009;29(4):1061–1076. https://doi.org/10.1523/JNEUROSCI.1679-08.2009.
134. Clark BJ, Brown JE, Taube JS. Head direction cell activity in the anterodorsal thalamus requires intact supragenual nuclei. *J Neurophysiol*. 2012;108(10):2767–2784. https://doi.org/10.1152/jn.00295.2012.
135. Mehlman ML, Winter SS, Valerio S, Taube JS. Functional and anatomical relationships between the medial precentral cortex, dorsal striatum, and head direction cell circuitry. I. Recording studies. *J Neurophysiol*. 2019 Feb 1;121(2):350–370.

Further reading

1. Bulens C, Meerwaldt JD, Van der Wildt GJ. Effect of stimulus orientation on contrast sensitivity in Parkinson's disease. *Neurology*. 1988;38(1):76–81.
2. Bulens C, Meerwaldt JD, van der Wildt GJ, et al. Contrast sensitivity in Parkinson's disease. *Neurology*. 1986;36(8):1121–1125.
3. Pieri V, Diederich NJ, Raman R, et al. Decreased color discrimination and contrast sensitivity in Parkinson's disease. *J Neurol Sci*. 2000;172(1):7–11.
4. Regan D, Maxner C. Orientation-selective visual loss in patients with Parkinson's disease. *Brain*. 1987; 110(Pt 2):415–432.
5. Trick GL, Kaskie B, Steinman SB. Visual impairment in Parkinson's disease: deficits in orientation and motion discrimination. *Optom Vis Sci*. 1994;71(4):242–245.
6. Harris JP, Atkinson EA, Lee AC, et al. Hemispace differences in the visual perception of size in left hemiParkinson's disease. *Neuropsychologia*. 2003;41(7):795–807.
7. Harris MG, Carre G. Is optic flow used to guide walking while wearing a displacing prism? *Perception*. 2001;30(7):811–818.
8. Owen AM, Beksinska M, James M, et al. Visuospatial memory deficits at different stages of Parkinson's disease. *Neuropsychologia*. 1993;31(7):627–644.
9. Davidsdottir S, Wagenaar R, Young D, et al. Impact of optic flow perception and egocentric coordinates on veering in Parkinson's disease. *Brain*. 2008;131(Pt 11):2882–2893. https://doi.org/10.1093/brain/awn237.
10. Ren X, Salazar R, Neargarder S, et al. Veering in hemi-Parkinson's disease: primacy of visual over motor contributions. *Vision Res*. 2015;115(Pt A):119–127. https://doi.org/10.1016/j.visres.2015.08.011.
11. Young DE, Wagenaar RC, Lin CC, et al. Visuospatial perception and navigation in Parkinson's disease. *Vision Res*. 2010;50(23):2495–2504. https://doi.org/10.1016/j.visres.2010.08.029.
12. Stein BE, Stanford TR. Multisensory integration: current issues from the perspective of the single neuron. *Nat Rev Neurosci*. 2008;9(4):255–266. https://doi.org/10.1038/nrn2331.
13. Schlicht EJ, Schrater PR. Impact of coordinate transformation uncertainty on human sensorimotor control. *J Neurophysiol*. 2007;97(6):4203–4214. https://doi.org/10.1152/jn.00160.2007.
14. Bronstein AM, Rudge P. Vestibular involvement in spasmodic torticollis. *J Neurol Neurosurg Psychiatry*. 1986;49(3):290–295.
15. Stell R, Bronstein AM, Marsden CD. Vestibulo-ocular abnormalities in spasmodic torticollis before and after botulinum toxin injections. *J Neurol Neurosurg Psychiatry*. 1989;52(1):57–62.
16. Diamond SG, Markham CH, Baloh RW. Ocular counterrolling abnormalities in spasmodic torticollis. *Arch Neurol*. 1988;45(2):164–169.
17. Diamond SG, Markham CH, Baloh RW. Vestibular involvement in spasmodic torticollis: an old hypothesis with new data from otolith testing. *Adv Oto Rhino Laryngol*. 1988;42:219–223.
18. Huygen PL, Verhagen WI, Van Hoof JJ, et al. Vestibular hyperreactivity in patients with idiopathic spasmodic torticollis. *J Neurol Neurosurg Psychiatry*. 1989;52(6):782–785.
19. Anastasopoulos D, Bhatia K, Bisdorff A, et al. Perception of spatial orientation in spasmodic torticollis. Part I: the postural vertical. *Mov Disord*. 1997;12(4):561–569. https://doi.org/10.1002/mds.870120413.
20. Bray A, Subanandan A, Isableu B, et al. We are most aware of our place in the world when about to fall. *Curr Biol*. 2004;14(15):R609–R610. https://doi.org/10.1016/j.cub.2004.07.040.

CHAPTER

10

Multisensory flavor perception: A cognitive neuroscience perspective

Charles Spence

Crossmodal Research Laboratory, Department of Experimental Psychology, Oxford University, Oxford, United Kingdom

Introduction

Eating and drinking are among life's most enjoyable experiences and at the same time also among the most multisensory.[1–3] While the focus of much of the scientific research relating to flavor in recent years has been on modernist cuisine (i.e., on the scientific approach to the preparation of food and drink; e.g., Refs. 4,5), there is a growing realization that the pleasures of the table are as much about what is going on in the mind as what is going on in the mouth of the person who is eating or drinking.[6] This has led the neuroscientist Gordon Shepherd[7] to establish a new field of research known as "neurogastronomy"—essentially the study of the brain on flavor (though see Ref. 8).[a] Others, meanwhile, have chosen to stress the importance of systematically studying our food perception and behaviors under more naturalistic (i.e., ecologically valid) conditions than are typically possible in the scanner environment. The latter approach goes by the name of "gastrophysics."[6] What the research tends to agree on though is that all of the senses can potentially contribute to the experience of the taste/flavor of food and drink. What is more, these multisensory contributions are not only restricted to the sensory properties of the foodstuff itself but also extend to the sensory attributes of the packaging, plateware, cutlery, and even the environments in which we happen to eat and drink.

At the outset, one can ask in what ways the multisensory integration and crossmodal perceptual phenomena that apply to food and drink stimuli are similar to/different from what one finds when considering the integration of the other, nonchemical, sensory inputs (see the other chapters in this volume). To the extent that the same rules apply, this would potentially allow for the emerging insights from the relatively "easier-to-study" senses (namely vision and hearing, and to a lesser extent touch) to be extended to the typically

[a]Not forgetting the new and related field of "neuroenology."[9]

Multisensory Perception
https://doi.org/10.1016/B978-0-12-812492-5.00010-3

"harder-to-study" chemical senses. Early attempts to transfer our understanding across sensory domains have already provided a number of useful insights (e.g., see, Refs. 10–12 for some representative early examples). Of course, one might also want to know whether any of the insights from studies of the chemical senses have implications in the opposite direction too, that is for the way in which we think about the integration of the so-called "higher" spatial senses (e.g., Refs. 2,13). The answer here, too, is in the affirmative, though there are fewer examples.

Food and the brain

At the outset, it is important to note that the brain evolved to predict/track nutrients in the environment (e.g., Refs. 14–17). It has even been suggested that trichromatic color vision may have evolved to help our ancestors discriminate the ripe red fruits from among the dense green forest canopy (e.g., see Refs. 18–20).[b] As such, even if one is not interested in the brain's response to flavor stimuli per se, one should at least be aware, as the eminent British biologist J. Z. Young put it half a century ago now, that: "*No animal can live without food. Let us then pursue the corollary of this: namely, food is about the most important influence in determining the organization of the brain and the behavior that the brain organization dictates*" Young[22(p21)]. Beyond that, it is perhaps also worth noting that it is presumably no coincidence that, as once again pointed out by Young, the brain is situated close to the point of ingestion (in humans, the mouth) in pretty much all known creatures.[c]

Rather than having to stick a potential foodstuff into our mouths to know whether it is likely to be nutritious or not, our brains make predictions given the various food experiences that we have had previously. In fact, by 4 months of age, young infants are already learning to associate particular colors with specific tastes/flavors (Ref. 23; see also, Ref. 24 for a review). These flavor predictions, derived from what we see and smell orthonasally ("orthonasal" smell when we inhale or sniff can be contrasted with "retronasal" olfaction, volatile-rich air pulsed out from the back of the nose/throat when we swallow;[25]), but also sometimes from what we hear[26] and feel,[27] help to set the food-related expectations that tend to anchor the subsequent multisensory flavor experience when we eventually come to taste a food or drink item (see, Ref. 28 for a review).[d] If the tasting experience turns out to be close to what our expectations led us to believe was going to be the case then our perceptual experience tends to mirror our flavor expectations (determined in advance of tasting). If, however, what we taste turns out to be far removed from our expectations, then that is likely

[b]One other evolutionary argument that one sometimes hears is that the shift from four to two legs resulted in a reduction in the importance of olfaction relative to vision in our ancestors. However, researchers have recently provided evidence against this explanation of visual dominance and olfactory decline. In fact, the sense of smell in humans turns out to be remarkably good (see, Ref. 21 for an excellent recent reevaluation of the literature on olfaction in humans).

[c]As Young([22], p. 22) puts it: "*The fact that the brain and the mouth are both at the same end of the body may not be as trivial as it seems.*"

[d]Note also that we also have expectations concerning the fact that foods will not suddenly change their taste (see Refs. 29, 29a).

to result in a negatively valenced "disconfirmation of expectation" response instead, or else no influence of the expectations on the experience (e.g., Refs. 30–32). Now, quite how wide the separation between the expectation and the experience needs to be to trigger these various outcomes is understandably of great interest to many of those working in the field (both academically and commercially). Unfortunately, however, as yet, we do not have a good way of predicting the outcome.

It is important to remember here that how people respond to surprise (i.e., a noticeable discrepancy between the expected and the actual taste/flavor experience) in food and drink is very much context-dependent. Indeed, what might be pleasurable/amusing in the context of a modernist restaurant, say, may well be rather less pleasant for someone who finds themselves in the science lab or lying in the brain scanner (see, Refs. 33,34 for reviews).[e] In fact, many of the food and beverage companies who have attempted to capitalize on surprise in their product offerings (i.e., by having them look one way, but taste very different) have failed to sustain the interest of consumers into the longer term (see Ref. 35). Certainly, according to some researchers,[36] getting the perception of food and drink right (e.g., so as to avoid the danger of poisoning) is so important that one might expect crossmodal illusions to be much rarer in relation to the foods that we ingest (cf.[37]) than when studying interactions between the spatial senses of hearing, vision, and touch, say.

There are few stimuli that get our brains going like the sight and smell of our favorite foods when we are hungry (e.g., Refs. 38–40). For instance, in one positron emission tomographic study reported by Wang and colleagues, an average 24% increase in cerebral metabolism was documented in 12 food-deprived participants shown pictures of their favorite foods while they talked about eating them. Note also that the food in question was warmed to deliver the relevant orthonasal olfactory cues to the participant's nostrils and a Q-tip that had been dipped in the food was rubbed on their tongues to give them a taste of the food too. The increases in brain metabolism reported in the study were particularly pronounced in the orbitofrontal cortex (OFC), the anterior insula, and the superior temporal regions. Killgore and Yurgelun-Todd[41] have shown that an individual's body mass provides an accurate predictor of orbitofrontal brain activity during the visual presentations of high-calorie foods.

Perhaps surprisingly, the neural activation seen in response to the anticipation of food is typically much greater than when actually consuming it (e.g., Refs. 42,43). According to an extensive recent review of the functional magnetic resonance imaging (fMRI) literature on eating by Chen, Papies, and Barsalou,[44] a core brain network underlies our various eating behaviors including a ventral reward pathway and a dorsal control pathway. These authors have argued that diverse fMRI data are broadly consistent with the view that neural responses to food cues (including food pictures) use the same core eating network as when engaged in eating itself. Based on the literature on grounded cognition (e.g., Refs. 45,46), their argument is that food cues activate eating simulations that produce reward predictions about a perceived food and potentially motivate its consumption.

Malik, McGlone, Bedrossian, and Dagher[47] have reported that neural responses to images of appetizing foods shown to the participants in their fMRI study were modulated by levels

[e]Notice that surprise in food is particularly hard to study using neuroscience techniques, given the difficulty associated with trying to "surprise" people repeatedly with food (and so acquire sufficient data to analyze noisy brain signals).

of circulating ghrelin (a peptide hormone secreted by the gut and known as the "hunger hormone"). The effects of ghrelin, administered intravenously, on the neural responses seen in the OFC and amygdala were correlated with self-rated hunger ratings, the suggestion here being that metabolic signals such as ghrelin may bias us toward food stimuli by enhancing the value placed on food-related cues. Meanwhile, leptin, the so-called "satiety hormone" (made by adipose cells), which helps to regulate the energy balance by inhibiting hunger (working in opposition to ghrelin), has also been shown to modulate the brain's response to appetizing images of food (see Ref. 48).

Another relevant finding to have emerged from the research conducted over the last decade or so is that our attention is directed preferentially to images of food. In fact, we appear to be especially drawn to (images of) high-fat (that is, energy-dense) foods (e.g., Ref. 17; see also Refs. 49,50). Our visual attention is also captured preferentially by those foods that we like.[51,52] And, as one might have guessed, attentional capture by images of food is also more pronounced in those participants who are hungry (e.g., Ref. 53; see also Refs. 54,55). By contrast, attentional capture is reduced for images of those foods that have just been eaten to satiety.[56,57] Finally here, Nummenmaa, Hietanen, Calvo, and Hyönä[58] have demonstrated that the capture of visual attention by images of food is contingent on a participant's body mass index.[f] According to Chen et al.,[44] many of the just-mentioned neuroimaging results can be accounted for by the modulation of a subset of the areas in the so-called core eating network.

Before concluding this section, it is perhaps worth drawing attention to the fact that the majority of research on visual influences on brain activity and over other physiological responses (such as salivation; see, Ref. 60 for a review) have been conducted with participants viewing images of foods presented on a computer monitor, rather than using real food stimuli. Similarly, those studies that have attempted to assess neural markers of particular food choices/behaviors have also typically often involved virtual food selection tasks (e.g., see Ref. 61). This may turn out to be important, at least given suggestive findings implying that the physiological/neurophysiological responses seen may sometimes differ as a function of whether a participant actually has any real chance of eating (or at least believes that they will get to eat) the foods that they are viewing/evaluating (e.g., Ref. 62; see also Ref. 60). Nevertheless, taken together, the results reviewed in this section highlight the continuous interplay between food cues and observer states and traits that will determine just how attention-capturing, and hedonically pleasing, foods (or, as is more often the case in these experimental studies, images of foods) are.[g]

[f]Researchers have also demonstrated increased resting state brain activity in the oral—somatosensory cortex of obese individuals.[59]

[g]Although, as yet, less well-studied, there has also been some intriguing neuropsychological research documenting the emergence of "Gourmand syndrome"—identified as a preoccupation with food and a preference for quality food experiences—following right anterior brain lesions.[63] Meanwhile, Camus, Halelamien, and Plassmann, et al.[64] have reported that repetitive transcranial magnetic stimulation, another form of neural disruption, over the right dorsolateral prefrontal cortex decreases valuations during food choices.

Multisensory flavor perception: gustatory and olfactory inputs

Researchers disagree on quite how flavor perception should be defined and, in particular, which senses are constitutive of flavor versus which are merely modulatory (see, Ref. 65 for a review). However, the two senses that everyone would seem to agree as being constitutive of flavor are taste (or gustation) and retronasal olfaction. Note here that only "retronasal" smell is constitutive of flavor, while orthonasal olfaction helps set our flavor expectations. The taste papillae that are grouped together in the taste buds distributed over the anterior surface of the tongue detect sapid molecules giving rise to the perception of the so-called basic tastes of sweet, sour, bitter, salty, and umami[66].[h] Gustation appears to rely on distributed neural processing (e.g., Ref. 71). Intriguing research in rats suggests that the various basic tastes may actually be coded spatiotopically in primary taste cortex.[72]

It has been widely reported that somewhere in the region of 70%–95% of what most people commonly think of as "taste" really comes not from the taste buds but rather from information detected by the receptors in the olfactory epithelium instead. While coming to a precise estimate of olfaction's role in flavor perception is undoubtedly difficult if not impossible (see, Ref. 73 on this theme), everyone would, at least, seem to agree that olfaction is the most important input as far as the constitutive flavor senses are concerned. It is, after all, olfaction that gives rise to the meaty, fruity, herbal, burnt, floral notes in food and drink. But the question remains as to why people should be confused about the source of information (retronasal smell) giving rise to their taste sensations? The phenomenon of "oral referral," whereby the information coded by the olfactory receptors in the nasal mucosa is mislocalized so as to be perceived as originating from the mouth, may help explain why most of us remain unaware of the olfactory contribution to taste[74].[i] Some have been tempted to refer to this as a kind of ventriloquist illusion in the world of flavor. What is clear is that the intimate link between gustatory and olfactory information processing (e.g., Refs. 75,76) is unlikely to be replicated for any other pairing of senses.

What is less often commented on, though, is the similarity in the orthonasal and retronasal olfactory experience. In fact, despite the fact that some researchers have wanted to argue that orthonasal and retronasal olfaction represent two distinct senses of smell,[25] the evidence suggests that people are able to match olfactory stimuli delivered by the two different routes with surprisingly little difficulty under most everyday situations.[77,78]

Evidence that orthonasal olfactory and gustatory cues are subject to multisensory integration comes from research conducted by Dalton et al.,[10]; see also Ref. 79). The participants in Dalton et al.'s studies had to decide which bottle from two pairs of bottles contained the

[h]Note here that this is not an exhaustive list. There are probably a few more basic tastes, such as the taste associated with certain fatty acids (Ref. 67; see also Ref. 68). There are also other stimuli that researchers have a hard time classifying as tastes, flavors, aromas, and/or trigeminal stimulants. For instance, some researchers have wanted to suggest that metal salts such as ferrous sulfate can give rise to a metallic taste, whereas others suspect that metallic might (also) be a flavor (see Ref. 69). Compounds such as 1-menthol (the principal flavor compound in mint) also appear to stimulate receptors in the oral cavity as well as temperature receptors (cooling), and a distinctive minty aroma.[70]

[i]Notice how, when one's nose is blocked, people mostly say that food loses its taste. In most case, however, the taste buds are working just fine, it is the olfactory input that has been removed from the experience due to blockage of the nasal airways. However, we seem mostly unaware of this.

almond-cherry like scent of benzaldehyde by sniffing them. The scent had been added to just one of the four bottles. The concentration of the olfactant was varied on a trial-by-trial basis to home in on each participant's detection threshold. Surprisingly, when the participants performed this task while holding a subthreshold solution of saccharin in their mouths (i.e., a solution that had no discernible taste or smell), the cherry-almond smell was perceived as significantly more intense relative to a baseline condition in which a tasteless water solution was held in the mouth instead (see Fig. 10.1). By contrast, holding a subthreshold solution of monosodium glutamate on the tongue did not give rise to any such change in the perceptibility of the benzaldehyde aroma. Such results have been taken to suggest that tastants and odorants are integrated in a stimulus combination-specific manner.

Similar results have subsequently been reported in a number of other studies. For instance, one group of researchers demonstrated a 50% lowering of the olfactory threshold—i.e., complete additivity—in the majority of participants when the relevant gustatory and olfactory stimuli were presented simultaneously.[80] Importantly, though, similar results were reported regardless of whether the odor was delivered orthonasally or retronasally (only the latter, remember, being a constitutive part of our multisensory flavor experiences). Moving the experimental situation even closer to everyday life, similar effects have now been reported with participants tasting actual flavored solutions (see Ref. 81; see, Ref. 24 for a review).

Olfactory–gustatory interactions are interestingly different from what is seen between, for example, the spatial senses. On the one hand, regular coexposure of specific olfactory and gustatory stimuli leads to the olfactory stimulus taking on the taste qualities.[82] Some have even been tempted to suggest that we are all synesthetic as far as smell–taste confusions.[83] Others, however, have argued against this notion.[84] Cross-cultural research suggests that our brains learn to combine those tastes and smells that regularly cooccur in the foods that we grow up with. The underlying idea here is that while everyone's brain may use the same rules to combine the various sensory inputs to deliver multisensory flavor perception,

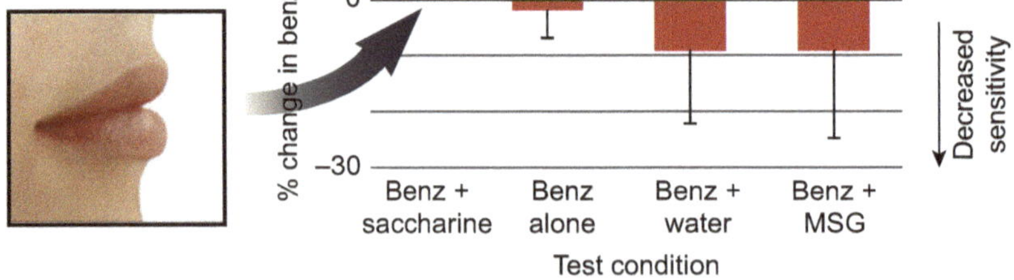

FIGURE 10.1 Multisensory interactions between congruent orthonasal olfactory and gustatory stimuli. The figure shows the results of a series of experiments by Dalton et al.,[10] showing the integration of congruent, but not of incongruent, orthonasal olfactory and gustatory cues. Congruency here determined by prior food experiences. Benz: benzaldehyde.

the particular combinations of tastants and olfactory stimuli that lead to multisensory enhancement (or suppression, when the taste and smell do not match; see e.g., Ref. 85) depends on the combination of ingredients, and hence of sensory cues, that tend to cooccur in the cuisine of the region where people have grown up. Such learning apparently starts in utero (see Ref. 86; and, Ref. 87 for a review). Schaal and his colleagues have demonstrated that neonates whose mothers consumed anise-flavored food during pregnancy are more likely to orient toward the smell of anise after birth, while elsewhere it has been shown that young children are more likely to eat carrots if their mothers happened to drink carrot-flavored milk during pregnancy.

Oral–somatosensory contributions to multisensory flavor perception

The trigeminal sense, giving rise to burning and cooling sensations (e.g., of chilli and menthol, respectively), is also considered a constitutive, if not always a necessary, component of flavor perception (Ref. 88; see, Ref. 65 for a review). However, it is important to note that nociceptive stimuli such as capsaicin (of chilli pepper) not only stimulate the trigeminal nerve but also the glossopharyngeal and vagus nerves as well.[89] Intriguingly, beyond the well-known mutual suppression of one taste by another (e.g., Refs. 90,91), fMRI research has demonstrated that carbonation (e.g., in a fizzy drink) also suppresses the brain's response to sweetness (especially to sucrose, and less to artificial sweeteners).[92]

To date, there has been far less research on the oral–somatosensory (or mouthfeel) contributions to flavor perception, relative to what one finds for other areas (senses). In part, this is presumably because of the difficulty associated with delivering carefully calibrated sets of experimental stimuli, where all that varies are the texture/oral somatosensory properties of the samples (see, Ref. 27 for a review). Indeed, one of the problems with early research in this area related to difficulties associated with disentangling the effects of any change in viscosity on the release of aromatic volatiles from the surface of a food/liquid (i.e., physicochemical interactions;[93,94] from any impact attributable to multisensory neural integration effects). In recent years, though, researchers have managed to deliver retronasal aromas and textured substrates separately to the mouths of their participants. By so doing, it has been possible to demonstrate the crossmodal interactions taking place between olfaction and taste, such that the addition of a "creamy" aroma, say, may result in people asserting that the substance that they are evaluating in their mouth is thicker/creamier than might otherwise be the case (e.g., Refs. 95,96). De Araujo and Rolls[97] have also conducted the relevant neuroimaging research detailing the neural representation of food texture and oral fat.

Although a detailed discussion is beyond the scope of the present review, it is also interesting to note how the felt texture of food/plateware in the hand can influence multisensory flavor perception (e.g., Refs. 98–103). So, for instance, Biggs et al. were able to show that people will rate ginger biscuits as tasting more gingery when served from a plate having a rough texture than from a smooth plate instead (see chapter by Spence and Sathian, this volume). Meanwhile, Kampfer, Leischnig, Ivens, and Spence[104] recently demonstrated that people rated a can of soft drink and a chocolate taken from an assortment box as tasting better when a small weight was added to the bottom of the can/box.

Visual contributions to multisensory flavor perception

Perhaps the most salient visual attribute of food and drink is its color. In fact, since the first study by the British chemist Moir[105] showing that changing the color of food changed what people had to say about the flavor, more than 200 studies have been published (see, Refs. 106,107 for reviews). For practical reasons, the majority of this research has been conducted on colored beverages, as it is simply much easier to control the stimuli (i.e., liquids are easier to color than solid foods like cakes; see Ref. 108). The majority, but by no means all, of the research shows that changing the color of food or drink can change the rated taste/flavor intensity, as well as the flavor identified. There is even evidence to suggest that taste thresholds can be influenced by the presence versus absence of color.[109] The question of how much of color's effect on suprathreshold ratings/identification responses should be attributed to more decisional, as opposed to interactions operating at a more perceptual, level is a topic that has interested researchers for decades (see, Ref. 110 on this). Surprisingly, instructions to ignore the color of a drink (e.g., because it is likely misleading concerning the drink's flavor) do not seem to exert much of an influence over visual dominance effects (e.g., see, Refs. 111–113 for a couple of representative examples).

The magnitude of color's influence on flavor identification has been shown to vary both as a function of a person's age (see, Ref. 24 for a review) and their taster status.[113] Some people, known as supertasters, have far more taste buds on their tongues than nontasters, in some cases up to 14 times as many (Ref. 114; though see also Ref. 115). While supertasters tend to be more sensitive to bitter tastes and to oral–somatosensory textures, such as the fattiness of salad dressings and ice cream,[116] Zampini and his colleagues[113] also showed that supertasters (or at least those who were more sensitive to propylthiouracil, one of the standard chemicals used to test for taster status) were less influenced by the presence of an inappropriate color in a range of fruit-flavored drinks.

The impact of color has been shown to depend on the taste/flavor that the taster normally associates with foods of that color.[117,118] That said, as Velasco et al.[35] have shown recently, some of these patterns (such as the association between the color red and sweetness) appear to be present regardless of the continent from which the participant comes from (see Table 10.1).

To date, there has only been limited neuroimaging research around the topic of the crossmodal influences of color on flavor perception (see Ref. 119; see also Ref. 32). For instance, Österbauer et al. reported a superadditive neural response in the OFC following the presentation of congruent (as compared with incongruent) combinations of visual and olfactory stimuli (just imagine a strawberry aroma being paired with the color red vs. with turquoise instead). Elsewhere, researchers have demonstrated that when people are presented with food-relevant odors that they cannot identify immediately, they often recruit visual cortex, as if trying to generate an appropriate visual mental image that may help them to identify the likely source of the odor (Refs. 120,121; see also, Ref. 122 and chapter by Lacey and Sathian, this volume, on the topic of crossmodal mental imagery).

A growing body of research conducted over the last 5 years or so has started to demonstrate how the background color against which a food stimulus is seen can also impact people's taste/flavor ratings too. So, for example, in one of the first studies to be published in this

TABLE 10.1 Summary of the percentage of color responses to the question "Which drink looks sweetest?" as a function of the region from which the participant originated in a study conducted at London's Science Museum. The participants saw six different colored drinks on the screen. The "N" column indicates the number of participants from each region.

	Color						
Region	**Blue**	**Green**	**Orange**	**Purple**	**Red**	**Yellow**	*N*
Africa	21.62	4.05	9.46	18.92	43.24	2.70	74
Asia	17.03	3.47	6.94	28.39	37.22	6.94	317
Europe	20.94	1.87	8.00	22.89	42.03	4.26	1337
North America	28.61	1.77	5.31	11.21	48.08	5.01	339
Oceania	26.67	2.00	4.67	19.33	41.33	6.00	150
South America	16.51	0.00	5.50	22.02	51.38	4.59	109
United Kingdom	32.58	1.27	5.48	15.64	39.09	5.95	2993
None	0.00	0.00	0.00	0.00	66.67	33.33	3
Total	27.81	1.62	6.22	18.21	40.68	5.47	5322

Table reprinted with permission from Velasco C, Michel C, Youssef J, et al., Colour-taste correspondences: designing food experiences to meet expectations or to surprise. Int J Food Des 2016;1:83–102.

area, Piqueras-Fiszman, Alcaide, Roura, and Spence[123] conducted a within-participants experiment at the Alicia Foundation in Spain in which people were shown to rate a pinkish-red strawberry mousse as tasting 7% sweeter, 13% more flavorful, and 9% more liked when tasted from a round white plate than from a round black plate instead. In the years since this study was published, several further studies have been reported that have come to much the same conclusion (e.g., Refs. 124–126). Elsewhere, it has been reported that red plateware leads to a reduction in people's consumption of those foods that they consider to be less healthy (see Refs. 127–129; see, Ref., 130 for a review).

Beyond the color of the food/drink, or the plateware/glassware against which it is presented, a few researchers have also assessed the impact of changing the color of the environment on people's tasting experiences.[131,132] So, for example, Spence et al.[132] tested nearly 3500 people in a study in which the latter had to rate a glass of red wine (served in a black tasting glass so as to completely obscure its visual appearance) while under normal white lighting, while under red ambient illumination, and while under green lighting. The results showed a near 15% mean change in people's rating of the fruitiness/freshness of one and the same wine (a Rioja) simply as a function of the change in the color of the ambient lighting.

Elsewhere, researchers have also reported that the brightness of the lighting can bias our behavior around spicy food,[133] and the consumption of bitter-tasting coffee (Ref. 134; see also Ref. 135). Taken together, the research that has been published over the last 80 years or so clearly indicates that color plays an important role in setting our taste/flavor expectations and hence modulating our experience of food and drink. While everyone is likely influenced by the appearance properties of foods, specific taste/flavor associations that we hold with different food colors differ somewhat by region/culture (see Refs. 117,118), and the impact of color most definitely depends on the context in which the color is seen (i.e., is the color in the product, in the plateware, in the environment, or in the product packaging). Before closing this section, it is worth noting that other visual appearance cues likely also matter though there has been far less research on turbidity/glossiness to date.[136]

Auditory contributions to multisensory flavor perception

Described by some as "the forgotten flavor sense" (see Ref. 6), there is growing evidence that what we hear when (and just before) we eat and drink can exert a profound influence over our multisensory tasting experiences (Ref. 137; see, Ref. 26 for a review). For example, the perceived crispness/crunchiness of potato chips and apples can be modified simply by filtering or boosting the consumer-produced sounds of biting/mastication (Refs. 37,138; see also Ref. 113).[j] Relevant to the themes of this chapter, this approach to the study of multisensory flavor perception emerged out of basic research on the parchment skin illusion (e.g., Refs. 141,142), whereby people's experience of the feel of a textured surface can be changed simply by changing the sound that they hear.

However, beyond the sound of our interaction with food while eating and drinking, it is important to note that the sounds of preparation and even the sound of the packaging can set certain expectations about the taste experience to come. These expectations, in turn, can modify our subsequent perception/ratings[6] everything from the sizzle of the steak on the hotplate (Ref. 143; see also Ref. 144), through the rattle of the crisp packet,[145] and the sound of the coffee machine. In some of our most recent research (see Ref. 137), we investigated the impact of the sound of a bottle closure on the rated quality of wine. 140 people were given four glasses of wine to taste. However, before tasting each glass of wine, the participants heard the sound of either a cork being pulled from a wine bottle or else the sound of a screw-cap bottle being opened. In fact, which of the two wines was presented with which opening sound was counterbalanced across participants. Nevertheless, the results demonstrated that the British participants tested in this experiment rated the Argentinian red wines as up to 15% higher in quality when tasted after hearing the pop of the cork.[k]

One stage further removed from the food and drink itself, the background noise levels in the places in which we eat and drink can also modify our taste perception. So, for example, researchers have documented a suppression of sweet and salt perception and an enhancement in the perception of umami when participants are exposed to loud white noise[146,147] or airplane engine noise.[148] On the opposite side, though, music and soundscapes can also be used to enhance the sensory-discriminative and/or hedonic attributes of food (Ref. 149; see, Ref. 26 for a review). This is a whole new area of research that is known as "sonic seasoning." One question that is of particular interest currently is to determine just how early

[j]One question that deserves some serious scientific consideration is why it is that no one likes a soggy crisp (or potato chip for those in the United Kingdom; see Ref. 139). Indeed, some of the latest evidence has suggested that noisy foods may retain their flavor for longer (i.e., the sonic element would seem to counteract the effects of adaptation (see Ref. 140).

[k]Subsequent questioning revealed that the majority of the participants preferred wine from cork-stoppered bottles.

in neural information processing the effect of sonic seasoning operates. Can playing putatively "sweet,"[149] "spicy,"[31] or "creamy" music[150] influence the neural processing of incoming food stimuli at the level of the relevant primary sensory areas (for some preliminary data in this regard, see Ref. 151; cf. 152)?

Another important topic for neuroimaging research in the years to come will be to assess the extent to which attention can be used to account for the behavioral/neural changes associated with labeling, sonic seasoning, etc. Certainly, there is already evidence that selective attention to a sensory modality can modulate the processing of gustatory, olfactory, and flavor stimuli (e.g., Refs. 153−155). One popular suggestion is that music can be used to draw people's attention to something in their tasting experience and so make it more salient.[156] Of course, crossmodal influences may operate in the opposite direction too (though see Ref. 157).

Conclusions

There can be little doubting the fact that our understanding of multisensory flavor perception has come a long way in recent years, due to the insights emerging from the study of the spatial senses (e.g., audition, vision, and touch) being extended to the harder-to-study chemical senses. That said, researchers have yet to come to any kind of consensus regarding which senses are constitutive of multisensory flavor perception and which other senses merely modulate the experience.[65] One of the important distinctions to have emerged is between flavor expectations on the one hand and the multisensory experience of flavor on the other (see Refs. 2,28 for reviews). Gustation, retronasal olfaction, and trigeminal/oral-somatosensory cues are often considered to be constitutive of flavor. By contrast, visual, orthonasal olfactory, auditory, and nonoral tactile cues play an important role in helping set our flavor expectations, and hence anchoring the flavor experience that may follow. It is clear that different combinations of senses are involved in the various stages of our interaction with food and drink and different neural substrates are involved too.[7,43,44,158−160] The latest neuroimaging has started to reveal the many factors determining which complex (i.e., real-world) food stimuli we attend to. Thus far, a number of insights have emerged from the research, showing that the brain's response to both the sensory-discriminative and hedonic aspects of food depends on many factors including a person's hunger state and body mass, as well as their own food preferences (see Ref. 44 for a review). In the future, the hope for many is that our growing understanding of how the human brain processes flavor stimuli may help provide interventions that may in some small way tackle the growing obesity crisis (see Ref. 7).

References

1. Brillat-Savarin JA. Physiologie du goût [The philosopher in the kitchen/the physiology of taste]. In: Meline JP, ed. *Bruxelles. Translated by A. Lalauze (1884), A Handbook of Gastronomy*. London, UK: Nimmo & Bain; 1835.
2. Spence C. Multisensory flavour perception. *Cell*. 2015;161:24−35.
3. Spence C, Piqueras-Fiszman B. *The Perfect Meal: The Multisensory Science of Food and Dining*. Oxford, UK: Wiley-Blackwell; 2014.
4. Myhrvold N, Young C. *Modernist Cuisine. The Art and Science of Cooking*. La Vergne, TN: Ingram Publisher Services; 2011.

5. Youssef J. *Molecular Cooking at Home: Taking Culinary Physics Out of the Lab and into Your Kitchen*. London, UK: Quintet Publishing; 2013.
6. Spence C. *Gastrophysics: The New Science of Eating*. London, UK: Viking Penguin; 2017.
7. Shepherd GM. *Neurogastronomy: How the Brain Creates Flavor and Why it Matters*. New York, NY: Columbia University Press; 2012.
8. Spence C. Book review: 'Neurogastronomy: how the brain creates flavor and why it matters' by Gordon M. Shepherd. *Flavour*. 2012;1:21.
9. Shepherd GM, Neuroenology. How the brain creates the taste of wine. *Flavour*. 2015;4:19.
10. Dalton P, Doolittle N, Nagata H, et al. The merging of the senses: integration of subthreshold taste and smell. *Nat Neurosci*. 2000;3:431–432.
11. Small DM, Jones-Gotman M, Zatorre RJ, et al. Flavor processing: more than the sum of its parts. *Neuroreport*. 1997;8:3913–3917.
12. Zampini M, Spence C. The role of auditory cues in modulating the perceived crispness and staleness of potato chips. *J Sensory Sci*. 2004;19:347–363.
13. Small DM. Crossmodal integration – insights from the chemical senses. *Trends Neurosci*. 2004;27:120–123.
14. Allen JS. *The Omnivorous Mind: Our Evolving Relationship with Food*. London, UK: Harvard University Press; 2012.
15. Spence C, Okajima K, Cheok AD, et al. Eating with our eyes: from visual hunger to digital satiation. *Brain Cogn*. 2016;110:53–63.
16. Tang DW, Fellows LK, Dagher A. Behavioral and neural valuation of foods is driven by implicit knowledge of caloric content. *Psych Sci*. 2014;25:2168–2176.
17. Toepel U, Knebel J, Hudry J, et al. The brain tracks the energetic value in food images. *Neuroimage*. 2009;44:967–974.
18. Bompas A, Kendall G, Sumner P. Spotting fruit versus picking fruit as the selective advantage of human colour vision. *I-Perception*. 2013;4:84–94.
19. Sumner P, Mollon JD. Catarrhine photopigments are optimized for detecting targets against a foliage background. *J Exp Biol*. 2000;203:1963–1986.
20. Sumner P, Mollon JD. Did primate trichromacy evolve for frugivory or folivory? In: Mollon JD, Pokorny J, Knoblauch K, eds. *Normal and Defective Colour Vision*. Oxford, UK: Oxford University Press; 2003:21–30.
21. McGann JP. Poor human olfaction is a 19th-century myth. *Science*. 2017;356. eaam7263.
22. Young JZ. Influence of the mouth on the evolution of the brain. In: Person P, ed. *Biology of the Mouth: A Symposium Presented at the Washington Meeting of the American Association for the Advancement of Science, 29–30 December 1966*. Washington, DC: American Association for the Advancement of Science; 1968:21–35.
23. Reardon P, Bushnell EW. Infants' sensitivity to arbitrary pairings of color and taste. *Infant Behav Dev*. 1988;11:245–250.
24. Spence C. The development and decline of multisensory flavour perception. In: Bremner AJ, Lewkowicz D, Spence C, eds. *Multisensory Development*. Oxford, UK: Oxford University Press; 2012:63–87.
25. Rozin P. "Taste-smell confusions" and the duality of the olfactory sense. *Percept Psychophys*. 1982;31:397–401.
26. Spence C. Eating with our ears: assessing the importance of the sounds of consumption to our perception and enjoyment of multisensory flavour experiences. *Flavour*. 2015;4:3.
27. Spence C, Piqueras-Fiszman B. Oral-somatosensory contributions to flavor perception and the appreciation of food and drink. In: Piqueras-Fiszman B, Spence C, eds. *Multisensory Flavor Perception: From Fundamental Neuroscience through to the Marketplace*. Duxford, CB: Elsevier; 2016:59–79.
28. Piqueras-Fiszman B, Spence C. Sensory expectations based on product-extrinsic food cues: an interdisciplinary review of the empirical evidence and theoretical accounts. *Food Qual Prefer*. 2015;40:165–179.
29. Woods AT, Poliakoff E, Lloyd DM, et al. Flavor expectation: the effects of assuming homogeneity on drink perception. *Chemosens Percept*. 2010;3:174–181.

29a. Spence, C. On the changing colour of food & drink. *International Journal of Gastronomy & Food Science*. (in press). https://doi.org/10.1016/j.ijgfs.2019.100161

30. Yeomans M, Chambers L, Blumenthal H, et al. The role of expectancy in sensory and hedonic evaluation: the case of smoked salmon ice-cream. *Food Qual Prefer*. 2008;19:565–573.
31. Wang Q(J), Keller S, Spence C. Sounds spicy: enhancing the evaluation of piquancy by means of a customised crossmodally congruent soundtrack. *Food Qual Prefer*. 2017;58:1–9.

32. Skrandies W, Reuther N. Match and mismatch of taste, odor, and color is reflected by electrical activity in the human brain. *J Psychophysiol*. 2008;22:175–184.
33. Piqueras-Fiszman B, Spence C. Sensory incongruity in the food and beverage sector: art, science, and commercialization. *Petits Propos Culinaires*. 2012a;95:74–118.
34. Schifferstein HNJ, Spence C. Multisensory product experience. In: Schifferstein HNJ, Hekkert P, eds. *Product Experience*. London, UK: Elsevier; 2008:133–161.
35. Velasco C, Michel C, Youssef J, et al. Colour-taste correspondences: designing food experiences to meet expectations or to surprise. *Int J Food Des*. 2016;1:83–102.
36. Koza BJ, Cilmi A, Dolese M, et al. Color enhances orthonasal olfactory intensity and reduces retronasal olfactory intensity. *Chem Senses*. 2005;30:643–649.
37. Zampini M, Spence C. Modifying the multisensory perception of a carbonated beverage using auditory cues. *Food Qual Prefer*. 2005;16:632–641.
38. Ohla K, Toepel U, le Coutre J, et al. Visual-gustatory interaction: orbitofrontal and insular cortices mediate the effect of high-calorie visual food cues on taste pleasantness. *PLoS One*. 2012;7(3):e32434.
39. Simmons WK, Martin A, Barsalou LW. Pictures of appetizing foods activate gustatory cortices for taste and reward. *Cerebr Cortex*. 2005;15:1602–1608.
40. Wang GJ, Volkow ND, Telang F, et al. Exposure to appetitive food stimuli markedly activates the human brain. *Neuroimage*. 2004;212:1790–1797.
41. Killgore WD, Yurgelun-Todd DA. Body mass predicts orbitofrontal activity during visual presentations of high-calorie foods. *Neuroreport*. 2005;16:859–863.
42. Small DM, Gerber JC, Mak YE, et al. Differential neural responses evoked by orthonasal versus retronasal odorant perception in humans. *Neuron*. 2005;47:593–605.
43. Small DM, Veldhuizen MG, Felsted J, et al. Separable substrates for anticipatory and consummatory food chemosensation. *Neuron*. 2008;57:786–797.
44. Chen J, Papies EK, Barsalou LW. A core eating network and its modulations underlie diverse eating phenomena. *Brain Cogn*. 2016;110:20–42.
45. Barsalou LW. Grounded cognition. *Annu Rev Psychol*. 2008;59:617–645.
46. Papies EK, Barsalou LW. Grounding desire and motivated behavior: a theoretical framework and review of empirical evidence. In: Hofmann W, Nordgren LF, eds. *The Psychology of Desire*. New York, NY: Guilford Press; 2015:36–60.
47. Malik S, McGlone F, Bedrossian D, et al. Ghrelin modulates brain activity in areas that control appetitive behavior. *Cell Metabol*. 2008;7:400–409.
48. Rosenbaum M, Sy M, Pavlovich K, et al. Leptin reverses weight loss-induced changes in regional neural activity responses to visual food stimuli. *J Clin Investig*. 2008;118:2583–2591.
49. Harrar V, Toepel U, Murray M, et al. Food's visually-perceived fat content affects discrimination speed in an orthogonal spatial task. *Exp Brain Res*. 2011;214:351–356.
50. Van der Laan LN, de Ridder DTD, Viergever MA, et al. The first taste is always with the eyes: a meta-analysis on the neural correlates of processing visual food cues. *Neuroimage*. 2011;55:296–303.
51. Bielser ML, Crézé C, Murray MM, et al. Does my brain want what my eyes like? – how food liking and choice influence spatio-temporal brain dynamics of food viewing. *Brain Cogn*. 2016;110:64–73.
52. Di Pellegrino G, Masgarelli S, Mengarelli F. Food pleasantness affects visual selective attention. *Quart J Exp Psych*. 2011;64:560–571.
53. Piech RM, Pastorino MT, Zald DH. All I saw was the cake. Hunger effects on attentional capture by visual food cues. *Appetite*. 2010;54:579–582.
54. Brignell C, Griffiths T, Bradley BP, et al. Attentional and approach biases for pictorial food cues. Influence of external eating. *Appetite*. 2009;52:299–306.
55. Radel R, Clement-Guillotin C. Evidence of motivational influences in early visual perception: hunger modulates conscious access. *Psych Sci*. 2012;23:232–234.
56. Garcia-Burgos D, Lao J, Munsch S, et al. Visual attention to food cues is differentially modulated by gustatory-hedonic and post-ingestive attributes. *Food Res Int*. 2017;97:199–208.
57. Mogg K, Bradley BP, Hyare H, et al. Selective attention to food-related stimuli in hunger: are attentional biases specific to emotional and psychopathological states, or are they also found in normal drive states? *Behav Res Ther*. 1998;36:227–237.

58. Nummenmaa L, Hietanen JK, Calvo MG, et al. Food catches the eye but not for everyone: a BMI-contingent attentional bias in rapid detection of nutriments. *PLoS One*. 2011;6(5):e19215.
59. Wang GJ, Volkow ND, Felder C, et al. Enhanced resting state activity of the oral somatosensory cortex in obese subjects. *Neuroreport*. 2002;13:1151–1155.
60. Spence C. Mouth-watering: the influence of environmental and cognitive factors on salivation and gustatory/flavour perception. *J Texture Stud*. 2011;42:157–171.
61. Toepel T, Bielser ML, Forde C, et al. Brain dynamics of meal size selection in humans. *Neuroimage*. 2015;113:133–142.
62. Medic N, Ziauddeen H, Forwood SE, et al. The presence of real food usurps hypothetical health value judgment in overweight people. *eNeuro*. 2016;3(2). ENEURO.0025-16.2016.
63. Regard M, Landis T. "Gourmand syndrome": eating passion associated with right anterior lesions. *Neurol Now*. 1997;48:1185–1190.
64. Camus MC, Halelamien N, Plassmann H, et al. Repetitive transcranial magnetic stimulation over the right dorsolateral preforntal cortex decreases valuations during food choices. *Eur J Neurosci*. 2009;30:1980–1988.
65. Spence C, Smith B, Auvray M. Confusing tastes and flavours. In: Stokes D, Matthen M, Biggs S, eds. *Perception and its Modalities*. Oxford, UK: Oxford University Press; 2015:247–274.
66. Breslin P, Huang L. Human taste: peripheral anatomy, taste transduction, and coding. *Taste Smell*. 2006;63:152–190.
67. Chale-Rush A, Burgess JR, Mattes RD. Multiple routes of chemosensitivity to free fatty acids in humans. *Am J Physiol Gastrointest Liver Physiol*. 2007;292:G1206–G1212.
68. Lapis TJ, Penner MH, Lim J. Humans can taste glucose oligomers independent of the hT1R2/hT1R3 sweet taste receptor. *Chem Senses*. 2016;41:755–762.
69. Lawless HT, Schlake S, Smythe J, et al. Metallic taste and retronasal smell. *Chem Senses*. 2004;29:25–33.
70. Nagata H, Dalton P, Doolittle N, et al. Psychophysical isolation of the modality responsible for detecting multimodal stimuli: a chemosensory example. *J Exp Psychol Hum Percept Perform*. 2005;31:101–109.
71. Simon SA, de Araujo IE, Gutierrez R, et al. The neural mechanisms of gustation: a distributed processing code. *Nat Rev Neurosci*. 2006;7:890–901.
72. Chen X, Gabitto M, Peng Y, et al. A gustotopic map of taste qualities in the mammalian brain. *Science*. 2011;333:1262–1266.
73. Spence C. Just how much of what we taste derives from the sense of smell? *Flavour*. 2015;4:30.
74. Spence C. Oral referral: mislocalizing odours to the mouth. *Food Qual Prefer*. 2016;50:117–128.
75. Veldhuizen MG, Nachtigal D, Teulings L, et al. The insular taste cortex contributes to odor quality coding. *Front Hum Neurosci*. 2010;4:58.
76. Veldhuizen MG, Shepard TG, Wang MF, et al. Coactivation of gustatory and olfactory signals in flavour perception. *Chem Senses*. 2010;35:121–133.
77. Pierce J, Halpern BP. Orthonasal and retronasal odorant identification based upon vapor phase input from common substances. *Chem Senses*. 1996;21:529–543.
78. Voirol E, Daget N. Comparative study of nasal and retronasal olfactory perception. *Lebensm-Wiss & Technol*. 1986;19:316–319.
79. Davidson JM, Linforth RST, Hollowood TA, et al. Effect of sucrose on the perceived flavor intensity of chewing gum. *J Agric Food Chem*. 1999;47:4336–4340.
80. Pfeiffer JC, Hollowood TA, Hort J, et al. Temporal synchrony and integration of sub-threshold taste and smell signals. *Chem Senses*. 2005;30:539–545.
81. Delwiche J, Heffelfinger AL. Cross-modal additivity of taste and smell. *J Sens Stud*. 2005;20:137–146.
82. Stevenson RJ, Boakes RA. Sweet and sour smells: learned synaesthesia between the senses of taste and smell. In: Calvert GA, Spence C, Stein BE, eds. *The Handbook of Multisensory Processing*. Cambridge, MA: MIT Press; 2004:69–83.
83. Stevenson RJ, Tomiczek C. Olfactory-induced synesthesias: a review and model. *Psychol Bull*. 2007;133:294–309.
84. Auvray M, Spence C. The multisensory perception of flavor. *Conscious Cognit*. 2008;17:1016–1031.
85. De Araujo IET, Rolls ET, Kringelbach ML, et al. Taste-olfactory convergence, and the representation of the pleasantness of flavour, in the human brain. *Eur J Neurosci*. 2003;18:2059–2068.
86. Schaal B, Marlier L, Soussignan R. Human foetuses learn odours from their pregnant mother's diet. *Chem Senses*. 2000;25:729–737.

87. Schaal B, Durand K. The role of olfaction in human multisensory development. In: Bremner AJ, Lewkowicz D, Spence C, eds. *Multisensory Development*. Oxford, UK: Oxford University Press; 2012:29–62.
88. Viana F. Chemosensory properties of the trigeminal system. *ACS Chem Neurosci*. 2011;2:38–50.
89. Rentmeister-Bryant H, Green BG. Perceived irritation during ingestion of capsaicin or piperine: comparison of trigeminal and non-trigeminal areas. *Chem Senses*. 1997;22:257–266.
90. Breslin PAS, Beauchamp GK. Salt enhances flavor by suppressing bitterness. *Nature*. 1997;387:563.
91. Suwankanit C, Dermiki M, Kennedy OB, et al. Umami: suppressed by all other tastes but itself an enhancer of salty and sweet perception. In: *Poster presented at 10th Pangborn Sensory Science Symposium (11–15th August 2013, Rio de Janeiro, Brazil); 2013*. 2013.
92. Di Salle F, Cantone E, Savarese MF, et al. Effect of carbonation on brain processing of sweet stimuli in humans. *Gastroenterol*. 2013;145:537–539. e1-e3.
93. Christensen CM. Effects of taste quality and intensity on oral perception of viscosity. *Percept Psychophys*. 1980;28:315–320.
94. Christensen CM. Effects of solution viscosity on perceived saltiness and sweetness. *Percept Psychophys*. 1980;28:347–353.
95. Bult JHF, de Wijk RA, Hummel T. Investigations on multimodal sensory integration: texture, taste, and ortho- and retronasal olfactory stimuli in concert. *Neurosci Lett*. 2007;411:6–10.
96. Roudnitzky N, Bult JHF, de Wijk RA, et al. Investigation of interactions between texture and ortho- and retronasal olfactory stimuli using psychophysical and electrophysiological approaches. *Behav Brain Res*. 2011;216:109–115.
97. De Araujo IE, Rolls ET. Representation in the human brain of food texture and oral fat. *J Neurosci*. 2004;24:3086–3093.
98. Barnett-Cowan M. An illusion you can sink your teeth into: haptic cues modulate the perceived freshness and crispness of pretzels. *Perception*. 2010;39:1684–1686.
99. Biggs L, Juravle G, Spence C. Haptic exploration of plateware alters the perceived texture and taste of food. *Food Qual Prefer*. 2016;50:129–134.
100. Piqueras-Fiszman B, Spence C. The influence of the feel of product packaging on the perception of the oral-somatosensory texture of food. *Food Qual Prefer*. 2012;26:67–73.
101. Tu Y, Yang Z, Ma C. Touching tastes: the haptic perception transfer of liquid food packaging materials. *Food Qual Prefer*. 2015;39:124–130.
102. Van Rompay TJL, Finger F, Saakes D, et al. "See me, feel me": effects of 3D-printed surface patterns on beverage evaluation. *Food Qual Prefer*. 2017;62:332–339.
103. Van Rompay TJL, Groothedde S. The taste of touch: Enhancing saltiness impressions through surface texture design. *Food Qual Prefer*. 2019;73:248–254. https://doi.org/10.1016/j.foodqual.2018.11.003.
104. Kampfer K, Leischnig A, Ivens BS, Spence C. Touch-taste-transference: assessing the effect of the weight of product packaging on flavor perception and taste evaluation. *PLoS One*. 2017;12(10).
105. Moir HC. Some observations on the appreciation of flavour in foodstuffs. *J. Soc. Chem. Ind*. 1936;14:145–148.
106. Spence C. On the psychological impact of food colour. *Flavour*. 2015;4:21.
107. Spence C, Levitan C, Shankar MU, et al. Does food color influence taste and flavor perception in humans? *Chemosens Percept*. 2010;3:68–84.
108. Spence C. Visual contributions to taste and flavour perception. In: Scotter M, ed. *Colour Additives for Food and Beverages*. Cambridge, UK: Woodhead Publishing; 2015e:189–210.
109. Maga JA. Influence of color on taste thresholds. *Chem Senses Flavor*. 1974;1:115–119.
110. Hidaka S, Shimoda K. Investigation of the effects of color on judgments of sweetness using a taste adaptation method. *Multisensory Res*. 2014;27:189–205.
111. Parr WV, White KG, Heatherbell D. The nose knows: influence of colour on perception of wine aroma. *J Wine Res*. 2003;14:79–101.
112. Zampini M, Sanabria D, Phillips N, et al. The multisensory perception of flavor: assessing the influence of color cues on flavor discrimination responses. *Food Qual Prefer*. 2007;18:975–984.
113. Zampini M, Wantling E, Phillips N, et al. Multisensory flavor perception: assessing the influence of fruit acids and color cues on the perception of fruit-flavored beverages. *Food Qual Prefer*. 2008;19:335–343.
114. Miller IJ, Reedy DP. Variations in human taste bud density and taste intensity perception. *Physiol Behav*. 1990;47:1213–1219.

115. Garneau NL, Nuessle TM, Sloan MM, et al. Crowdsourcing taste research: genetic and phenotypic predictors of bitter taste perception as a model. *Front Integr Neurosci*. 2014;8:33.
116. Eldeghaidy S, Marciani L, McGlone F, et al. The cortical response to the oral perception of fat emulsions and the effect of taster status. *J Neurophysiol*. 2011;105:2572–2581.
117. Shankar MU, Levitan C, Spence C. Grape expectations: the role of cognitive influences in color-flavor interactions. *Conscious Cogn*. 2010;19:380–390.
118. Shankar M, Simons C, Shiv B, et al. An expectations-based approach to explaining the influence of color on odor identification: the influence of degree of discrepancy. *Atten Percept Psychophys*. 2010;72:1981–1993.
119. Österbauer RA, Matthews PM, Jenkinson M, et al. Color of scents: chromatic stimuli modulate odor responses in the human brain. *J Neurophysiol*. 2005;93:3434–3441.
120. Qureshy A, Kawashima R, Imran MB, et al. Functional mapping of human brain in olfactory processing: a PET study. *J Neurophysiol*. 2000;84:1656–1666.
121. Royet JP, Koenig O, Gregoire MC, et al. Functional anatomy of perceptual and semantic processing for odors. *J Cogn Neurosci*. 1999;11:94–109.
122. Spence C, Deroy O. Crossmodal mental imagery. In: Lacey S, Lawson R, eds. *Multisensory Imagery: Theory and Applications*. New York, NY: Springer; 2013:157–183.
123. Piqueras-Fiszman B, Alcaide J, Roura E, et al. Is it the plate or is it the food? Assessing the influence of the color (black or white) and shape of the plate on the perception of the food placed on it. *Food Qual Prefer*. 2012;24:205–208.
124. Piqueras-Fiszman B, Giboreau A, Spence C. Assessing the influence of the color of the plate on the perception of a complex food in a restaurant setting. *Flavour*. 2013;2:24.
125. Stewart PC, Goss E. Plate shape and colour interact to influence taste and quality judgments. *Flavour*. 2013;2:27.
126. Tu Y, Yang Z, Ma C. The taste of plate: how the spiciness of food is affected by the color of the plate used to serve it. *J Sens Stud*. 2016;31:50–60.
127. Bruno N, Martani M, Corsini C, et al. The effect of the color red on consuming food does not depend on achromatic (Michelson) contrast and extends to rubbing cream on the skin. *Appetite*. 2013;71:307–313.
128. Genschow O, Reutner L, Wanke M. The color red reduces snack food and soft drink intake. *Appetite*. 2012;58:699–702.
129. Reutner L, Genschow O, Wänke M. The adaptive eater: perceived healthiness moderates the effect of the color red on consumption. *Food Qual Prefer*. 2015;44:172–178.
130. Spence C. Background colour & its impact on food perception & behaviour. *Food Qual Prefer*. 2018;68:156–166.
131. Oberfeld D, Hecht H, Allendorf U, et al. Ambient lighting modifies the flavor of wine. *J Sens Stud*. 2009;24:797–832.
132. Spence C, Velasco C, Knoeferle K. A large sample study on the influence of the multisensory environment on the wine drinking experience. *Flavour*. 2014;3:8.
133. Xu AJ, Labroo AA. Incandescent affect: turning on the hot emotional system with bright light. *J Consum Psychol*. 2014;24:207–216.
134. Gal D, Wheeler SC, Shiv B, 2007, unpublished manuscript. Cross-modal influences on gustatory perception. Available at: SSRN: http://ssrn.com/abstract=1030197.
135. Biswas D, Szocs C, Chacko R, et al. Shining light on atmospherics: how ambient light influences food choices. *J Mark Res*. 2017;54(1):111–123.
136. Murakoshi T, Masuda T, Utsumi K, et al. Glossiness and perishable food quality: visual freshness judgment of fish eyes based on luminance distribution. *PLoS One*. 2013;8(3):e58994.
137. Spence C, Wang QJ. Assessing the impact of closure type on wine ratings and mood. *Beverages*. 2017;3:52. https://doi.org/10.3390/beverages3040052.
138. Demattè ML, Pojer N, Endrizzi I, et al. Effects of the sound of the bite on apple perceived crispness and hardness. *Food Qual Prefer*. 2014;38:58–64.
139. Weiss G. Why is a soggy potato chip unappetizing? *Science*. 2001;293:1753–1754.
140. Luckett CR, Meullenet JF, Seo HS. Crispness level of potato chips affects temporal dynamics of flavor perception and mastication patterns in adults of different age groups. *Food Qual Prefer*. 2016;51:8–19.
141. Guest S, Catmur C, Lloyd D, et al. Audiotactile interactions in roughness perception. *Exp Brain Res*. 2002;146:161–171.
142. Jousmäki V, Hari R. Parchment-skin illusion: sound-biased touch. *Curr Biol*. 1998;8:869–872.

143. Wheeler E. *Tested Sentences that Sell*. New York, NY: Prentice & Co. Hall; 1938.
144. Gopnik A. *The Table Comes First: Family, France, and the Meaning of Food*. London, UK: Quercus; 2012.
145. Spence C, Shankar MU, Blumenthal H. 'Sound bites': auditory contributions to the perception and consumption of food and drink. In: Bacci F, Melcher D, eds. *Art and the Senses*. Oxford, UK: Oxford University Press; 2011:207–238.
146. Ferber C, Cabanac M. Influence of noise on gustatory affective ratings and preference for sweet or salt. *Appetite*. 1987;8:229–235.
147. Woods AT, Poliakoff E, Lloyd DM, et al. Effect of background noise on food perception. *Food Qual Prefer*. 2011;22:42–47.
148. Yan KS, Dando R. A crossmodal role for audition in taste perception. *J Exp Psychol Hum Percept Perform*. 2015;41:590–596.
149. Crisinel AS, Cosser S, King S, et al. A bittersweet symphony: systematically modulating the taste of food by changing the sonic properties of the soundtrack playing in the background. *Food Qual Prefer*. 2012;24:201–204.
150. Reinoso Carvalho F, Wang Q(J), Van Ee R, et al. "Smooth operator": music modulates the perceived creaminess, sweetness, and bitterness of chocolate. *Appetite*. 2017;108:383–390.
151. Callan A, Callan D, Ando H. *Differential effects of music and pictures on taste perception –an fMRI study*. Poster Presented at the Annual Meeting of the International Multisensory Research Forum. June, 14–17th June, Toronto, CA. 2018.
152. Woods AT, Lloyd DM, Kuenzel J, et al. Expected taste intensity affects response to sweet drinks in primary taste cortex. *Neuroreport*. 2011;22:365–369.
153. Ashkenazi A, Marks LE. Effect of endogenous attention on detection of weak gustatory and olfactory flavors. *Percept Psychophys*. 2004;66:596–608.
154. Spence C, Kettenmann B, Kobal G, et al. Selective attention to the chemosensory modality. *Percept Psychophys*. 2000;62:1265–1271.
155. Forster S, Spence C. "What smell?" Temporarily loading visual attention induces a prolonged loss of olfactory awareness. *Psych Sci*. 2018;29(10):1642–1652.
156. Spence C, Wang Q(J). Wine & music (II): can you taste the music? Modulating the experience of wine through music and sound. *Flavour*. 2015;4:33.
157. Spence C. On the relationship(s) between colour and taste. *Exp Psychol*. 2019;66:99–111.
158. Small DM. Flavor is in the brain. *Physiol Behav*. 2012;107:540–552.
159. Spence C. The neuroscience of flavor. In: Piqueras-Fiszman B, Spence C, eds. *Multisensory Flavor Perception: From Fundamental Neuroscience through to the Marketplace*. Oxford, UK: Elsevier; 2016:235–248.
160. Verhagen JV, Engelen L. The neurocognitive bases of human multimodal food perception: sensory integration. *Neurosci Biobehav Rev*. 2006;30:613–650.

CHAPTER

11

Audiovisual crossmodal correspondences: behavioral consequences and neural underpinnings

Charles Spence[1], *K. Sathian*[2,3]

[1]Crossmodal Research Laboratory, Department of Experimental Psychology, Oxford University, Oxford, United Kingdom; [2]Department of Neurology, Neural & Behavioral Sciences, and Psychology, Pennsylvania State University, Hershey, PA, United States; [3]Departments of Neurology and Psychology, Emory University, Atlanta, GA, United States

Introduction

Crossmodal correspondences have been defined as a tendency for a feature, attribute, dimension, or stimulus in one sensory modality, either physically present or merely imagined, to be matched (or associated) with a feature, attribute, dimension, or stimulus in another sensory modality (see Refs. 1,2). The last few years have seen an explosive growth of interest in the study of the crossmodal correspondences. While this is undoubtedly a good thing for the field, it does make any attempt at a comprehensive review all but impossible (at least within the space constraints that are inherent in an edited volume such as this). The focus here will therefore be on trying to understand what has been learnt from recent studies of audiovisual correspondences in neurologically normal adult human participants. Audiovisual correspondences are the most extensively studied kind of multisensory pairing that have been examined to date. There has, as yet, been rather less interest in the study of crossmodal correspondences involving the sense of touch/haptics; it is hoped that future work will address this gap. Given the space limitations, this review will stay away from those

Multisensory Perception
https://doi.org/10.1016/B978-0-12-812492-5.00011-5

studies documenting crossmodal correspondences involving the chemical senses (including taste, smell, and the common chemical sense associated with activation of nasal or oral nociceptors[a]; e.g., see Refs. 6–8 for early work). The interested reader is directed to reviews documenting the many crossmodal correspondences between, for example, olfaction and sound,[9] taste/flavor and sound/music,[10,11] and between taste/smell/flavor stimuli and visually perceived dimensions, such as color and shape (e.g., Refs. 12,13). The questions of how, and when, the various correspondences are acquired, or emerge, during the course of human development, while fascinating in their own right, once again fall beyond the scope of this chapter, especially given the amount of recent literature on this question (see Ref. 14 for a review). Meanwhile, those interested in historical aspects of the study of the correspondences are directed to the earlier reviews provided by Marks[15] and Spence.[2]

From a theoretical viewpoint, it can be argued that our understanding of the crossmodal correspondences, developed by studying one pair of senses (here, audition and vision), is likely to apply equally well to any other pair of senses that one should be interested in. A wide range of audiovisual correspondences has been documented to date, ranging from correspondences between simple features, semantic correspondences, and more complex correspondences involving say classical music and paintings (e.g., Ref. 16,16a). This review will be restricted to a consideration of those correspondences between simple stimuli that do not have any particular semantic/esthetic meaning/association.

Kinds of audiovisual crossmodal correspondence and their effects on behavior

According to Spence,[2] it is possible to distinguish between several qualitatively different types of crossmodal correspondence. These include statistical, structural, semantic (or perhaps better said, linguistic; see Ref. 17), and possibly also affective (or emotional) correspondences. The suggestion is that these qualitatively different kinds of crossmodal correspondences might have distinct developmental trajectories and neural substrates, and perhaps even different consequences for perception, behavior, etc. For instance, according to the statistical account, crossmodal correspondences can be thought of as the internalization of the statistical regularities of the environment.[1,2] According to the structural (innate) account, we are born with systems for arranging stimuli in terms of (e.g.) magnitude/intensity.[b] One might think of polar correspondences (e.g., Refs. 20–22) as exemplifying structural correspondences. Linguistic correspondences refer to stimuli that may be related because of the similar terms we sometimes use to describe stimuli presented in the different senses, such as the words "high" and "low" to describe both pitch and spatial elevation in English (e.g., Refs. 23–25; though see also Ref. 26). It is worth stressing here that these various explanations should not be thought of as mutually exclusive, and indeed, numerous studies have

[a] See Hubbard Jones,[3] Cain,[4] and Silver[5] on the common chemical sense—namely the response to chemical irritants such as capsaicin in chilli, piperine in black pepper, or the carbon dioxide in a fizzy drink.

[b] Some researchers have seemingly been driven to the innate account when they cannot point to an environmental regularity as the source of the correspondence and/or when the correspondence emerges very early in development (e.g., see Refs. 18,19).

demonstrated that different combinations of correspondences appear capable of explaining some part of the variance in people's pairing responses.

Spence[2] only briefly mentioned the affective/emotional mediation of crossmodal correspondences. However, some of the research that has been published since then has really started to emphasize that some crossmodal correspondences are mediated, at least in part, by emotion.[27] And while the affective account might well be more relevant when it comes to more complex, esthetically meaningful stimuli, it is worth noting that even simple stimuli, such as squiggly lines and tones, are associated with specific affective value.[28–35] Thus, it may be appropriate to consider the affective modulation of crossmodal correspondences as just another dimension influencing behavioral reactions to the stimuli concerned.

Over the last half century or so, many laboratory studies have demonstrated the behavioral consequences of crossmodal correspondences for human performance when responding to multisensory stimuli. Typically, the speed and/or accuracy of participants' behavioral responses are compared on those trials in which the auditory and visual stimuli align in terms of some putative correspondence and trials where the stimulus mapping has been reversed. This sort of approach has, for instance, been used in innumerable speeded classification studies (e.g., Refs. 36–43; see Refs. 44, for a review). However, essentially the same results (in terms of crossmodal correspondences influencing people's speeded performance) have been obtained in those studies that have used a variant of the Implicit Association Test (e.g., Refs. 45; see also Refs. 46–48).[c] Similarly, some years ago, Miller[49] demonstrated that performance on the redundant target effect paradigm is also affected by the correspondence between the audiovisual stimuli on bimodal target trials (see also Ref. 50).

In fact, the Colavita visual dominance effect (see Ref. 51) is one of the few speeded behavioral tasks where crossmodal correspondences appear not to influence human performance. In a typical paradigm, people have to respond to a random sequence of unimodal auditory, unimodal visual, and rarer bimodal audiovisual targets as rapidly as possible (see Refs. 52, for a review). The oft-reported result from such studies is that participants sometimes fail to respond to the auditory stimulus on the bimodal target trials (this is the Colavita visual dominance effect). Recently, Stekelenburg and Keetels[53] assessed whether the magnitude of this effect would be influenced by the corresponding versus noncorresponding relationship between auditory pitch and visual size on the bimodal target trials (high pitch corresponding with smaller objects and low pitch with larger objects). However, no such modulatory effect was found. That said, it should be noted that these authors also failed to find reaction time (RT) differences related to the crossmodal correspondence.

A number of researchers have now gone further in terms of demonstrating crossmodal interactions in a variety of unspeeded psychophysical tasks (e.g., Refs. 54–57). So, for instance, Parise and Spence[56] reported that spatial and temporal ventriloquism effects are enhanced for crossmodally congruent pairs of auditory and visual stimuli as compared with pairs of incongruent stimuli (see Fig. 11.1). Meanwhile, Orchard-Mills et al.[58] demonstrated that the crossmodal correspondence between auditory pitch and visual elevation (high pitch corresponds

[c] The IAT results are particularly intriguing because only one stimulus is presented on each trial. This means that any influence of crossmodal correspondences on performance in this paradigm cannot be attributed to multisensory perception as all information processing is unisensory.

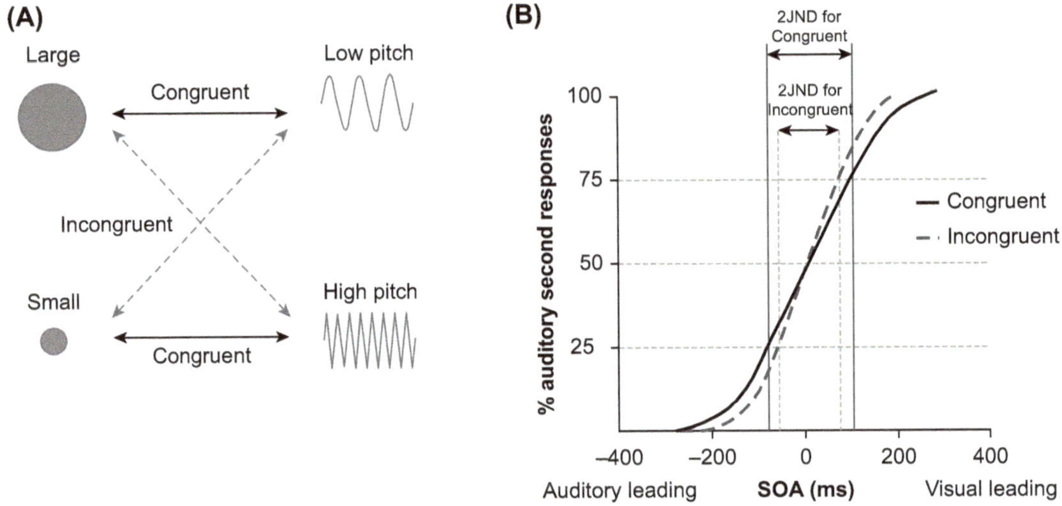

FIGURE 11.1 (A) The crossmodal correspondence between visual size and auditory pitch used in one of Parise and Spence's[56] temporal order judgment (TOJ) studies. (B) The results (of their Experiment 1) revealed that participants found it harder to correctly judge the temporal order of the visual and auditory stimuli (i.e., the just noticeable difference (JND) was significantly higher) when they were crossmodally congruent than when they were crossmodally incongruent. A similar pattern of results was also obtained in a second study in which the pitch and waveform of the sound and the visual features of curvilinearity and the magnitude of the angles of regular shapes were varied.

with high elevation and low pitch with low elevation) also modulates the temporal ventriloquism effect (see Refs. 59, for a similar claim concerning the pitch-size correspondence; though see Ref. 60). Parise and Spence showed that the pitch-size crossmodal correspondence modulated the spatial ventriloquism effect as well. Elsewhere, Sweeny et al.[57] reported that heard speech sounds affect the perceived shape of briefly presented visual stimuli. In particular, ovals, whose aspect ratio (relating width to height) varied on a trial-by-trial basis, were rated as looking taller when a /woo/ sound (typically associated with vertical elongation of the mouth) was presented, and as wider when a /wee/ sound (typically associated with horizontal elongation of the mouth) was presented instead. Furthermore, the corresponding speech sounds enhanced adaptation effects to vertically versus horizontally elongated ovals, as indexed by the rating of a subsequently presented symmetric shape,[57] suggesting that the neural processing of these sounds and shapes interacts. Such findings therefore add to a growing body of scientific evidence demonstrating that audiovisual correspondences can give rise to perceptual (in addition to decisional) effects.[61]

Assessing the impact of crossmodal correspondences on other aspects of cognition

While most of the research on the crossmodal correspondences that has been published to date has tended to focus on any perceptual/behavioral effects that may exist, some researchers have now started to turn their attention toward the question of whether working memory may also be sensitive to the influence of crossmodal correspondences.[62] The answer, at least from this one study, would appear to be in the affirmative. In particular, Brunetti and his colleagues found that performance on an audiovisual two-back task was significantly affected by the pitch-shape and pitch-elevation correspondences.[d] The influence of crossmodal correspondences on mental imagery is another area that is starting to open up to empirical investigation too.[63] Furthermore, beyond their influence on perception, in the coming years, there is likely to be an expanding area of research around the effect of the correspondences on action (e.g., speech production/vocalization[57,64,65] and gesture[66]).

Do crossmodal correspondences occur early or late in human information processing?

An important issue that various authors have addressed concerns the stage(s) in information processing at which the effects of crossmodal correspondences occur. Relevant to this issue is the question of the automaticity of such effects (see Ref. 67 for a review). When thinking about this question, it is worth remembering that multiple criteria are used by those working in the field to assess automaticity. These include the goal-independence criterion; the nonconscious criterion; the load-insensitivity criterion; and the speed criterion.[68] According to Spence and Deroy's review,[67] the evidence would appear to support the claim that crossmodal correspondences are automatic according to several but not all of the commonly used criteria. One of the lines of evidence that tends to support the view of automaticity comes from studies showing that crossmodal correspondences affect the spatial distribution of exogenous attention. So, for instance, Chiou and Rich[69] conducted a number of experiments demonstrating that the presentation of a task-irrelevant (i.e., spatially nonpredictive) relatively high (vs. low-)-pitched sound before the onset of a visual target from above (vs. below) fixation gives rise to an exogenous shift of spatial attention upward (vs. down). Similar results have also been reported by Mossbridge et al.[70] and by Fernández-Prieto and Navarra.[71]

The idea that crossmodal correspondences are automatic is also supported by a recent study in the domain of sound symbolism, a type of crossmodal correspondence where visual shapes are associated with particular sounds comprising words or nonwords (more on this later). This study, by Hung et al.,[72] used the technique of continuous flash suppression to block rapid conscious perception of visual shapes and showed that the shapes emerged into consciousness sooner when paired with congruent as compared with incongruent words/nonwords. Consistent with this, Ković et al.[73] reported neural evidence, in event-related potentials (ERPs) recorded by electroencephalography (EEG), of a sensitivity to sound-symbolic associations that emerged as early as 140–180 ms after stimulus onset over occipital electrodes, along with a later parietal effect between 340 and 520 ms (this study did not report source localization for the ERP waveforms).

The findings from other studies favor later effects of crossmodal correspondences, although these later effects are not necessarily in conflict with the findings reviewed above suggesting that the correspondences themselves might be automatic and could also trigger

[d] There is perhaps a relevant link to the IAT results mentioned earlier,[45] because the latter task presumably requires the participant to maintain the relevant response mappings in working memory.

earlier effects. However, the finding of Stekelenberg and Keetels,[53] cited above, that the pitch-size correspondence did not modulate the Colavita effect, supports the view that the modulatory effects of crossmodal correspondences emerge relatively late in human information processing. Stekelenburg and Keetels, in their study,[53] also recorded EEG responses to both congruent and incongruent audiovisual stimulus pairs. In keeping with their behavioral results, these researchers found relatively late ERP crossmodal congruency effects: a frontal effect at around 400–550 ms and a later occipitoparietal effect in the 690–800 ms time window. Neural sources for the earlier of these two effects were localized to the anterior cingulate and premotor cortex, while the later effect originated in the premotor cortex, inferior parietal lobule, and the posterior middle temporal gyrus. Bien et al.[74] used a combination of behavioral, transcranial magnetic stimulation (TMS), and ERP methods to demonstrate the involvement of cortex around the right intraparietal sulcus in the pitch-size correspondence, at a latency of around 250 ms, i.e., somewhat earlier than reported in the study of Stekelenberg and Keetels.[53] Relatedly, Sadaghiani et al.[25] demonstrated behaviorally that rising/falling pitch influenced the perceived direction of visual motion along the vertical axis, with a corresponding, albeit weak, neural effect as measured with functional magnetic resonance imaging (fMRI) in superior parietal cortex. Revill et al.[75] also using fMRI, found stronger activation in left superior parietal cortex, when listening to words in a foreign (uncomprehended) language, for sound-symbolic words relative to words that were not sound-symbolic; furthermore, fractional anisotropy (a measure of white matter organization) in the superior longitudinal fasciculus, measured with diffusion tensor imaging, was related to individual behavioral sensitivity to the sound-symbolic effect. The parietal loci of the fMRI effects in the studies of Sadaghiani et al.[25] and Revill et al.[75] are more in keeping with later rather than earlier effects, although it is worth pointing out that such studies do not lend themselves well to investigating the timing of neural processes.

Three recent fMRI studies examined the neural basis of crossmodal correspondences, using presentations of audiovisual stimuli that were either congruent or incongruent for the correspondence of interest in each study. In one of these studies, directed at the pitch-elevation crossmodal correspondence,[48] an event-related design and a one-back comparison yielded congruency-related activations (with stronger activity for the congruent as compared with the incongruent condition) in inferior frontal and insular cortex bilaterally and in the right frontal eye field and inferior parietal cortex. These activations were considered to be most consistent with the involvement of multisensory attentional processes. Although there was no significant behavioral congruency effect during scanning, the IAT was used to demonstrate a behavioral congruency effect in the same participants outside the scanner. Meanwhile, another study looked at the sound-symbolic correspondence between two visual shapes having pointed/rounded protuberances and the auditory nonwords "keekay"/"lomo," respectively.[47] Congruent and incongruent trials were blocked. When participants attended to the auditory stimuli to make a two-alternative forced-choice speeded classification, a behavioral congruency effect was found during scanning for both accuracy and RT, in addition to an IAT congruency effect outside the scanner. Activations, in this case stronger for the incongruent than congruent condition, were found in the left middle frontal gyrus and in multiple parietal cortical foci of both hemispheres. These results were thought to reflect either phonological processing or to be related to multisensory attention. In a third study,[76] participants were asked to attend to pairings of auditory nonwords and visual shapes to detect rare targets that could be in either modality. Incongruent stimuli in this study, relative to congruent stimuli, activated dorsolateral prefrontal cortex bilaterally. Thus, all these fMRI studies, despite their differences in design and the type of congruency effects that were observed, appear to support high-level and therefore presumably late effects of crossmodal correspondences, consistent with the studies reviewed in the preceding paragraph. However, these recent fMRI studies relied on congruency manipulations, which may preferentially target mechanisms of attention,[77] and thus potentially favor finding high-level neural effects. One of these studies also provided evidence for lower-level sensory effects,[76] observing greater differential activation in auditory cortex for "rounded" nonwords (analogous to "lomo") relative to "pointed" nonwords (analogous to "keekay") on congruent compared with incongruent trials, as well as a correlation between the magnitude of the congruency effect measured behaviorally and by fMRI in occipitotemporal visual cortex.

Elevation as a fundamental organizational dimension for many crossmodal correspondences

Elevation appears to be a dominant/strong dimension against which other modalities/classes of sensory/semantic/conceptual information are mapped (see Refs. 78–81). For instance, the crossmodal correspondence between auditory pitch and visual elevation is a robust one that has been the subject of extensive study.[2] Some have argued, though, that this correspondence may not be actually a sensory one but rather more semantic (i.e., cognitive), because the same word "high" is used polysemously for elevation and pitch.[82] The axis along which elevation judgments affect this correspondence has also been the subject of empirical research: the direction of frequency sweeps (ascending or descending) served as an effective cue for visuospatial attention (at high vs. low loci), but only along the vertical axis and not when the head was tilted with respect to the body.[70] Manipulations of observers' body postural orientations (upright vs. lying on one side), however, revealed that the pitch-elevation correspondence has components along the long axis of the body as well as the gravitational vertical.[83]

To further complicate matters, the pitch of a sound has been shown to influence people's judgments of the localization of its source: people are biased in their localization of the elevation and, to a lesser degree, azimuthal position of sounds[84–88]; there is a strong statistical association between high pitch and high elevation out there in the environment[85]; and the property of elevation is a modality-independent one that can be assessed via multiple senses, including vision, hearing, and touch. Thus, the question arises as to whether the crossmodal correspondence between auditory pitch and visual elevation might be mediated through auditory elevation, which obviously is colocalized with visual elevation for a given object.

To assess this possibility, Jamal et al.[89] measured performance in a speeded classification task during attention to auditory pitch, auditory elevation, or visual elevation. They found evidence for the independent existence of three correspondences: the crossmodal correspondence between auditory pitch and visual elevation, the intramodal correspondence between auditory pitch and auditory elevation, and the crossmodal spatial correspondence between auditory and visual elevation. Furthermore, there was a complex set of modulatory interactions between these correspondences when classifying auditory stimuli (but not visual elevation) as high or low. In a nutshell, the intramodal correspondence between pitch and auditory elevation modulated both the crossmodal correspondences regardless of which auditory attribute was attended, whereas the crossmodal correspondences modulated the intramodal correspondence only when attending to the attribute common to the two correspondences under consideration (i.e., pitch in the case of the crossmodal pitch-elevation correspondence; auditory elevation in the case of the crossmodal spatial correspondence). According to Jamal et al.,[89] these results therefore suggest that statistical learning of environmental cues is important in the development of at least these correspondences.

On the relative versus absolute nature of crossmodal correspondences

The general assumption in the literature would appear to be that crossmodal correspondences are more "relative" than "absolute" (thus contrasting with the absolute mappings seen in the case of synesthesia; see Ref. 90). Thus, it is the larger of two circles that will be matched with the lower-pitched of two sounds (i.e., the actual pitch and size of the auditory and visual stimuli do not seem to matter too much). What this means, in practice, is that the experimenter working on the correspondences normally does not need to worry about precisely calibrating the experimental stimuli in the two modalities; just as long as there is a relative difference in each modality (say a clear difference in pitch and a perceptible difference in size or elevation), an effect of the crossmodal correspondences (congruent vs. incongruent) on performance will likely be obtained.[90a]

Support for the claim that correspondences are relative phenomena comes from findings such as those reported by Gallace and Spence.[39] These researchers conducted a series of four speeded classification studies in which the crossmodal correspondence between auditory pitch and visual size was found to modulate performance when the stimuli were varied randomly on a trial-by-trial basis (Experiment 1), but not when the stimulus mapping was fixed on a block-by-block basis instead (Experiment 2; see Refs. 37,58, for a similar pattern of results).[e] Furthermore, other researchers have since demonstrated that a given stimulus may be treated as either high or low depending on the specific context in which it happens to be presented (see Ref. 92).[f] So, for instance, when low-, medium-, and high-pitched sounds were presented in a random sequence on successive trials, the medium stimulus was shown to act like a "low" stimulus if it followed a high stimulus on the preceding trial, but like a "high" stimulus if the preceding stimulus had been a "high" tone instead.[92] However, that said, other researchers have argued that extreme stimuli may sometimes give rise to

[e] There is an interesting link here to the separate literature on integral versus separable perceptual dimensions (see Ref. 91). The fact that Gallace and Spence[39] only observed a behavioral effect of the crossmodal correspondence under those conditions in which the crossmodal stimulus mapping varied on a trial-by-trial basis could be taken to support the view that auditory pitch and visual size are separable perceptual dimensions, albeit susceptible to integration under some circumstances. Ben-Artzi and Marks[36] demonstrated Garner interference effects for the pitch-elevation crossmodal correspondence, indicating that attention cannot be entirely selectively allocated to the auditory pitch and visual elevation dimensions, but this approach has not as yet been used to examine other crossmodal correspondences.

[f] Note that the visual brightness—haptic size and auditory pitch—visual size/elevation correspondences also exhibit the same relative properties.[69,93].

correspondences that can be reliably ranked in the absence of any comparison stimulus (see Refs. 36,67,74,94).[g]

Relevant here, Guzman-Martinez et al.[54] found that certain correspondences operate under seemingly absolute constraints. The participants in the study had to adjust the auditory amplitude modulation (AM) rate and pitch of a sound until it perceptually "matched" the visual spatial frequency of Gabor displays having one of three different spatial frequencies. Interestingly, the participants consistently matched a specific auditory AM rate to each of the visual spatial frequencies. These matches persisted when the observer had to determine an auditory AM rate that matched only a single visual spatial frequency (the same result was also obtained between tactile AM rate and visual spatial frequency). These results led Guzman-Martinez et al.[54] to conclude that an absolute crossmodal mapping between visual spatial frequency and auditory AM rate does indeed exist. Importantly, however, such claims have only been partially confirmed by subsequent empirical research. In particular, when Orchard-Mills et al.[95] tried to replicate Guzman-Martinez et al.'s findings, they did indeed obtain a linear relationship between AM auditory noise and visual spatial frequency (so pointing toward an absolute mapping). That said, other data from their study suggested that this crossmodal interaction is flexible and based on relatively unspecific frequencies: e.g., the evidence that correspondence effects also appear with variable pairs of visual and auditory frequencies leaves room for there being both absolute constraints and some degree of relative context-dependent flexibility (within those constraints). As mentioned earlier, perhaps the degree of absoluteness of any given correspondence will depend on the extent to which the sensory estimates pick up on the same, or analogous, property out there in the environment.[h] Relative correspondences may turn out to be a feature of those correspondences involving prothetic (size or intensity) and metathetic continua (such as pitch) where the stimuli can be organized without being categorized. By contrast, absolute correspondences may be more a feature of those continua or classes of stimuli that can be given a categorical label (such as a hue, or basic taste property). In this regard, in might be interesting to investigate pitch-based correspondences in those nonsynesthetes who experience absolute pitch (and who are presumably able to give a categorical label to the sounds that they hear).

Sound symbolism and crossmodal correspondences

Sound symbolism has been referred to a few times already in this chapter. Examples of sound symbolism[96–99] can be thought of as constituting a subset of all crossmodal correspondences (see Refs. 2,100; see also Ref. 101)[i]: specifically, those correspondences in which speech sounds are associated with object properties (typically assessed visually). In terms of the visual properties that may be associated with speech sounds, the focus of most of the research

[g] Should the pitch-elevation association reflect a statistical correspondence, then the data from Parise et al.'s study[85] might be taken to suggest that the correspondence would not be documented, or might even be reversed, should either or both of the auditory stimuli be in the very high audible frequency range.

[h] Indeed, perhaps the degree of absolute/relative value expressed by a particular crossmodal correspondence may correlate with how surprising we find it to be on first hearing about it.

[i] Sound symbolism refers to the nonarbitrary mappings that exist between phonetic properties of speech sounds and their meaning.

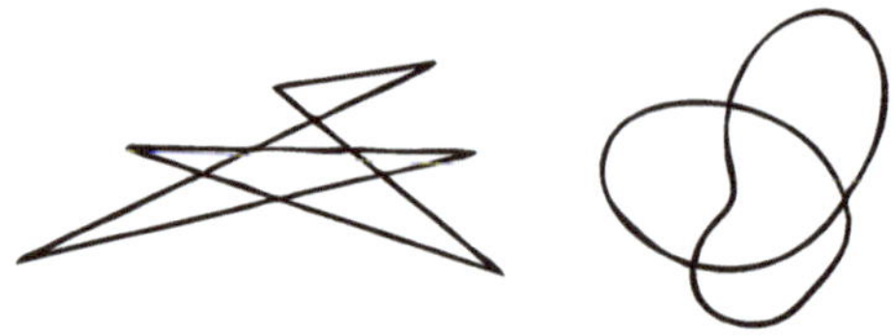

FIGURE 11.2 The pair of visual stimuli first introduced by Köhler[102,103] to demonstrate the phenomenon of shape symbolism. The majority of participants say that the word "takete" better fits with the angular shape shown on the left, whereas the word "maluma" (or "baluma" in the original 1929 paper) fits better with the rounded shape on the right. Even those individuals from cultures without any written language appear to exhibit this effect (see Refs. 105; though see also Ref. 106). Some of the only exceptions to this generalization would appear to be those suffering from damage to the angular gyrus[104] (see also Refs. 107, on the issue of brain damage and the correspondences) and those individuals with autism spectrum disorder[108–110] who are less sensitive to this kind of correspondences than the general population.

that has been published to date has been on shape and size symbolism. These two examples of sound-symbolic correspondence were first popularized by the work of Köhler[102,103] on what has come to be known as the bouba/kiki effect[104] relating these nonword sounds to angularity versus roundedness of visual shapes (see Fig. 11.2) and by Sapir[111] in his work on the mil/mal effect published, coincidentally, in the same year. The latter effect represents a crossmodal correspondence between the size of the mouth when uttering particular vowel sounds and the relative size implied by the adjectives incorporating these sounds. It is somewhat surprising how much of the subsequent research that has been published in this area has failed to go beyond the basic stimuli first introduced by Köhler[101] and Sapir,[110] almost a century ago. Thus, it is difficult to know what aspect(s) of these visual stimuli, in particular, is/are actually driving the correspondence. In fact, one thing that, at this point, remains relatively unclear is the precise nature of the relationship tying sound symbolism to the other crossmodal correspondences. Further research is undoubtedly needed in this area to figure out the ways in which these phenomena are similar versus different. And while we are a long way from having any kind of unifying model encompassing both sound symbolism and crossmodal correspondences, Sidhu and Pexman[100] have at least made a start recently by outlining five key mechanisms of sound-symbolic association.

As a consequence, some investigators have been working with radial frequency patterns, as they provide a systematic means of varying the shape properties of stimuli to assess the underlying drivers of any crossmodal sound-symbolic associations (e.g., see Ref. 112). The results of this research have so far revealed that people are more likely to match radial frequency patterns to the nonword "kiki" than to the nonword "bouba" when the value of each factor (frequency, amplitude, and spikiness) is increased (see Fig. 11.3). Elsewhere, Knöferle et al.[113] investigated the acoustic/phonetic underpinnings of both size and shape symbolism using carefully constructed sets of auditory stimuli. The participants who took part in Knöferle et al.'s study had to judge whether a given consonant–vowel speech sound was large or small, angular or rounded, using a size or shape scale. The results of the latter study demonstrated that size and shape symbolism are not induced by a common underlying mechanism, but rather are distinctly affected by the acoustic (i.e., by the spectrotemporal) properties of speech sounds. So, for example, sound symbolism for size was found to be influenced by formants F1 and F2 and particularly their interaction, as well as by duration.

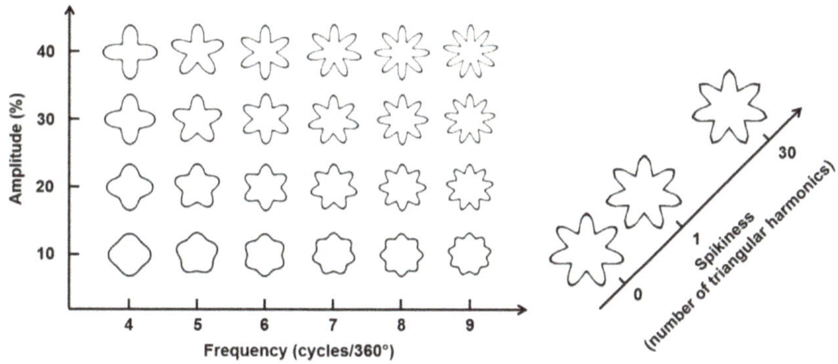

FIGURE 11.3 Radial frequency (RF) patterns used in Chen et al.'s study[112] of the sound symbolic bouba/kiki effect. Three attributes of the visual stimuli were manipulated: frequency, amplitude, and the spikiness of the sinusoidal modulations making up the shape's circumference of these outline shapes.

By contrast, sound symbolism for shape was predicted by formants F2 and F3. Such findings therefore portray sound symbolism as a process that is not based merely on broad categorical contrasts, such as round/unround and front/back vowels. Rather, it would appear that people appear to base their judgments on specific sets of acoustic cues, extracted from speech sounds, which vary across the dimensions being judged. Consistent with this suggestion, List et al.,[114] using analyses of dissimilarity matrices for 537 nonwords varying parametrically in their perceptual ratings of pointedness or roundedness, found that certain acoustic properties of the nonwords, for instance, their spectral composition as indexed by the fast Fourier transform and spectral tilt, and measures of their amplitude profile, such as the speech envelope and shimmer (amplitude variability over time), were related to their perceptual ratings. The systematic relationships to acoustic parameters found across these studies offer the potential for a deeper neural understanding of sound symbolism than is currently possible.

So, does sound symbolism represent but one class of crossmodal correspondence (as suggested by Spence[2]) or is it instead different in kind from all the rest? On the one hand, Parise and Spence's[45] IAT studies revealed similar magnitudes of effects for both sound-symbolic and nonsound-symbolic correspondences (see Fig. 11.4). Also relevant to answering the question, Lacey et al.[46] reported that synesthesia appears to strengthen sound-symbolic crossmodal correspondences (specifically shape symbolism), while pitch-size and pitch-elevation correspondences were not statistically different between the synesthetic and nonsynesthetic participant groups they tested. This might suggest that, at least for synesthetes, sound-symbolic correspondences are different from other kinds of crossmodal correspondences (see below). Bankieris and Simner[115] suggested that sound symbolism and synesthesia may be in some way connected. They based this claim on evidence from their study suggesting that synesthetes tend to be more sensitive to sound-symbolic correspondences in other languages than are nonsynesthetes. However, while sound-symbolic correspondences likely have different neural substrates than do other classes of correspondences, as suggested by some of the studies reviewed earlier, it is equally likely that various other kinds of correspondences (e.g., perhaps the quintessential case of structural correspondences, see Spence[2]) will

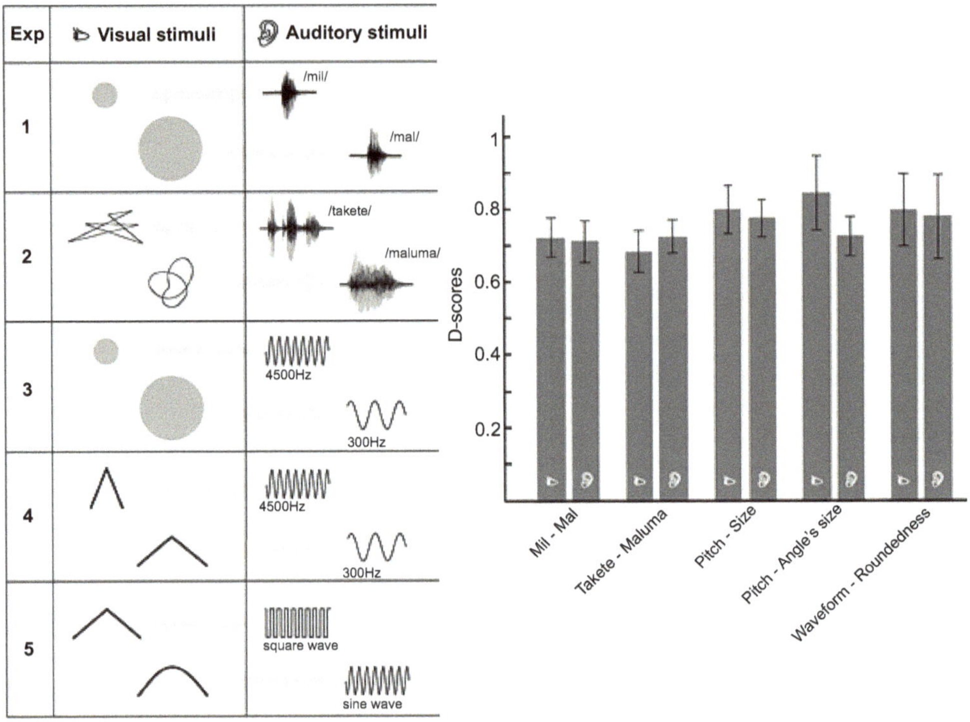

FIGURE 11.4 Stimuli and results of the five IAT experiments reported by Parise and Spence.[45] The *light gray double-headed arrows* in the left panel connect compatible audiovisual pairs of stimuli. Intriguingly, the magnitude of the IAT (shown on the right) did not differ significantly as a function of whether the crossmodal correspondence under investigation happened to be sound symbolic or not.

also turn out to have varying neural signatures. Unfortunately, as yet, we are unaware of any relevant empirical studies having tackled this important question directly (i.e., discriminating the neural substrates underlying different classes of correspondence).

Crossmodal correspondences and synesthesia

Over the years, there has been a long-standing tendency to want to link the crossmodal correspondences or at least the more surprising of them (such as the pitch-size correspondence), to synesthesia proper (see also Chapters 12 and 13). One way in which this occurs is when researchers refer to crossmodal correspondences as "synesthetic correspondences" (e.g., Refs. 19,24,41,42,116) or as "synesthetic associations."[59] Some researchers, though, have gone further in describing crossmodal correspondences as a weak form of synesthesia[117,118] (see also Refs. 31,119,120).[j] Others, meanwhile, have argued vociferously against

[j] Specifically, according to Martino and Marks (Ref. 118, p. 61), synesthesia comes in two kinds: "*Strong* synesthesia *is characterized by a vivid image in one sensory modality in response to stimulation in another one.* *Weak* synesthesia *is characterized by cross-sensory correspondences expressed through language, perceptual similarity, and perceptual interactions during information processing.*"

any such conflation of the two phenomena (see Refs. 14,89,121,122 for reviews), interesting though both undoubtedly are in their own right.[k]

It is worth stressing here the many important differences separating synesthesia from the crossmodal correspondences. So, for instance, while the synesthetic relation between inducer and concurrent tends to be fixed across the lifetime in synesthetes, many crossmodal correspondences update as a function of the changing statistics of the environment or rather of the environmental stimulation that we happen to be exposed to (e.g., Ref. 124). That said, it is interesting to note how recent research suggests that synesthesia may actually follow something of a similar pattern. For instance, children in primary school who experience letter-color synesthesia demonstrated an increase in the number of their synesthetic associations over a year of follow-up.[125] Intriguingly, the associations appear to be tuned according to the statistics of the environment (including such factors as the frequency of letter usage, semantic, and ordinal associations), resulting in biases for more similarly shaped and similar-sounding letters to have similar colors associated with them.[126]

Another important difference is that in synesthetes the relation between inducer and concurrent tends to be absolute, whereas many of the most frequently studied of typical crossmodal correspondences are relative in nature—that is, it is the higher-pitched of two sounds that will be matched with the smaller of two circles, for example.[89] Furthermore, while synesthesia, by definition, gives rise to an idiosyncratic conscious concurrent (at least according to Grossenbacher and Lovelace's[127] definition), the typical correspondences seem to be characterized by a feeling that operates without the need for conscious concurrents. That said, some commentators have recently started to question the need for synesthetic concurrents to be conscious (see Ref. 128), and one might wonder whether those with particularly vivid crossmodal mental imagery might not also experience some kind of concurrent based on the correspondences.[129,130] In this context, it is relevant to note that synesthesia has been associated with enhanced vividness of visual imagery[131] and greater use of mental imagery.[132]

Given these, and the many other differences between synesthesia and the correspondences, one might reasonably expect that the neural substrates associated with the two phenomena would also differ; however, the limited evidence gathered to date does not permit a clear answer to this question. For instance, a TMS study designed to interfere with crossmodal correspondences and one aimed at knocking out synesthetic concurrents were both applied to parietal cortex[74,133] (see Ref. 134 for a discussion). Specifically, Bien and her colleagues[74] attempted to disrupt the function of cortex in the right intraparietal sulcus using TMS over the parietal site P4 (an electrode location in the standard clinically used EEG electrode montage) and temporarily eliminated the effect of the crossmodal correspondence between auditory pitch and visual size on the spatial ventriloquism effect. Esterman and colleagues[133] targeted TMS to the right angular gyrus at the junction of the intraparietal and transverse occipital sulci, using a neuronavigation system that allowed precise targeting of this anatomical locus in each participant and found that the Stroop effect typically elicited by the synesthetically induced color was eliminated by TMS (see also Ref. 135).

[k] Although both phenomena often occur crossmodally, it should be remembered that by far the most common type of synesthesia is the unimodal color–grapheme variety, experienced by nearly 70% of synesthetes, according to Day.[123].

Unfortunately, due to the differences in the methods, the sites in the two studies cannot be directly compared, precluding judgment on whether the same or different processes were involved in the two studies.

Those who wish to connect the phenomena of synesthesia with the crossmodal correspondences often point to the similarity between certain synesthetic concurrents and the correspondences that have been demonstrated in nonsynesthetes (e.g., see Refs. 136,137). At this point, it is perhaps worth remembering that synesthetes presumably experience crossmodal correspondences just like nonsynesthetes.[1] Perhaps the more interesting question here, though, is whether they are more sensitive to the effect of certain crossmodal correspondences than are nonsynesthetes. One might, a priori, expect stronger crossmodal correspondences for synesthetes in the domain, or modalities, in which they experience synesthesia (i.e., between the modalities of the inducer and of the concurrent). Interestingly, though, stronger crossmodal correspondences, as indexed by the strength of the IAT congruency effect, were found in synesthetes compared with nonsynesthetes for sound-symbolic crossmodal correspondences (between rounded/pointed visual shapes and the auditory nonwords "lomo" and "keekay"), but not for the lower-level crossmodal correspondences tested, i.e., pitch-elevation and pitch-size.[46] This suggests that indeed synesthesia may be associated with a greater tendency to form sound-symbolic correspondences, sound symbolism being a domain outside those of the types of synesthesia reported by the participants. It is worth noting that almost all the synesthetes in this study were associators (whose synesthetic experiences are reported to occur in the "mind's eye"); thus, it remains an open question if projectors (whose synesthetic experiences are reported to occur out in the world where the inducing stimuli are perceived) might show stronger low-level correspondences in addition to sound-symbolic ones.[142]

Conclusions

Beyond the spatiotemporal factors that have, in recent decades, been considered central to solving the crossmodal binding problem, it is clear that the crossmodal correspondences, as but one example of "the unity effect" (see Refs. 143, for a review), offer some potentially relevant weak constraints on crossmodal binding too. However, moving forward, it is clear that more cognitive neuroscience data will be needed on the neural underpinnings of the putatively different classes of crossmodal correspondences, on sound symbolism, and on synesthesia. Developmental and comparative perspectives will likely also provide useful information here.

Many crossmodal correspondences have now been documented between more complex stimuli, such as, for example, classical music and paintings (e.g., Refs. 16,16a; see also Ref. 27). However, a careful analysis of such research and the relationship between such high-level or semantically (esthetically) meaningful correspondences and the correspondences between simpler dimensions that have been reviewed here will have to wait for another day. That said, and as suggested at the start of this piece, our theoretical

[1] Note that we are not suggesting here that there are not important individual differences in the strength of crossmodal correspondences, but rather that those individual differences may have more to do with the notion of consensuality and metacognition (see Refs. 138–140), than of "degree of synesthetic tendencies.[127,141]

understanding of the crossmodal correspondences from the better-studied pairs of senses (such as the audiovisual pairing reviewed here) will likely also apply to those correspondences between any other pair of modalities as well. Similarly, we believe that our understanding of crossmodal correspondences involving complex stimuli will be furthered by our growing understanding on the underlying nature of the correspondences that connect simple stimuli.

In closing, it is important to stress that the focus of this literature review has primarily been on data collected from WEIRDos—that is, young participants from Western, educated, industrialized, rich, and democratic countries, mostly studying undergraduate psychology.[144,145] As such, it leaves unanswered many intriguing questions concerning possible cross-cultural similarities and differences that may affect audiovisual crossmodal correspondences in general, or sound symbolism in particular (see Refs. 105,106,112,146,147, for some preliminary evidence on this theme). Similarly, there has not been space in this review to deal in any depth with the developmental (see Refs. 14, for a review) or cross-species angles on the topic (see Ref. 148 for a review). It will, for instance, be interesting to compare the different developmental trajectories for the acquisition of different classes of correspondence (e.g., Refs. 21,149). Certainly, the question of which classes of crossmodal correspondence are experienced cross-species remains another intriguing question for future research (cf. Refs. 150, for some preliminary data on this issue from dogs). The data that have been published to date suggest that chimpanzees many be sensitive to some of the same correspondences as humans (e.g., Refs. 18,151).[m] One oft-mentioned example from the comparative biology literature is the lowering of the pitch of animal vocalizations to sound "bigger" (what is called "dishonest signaling").[153] Ultimately though, it is to be hoped that the rapidly expanding body of research on the crossmodal correspondences will, in the years to come, help our theorizing concerning their underpinning nature.

References

1. Parise CV. Crossmodal correspondences: standing issues and experimental guidelines. *Multisensory Res.* 2016;29:7–28.
2. Spence C. Crossmodal correspondences: a tutorial review. *Atten Percept Psychophys.* 2011;73:971–995.
3. Hubbard Jones M. A study of the 'common chemical sense'. *Am J Psychol.* 1954;67:696–699.
4. Cain WS. Olfaction and the common chemical sense: some psychophysical contrasts. *Sensory Process.* 1976;1:57–67.
5. Silver W. The common chemical sense. In: Finger TE, Silver WL, eds. *Neurobiology of Taste and Smell.* New York: John Wiley & Sons; 1987:65–87.
6. Belkin K, Martin R, Kemp SE, et al. Auditory pitch as a perceptual analogue to odor quality. *Psychol Sci.* 1997;8:340–342.
7. Gilbert AN, Martin R, Kemp SE. Cross-modal correspondence between vision and olfaction: the color of smells. *Am J Psychol.* 1996;109:335–351.
8. Kemp SE, Gilbert AN. Odor intensity and color lightness are correlated sensory dimensions. *Am J Psychol.* 1997;110:35–46.
9. Deroy O, Crisinel AS, Spence C. Crossmodal correspondences between odors and contingent features: odors, musical notes, and geometrical shapes. *Psychon Bull Rev.* 2013;20:878–896.
10. Knoeferle KM, Woods A, Käppler F, et al. That sounds sweet: using crossmodal correspondences to communicate gustatory attributes. *Psychol Mark.* 2015;32:107–120.

[m] von Hornbostel[152] even argued for the amodal detection of brightness in goldfish!

11. Knöferle KM, Spence C. Crossmodal correspondences between sounds and tastes. *Psychon Bull Rev.* 2012;19:992–1006.
12. Spence C, Deroy O. Tasting shapes: a review of four hypotheses. *Theor Histor Sci.* 2013;10:207–238.
13. Spence C, Wan X, Woods A, et al. On tasty colours and colourful tastes? Assessing, explaining, and utilizing crossmodal correspondences between colours and basic tastes. *Flavour.* 2015;4:23.
14. Deroy O, Spence C. Learning 'arbitrary' crossmodal correspondences: staying away from neonatal synaesthesia. *Neurosci Biobehav Rev.* 2013;37:1240–1253.
15. Marks L. *The Unity of the Senses: Interrelations Among the Modalities.* New York: Academic Press; 1978.
16. Albertazzi L, Canal L, Micciolo R. Cross-modal association between materic painting and classical Spanish music. *Front Psychol.* 2015;6:424.

16a. Spence, C. (in press). Assessing the role of emotional mediation in explaining crossmodal correspondences involving musical stimuli. *Multisensory Research.*

17. Walker P. Cross-sensory correspondences and cross talk between dimensions of connotative meaning: visual angularity is hard, high-pitched, and bright. *Atten Percept Psychophys.* 2012;74:1792–1809.
18. Ludwig VU, Adachi I, Matzuzawa T. Visuoauditory mappings between high luminance and high pitch are shared by chimpanzees (*Pan troglodytes*) and humans. *Proc Nat Acad Sci USA.* 2011;108:20661–20665.
19. Walker P, Bremner JG, Mason U, et al. Preverbal infants' sensitivity to synesthetic cross-modality correspondences. *Psychol Sci.* 2010;21:21–25.
20. Chang S, Cho YS. Polarity correspondence effect between loudness and lateralized response set. *Front Psychol.* 2015;6:683.
21. Gardner H. Metaphors and modalities: how children project polar adjectives onto diverse domains. *Child Dev.* 1974;45:84–91.
22. Proctor RW, Cho YS. Polarity correspondence: a general principle for performance of speeded binary classification tasks. *Psychol Bull.* 2006;132:416–442.
23. Martino G, Marks LE. Perceptual and linguistic interactions in speeded classification: tests of the semantic coding hypothesis. *Perception.* 1999;28:903–923.
24. Martino G, Marks LE. Cross-modal interaction between vision and touch: the role of synesthetic correspondence. *Perception.* 2000;29:745–754.
25. Sadaghiani S, Maier JX, Noppeney U. Natural, metaphoric, and linguistic auditory direction signals have distinct influences on visual motion processing. *J Neurosci.* 2009;29:6490–6499.
26. Parkinson C, Kohler PJ, Sievers B, et al. Associations between auditory pitch and visual elevation do not depend on language: evidence from a remote population. *Perception.* 2012;41:854–861.
27. Palmer SE, Schloss KB, Xu Z, et al. Music-color associations are mediated by emotion. *Proc Nat Acad Sci USA.* 2013;110:8836–8841.
28. Bozzi P, Flores D'Arcais G. Ricerca sperimentale sui rapporti intermodali fra qualità espressive [Experimental research on the intermodal relationships between expressive qualities]. *Arch Psicol Neurol Psichiatr.* 1967;28:377–420.
29. Collier GL. Affective synaesthesia: extracting emotion space from simple perceptual stimuli. *Motiv Emot.* 1996;20:1–32.
30. Cowles JT. An experimental study of the pairing of certain auditory and visual stimuli. *J Exp Psychol.* 1935;18:461–469.
31. Hubbard TL. Synesthesia-like mappings of lightness, pitch, and melodic interval. *Am J Psychol.* 1996;109:219–238.
32. Janković D. Cross-modal nature of evaluative meaning. In: Galmonte A, Actis-Grosso R, eds. *Different Psychological Perspectives on Cognitive Processes: Current Research Trends in Alps-Adria Region.* Newcastle Upon Tyne: Cambridge Scholars Publishing; 2014:58–75.
33. Lindauer MS. The meanings of the physiognomic stimuli taketa and maluma. *Bull Psychon Soc.* 1990;28:47–50.
34. Lundholm H. The affective tone of lines: experimental researches. *Psychol Rev.* 1921;28:43–60.
35. Poffenberger AT, Barrows BE. The feeling value of lines. *J Appl Psychol.* 1924;8:187–205.
36. Ben-Artzi E, Marks LE. Visual-auditory interaction in speeded classification: role of stimulus difference. *Percept Psychophys.* 1995;57:1151–1162.
37. Bernstein IH, Edelstein BA. Effects of some variations in auditory input upon visual choice reaction time. *J Exp Psychol.* 1971;87:241–247.

38. Evans KK, Treisman A. Natural cross-modal mappings between visual and auditory features. *J Vis.* 2010;10(1):6.1–6.12.
39. Gallace A, Spence C. Multisensory synesthetic interactions in the speeded classification of visual size. *Percept Psychophys.* 2006;68:1191–1203.
40. Melara RD. Dimensional interaction between color and pitch. *J Exp Psychol Hum Percept Perform.* 1989;15:69–79.
41. Melara RD. Similarity relations among synesthetic stimuli and their attributes. *J Exp Psychol Hum Percept Perform.* 1989;15:212–231.
42. Melara RD, O'Brien TP. Interaction between synesthetically corresponding dimensions. *J Exp Psychol Gen.* 1987;116:323–336.
43. Patching GR, Quinlan PT. Garner and congruence effects in the speeded classification of bimodal signals. *J Exp Psychol Hum Percept Perform.* 2002;28:755–775.
44. Marks LE. Cross-modal interactions in speeded classification. In: Calvert GA, Spence C, Stein BE, eds. *Handbook of Multisensory Processes.* Cambridge, MA: MIT Press; 2004:85–105.
45. Parise CV, Spence C. Audiovisual crossmodal correspondences and sound symbolism: an IAT study. *Exp Brain Res.* 2012;220:319–333.
46. Lacey S, Martinez M, McCormick K, et al. Synesthesia strengthens sound-symbolic cross-modal correspondences. *Eur J Neurosci.* 2016;44:2716–2721.
47. McCormick K, Lacey S, Stilla R, et al. Neural basis of crossmodal correspondence between auditory pseudowords and visual shapes. *bioRxiv;* 2018. https://doi.org/10.1101/478347.
48. McCormick K, Lacey S, Stilla R, et al. Neural basis of the crossmodal correspondence between auditory pitch and visuospatial elevation. *Neuropsychologia.* 2018;112:19–30.
49. Miller JO. Channel interaction and the redundant targets effect in bimodal divided attention. *J Exp Psychol Hum Percept Perform.* 1991;17:160–169.
50. Makovac E, Gerbino W. Sound-shape congruency affects the multisensory response enhancement. *Vis Cogn.* 2010;18:133–137.
51. Colavita FB. Human sensory dominance. *Percept Psychophys.* 1974;16:409–412.
52. Spence C, Parise C, Chen YC. The Colavita visual dominance effect. In: Murray MM, Wallace M, eds. *Frontiers in the Neural Bases of Multisensory Processes.* Boca Raton, FL: CRC Press; 2011:523–550.
53. Stekelenburg JJ, Keetels M. The effect of synesthetic associations between the visual and auditory modalities on the Colavita effect. *Exp Brain Res.* 2016;234:1209–1219.
54. Guzman-Martinez E, Ortega L, Grabowecky M, et al. Interactive coding of visual spatial frequency and auditory amplitude-modulation rate. *Curr Biol.* 2012;22:383–388.
55. Maeda F, Kanai R, Shimojo S. Changing pitch induced visual motion illusion. *Curr Biol.* 2004;14:R990–R991.
56. Parise C, Spence C. 'When birds of a feather flock together': synesthetic correspondences modulate audiovisual integration in non-synesthetes. *PLoS One.* 2009;4(5):e5664.
57. Sweeny TD, Guzman-Martinez E, Ortega L, et al. Sounds exaggerate visual shape. *Cognition.* 2012;124:194–200.
58. Orchard-Mills E, Van der Burg E, Alais D. Crossmodal correspondence between auditory pitch and visual elevation affects temporal ventriloquism. *Perception.* 2016;45:409–424.
59. Parise C, Spence C. Synesthetic congruency modulates the temporal ventriloquism effect. *Neurosci Lett.* 2008;442:257–261.
60. Keetels M, Vroomen J. No effect of synesthetic congruency on temporal ventriloquism. *Atten Percept Psychophys.* 2011;73:209–218.
61. Spence C, Deroy O. Hearing mouth shapes: sound symbolism and the reverse McGurk effect. *I-Perception.* 2012;3:550–552.
62. Brunetti R, Indraccolo A, Mastroberardino S, et al. The impact of cross-modal correspondences on working memory performance. *J Exp Psychol Hum Percept Perform.* 2017;43:819–831.
63. Lowe ML, Haws KL. Sounds big: the effects of acoustic pitch on product perceptions. *J Market Res.* 2017;54(2):331–346.
64. Parise CV, Pavani F. Evidence of sound symbolism in simple vocalizations. *Exp Brain Res.* 2011;214:373–380.
65. Vainio L, Tiainen M, Tiippana K, et al. Sharp and round shapes of seen objects have distinct influences on vowel and consonant articulation. *Psychol Res.* 2017;81:827–839.
66. Salgado-Montejo A, Marmolejo-Ramos F, Alvarado JA, et al. Drawing sounds: representing tones and chords spatially. *Exp Brain Res.* 2016;234:3509–3522.

67. Spence C, Deroy O. How automatic are crossmodal correspondences? *Conscious Cognit.* 2013;22:245–260.
68. Moors A, De Houwer J. Automaticity: a theoretical and conceptual analysis. *Psychol Bull.* 2006;132:297–326.
69. Chiou R, Rich AN. Cross-modality correspondence between pitch and spatial location modulates attentional orienting. *Perception.* 2012;41:339–353.
70. Mossbridge JA, Grabowecky M, Suzuki S. Changes in auditory frequency guide visual-spatial attention. *Cognition.* 2011;121:133–139.
71. Fernández-Prieto I, Navarra J. The higher the pitch the larger its crossmodal influence on visuospatial processing. *Psychol Music.* 2017;45(5):713–724.
72. Hung SM, Styles SJ, Hseih PJ. Can a word sound like a shape before you have seen it? Sound-shape mapping prior to conscious awareness. *Psychol Sci.* 2017;28(3):263–275.
73. Ković V, Plunkett K, Westermann G. The shape of words in the brain. *Cognition.* 2010;114:19–28.
74. Bien N, ten Oever S, Goebel R, et al. The sound of size: crossmodal binding in pitch-size synesthesia: a combined TMS, EEG, and psychophysics study. *Neuroimage.* 2012;59:663–672.
75. Revill KP, Namy LL, DeFife LC, et al. Cross-linguistic sound symbolism and crossmodal correspondence: evidence from fMRI and DTI. *Brain Lang.* 2014;128:18–24.
76. Peiffer-Smadja N, Cohen L. The cerebral bases of the bouba-kiki effect. *Neuroimage.* 2019;186:679–689.
77. Noppeney U. Characterization of multisensory integration with fMRI. In: Murray MM, Wallace MT, eds. *The Neural Bases of Multisensory Processes.* Boca Raton, FL: CRC Press/Taylor and Francis; 2012:233–252.
78. Cian L. Verticality and conceptual metaphors: a systematic review. *J Assoc Consum Res*; 2017. http://www.journals.uchicago.edu/doi/abs/10.1086/694082.
79. Deroy O, Fernandez-Prieto I, Navarra J, et al. Unravelling the paradox of spatial pitch. In: Hubbard TL, ed. *Spatial Biases in Perception and Cognition.* Cambridge, UK: Cambridge University Press; 2018:77–93.
80. Lynott D, Coventry K. On the ups and downs of emotion: testing between conceptual-metaphor and polarity accounts of emotional valence–spatial location interactions. *Psychon Bull Rev.* 2014;21:218–226.
81. Meier BP, Robinson MD. Why the sunny side is up: association between affect and vertical position. *Psychol Sci.* 2004;15:243–247.
82. Walker L, Walker P, Francis B. A common scheme for cross-sensory correspondences across stimulus dimensions. *Perception.* 2012;41:1186–1192.
83. Carnevale MJ, Harris LR. Which direction is up for a high pitch? *Multisensory Res.* 2016;29:113–132.
84. Mudd SA. Spatial stereotypes of four dimensions of pure tone. *J Exp Psychol.* 1963;66:347–352.
85. Parise CV, Knorre K, Ernst MO. Natural auditory scene statistics shapes human spatial hearing. *Proc Nat Acad Sci USA.* 2014;111:6104–6108.
86. Pratt CC. The spatial character of high and low tones. *J Exp Psychol.* 1930;13:278–285.
87. Roffler SK, Butler RA. Factors that influence the localization of sound in the vertical plane. *J Acoust Soc Am.* 1968;43:1255–1259.
88. Stumpf K. *Tonpsychologie I [Psychology of the Tone].* Leipzig: Hirzel; 1883.
89. Jamal Y, Lacey S, Nygaard L, et al. Interactions between auditory elevation, auditory pitch and visual elevation during multisensory perception. *Multisensory Res.* 2017;30:287–306.
90. Deroy O, Spence C. Weakening the case for 'weak synaesthesia': why crossmodal correspondences are not synaesthetic. *Psychon Bull Rev.* 2013;20:643–664.
90a. Spence C. On the relative nature of (pitch-based) crossmodal correspondences. *Multisens Res.* 2019;32:235–265.
91. Garner WR. Asymmetric interactions of stimulus dimensions in perceptual information processing. In: Tighe TJ, Shepp BE, eds. *Perception, Cognition, and Development: Interactional Analyses.* Hillsdale, NJ: Erlbaum; 1983:1–37.
92. Brunetti R, Indraccolo A, Del Gatto C, et al. Are crossmodal correspondences absolute or relative? Context effects on speeded classification. *Atten Percept Psychophys.* 2018;80(2):527–534.
93. Walker L, Walker P. Cross-sensory mapping of feature values in the size–brightness correspondence can be more relative than absolute. *J Exp Psychol Hum Percept Perform.* 2016;42:138–150.
94. Marks LE, Szczesiul R, Ohlott P. On the cross-modal perception of intensity. *J Exp Psychol Hum Percept Perform.* 1986;12:517–534.
95. Orchard-Mills E, Van der Burg E, Alais D. Amplitude-modulated auditory stimuli influence selection of visual spatial frequencies. *J Vis.* 2013;13:6.
96. Blasi DE, Wichmann S, Hammarström H, et al. Sound-meaning association biases evidenced across thousands of languages. *Proc Nat Acad Sci USA.* 2016;113:10818–10823.

97. Hinton L, Nichols J, Ohala JJ, eds. *Sound Symbolism*. Cambridge: Cambridge University Press; 1994.
98. Lockwood G, Dingemanse M. Iconicity in the lab: a review of behavioral, developmental, and neuroimaging research into sound-symbolism. *Front Psychol*. 2015;6:1246.
99. Nuckolls J. The case for sound symbolism. *Annu Rev Anthropol*. 2003;28:225–252.
100. Sidhu DM, Pexman PM. Five mechanisms of sound symbolic association. *Psychon Bull Rev*. 2018;25:1619–1643.
101. Walker P. Cross-sensory correspondences and symbolism in spoken and written language. *J Exp Psychol Learn Mem Cogn*. 2016;42:1339–1361.
102. Köhler W. *Gestalt Psychology*. New York: Liveright; 1929.
103. Köhler W. *Gestalt Psychology: An Introduction to New Concepts in Modern Psychology*. New York: Liveright Publication; 1947.
104. Ramachandran VS, Hubbard EM. Hearing colors, tasting shapes. *Sci Am*. 2003;288(May):43–49.
105. Bremner A, Caparos S, Davidoff J, et al. Bouba and Kiki in Namibia? A remote culture make similar shape-sound matches, but different shape-taste matches to Westerners. *Cognition*. 2013;126:165–172.
106. Styles SJ, Gawne L. When does maluma/takete Fail? Two key failures and a meta-analysis suggest that phonology and phonotactics matter. *I-Perception*. 2017;8:1–17.
107. Nahm FKD, Tranel D, Damasio H, et al. Cross-modal associations and the human amygdala. *Neuropsychologia*. 1993;31:727–744.
108. Oberman LM, Ramachandran VS. Preliminary evidence for deficits in multisensory integration in autism spectrum disorders: the mirror neuron hypothesis. *Soc Neurosci*. 2008;3:348–355.
109. Occelli V, Esposito G, Venuti P, et al. The takete-maluma phenomenon in autism spectrum disorders. *Perception*. 2013;42:233–241.
110. Ramachandran VS, Oberman LM. Broken mirrors. *Sci Am*. 2006;295(5):62–69.
111. Sapir E. A study in phonetic symbolism. *J Exp Psychol*. 1929;12:225–239.
112. Chen YC, Huang PC, Woods A, et al. When "Bouba" equals "Kiki": cultural commonalities and cultural differences in sound-shape correspondences. *Sci Rep*. 2016;6:26681.
113. Knöferle K, Li J, Maggioni E, Spence C. What drives sound symbolism? Different acoustic cues underlie sound-size and sound-shape mappings. *Sci Rep*. 2017;7(1):5562. https://doi.org/10.1038/s41598-017-05965-y.
114. List SM, McCormick K, Lacey S, et al. Acoustic and visual stimulus parameters underlying sound symbolic crossmodal correspondences. *bioRxiv*; 2019. https://doi.org/10.1101/517581.
115. Bankieris K, Simner J. What is the link between synaesthesia and sound symbolism? *Cognition*. 2015;136:186–195.
116. Braaten R. Synesthetic correspondence between visual location and auditory pitch in infants. In: *Paper Presented at the Annual Meeting of the Psychonomic Society*. 1993.
117. Marks LE, Mulvenna CM. Synesthesia, at and near its borders. *Front Psychol*. 2013;4:651.
118. Martino G, Marks LE. Synesthesia: strong and weak. *Curr Dir Psychol Sci*. 2001;10:61–65.
119. Berman G. Synesthesia and the arts. *Leonardo*. 1999;32:15–22.
120. Wicker FW, Holahan CK. Analogy training and synesthetic phenomena. *J Gen Psychol*. 1978;98:113–122.
121. Deroy O, Spence C. Lessons of synaesthesia for consciousness: learning from the exception, rather than the general. *Neuropsychology*. 2016;88:49–57.
122. Deroy O, Spence C. Questioning the continuity claim: what difference does consciousness make? In: Deroy O, ed. *Sensory Blending: New Essays on Synaesthesia*. Oxford: Oxford University Press; 2017:191–214.
123. Day S. Some demographic and socio-cultural aspects of synesthesia. In: Robertson LC, Sagiv N, eds. *Synesthesia: Perspectives from Cognitive Neuroscience*. New York: Oxford University Press; 2005:11–33.
124. Ernst MO. Learning to integrate arbitrary signals from vision and touch. *J Vis*. 2007;7(5):7.1–7.14.
125. Simner J, Harrold J, Creed H, et al. Early detection of markers for synaesthesia in childhood populations. *Brain*. 2009;132:57–64.
126. Watson MR, Akins KA, Spiker C, et al. Synesthesia and learning: a critical review and novel theory. *Front Hum Neurosci*. 2014;8:98.
127. Grossenbacher PG, Lovelace CT. Mechanisms of synesthesia: cognitive and physiological constraints. *Trends Cogn Sci*. 2001;5:36–41.
128. Deroy O, ed. *Sensory Blending: On Synaesthesia & Related Phenomena*. Oxford, UK: Oxford University Press; 2017.
129. Rader CM, Tellegen A. An investigation of synesthesia. *J Personal Soc Psychol*. 1987;52:981–987.
130. Spence C, Deroy O. Crossmodal mental imagery. In: Lacey S, Lawson R, eds. *Multisensory Imagery: Theory and Applications*. New York, NY: Springer; 2013:157–183.

131. Meier B, Rothen N. Graphene-color synaesthesia is associated with a distinct cognitive style. *Front Psychol.* 2013;4. https://doi.org/10.3389/fpsyg.2013.00632.
132. Chun CA, Hupé JM. Are synesthetes exceptional beyond their synesthetic associations? A systematic comparison of creativity, personality, cognition and mental imagery in synesthetes and controls. *Br J Psychol.* 2016;107:397–418.
133. Esterman M, Verstynen T, Ivry RB, et al. Coming unbound: disrupting automatic integration of synesthetic color and graphemes by transcranial magnetic stimulation of the right parietal lobe. *J Cogn Neurosci.* 2006;18:1570–1576.
134. Spence C, Parise CV. The cognitive neuroscience of crossmodal correspondences. *I-Perception.* 2012;3:410–412.
135. Muggleton N, Tsakanikos E, Walsh V, et al. Disruption of synaesthesia following TMS of the right posterior parietal cortex. *Neuropsychologia.* 2007;45:1582–1585.
136. Cohen Kadosh R, Henik A, Walsh V. Small is bright and big is dark in synaesthesia. *Curr Biol.* 2007;17:R834–R835.
137. Simner J, Ward J, Lanz M, et al. Non-random associations of graphemes to colours in synaesthetic and non-synaesthetic populations. *Cogn Neuropsychol.* 2005;22:1069–1085.
138. Chen YC, Huang PC, Woods A, et al. I know that "Kiki" is angular: the metacognition underlying sound-shape correspondences. *Psych Bull Rev.* 2019;26:261–268.
139. Koriat A. Another look at the relationship between phonetic symbolism and the feeling of knowing. *Mem Cogn.* 1976;4:244–248.
140. Koriat A. Subjective confidence in one's answers: the consensuality principle. *J Exp Psychol Learn Mem Cogn.* 2008;34:945–959.
141. Marks LE. For hedgehogs and foxes: individual differences in the perception of cross-modal similarity. In: Ljunggren G, Dornic S, eds. *Psychophysics in Action.* Berlin: Springer Verlag; 1989:55–65.
142. Ward J. Synaesthesia. In: Stein BE, ed. *The New Handbook of Multisensory Processes.* Cambridge, MA: MIT Press; 2012:319–333.
143. Chen YC, Spence C. Assessing the role of the 'unity effect' on multisensory integration: a review. *Front Psychol.* 2017;8:445.
144. Arnett JJ. The neglected 95%: why American psychology needs to become less American. *Am Psychol.* 2008;63:602–614.
145. Henrich J, Heine SJ, Norenzayan A. The weirdest people in the world? *Behav Brain Sci.* 2010;33:61–135.
146. Osgood CE, May W, Miron M. *Cross-Cultural Universals of Affective Meaning.* Urbana: University of Illinois Press; 1975.
147. Walker R. The effects of culture, environment, age, and musical training on choices of visual metaphors for sound. *Percept Psychophys.* 1987;42:491–502.
148. Ratcliffe VF, Taylor AM, Reby D. Cross-modal correspondences in non-human mammal communication. *Multisensory Res.* 2016;29:49–91.
149. Marks LE, Hammeal RJ, Bornstein MH. Perceiving similarity and comprehending metaphor. *Monogr Soc Res Child Dev.* 1987;52(215):1–102.
150. Faragó T, Pongrácz P, Miklósi Á, et al. Dogs' expectation about signalers' body size by virtue of their growls. *PLoS One.* 2010;5(12):e15175.
151. Ravignani A, Sonnweber R. Chimpanzees process structural isomorphisms across sensory modalities. *Cognition.* 2017;161:74–79.
152. von Hornbostel EM. Über Geruchshelligkeit [On smell brightness]. *Pflugers Arch für Gesamte Physiol Menschen Tiere.* 1931;227:517–538.
153. Fitch WT. The evolution of speech: a comparative review. *Trends Cogn Sci.* 2000;4:258–267.

CHAPTER

12

How do crossmodal correspondences and multisensory processes relate to synesthesia?

David Brang[1], *Vilayanur S. Ramachandran*[2]

[1]University of Michigan, Ann Arbor, MI, United States; [2]Department of Psychology and Center for Brain and Cognition, University of California, San Diego, CA, United States

Introduction

Synesthesia is an automatic and involuntary phenomenon in which one modality evokes activation in a second, typically unrelated sensory or cognitive modality, resulting in the experience of atypical qualia (see Chapter 13). For example, in one of the best-studied forms known as grapheme-color synesthesia, viewing numbers or letters elicits the experience of a specific color[2] (Fig. 12.1). Synesthetic experiences are consistent across decades and typically begin early in life.[2–4] While this phenomenon can occur due to brain damage or altered states of consciousness[5] (for review see Refs. 6,7), synesthesia is usually considered within a genetic framework.[2–4,8–10] After being brushed under the carpet for over a century, there was a renaissance of interest in this alluringly quirky phenomenon in the late 1990s, after we and a handful of other researchers rescued it from oblivion. In particular, we showed that it was a robust sensory process that could be studied in detail and that it had vast implications for understanding a wider range of cognitive processes than people had imagined, including creativity, metaphor, sound symbolism (e.g., association of the auditory pseudowords bouba/kiki with ameboid and jagged shapes, respectively), memory (see Chapter 14), e.g., enhanced memory for phone numbers[9] and visual eidetic memory,[11] and psycholinguistics,[12] among others. In fact, there has been such a tremendous resurgence of investigations (over 23 books and hundreds of peer-reviewed articles) that it is hard to keep track of it all.

The neural substrates that support some forms of synesthesia have been studied using both psychophysical tests[9,10] and neuroimaging techniques.[13–19] When viewing black letters or numbers, individuals with grapheme-color synesthesia show enhanced activity within posterior inferior temporal color regions,[15,17,20] consistent with subjective reports from some of these individuals that synesthetic colors can be experienced as perceptually real sensations.

Multisensory Perception
https://doi.org/10.1016/B978-0-12-812492-5.00012-7

FIGURE 12.1 Number–color associations for one of our synesthetes. *From Fig. 1 of Ramachandran VS, Brang D. Synesthesia.* Scholarpedia. *2008;3(6):3981. Reproduced with permission.*

Especially credible models are those that rely on both the anatomical pathways and the timing of activation in these networks in their explanation of the physiological basis for synesthesia. In the cross-activation model, researchers have argued that an excess of connectivity within pathways connecting regions of the brain involved in the synesthetic experience give rise to these crossmodal sensations,[9] potentially due to less neural pruning during development.[21] Studies have provided some support for this theory in that grapheme-color synesthetes show altered patterns of gray matter density and connectivity within inferior temporal lobe regions, close to areas responsible for early perceptual color processing,[22–24] in addition to co-activation of grapheme and color regions when viewing achromatic letters.[17]

Updated models examining the neurophysiological bases for the experience of synesthesia have argued for two stages in the process.[25] First, in line with the cross-activation model, synesthetic experiences are generated through direct sensory interactions (e.g., grapheme areas in the fusiform gyrus directly activate neighboring color areas).[15,17] Second, the real and synesthetic percepts (numbers and colors) are subsequently bound into conscious experiences through "hyper-binding" mechanisms in parietal and frontal lobes.[25–27] One critical prediction of this model that has recently been verified is that synesthetic activation can occur without conscious awareness, which could be due to activity failing to reach the second stage.[28]

This model better explains the full range of neuroimaging results from studies of synesthesia (for a meta-analysis see Ref. 29) that show broader differences in functional activity between synesthetes and controls, including regions in the inferior temporal lobes, parietal lobes, and frontal regions.[30,31] Indeed, this two-stage account is necessary to explain evidence that transient interference with the functioning of parietal regions can weaken the experience of synesthetic concurrents, as demonstrated with transcranial magnetic stimulation (TMS).[32,33] Taken collectively, this more recent evidence suggests that synesthesia operates both through direct communication between the senses and the integration of information at multisensory hubs in frontal and parietal areas, with individual differences in these networks potentially accounting for differences between projector and associator synesthetes (projectors experience the atypical sensory experience at the locus of the triggering stimulus, whereas associators report such experience in the mind's eye).

If the goal of cognitive neuroscience is to explain complex perceptual and cognitive phenomena in terms of structure (as happened for DNA and the genetic code), then synesthesia may be a prime example of this application. Over the past 2 decades, behavioral research has succeeded in demonstrating the validity of synesthetes' subjective reports[9,19,34]; accumulating neuroimaging evidence has increasingly illuminated the physiological basis for synesthesia; and genetic research clarifies the hereditary nature of these experiences[35], confirming Galton's speculations. Nevertheless, the broader relationship between synesthesia and cognitive processes present in all individuals remains a matter of continued debate. In particular, interest in this relationship was largely catalyzed by Ramachandran and Hubbard's suggestion that sound symbolism (*bouba–kiki*) may serve as a probe to explore the link between synesthesia and higher-level cognition. Though Ramachandran and Hubbard also noted that sound–symbolic associations "ring true" and are not arbitrary, whereas, the link between the number 5 and the color red seems to be so, yet deeper analysis from our lab and others shows that there may indeed be hidden reasons for even this seemingly random association; see Ref. 36. Extensive models have been proposed and research conducted to identify a link from synesthetic experiences to some form of related cognitive or perceptual processes in typically developing individuals.[37–41] Indeed, many crossmodal experiences are present in nonsynesthetes and it is possible that synesthesia developed as a logical outgrowth from the underlying common neural mechanisms.

Nevertheless, researchers (including the present authors) have largely neglected the possibility that synesthesia is a phenomenon unto itself and not related to any nonsynesthetic process. However, while it is possible that synesthesia is an isolated and unique cognitive phenomenon, this fails intuitions about the emergence of synesthetic experiences[8] and such a result would run counter to the motivations of researchers to identify the relevance of synesthesia to broader cognitive processes. Only by examining synesthesia's relationship to other more common crossmodal phenomena and cognitive processes can we determine whether synesthesia is merely a curiosity, a psychological teratoma, or whether synesthesiologists are paleontologists of the mind, who unearth neural fossils that can help us understand more complex enigmatic mental functions like metaphor, embodied cognition, and mental calendars (see last section of this chapter). Such functions have remained elusive precisely because most scholars ignore the many erratic evolutionary trajectories that culminated in the multiple arrays of crude mechanisms that constitute human cognition. Evolution prefers crude shortcuts that are merely adequate rather than optimal solutions to environmental challenges ("hyperadaptationism"). An added advantage is that many crude shortcuts are easier to evolve than a single sophisticated one (since evolution has no foresight), secondly deploying multiple shortcuts also buys you immunity from damage and tolerance for noisy sensory inputs. As an analogy, think of two inebriated individuals ready to cross the street: each should have considerable difficulty in crossing, but by leaning on each other they manage to stumble across safely.

Thus, even if the phenomenology (and therefore underlying physiology) is utterly different, understanding one (either synesthetic or related cognitive/perceptual phenomena) can enrich our understanding of the other. Indeed, we have previously suggested (and alluded to above) that one could go all the way up to even link the experiences of synesthetes to the genesis of metaphor in literature.[9,10] Following this trajectory, research has examined numerous models proposing how synesthesia might be related to other nonsynesthetic processes that show phenomenological similarities. In this chapter we address two main issues of interest: (1) Is synesthesia related to crossmodal experiences present in all individuals, both phenomenologically and mechanistically, possibly indicating that it exists along a continuum of normal crossmodal correspondences? (2) How do we reconcile these models of synesthetic crossmodal correspondences in the context of other cognitive and perceptual differences in synesthetes?

Crossmodal correspondences

All individuals demonstrate the ability to combine information among their sensory systems. The field of multisensory perception has rigorously demonstrated that auditory, visual, and tactile signals are integrated at both cortical and subcortical sites, leading to enhanced sensory processing and more statistically reliable representations of the natural environment (for a review see Ref. 42). Research in this area emphasizes that perceptual integration of these sensory signals is optimal when the information is matched along certain sensory dimensions (e.g., temporally or spatially coincident,[42–44] see Chapter 3). Whereas multisensory perception reflects the ability for sensory information from separate modalities to be combined in a perceptual manner, crossmodal correspondences capitalize on unifying dimensions that exist among the sensory modalities (allowing the abstraction of a common denominator between the two).

Crossmodal correspondences are defined as pairs of associations between two sensory or cognitive processes that are generally agreed upon by most individuals within a population[45] (see Chapter 11). For example, one of the earliest demonstrations of crossmodal correspondences was found between linguistic nonwords and the angular/round dimension of visual shapes. Specifically, when individuals are instructed to match the nonwords "bouba" and "kiki" to two shapes differing in their angular/roundness dimensions (see Fig. 12.2), individuals reliably report that "kiki" is associated with the more angular of the two images, whereas "bouba" is associated with the rounder of the shapes.[46,47] Ramachandran and Hubbard[9,10] found that Tamil speakers in South India chose the same correspondences and suggested that the *bouba–kiki* effect has relevance to understanding other phenomena such as synesthesia, metaphor, and the evolution of

FIGURE 12.2 Schematic illustrations of (left) bouba/maluma or (right) takete/kiki of the types used by Ramachandran and Hubbard 2001.

words in protolanguage (see introductory chapter 1 by Ramachandran, Marcus, and Chunharas, this volume). Researchers have subsequently argued that this sound symbolism develops due to the temporal pattern of the sound as well as the shape of the mouth made during speech production (suggesting the import of a naturally acquired, statistically learned congruence or "resonance" between or across the two modalities for these objects).[9,48] Ramachandran and Hubbard[9,10] have also argued, however, that the property of jaggedness might be abstracted across the "kiki" sound and jagged shape (e.g., based on Fourier components) and not necessarily because of a Hebbian link between the two. After all, we do not often see jagged objects that say "kiki" which is different from "dog" being paired with "bow wow" (or indeed even "wurf wurf") which is a purely arbitrary link for the most part: there is nothing doglike about the sound "wurf wurf" or "bow wow," though a silhouette of a dog will evoke "arf arf" because of associative learning (notice this is the Sapir–Whorf debate in disguise).

Dozens of crossmodal correspondences have been identified to date (for a review see Ref. 45), showing relationships among many sensory modalities and concepts. For example, extensive research has demonstrated that children and adults match auditory loudness with visual brightness along a similar dimension, such that bright stimuli are associated with louder sounds than dark stimuli.[49] It has been suggested that this form of crossmodal correspondence is supported by individuals' understanding of amodal stimulus dimensions such as intensity, with both loud and bright stimuli falling on the upper range of this mapping.[50] Similar mappings have been identified between auditory pitch and visual elements of spatial elevation, object size, and object brightness. Specifically, high-pitched sounds are associated with high spatial elevations,[51] small objects,[52] as well as lighter colors.[53] Several of these findings have been replicated in both children and adults.[53] While the mapping between these associations has been argued to be due to amodal dimensional agreement, this is not the only possibility.[45]

Several forms of crossmodal correspondences exist, but the boundaries delineating each variety are a matter of continued debate. Spence[45] argues for three categories of crossmodal correspondences: (1) Statistical correspondences—cross-sensory mappings defined purely according to statistical co-occurrence in the natural environment (e.g., a child refers to a dog as "bow wow" based on the sound the dog makes, and one can quickly learn that larger objects tend to produce louder sounds[54]); (2) Structural correspondences—relationships between the sensory modalities engendered by the structure of the brain (e.g., it has been argued that pitch and size relationships, such as a small ball making a high-pitched squeak, cannot be attributed to learning as these associations are present in very young children[52]); (3) Semantically mediated correspondences—abstract dimensions that unify pairs of sensory stimuli (e.g., the dimension of intensity permeates all the sensory systems, so that loud sounds are associated with both bright lights or strong flavors). According to this view, both statistical and semantic crossmodal correspondences would reflect learned associations, but semantic links would reflect similar conceptual or amodal representations across the senses, whereas statistical associations arise through spontaneous associations through experience in the environment. We would add one dimension of sound–symbolic effects as a fourth to Spence's list, as they may be based on abstracting the common denominator (e.g., both visual "kiki" and auditory "kiki" have sudden inflections).

Many crossmodal correspondences have the potential to be grouped into more than one of these categories, as the defining features of the categories are not wholly independent. Most broadly, these categories differ in how an association was established (through statistical learning, innate brain structure, or via semantic links); yet the origin of an association could plausibly occur from more than one of these causes. For example, observational research has verified that the size of an object/animal will predict the loudness of the sound generated during vocalization or on collision with another object (e.g., Ref. 55); simply stated, larger objects produce louder sounds. This elementary association can reflect easily acquired statistical information about the world (one only needs a few instances of hearing a small vs. large animal generate a sound to understand the relationship between size and sound intensity) as well as amodal dimensions of intensity (large and loud both fall on the more salient sides of their respective dimensions). Thus, understanding of how each association is acquired in a mechanistic sense is needed to discriminate the two. Nevertheless, preliminary evidence suggests that different psychological rules apply to each of the crossmodal correspondence categories. Semantically mediated crossmodal correspondences tend to be relative and nonabsolute,[56] such that individuals scale their crossmodal preferences according to the range that they are presented with. For example, the specific tone associated with a white object will typically be the highest-pitched tone from the given options, which in other contexts might be associated with a less bright stimulus if more options are presented (for a review see Ref. 45). Conversely, statistically learned relationships between the senses may operate as stronger crossmodal correspondences and typically do not show relative mappings.[57] For example, roughness and dryness perception during tactile exploration are associated with specific patterns of auditory noise in a nonrelative manner: loud noise elicits stable but large levels of bias in the degree to which a sound changes the experience of a tactile sensation.[58,59] Statistically learned behaviors like these probably occur frequently in the environment, such as our exposure to both hearing *and* feeling our skin making contact with itself and other objects. Nevertheless, the origin of shared relations between many crossmodal correspondences remains imperfectly understood, but individuals nonetheless show remarkably reliable associations suggesting some common mapping, dimension, or learned statistical representation.

Phenomenological similarities between synesthesia and crossmodal correspondences: grapheme-color, sound-color, and number-form synesthesias

Crossmodal correspondences have long been considered a "weak" form of synesthesia[60] implying that synesthesia exists at the end of the continuum of crossmodal correspondences (e.g., Ref. 61). Research examining the categories, perceptual consequences of, and neural mechanisms underlying crossmodal correspondences is still an active field of research, and the resurgence of interest in synesthesia has highlighted the presumed relationship of this phenomenon with crossmodal correspondences. Indeed, many researchers point to optimism that research into either crossmodal correspondences or synesthesia will benefit understanding of the other,[14,41,60,62] particularly in terms of shared neural mechanisms. However, only limited research to date has provided evidence demonstrating that synesthesia and crossmodal correspondences share similar neural mechanisms or arise through a similar selection process in the brain. Such a finding is necessary to understand the relationship between these phenomena in order to better understand multisensory processes in general. Here we examine the models that have been proposed to describe how these phenomena may be linked, the studies that have tested these models, and alternative explanations for findings in these studies (see also Chapter 13).

Part of the allure in grouping synesthesia with crossmodal correspondences may come from the large number of similarities between the two phenomena. Indeed, the primary qualification linking these two experiences is that both reflect the mapping of information across separate sensory modalities, with the specific associations appearing to be both somewhat arbitrary, but also generally intuitive. Under the belief that synesthesia and crossmodal correspondences share a similar physiological basis, researchers have argued that they should follow the same sets of rules in defining the observed crossmodal mappings (e.g., Ref. 63). This qualification poses an initial challenge for synesthesia: while crossmodal correspondences are consistent both over time for the same individual[64] as well as across individuals in the general population, synesthetic associations have been long believed to be random and arbitrary mappings between the senses. However, research over the last 15 years has established that some elements of synesthetes' associations are predictable and reliable across synesthetes, and most importantly, these mappings are in agreement with crossmodal correspondences experienced by nonsynesthetes.[65]

One point that is not under contention is that both synesthetic associations and crossmodal correspondences must be acquired through developmental maturation (as a synesthete is not born recognizing the letter 2, let alone that that 2 is yellow), suggesting that research into how synesthetic associations develop is a potentially fruitful area in understanding its relationship with crossmodal correspondences. Research on children with synesthesia indicates that grapheme-color associations evolve over time, as demonstrated by longitudinal tracking of their reliability from ages 6–11 years,[66,67] later becoming highly stable associations in adulthood.[68,69] Indeed, it is accepted that the specific synesthetic associations one develops are driven by statistical experiences in the world. For example, one's native language exerts strong influences based on prominent linguistic elements, such as color words (e.g., the letter R is most often associated with the color red for synesthetes[65]). Furthermore, one's experience with letters themselves will bias the perceived colors, such that letters and numbers more frequently encountered in English evoke more luminant synesthetic colors (e.g., vowels are typically evocative of bright colors, whereas X and Z typically trigger black or dark purple).[13,70,71] Indeed, preliminary research has demonstrated that early childhood exposure to synesthetic pairings (such as colored refrigerator magnets) can significantly bias synesthetes' associations throughout their lifetime.[72] Furthermore, similarly shaped English letters show similar colors[73–75] and similar shape-color or phoneme-color similarities have been found in other languages as well,[76–78] indicating that the specific synesthetic associations one experiences are highly tailored and meaningful pairings. Additional evidence for the protracted nature of synesthesia's development comes from research on second-language acquisition: nonmeaningful graphemes tend to be colorless to synesthetes, but through extensive experience with the graphemes (most commonly observed when one is learning a new language) synesthetic colors can develop over time.[74,79–81] Indeed, the transfer of associations, such as color, from a synesthetes' preexisting set of associations to the novel graphemes of a new language can occur in a systematic way, based on shape or phonemic similarity.[80] For this specific form of synesthesia at least, research thus demonstrates that synesthetic associations between graphemes and colors occur through systematic experience in the environment and are not random pairings, similar to the manner in which statistical mappings occur in natural crossmodal correspondences.

The next qualifying element for models proposing similar physiological bases between these two phenomena is whether there is any similarity in the associations synesthetes make relative to those that nonsynesthetes experience as crossmodal correspondences. Research strongly supports this view as demonstrated across several forms of synesthesia. Grapheme-color synesthesia has been studied extensively in this context, with the finding that nonsynesthetes who are instructed to identify a color that is best associated with a particular number or letter will generate associations that are consistent with those experienced by synesthetes.[65] For example, similar to the most common associations experienced by synesthetes, nonsynesthetes paired red with A, blue with B, yellow with C, black with X and Z, etc., demonstrating similar cognitive assessments in assigning these associations.[65]

Examining how crossmodal correspondences associations relate to other forms of synesthesia, extensive work has additionally been conducted on sound–color synesthesia, starting with reports from Marks[82] that a similar mapping between sounds and colors exists between synesthetes and nonsynesthetes. This finding has been replicated and extended, such that high frequency sounds generate brighter synesthetic colors than do low frequency sounds, and this pattern matches those obtained from nonsynesthetic individuals.[83] Marks argued, influentially, that these mappings were universal across synesthetes and nonsynesthetes, reflecting common processes in both groups.[82] While more limited research has been conducted on sound–color synesthesia relative to grapheme-color synesthesia, several studies have examined the relationship between synesthetically evoked colors and those of crossmodal correspondences to the same sounds. Pitch-color translations are one such dimension that is consistent across synesthetic associations and nonsynesthetic crossmodal correspondences, such that both groups experience a reliable mapping of low-pitched sounds with dark colors and high-pitched sounds with bright colors.[53,62,82–86] More recent examinations have replicated and expanded this finding[62] and argue that since crossmodal dimensions are shared across synesthetes and nonsynesthetes these experiences likely use the same mechanisms in both groups. However, it is important to note that it is equally plausible that the causality is reversed, such that the specific synesthetic experiences one possesses could be based on preexisting associations made through crossmodal correspondences. This view is supported by findings that crossmodal correspondences are present in children,[53] yet the specific synesthetic associations one experiences do not solidify until late childhood or early adolescence.[67,87]

Phenomenological similarities between synesthesia and crossmodal correspondences: sequence-space synesthesias and synesthetic mental calendars

Many forms of synesthesia exist, and with each form comes the potential that there are shared neural mechanisms that evoke a related crossmodal correspondence in nonsynesthetes. The cases above present data on two of the most canonical forms of synesthesia, but other variants have also received much attention in this context, and they, by their nature, blur the boundary between synesthesia and crossmodal correspondences. One type of synesthesia that fits this more ambiguous territory between synesthetic experiences and typical nonsynesthetic mappings is sequence-space synesthesias (also known as time-space, number-form, and calendar synesthesias), which was first reported in the late 19th century by Francis Galton as numbers occupying particular regions of space (Fig. 12.3).[88]

Prevalence rates vary, but estimates range from 5% to 15% of the population as experiencing one form of spatial-sequence synesthesia.[89] The high prevalence of this condition and the difficulty in establishing a cutoff between individuals who could simply imagine a

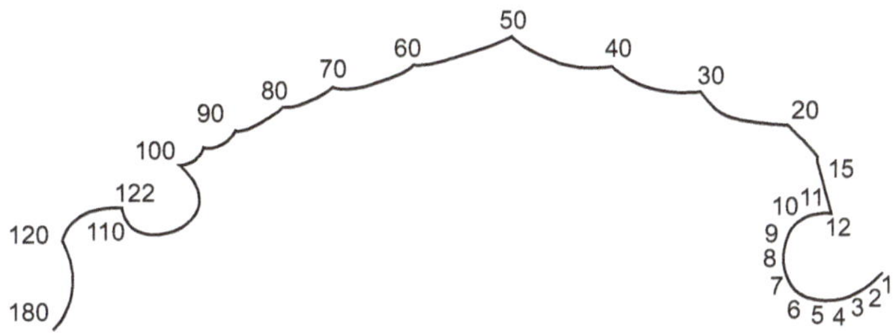

FIGURE 12.3 Number-line generated by one of Francis Galton's synesthetic subjects in 1880. *From Fig. 4 of Ramachandran VS, Brang D. Synesthesia.* Scholarpedia. *2008;3(6):3981. Reproduced with permission.*

number-line instead of automatically experiencing a synesthetic one has caused debate whether this is simply a crossmodal correspondence in itself and a form of synesthesia at all,[90] particularly as no sensory concurrent sensations are evoked (sequences are mapped in space). However, neuroimaging research, behavioral findings, and consistency of self-reported sensations have all pointed to this as a veridical form of synesthesia.[71,91–94] Indeed, individuals with spatial-sequence synesthesia are more likely to have other forms of synesthesia as well, indicating a strong link in the cluster or family of synesthesias.[94,95]

Substantial evidence demonstrates that individuals in the general population experience some form of (implicit) mental number line, with evidence coming largely from the SNARC effect (Spatial-Numerical Association of Response Codes).[90] In this effect, subjects are instructed to make task-irrelevant judgments about numbers using either their left or right hand (e.g., respond with your left hand if the number is odd, and right hand if the number is even). Even though the task has nothing to do with the magnitude of the numbers, subjects will respond significantly faster with their left hand to small numbers and with their right hand to large numbers, indicative of an implicit number line running from the left to the right. Identifying a similar pattern of experiences between synesthetes and typical crossmodal correspondences, the majority of Western individuals experience implicit number lines running from left to right, and indeed the majority of Western synesthetes also experience synesthetic number-forms running left to right.[94] However, as can be seen in Fig. 12.3 depicting the number line produced by a British synesthete in 1880, this is not always the case.[88] While no explanation for the leftward direction of this number line was provided, the synesthete himself inferred that the path of the numbers 1–10 followed that of a clock face:

> You will observe that the first part of the diagram roughly follows the arrangement of figures on a clock face, and I am inclined to think that may have been in part the unconscious source of it, but I have always been utterly at a loss to account for the abrupt change at 10 and again at 12.[88]

When nonsynesthetic individuals are asked to visualize an annual calendar, most report a vague indistinct rectangular grid in front of them on the parallel plane. But, amazingly, a small proportion of the population "literally" sees a black or colorful outlined circle floating in space often with the months clearly labeled with a specific font! It may even rest on a horizontal plane like a Hula-Hoop—with their chest as a tangent. In the variant of sequence-space synesthesia referred to as time-space synesthesia, units of time, such as the months of the year or days of the week, are mapped into a spatial landscape, following well-defined paths.[96]

Research from our lab cataloged the types of forms time-space synesthetes' experience, demonstrating a variety of shapes ranging from basic straight lines to circles and complex 3D landscapes.[97] Similar to other forms of synesthesia, time-space calendars are remarkably consistent over time for a given synesthete[96] and there is similarity of the spatial arrangements across synesthetes.[97] For example, the months of the year typically follow the pattern of a circle as if to mimic a clock face.[97] Interestingly, however, there is substantial variability across synesthetes in the specific location of months on the circle (e.g., the location of January is not always at the top of the circle).[98] Further demonstrating statistical regularities in the pattern of experiences, the clockwise or counterclockwise spatial arrangement of these sequences tends to follow handedness, with right-handed synesthetes showing clockwise arrangements, while left-handed synesthetes show counterclockwise arrangements.[99] Nevertheless, when comparing synesthetes' calendar spatial forms to those that nonsynesthetes generate, the circular nature of synesthetic calendars is the most salient difference between the groups.[97] Indeed, as mentioned above, nonsynesthetes most often produce calendars that follow the spatial arrangement of a single linear placement or a grid of rows, similar to that which would be displayed on a wall calendar.

While the specific patterns of associations found in synesthetes do not fully mirror the mappings made by nonsynesthetes, substantial research has demonstrated that time is treated in a spatial manner by individuals in the general population, extending the link to synesthesia. Extensive work from Lakoff and colleagues (e.g., Ref. 100) indicates that all individuals think about time (at least implicitly) in a spatial manner. Experimental studies have similarly demonstrated that thinking about space and spatial processes affect our reasoning about time.[101–105] Furthermore, damage to right parietal regions that deal with spatial mapping can affect how individuals think about time.[106] These studies demonstrate the pervasiveness of spatial-temporal crossmodal correspondences in language and thought (e.g., a past event is behind me) and also in ways that affect behavior,[101,103,104] and neural responses.[107]

To establish the "reality" of these calendars we recently introduced a novel "recite alternate months backward" task.[108,109] Most of us do this algorithmically (and therefore, slowly) taking twice the time for backward as forward. This difference in speed is reduced significantly in time-space synesthetes (by about 50%), as is their overall speed. This is presumably because they are just reading off the calendar perceptually "displayed" in front of them; the fact that they moved their eyes and index finger along the rim of the calendar, and, when asked about events in (say) July they spontaneously shifted their gaze to look at its location, supports this view.

We have additionally observed in a recent synesthete that, remarkably, the calendar was vertical with a strange L-shaped ribbon tilted in 3D. For this individual, the months were clearly marked with color splotches delineated by thin black outlines. In half of the months the calendar remained stuck in front when she looked to the right and then the left side of the ribbon became indistinct, making it hard to retrieve episodic memories from those months (one of the most striking examples of embodied cognition that we know of). It was as though she was looking at an actual physical calendar rather than an abstract internal representation.

The most compelling results observed[108,109] were from "projector" time-space synesthetes (~30% of individuals) who reported "seeing" the lines delineating the calendar's boundaries as well as boundaries between months (e.g., see Fig. 12.4). Those who reported that the experience was in their "mind's eye", i.e., "associators", nonetheless reported that these experiences were more "intrusive" than regular mental images (e.g., you think of glass shoes

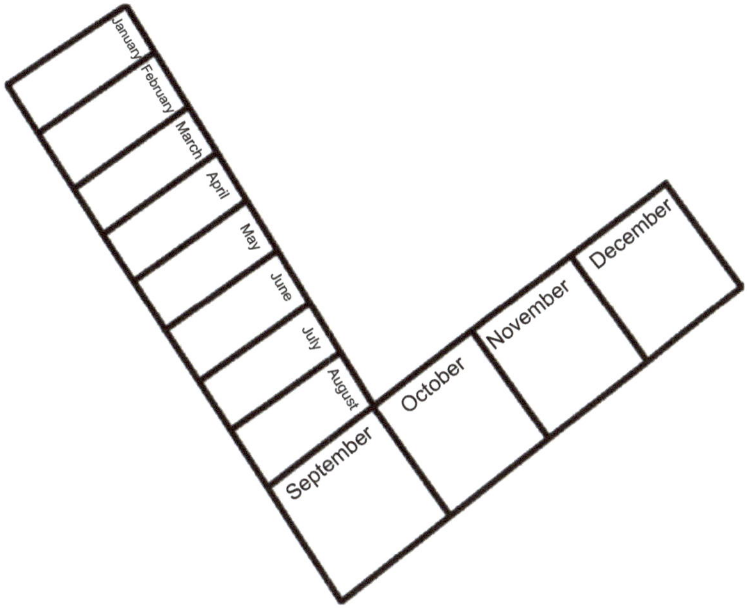

FIGURE 12.4 Calendar form from a synesthete. *From Fig. 1 of Ramachandran V, Seckel E. Synesthetic colors induced by graphemes that have not been consciously perceived.* Neurocase. *2015;21(2):216–219. Reproduced with permission.*

when thinking of Cinderella, but it is not obligatory, and you can switch to thinking of the prince, the pumpkin, the chariot, etc.). Three additional effects of these calendars are worthy of note:

First, some synesthetes are able to adopt an allocentric view of their calendar, such as is the case for one of our synesthetes who experiences a U shape. When she changes her vantage point, the fonts defining the months all became mirror reversed and the subject registered surprise saying "I've never tried this before—why are the letters mirror-reversed? Oh I see why now, etc."

Second, we wondered to what extent do the lines defining the calendar—whether straight (as in the shape the phenomenon of M) or U-shaped, that are obviously generated top-down—mimic the activity of early visual areas stimulated bottom-up? For instance, it is known[110] that if a loop of thin black wire is moved in front of the dynamic visual noise or snow on a TV screen the twinkling dots inside the loop coalesce into clusters that move slower. Surprisingly, at least three of four subjects could get the same effect by simply projecting their mental calendar on the TV static! On the other hand, if they imagined (say) a *regular* calendar or a square, no such clustering or adhesive movement was seen.

The last observation was the most intriguing. When synesthete H. projected or superimposed her calendar (L-shaped) on top of art papers (such as MacKay's rays) or printed stripes, the lines defining the calendar became bent or distorted in a manner seen in geometric optical illusions such as the Hermann grid illusion. This is the first example we know of where an entirely internally generated image of a line can be subjected to distortion by a real external image—where fantasy yields to reality.

The question emerges: what is the neural locus of the calendar, whether indistinct as in nonsynesthetes or crisply defined and vivid as in those with time-space synesthesia? We propose a theory based on 3 bits of information. First, there are strong hints from our lab (VSR) that (1) the left angular gyrus is involved in sequencing and sequential mapping: dyslexics with damage to this area have problems with sequencing (e.g., "pig" will be written as "gip") and have difficulty in keeping track of appointments. Adults with similar damage develop Gerstmann syndrome, characterized by left–right confusion and problems with arithmetic and writing. In both cases the problem lies in *mapping*, e.g., of time to space or a numerical quantity to space. (2) The hippocampal place cells give you your own GPS to locate yourself in a novel environment whereas the associated grid cells in parahippocampal areas enable you to trace out your trajectory in space-time—essential for using one's mental calendar. (3) The inferior longitudinal fasciculus (a dominant pathway present in all individuals) clearly connects the angular gyrus with the place and grid cells. Add to the equation the fact that there is also a vestibular input to the hippocampus and you have all the facts clicking into place—a system of interconnected neuronal circuits that gives you the ability to construct a calendar in your brain, allowing you to navigate space and time while remaining firmly anchored in your body.

Phenomenological similarities between synesthesia and crossmodal correspondences: less well-studied varieties of synesthesia

While the variants of synesthesia described above have received the majority of research attempting to bridge the associations present in crossmodal correspondences and synesthesia, limited research exists into a myriad of other less common forms of the condition as well. One particularly interesting form of synesthesia that we identified in 2007 was tactile-emotion synesthesia, in which tactile textures evoked profound emotional induction for two individuals with the condition.[111] Much like other forms of synesthesia, these experiences were extremely consistent over time, automatic, unable to be ignored, and highly specific (e.g., denim evoked disgust, a specific type of ribbed plastic caused laughter and joy, and the exterior surface of a medicinal gel cap elicited feelings of jealousy). While never studied systematically, many of the textures that yielded either strongly positive or negative emotions were generally consistent across the synesthetes and in line with descriptions of the preferences for the textures reported by nonsynesthetic participants.

One final form of synesthesia that is worth highlighting here is ordinal linguistic personification, which is defined as the consistent and strong association of genders and/or personality traits for numbers and letters.[112] This is an understudied form of synesthesia that is often related to the experience of grapheme-color synesthesia, with clear links to crossmodal correspondences. It was first reported in the 19th century by Flournoy, who included a vivid description from one of his synesthetic subjects (taken from Ref. 112):

> 1, 2, 3 are children [who] play together. 4 is a good peaceful woman, absorbed by down-to-earth occupations...5 is a young man, ordinary and common in his tastes and appearance, but extravagant and self-centered. 6 is a young man...polite, gentle,...average intelligence; orphan. 7 is a bad sort, although brought up well; spiritual, extravagant, gay, likeable; capable of very good actions on occasion...8 is a very dignified lady, who acts appropriately... She is the wife of 9 [who is] self-centered, maniacal, grumpy, endlessly reproaching his wife for one thing or another (Ref. 113, pp. 219–220).

Indeed, while these reports sound fantastical, Simner, Gärtner, and Taylor[66] demonstrated a similar pattern of responses between synesthetes and nonsynesthetes who were instructed to generate personality traits for graphemes. Specifically, as the luminance of synesthetic colors is predicted by the frequency of the letters in grapheme-color synesthesia, in this variant of synesthesia, letter frequency predicted personality traits: both synesthetes and nonsynesthetes alike ascribed more agreeable and less neurotic personality traits to high frequency letters, pointing again to overlapping sets of associations present in synesthesia and crossmodal correspondences.

How do these models account for other cognitive and perceptual differences present in synesthetes?

While the data presented above are at least preliminarily consistent with models indicating that synesthesia is an outgrowth of normal crossmodal or multisensory processes, several pieces of evidence are only partially consistent with this model and there are a significant number of gaps that will require additional research to understand this relationship. The collective wisdom emerging from these data demonstrates that synesthetes show nonrandom mappings in their synesthetic associations and that these associations are phenomenologically similar to crossmodal correspondences in nonsynesthetes. However, these data rely largely on identifying consistent mappings between the two populations, which is correlative in nature and does not explicitly require that the mechanisms underlying these two phenomena be the same. While there is preliminary evidence that synesthetes show enhanced processing for a specific form of crossmodal correspondence, namely sound symbolism,[61] future research is needed to understand whether this generalizes to all forms of synesthesia. Furthermore, for models to provide robust and mechanistic explanations of synesthesia and its relevance to other crossmodal or multisensory processes, they should be able to account for at least some of the myriad of other perceptual differences that have been observed in synesthetes relative to nonsynesthetes.

There is an abundance of research indicating that synesthetes show perceptual and cognitive differences from the general population, besides the simple experience of one sensory modality evoking percepts in a second system. In the perceptual domain, synesthetes show altered processing of basic sensory stimuli in areas related to their synesthetic experiences. For example, grapheme-color synesthetes show enhanced discrimination abilities for colors[114] and a larger McCollough effect[115] even though colors in themselves do not evoke any synesthetic experience. Researchers have posited that this change in sensory processing in areas related to the synesthetic experience could either reflect bidirectional patterns of activity or enhanced perceptual gain in synesthetes (e.g., Ref. 38). Indeed, this latter view is supported by evidence that basic visual cortical processes are altered in synesthesia, such that grapheme-color synesthetes show altered visual evoked potentials to sinusoidal gratings that do not evoke synesthetic colors; these potentials occur in primary visual areas within 70 ms.[116] Consistent with these findings of enhanced visual cortical activity in synesthetes, Terhune and colleagues[117] reported a striking difference between synesthetes and nonsynesthetes in visual cortical excitability[118] using TMS. Specifically, TMS can be used to evoke visual phosphenes by inducing electromagnetic current into early visual areas, with the amount of energy required to evoke a phosphene indicating visual cortical excitability. Here, the authors demonstrated that synesthetes required 300% less magnetic energy to evoke visual phosphenes, indicating extremely high resting levels of visual cortical activity in synesthetes,[117] in line with the findings of other perceptual differences in synesthetes discussed above. In the cognitive domain, several studies have demonstrated that grapheme-color synesthetes show enhanced memory for letters and numbers, mirroring synesthetes' subjective reports that synesthetic associations serve as an extra dimension of information that is encoded, leading to facilitated memory.[9,119] As alluded to earlier, time-space synesthetes have reported similar benefits specific to their synesthetic experiences, with better memory for both past and future dates/events,[120] improved spatial working memory,[97,120] as well as increased spatial processing abilities including mental rotation.[120,121] Relatedly, one of the best-replicated findings within the synesthesia literature is that synesthetes show enhanced imagery capacities.[11,122] While any model examining the relationship between synesthesia and crossmodal correspondences, particularly those that suggest synesthesia is an enhancement of crossmodal correspondences, should also be able to explain other differences in synesthetes as well, it is unclear at present how these perceptual and cognitive differences can be integrated with current models.

Relationships between crossmodal correspondences and synesthesia still requiring clarification

Spence and colleagues have argued extensively about the differences between synesthesia and crossmodal correspondences. A review dedicated to this subject highlighted that crossmodal correspondences should be "acquired, malleable, relative, and transitive".[63] However, several of these qualifications are either consistent with synesthesia, or inconsistent with forms of crossmodal correspondences, establishing these limits as overly restrictive. Specifically, as described above, synesthetic associations are at least partially acquired, as evidenced by the fact that the pairings change throughout development[87] and can transfer to newly learned languages[73–77]; though this does not overshadow the more obvious fact that they remain largely stable. Furthermore, in some contexts synesthetic associations are context-dependent, in that the color evoked by a given grapheme can be modulated by the context of the word and surrounding letters or the attentional focus of the individual in the case of Navon figures.[9] Finally, the requirement that crossmodal correspondences be relative in nature is inconsistent with several established types of crossmodal correspondences that show *nonrelative* mappings.[58] Nevertheless, several important differences do exist between synesthesia and crossmodal correspondences that question whether they share a similar physiological basis.

One of the most significant differences between these two phenomena is that while crossmodal correspondences are defined as a mapping between two sensory modalities leading to cognitive or perceptual *biases*, synesthesia is a crossmodal mapping that results in altered *conscious experiences*. For example, while a nonsynesthete may conceptually understand that a high-pitched tone is associated with a yellow flash of light, for a synesthete, the high-pitched tone could evoke the conscious visual experience of a yellow ameboid-shaped object. Some authors have argued that this difference is attributable solely to the graded nature of these experiences, indicating that consciousness occurs at one end of the distribution of crossmodal correspondences,[60] with crossmodal correspondences as a form of "weak synesthesia". Others, however (see Refs. 45,63), argue that that consciousness should not be part of this scale. Indeed, the latter researchers highlight that if synesthesia and crossmodal correspondences exist as extremes on a linear gradient, we should expect there to be more quasi-conscious experiences existing between the two extremes (whereas conscious experiences seem largely all-or-none in nature). Synesthesia can occur even unconsciously[28,123] indicating that consciousness is not a necessary element of the phenomenon. In further support of this view, sensory deprivation,[1] brain damage,[124] and drug use[5,7] can lead to the spontaneous experience of synesthesia, which likely only affects the conscious experience of cross-sensory percepts without strengthening the underlying crossmodal correspondences. Drug-induced synesthesias are the best studied of these instances, and research indicates a strong role of serotonergic receptors in inducing synesthetic experiences (for reviews see Refs. 6,125). We provided the first evidence for this model describing synesthetes who experienced a temporary loss of their synesthesia through modulation of serotonin (5HT-2A) receptor activation.[5] Early reports from Simpson and McKellar[126] indicated that synesthetes under the influence of mescaline (which influences 5HT-2A receptors) perceived heightened intensity of their native synesthetic experiences, in addition to temporary experiences of new forms of synesthesia as well. Of important note, preliminary reports do indicate that with extreme training of synesthetic associations (letter-color pairings), some nonsynesthetes have reported developing conscious synesthetic qualia,[127] leaving in place the possibility that consciousness exists at the end of a continuum of crossmodal associations.

Examining other critical potential differences between crossmodal correspondences and synesthesia, the latter is defined by extremely specific associations[88,98] whereas crossmodal correspondences tend to encompass general preferences or relativistic associations.[45] Synesthetes will often labor to match an external color to the precise shade that is synesthetically experienced, stating that the specific hue never feels precise enough or truly indicative of their experience. More generally, crossmodal correspondences are broadly bidirectional in nature (e.g., a loud sound is associated with a bright visual stimulus, and vice-versa[60,128,129]) but synesthetic associations are in most cases unidirectional (e.g., 7 would induce a sensation of the color green, but green fails to induce the conscious sensation of the number 7[9]); however, some evidence nonetheless suggests that synesthesia may be unconsciously bidirectional.[130,131] Finally, as pointed out by Martino and Marks,[60] one critical difference is that crossmodal correspondences are consciously experienced as associative links, whereas synesthesia is a literal and visceral link. Indeed, synesthetes experience extreme discomfort in seeing an incongruent synesthetic association (e.g., a green 2 for a synesthete who experiences blue for the number 2)[9,132,133] whereas nonsynesthetes experience mild slowing of reaction times or slightly impaired perceptual processing in response to incongruent crossmodal correspondences.[134]

Finally, arguments have been put forward by several groups that if synesthesia is the outgrowth of either multisensory processing or crossmodal correspondences, then synesthetes should show above average performance on other multisensory or crossmodal tasks.[38,41,45,62] As described in the introduction, interactions among our sensory systems are not unique to synesthetes and arise in the form of both multisensory interactions (reflecting synergistic effects resulting from the joint stimulation of two modalities) and crossmodal abstractions (reflecting cross-dimensional mappings among the senses). Research has not provided conclusive support for the hypothesis that synesthesia is an extension of more common multisensory perceptual interactions. In particular, three studies have examined this model using tests of multisensory integration between modalities unrelated to the synesthetic experience for the participants and found conflicting results of either increased[38] or decreased[135] multisensory integration, or no differences between the groups whatsoever.[136] These initial studies examined enhanced multisensory processing through a broad lens, with the presumption that synesthesia might reflect hyperactivation of a general multisensory hub in the brain. In hindsight, this model of generalized enhancements in multisensory processes in synesthetes is inconsistent with contemporary models of multisensory processing. While there do exist convergence zones in the cortex that are important to multisensory integration, several such areas exist, including the posterior superior temporal sulcus, intraparietal sulcus, angular gyrus, frontal regions, as well as subcortical and thalamic areas (for a review see Ref. 42), with no evidence suggesting that enhanced processing in one region uniformly enhances all forms of multisensory integration. Indeed, each multisensory process is governed by a very specific network of brain areas and neural processes, including phase-resetting, spike rates, power changes in intrinsic oscillations, different anatomical or functional networks, and variable levels of attention, among other factors.[137–141] All of this is to highlight that multisensory integration is not a unitary phenomenon. Thus, the varied anatomical bases underlying different forms of multisensory integration suggest that if synesthetes were to show enhanced multisensory integration abilities, they would likely be restricted to a single process or a specific network related to the interactions present in their synesthetic experiences (which no research has examined in detail to date).

In the domain of crossmodal correspondences, however, recent research has extended these efforts. Examining whether synesthetes show increased strength on one or more common crossmodal correspondences, Lacey, Martinez, McCormick, and Sathian[61] used the implicit association test to compare synesthetes' and nonsynesthetes' biases for crossmodal pairings occurring between three distinct dimensions: auditory pitch and visual elevation, auditory pitch and visual size, and sound–symbolic combinations. Critically, each of these features has been repeatedly demonstrated to show a consistent mapping in the general population (for a review see Ref. 45) and the paradigm used was previously validated in the examination of crossmodal correspondences.[129] Lacey, Martinez, McCormick, and Sathian[61] observed greater congruency effects in synesthetes compared to controls for only one of the tested crossmodal correspondences: sound–symbolic pseudoword-to-shape pairings (of the bouba/kiki variety; e.g., Fig. 12.2), but not pitch-elevation or pitch-size mappings. The authors interpreted these results to mean that the shared mechanisms underlying synesthesia and crossmodal correspondences occurred postperceptually (given the higher-order cognitive mapping between speech sounds and shapes required for sound symbolism), as no differences were observed with presumably lower-level perceptual associations between auditory pitch and visual elevation or auditory pitch and visual size. An alternative possibility is that synesthetes in general may show enhanced statistical learning behaviors, of the type that support sound symbolism[61]; this model would be consistent with synesthetes' maintenance of consistent associations between arbitrary sensory percepts as well as repeated findings of enhanced memory abilities in synesthetes (e.g., Ref. 119).

One significant caveat to studies examining the relationship between synesthesia and crossmodal correspondences is that research has been restricted to adult synesthetes who have had their synesthetic associations for decades. This late-life examination limits investigation of the independence of synesthetic associations from crossmodal correspondences and instead only demonstrates that for those who already have crossmodal correspondences, synesthetic associations will tend to match them. Thus, studies of adult synesthetes cannot address the qualitative nature of synesthesia before one acquires crossmodal correspondences. Indeed, it is possible that synesthetic correspondences tend to match crossmodal correspondences simply due to both groups having similar experiences within their environment (and as pointed out before, synesthetic associations are modified in line with these environmental experiences; see Refs. 13,71,72). In this view, the fact that synesthetic experiences tend to match crossmodal correspondences is entirely an emergent property; synesthetes need to associate something with their aberrant sensory activation, so it is logical that the brain would use the associations (crossmodal correspondences) that are already present. As a test case to this model, what happens to a synesthete's crossmodal correspondences if their synesthesia encompasses the same types of associations, and can they be dissociated? For example, in tone-color synesthesia, simple tones evoke vivid and colorful flashes of light. While many synesthetes experience a mapping between pitch and lightness, such that lower frequency tones evoke darker colors, this is by no means the rule for all synesthetes. As a thought experiment, for a synesthete who experiences low-frequency tones as bright colors and high-frequency tones as dark colors, would this individual show any remnants of the typical tone-brightness mappings observed in typical crossmodal correspondences, perhaps using the methods described in Ref. 61? In this test case, if the synesthete's crossmodal correspondences were completely overwritten by their synesthetic associations (even when tested implicitly), it might suggest a similar mechanism underlying the two types of associations. However, if the individual experienced different tone-color mappings according to context (with the synesthetic mappings dominating during the perceptual experience of a tone and crossmodal correspondences dominating during conceptual processing of tone-color relationships) or reported a duality of experiences for crossmodal correspondences and synesthetic associations, this would suggest independence across the processes.

Conclusions

The phenomenological similarities between synesthesia and crossmodal correspondences are apparent. What is undeniable is that, were a synesthete to report that a 5000 Hz tone evokes a small bright flash of light in her upper visual field, this association would "make sense" to nonsynesthetes even though they did not experience the subjective flash of light themselves. Research seeking to understand the relationship between these two processes, however, remains in a nascent phase, leaving several possibilities as to why synesthetic associations and crossmodal correspondences qualitatively tend to match with one another, whether synesthesia and crossmodal correspondences exist along a linear gradient relating the two, the neural bases (including any overlap) across these two phenomena, and why crossmodal correspondences would be experienced as stronger and qualitatively different percepts for synesthetes. Here we summarize four main models linking synesthesia and crossmodal correspondences:

(1) Linear Gradient Model: Crossmodal correspondences operate along a continuum, with synesthesia at one extreme. This provides a descriptive account, without proposing specific mechanisms that could support both processes. The all-or-none, qualitative nature of synesthesia proposes a challenge for linear gradient accounts.

(2) Inconvenient, Independence Model: Synesthesia and crossmodal correspondences are derived from unique biological bases, and the shared mapping of associations (e.g., a synesthete may experience bright colors evoked by high-pitched tones just as nonsynesthetes *associate* bright colors with high-pitched tones) is an emergent property from both processes capitalizing on the statistics and information available in the environment. This model proposes that any surface-level similarities between synesthesia and crossmodal correspondences are merely due to both phenomena picking up on natural statistics in the environment as well as using nonarbitrary (amodal and intrinsic) mappings to link sensory information.

(3) Synesthetes as Expert Statistical Learners Model: Some aspects of crossmodal correspondences require learning to solidify the relationships between arbitrary and nonarbitrary sensory information. This model suggests that synesthetes are experts in statistical learning, capable of identifying relationships between arbitrary or nonarbitrary items and consolidating them into stable percepts that can last decades, explaining the stronger associations present in synesthetes for some crossmodal correspondences, as synesthetes would be better at maintaining natural associations in general. Consistent with this model, synesthetes report enhanced abilities in learning new languages, and their synesthetic associations quickly transfer across languages. Furthermore, if this ability to find and consolidate random or nonarbitrary associations were to go into overdrive, it could explain why some synesthetes report seemingly random associations for genders and personalities with graphemes (as occurs in ordinal linguistic personification) or the experience of multiple colors for the same letters and numbers (see the splotchy colors present in the numbers 7 and 8 in Fig. 12.1).

(4) Local and Specific Model: This model proposes that the mechanisms underlying synesthesia will neither affect nor relate to the processing of crossmodal correspondences, except if there is overlap in the neural mechanisms supporting a specific crossmodal correspondence and the specific brain areas that are involved in generating a synesthetic association. For example, a tone-color synesthete, but not a grapheme-color synesthete, would show enhanced crossmodal correspondences for tone-brightness associations.

References

1. Armel KC, Ramachandran VS. Acquired synesthesia in retinitis pigmentosa. *Neurocase*. 1999;5(4):293–296.
2. Baron-Cohen S, Burt L, Smith-Laittan F, Harrison J, Bolton P. Synaesthesia: prevalence and familiality. *Perception*. 1996;25(9):1073–1079.
3. Asher JE, Lamb JA, Brocklebank D, et al. A whole-genome scan and fine-mapping linkage study of auditory-visual synesthesia reveals evidence of linkage to chromosomes 2q24, 5q33, 6p12, and 12p12. *Am J Hum Genet*. 2009;84(2):279–285.
4. Ward J, Simner J. Is synaesthesia an X-linked dominant trait with lethality in males? *Perception*. 2005;34(5):611–623.
5. Brang D, Ramachandran VS. Psychopharmacology of synesthesia; the role of serotonin S2a receptor activation. *Med Hypotheses*. 2008;70(4):903–904.
6. Luke DP, Terhune DB. The induction of synaesthesia with chemical agents: a systematic review. *Front Psychol*. 2013;4.
7. Sinke C, Halpern JH, Zedler M, Neufeld J, Emrich HM, Passie T. Genuine and drug-induced synesthesia: a comparison. *Conscious Cognit*. 2012;21(3):1419–1434.

8. Brang D, Ramachandran VS. Survival of the synesthesia gene: why do people hear colors and taste words? *PLoS Biol*. 2011;9(11):e1001205.
9. Ramachandran VS, Hubbard EM. Synaesthesia — a window into perception, thought and language. *J Conscious Stud*. 2001;8(12):3—34.
10. Ramachandran VS, Hubbard EM. Psychophysical investigations into the neural basis of synaesthesia. *Proc R Soc Lond B Biol Sci*. 2001;268(1470):979—983.
11. Brang D, Ramachandran VS. Visual field heterogeneity, laterality, and eidetic imagery in synesthesia. *Neurocase*. 2010;16(2):169—174.
12. Simner J, Synaesthesia WJ. The taste of words on the tip of the tongue. *Nature*. 2006;444(7118):438.
13. Beeli G, Esslen M, Jäncke L. Time course of neural activity correlated with colored-hearing synesthesia. *Cerebr Cortex*. 2007;18(2):379—385.
14. Brang D, Edwards L, Ramachandran V, Coulson S. Is the sky 2? Contextual priming in graphene-color synaesthesia. *Psychol Sci*. 2008;19(5):421—428.
15. Brang D, Hubbard EM, Coulson S, Huang M, Ramachandran VS. Magnetoencephalography reveals early activation of V4 in graphene-color synesthesia. *Neuroimage*. 2010;53(1):268—274.
16. Goller AI, Otten LJ, Ward J. Seeing sounds and hearing colors: an event-related potential study of auditory—visual synesthesia. *J Cogn Neurosci*. 2009;21(10):1869—1881.
17. Hubbard EM, Arman AC, Ramachandran VS, Boynton GM. Individual differences among graphene-color synesthetes: brain-behavior correlations. *Neuron*. 2005;45(6):975—985.
18. Nunn JA, Gregory LJ, Brammer M, et al. Functional magnetic resonance imaging of synesthesia: activation of V4/V8 by spoken words. *Nat Neurosci*. 2002;5(4):371—375.
19. Palmeri TJ, Blake R, Marois R, Flanery MA, Whetsell W. The perceptual reality of synesthetic colors. *Proc Natl Acad Sci*. 2002;99(6):4127—4131.
20. Sperling JM, Prvulovic D, Linden DE, Singer W, Stirn A. Neuronal correlates of colour-graphemic synaesthesia: Afmri study. *Cortex*. 2006;42(2):295—303.
21. Maurer D. Neonatal synesthesia: implications for the processing of speech and faces. In: *Developmental Neurocognition: Speech and Face Processing in the First Year of Life*. Springer; 1993:109—124.
22. Jäncke L, Beeli G, Eulig C, Hänggi J. The neuroanatomy of graphene—color synesthesia. *Eur J Neurosci*. 2009;29(6):1287—1293.
23. Rouw R, Scholte HS. Increased structural connectivity in graphene-color synesthesia. *Nat Neurosci*. 2007;10(6):792—797.
24. Weiss PH, Kalckert A, Fink GR. Priming letters by colors: evidence for the bidirectionality of graphene—color synesthesia. *J Cogn Neurosci*. 2009;21(10):2019—2026.
25. Hubbard EM, Brang D, Ramachandran VS. The cross — activation theory at 10. *J Neuropsychol*. 2011;5(2):152—177.
26. Hubbard E. Neurophysiology of synesthesia. *Curr Psychiatr Rep*. 2007;9(3):193—199.
27. Robertson LC. Binding, spatial attention and perceptual awareness. *Nat Rev Neurosci*. 2003;4(2):93—102.
28. Ramachandran V, Seckel E. Synesthetic colors induced by graphemes that have not been consciously perceived. *Neurocase*. 2015;21(2):216—219.
29. Rouw R, Scholte HS, Colizoli O. Brain areas involved in synaesthesia: a review. *J Neuropsychol*. 2011;5(2):214—242.
30. Rouw R, Scholte HS. Neural basis of individual differences in synesthetic experiences. *J Neurosci*. 2010;30(18):6205—6213.
31. Weiss PH, Fink GR. Graphene-colour synaesthetes show increased grey matter volumes of parietal and fusiform cortex. *Brain*. 2009;132(1):65—70.
32. Esterman M, Verstynen T, Ivry RB, Robertson LC. Coming unbound: disrupting automatic integration of synesthetic color and graphemes by transcranial magnetic stimulation of the right parietal lobe. *J Cogn Neurosci*. 2006;18(9):1570—1576.
33. Muggleton N, Tsakanikos E, Walsh V, Ward J. Disruption of synaesthesia following TMS of the right posterior parietal cortex. *Neuropsychologia*. 2007;45(7):1582—1585.
34. Ward J, Jonas C, Dienes Z, Seth A. Graphene-colour synaesthesia improves detection of embedded shapes, but without pre-attentive 'pop-out' of synaesthetic colour. *Proc R Soc Lond B Biol Sci*. 2010;277(1684):1021—1026.

35. Tilot AK, Kucera KS, Vino A, Asher JE, Baron-Cohen S, Fisher SE. Rare variants in axonogenesis genes connect three families with sound–color synesthesia. *Proc Natl Acad Sci*. 2018;115(12):3168–3173.
36. Root NB, Rouw R, Asano M, et al. Why is the synesthete's "A" red? Using a five-language dataset to disentangle the effects of shape, sound, semantics, and ordinality on inducer–concurrent relationships in graphene-color synesthesia. *Cortex*. 2018;99:375–389.
37. Bien N, ten Oever S, Goebel R, Sack AT. The sound of size: crossmodal binding in pitch-size synesthesia: a combined TMS, EEG and psychophysics study. *Neuroimage*. 2012;59(1):663–672.
38. Brang D, Williams LE, Ramachandran VS. Graphene-color synesthetes show enhanced crossmodal processing between auditory and visual modalities. *Cortex*. 2012;48(5):630–637.
39. Ludwig J, Sanbonmatsu L, Gennetian L, et al. Neighborhoods, obesity, and diabetes—a randomized social experiment. *N Engl J Med*. 2011;365(16):1509–1519.
40. Mulvenna CM, Walsh V. Synaesthesia: supernormal integration? *Trends Cogn Sci*. 2006;10(8):350–352.
41. Sagiv N, Ward J. Crossmodal interactions: lessons from synesthesia. *Prog Brain Res*. 2006;155:259–271.
42. Stein BE, Stanford TR. Multisensory integration: current issues from the perspective of the single neuron. *Nat Rev Neurosci*. 2008;9(4):255–266.
43. Radeau M, Bertelson P. Auditory-visual interaction and the timing of inputs. *Psychol Res*. 1987;49(1):17–22.
44. Stein BE, Meredith MA, Wolf S. *The Merging of the Senses*. Cambridge, MA: MIT Press; 1993.
45. Spence C. Crossmodal correspondences: a tutorial review. *Atten Percept Psychophys*. 2011;73(4):971–995.
46. Köhler W. *Gestalt Psychology*. New York, NY: Liveright; 1929.
47. Köhler W. *Gestalt Psychology: The Definitive Statement of the Gestalt Theory*. New York: Liveright; 1947.
48. Ramachandran VS, Hubbard EM. Hearing colors, tasting shapes. *Sci Am*. 2003;288(5):52–59.
49. Stevens JC, Marks LE. Cross-modality matching of brightness and loudness. *Proc Natl Acad Sci*. 1965;54(2):407–411.
50. Krantz DH. A theory of magnitude estimation and cross-modality matching. *J Math Psychol*. 1972;9(2):168–199.
51. Mudd S. Spatial stereotypes of four dimensions of pure tone. *J Exp Psychol*. 1963;66(4):347.
52. Mondloch CJ, Maurer D. Do small white balls squeak? Pitch-object correspondences in young children. *Cognit Affect Behav Neurosci*. 2004;4(2):133–136.
53. Simpson RH, Quinn M, Ausubel DP. Synesthesia in children: association of colors with pure tone frequencies. *J Genet Psychol*. 1956;89(1):95–103.
54. Coward SW, Stevens CJ. Extracting meaning from sound: Nomic mappings, everyday listening, and perceiving object size from frequency. *Psychol Rec*. 2004;54(3):349–364.
55. Grassi M. Do we hear size or sound? Balls dropped on plates. *Atten Percept Psychophys*. 2005;67(2):274–284.
56. Gallace A, Spence C. Multisensory synesthetic interactions in the speeded classification of visual size. *Percept Psychophys*. 2006;68(7):1191–1203.
57. Smith EL, Grabowecky M, Suzuki S. Auditory-visual crossmodal integration in perception of face gender. *Curr Biol*. 2007;17(19):1680–1685.
58. Guzman-Martinez E, Ortega L, Grabowecky M, Mossbridge J, Suzuki S. Interactive coding of visual spatial frequency and auditory amplitude-modulation rate. *Curr Biol*. 2012;22(5):383–388.
59. Jousmäki V, Hari R. Parchment-skin illusion: sound-biased touch. *Curr Biol*. 1998;8(6):R190–R191.
60. Martino G, Marks LE. Synesthesia: strong and weak. *Curr Dir Psychol Sci*. 2001;10(2):61–65.
61. Lacey S, Martinez M, McCormick K, Sathian K. Synesthesia strengthens sound-symbolic cross-modal correspondences. *Eur J Neurosci*. 2016;44(9):2716–2721.
62. Ward J, Huckstep B, Tsakanikos E. Sound-colour synaesthesia: to what extent does it use cross-modal mechanisms common to us all? *Cortex*. 2006;42(2):264–280.
63. Deroy O, Spence C. Why we are not all synesthetes (not even weakly so). *Psychon Bull Rev*. 2013;20(4):643–664.
64. Gilbert AN, Martin R, Kemp SE. Cross-modal correspondence between vision and olfaction: the color of smells. *Am J Psychol*. 1996:335–351.
65. Simner J, Ward J, Lanz M, et al. Non-random associations of graphemes to colours in synaesthetic and non-synaesthetic populations. *Cogn Neuropsychol*. 2005;22(8):1069–1085.
66. Simner J, Gärtner O, Taylor MD. Cross-modal personality attributions in synaesthetes and non-synaesthetes. *J Neuropsychol*. 2011;5(2):283–301.
67. Simner J, Harrold J, Creed H, Monro L, Foulkes L. Early detection of markers for synaesthesia in childhood populations. *Brain*. 2009;132(1):57–64.

68. Eagleman DM, Kagan AD, Nelson SS, Sagaram D, Sarma AK. A standardized test battery for the study of synesthesia. *J Neurosci Methods*. 2007;159(1):139–145.
69. Rothen N, Seth AK, Witzel C, Ward J. Diagnosing synaesthesia with online colour pickers: maximising sensitivity and specificity. *J Neurosci Methods*. 2013;215(1):156–160.
70. Kadosh RC, Henik A, Walsh V. Small is bright and big is dark in synaesthesia. *Curr Biol*. 2007;17(19):R834–R835.
71. Smilek D, SA Carriere J, Dixon MJ, Merikle PM. Graphene frequency and color luminance in graphene-color synaesthesia. *Psychol Sci*. 2007;18(9):793–795.
72. Witthoft N, Winawer J. Learning, memory, and synesthesia. *Psychol Sci*. 2013;24(3):258–265.
73. Brang D, Rouw R, Ramachandran VS, Coulson S. Similarly shaped letters evoke similar colors in graphene–color synesthesia. *Neuropsychologia*. 2011;49(5):1355–1358.
74. Jürgens U, Nikolić D. IdeaesthesIa: conceptual processes assign similar colours to similar shapes. *Transl Neurosci*. 2012;3(1):22–27.
75. Watson MR, Akins KA, Enns JT. Second-order mappings in graphene–color synesthesia. *Psychon Bull Rev*. 2012;19(2):211–217.
76. Asano M, Yokosawa K. Synesthetic colors are elicited by sound quality in Japanese synesthetes. *Conscious Cognit*. 2011;20(4):1816–1823.
77. Asano M, Yokosawa K. Synesthetic colors for Japanese late acquired graphemes. *Conscious Cognit*. 2012;21(2):983–993.
78. Shin E-h, Kim C-Y. Both "나" and "な" are yellow: cross-linguistic investigation in search of the determinants of synesthetic color. *Neuropsychologia*. 2014;65:25–36.
79. Blair CD, Berryhill ME. Synesthetic graphene-color percepts exist for newly encountered Hebrew, Devanagari, Armenian and Cyrillic graphemes. *Conscious Cognit*. 2013;22(3):944–954.
80. Brang D, Ghiam M, Ramachandran VS. Impaired acquisition of novel graphene-color correspondences in synesthesia. *Front Hum Neurosci*. 2013;7.
81. Mroczko A, Metzinger T, Singer W, Nikolić D. Immediate transfer of synesthesia to a novel inducer. *J Vis*. 2009;9(12):25.
82. Marks LE. On colored-hearing synesthesia: cross-modal translations of sensory dimensions. *Psychol Bull*. 1975;82(3):303.
83. Marks LE. On associations of light and sound: the mediation of brightness, pitch, and loudness. *Am J Psychol*. 1974:173–188.
84. Hubbard TL. Synesthesia-like mappings of lightness, pitch, and melodic interval. *Am J Psychol*. 1996:219–238.
85. Riggs LA, Karwoski T. Synaesthesia. *Br J Psychol*. 1934;25(1):29–41.
86. Zigler MJ. Tone shapes: a novel type of synaesthesia. *J Gen Psychol*. 1930;3(2):277–287.
87. Simner J, Bain AE. A longitudinal study of graphene-color synesthesia in childhood: 6/7 years to 10/11 years. *Front Hum Neurosci*. 2013;7.
88. Galton F. *Visualised Numerals*. Nature Publishing Group; 1880.
89. Seron X, Pesenti M, Noel M-P, Deloche G, Cornet J-A. Images of numbers, or "When 98 is upper left and 6 sky blue". *Cognition*. 1992;44(1):159–196.
90. Dehaene S, Bossini S, Giraux P. The mental representation of parity and number magnitude. *J Exp Psychol Gen*. 1993;122(3):371.
91. Grossenbacher PG, Lovelace CT. Mechanisms of synesthesia: cognitive and physiological constraints. *Trends Cogn Sci*. 2001;5(1):36–41.
92. Hubbard EM, Piazza M, Pinel P, Dehaene S. Interactions between number and space in parietal cortex. *Nat Rev Neurosci*. 2005;6(6):435–448.
93. Price MC, Mentzoni RA. Where is January? The month-SNARC effect in sequence-form synaesthetes. *Cortex*. 2008;44(7):890–907.
94. Sagiv N, Simner J, Collins J, Butterworth B, Ward J. What is the relationship between synaesthesia and visuospatial number forms? *Cognition*. 2006;101(1):114–128.
95. Novich S, Cheng S, Eagleman DM. Is synaesthesia one condition or many? A large-scale analysis reveals subgroups. *J Neuropsychol*. 2011;5(2):353–371.
96. Smilek D, Callejas A, Dixon MJ, Merikle PM. Ovals of time: time-space associations in synaesthesia. *Conscious Cognit*. 2007;16(2):507–519.

97. Brang D, Teuscher U, Ramachandran VS, Coulson S. Temporal sequences, synesthetic mappings, and cultural biases: the geography of time. *Conscious Cognit*. 2010;19(1):311–320.
98. Eagleman DM. The objectification of overlearned sequences: a new view of spatial sequence synesthesia. *Cortex*. 2009;45(10):1266–1277.
99. Brang D, Teuscher U, Miller LE, Ramachandran VS, Coulson S. Handedness and calendar orientations in time–space synaesthesia. *J Neuropsychol*. 2011;5(2):323–332.
100. Lakoff G, Johnson M. *Metaphors We Live by*. University of Chicago press; 2008.
101. Boroditsky L. Metaphoric structuring: understanding time through spatial metaphors. *Cognition*. 2000;75(1):1–28.
102. Boroditsky L, Ramscar M. The roles of body and mind in abstract thought. *Psychol Sci*. 2002;13(2):185–189.
103. Casasanto D, Boroditsky L. Time in the mind: using space to think about time. *Cognition*. 2008;106(2):579–593.
104. Gentner D, Imai M, Boroditsky L. As time goes by: evidence for two systems in processing space→ time metaphors. *Lang Cogn Process*. 2002;17(5):537–565.
105. Núñez RE, Motz BA, Teuscher U. Time after time: the psychological reality of the ego-and time-reference-point distinction in metaphorical construals of time. *Metaphor Symbol*. 2006;21(3):133–146.
106. Bonato M, Saj A, Vuilleumier P. Hemispatial neglect shows that "before" is "left". *Neural Plast*. 2016;2016.
107. Teuscher U, McQuire M, Collins J, Coulson S. Congruity effects in time and space: behavioral and ERP measures. *Cogn Sci*. 2008;32(3):563–578.
108. Ramachandran V, Chunharas C, Marcus Z. Hypothesis concerning embodied calendars: a case study of number form, color spreading, and taste-color synaesthesia. *Med Hypotheses*. 2016;94:58–62.
109. Ramachandran VS, Vajanaphanich M, Chunharas C. Calendars in the brain; their perceptual characteristics and possible neural substrate. *Neurocase*. 2016;22(5):461–465.
110. MacKay DM. Visual stability and voluntary eye movements. In: *Central Processing of Visual Information A: Integrative Functions and Comparative Data*. Springer; 1973:307–331.
111. Ramachandran VS, Brang D. Tactile-emotion synesthesia. *Neurocase*. 2008;14(5):390–399.
112. Simner J, Holenstein E. Ordinal linguistic personification as a variant of synesthesia. *J Cogn Neurosci*. 2007;19(4):694–703.
113. Flournoy T. *Des phénomènes de synopsie (audition colorée) photismes, schèmes visuels, personnifications*. Alcan; 1893.
114. Banissy MJ, Walsh V, Ward J. Enhanced sensory perception in synaesthesia. *Exp Brain Res*. 2009;196(4):565–571.
115. Ramachandran V, Marcus Z. Synesthesia and the McCollough effect. *i-Perception*. 2017;8(3), 2041669517711718.
116. Barnett KJ, Foxe JJ, Molholm S, et al. Differences in early sensory-perceptual processing in synesthesia: a visual evoked potential study. *Neuroimage*. 2008;43(3):605–613.
117. Terhune DB, Tai S, Cowey A, Popescu T, Kadosh RC. Enhanced cortical excitability in graphene-color synesthesia and its modulation. *Curr Biol*. 2011;21(23):2006–2009.
118. Terhune DB, Murray E, Near J, Stagg CJ, Cowey A, Cohen Kadosh R. Phosphene perception relates to visual cortex glutamate levels and covaries with atypical visuospatial awareness. *Cerebr Cortex*. 2015;25(11):4341–4350.
119. Rothen N, Meier B, Ward J. Enhanced memory ability: insights from synaesthesia. *Neurosci Biobehav Rev*. 2012;36(8):1952–1963.
120. Simner J, Mayo N, Spiller M-J. A foundation for savantism? Visuo-spatial synaesthetes present with cognitive benefits. *Cortex*. 2009;45(10):1246–1260.
121. Brang D, Miller LE, McQuire M, Ramachandran VS, Coulson S. Enhanced mental rotation ability in time-space synesthesia. *Cogn Process*. 2013;14(4):429–434.
122. Spiller MJ, Jonas CN, Simner J, Jansari A. Beyond visual imagery: how modality-specific is enhanced mental imagery in synesthesia? *Conscious Cognit*. 2015;31:73–85.
123. Ramachandran VS, Seckel E. Graphemes evoke synesthetic colors even before they are consciously recognized. *Perception*. 2011;40(4):490–492.
124. Ro T, Farnè A, Johnson RM, et al. Feeling sounds after a thalamic lesion. *Ann Neurol*. 2007;62(5):433–441.
125. Brogaard B. Serotonergic hyperactivity as a potential factor in developmental, acquired and drug-induced synesthesia. *Front Hum Neurosci*. 2013;7.
126. Simpson L, McKellar P. Types of synaesthesia. *Br J Psychiatry*. 1955;101(422):141–147.
127. Bor D, Rothen N, Schwartzman DJ, Clayton S, Seth AK. Adults can be trained to acquire synesthetic experiences. *Sci Rep*. 2014;4:7089.

128. Martino G, Marks LE. Perceptual and linguistic interactions in speeded classification: tests of the semantic coding hypothesis. *Perception*. 1999;28(7):903–923.
129. Parise CV, Spence C. Audiovisual crossmodal correspondences and sound symbolism: a study using the implicit association test. *Exp Brain Res*. 2012;220(3–4):319–333.
130. Kadosh RC, Kadosh KC, Henik A. The neuronal correlate of bidirectional synesthesia: a combined event-related potential and functional magnetic resonance imaging study. *J Cogn Neurosci*. 2007;19(12):2050–2059.
131. Niessen E, Fink GR, Schweitzer L, Kluender N, Weiss PH. Implicit interactions between number and space in digit-color synesthesia. *Cortex*. 2015;64:225–234.
132. Hochel M, Milán EG, Martín JM, et al. Congruence or coherence? Emotional and physiological responses to colours in synaesthesia. *Eur J Cogn Psychol*. 2009;21(5):703–723.
133. Ward J. Emotionally mediated synaesthesia. *Cogn Neuropsychol*. 2004;21(7):761–772.
134. Parise CV, Spence C. 'When birds of a feather flock together': synesthetic correspondences modulate audiovisual integration in non-synesthetes. *PLoS One*. 2009;4(5):e5664.
135. Neufeld J, Sinke C, Zedler M, Emrich HM, Szycik GR. Reduced audio–visual integration in synaesthetes indicated by the double-flash illusion. *Brain Res*. 2012;1473:78–86.
136. Whittingham KM, McDonald JS, Clifford CW. Synesthetes show normal sound-induced flash fission and fusion illusions. *Vis Res*. 2014;105:1–9.
137. Driver J, Spence C. Multisensory perception: beyond modularity and convergence. *Curr Biol*. 2000;10(20):R731–R735.
138. Foxe JJ, Schroeder CE. The case for feedforward multisensory convergence during early cortical processing. *Neuroreport*. 2005;16(5):419–423.
139. Kayser C, Logothetis NK. Do early sensory cortices integrate cross-modal information? *Brain Struct Funct*. 2007;212(2):121–132.
140. Watkins S, Shams L, Josephs O, Rees G. Activity in human V1 follows multisensory perception. *Neuroimage*. 2007;37(2):572–578.
141. Watkins S, Shams L, Tanaka S, Haynes J-D, Rees G. Sound alters activity in human V1 in association with illusory visual perception. *Neuroimage*. 2006;31(3):1247–1256.

CHAPTER

13

Synesthesia: the current state of the field

Jamie Ward, Julia Simner

School of Psychology, University of Sussex, Falmer, Brighton, United Kingdom

Synesthesia (whose British spelling is *synaesthesia*) is a condition that affects a relatively small percentage of the population but causes remarkable differences in how these individuals perceive the world. *Synesthetes* (i.e., people with synesthesia) experience their environment with unusual additions of colors, tastes, smells, touch, and other sensations. For example, someone with synesthesia might see colors in the visual field when they hear music or when they eat certain foods, or they might hear sounds when watching (silently) moving objects. These unusual experiences—although very different to those of the average person—form part of the normal everyday perceptual life of people with synesthesia, and we shall see below that these sensations are tied to subtle differences in the structure and function of synesthetes' brains. We know that the condition affects *at least* 4.4% of the population[1] but the true prevalence is unknown because synesthesia has many different manifestations (see below) and because epidemiological studies have yet to screen for all variants comprehensively. One reason that synesthesia could exist in high numbers without great public awareness is that people with synesthesia tend to assume that their sensations are everyday and pedestrian—so normal that they do not warrant mention. This in turn leads to surprise both from synesthetes and nonsynesthetes, that each other exists in a different perceptual world. The focus of this chapter is *developmental* or *congenital synesthesia*; instances when synesthesia emerges from an early age, apparently predisposed in some way from birth. Rarer instances have been reported, however, of *acquired synesthesia*, which is the emergence of synesthesia in later life, and in a way that is triggered by some life event such as the onset of disease or pathology.[2]

Synesthesia is often described as a "condition" because it affects a small number of people and has a presumed biological origin but it is also alternatively referred to as a "trait," particularly if the synesthete in question feels no debilitating effects from his or her experiences whatsoever. But this introduces a key characteristic of synesthesia: that it is a multivariant condition with many different ways of manifesting itself from one synesthete to the next. Most synesthetes experience synesthesia in a way that could be described as a benign alternative form

Multisensory Perception
https://doi.org/10.1016/B978-0-12-812492-5.00013-9

of perception. However, a small number of synesthetes experience considerable difficulties tied to their synesthesia, either because the sensations themselves are unpleasant or intrusive in some way (e.g., when synesthete JIW hears certain words, his mouth is flooded with the unpleasant taste of earwax[3]) or because having synesthesia might be accompanied by other medical conditions. For example, three recent studies have now suggested a link between synesthesia and autism spectrum conditions (ASCs). Two of these studies found that people with ASC were far more likely to have synesthesia than the average person.[4,5] This does not mean, conversely, that all people with synesthesia will also have autism: the overwhelming majority will not. But there are higher rates of synesthesia among people with ASC than we would otherwise expect to find in the wider population. Nonetheless, our own study[6] has recently shown that synesthesia does not appear to be linked to autism per se, but specifically to the gifts (or "savant skills") that sometimes accompany autism.[7]

Definition and diagnosis

At the time of writing this chapter, more than 100 different types of synesthesia had been reported in the science literature, either empirically studied in detail or anecdotally noted.[8] Because synesthesia is best understood by giving examples of its many different manifestations, we start with brief snapshots of some well-known variants of synesthesia and their definitions. One relatively well-studied form of synesthesia is *sound-color synesthesia*. We point out that here and throughout this chapter, we follow the scientific convention of placing the trigger for synesthesia (here, sound) before the additional sensation experienced by the synesthete (in this example, color). So for sound-color synesthetes, hearing sounds causes unusual sensations of color. For example, the sound of middle C on a piano keyboard might evoke a temporary shimmering patch of redness in the visual field. Or the sound of violins might be a deep purple. It is important to note that the synesthetic color does not *replace* the usual auditory sensation of sound; so the synesthete would still hear the note, but its sound would be accompanied by an unusual visual perception of color. This example illustrates why synesthesia is often described as "a merging of the senses": one sense (here sound; which in scientific parlance is "the *inducer*") crosses or merges with another (here vision; known conventionally as "the *concurrent*").

There are a number of different types of synesthesia which cause a merging of the senses. One well-documented example was of "The Man who Tasted Shapes."[9] This synesthete, identified in the case study as Michael Watson, felt unusual touch sensations of texture in his hands that were triggered by the taste of food in his mouth. The taste of chicken, for example, was a prickly, pointed sensation, like touching a bed of nails. When the food had a particularly intense flavor, the feeling would sweep down Michael Watson's arm into his hands, where he could feel its synesthetic shape, as well as its texture, weight, and even temperature—as if he were actually grasping something on his palm.

In other cases of synesthesia, the sensations are triggered by more abstract information, such as thinking, reasoning, or other higher-order functions. Synesthete JIW, for example, tastes language; that is, he experiences a flood of synesthetic taste in his mouth when he reads, speaks, hears, or even thinks about words. Each word in English fills his mouth with a different flavor, so when James hears the name "Phillip," for example, his mouth is flooded with the taste of bitter oranges. The word "society" fills his mouth with onions. And the word "audience" tastes of tinned peas. These are phantom flavors, triggered by unusual activity in JIW's brain,[10] but the tastes themselves are as "real" as any flavors of food one might put in the mouth—although there is nothing in JIW's mouth when he tastes them. His synesthetic tastes are triggered by words because JIW has *lexical-gustatory synesthesia*. Language is the trigger not only for synesthetic tastes but for a range of other synesthetic sensations too. One study found that 88% of the synesthesias detected in a large sample of people screened for synesthesia were triggered by language in some way (e.g., triggered by words, by letters, by days of the week, by proper names).[1] Indeed, perhaps the best-studied variant of synesthesia is that of colors triggered by letters and digits. Collectively, these triggers are known as "graphemes" and so this variant is called grapheme-color synesthesia. For grapheme-color synesthetes, each letter or digit has its own specific hue. So A might be crimson red, 7 might be pale green, Z might be deep yellow, and so on. For some grapheme-color synesthetes, the colors are projected externally on the text (and these synesthetes are called "projectors"), while for others the color experience is on an "inner screen" or the colors are simply known (and these synaesthetes are called "associators").[11] There are around one or two grapheme-color synesthetes per 100 people in the population at large, and below we shall see that the condition emerges midchildhood, which is the time when children first acquire literacy (in the United Kingdom at least where the study took place[1]).

In addition to colors and tastes, other synesthetes experience smells, shapes, touch, movements, textures, and a range of other sensations. Synesthetic hearing has been reported in a number of recent articles describing the link between sound and movement.[12,13] The synesthetes identified in these studies hear movement or flashes of light as faint sounds such as buzzing, whooshes, or humming. These phantom sounds are particularly tied to the rhythm of the movement that triggers it, so a rhythmic flashing light would trigger the same rhythm of sound, synced in time to the visual inducer. With regard to touch, some synesthetes experience tactile sensations on their own body when seeing other people being touched. This has been termed mirror-touch synesthesia and affects about 1.6% of the population.[14]

Some types of synesthesia, such as mirror-touch and hearing-motion synesthesia, appear to be closely related to processes found in neurotypical people (e.g., embodied empathy, auditory imagery). This poses the question of why we would consider grouping them together with other kinds of synesthesia such as grapheme-color and lexical-gustatory.[15] For all these manifestations to be "synesthesia," it would be important to consider carefully what the definition of synesthesia might need to be.[16] A simple "merging of the senses" does not suffice as a definition because we have seen that many synesthesias have inducers (i.e., triggers) that are *not sensory* (e.g., some synesthesias are triggered by higher-order processes such as thinking or using language). Furthermore, not just inducers but even synesthetic concurrents can be nonsensory. In *sequence-personality synesthesia*, for example, ordered sequences like letters of the alphabet or days of the week take on concurrents that are personality traits (e.g., the synesthete might view the letter A as a bossy mother). Personality traits are, again, abstract constructs of thinking rather than sensory perceptions per se and so do not fit well either into any notion of synesthesia as a "merging of senses." Whether or not these should be included as variants of synesthesia has been debated, but all these different experiences do nonetheless reflect what synesthesia is at its very core: the bringing together of qualities that the average person does not usually experience together.

The wide range of conditions that fall under the umbrella term of "synesthesia" also poses a challenge for how to test for synesthesia, and there are several ways we might hypothesize doing this. We shall see below that grapheme-color synesthesia has known neural correlates.[17] But less is known about other types of synesthesia and scanning is not used as a diagnostic tool, not least because magnetic resonance imaging (MRI) is time-consuming and costly. Equally, although synesthesia is likely to be inherited via genetic transmission,[18,19] there is no genetic test either; our knowledge here is too nascent to effectively guide us to diagnosis in this way, and so an alternative diagnostic method is required. For all these reasons, the "gold standard" test for synesthesia is not a biological one, but a behavioral one. It rests on a widely established test that looks for one key behavioral feature of synesthesia, which is the fact that synesthetic experiences are relatively stable over time. So, a synesthete with colored letters will tend to associate the same letter with the same color over time (e.g., if A is synesthetically blue for any particularly synesthete, it will always tend to be blue for that person). In contrast, someone *without* synesthesia may not only find the idea of colored letters somewhat alien but would pair letters to colors only inconsistently, even if they were highly incentivized or trying effortfully.[3] This difference between consistent synesthetes and *in*consistent *non*synesthetes forms the "consistency test/test of genuineness," the widely used diagnostic for synesthesia. Synesthetes are presented with a list of triggering stimuli (e.g., a list of letters, for a grapheme-color synesthete) to which they must give their synesthetic associations first once, and then in a surprise retest. The responses are compared over time for each stimulus (e.g., is the same color given over time for the letter A?) and only those scoring significantly more consistently than controls are deemed genuine synesthetes.

The test and retest involved in recognizing synesthesia were traditionally separated by a number of weeks, months, or even years. One study showed that the synesthetic tastes triggered by words in a lexical-gustatory synesthete were 100% consistent in a surprise retest over approximately 30 years.[20] Clearly, however, practical considerations limit the amount of time that can be taken up with a test to diagnose synesthesia. Contemporary tests now rely on a short 15-minute session that can test and retest each letter (or other inducer) up to 3 times within a single test. In these tests, such as the widely used *Synesthesia Battery*,[21] each trigger stimulus is presented three times in a random order and allocated a color on each occasion. The synesthete selects his or her colors from a digital color-palette and the scoring takes into account the numerical distance in color-space between colors for the same letter (e.g., the three colors for the letter A). When the distances are particularly short, this means the color selection was highly consistent, and the participant is likely to be a synesthete. The use of such speedy tests, and their availability online, has revolutionized the study of synesthesia by allowing researchers to quickly demonstrate the genuineness of their case studies in a fast, robust, and overall elegant way. One important drawback to the use of consistency as the central diagnostic for synesthesia is the possibility that consistency is not, in fact, a defining feature of synesthesia at all. While it is true that many synesthetes are highly consistent, and while we do have a tool that will assess consistency, ruling out synesthetes that fail in consistency risks falling foul of Maslow's Law of the Instrument.[22] Maslow famously described that there is a temptation to treat everything as if it were a nail, if the only tool we have is a hammer. It is completely possible that some synesthetes are actually *in*consistent; if so, they have become sidelined in modern synesthesia research because the test of consistency is such a compelling tool.[16]

We turn now to the question of where synesthetic associations might derive from and we consider this question from a number of different angles. We look at the neurological underpinnings of the condition by considering the ways in which the brains of people with synesthesia are different—in both their structure and in their function—to the brain of people without synesthesia. We look too at the roots of synesthesia in early childhood. We ask how these unusual associations might have developed over the early life span, both in their brain structure and in the ways in which particular inducers come to be paired with particular concurrents. Why might A be red for any given synesthete? Why might "society" taste of onions, or the pitch of middle C on a piano evoke a certain shape, texture, or other association? In the following sections, we will explore these questions.

The neural basis of synesthesia

Neuroscientific theories of synesthesia have centered on the notion of atypical (normally excessive) connectivity. This is unsurprising given that synesthesia itself is defined in terms of an atypical connection between stimuli and experiences. However, the precise underlying mechanisms that give rise to this are not fully understood and remain a source of controversy.

Neurodevelopmental accounts of synesthesia

The most influential neurodevelopmental account of synesthesia is the Neonatal Synesthesia Hypothesis (or Infantile Synesthesia Hypothesis) originally put forward by Daphne Maurer and colleagues (23,24). Put simply, the idea behind the theory is that all human infants are synesthetes and most people lose this ability during development (becoming adult nonsynesthetes) but a few retain this ability (becoming adult synesthetes). The evidence for the theory comes from several different observations:

1) *Increased connectivity during infancy.* Synaptic density is greatest soon after birth with synaptic density in sensory cortical regions decreasing toward adult levels earlier than in other regions.[25] Glucose metabolism, a sign of functional synaptic activity (rather than amount of synapses), also shows an early peak and fall.[26] In particular, there is anatomical evidence of pathways from auditory to visual cortex that are normally reduced or removed during development.[27,28]
2) *Less domain specificity during infancy.* Cortical regions are far less specialized during infancy and, in particular, may respond more strongly to multiple sensory modalities relative to older children or adults. For example, regions normally specialized for spoken language respond more strongly to visual inputs early in life.[29]
3) *Presence of synesthetic-like correspondences in early life.* For example, 3- to 4-month-old infants will orient toward high and pointed shapes when played a high-pitched tone and will orient toward low and rounded shapes when played a low-pitched tone.[30] This has been taken as evidence that these correspondences are innate rather than learned.

In this theory, the triggering mechanism that causes individuals to develop either along a synesthetic or nonsynesthetic trajectory is assumed to be genetic. One candidate mechanism is apoptosis, or programmed cell death, which is known to be under genetic control.[31] Under this account, genetic differences in synesthetes cause certain pathways to be retained that tend to be lost in most other people. Another candidate mechanism is synaptogenesis which is under genetic control,[32] although synaptic elimination itself is more strongly driven by the environment 33 Bargary and Mitchell[34] discuss a number of other possible genetic mechanisms that would result in atypical connectivity including polymorphic variations in molecules that support axon guidance or that normally create borders between adjacent regions.

The Neonatal Synesthesia Hypothesis is attractive because it attempts to account for a wide range of data and is grounded in a plausible account of brain development. However, it has attracted criticisms and controversy 35. First of all, the nature of the evidence is all indirect: nobody has ever studied synesthesia in people this young. (Although it is not possible to diagnose synesthesia in people this young, one can test people "at risk" of synesthesia and then classify them later in life as is done elsewhere.[36]) Some types of synesthesia, such as grapheme color, appear to emerge in midchildhood (6–10 years) rather than infancy 37. However, it remains possible that other kinds of synesthesia are in place before this or that differences in brain wiring predate the emergence of synesthesia. Crossmodal correspondences have been argued to be irrelevant to the debate because they are pervasive rather than a unique characteristic of synesthetes (e.g., Ref. 35) and because some correspondences that were believed to be innate (because they supposedly do not exist in the environment) have been shown to reflect statistical regularities in the environment[38] (see chapters in this volume by Spence and Sathian; Brang and Ramachandran).

Evidence of functional and structural brain differences in adults

This subsection considers a number of questions relating to the adult synesthetic brain. Where in the adult brain are the inducer and concurrent connected together? What is the nature of the connectivity (e.g., functional vs. structural)? Are there atypical features of the synesthete's brain more widely?

What is the nature of the connectivity (e.g., functional vs. structural)?

One of the most enduring debates is whether atypical connectivity in synesthetes' brains is functional in nature (e.g., disinhibited feedback between regions) or structural (e.g., the amount or pattern of dendritic and/or axonal connections).

Grossenbacher and Lovelace[39] were the original proponents of the disinhibited feedback model. Under this account, both synesthetes and nonsynesthetes possess a similar set of structural pathways between modalities but links between the inducer and concurrent, normally inhibited in nonsynesthetes (preventing one from eliciting the other), would be disinhibited in synesthetes. Evidence in support of this kind of mechanism in the nonsynesthetic brain comes from studies of multisensory plasticity following sensory deprivation such that, for instance, sounds and touch may come to activate visual cortex following blindfolding or blindness.[40] The relative rapidity of this mechanism is more consistent with functional disinhibition (although structural changes may follow). This mechanism also thus offers a reasonable account of acquired synesthesia which is typically triggered by some form of sensory loss 2.

The disinhibited feedback model has tended to have been less accepted in the case of developmental synesthesia for which there is convincing evidence of structural brain differences. At least, one can say with conviction that disinhibited feedback cannot be the sole mechanism at play in developmental synesthesia, even if it is hard to rule out completely. Diffusion tensor imaging (DTI) provides a measure of the degree of organization of white matter, and this has been shown to be increased, in several regions, in grapheme-color synesthesia[17] and music-color synesthesia.[41] But note that the reverse pattern (less white matter organization) was observed in several regions for ordinal linguistic personification.[42] The method of voxel-based morphometry has typically been used to assess gray matter density (although it can be applied to white matter too): where gray matter density reflects the concentration of cell bodies and also synapses. Increases in gray matter density have been found in regions such as area V4 and parietal cortex in grapheme-color synesthesia (see Ref. 40) and somatosensory cortex in mirror-touch synesthesia.[43]

Where in the adult brain are the inducer and concurrent connected together?

Taken together with the previous debate, Bargary and Mitchell[34] outlined a 2×2 space of possible neural models depending on whether the connections are direct/indirect X structural/functional. This section will focus on whether the inducer and concurrent are connected directly or indirectly to each other.

Cytowic[44] initially speculated that synesthesia is generated in the limbic lobe (hippocampus, amygdala, etc.), and that this is where the inducer and concurrent connect together. In his view, synesthesia was an atavism, i.e., a throwback to the evolutionary past and a more primitive way of sensing. Although these regions do indeed process inputs from multiple senses,[45] there is no consistent evidence from structural and functional neuroimaging for a key role of the limbic system in synesthesia.[17]

An alternative viewpoint is that the brain regions that support processing of the inducer and concurrent connect together directly. The Cross-Activation Theory, or Adjacency Theory, is a good example of such a theory.[46,47] This theory argues that inducers and concurrents that are anatomically neighboring each other are more likely to connect together than those that do not: for instance, the high prevalence of grapheme-color synesthesia is assumed to reflect the anatomical proximity of color-selective and grapheme-selective regions in the visual ventral stream. This could also account for other associations. Ward, Simner, and Auyeung[48] noted that whereas the synesthetic color of words depends on their spelling pattern, the synesthetic taste of words depends on their sound pattern. It is hard to imagine what statistical regularities in the environment could underpin this trend, but regularities in brain topology (e.g., flavor and speech being anterior, vision posterior) would account for this.

Is there direct evidence in support of the Adjacency Theory from neuroimaging? Rouw and Scholte[49] examined white matter connectivity in grapheme-color synesthetes using DTI. There was evidence for increased white matter organization in inferior temporal cortex particularly for projector synesthetes. This was near to regions involved in processing the inducer and concurrent, although somewhat anterior to it, but has been interpreted as providing support for the Adjacency Theory.[46] Using a measure of functional rather than structural connectivity (dynamic causal modeling of functional MRI (fMRI) data), van Leeuwen, den Ouden, and Hagoort[50] found that direct cross activation between regions processing the inducer and concurrent offered the best account of projector grapheme-color synesthetes, whereas indirect cross activation (via parietal cortex) offered the best account of associator grapheme-color synesthetes. In summary, while there is neuroimaging evidence that inducers and concurrents can directly connect together in grapheme-color synesthetes, it may not be the case for all of them and there is little evidence on this issue for other kinds of synesthesia.

Finally, there is a more radical answer to the question: where in the adult brain are the inducer and concurrent connected together? Namely, that instead of assuming two sets of neurons (e.g., a set coding graphemes and a set coding colors) that connect together, there may be a single population of neurons that code multiple features. This is in the spirit of the Neonatal Synesthesia Hypothesis suggesting that synesthetes undergo less of a domain-specific reorganization during development, or what Baron-Cohen et al.[51] referred to as a "breakdown of modularity." For some synesthetes, all sounds trigger vision and one could reasonably refer to them as having a visual modality and an audiovisual modality (i.e., without evoking the notion of connectivity) instead of assuming separate auditory and visual modalities with atypical connectivity.[52] Ultimately, strong evidence for this would come from single-cell electrophysiology in synesthetes or, less strongly, from multivariate pattern analysis in fMRI.

Are there atypical features of the synesthete's brain more widely?

At one extreme, one could imagine that the brains of synesthetes differ in some very narrow way (e.g., connectivity between grapheme and color processing regions) and, at the other extreme, they may differ in terms of global brain architecture (e.g., hyperconnectivity across all regions). Neither of these extreme views is justified by the available data but, instead, there appears to be a middle ground in which there are multiple regional differences in the brains of synesthetes. For instance, grapheme-color synesthetes have increased gray matter density in the intraparietal sulcus.[17] While this may be functionally relevant to synesthesia itself,[50] its impact is likely to extend beyond that. Hänggi, Wotruba, and Jäncke[53] used graph network theory to demonstrate that these parietal regions, in synesthetes, have a higher degree centrality, i.e., they act as key hubs to many brain regions. In addition to parietal cortex, there is evidence for structural differences (increased gray matter) in motor/premotor regions, and of increased prefrontal activity when processing inducers.[17]

In short, one could perhaps think of synesthesia as a symptom of a particular kind of brain organization. But the best ways of characterizing this brain organization and its implications—e.g., for cognition, other neurodevelopmental conditions—remain to be elucidated. In the section below, we will discuss how synesthesia is related to differences in cognition. For instance, one idea is that the special brain architecture of synesthetes enables them to better connect ideas together and, hence, be more creative.[47] In this view, it is not the synesthesia (e.g., having colors for numbers) itself that is important, but it is the particular brain organization that predisposes to synesthesia. Synesthesia becomes what Ward[54] has referred to as a "colorful sideshow" to the main event (whatever that may turn out to be). A related question is how the wider brain differences present in synesthesia are linked to those present in other developmental conditions such as autism and schizophrenia. In particular, recent studies showing an increased comorbidity of synesthesia and autism suggest that there are likely to be shared mechanisms in terms of brain organization and/or genetics 5.

Where do synesthetic associations come from?

One question commonly asked in synesthesia research, and indeed of synesthetes themselves, is why certain inducers (triggers) come to be paired with certain concurrents (sensations). What governs the types of associations that arise in synesthesia and is there any sense or pattern in these pairings? Put simply, why might the letter A be red (say) for any given synesthete, and not green, blue, purple or another color altogether? It is important to clarify first that synesthetes tend to differ one from the next in the makeup of their synesthetic associations. This means that synesthetes tend to disagree, one-on-one, in the colors of their alphabets, the tastes of their words, the shape of their tastes, and so on. But when *large groups* of synesthetes are compared together, surprising patterns emerge. Two studies in 2005 compared the colors of letters in large cohorts of grapheme-color synesthetes and found that the letter A tended to be red significantly more often than chance would predict, that B tended to be blue, C tended to be yellow, and so on.[55,56] When each letter was inspected, it was possible to detect a "prototypical alphabet" in that each letter tended to have "its own color," at higher than chance levels. There was still considerable variation from one person to the next: so although A tended to be red it could in fact be any color at all when every synesthete was considered. But there were significant underlying trends causing unity across synesthetes as a group. And this prototypical coloring of the alphabet yielded surprising structure, which could be described as a series of unconscious "rules" the synesthetes were following, even without knowing it. So, for example, letters that appear with high frequency in the English language (e.g., A, S) tend to be paired with high-frequency color terms (e.g., red, yellow), while less frequent letters paired with less frequent colors (e.g., Q = purple). These two early papers spawned a subfield of synesthesia research which seeks to understand the structure underlying synesthetic pairings. Later papers, for example, showed that the frequency of the inducer was also tied to the luminance and saturation of synesthetic colors (e.g., high-frequency letters tend to take colors that are more saturated[57]). Other studies showed systematicity in other forms of synesthesia: synesthetic colors triggered by music become more luminant as the pitch of the musical note gets higher[58]; and synesthetic tastes draw on the underlying phonological structure of food names (e.g., the taste of peach might be triggered by words like "beach" and "reach"[3]). There are even underlying biases linking synesthetic colors to temporary mood states: the colors of letters are more luminant when grapheme-color synesthetes are in positive mood states.[59]

Above, we have referred to the systematicity found in synesthetic associations as a system of "rules." We should be clear, however, that these rules are unconscious: synesthetes are themselves often unaware of where their sensations come from and it takes a great deal of "detective work" by researchers to uncover the rules beneath the surface. Related to this is the fact that although these rules provide elements of structure, synesthetic associations nonetheless appear chaotic and somewhat random on the surface. So, for example, there is nothing immediately obvious about why the word "shuttlecock" triggers the taste of egg whites for JIW. However, detailed examination of the system of tastes as a whole reveals that word in English containing the phoneme / ʃ / ("sh") significantly tend to taste of eggs for JIW, likely via the phoneme structure of the semantically related word "shell."[3]

One interesting facet of this body of research is the finding that even nonsynesthetes internalize cross-sensory rules, and that these rules are often same rules found in synesthesia. Collectively, these rules in nonsynesthetes are typically referred to as crossmodal correspondences[60] (see also chapters in this volume by Spence and Sathian; Brang and Ramachandran). Consider, for example, the finding that synesthetic colors from musical notes are more luminant when the pitch of the note is higher. A key finding has been that even nonsynesthetes associate colors to music, and that they too follow the same unconscious bias. In our study, we asked nonsynesthetes to listen to individual piano notes and to select a color for each, simply from intuition.[58] Even though our nonsynesthete subjects thought they were guessing, they in fact provided systematic choices; just like synesthetes, their colors were more luminant as the piano note got higher. The same mirroring between synesthetes and controls has been found in other forms of synesthesia too—all people have intuitive preferences for the colors of letters, and there is a greater than chance similarity between the colors of synesthetes and nonsynesthetes (e.g., both groups feel that A is red, and this association is picked at greater than chance levels[55,56]).

A small number of studies have also shown that synesthetes sometimes develop their synesthetic associations by internalizing some type of association from the environment. Hancock[61] presented a case study of grapheme-color synesthetic twins whose colored letters mapped onto a colored alphabet toy they used as a younger child. This toy is one of a category of educational materials, also including alphabet books and ABC posters, which explicitly color-code letters. In some instances at least, these serve as the template for what later becomes internalized as synesthesia. A larger-scale study showed that this tendency can be observed in other synesthetes too.[62] And Mankin and colleagues have recently shown that *all people* internalize the colors of educational ABC materials.[63] In that study, large numbers of people from the general population were asked to provide their colors for letters, while a second group were asked to complete phrases of the type "A is for …" (e.g., A is for apple). A third group of subjects were required to provide the most common colors of these objects (e.g., apples tend to be red). When these three bodies of data were brought together, there was clear evidence that the colors chosen for letters were based on the information gathered from the other groups. So, A tended to be red, because A tends to be for apple, and because apples tend to be red. These internalized letter-colors are likely to form the basis of synesthetic colors too, given that synesthesia often mirrors the unconscious intuitions of nonsynesthetes.

It seems clear then that synesthetic associations are not random, and that they derive in many cases from unconscious biases found in the greater population. Some of these reflect more abstract crossmodal correspondences (high pitch notes mapping to more luminant colors), while others are more firmly based on environmental influences—both direct influences (e.g., alphabet toys) and indirect influences (e.g., the frequency of letters in the linguistic environment). It is also likely that the colors of grapheme-color synesthetes are constructed to some extent around literacy acquisition: letters cannot become colored until the synesthete has first learned those letters as a child. Indeed, our own study showed the slow consolidation of colored graphemes in synesthetic children between the ages of 6 and 10 years.[37,64] Another study, however, showed that some synesthesia emerges *before* letters are fully acquired. In one study, nonsynesthetic children as young as 3 years old were asked to provide colors for letters.[65] Even before these children could recognize the alphabet, they already had shared intuitive preferences for the coloring of some letters. In particular, the letters X and O were consistently colored black and white, respectively, at higher than chance levels even. These color-pairings (X-black; O-white) are also found in the prototypical associations of synesthetes. This suggests that at least some element of grapheme-color synesthesia may lie with basic shape perception rather than higher-order letter identification. This conclusion is supported by studies with adult grapheme-color synesthetes which show similarities in the colors of letters with the same shape. For example, letters such as "d" and "b" are likely to be closer in color than letters such as "d" and "j."[66] When taken with the evidence above, it seems that synesthetic colors for letters depend not only on literacy acquisition (i.e., learning letters; and using "A is for apple" educational techniques) but might also stem back further into much earlier childhood, consistent with the Neonatal Synesthesia Hypothesis.

A final way to consider where synesthetic associations might come from is to think in genetic terms. Synesthesia appears to pass down through families.[67] And the way synesthesia is passed down suggests genetic inheritance rather than something taught from one generation to the next. For example, parents with one type of synesthesia can have children with a different type altogether[48] and grapheme-color associations within families are no more similar than those across families.[67] Third and most conclusively, two studies have attempted to link the trait of synesthesia directly to the genes. Asher et al.[18] conducted a whole-genome scan of synesthetic families with colored graphemes and music, and a second study[19] considered families with colored letters, numbers, and other linguistic sequences (colored days and months). These studies pinpointed five separate chromosomal regions on which the gene(s) for these forms of synesthesia are likely to lie. The numerous chromosomal hotspots identified suggest that synesthesia is likely to have a complex genetic inheritance, with many genes contributing to single variants or indeed vice versa, with many variants related to the same gene. One study at least has suggested that different phenotypes may cluster together under different genetic causes. Novich et al.[68] looked at the types of synesthesias that are found coexisting within single individuals and found five distinct clusters of synesthesias (e.g., colors triggered by sequences clustered differently to synesthesias triggering tastes). It appears then that synesthetic associations may not only derive from postbirth events such as literacy acquisition and environmental submersion but are also likely to be predetermined with some degree of complexity at birth.

The cognitive profile of synesthesia

We saw above that synesthetes not only differ from one to the next in their individual pairings of inducers and concurrents but that they also show similarities as a group. Colors for letters, for example, are wide ranging and varied, but synesthetes nonetheless tend to follow a set of unconscious biases in the formation of these associations, which creates similarities among them. But there are other ways, too, that synesthetes act similarly to each other. Synesthetes have a shared cognitive-perceptual profile, beyond that of the synesthesia itself. In this section, we review several areas in which synesthetes, as a group, are different from non-synesthetes and what this might reveal about the nature of synesthesia itself.

Perception

People with synesthesia not only have unusual percept-like experiences but they also show differences in their processing of sensory stimuli. We will first consider unimodal stimuli which have revealed a pattern of relatively circumscribed abilities and then consider multisensory stimuli for which the literature is less consistent.

For visual stimuli, both grapheme-color synesthetes and sequence-space synesthetes show enhanced abilities for discrimination of color and also of shape varying between a perfect and imperfect circle.[69] This is not found in all tests of visual perception; for instance, grapheme-color synesthetes appear to have normal[69] or worse[70] perception of visual motion. There is also evidence that they have enhanced processing of spatial frequency gratings, for which early visual cortex (e.g., V1) is highly sensitive. They find mid-to-high spatial frequency stimuli as more aversive,[71] and they show higher amplitude visual-evoked potentials in response to such stimuli on electroencephalography.[72] Paired presentation of grating stimuli with colors (e.g., horizontal stripes + red; vertical stripes + green) leads to longer lasting color-opponent after effects in grapheme-color synesthetes, i.e., a greater McCollough effect.[73] This is assumed to reflect cortical processes (because orientation tuning is not found before V1) and, indeed, retinal after-images are not prolonged in this group. Stimulation of the visual cortex of grapheme-color synesthetes (with transcranial magnetic stimulation) results in phosphenes at lower stimulation intensities than controls, implying that the visual cortex is intrinsically more excitable.[74]

Far less is known about nonvisual modalities, but there is evidence of selective abilities here too. In the auditory domain, it has been suggested that perfect pitch is linked to synesthesia.[75] People with mirror-touch synesthesia have enhanced tactile spatial acuity (but not enhanced color discrimination), whereas grapheme-color synesthetes show the reverse profile.[76] Across all sensory modalities, grapheme-color synesthetes report more subjective sensory sensitivity on questionnaire measures designed to measure these traits in people with autism.[71]

Given that synesthesia is often defined as a multisensory phenomenon and given the evidence for enhanced unimodal perceptual abilities, one might expect synesthetes to also show enhanced multisensory perception—but here the evidence is mixed. The synesthetic color of audiovisual speech is driven by the integrated percept in the case of the McGurk illusion.[77] However, overall susceptibility to the McGurk illusion is lower and synesthetes are less likely to use visual cues to identify speech presented against noise.[78] In the double flash illusion, a double beep and a single flash can sometimes result in the perception of a double flash. Whereas one study found this to be reduced in synesthetes,[79] another suggested it was increased.[80] The only other evidence in favor of enhanced multisensory perception in synesthetes comes from sound symbolism: for instance, the tendency to think of "bouba" as round and "kiki" as pointed. Synesthetes are better at guessing the meaning of foreign words with a sound symbolic component[81] and show more interference, in reaction time tests, to incongruent sound-shape correspondences but not for other kinds of crossmodal correspondence[82] (see also chapters in this volume by Spence and Sathian; Brang and Ramachandran).

Imagery

With regard to visual imagery, a distinction is frequently drawn between object-based visual imagery (e.g., the vividness and detail of imagined objects and scenes) versus spatial imagery (e.g., the ability to navigate around and manipulate images).[83] Both grapheme-color synesthetes and sequence-space synesthetes consistently report higher object-based visual imagery 84,85. This extends into their everyday cognition, for instance, their autobiographical memories have subjectively more sensory qualities.[86] The evidence for enhanced spatial imagery is more mixed.[85] With regard to tests of mental rotation, which is thought to tap the spatial component of imagery, Havlik, Carmichael, and Simner[87] show that people with sequence-space synesthesia are better at this task but it depends on whether their spatial form is perceived externally. In other sensory modalities, synesthetes also have enhanced mental imagery on both self-report and objective measures and this is increased if their synesthesia involves that modality.[88] In sum, people with synesthesia report rich mental imagery in at least some domains. While this may be influenced by the types of synesthesia they have, it is not narrowly limited to their inducers and concurrents. For example, a grapheme-color synesthete does not just have rich imagery of letters/numbers but also for faces, scenes, and such like. The same pattern is broadly true when we consider memory (see also chapter by Gosavi and Hubbard, this volume).

Memory

One might expect there to be a relationship between synesthesia and memory insofar as there is a general tendency for stimuli that are richly encoded to be better remembered. Thus, the colors induced by words and numbers (in a grapheme-color synesthete) would provide an additional cue, not available to others, for remembering. However, this theory predicts that any memory advantage would be narrowly limited to inducing stimuli, which is not the case. In tests of episodic memory, synesthetes have enhanced memory for both verbal and visual stimuli but the enhancement tends to be greatest for visual stimuli.[89] This can occur for meaningless stimuli such as fractals, which are hard to recode verbally.[90] In a task of associating pairs of fractal images together, synesthetes show a distinct pattern of increased and diminished fMRI brain activity (relative to controls) when recognizing and retrieving the associate, respectively.[91] But there were no differences in regions such as the hippocampus that are more traditionally linked to memory functioning. When memory for colors is directly contrasted against memory for digits[92,93] or shapes,[94] keeping general task demands constant (e.g., associating a stimulus with a location), then colors tend to outperform other stimulus features. Thus, although visual memory is overall enhanced, this may be increased further when color is the relevant feature.

In summary, the memory advantage of synesthetes has parallels with the patterns described for both perception and imagery. Namely, that cognitive enhancement is not narrowly limited to stimuli participating in the synesthesia. This is explained in terms of the more general cognitive style, noted above, to be "visual thinkers" (or, more generally, to think using sensory images). We assume that the root of this lies in the sensitivity of early sensory processing together with a distinct set of brain connectivity patterns.

Art, personality, and creativity

It has been observed that both the hobbies and occupations of synesthetes are skewed toward the creative industries, relative to national estimates of numbers employed in that sector, in both Australia[55] and the United Kingdom.[95] Moreover, the prevalence of grapheme-color synesthesia is higher in arts students compared with those from other subjects[96] and synesthetes (as a mixed group) are more likely to be engaged in visual arts as a hobby.[55,95] This difference may be underpinned by some of the cognitive differences already noted (e.g., a tendency to think in images, increased sensitivity to color), and it may also relate to differences in personality linked to synesthesia. One of the most consistent findings, found across different types of synesthesia and in samples that are not self-referred, is that synesthetes report greater "openness to experience" (e.g., active imagination, intellectual curiosity) on questionnaire measures of the "big five" personality traits.[97,98]

With regard to creativity, Ramachandran and Hubbard[99] suggest that the kinds of brain changes that give rise to synesthesia may also give rise to creativity as a cognitive ability. While this explanation remains untested, there is evidence that synesthetes score higher on at least some psychometric measures of creativity[95,98] and may have generally higher intelligence.[97]

Cognitive weaknesses

Although most research has focused on enhanced cognitive abilities in synesthesia, it is possible that this pattern is offset with particular weaknesses that have yet to be fully explored. When synesthetes are asked, open-endedly, to report their strengths and weaknesses, then Rich et al. (2005) found several areas of weakness that were reported significantly more than for controls, and these included coordination/balance/sports, and sense of direction. Any weaknesses are hard to characterize in precise cognitive terms, but some could involve problems in linking external sensory information with motor processes (and related functions such as proprioception). The strengths that synesthetes reported, more than controls, included many of the things discussed so far (memory, art, languages), but the sample was split in terms of mathematics: some listed it as a distinct strength, and others as a distinct weakness. Mathematics involves a range of cognitive processes ranging from an intuitive processing of quantity to far more abstract concepts such as algebra,[100] and it is unclear exactly where the relative strengths and/or weaknesses may lie (and indeed, how much weight to give to these self-reports). But in one behavioral study, at least, sequence-space synesthetes were objectively slower at simple (single digit) mental arithmetic compared with controls, with a particular weakness in multiplication (over division, subtraction, and addition). Because multiplication relies heavily on verbal memorization, one interpretation is that synesthetes use different strategies to controls (i.e., relying too heavily on their spatial projection of numbers as opposed to stored verbal facts, even where this was not appropriate for the task.[101]

Future directions

Research into synesthesia has made substantial progress over the last 20 years. Synesthesia is a unique and fascinating alternative way of perceiving the world, and one that is linked to a distinct cognitive and neural profile beyond having "extra" sensations. To some extent, the current body of work has identified many of the key pieces of the jigsaw puzzle, but has not yet fitted them together. For instance, we know that synesthesia has a genetic component and we know that the brains of synesthetes differ. But we do not know how these two facts are related; i.e., what role these genes play in neurotypical and synesthetic brain development and function. Nor do we understand how this process would differ depending on the type of synesthesia that is manifested (e.g., grapheme-color, mirror-touch, lexical-gustatory). Very little is known about how synesthesia changes over the life span, particularly in the early years in which synesthesia first emerges, and in which normal multisensory perception matures. Similarly, we do not know whether the cognitive profile linked to synesthesia is a necessary outcome of having this condition (two sides of the same coin) or a correlational one. The answer to these questions will be an interdisciplinary endeavor involving (minimally) psychologists, neuroscientists, and geneticists.

References

1. Simner J, Mulvenna C, Sagiv N, et al. Synesthesia: the prevalence of atypical cross-modal experiences. *Perception*. 2006;35:1024–1033.
2. Afra M, Funke M, Matsuo F. Acquired auditory-visual synesthesia: a window to early cross-modal sensory interactions. *Psychol Res Behav Manag*. 2009;2:31–37.
3. Ward J, Simner J. Lexical-gustatory synesthesia: linguistic and conceptual factors. *Cognition*. 2003;89:237–261.
4. Baron-Cohen S, Johnson D, Asher J, et al. Is synesthesia more common in autism? *Mol Autism*. 2013;4(1):40.
5. Neufeld J, Roy M, Zapf A, et al. Is synesthesia more common in patients with Asperger syndrome? *Front Hum Neurosci*. 2013;7.
6. Hughes J, Simner J, Baron-Cohen S, Treffert D, Ward J. Is synesthesia more prevalent in autism spectrum conditions? only where there is prodigious talent. *Multisensory Res*. 2017;30:391–408.
7. Frith U. *Autism: Explaining the Enigma*. Oxford: Blackwell; 1989.
8. Eagleman DM, Cytowic RE. *Wednesday Is Indigo Blue*. Boston, MA: MIT Press; 2009.
9. Cytowic RE. *The Man Who Tasted Shapes*. London: Abacus Books; 1993.
10. Jones C, Gray MA, Minati L, Simner J, Critchley HD, Ward J. The neural basis of illusory gustatory sensations: two rare cases of lexical-gustatory synesthesia. *J Neuropsychol*. 2011;5:243–254.
11. Dixon MJ, Smilek D, Merikle PM. Not all synaesthetes are created equal: projector vs. associator synaesthetes. *Cognit Affect Behav Neurosci*. 2004;4:335–343.
12. Saenz M, Koch C. The sound of change: visually-induced auditory synesthesia. *Curr Biol*. 2008;18:R650–R651.
13. Fassnidge C, Marcotti CC, Freeman E. A deafening flash! Visual interference of auditory signal detection. *Conscious Cognit*. 2017;49:15–24.
14. Banissy MJ, Kadosh RC, Maus GW, Walsh V, Ward J. Prevalence, characteristics and a neurocognitive model of mirror-touch synesthesia. *Exp Brain Res*. 2009;198(2–3):261–272.
15. Rothen N, Meier B. Why vicarious experience is not an instance of synesthesia. *Front Hum Neurosci*. 2013;7.
16. Simner J. Defining synesthesia. *Br J Psychol*. 2012;103:1–15.
17. Rouw R, Scholte HS, Colizoli O. Brain areas involved in synesthesia: a review. *J Neuropsychol*. 2011;5:214–242.
18. Asher JE, Lamb JA, Brocklebank D, et al. A whole-genome scan and fine-mapping linkage study of auditory-visual synesthesia reveals evidence of linkage to chromosomes 2q24, 5q33, 6p12, and 12p12. *Am J Hum Genet*. February 2009;84(2):279–285.
19. Tomson SN, Avidan N, Lee K, et al. The genetics of colored sequence synesthesia: suggestive evidence of linkage to 16q and genetic heterogeneity for the condition. *Behav Brain Res*. 2011;223(1):48–52.
20. Simner J, Logie RH. Synesthetic consistency spans decades in a lexical-gustatory synesthete. *Neurocase*. 2007;13:358–365.
21. Eagleman DM, Kagan AD, Nelson SS, Sagaram D, Sarma AK. A standardized test battery for the study of synesthesia. *J Neurosci Methods*. 2007;159:139–145.
22. Marslow AH. *The Psychology of Science*. New York: Harper and Row; 1966.
23. Maurer D, Maurer C. *The World of the Newborn*. New York: Basic Books; 1988.
24. Maurer D, Mondloch CJ. The infant as synesthete? *Atten Perform*. 2006;XXI:449–471.
25. Huttenlocher PR, Dabholkar AS. Regional differences in synaptogenesis in human cerebral cortex. *J Comp Neurol*. 1997;387:167–178.
26. Chugani HT, Phelps ME, Mazziotta JC. Positron emission tomography study of human brain functional development. *Ann Neurol*. 1987;22:487–497.
27. Falchier A, Clavagnier S, Barone P, Kennedy H. Anatomical evidence of multimodal integration in primate striate cortex. *J Neurosci*. 2002;22:5749–5759.
28. Dehay C, Bullier J, Kennedy H. Transient projections from the fronto-parietal and temporal cortex to areas 17, 18, 19 in the kitten. *Exp Brain Res*. 1984;57:208–212.

29. Tzourio-Mazoyer N, De Schonen S, Crivello F, Reutter B, Aujard Y, Mazoyer B. Neural correlates of woman face processing by 2-month-old infants. *Neuroimage*. 2002;15(2):454–461.
30. Walker P, Bremner JG, Mason U, et al. Preverbal infants' sensitivity to synesthetic cross-modality correspondences. *Psychol Sci*. 2010;21(1):21–25.
31. Baron-Cohen S. Is there a normal phase of synesthesia in development? *Psyche*. 1996;2.
32. Bourgeois J-P, Jastreboff P, Rakic P. Synaptogenesis in the visual cortex of normal and pretern monkeys: evidence for intrinsic regulation of synaptic overproduction. *Proc Natl Acad Sci U S A*. 1989;86:4297–4301.
33. Blakemore C, Garey LJ, Vital-Durand F. The physiological effects of monocular deprivation and their reversal in the monkey's visual cortex. *Physiology*. 1978;283:223–262.
34. Bargary G, Mitchell KJ. Synesthesia and cortical connectivity. *Trends Neurosci*. 2008;31:335–342.
35. Deroy O, Spence C. Are we all born synesthetic? Examining the neonatal synesthesia hypothesis. *Neurosci Biobehav Rev*. 2013;37(7):1240–1253.
36. Lloyd-Fox S, Blasi A, Elwell CE, Charman T, Murphy D, Johnson MH. Reduced neural sensitivity to social stimuli in infants at risk for autism. *Proc R Soc B Biol Sci*. 2013;280:20123026.
37. Simner J, Bain AE. A longitudinal study of grapheme-color synesthesia in childhood: 6/7 years to 10/11 years. *Front Hum Neurosci*. 2013;7.
38. Parise CV, Knorre K, Ernst MO. Natural auditory scene statistics shapes human spatial hearing. *Proc Natl Acad Sci U S A*. 2014;111(16):6104–6108.
39. Grossenbacher PG, Lovelace CT. Mechanisms of synesthesia: cognitive and physiological constraints. *Trends Cogn Sci*. 2001;5:36–41.
40. Pascual-Leone A, Amedi A, Fregni F, Merabet LB. The plastic human brain cortex. *Annu Rev Neurosci*. 2005;28:377–401.
41. Zamm A, Schlaug G, Eagleman DM, Loui P. Pathways to seeing music: enhanced structural connectivity in colored-music synesthesia. *Neuroimage*. 2013;74:359–366.
42. Simner J, Rehme MK, Carmichael DA, et al. Social responsiveness to inanimate entities: altered white matter in a 'social synesthesia. *Neuropsychologia*. 2016;91:282–289.
43. Holle H, Banissy MJ, Ward J. Functional and structural brain correlates of mirror-touch synesthesia. *Neuroimage*. 2013;83:1041–1050.
44. Cytowic RE. *Synesthesia: A Union of the Senses*. New York: Springer; 1989.
45. Quiroga RQ, Kraskov A, Koch C, Fried I. Explicit encoding of multimodal percepts by single neurons in the human brain. *Curr Biol*. 2009;19:1308–1313.
46. Hubbard EM, Brang D, Ramachandran VS. The cross-activation theory at 10. *J Neuropsychol*. 2011;5:152–177.
47. Ramachandran VS, Hubbard EM. Synesthesia – a window into perception, thought and language. *J Conscious Stud*. 2001;8:3–34.
48. Ward J, Simner J, Auyeung V. A comparison of lexical-gustatory and graphene-colour synesthesia. *Cogn Neuropsychol*. 2005;22:28–41.
49. Rouw R, Scholte HS. Increased structural connectivity in graphene-color synesthesia. *Nat Neurosci*. 2007;10:792–797.
50. van Leeuwen TM, den Ouden HEM, Hagoort P. Effective connectivity determines the nature of subjective experience in graphene-color synesthesia. *J Neurosci*. 2011;31(27):9879–9884.
51. Baron-Cohen S, Harrison J, Goldstein LH, Wyke M. Coloured speech perception: is synesthesia what happens when modularity breaks down? *Perception*. 1993;22:419–426.
52. Ward J. Visual music in arts and minds: explorations with synesthesia. In: Bacci F, Melcher D, eds. *Art and the Senses*. Oxford: Oxford University Press; 2011.
53. Hänggi J, Wotruba D, Jäncke L. Globally altered structural brain network topology in graphene-color synesthesia. *J Neurosci*. 2011;31(15):5816–5828.
54. Ward J. *The Frog Who Croaked Blue: Synesthesia and the Mixing of the Senses*. London: Routledge; 2008.
55. Rich AN, Bradshaw JL, Mattingley JB. A systematic, large-scale study of synesthesia: implications for the role of early experience in lexical-colour associations. *Cognition*. 2005;98:53–84.
56. Simner J, Lanz M, Jansari A, et al. Non-random associations of graphemes to colours in synesthetic and normal populations. *Cogn Neuropsychol*. 2005;22:1069–1085.
57. Beeli G, Esslen M, Jäncke L. Frequency correlates in graphene-color synesthesia. *Psychol Sci*. 2007;18:788–792.

58. Ward J, Huckstep B, Tsakanikos E. Sound-colour synesthesia: to what extent does it use cross-modal mechanisms common to us all? *Cortex*. 2006;42:264–280.
59. Kay CL, Carmichael D, Ruffell HE, Simner J. Colour fluctuations in graphene-colour synesthesia: the effect of clinical and non-clinical mood changes. *Br J Psychol*. 2015;106(3):487–504.
60. Spence C. Crossmodal correspondences: a tutorial review. *Atten Percept Psychophys*. 2011;73(4):971–995.
61. Hancock P. Monozygotic twins colour-number association:A case study. *Cortex*. 2006;42:147–150.
62. Witthoft N, Winawer J, Eagleman DM. Prevalence of learned graphene-color pairings in a large online sample of synesthetes. *PLoS One*. 2015;10(3).
63. Mankin JL, Simner J. A is for apple: the role of letter-word associations in the development of graphene-colour synesthesia. *Multisensory Res*. 2017;30:409–446.
64. Simner J, Harrold J, Creed H, Monro L, Foulkes L. Early detection markers for synesthesia in childhood populations. *Brain*. 2009;132:57–64.
65. Spector F, Maurer D. The colors of the alphabet: naturally-biased associations between shape and color. *J Exp Psychol Hum Percept Perform*. 2011;37(2):484–495.
66. Brang D, Rouw R, Ramachandran VS, Coulson S. Similarly shaped letters evoke similar colors in graphene-color synesthesia. *Neuropsychologia*. 2011;49(5):1355–1358.
67. Barnett KJ, Finucane C, Asher JE, et al. Familial patterns and the origins of individual differences in synesthesia. *Cognition*. 2008;106:871–893.
68. Novich S, Cheng S, Eagleman DM. Is synesthesia one condition or many? A large-scale analysis reveals subgroups. *J Neuropsychol*. 2011;5:353–371.
69. Ward J, Rothen N, Chang A, Kanai R. The structure of inter-individual differences in visual ability: evidence from the general population and synesthesia. Vision Res. (in press).
70. Banissy MJ, Muggleton N, Tester V, et al. Synesthesia for color is linked to improved color perception, but reduced motion perception. *Psychol Sci*. 2013;24(12):2390–2397.
71. Ward J, Hoadley C, Hughes JEA, et al. Atypical sensory sensitivity as a shared feature between synesthesia and autism. *Sci Rep*. 2017;7.
72. Barnett KJ, Foxe JJ, Malholm S, et al. Differences in early sensory-perceptual processing in synesthesia: a visual evoked potential study. *Neuroimage*. 2008;15:605–613.
73. Ramachandran VS, Marcus Z. Synesthesia and the McCollough effect. *IPerception*. 2017;8(3), 2041669517711718.
74. Terhune DB, Tai S, Cowey A, Popescu T, Kadosh RC. Enhanced cortical excitability in graphene-color synesthesia and its modulation. *Curr Biol*. 2011;21(23):2006–2009.
75. Gregersen PK, Kowalsky E, Lee A, et al. Absolute pitch exhibits phenotypic and genetic overlap with synesthesia. *Hum Mol Genet*. 2013;22(10):2097–2104.
76. Banissy MJ, Walsh V, Ward J. Enhanced sensory perception in synesthesia. *Exp Brain Res*. 2009;196(4):565–571.
77. Bargary G, Barnett KJ, Mitchell KJ, Newell FN. Colored-speech synesthesia is triggered by multisensory, not unisensory, perception. *Psychol Sci*. 2009;20(5):529–533.
78. Sinke C, Neufeld J, Zedler M, et al. Reduced audiovisual integration in synesthesia evidence from bimodal speech perception. *J Neuropsychol*. 2014;8(1):94–106.
79. Neufeld J, Sinke C, Zedler M, Emrich HM, Szycik GR. Reduced audio-visual integration in synaesthetes indicated by the double-flash illusion. *Brain Res*. 2012;1473:78–86.
80. Brang D, Williams LE, Ramachandran VS. Graphene-color synaesthetes show enhanced crossmodal processing between auditory and visual modalities. *Cortex*. 2012;48(5):630–637.
81. Bankieris K, Simner J. What is the link between synesthesia and sound symbolism? *Cognition*. March 2015;136:186–195.
82. Lacey S, Martinez M, McCormick K, Sathian K. Synesthesia strengthens sound-symbolic cross-modal correspondences. *Eur J Neurosci*. 2016;44(9):2716–2721.
83. Blajenkova O, Kozhevnikov M, Motes MA. Object-spatial imagery: new self-report imagery questionnaire. *Appl Cognit Psychol*. 2006;20(2):239–263.
84. Mealor AD, Simner J, Rothen N, Carmichael DA, Ward J. Different dimensions of cognitive style in typical and atypical cognition: new evidence and a new measurement tool. *PLoS One*. 2016;11(5):e0155483.
85. Price MC. Spatial forms and mental imagery. *Cortex*. 2009;45(10):1229–1245.
86. Chin T, Ward J. Synesthesia is linked to more vivid and detailed content of autobiographical memories and less fading of childhood memories. Memory. (in press).

87. Havlik AM, Carmichael DA, Simner J. Do sequence-space synaesthetes have better spatial imagery skills? Yes, but there are individual differences. *Cogn Process*. 2015;16(3):245–253.
88. Spiller MJ, Jonas CN, Simner J, Jansari A. Beyond visual imagery: how modality-specific is enhanced mental imagery in synesthesia? *Conscious Cognit*. 2015;31:73–85.
89. Rothen N, Meier B. Graphene-colour synesthesia yields an ordinary rather than extraordinary memory advantage: evidence from a group study. *Memory*. 2010;18(3):258–264.
90. Ward J, Hovard P, Jones A, Rothen N. Enhanced recognition memory in graphene-color synesthesia for different categories of visual stimuli. *Front Psychol*. 2013:4.
91. Pfeifer G, Ward J, Chan D, Sigala N. Representational account of memory: insights from aging and synesthesia. *J Cogn Neurosci*. 2016;28(12):1987–2002.
92. Yaro C, Ward J. Searching for Shereshevskii: what is superior about the memory of synaesthetes? *Q J Exp Psychol*. 2007;60:682–696.
93. Terhune DB, Wudarczyk OA, Kochuparampil P, Kadosh RC. Enhanced dimension-specific visual working memory in graphene-color synesthesia. *Cognition*. 2013;129(1):123–137.
94. Pritchard J, Rothen N, Coolbear D, Ward J. Enhanced associative memory for colour (but not shape or location) in synesthesia. *Cognition*. 2013;127(2):230–234.
95. Ward J, Thompson-Lake D, Ely R, Kaminski F. Synesthesia, creativity and art: what is the link? *Br J Psychol*. 2008;99:127–141.
96. Rothen N, Meier B. Higher prevalence of synesthesia in art students. *Perception*. 2010;39(5):718–720.
97. Rouw R, Scholte HS. Personality and cognitive profiles of a general synesthetic trait. *Neuropsychologia*. 2016;88:35–48.
98. Chun CA, Hupe JM. Are synaesthetes exceptional beyond their synesthetic associations? A systematic comparison of creativity, personality, cognition, and mental imagery in synaesthetes and controls. *Br J Psychol*. 2016;107(3):397–418.
99. Ramachandran VS, Hubbard EM. Hearing colors, tasting shapes. *Sci Am*. 2003;April:52–59.
100. Butterworth B. *The Mathematical Brain*. London: Macmillan; 1999.
101. Ward J, Sagiv N, butterworth B. The impact of visuo-spatial number forms on simple arithmetic. *Cortex*. 2009;45(10):1261–1265.

CHAPTER

14

How synesthesia may lead to enhanced memory

Radhika S. Gosavi, Edward M. Hubbard

Department of Educational Psychology, University of Wisconsin–Madison, Madison, WI, United States

Introduction

Synesthesia is a benign perceptual/cognitive variant in which stimulation in one sensory or cognitive stream leads to associated experiences in a second, unstimulated stream[1,2] (see Chapter 13). For example, in one of the most common forms of synesthesia, grapheme-color synesthesia, letters or numbers are perceived as if viewed through a colored overlay.[3,4] In another common form, ordinal linguistic personification, numbers, days of the week, and months of the year evoke personalities.[5,6] In sequence-space synesthesia, numbers, months of the year, days of the week, and other ordered sequences are experienced as corresponding with precise locations in space, such as a three-dimensional view of a year as a map.[7–9] "Colored hearing," which includes auditory word-color and music-color synesthesia,[10–12] is one of the few types that involves prototypical cross-modal linkages. Although this form of synesthesia is often presented as a paradigmatic example of synesthesia, it is less common than many of the previously mentioned forms that involve cross-stream, but within-modality, associations.

Although it is often referred to as a "neurological condition," synesthesia is not listed in either the DSM-IV or the ICD classification,[13] as it generally does not interfere with normal daily functioning. Indeed, most synesthetes report that their experiences are neutral or even pleasant.[14,15] Rather, we tend to think of synesthesia as being similar to color blindness or perfect pitch, a difference in perceptual experience. We sometimes refer to it as a neurological condition to reflect the brain basis of this perceptual difference, but this should not be taken to indicate that synesthesia typically yields functional impairments. The unusual reports of synesthetes may lead clinicians to think of synesthesia as a symptom of a psychiatric disorder, and indeed, the DSM-V revisions included discussions of including synesthesia as part of a differential diagnosis for other psychiatric conditions.[16] Its high prevalence rate

Multisensory Perception
https://doi.org/10.1016/B978-0-12-812492-5.00014-0

means that congenital synesthesia sometimes may be found in patients who present with psychiatric conditions. Finally, despite this general pattern, emerging evidence suggests that synesthesia may be associated with a heightened rate of radiological markers for multiple sclerosis[17] and anxiety disorders.[18]

Since the rediscovery of synesthesia in the mid-1980s,[19,20] much of synesthesia research has focused on demonstrating the reality of synesthetes' experiences[21–24] and exploring the neural substrates that lead to synesthetic experiences.[25,26] Fortunately, as these foundational issues have become more settled,[1] synesthesia research has expanded to address a much wider variety of topics. Synesthesia research has explored the impact of other life experiences on synesthesia,[27] as well as the impact of synesthesia on perceptual and cognitive functions[3,23,24] and on possible comorbidities between synesthesia and other cognitive and neural traits.[17,18,28] In this chapter, we focus on the impact of synesthesia on memory, reviewing over 50 years of research on this topic, and building on our own recent investigations of how synesthesia impacts multiple memory processing stages.

Synesthesia and long-term memory

While the reality of synesthetic experience along with its neural and perceptual basis has been more thoroughly investigated, less is known about how synesthetic associations shape cognitive faculties. For instance, synesthetic associations have been shown to impact synesthetes' memory abilities, though the underlying neural mechanisms remain unclear.[29] One particularly fascinating case study is of Luria's famous mnemonist, "S".[30] Luria reported that S's memory was essentially limitless, and he showed very little decay even over decades (indeed, S complained that he could not forget). S's memory was driven, in part, by his fivefold synesthesia (in which all of his senses were linked), and allowed him to recall speeches in full and memorize complex mathematical equations with ease. Interestingly, at times S missed the semantic structure (i.e., the meaning) of the information due to his heightened attention to detail. One limitation of Luria's case study is that it is difficult to distinguish whether S's spectacular memory was based on his associations or on strategies, such as his use of the method of loci. More modern case studies involving synesthetes with spectacular memory, such as "C"[31] and "JS",[32] used more targeted experimental methods to demonstrate that their subjects' memory enhancement was due to their synesthetic associations.

Going beyond case studies, group studies have shown small but consistent performance advantages for synesthetes, particularly on tests of long-term memory.[29,33,34] These investigations have demonstrated enhancements in the visuospatial and verbal memory domains. For instance, synesthetes who experience colors for verbally presented words demonstrate better memory recall than nonsynesthetes on verbal recall tasks,[35] while grapheme-color synesthetes outperform control subjects on tasks of long-term memory for alphanumeric digits.[31,36] Crucially, the latter studies also manipulated the congruency of the presented stimuli with the subjects' synesthetic associations. In the neutral condition, synesthetes were presented with black or gray digits, while in the congruent condition, digits were presented in the same color that each synesthete associates with that specific digit (e.g., *1* would be presented in red for a synesthete that associates the number *1* with red). In both conditions, synesthetes recalled more digits than nonsynesthetes.

Most interesting, however, is the finding that although synesthesia can aid memory, it can also be a hindrance if participants are presented with incongruent stimuli in which the colors of the stimuli do not match an individual's preexisting synesthetic associations (e.g., *1* would be presented in blue for a synesthete that associates the number *1* with red). In general, synesthetes show poorer recall of incongruent stimuli than congruent or neutral stimuli,[29] though this is not always the case. Radvansky et al.[37] found that synesthetes outperform nonsynesthetes on a memory task in which words were presented in colors that were neutral, congruent, or incongruent colors to their associations, although there was a reduction in performance noted for the incongruent condition. It has been suggested that differences in experimental design may account for this discrepancy in the literature, as other properties of synesthesia (such as bidirectionality or overall enhanced color processing) may be differentially tested by specific memory tasks.[35,37–39] Additionally, it is important to note that memory enhancement is only observed when synesthetes are presented with synesthesia-inducing (e.g., graphemes for grapheme-color synesthetes) as opposed to noninducing (e.g., symbols) stimuli.[37,40,41]

Beyond long-term memory

To more accurately assess the sources of synesthetic memory advantages, it is worth examining the classic "multistore model" or "modal model" of memory. Atkinson and Shiffrin[42] proposed three distinct memory stores: sensory (iconic) memory, short-term (working) memory, and long-term memory. This model provided an account for memory processing from sensory to long-term memory and how information moves through these different stores. The transfer of information through the various stages of memory raises the question: are synesthetic advantages in long-term memory derived from advantages in earlier stores of memory processing?

A recent study investigated if the relationship between perception and visual associative memory is continuous rather than modular in synesthetes and older adults.[43] This study found that synesthetes had greater neural activity in the early visual regions compared to nonsynesthetes in a delayed pair-associate retrieval task: participants had to decide if black-and-white images (which did not induce synesthetic experiences) were correctly paired with a previous image, in a set of four similar and four dissimilar pairs of images. Synesthetes were faster to learn the paired associates and showed greater neural activity in the early visual regions during retrieval, than nonsynesthetes. This study concluded that the observed memory advantage for synesthetes was driven by their enhanced visual sensitivity and visual perceptual abilities. Furthermore, there were no functional differences found between the medial temporal lobe and sensory cortices for memory and perception processes. Thus, Pfeifer et al. suggested that mnemonic mechanisms take place in a distributed network of areas along the ventral visual stream and are dependent on the stimulus. This account of memory suggests that visual stimuli are processed in a continuous, rather than a segmented, manner from perception to memory along the ventral visual stream.[44]

Testing earlier stages of memory in synesthesia

Our research group conducted a series of three studies to investigate the impact of synesthesia on iconic, working, and long-term memory. The aim of these studies was to test if synesthetic advantages in long-term memory derive from advantages in earlier stages of memory processing. First, we explored the capacity of iconic memory in grapheme-color synesthetes[45] using the well-established Sperling partial report paradigm[46] (Fig. 14.1). On each trial, a fixation cross was displayed in the center of the screen, followed by an array of either 3 × 3 or 4 × 3 randomized letters. Modeled after the original Sperling paradigm, participants heard a low, medium, or high tone after a variable delay (0, 150, 300, 500, and 1000 ms). Based on the tone, participants were instructed to report the corresponding portion of the letter array. If they heard a low tone, they were instructed to report the bottom row, a medium tone, the middle row, and a high tone, the top row. The participants therefore were not aware of which row they had to report until the presentation of the tone. They were given 2000 ms to recall the appropriate line, and the experiment advanced to the next trial once the response duration ended. Accuracy was calculated for each trial based on the number of correct letters identified. The whole report capacity, or number of items available in memory, was calculated based on the performance accuracy on the cued row. The results indicate that synesthetes have an overall advantage in iconic memory storage compared to nonsynesthetes, as capacity estimates were higher for the synesthetes than for the nonsynesthetes (Fig. 14.2). Furthermore, capacity for the synesthetes after a 500 ms delay (4.18 items) was almost identical to capacity for the nonsynesthetes immediately after the offset of the array (4.31 items). When the capacity estimates were broken up by the size of the letter array (9 or 12 letters) we found that capacity differences between groups became greatest when they were presented with higher memory loads. Beyond the relationship between synesthesia and memory, this study informs the ongoing debate about the basis of iconic memory. Our findings suggest that the perceptual and memory processes may be working in conjunction with each other and may not be mutually exclusive.

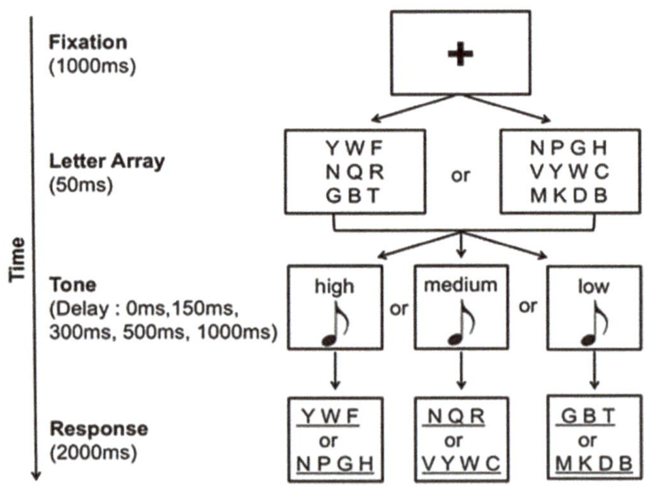

FIGURE 14.1 The original partial report paradigm replicated. Participants were presented an array of 9 or 12 randomized letters of the alphabet. After a variable delay, they were presented with a tone (high, medium, or low) and then had 2000 ms to report the letters in the appropriate row.

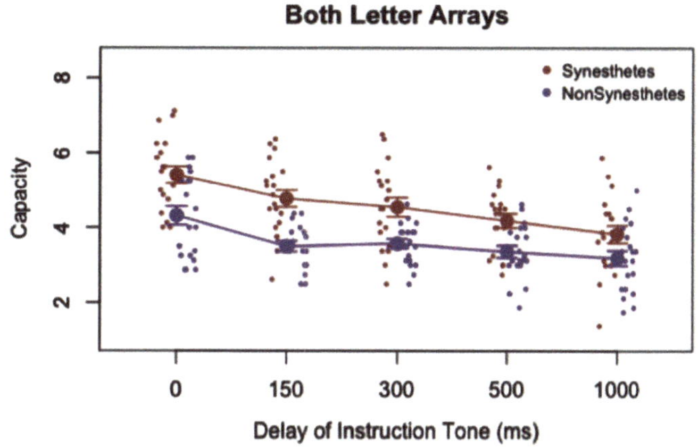

FIGURE 14.2 Iconic memory capacity estimates for synesthetes and nonsynesthetes when presented with letters. Capacity estimates collapsed across the 9 and 12 letter arrays. Red = synesthetes, blue = nonsynesthetes. *Large dots* indicate group means, while *small dots* represent individual participants. *Error bars* indicate the standard error of the mean. *Asterisks* above each panel indicate delays with significant group differences.

Next, we tested working memory performance in synesthetes and nonsynesthetes using a novel adaptation of the standard change detection paradigm.[47] In one of the only studies testing working memory in grapheme-color synesthetes to date, Terhune et al.[39] used n-back tasks in which participants were presented with a synesthesia-inducing or noninducing grapheme. Stimuli were presented in colors that were congruent with or incongruent with a given synesthete's grapheme-color associations. Participants were instructed to maintain the color of the grapheme or other information about the grapheme in memory and had to indicate if the stimulus properties matched the stimulus presented two or three trials previously. Results showed that synesthetes, compared to nonsynesthetes, displayed superior working memory performance while holding color information in memory, but not while holding grapheme information in memory across all types of stimuli. Terhune and colleagues[39] conclude that this result demonstrates enhancements in dimension-specific information (in this study, color) for synesthetes compared to nonsynesthetes.

To extend Terhune et al.'s[39] findings, we used a standard working memory change detection task in which participants are instructed to report whether a single target stimulus matched or did not match the stimulus at the same location in original memory array.[48] The experiment was divided into four conditions in a 2 (attention) × 2 (stimulus) design (Fig. 14.3). In the "attend-to-number" condition, participants were instructed to attend to the numbers. In the "attend-to-color" condition, participants were instructed to attend to the colors. The attentional conditions were blocked and their order was counterbalanced across participants. Within each attentional condition, the neutral baseline condition (color patches in the attend-to-color condition and black numerals in the attend-to-number condition) was always presented first, followed by the colored numbers (congruent/incongruent) condition. In the attend-to-color baseline condition, color patches were presented in colors that matched the synesthetes' number-color associations. Each colored number trial consisted of memory arrays composed of either congruent or

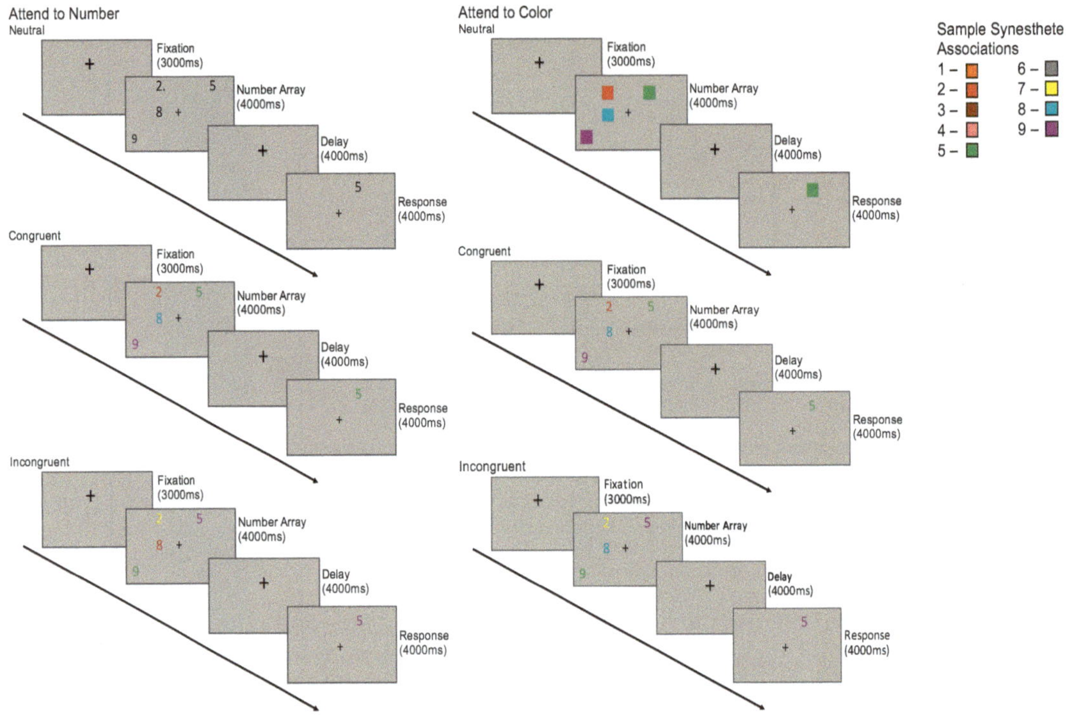

FIGURE 14.3 Participants attended to the number or colors in a change detection task. Stimuli were presented in neutral (black), or in colors that were congruent or incongruent to the synesthete's associations. Two, four, or eight number stimuli appeared in the array (this schematic only displays four numbers).

incongruently colored numbers. The congruent and incongruent trials were randomly intermixed. Congruent numbers were presented in colors that matched each synesthete's associations, and incongruently colored numbers were presented using reshuffled number-color pairings from within each synesthete's reported associations. The unattended dimension was always consistent (i.e., the color of the number never changed when the participant was attending to numbers). Nonsynesthetic controls were presented with the same set of stimuli as their respective matched synesthete.

Each trial started with a fixation cross in the center of the screen, followed by an array of two, four, or eight numbers on a gray background. The items were randomly distributed within cells of a 5×5 grid with a fixation cross displayed in the center cell (4000 ms). Next, there was a delay period (4000 ms), followed by the presentation of one of the previously presented numbers. Participants were asked to report whether a target stimulus matched the attended dimension (color/letter) of the stimulus presented at that location during the study period.

We calculated accuracy and capacity estimates in this study, as together they provide a deeper understanding of the working memory mechanisms at play. Overall, the results of this study reinforce previous findings, indicating that synesthetes have enhanced memory across stimuli while attending to either numbers or colors (Fig. 14.4). By analyzing performance accuracy and working memory capacity while participants attended to numbers and colors, we found that synesthetes had an advantage in both conditions. Interestingly, beyond four items,

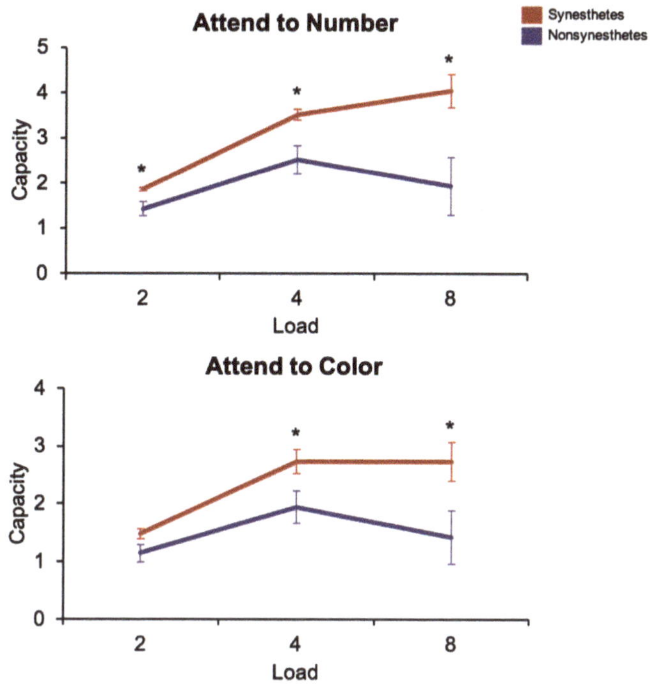

FIGURE 14.4 Working memory capacity estimates when synesthetes and nonsynesthetes were presented with neutral stimuli and were attending to number and color. Error bars represent standard error of the mean. Asterisks indicate loads with significant group differences.

our estimate of capacity for synesthetes actually *increased* as the load (number of items presented) increased while it *decreased* for nonsynesthetes. We observed a statistically significant interaction between load and group, such that the capacity for nonsynesthetes began to stabilize or decline after a load of four items, whereas the capacity for synesthetes was increasing even at a load of eight items. This finding demonstrated that while nonsynesthetes reach their capacity limit in working memory after four items, the capacity limit for synesthetes is likely beyond the limit imposed in this study. Performance and capacity differences across the two groups were largest at the highest load of eight items, a finding consistent with the results of the first study. Indeed, the effect of load on capacity was dependent on group, such that synesthetes and nonsynesthetes have a different relationship between load and capacity (synesthetes' capacity increases while nonsynesthetes' decreases). This suggests that grapheme-color associations are most beneficial to a synesthete's memory performance when working memory resources are stressed. Further investigations with higher working memory loads are required to test the working memory capacity limits for synesthetes.

In the third study, we replicated a previous long-term memory study[49] to contextualize results from the previous two studies within the existing literature on the impact of grapheme-color synesthesia on long-term memory (Fig. 14.5). By performing a meticulous extended replication of the original Smilek et al.[49] study, we hoped to gain further insight into stimulus congruency effects for synesthetes and their impact on long-term memory. Previous studies that have investigated long-term memory advantages for synesthetes have yielded variable results, particularly regarding the use of incongruent stimuli. Smilek et al.[49] reported that their synesthete, "C", had superior memory when letters were presented

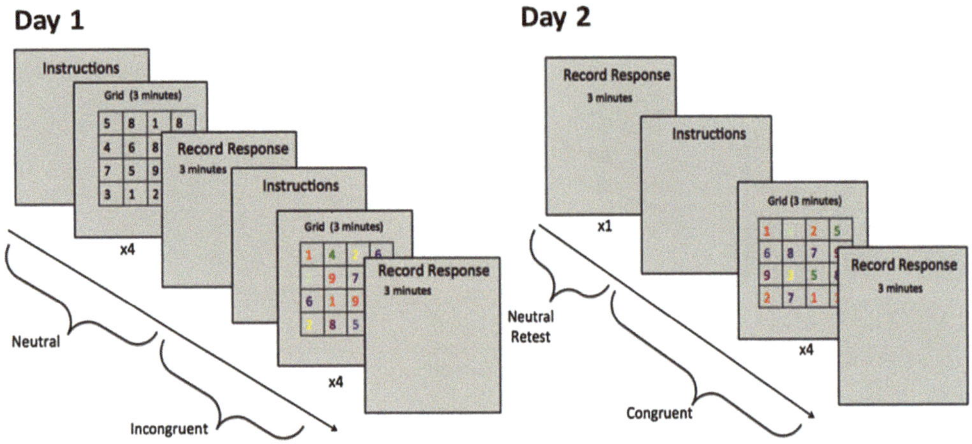

FIGURE 14.5 On Day 1 of testing, participants studied and were tested on the neutral and incongruent grid four times. On Day 2 of testing, participants first were asked to record their recollection of the neutral grid from day 1 of testing. Next, they studied and were tested on the congruent grid four times.

in either congruent or neutral colors, but dramatically impaired memory when letters were presented in incongruent colors. However, as noted by Yaro and Ward,[35] Smilek and colleagues[49] only reported results from the first learning period, when in fact the participants were given four learning periods. C's inability to learn how to use an appropriate strategy on the first attempt could have produced the drastic drop in memory recall when presented with incongruent stimuli. To overcome the shortcomings in the design of the original paradigm, subsequent studies have modified various aspects of the paradigm including the number of items presented (27 digits instead of 50), the number of learning periods (2 instead of 4), the retention period (2–3 weeks instead of 2 days), and the analysis methods (collapsing across learning periods instead of analyzing the first learning period).[35,50]

In our exact replication of the paradigm used by Smilek et al.,[49] participants were presented with a 10×5 grid with 50 randomly generated digits. There were three different matrices presented with neutral, incongruent, and congruent colored stimuli over 2 days of testing. On the first day, participants were presented with the neutral and incongruent colored matrices (four times per matrix), with a 3-minute study and a 3-minute free report period each time. On the second day of testing, the participants were first retested on the matrix that they were presented with on the first day of testing with neutrally colored, black-and-white, stimuli (Neutral Retest). Next, participants were presented with the congruent colored matrix four times, with a 3-minute study and a 3-minute report period each time. Nonsynesthetic controls were presented stimuli with the same color associations as their respective age and gender-matched synesthete.

Our exact replication of the Smilek et al.[49] long-term memory paradigm, with an expanded sample size, leads to a few interesting conclusions (Fig. 14.6). First, we replicated the original Smilek et al.[49] finding that synesthetes have enhanced recall relative to nonsynesthetes for black digits, particularly after a long delay period (Fig. 14.6A). In the neutral condition, we found that the greatest and only statistically significant advantage for synesthetes was at the retest time point, which occurred 48 h after the first session of testing (Fig. 14.6B). Forgetting, as assessed by the drop in accuracy between the last testing point in the first session (N4)

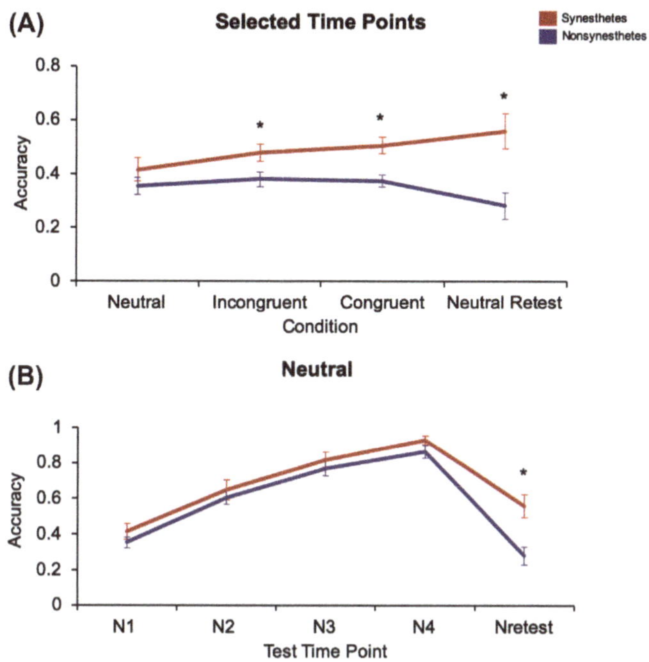

FIGURE 14.6 A comparison of long-term memory accuracies for synesthetes (blue) and nonsynesthetes (red). (A) The time points that were selected for analysis in the Smilek et al. paper. (B) Accuracies at the four test time points of the neutral grid on day 1 of testing (N1–N4), and the one test time point on day 2 of testing (Nretest). Error bars represent standard error of the mean. Asterisks indicate testing time points with significant group differences.

and the retest point, was smaller for synesthetes than for nonsynesthetes. In line with previous findings, this result implies that synesthetes forget less information over a long-term delay period.[29,49,51,52] We did not replicate the original Smilek et al.[49] finding that memory recall for synesthetes was significantly worse than for nonsynesthetes when they were presented with graphemes in colors that were incongruent to their associations. Instead, we found that synesthetes have a nonsignificant numerical advantage in the incongruent condition. While a significant memory advantage was not revealed for synesthetes across all conditions, we did find a numerical advantage for synesthetes across all testing time points in all conditions, a finding in line with previous studies.[21]

Collectively, these three studies paint a much richer picture of the mechanisms of synesthetic memory by testing across all three banks of memory storage in grapheme-color synesthetes and nonsynesthetes.

Load manipulations

These recent studies by our group make a key contribution to the synesthesia literature by introducing load manipulations into the experimental study design. Synesthesia researchers have not systematically varied load, as is commonly done in memory research, to find the parameters under which synesthetic memory advantages appear. Additionally, it is critical to survey the experimental design of studies investigating long-term memory in synesthetes.

To date, there have been several studies that test long-term memory in synesthesia that each utilize a different number of stimuli. However, no other existing studies have systematically varied load as a factor in their experimental paradigms. Indeed, we suggest that this might be an important factor in better understanding the discrepancies in the current literature on synesthesia and memory. Yaro and Ward[35] used a digit and matrix task which presented participants with a grid of 27 numbers from 0 to 9, and manipulated the color congruency of each set. This experiment did not find a significant memory advantage for synesthetes, and did not find an effect of congruence on memory performance. Smilek et al.[49] used a grid of 50 digits and found memory advantages for synesthetes compared to nonsynesthetes in the congruent and neutral conditions but not in the incongruent condition. Radvansky et al.[37] showed a significant memory advantage in synesthetes when stimuli were incongruent with their specific synesthetic associations using lists of words. The differences in findings between the aforementioned papers are consistent with the findings of our studies: Yaro and Ward[35] utilized a smaller load and did not find a significant memory advantage in synesthetes, while Smilek et al.,[49] and Radvansky et al.[37] utilized a larger load and did find an advantage. Similarly, in our studies, we found that the advantage in memory for synesthetes over nonsynesthete controls increased as the load increased.

The dual coding model

One possible explanation for the memory advantage seen in synesthesia relies on the dual coding model of memory.[53]This model proposes that having both a visual and verbal code leads to enhanced memory and learning. These two codes are created at the encoding stage and then can be called upon at the retrieval stage. Such an experience has been evoked in nonsynesthetic individuals by asking participants to associate mental images with meaningful words.[54] Paivio[53] also found that participants had higher recall for concrete words which evoked an image compared to abstract words that did not. Building on this account, Yaro and Ward[35] found a long-term memory advantage for synesthetes compared to nonsynesthetes when presented with words and digits. They suggest that this finding shows that dual coding is an automatic process in grapheme-color synesthesia as a color is automatically and consistently evoked every time a grapheme-color synesthete views a grapheme, whereas nonsynesthetes encode only the grapheme (Fig. 14.7). In this account, the dual coding mechanism produces enhanced memory recall for synesthetes due to their ability to access a greater amount of information about the stimulus via their synesthetic color associations.[29]

Building on this theoretical framework, a recent study[34] has put forth a modified dual coding account, which proposes that synesthetes may exhibit a memory advantage only when the "synesthetically linked information is more memorable than the presented stimuli" (p. 26). This study tested if synesthetic colors enhance serial position recall for structured (ascending and descending) and unstructured color and digits sequences. Synesthetes only showed a benefit for structured color sequences, while nonsynesthetes did not show a performance difference between any conditions. To explain these results, Teichmann and colleagues[34] proposed that synesthetes were only able to excel in memory for the color sequences by using their grapheme-color associations because the secondary information (digits) was more useful for the task than the primary information (presented colors). Interestingly, this recent finding that semantic structure aids memory for colors contradicts Luria's early finding, which demonstrated that S missed the semantic structure for the perceptual

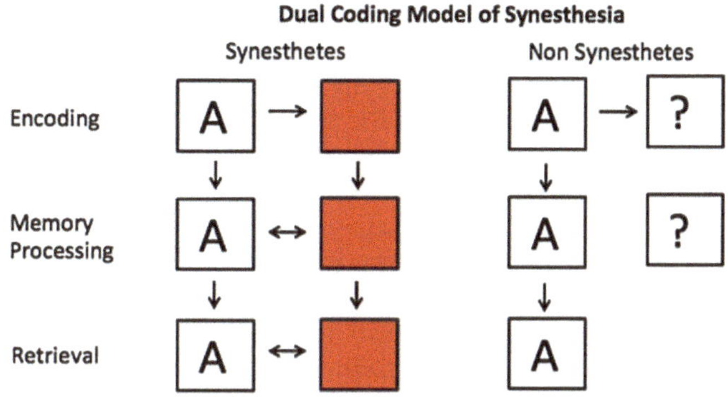

FIGURE 14.7 A schematic of the dual coding theory of synesthetic memory.

structure. This example of inconsistent reports in the literature illustrates the necessity to develop new paradigms and models with the goal of building a greater understanding of the underlying cognitive mechanisms of synesthesia.

The recoding model of memory

To better account for synesthetic memory advantages, we have developed a novel recoding model of memory. Grapheme-color synesthetes are able to encode stimuli in the form of synesthetic colors, while nonsynesthetes cannot. This ability to encode stimuli in multiple modalities may lead to the higher recall accuracy of synesthetes. Support for this dual coding mechanism has been shown in typical individuals as well. While the theoretical account of the dual coding model can partially explain the synesthetic memory advantage by proposing a mechanism for *why* this advantage arises, it cannot fully explicate the results found in this study. The original dual coding model further prompts the question: can this overall advantage in memory be driven only by this automatic, yet passive, dual code? To further understand the mechanisms that are driving this memory advantage for synesthetes across all banks of memory, we propose an extension to the dual coding theory. This modified version builds on the original theory and accounts for the results in this study by explaining *how* a synesthetic advantage can be seen through all three banks of memory. Considering recoding as the active component in the retention of information can start to explain how the presence of an automatic dual code can lead to a sustained memory advantage.

Recoding is a process by which one form of information is translated to another form.[55–57] Some studies have suggested that recoding can indeed have an impact on grapheme-color synesthetes' memory.[37] In a study testing synesthetic memory for words, Radvansky et al.[37] manipulated study material perceptually by using the von Restorff isolation effect.[58] In this manipulation, participants were presented with a list of words, in which only one word was colored while others were in black ink. Consistent with the previously known effect, they found that memory for the colored word was higher in nonsynesthete controls. This, however, was not the case for the synesthetes, who did not exhibit a von Restorff effect: their recall was not significantly better for the colored words than the words in black.

Radvansky and colleagues[37] proposed that this outcome could be attributed to the fact that there are synesthetic colors associated to all the words in the list in addition to the colored printed words. Thus, the advantage present for the colored words in nonsynesthetes was diminished for synesthetes due to the automatic translation of the presented grapheme information into color codes. While there are some negative consequences for the recoding of information, as illustrated above, this discussion focuses on the positives, which could explain the mechanisms behind the memory advantages observed for synesthetes.

Synesthetes are able to easily recode information due to the overlearned nature of their associations between graphemes and colors.[59,60] Overlearned associations are associations that have been practiced to the point of mastery and automaticity.[61,62] Previous studies have suggested that the consistent, automatic, and bidirectional quality of synesthetic associations leads to overlearned associations between graphemes and colors that impact a synesthete's cognitive system.[59,60] Many synesthetes report that they have had consistent color associations for their graphemes (e.g., the color red is associated with the number 1) "as long as [they] can remember".[20,63] Moreover, this consistency is noted for specific synesthesia-inducing stimuli (e.g., graphemes for a grapheme-color synesthete) such that when presented with the same stimuli (e.g., letters, numbers), the same synesthetic experiences are evoked (e.g., colors).[64] Synesthetic associations are automatic; the percept of a grapheme involuntarily evokes the experience of color in a synesthete.[65–68] To test automaticity, researchers have conducted experiments using a synesthetic Stroop test,[69] in which synesthetes have to name the color of a presented letter. The letters presented are either colored congruently (e.g., a blue "B" when the synesthetic color for "B" is blue) or incongruently (e.g., a red "B" when the synesthetic color for "B" is blue) with the synesthete's grapheme-color associations.[70,71] Synesthetes are slower to identify the color of the letter when the letter is presented in a color incongruent with their associations, further supporting the view of grapheme-color synesthesia as automatic in nature.[41,72,73] For many years, researchers believed synesthesia to be a strictly unidirectional phenomenon, and any reports of bidirectionality were overlooked.[74,75] In the context of grapheme-color synesthesia, associations are considered bidirectional if a grapheme triggers the experience of a color (e.g., *A* triggers red), and the same color triggers that same grapheme (e.g., red triggers *A*) for a given synesthete.[76] Notably, synesthesia appears to be unidirectional at the level of conscious experience, such that synesthetes will deny a conscious bidirectional experience when asked, but bidirectionality of associations has been observed at the implicit cognitive level.[77] Basing these overlearned associations on consistent, automatic, and bidirectional synesthetic associations allows synesthetes to recode information in a nonarbitrary code that nonsynesthetes cannot. Building on this automatic dual code, the recoding of presented information serves as the mechanistic vehicle through which synesthetes are able to achieve an advantage over nonsynesthetes in the domain of memory.

While this recoding model accounts for synesthetic advantages across multiple stages of memory and starts to explain how the dual coding model may be driving memory advantages for synesthetes, additional studies are needed to test the predictions that this extended model makes. The extension considers the synesthete's ability to efficiently recode information as an active component based on the overlearned and bidirectional nature of synesthetic associations. Synesthetes are able to use their overlearned associations to effectively transform information and thus remember it with higher fidelity. This model would predict that the ability to recode information might be mediated by the type of stimuli presented. Synesthetes may need to expend more resources to recode information presented in colors that are incongruent with their associations than those that are congruent with their associations. While synesthetes perform with higher accuracy when presented with congruent and incongruent stimuli, the differences in recoding across stimulus types might be observed in the reaction time measures. For example, if a stimulus is presented in a color that is incongruent with the synesthetic association, it may take longer to recode than a stimulus that is presented in a congruent color initially. Additionally, recoding may be less successful for higher memory loads than lower loads. Along with figuring out what the maximum capacity is, future studies that use higher item loads may reveal how synesthetic recoding abilities are affected when memory resources are stressed.

In contrast to nonsynesthetes who rehearse the same codes, the extended model also suggests that synesthetes may be able to restudy the information in a spaced and customized manner by recoding information efficiently. This self-generated restudy can potentially be more useful for synesthetes as it is based on their overlearned associations. Over 100 years of memory research has demonstrated a "spacing effect"—for constant study time, study that is spaced out over multiple time points leads to better retention than study massed into a single study session.[78,79] We suggest that a similar recoding of information happens at multiple time points in synesthetes, leading to a natural spacing effect. We can test this prediction moving forward by conducting additional studies to investigate if synesthetes show a larger spacing effect than nonsynesthetes. If synesthetes outperform nonsynesthetes on a paradigm designed to investigate spacing, it could lend support to the notion that synesthetes may indeed be using active recoding as a mechanism, thereby leading to the observed memory enhancements. Additionally, a varied stimulus set with congruency manipulations, and synesthesia-inducing and noninducing graphemes, can be used to further parse the nature of this effect. In this manner, we can begin to understand the role that recoding may play in the memory processing mechanisms of synesthetes.

Conclusions

In summary, the results of the studies presented in this chapter indicate that synesthetic advantages previously reported in long-term memory may indeed stem from advantages in earlier stages of memory, including iconic and working memory stages. While the recoding model of memory accounts for the results in three studies presented in this chapter and begins to explain how the dual coding model might be driving memory advantages for synesthetes, additional studies are needed to test the predictions that this model makes. This model would predict that the ability to recode information might be mediated by the type of stimuli presented. For this reason, future studies should use various types of nonsynesthesia-inducing stimuli to test memory performance. In this manner, we can begin to understand the role that recoding may play in the memory processing mechanisms of synesthetes.

Researchers have considered synesthesia to be a model system for multisensory integration (MSI) more broadly. The process of MSI combines distributed sensory information to form one, unified sensory message.[80] Studies have also shown that MSI improves performance on a range of tasks, from basic perceptual tasks such as low-level target detection, to a variety of cognitive tasks.[81–84] While researchers have begun investigating whether and how semantically congruent multisensory information automatically enhances long-term object memory in nonsynesthetes (see Chapter 6), these findings leave open a number of other memory-related questions. For example, what are the cognitive mechanisms involved in early sensory memory processes, and how do they differ between nonsynesthetes and synesthetes? Does a similar multisensory enhancement occur in working memory capacity? Furthermore, the ability to integrate information from multiple senses is highly dependent on the properties of the stimulus presented. As Matusz et al. note, current investigations have been confined to the use of semantically meaningful stimuli. However, simple stimuli, such as flashes and beeps, have formed the basis of many experimental studies and have been instrumental in revealing important findings about the organization of basic sensory pathways.[85,86] More recently, studies have started using and advocating for the use of more complex stimuli, which mimic naturally present environmental stimuli and are more semantically meaningful.[87] The use of these different stimuli raises the question: are the sensory modalities linked differently based on stimulus properties? To understand the interaction of cross-sensory associations and cognitive mechanisms at a deeper level, future studies should further investigate the impact of MSI on memory mechanisms in individuals without synesthesia.[83,88]

Finally, the impact of synesthesia on higher cognitive functions has several larger implications, including the potential advantages of synesthetic and multisensory associations in education. A better understanding of mechanisms of synesthesia will improve our understanding of the links between learning and synesthesia, and may provide new insights into beneficial educational methods for all students. Studying the interference patterns of synesthesia could shape new and innovative educational techniques in the future by determining which associations need to be eliminated or prioritized in teaching methods. For example, a teacher could use specific colors to reduce interference with a child's synesthetic associations, or could simply avoid attempts to color code in classrooms with synesthetes. Additionally, teaching synesthetic associations to nonsynesthetes would open up a new realm of questions about the relationship between learning and synesthesia. If the associations made in synesthesia are beneficial for general cognitive functioning and are shaped through explicit learning, it may be practical to implement these concepts into current teaching methods. In this way, all students may reap benefits of this interesting phenomenon.

References

1. Simner J, Hubbard EM. *Oxford Handbook of Synesthesia*. Oxford University Press; 2013.
2. Ward J. Synesthesia. *Annu Rev Psychol*. 2013;64(1):49–75. https://doi.org/10.1146/annurev-psych-113011-143840.
3. Hubbard EM, Ramachandran VS. Neurocognitive mechanisms of synesthesia. *Neuron*. 2005;48(3):509–520. https://doi.org/10.1016/j.neuron.2005.10.012.
4. Rich AN, Mattingley JB. Anomalous perception in synaesthesia: a cognitive neuroscience perspective. *Nat Rev Neurosci*. 2002. https://doi.org/10.1038/nrn702.
5. Simner J, Hubbard EM. Variants of synesthesia interact in cognitive tasks: evidence for implicit associations and late connectivity in cross-talk theories. *Neuroscience*. 2006. https://doi.org/10.1016/j.neuroscience.2006.08.018.

6. Amin M, Olu-Lafe O, Claessen LE, et al. Understanding grapheme personification: a social synaesthesia? *J Neuropsychol*. 2011. https://doi.org/10.1111/j.1748-6653.2011.02016.x.
7. Galton F. Visualised numerals. *Nature*. 1880;21:252–256. https://doi.org/10.1038/021494e0.
8. Seron X, Pesenti M, Noël MP, Deloche G, Cornet JA. Images of numbers, or "when 98 is upper left and 6 sky blue. *Cognition*. 1992. https://doi.org/10.1016/0010-0277(92)90053-K.
9. Sagiv N, Simner J, Collins J, Butterworth B, Ward J. What is the relationship between synaesthesia and visuo-spatial number forms? *Cognition*. 2006. https://doi.org/10.1016/j.cognition.2005.09.004.
10. Baron-Cohen S, Harrison J, Goldstein LH, Wyke M. Coloured speech perception: is synaesthesia what happens when modularity breaks down? *Perception*. 1993;22(4):419–426. https://doi.org/10.1068/p220419.
11. Paulesu E, Harrison J, Baron-cohen S, et al. The physiology of coloured hearing. A PET activation study of colour-word synaesthesia. *Brain*. 1995. https://doi.org/10.1093/brain/118.4.1073.
12. Nunn JA, Gregory LJ, Brammer M, et al. Functional magnetic resonance imaging of synesthesia: activation of V4/V8 by spoken words. *Nat Neurosci*. 2002. https://doi.org/10.1038/nn818.
13. Parmelee DX. DSM-IV made easy: the clinician's guide to diagnosis. *J Child Fam Stud*. 1996. https://doi.org/10.1007/BF02234682.
14. Robertson LC, Sagiv N, Day S. Some demographic and socio-cultural aspects of synesthesia. In: *Synesthesia: Perspectives from Cognitive Neuroscience*. 2005.
15. Simner J. Synesthesia: perspectives from cognitive neuroscience. *Perception*. 2005. https://doi.org/10.1068/p3410rvw.
16. Skodol AE, Clark LA, Bender DS, et al. Proposed changes in personality and personality disorder assessment and diagnosis for DSM-5 part I: description and rationale. *Personal Disord Theory, Res Treat*. 2011. https://doi.org/10.1037/a0021891.
17. Simner J, Carmichael DA, Hubbard EM, Morris Z, Lawrie SM. Rates of white matter hyperintensities compatible with the radiological profile of multiple sclerosis within self-referred synesthete populations. *Neurocase*. 2015. https://doi.org/10.1080/13554794.2014.892625.
18. Carmichael DA, Smees R, Shillcock RC, Simner J. Is there a burden attached to synaesthesia? Health screening of synaesthetes in the general population. *Br J Psychol*. 2018.
19. Baron-Cohen S, Wyke MA, Binnie C. Hearing words and seeing colours: an experimental investigation of a case of synaesthesia. *Perception*. 1987;16(6):761–767. https://doi.org/10.1068/p160761.
20. Cytowic RE. Synesthesia: a union of the senses. *Neurology*. 1989;39(10):1413. https://doi.org/10.1212/WNL.39.10.1413.
21. Ramachandran VS, Hubbard EH. Synaesthesia—a window into perception, thought and language. *J Conscious Stud*. 2001;8(12):3–34. https://doi.org/10.1111/1468-0068.00363.
22. Ramachandran VS, Hubbard EM. Psychophysical investigations into the neural basis of synaesthesia. *Proc R Soc B Biol Sci*. 2001;268(1470):979–983. https://doi.org/10.1098/rspb.2000.1576.
23. Palmeri TJ, Blake R, Marois R, Flanery MA, Whetsell W. The perceptual reality of synesthetic colors. *Proc Natl Acad Sci*. 2002;99(6):4127–4131. https://doi.org/10.1073/pnas.022049399.
24. Kim CY, Blake R. Revisiting the perceptual reality of synesthetic color. In: *Oxford Handbook of Synesthesia*. 2013.
25. Hubbard EM, Brang D, Ramachandran VS. The cross-activation theory at 10. *J Neuropsychol*. 2011;5(2):152–177. https://doi.org/10.1111/j.1748-6653.2011.02014.x.
26. Hubbard EM. Synesthesia and functional imaging. In: *Oxford Handbook of Synesthesia*. 2013.
27. Hancock P. Synesthesia, alphabet books, and fridge magnets. In: *Oxford Handbook of Synesthesia*. 2013:83–99.
28. Wendler E. Leakings, drafts, and magical thinking: synaesthesia, creativity, and obsessive-compulsive disorder - is there a link? *J Transpers Psychol*. 2017.
29. Rothen N, Meier B, Ward J. Enhanced memory ability: insights from synaesthesia. *Neurosci Biobehav Rev*. 2012;36(8):1952–1963. https://doi.org/10.1016/j.neubiorev.2012.05.004.
30. Luria A. *The Mind of a Mnemonist*. 1968.
31. Smilek D, Dixon MJ, Cudahy C, Merikle PM. Synaesthetic photisms influence visual perception. *J Cogn Neurosci*. 2001;13(7):930–936. https://doi.org/10.1162/089892901753165845.
32. Brang D, Ramachandran VS. Visual field heterogeneity, laterality, and eidetic imagery in synesthesia. *Neurocase*. 2010;16(2):169–174. https://doi.org/10.1080/13554790903339645.
33. Meier B, Rothen N. Developing synaesthesia: a primer. *Front Hum Neurosci*. 2015;9(April):2014–2015. https://doi.org/10.3389/fnhum.2015.00211.

34. Teichmann AL, Nieuwenstein MR, Rich AN. Digit-color synaesthesia only enhances memory for colors in a specific context: a new method of duration thresholds to measure serial recall. *J Exp Psychol Hum Percept Perform.* 2017;43(8):1494–1503. https://doi.org/10.1037/xhp0000402.
35. Yaro C, Ward J. Searching for Shereshevskii: what is superior about the memory of synaesthetes? *Q J Exp Psychol.* 2007;60(5):681–695. https://doi.org/10.1080/17470210600785208.
36. Green J a K, Goswami U. Synesthesia and number cognition in children. *Cognition.* 2008;106(January):463–473. https://doi.org/10.1016/j.cognition.2007.01.013.
37. Radvansky GA, Gibson BS, McNerney MW. Synesthesia and memory: color congruency, von Restorff, and false memory effects. *J Exp Psychol Learn Mem Cogn.* 2011;37(1):219–229. https://doi.org/10.1037/a0021329.
38. Banissy MJ, Tester V, Muggleton NG, et al. Synesthesia for color is linked to improved color perception but reduced motion perception. *Psychol Sci.* 2013;24(12):2390–2397. https://doi.org/10.1177/0956797613492424.
39. Terhune DB, Wudarczyk OA, Kochuparampil P, Cohen Kadosh R. Enhanced dimension-specific visual working memory in grapheme-color synesthesia. *Cognition.* 2013;129(1):123–137. https://doi.org/10.1016/j.cognition.2013.06.009.
40. Chiou R, Rich AN. The role of conceptual knowledge in understanding synesthesia: evaluating contemporary findings from a "hub-and-spokes" perspective. *Front Psychol.* 2014. https://doi.org/10.3389/fpsyg.2014.00105.
41. Dixon MJ, Smilek D, Cudahy C, Merikle PM. Five plus two equals yellow. *Nature.* 2000;406(6794):365. https://doi.org/10.1038/35019148.
42. Atkinson RC, Shiffrin RM. Human memory: a proposed system and its control processes. *Psychol Learn Motiv.* 1968;2:89–195.
43. Pfeifer G, Ward J, Chan D, Sigala N. Representational account of memory: insights from aging and synesthesia. *J Cogn Neurosci.* 2016.
44. Bussey TJ, Saksida LM. Memory, perception, and the ventral visual-perirhinal-hippocampal stream: thinking outside of the boxes. *Hippocampus.* 2007;17(9):898–908.
45. Gosavi RS, Hubbard EM. A colorful advantage in iconic memory. *Cognition.* 2019;187:32–37. https://doi.org/10.1016/j.cognition.2019.02.009.
46. Sperling G. The information available in brief visual presentations. *Psychol Monogr Gen Appl.* 1960;74(11):1.
47. Gosavi R.S. and Hubbard E.M., The impact of grapheme-color synesthesia on working memory, in prep.
48. Luck SJ, Vogel EK. The capacity of visual working memory for features and conjunctions. *Nature.* 1997;390(6657):279–284. https://doi.org/10.1038/36846.
49. Smilek D, Dixon MJ, Cudahy C, Merikle PM. Synesthetic color experiences influence memory. *J Cogn Neurosci.* 2001;13(7):930–936.
50. Rothen N, Meier B. Do synesthetes have a general advantage in visual search and episodic memory? A case for group studies. *PLoS One.* 2009;4(4). https://doi.org/10.1371/journal.pone.0005037.
51. Deroy O, Spence C. Why we are not all synesthetes (not even weakly so). *Psychon Bull Rev.* 2013;20(4):643–664. https://doi.org/10.3758/s13423-013-0387-2.
52. Mills CB, Innis J, Westendorf T, Owsianiecki L, McDonald A. Effect of a synesthete's photisms on name recall. *Cortex.* 2006;42(2):155–163. https://doi.org/10.1016/S0010-9452(08)70340-X.
53. Paivio A. Mental imagery in associative learning and memory. *Psychol Rev.* 1969;76(3):241–263. https://doi.org/10.1037/h0027272.
54. Ishai A, Sagi D. Visual imagery facilitates visual perception: psychophysical evidence. *J Cogn Neurosci.* 1997;9(4):476–489.
55. Cruse D, Clifton C. Recoding strategies and the retrieval of information from memory. *Cogn Psychol.* 1973;4(2):157–193. https://doi.org/10.1016/0010-0285(73)90010-8.
56. Loftus GR, Loftus EF. *Human Memory: The Processing of Information.* Psychol Press; 1976.
57. Baddeley A. Working memory. *Science.* 1992;255(ii):556–559. https://doi.org/10.4249/scholarpedia.3015.
58. Von Restorff H. Über die wirkung von bereichsbildungen im spurenfeld. *Psychol Forsch.* 1933;18(1):299–342.
59. Berteletti I, Hubbard EM, Zorzi M. Implicit versus explicit interference effects in a number-color synesthete. *Cortex.* 2010;46(2):170–177. https://doi.org/10.1016/j.cortex.2008.12.009.
60. Newell FN, Mitchell KJ. Multisensory integration and cross-modal learning in synaesthesia: a unifying model. *Neuropsychologia.* 2016;88:140–150. https://doi.org/10.1016/j.neuropsychologia.2015.07.026.
61. Krueger WCF. The effect of overlearning on retention. *J Exp Psychol.* 1929;12(1):71–78. https://doi.org/10.1037/h0072036.
62. Rohrer D, Taylor K, Pashler H, Wixted JT, Cepeda NJ. The effect of overlearning on long-term retention. *Appl Cognit Psychol.* 2005;19(3):361–374. https://doi.org/10.1002/acp.1083.

63. Simner J, Bain AE. A longitudinal study of grapheme-color synesthesia in childhood: 6/7 years to 10/11 years. *Front Hum Neurosci*. 2013;7. https://doi.org/10.3389/fnhum.2013.00603.
64. SVARTDAL F, IVERSEN T. Consistency in synesthetic experience to vowels and consonants: five case studies. *Scand J Psychol*. 1989. https://doi.org/10.1111/j.1467-9450.1989.tb01084.x.
65. Linkovski O, Akiva-Kabiri L, Gertner L, Henik A. Is it for real? Evaluating authenticity of musical pitch-space synesthesia. *Cogn Process*. 2012;13(1 suppl). https://doi.org/10.1007/s10339-012-0498-0.
66. Lupiáñez J, Callejas A. Automatic perception and synaesthesia: evidence from colour and photism naming in a stroop-negative priming task. *Cortex*. 2006;42(2):204–212. https://doi.org/10.1016/S0010-9452(08)70345-9.
67. Mattingley JB. Attention, automaticity, and awareness in synesthesia. *Ann N Y Acad Sci*. 2009;1156:141–167. https://doi.org/10.1111/j.1749-6632.2009.04422.x.
68. Sinke C, Halpern JH, Zedler M, Neufeld J, Emrich HM, Passie T. Genuine and drug-induced synesthesia: a comparison. *Conscious Cognit*. 2012;21(3):1419–1434. https://doi.org/10.1016/j.concog.2012.03.009.
69. Stroop JR. Studies of interference in serial verbal reactions. *J Exp Psychol*. 1935;18(6):643–662. https://doi.org/10.1037/h0054651.
70. Colizoli O, Murre JMJ, Scholte HS, Rouw R. Creating colored letters: familial markers of grapheme–color synesthesia in parietal lobe activation and structure. *J Cogn Neurosci*. 2017;29(7):1–14. https://doi.org/10.1162/jocn_a_01105.
71. Gertner L, Henik A, Reznik D, Cohen Kadosh R. Implications of number-space synesthesia on the automaticity of numerical processing. *Cortex*. 2013;49(5):1352–1362. https://doi.org/10.1016/j.cortex.2012.03.019.
72. Colizoli O, Murre JMJ, Rouw R. Defining (trained) grapheme-color synesthesia. *Front Hum Neurosci*. 2014;8:368. https://www.frontiersin.org/article/10.3389/fnhum.2014.00368.
73. Mattingley JB, Rich AN, Yelland G, Bradshaw JL. Unconscious priming eliminates automatic binding of colour and alphanumeric form in synaesthesia. *Nature*. 2001;410(6828):580–582. https://doi.org/10.1038/35069062.
74. Baron-Cohen S. Is there a normal phase of synaesthesia in development. *Psyche*. 1996;2(27):223–228. http://journalpsyche.org/files/0xaa3f.pdf.
75. Marks LE. On colored-hearing synesthesia: cross-modal translations of sensory dimensions. *Psychol Bull*. 1975;82(3):303–331. https://doi.org/10.1037/0033-2909.82.3.303.
76. Cohen-Kadosh R, Henik A. Numbers, synesthesia, and directionality. In: *Oxford Handbook of Synesthesia*. 2013.
77. Kadosh RC, Sagiv N, Linden DEJ, Robertson LC, Elinger G, Henik A. When blue is larger than red: colors influence numerical cognition in synesthesia. *J Cogn Neurosci*. 2005. https://doi.org/10.1162/089892905774589181.
78. Carpenter SK, Cepeda NJ, Rohrer D, Kang SH, Pashler H. Using spacing to enhance diverse forms of learning: review of recent research and implications for instruction. *Educ Psychol Rev*. 2012;24(3):369–378.
79. Cepeda NJ, Vul E, Rohrer D, Wixted JT, Pashler H. Spacing effects in learning: a temporal ridgeline of optimal retention: research article. *Psychol Sci*. 2008. https://doi.org/10.1111/j.1467-9280.2008.02209.x.
80. Calvert G, Spence C, Stein BE. *The Handbook of Multisensory Processes*. In: *Handb Multisensory Process*. 2004:933 (nicht verfügbar?).
81. Botta F, Santangelo V, Raffone A, Sanabria D, Lupiáñez J, Belardinelli MO. Multisensory integration affects visuo-spatial working memory. *J Exp Psychol Hum Percept Perform*. 2011;37(4):1099–1109. https://doi.org/10.1037/a0023513.
82. Cichy RM, Teng S. Resolving the neural dynamics of visual and auditory scene processing in the human brain: a methodological approach. *Philos Trans R Soc B Biol Sci*. 2017;372(1714):20160108. https://doi.org/10.1098/rstb.2016.0108.
83. Quak M, London RE, Talsma D. A multisensory perspective of working memory. *Front Hum Neurosci*. 2015;9. https://doi.org/10.3389/fnhum.2015.00197.
84. Stein BE, Meredith MA. *The Merging Senses*. 1993.
85. Hartline HK. The receptive fields of optic nerve fibers. *Am J Physiol*. 1940;130:690–699.
86. Hubel DH, Wiesel TN. Receptive fields, binocular interaction and functional architecture in the cat's visual cortex. *J Physiol*. 1962;160(1):106–154. https://doi.org/10.1113/jphysiol.1962.sp006837.
87. Felsen G, Dan Y. A natural approach to studying vision. *Nat Neurosci*. 2005;8(12):1643–1646. https://doi.org/10.1038/nn1608.
88. Lehmann S, Murray MM. The role of multisensory memories in unisensory object discrimination. *Cogn Brain Res*. 2005. https://doi.org/10.1016/j.cogbrainres.2005.02.005.

SECTION III

Clinical applications

CHAPTER

15

Task-selectivity in the sensory deprived brain and sensory substitution approaches for clinical practice: evidence from blindness

Benedetta Heimler[1], *Amir Amedi*[1,2]

[1]**Department of Medical Neurobiology, Institute for Medical Research Israel-Canada, Faculty of Medicine, Hebrew University of Jerusalem, Hadassah Ein-Kerem, Jerusalem, Israel;**
[2]**Department of Cognitive Science, Faculty of Humanities, Hebrew University of Jerusalem, Hadassah Ein-Kerem, Jerusalem, Israel**

Introduction

To what extent is the human brain already defined at birth, and alternatively, how much does the brain change in response to cognitive and sensory experiences? This question has inflamed philosophical debate for centuries, and even today, is still one of the most crucial questions in cognitive neuroscience. In general terms, the most accepted notion that has prevailed over decades is that functional brain specializations arise from evolutionary programming that developed through natural selection. This conclusion was supported by the repeatedly observed anatomical consistency of brain specializations across individuals, not only concerning the division of sensory labor (e.g., the division of the brain into visual, auditory, or somatosensory regions) but even within specific sensory cortices (e.g., the division in the visual cortex between retinotopic mapping in early visual cortices and selectivity for specific visual categories such as faces or body images in higher-order visual cortices).

But then, within the natural selection framework, what is the role of cognitive and sensory experiences? Do they play a role only for evolutionary purposes, i.e., on a time span of thousands of years, or do they also affect brain specializations in the time span of the life of an organism? For many decades, the main assumption in this matter was that cognitive and sensory experiences do play a crucial role during human life, but their role is strongly

Multisensory Perception
https://doi.org/10.1016/B978-0-12-812492-5.00015-2

constrained by the closure of critical/sensitive periods of development (i.e., an optimal time window in early infancy when the development of a particular sensory system should be pursued in order for the corresponding brain specializations to develop in a typical manner[1]). In other words, the influence of sensory experiences was classically considered to be very minimal during adulthood: if a specific sensory system did not develop during critical periods early in life, it would never properly develop to a level comparable to the control population.[1–7] These conclusions strongly rely on the seminal work by Hubel and Wiesel on kittens visually deprived in one eye at different times after birth and for different time periods.[2] Specifically, Hubel and Wiesel showed that even short periods of monocular deprivation of a few days permanently affected cortical physiology. They also showed that after 3 months of monocular deprivation, deficits in cortical physiology and atrophy in crucial regions of the visual system persisted for years, even though, behaviorally, vision partially recovered in the deprived eye.[2] Later studies on patients who recovered vision during adulthood corroborated these seminal findings by further documenting the lack of a proper development of visual-related brain specializations, even after years of therapy aimed at visual restoration.[8–12]

However, this classical view has been called into question during the last decades, due to accumulating evidence highlighting that the brain still retains a considerable amount of plasticity during adulthood that can be triggered via specific training regimens.[13–18] Numerous studies of this kind showed the remarkable benefits of specific training for the efficient (re-) wiring of the brain in several domains: for the recovery of higher-order abilities such as language, memory, and executive functions mainly in the aging brain[19–24]; for the improvement of specific sets of sensory/cognitive abilities either as a consequence of specific acquired expertise (e.g., in musicians[25–30]) or as a consequence of partial lesions[31,32] or of lesions occurring during adulthood.[33,34] This body of studies then unraveled the lifelong capacity for plasticity in the brain and the crucial role of specific training programs for efficiently triggering it. However, from this perspective, it is still posited that each specific ability/sensory-specific function must be at least partially experienced during critical periods early in development in order for the corresponding brain specialization to emerge (and later in life to be further modifiable by training). In other words, the currently accepted view still postulates that sensory brain regions are strictly sensory-specific in nature (i.e., determined by evolution), and that there is an unalterable link between a given brain sensory region and a sensory-specific computation/cognitive task which must be established during critical periods via sensory-specific experiences.

In this chapter, we will review a series of studies conducted in the last few decades with sensory-deprived adults and specifically, with congenitally blind adults, challenging the classic assumptions on the factors driving sensory brain organization in relation to critical periods of development (see companion chapter by Lomber et al., this volume, for a comprehensive review of studies on congenital and acquired deafness). We will focus in particular on the (re)-organization occurring within higher-order visual cortices deprived of their natural sensory input (vision), describing the mechanisms we propose as underlying the emergence of their organization and redefining the assumptions regarding critical periods based on the available results. We will then discuss the implications of these results and novel frameworks for rehabilitative approaches to sensory restoration with a special emphasis on sensory substitution devices (SSDs) and the role of specific training for reshaping/rewiring brain sensory functions. Before diving into this exciting literature, we will introduce the concept of sensory substitution, as we will refer to this throughout the chapter.

Sensory substitution devices

SSDs aim at conveying the information typically delivered by one sensory modality (e.g., vision) to another sensory modality (e.g., audition or touch) via predetermined algorithms that can be learned by the users through specific training programs.[35–38] At first, the concept of sensory substitution seems very intuitive. For instance, everybody relies on different sensory modalities when visual information is unavailable: we rely on touch when searching for our wallet in our backpack and we rely on our audition to know, for instance, if there are people in a room when the door is closed. What differentiates SSDs from these other strategies is the structured training that users need to undertake in order to learn to interpret the SSD-specific algorithm. The first structured sensory substitution system was probably Braille reading. This technique was originally developed at the beginning of the 19th century by Barbier as a means of writing and reading in the dark for the French military in the Napoleonic era, and then revised by Louis Braille to enable the blind to read by substituting visual letters with tactile dot patterns coding for the letters.[39] This approach was further enhanced in the early 1950s with the development of automatic text-to-braille converters such as the Optacon.[40] A highly interesting effort which is often neglected historically, was the Elektroftalm that attempted for the first time to convey composite visual information, i.e., to electronically transform a visual image into auditory (late 1890s) and tactile (1950s) stimulation using one or several sensors.[41] These early attempts led to the more organized and methodologically sound attempts in the 1970s, by Paul Bach-y-Rita, who is considered the pioneer of the extensive use of sensory substitution for research aimed at visual rehabilitation for the blind population. Bach-y-Rita built his framework for SSD research based on the accepted idea that visual perception mainly takes place in our brain rather than in our eyes.[42] He points out that visual information travels from the retina to the brain in the form of electrical and chemical pulses, and it is the brain that interprets this information as vision.[42] The perception of an image requires much more from the brain than a simple image analysis. Visual perception is based on memory, learning, and interpretation of contextual, cultural, and social factors.[42–44] Thus, he suggests that SSDs are the perfect tools to unravel the extent to which the eyes are essential to vision and the ears are essential to audition.[42] This research question might seem absurd, but it is indeed the problem posed by sensory substitution. Can Braille reading be qualified as vision? Or is it rather a tactile experience that replaces vision?

The biggest challenge that Bach-y-Rita needed to face when building his SSD concerned the choice of the sensory modality he would be using to convey visual information to blind users. Indeed vision has special properties, such as its high informational capacity (or bandwidth), and its capacity for parallel processing, that exceeds those of the other senses.[45] Bach-y-Rita chose to use touch to convey complex visual images, probably because of the possibility to convey tactile information in parallel and the quite obvious skin/retina analogue related to the topographic representation of spatial locations on both sensory organs.[46] The visual-to-tactile SSD that Bach-y-Rita developed is known as "Tactile Vision

Sensory Substitution" (TVSS).[36] He used a camera to capture images and then transmitted them to an electrode grid positioned on the backs of the users, to stimulate cutaneous receptors. Case studies conducted in Bach-y-Rita's laboratory demonstrated that after extensive training, congenitally blind participants were able to make judgments of distance, grab objects in motion, and even recognize novel objects. Later, this device was adapted to stimulate the tongue. The reason for the choice of the tongue rather than the skin of the back is twofold: firstly, the tongue is embedded in a wet milieu, making possible to use much safer microcurrents for stimulation, and secondly the tongue is a much more sensitive sensory organ with a much higher density of receptors than the back, thus providing much better spatial resolution. However, most research with SSDs, nowadays, relies on the auditory rather than on the tactile modality to convey the missing visual information. This is because users of visual-to-auditory SSD do not need specific materials (i.e., vibrators) to be able to use the device (Fig. 15.1), thus making it a more convenient and easy-to-disseminate approach. In addition, from a more theoretical perspective, the auditory system, compared to the skin,

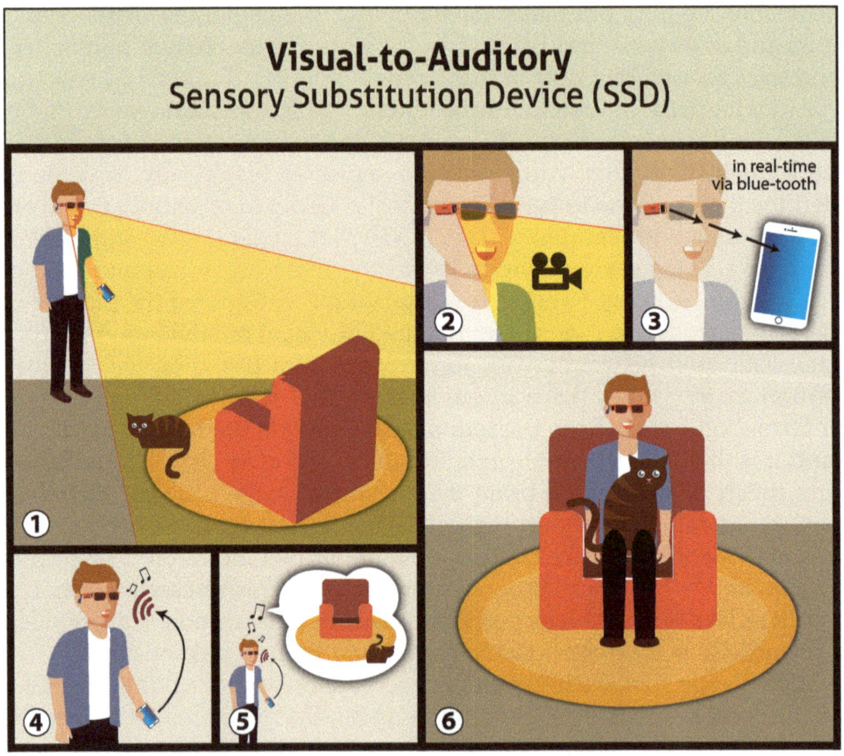

FIGURE 15.1 Current visual-to-auditory sensory substitution devices (SSDs). (1–3) Visual images are captured via a camera which is connected via Bluetooth to an app on the smartphone. (4) The app contains the SSD algorithm which transforms the visual image into an auditory soundscape maintaining all the basic features of the visual scene (i.e., shape, size, color, and location of objects). (5–6) After specific training, the users learn to interpret the SSD algorithm and are able to understand the scene in front of them and efficiently and independently interact with the external environment.

provides a higher informational capacity (bandwidth) to convey visual information to the brain.[46] Indeed, the informational capacity of the human eye has been estimated to be around 4.3×10^6 bits per second.[47] This is four orders of magnitude greater bandwidth than the estimated bandwidth of the human fingertip—100 bits per second,[48] and of other areas of skin for which the bandwidth estimated was even lower, from 5 to 56 bits per second.[49] The information capacity of the human ear is the highest after vision, with a capacity of 10^4 bits per second.[50] In addition, audition has a much wider spectrum than vibration. These factors, in turn, allow the more rapid presentation of more detailed visual images in audition compared to touch. Examples of visual-to-auditory SSDs are the VOiCe[38] or the EyeMusic,[35] which transform visual images into what are called auditory soundscapes maintaining all the basic features of visual stimuli in a scene, such as their shapes and exact spatial locations (Fig. 15.1). Training with visual-to-auditory SSDs was shown to be effective in teaching blind users to perform a variety of "visual" tasks such as object recognition and localization,[51,52] as well as navigation in real and virtual environments,[53,54] among many other tasks[39,55] (Fig. 15.1). Additionally, visual-to-auditory SSDs have been successfully used for teaching inherently visual concepts to blind users such as color perception[35] or visual parsing[56] and were also proven effective to permit users to perform visual acuity tests at a level above the threshold for legal blindness.[57,58]

In this section, we provided an overview on the considerations behind the invention of SSDs and described the main results obtained in the context of visual rehabilitation. Throughout the rest of the chapter, we will discuss other uses of SSDs, for example, for uncovering the properties of our sensory brain organization and for maximizing sensory restoration outcomes.

Crossmodal plasticity in cases of sensory deprivation

The term crossmodal plasticity generally refers to the recruitment of a deprived region of sensory cortex (e.g., the visual cortex in case of blindness or the auditory cortex in case of deafness) by the intact sensory modalities (e.g., audition or vision). This notion emerged from seminal studies reporting high metabolic/electroencephalographic (EEG) activity in the deprived sensory cortices of adults who became either blind or deaf early in life. The results suggested that deprived cortices in blind adults were not silent but were activated by a variety of tactile and auditory stimuli.[59–64] Similar results were also obtained in the deprived auditory cortices of deaf adults by visual inputs.[65–68] These intriguing results prompted a series of studies aiming at investigating the organizational properties of these crossmodal activations which will be discussed in the next sections.

Task-selective sensory-independent organization in the deprived higher-order "visual" cortices

Perhaps the most groundbreaking result in the past decades regarding the properties of crossmodal activations in the deprived sensory cortices is the finding that most of the known category-selective regions in the deprived higher-order visual cortices maintain their category-selective functionality (e.g., to process objects, letters or numbers), albeit recruited by the spared sensory modalities (task-selective sensory-independent [TSSI] recruitment; see for reviews[69,70]). These results have been obtained in studies using functional magnetic resonance imaging (fMRI), conducted mainly with congenitally blind participants. Specifically, this body of studies reported TSSI recruitment by auditory and/or tactile inputs, respecting the broad division of labor between ventral (what) and dorsal (where) "visual" pathways[71] as well as specific category-selective specializations in both "visual" pathways, such as spatial localization[72,73]; motion detection[74]; tool and object perception[75]; reading[52,76]; number identification[77]; and body image perception.[78] This body of work ultimately suggests that the brain is organized along the lines of flexible task machinery, rather than sensory-specific machinery as classically conceived.[79] In addition, several of these studies documented, in congenitally blind participants who showed TSSI recruitment in their "visual" cortices, the preservation of functional connectivity patterns between specific category-selective "visual" regions and the other brain regions known to be involved in the same computation in the sighted brain.[77,78] For instance, it has been shown that number identification via audition recruited, in congenitally blind participants, the right number form area (NFA) in the ventral "visual" stream.[77] Additional analyses showed that these same participants had preserved functional connections between their NFA and other nodes that are known to be part of the numeral processing network in the control population, such as the right intraparietal sulcus.[77] Importantly, some of the most convincing results among the studies discussed above, including the latter work,[77] were obtained using visual-to-auditory SSDs. What is unique about using SSDs to investigate brain organization is that, during SSD training programs, users learn a new sensory pairing between a "visual" category (e.g., body shape recognition) and a sensory modality (e.g., audition) which had never been used to perform this specific task before.[52,75,77,78] Thus, the finding that TSSI recruitment of "visual" category-selective regions emerged in congenitally blind adults following relatively short SSD training showed that higher-order "visual" regions are incredibly flexible for the activating sensory modality. Indeed, the SSD training programs implemented in the studies reviewed above, reporting TSSI recruitment in the congenitally blind brain, lasted between 10 and 50 hours.[52,75,77,78] This observed flexibility refutes the idea, that in order for typical specializations to emerge, the pairing between a specific computation and a specific sensory input absolutely needs to take place during critical periods of development early in life.

Overall then, the classic account of the brain as strictly sensory-specific in nature cannot explain the set of findings documenting TSSI recruitment of deprived visual cortices in congenitally blind adults. Thus, if sensory-specific input does not drive the emergence of our sensory brain organization, what are the mechanisms underlying such organization? Based on all the available results, it was recently suggested that TSSI organization arises from a combination of two principles that are not mutually exclusive: a sensitivity to task-distinctive features that is invariant to the input sensory modality (e.g., body shape perception in the extrastriate body area [EBA] independent of the sensory modality used as input) and the preservation of large-scale anatomical and functional (partly innate?) connectivity patterns (e.g., the connections between EBA and all the other brain regions associated with body shape perception[80–82]).

Does task-selective and sensory-independent organization extend to higher-order auditory regions as well?

Is TSSI recruitment a general principle of brain (re)-organization, or is it specific to the (re)-organization of visual cortices? Unfortunately, when it comes to other deprived sensory cortices (e.g., auditory or somatosensory), evidence for TSSI recruitment is more limited compared to the results obtained with the congenitally blind population. Nonetheless, accumulating evidence from the deaf population supports the findings documented for the blind population, ultimately suggesting that TSSI organization extends beyond visual regions. In the case of congenital deafness, the most elegant evidence regarding the presence of TSSI recruitment of the deprived auditory cortices comes from a series of studies by Lomber's group with congenitally deaf cats showing that deprived higher-order auditory regions maintained their typical computations, albeit being activated by vision[83–85](see for details the chapter by Lomber et al. in this volume). Evidence in favor of TSSI recruitment in the deprived higher-order "auditory" cortices of deaf humans is less straightforward but is also beginning to accumulate. Indeed, for a long time, the only clear result supporting the task-selective recruitment of deprived auditory cortices in congenitally deaf adults concerned the processing of sign language.[86] It has been repeatedly shown with fMRI that sign language in early or native deaf signers recruits the same auditory regions typically recruited by spoken language both during sign production tasks[87–92] and during sign language comprehension tasks.[91,93–95] These neuroimaging results are corroborated by neuropsychological evidence on selective sign language impairments in deaf adults as a consequence of damage to the left auditory cortex (i.e., the cortex in which spoken language is processed), whereas sign language skills were unaltered after right hemisphere auditory damage.[96–99] All these results suggest that the language network maintains its distinctive large-scale properties independently of the sensory modality used as input (audition or vision). Importantly, these results also provide initial evidence suggesting that the two principles proposed above as underlying TSSI organization in the deprived visual brain (i.e., sensitivity to task-distinctive features and preserved connectivity[82]) might also extend to higher-order "auditory" regions in humans.[80] However, to confirm that these are general principles driving the emergence of the organization of higher-order sensory cortices, more evidence is needed. Apart from language-related activations, evidence for TSSI recruitment in the deprived human auditory cortex started to arise only recently. One EEG study suggests the presence of sensory-independent task-selectivity in early deaf adults for the automatic detection of changes in the environment,[100] a skill that has been primarily ascribed to the auditory system.[101,102] In this study, early deaf and normally hearing adults were tested in a visual mismatch negativity (vMMN) task. vMMN is a well-known electrophysiological marker of sensory expectancy, which is considered to underlie the automatic detection of visual changes in the environment.[103] Source-estimate localization analyses revealed that the early deaf adults, but not adults with normal hearing, recruited their auditory cortices when automatically detecting visual changes.[100] These results provide initial hints in favor of the maintenance of an automatic change-detection functionality within the deprived auditory cortex of early deaf adults.[100] However, this study does not provide conclusive evidence for task selectivity since electrophysiology has rather poor spatial localization and source estimates may not be entirely reliable. In addition, the authors of this study[100] did not test the auditory counterpart of the task, by asking hearing participants to automatically detect auditory changes and then comparing the resulting auditory MMN source estimate with the one reported for vMMN in the deaf population.

Recently, two conclusive pieces of evidence for sensory-independent task selectivity during perceptual tasks in deaf humans were put forward for visual rhythm sequence perception[104] and for facial identification.[105] Bola and colleagues[104] documented auditory cortex recruitment in congenitally deaf and hearing adults when discriminating visual or auditory rhythm sequences, respectively. In both sensory modalities (vision, audition), the activation for perception of rhythms peaked in the posterior and lateral part of the high-level auditory cortex, i.e., in the same anatomic (auditory) region independently of the sensory modality used as input. Similarly, Benetti and colleagues[105] showed that the region in the auditory cortex which in the control population responds to voice perception is involved in face identity processing in congenitally deaf adults (especially in the right hemisphere). Importantly, the same group also showed, in the same deaf participants, largely preserved connectivity patterns between this task-selective temporal region and occipital regions.[106] This latter result provides corroborative evidence suggesting that indeed preserved connectivity together with sensitivity to task-distinctive features[82] might underlie TSSI organization in higher-order "auditory" regions,[80] as in visual cortical regions (see above). Unfortunately, studies with the deaf population using SSDs to train the auditory-deprived brain to perceive auditory information through visual or tactile sensory channels are still missing. Implementing this approach, however, could allow investigation of the interesting question of whether the flexibility for the sensory modality triggering TSSI recruitment reported for category-selective regions in the deprived visual cortices also extends to the deprived auditory cortices.

Taken together, this body of works suggests that TSSI organization is a general principle characterizing the organization of higher-order sensory cortices, extending beyond visual cortices.

Does TSSI organization extend to deprived primary sensory cortices as well?

Unfortunately, data regarding the extent to which TSSI organization extends to primary sensory cortices are still quite controversial and not conclusive (for a review see[82]). One of the reasons behind the disparity of findings between primary sensory and higher-order cortices is that while there were clear hypotheses regarding the properties of TSSI recruitment in category-selective regions in higher-order sensory cortices, the hypotheses related to TSSI organization in primary sensory cortices appeared weak. Indeed, primary sensory cortices are the first relay of sensory information in the cerebral cortex and are known to compute basic analyses of sensory features. Thus, among all cortical regions, they are considered the most sensory-specific regions. What sensory-independent and task-selective computation could they perform if deprived of their natural sensory inputs? We propose, that in order to test whether TSSI organization can emerge in these cortices, instead of focusing on specific computations, one must focus on the main and large-scale organizational principle of primary sensory cortices, namely topographic mapping (e.g., retinotopy or tonotopy for visual and auditory primary sensory cortices, respectively). Is topographic mapping, or at least broad topographic division, maintained in the deprived primary sensory cortices? Very interestingly, recent studies demonstrated the maintenance of the large-scale functional connectivity patterns characterizing retinotopic and tonotopic biases in the congenitally blind[107] and congenitally deaf,[108] respectively. However, so far, the functional meaning of these preserved anatomical connections remains unknown. Crucially, these results are quite puzzling for the scientific community, as available results showing crossmodal recruitment of the deprived early sensory cortices, and mainly of the deprived early visual cortices, never hinted at any preserved functional topographic maps. Actually, accumulating evidence suggests "task-switching" in the deprived primary visual cortex toward higher cognitive functions such as language, verbal and episodic memory or numerical cognition,[109–117] focused attention,[118] and executive control.[119] These results are generally described as dramatically diverging from the predictions of TSSI brain organization. This is because such functions do not typically recruit early visual areas in sighted individuals (but see[120]), are not sensory in nature, and are not organized topographically. However, we suggest that functional topographic organizations might emerge independently of the input used, if the information provided carries core "retinotopic" features. This means, for example, that the known eccentricity bias characterizing classic retinotopic mapping might be conceptualized as a TSSI high versus low shape resolution bias. This would predict, for instance, the activation of foveally responsive regions for Braille reading (a task requiring high-resolution shape analyses) in the deprived primary visual cortex.[80] Interestingly, initial support for this prediction comes from the results obtained in the case study of patient S who experienced severe visual acuity reduction due to corneal opacification from the age of six years.[121] Using fMRI, the authors observed that in patient S classic foveally driven regions were recruited by Braille letters, while classic peripherally responsive regions were active during visual processing.[121] Given the low acuity of vision in patient S, this case study suggests, in line with our hypothesis, that the eccentricity bias may indeed be conceptualized as a sensory-independent high versus low shape resolution bias. However, these results were obtained in one participant only, who underwent normal visual development during critical periods. Future studies may further test these intriguing questions in congenitally blind participants, ultimately unraveling whether the whole brain is organized in a sensory-independent and task-selective manner, or if alternatively, there are indeed some constraints in the human brain with respect to specific sensory inputs.

Task-switching versus TSSI organization in higher-order "visual" cortices

It is important to note that many of the studies reporting task-switching plasticity toward higher-order cognitive tasks, such as verbal memory, semantic and syntactic processing of language, or mathematical reasoning, in the early visual cortices of congenitally blind adults, also reported extensive crossmodal recruitment for these tasks beyond these early regions, across higher-order "visual" regions.[109–112,116] These are the same regions for which TSSI recruitment has been shown. Then, how can these divergent findings be integrated together into a unified framework on sensory reorganization following blindness? Unfortunately, there are not many studies addressing this crucial question. A recent investigation by Kim et al.[122] tested the (re)-organization properties of the visual word form area (VWFA), a region in the "ventral" visual stream repeatedly described as TSSI and responsive to symbol-to-phoneme conversions[52,76,80] but that was also shown to be recruited by less specific linguistic tasks.[109–112] Specifically, Kim and colleagues showed that the VWFA was responsive to both Braille letters and the grammatical complexity of auditory sentences in congenitally blind adults, whereas in sighted adults it was activated only during reading of print and not auditory sentences.[122] The authors interpreted these results as evidence suggesting that the deprived visual cortex lost its selectivity to specific computations, supporting Bedny's proposal that the deprived visual cortex is pluripotent with the ability to take over a wide range of functions.[115] In other words, Bedny proposes that brain specializations are constrained neither to a specific sensory modality (i.e., the natural selection account) nor to specific sensory-independent computations (i.e., the TSSI account), but rather that they are only constrained by preexistent connectivity patterns and by experience during critical periods early in development.[115] Recently, she further refined her proposal by suggesting that the strongest weight to cortical repurposing is provided by experiences during critical periods rather than by connectivity biases.[123] Specifically, Kanjlia and colleagues tested congenitally blind, late blind, and sighted controls in mathematics and language-related tasks manipulating cognitive load (i.e., all tasks that have been shown to recruit the deprived visual cortex).[123] The authors also acquired resting-state data on the same participants.[123] Their results indicated that, while resting-state functional connectivity between the deprived visual cortex and the rest of the brain was similar in the two blind groups, regional specialization for mathematics and language as well as load-dependent activity across the deprived visual cortices was observed only in congenital blindness.[123] The authors concluded that there are critical periods for the repurposing of the visual pluripotent cortex, i.e., that experiences early in development play a crucial role in determining the properties of cortical specializations.[123]

However, there are numerous studies showing TSSI organization in the deprived higher-order visual cortices.[80] Thus, which of the two organizational principles, namely TSSI or task-switching, is more dominant in shaping the organization of these cortices? A recent study from our laboratory provides initial results in answering this crucial question.[124] Similarly to the work by Kim and colleagues,[122] Sigalev et al. used fMRI to examine the (re)-organization properties of VWFA in congenitally blind,[124]after training on reading letters via an SSD. After SSD training, in congenitally blind participants, the VWFA responded only to SSD-presented words and not during an auditory semantic task.[124] These results suggest that, with the appropriate training, TSSI organization may overcome task-switching plasticity.[124] These findings are not conclusive in this matter as the authors did not test VWFA recruitment by semantics before the SSD training. Nonetheless, this study suggests the interesting working hypothesis that there might be indeed some predispositions to specific sensory-independent computations in the higher-order visual cortices as suggested by the TSSI account for brain (re)-organization. Furthermore, it suggests that such predispositions might be somewhat (re)-awakened or strengthened by task-specific training, even if such training is relatively short compared to the lifelong experience following task-switching in a given region, and even if the training is undertaken during adulthood. Future studies could investigate this issue more systematically, for instance, by performing longitudinal studies during which blind participants are scanned in both task-switching and TSSI-eliciting tasks before and after SSD task-specific learning.

Beyond the notion of strictly sensory-specific critical periods

All the aforementioned results on TSSI organization highlight that experiences in a specific sensory modality are not essential for the related typical brain specializations to emerge. Indeed, several studies provided robust evidence in this direction by showing, for instance, that "visual" or "auditory" cortices can develop their typical category-selectivity specializations while being activated by an atypical sensory modality. Studies with SSDs further showed the incredible flexibility for the sensory modality inducing TSSI recruitment[52,75,77,78] (for more details see above). Then, how do these results reconcile with the dominant view on the emergence of sensory brain specializations positing an unalterable link between a specific specialized sensory region in the brain and a given sensory-specific function/computation that must be established early in development via sensory-specific experiences? What we would like to stress here is that this classic framework never entirely took into account that every cognitive function/computation has its own specific critical period of development.[125] In other words, we do not deny the existence of sensory-specific critical periods or that they play an important role in cortical development. Rather, we argue that, to achieve the full development of a given cortical area, two types of processes must occur during different critical periods: (1) the development of neural connections underlying proper sensory processing, during sensory critical periods and (2) the development of cognitive/computational units typical of that cortical area, during functional critical periods (e.g., language; object recognition, etc.; see also[70,82,126]). Within this framework, TSSI phenomena may be conceived as possible evidence in favor of the efficient development of a given cognitive/computational unit within its corresponding functional critical period, despite the different sensory modality tuning of that particular unit compared to the control population (see also[39]). Thus, available data on TSSI recruitment, and especially the TSSI results obtained with SSDs, suggest that the two types of critical periods might be independent from each other. Future studies could investigate whether one of the two types of critical periods is more predominant than the other in shaping the development of a given cortical area and related cortical network. Future studies may also investigate how critical periods and their related constraints interact with the proposed principles underlying sensory brain organization (i.e., task-selective distinctive features and preserved connectivity biases[82]). We think the best model to investigate these crucial questions is sensory restoration. In the following section, we will briefly discuss our proposed approach in this direction and we will put forward how we think the results reviewed in this chapter can maximize the outcomes of sensory restoration programs.

Specific multisensory training as a tool to maximize sensory restoration outcomes

The repeated findings of TSSI recruitment in the deprived sensory cortices together with findings showing TSSI recruitment after short-term training with SSDs (e.g.,[75,78]) pose crucial questions for sensory restoration. Indeed, if sensory cortices still maintain their typical computations while being so flexible for their activating sensory modality, can TSSI brain organization be exploited to maximize sensory restoration outcomes? Unfortunately, clear results from studies testing this crucial question are still lacking. In our opinion, this is due to the fact that the great majority of the evidence currently available on the efficacy of sensory restoration outcomes has been provided by studies on auditory restoration through cochlear implantation (see chapter by Lomber et al., this volume). Cochlear implants (CIs) are the most established invasive procedure for sensory restoration.[127] However, seminal studies with CI patients reported that patients with poor spoken language recovery had, prior to the surgery, high metabolic activity in response to visual stimulation in their deprived auditory cortex.[128,129] This, in turn, led to the dogma within sensory restoration practices that crossmodal recruitment in the deprived sensory cortex is a negative predictor of sensory restoration.[130] As a result, clinicians suggested that patients avoid, for instance, the learning of a sign language prior to the CI surgery, and the postsurgery training was encouraged to be undertaken in audition alone.[126] We propose, instead, based on results documenting TSSI recruitment, that training within sensory restoration programs should focus on specific computations and use a multisensory approach where the newly restored sensory modality is paired with a familiar and spared one (see also[70,82]).

Importantly, in line with our proposal, in the CI-related literature, evidence has started to document the higher efficacy of multisensory training programs (e.g., audiovisual) compared to unisensory ones (i.e., auditory only) for recovery of function on specific auditory tasks. For instance, exposure to audiovisual language rehabilitative training (speech-reading therapy, pairing sign language with spoken language) substantially improves auditory linguistic recovery compared to auditory-only training in CI patients.[70,126,131,132] Furthermore, a recent study showed that learning sign language boosts auditory linguistic recovery in early implanted deaf children.[133] Thus, this latter result suggests that the presence of TSSI recruitment (e.g., the recruitment of the classical language network by atypical sensory information, i.e., by vision instead of audition in the case of sign language[89,91]) can facilitate rather than impede the recovery of a given cognitive task/computation by the restored sensory modality.[82,126] Interestingly, recent evidence highlights the incredible potential of multisensory training for maximizing sensory restoration outcomes even for cognitive tasks/computations that most probably were not even learned during infancy.[134] Isaiah and colleagues (2014)[134] showed that in early deaf ferrets who received bilateral CIs in adulthood, an audiovisual specific training was more effective than an auditory one for recovering auditory localization abilities both at the neural and the behavioral level. This result, in turn, further suggests that binding inputs from different sensory modalities during task-oriented training programs, and especially the combination of a familiar modality (e.g., vision) with a novel, newly restored one (e.g., audition) might be a powerful way to restore efficient and task-specific sensory recovery, even in case of interventions occurring in adulthood.

We propose that a similar multisensory approach to training holds a lot of promise in the case of sight restoration as well, making SSDs tools with great potential in this context. Unlike auditory restoration and CI interventions, sight restoration still lacks a unified set of procedures. However, given the incredibly fast advances in biotechnology that characterize our era, we hope that, in the near future, sight restoration may also enjoy more standardized procedures and better expected outcomes. Thus, it is crucial to begin to prepare potential rehabilitation programs to further maximize such outcomes. Within this context, we propose it will be crucial to exploit the documented TSSI properties of "visual" cortices. Indeed, the few studies documenting sight restoration reported far from optimal recovery results in these patients.[8,135–138] However, these patients are mainly impaired on high-level visual tasks such as feature binding, object-background segregation, and perception of 3D shapes and faces.[136,137,139,140] Since these are precisely the visual abilities that can be learned using SSDs,[52] one logical step is the systematic implementation of multisensory training programs, where SSD input is paired with the restored visual modality to boost the recovery of specific computational tasks[82] (Fig. 15.2). Specifically, candidates for sight restoration might use SSDs prior to the intervention, and learn, for example, to perceive SSD-presented body-shapes, ultimately recruiting the EBA and its related network of processing[78] (i.e., TSSI recruitment). Then, after surgical sight restoration, the SSD stimulation can be paired with visual input, mediating two types of benefits. The familiar SSD input can help to better understand the newly restored visual input. For example, presenting a body shape both through an SSD and through vision simultaneously may help the patient to perceive fine details of the image or bind visual features into a coherent shape. Moreover, such pairing may facilitate a neural network's adaptability and thus allow it to efficiently process its typical sensory input. For instance, in the case of body shapes, it has been shown that SSD-presented body shapes

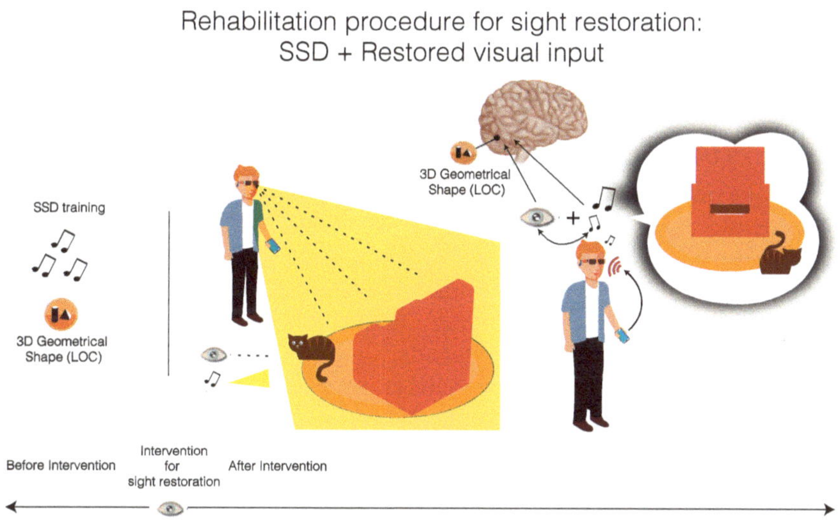

FIGURE 15.2 Proposed rehabilitative procedure for sight restoration: sensory substitution device (SSD) paired with restored visual input. This figure shows the visual-to-auditory SSD used to teach processing of visual objects (and the corresponding brain activations) as an example. The same approach could be implemented using a visual-to-tactile SSD as well as being applicable to many other cognitive/computational tasks. Left: Before the intervention for sight restoration, patients can be trained with SSDs to teach the brain to process (typically visual) specific tasks through a sensory modality (e.g., audition) that has never performed such a task, thus activating task-selective and sensory-independent (TSSI) regions and their related network (in the case depicted in the figure, SSD training on object recognition will trigger the activation of the lateral occipital complex (LOC) in the ventral "visual" stream, a TSSI region involved in 3D geometrical shape analyses). Right: After the intervention for sight restoration, patients can pair the newly acquired and developing visual input with a familiar sensory input (e.g., auditory SSD input). Medical or surgical visual restoration and SSDs could be used together to facilitate, strengthen, and complete the visual experience. This pairing aims to exploit TSSI brain organization, and we propose that it may eventually facilitate the adaptability of the visual cortex to process its typical sensory input (vision; Top).

recruit the EBA in a TSSI manner, and that this region is functionally connected to other regions typically involved in body shape processing in the blind population.[78,82] Given the documented flexibility in the activating sensory modality of TSSI cortical regions, pairing an SSD and visually presented body shapes may aid the visual cortex to tune toward specific visual inputs. A similar logic can be applied to basically all the known visual categories and perhaps even for more low-level computations (see, for instance, Fig. 15.2 where this approach is explained for object recognition in the lateral occipital complex [LOC]).

In short, we propose that to maximize the outcomes of sensory restoration procedures, TSSI recruitment of the deprived higher-order sensory cortices by atypical sensory modalities may be beneficial rather than detrimental for proper sensory recovery. We further suggest that the sensory flexibility documented in these cortices may be exploited during training programs for sensory recovery. Specifically, such training programs must be oriented to the recovery of specific tasks and must be multisensory in nature (i.e., pairing the restored sensory modality with a familiar one). We propose such multisensory training can both facilitate the understanding of novel stimuli as well as facilitate the task-selective recruitment of the newly restored sensory cortex by its natural sensory input.

SSD training as a tool to maximize visual recovery after partial visual loss

While the approach proposed above holds a lot of promise for the future, the patients that can currently benefit from it are very few (e.g., currently there are around 250 patients with retinal prosthesis worldwide). Nonetheless, there are currently over 300 million people worldwide suffering from various types of visual impairments that limit their function.[141] We would like to highlight here that SSDs can be excellent tools not only for those who are blind, but also for recovering function in cases of visual impairments and partial visual loss. SSDs carry the advantage of maintaining many features of visual images such as shape, size, and spatial locations of objects, ultimately making them extremely suitable to be paired and integrated with actual vision. In other words, we propose that visually impaired patients, in addition to blind people, may greatly benefit from SSD training programs aimed at visual rehabilitation and visual recovery. Following a similar logic to the one described in the section above, SSDs can be paired with the residual visual input and allow a more complete understanding of the degraded visual information. This in turn may yield enormous benefits for daily life activities, ultimately boosting the functional independence of this population.

In addition, training with SSDs can also be embedded within specific rehabilitation programs aimed at the visual recovery of patients suffering from visual impairment (Fig. 15.3). Indeed, accumulating evidence suggests that multisensory stimulation may be an effective rehabilitation method for visual impairments acquired in adulthood (e.g., after stroke[142]). The majority of these studies have been conducted with hemianopic patients, i.e., patients with an acquired lesion in the visual structures located in the early visual pathway behind the optic chiasm. Such lesions generally result in visual loss in up to one half of the visual field. This condition, in turn, results in many difficulties in daily life including reading, scanning scenes, and obstacle avoidance, especially relating to the affected portion of the visual field.[143] Several studies documented the greater benefit of multisensory than unisensory training with these patients by showing that the greatest improvement in

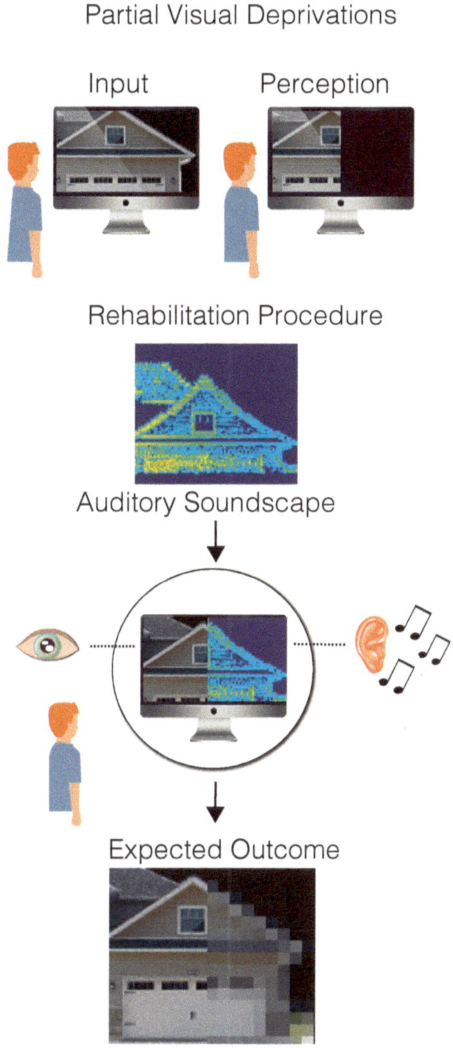

FIGURE 15.3 Proposed rehabilitative procedure for partial visual deprivations. Top: This figure depicts a case of a hemianopic patient who lost his vision in one half of the visual field as an example. The same approach can be used for many other types of visual impairment. Middle: We propose that the rehabilitative procedure should include computerized training programs where visual inputs (in this case objects) are presented together with visual-to-auditory SSD inputs. In this way, the patient is able to complete the missing visual information through the auditory soundscape, ultimately integrating together the information from the two sensory modalities into a unified percept. Bottom: Such multisensory training may lead to better recovery of vision.

visual performance in the affected portion of the visual field was achieved when a visual stimulus was presented together with a coincident sound[144–146] (see also chapter by Stein and Rowland, this volume). We believe that visual-to-auditory SSDs are very promising tools to use for the rehabilitation of these patients. Fig. 15.3 shows an example of such an approach where the auditory-to-visual SSD input is paired with the remaining visual input to boost visual recovery. Specifically, training programs may present stimuli on a computer screen (in the figure we show objects as an example). During training patients integrate the available visual information with the SSD input and achieve a full perception of the presented object. In this way, patients may be helped to recognize objects by being able to perceive them entirely. Furthermore, such pairing may facilitate a neural network's adaptability and thus allow it to efficiently recover the missing visual input, as discussed in an earlier section. Thus, we propose that this multisensory approach may further maximize the outcomes of rehabilitative procedures and functional visual recovery (Fig. 15.3).

Note that the case of hemianopic patients is used here as an example, but the proposed multisensory approach may be suited to implementation for many other types of visual impairments (e.g., peripheral visual impairment in retinitis pigmentosa or central visual loss in macular degeneration).

General conclusions

In this chapter, we reviewed the current literature on the (re)-organization of sensory cortices following sensory deprivation, mainly blindness. We suggest that the brain may be organized as a task-oriented rather than a sensory-oriented machine as classically conceived. Finally, we propose that this task-oriented and sensory-flexible organization may be exploited during sensory restoration programs to maximize sensory recovery, and more generally, that multisensory training pairing SSD and visual inputs holds great promise to maximize the outcomes of visual recovery. Throughout this chapter, we showed how studies with SSDs played a crucial role in unraveling the incredible sensory flexibility of our brain across the life span. Thus, we can conclude that all the work that was carried on with SSDs after the pioneering studies of Bach-y-Rita in the 1970s corroborated his intuition on the relevance to visual rehabilitation of the fact that we see with our brain rather than only with our eyes. Such a conclusion carries crucial implications for sensory recovery practices, ultimately unraveling novel exciting paths to maximize rehabilitation outcomes.

References

1. Knudsen EI. Sensitive periods in the development of the brain and behavior. *J Cogn Neurosci.* 2004;16(8):1412–1425. https://doi.org/10.1162/0898929042304796.
2. Hubel DH, Wiesel TN. The period of susceptibility to the physiological effects of unilateral eye closure in kittens. *J Physiol.* 1970;206(2):419–436.
3. Cynader M, Chernenko G. Abolition of direction selectivity in the visual cortex of the cat. *Science.* 1976;193(4252):504–505. https://doi.org/10.1126/science.941025.
4. Cynader M, Mitchell DE. Monocular astigmatism effects on kitten visual cortex development. *Nature.* 1977;270:177–178.

5. Daw NW. Critical periods: motion sensitivity is early in all areas. *Curr Biol.* 2009. https://doi.org/10.1016/j.cub.2009.02.048.
6. Daw NW. The foundations of development and deprivation in the visual system. *J Physiol.* 2009;587(12):2769–2773.
7. Knudsen EI. Early blindness results in a degraded auditory map of space in the optic tectum of the barn owl. *Proc Natl Acad Sci U S A.* 1988. https://doi.org/10.1073/pnas.85.16.6211.
8. Levin N, Dumoulin SO, Winawer J, Dougherty RF, Wandell BA. Cortical maps and white matter tracts following long period of visual deprivation and retinal image restoration. *Neuron.* 2010;65(1):21–31. https://doi.org/10.1016/j.neuron.2009.12.006.
9. Roder B, Ley P, Shenoy BH, Kekunnaya R, Bottari D. Sensitive periods for the functional specialization of the neural system for human face processing. *Proc Natl Acad Sci.* 2013. https://doi.org/10.1073/pnas.1309963110.
10. Grady CL, Mondloch CJ, Lewis TL, Maurer D. Early visual deprivation from congenital cataracts disrupts activity and functional connectivity in the face network. *Neuropsychologia.* 2014. https://doi.org/10.1016/j.neuropsychologia.2014.03.005.
11. Bottari D, Troje NF, Ley P, Hense M, Kekunnaya R, Röder B. Sight restoration after congenital blindness does not reinstate alpha oscillatory activity in humans. *Sci Rep.* 2016. https://doi.org/10.1038/srep24683.
12. Bottari D, Kekunnaya R, Hense M, Troje NF, Sourav S, Röder B. Motion processing after sight restoration: No competition between visual recovery and auditory compensation. *Neuroimage.* 2018. https://doi.org/10.1016/j.neuroimage.2017.11.050.
13. Recanzone GH, Schreiner CE, Merzenich MM. Plasticity in the frequency representation of primary auditory cortex following discrimination training in adult owl monkeys. *J Neurosci.* 1993. doi:papers://FAFC0638-5DD4-4A81-A69F-F8A54DFE70C3/Paper/p.11227.
14. Buonomano DV, Merzenich MM. Cortical plasticity: from synapses to maps. *Annu Rev Neurosci.* 1998. https://doi.org/10.1146/annurev.neuro.21.1.149.
15. Recanzone G, Dinse R, Merzenich MM, Jenkins WM, Grajski Ka, Dinse HR. Topographic reorganization of the hand representation in cortical area 3b of owl monkeys trained in a frequency-discrimination task. *J Neurophysiol.* 1992, 1597696.
16. Thomas C, Baker CI. Teaching an adult brain new tricks: a critical review of evidence for training- dependent structural plasticity in humans. *Neuroimage.* 2013. https://doi.org/10.1016/j.neuroimage.2012.03.069.
17. Draganski B, May A. Training-induced structural changes in the adult human brain. *Behav Brain Res.* 2008. https://doi.org/10.1016/j.bbr.2008.02.015.
18. Reetzke R, Xie Z, Llanos F, Chandrasekaran B. Tracing the trajectory of sensory plasticity across different stages of speech learning in adulthood. *Curr Biol.* 2018. https://doi.org/10.1016/j.cub.2018.03.026.
19. Cheng Y, Jia G, Zhang Y, et al. Positive impacts of early auditory training on cortical processing at an older age. *Proc Natl Acad Sci.* 2017. https://doi.org/10.1073/pnas.1707086114.
20. Anderson S, White-Schwoch T, Parbery-Clark A, Kraus N. Reversal of age-related neural timing delays with training. *Proc Natl Acad Sci.* 2013. https://doi.org/10.1073/pnas.1213555110.
21. Smith GE, Housen P, Yaffe K, et al. A cognitive training program based on principles of brain plasticity: results from the improvement in memory with plasticity-based adaptive cognitive training (IMPACT) study. *J Am Geriatr Soc.* 2009. https://doi.org/10.1111/j.1532-5415.2008.02167.x.
22. Mishra J, deVillers-Sidani E, Merzenich M, Gazzaley A. Adaptive training diminishes distractibility in aging across species. *Neuron.* 2014. https://doi.org/10.1016/j.neuron.2014.10.034.
23. de Villers-Sidani E, Alzghoul L, Zhou X, Simpson KL, Lin RCS, Merzenich MM. Recovery of functional and structural age-related changes in the rat primary auditory cortex with operant training. *Proc Natl Acad Sci.* 2010. https://doi.org/10.1073/pnas.1007885107.
24. Mahncke HW, Connor BB, Appelman J, et al. Memory enhancement in healthy older adults using a brain plasticity-based training program: a randomized, controlled study. *Proc Natl Acad Sci.* 2006. https://doi.org/10.1073/pnas.0605194103.
25. Lappe C, Trainor LJ, Herholz SC, Pantev C. Cortical plasticity induced by short-term multimodal musical rhythm training. *PLoS One.* 2011. https://doi.org/10.1371/journal.pone.0021493.
26. Pantev C, Lappe C, Herholz SC, Trainor L. Auditory-somatosensory integration and cortical plasticity in musical training. In: *Annals of the New York Academy of Sciences.* 2009. https://doi.org/10.1111/j.1749-6632.2009.04588.x.

27. Pantev C, Oostenveld R, Engelien A, Ross B, Roberts LE, Hoke M. Increased auditory cortical representation in musicians. *Nature*. 1998. https://doi.org/10.1038/33918.
28. Pantev C, Herholz SC. Plasticity of the human auditory cortex related to musical training. *Neurosci Biobehav Rev*. 2011. https://doi.org/10.1016/j.neubiorev.2011.06.010.
29. Herholz SC, Zatorre RJ. Musical training as a framework for brain plasticity: behavior, function, and structure. *Neuron*. 2012. https://doi.org/10.1016/j.neuron.2012.10.011.
30. Chen JL, Penhune VB, Zatorre RJ. Moving on time: brain network for auditory-motor synchronization is modulated by rhythm complexity and musical training. *J Cogn Neurosci*. 2008. https://doi.org/10.1162/jocn.2008.20018.
31. Zhou X, Merzenich MM. Intensive training in adults refines A1 representations degraded in an early postnatal critical period. *Proc Natl Acad Sci*. 2007. https://doi.org/10.1073/pnas.0707348104.
32. Pan Y, Zhang J, Cai R, Zhou X, Sun X. Developmentally degraded directional selectivity of the auditory cortex can be restored by auditory discrimination training in adults. *Behav Brain Res*. 2011. https://doi.org/10.1016/j.bbr.2011.08.033.
33. Kerr AL, Cheng SY, Jones TA. Experience-dependent neural plasticity in the adult damaged brain. *J Commun Disord*. 2011. https://doi.org/10.1016/j.jcomdis.2011.04.011.
34. Xerri C, Merzenich MM, Peterson BE, Jenkins WM. Plasticity of primary somatosensory cortex paralleling sensorimotor skill recovery from stroke in adult monkeys. *J Neurophysiol*. 1998. https://doi.org/10.1152/jn.1998.79.4.2119.
35. Abboud S, Hanassy S, Levy-Tzedek S, Maidenbaum S, Amedi A. EyeMusic: introducing a "visual" colorful experience for the blind using auditory sensory substitution. *Restor Neurol Neurosci*. 2014;32(2):247–257. https://doi.org/10.3233/RNN-130338.
36. Bach-Y-Rita P, Colliins CC, Saunders FA, White B, Scadden L. Vision substitution by tactile image projection. *Nature*. 1969;221(5184):963–964. https://doi.org/10.1038/221963a0.
37. Bach-y-Rita P. Tactile sensory substitution studies. *Ann N Y Acad Sci*. 2004;1013:83–91. https://doi.org/10.1196/annals.1305.006.
38. Meijer PBL. An experimental system for auditory image representations. *IEEE Trans Biomed Eng*. 1992;39(2):112–121. https://doi.org/10.1109/10.121642.
39. Maidenbaum S, Abboud S, Amedi A. Sensory substitution: closing the gap between basic research and widespread practical visual rehabilitation. *Neurosci Biobehav Rev*. 2014;41:3–15.
40. Goldish LH, Taylor HE. The optacon: a valuable device for blind persons. *N Outlook Blind*. 1974;68(2):49–56. https://eric.ed.gov/?id=EJ096181.
41. Starkiewicz W, Kuliszewski T. Progress report on the elektroftalm mobility aid. In: *Proceedings of the Rotterdam Mobility Research Conference*. New York: American Foundation for the Blind; 1965:27–38.
42. Bach-y-Rita P. *Brain Mechanisms in Sensory Substitution*. Academic Press.; 1972.
43. Bach-y-Rita P. Conservation of space and energy in the brain. *Restor Neurol Neurosci*. 1996;10(1):1–3.
44. Bach-y-Rita P, Tyler ME, Kaczmarek KA. Seeing with the brain. *Int J Hum Comput Interact*. 2003. https://doi.org/10.1207/S15327590IJHC1502_6.
45. Pasqualotto A, Proulx MJ. The role of visual experience for the neural basis of spatial cognition. *Neurosci Biobehav Rev*. 2012. https://doi.org/10.1016/j.neubiorev.2012.01.008.
46. Proulx MJ, Gwinnutt J, Dell'Erba S, Levy-Tzedek S, de Sousa A a, Brown DJ. Other ways of seeing: from behavior to neural mechanisms in the online "visual" control of action with sensory substitution. *Restor Neurol Neurosci*. 2015;34(1):29–44. https://doi.org/10.3233/RNN-150541.
47. Jacobson H. The informational capacity of the human eye. *Science*. 1951;113(2933):292–293.
48. Kokjer KJ. The information capacity of the human fingertip. *IEEE Trans Syst Man Cybern*. 1987. https://doi.org/10.1109/TSMC.1987.289337.
49. Schmidt RF. Somatovisceral sensibility. In: *Fundamentals of Sensory Physiology*. Springer Berlin Heidelberg; 1981:81–125.
50. Jacobson H. The informational capacity of the human ear. *Science*. 1950;112(2901):143–144.
51. Proulx MJ, Stoerig P, Ludowig E, Knoll I. Seeing "where" through the ears: effects of learning-by-doing and long-term sensory deprivation on localization based on image-to-sound substitution. *PLoS One*. 2008;3(3). https://doi.org/10.1371/journal.pone.0001840.

52. Striem-Amit E, Cohen L, Dehaene S, Amedi A. Reading with sounds: sensory substitution selectively activates the visual word form area in the blind. *Neuron.* 2012;76(3):640–652. https://doi.org/10.1016/j.neuron.2012.08.026.
53. Chebat D-R, Maidenbaum S, Amedi A. Navigation using sensory substitution in real and virtual mazes. *PLoS One.* 2015;10(6):e0126307.
54. Maidenbaum S, Buchs G, Abboud S, Lavi-Rotbain O, Amedi A. Perception of graphical virtual environments by blind users via sensory substitution. *PLoS One.* 2016;11(2):e0147501. https://doi.org/10.1371/journal.pone.0147501.
55. Ward J, Meijer P. Visual experiences in the blind induced by an auditory sensory substitution device. *Conscious Cognit.* 2010;19(1):492–500. https://doi.org/10.1016/j.concog.2009.10.006.
56. Reich L, Amedi A. "Visual" parsing can be taught quickly without visual experience during critical periods. *Sci Rep.* 2015;5:15359. https://doi.org/10.1038/srep15359.
57. Levy-Tzedek S, Riemer D, Amedi A. Color improves "visual" acuity via sound. *Front Neurosci.* 2014;8:1–7. https://doi.org/10.3389/fnins.2014.00358.
58. Striem-Amit E, Guendelman M, Amedi A. 'Visual' acuity of the congenitally blind using visual-to-auditory sensory substitution. Serino A, ed *PLoS One.* 2012;7(3):e33136. https://doi.org/10.1371/journal.pone.0033136.
59. Sadato N, Pascual-Leone A, Grafman J, et al. Activation of the visual cortex by Braille reading in blind subjects. *Nature.* 1996;380(6574):526. https://doi.org/10.1093/acprof:oso/9780198528999.003.0022.
60. Kujala T, Alho K, Paavilainen P, Summala H, Näätänen R. Neural plasticity in processing of sound location by the early blind: an event-related potential study. *Electroencephalogr Clin Neurophysiol Evoked Potentials.* 1992. https://doi.org/10.1016/0168-5597(92)90034-9.
61. Büchel C, Price C, Frackowiak RSJ, Friston K. Different activation patterns in the visual cortex of late and congenitally blind subjects. *Brain.* 1998. https://doi.org/10.1093/brain/121.3.409.
62. Weeks R, Horwitz B, Aziz-Sultan A, et al. A positron emission tomographic study of auditory localization in the congenitally blind. *J Neurosci Off J Soc Neurosci.* 2000.
63. Burton H, Snyder AZ, Conturo TE, Akbudak E, Ollinger JM, Raichle ME. Adaptive changes in early and late blind: a fMRI study of braille reading. *J Neurophysiol.* 2002. https://doi.org/10.1152/jn.00285.2001.
64. Sadato N, Pascual-Leone A, Grafman J, Deiber MP, Ibañez V, Hallett M. Neural networks for Braille reading by the blind. *Brain.* 1998. https://doi.org/10.1093/brain/121.7.1213.
65. Catalán-Ahumada M, Deggouj N, De Volder A, Melin J, Michel C, Veraart C. *High Metabolic Activity Demonstrated by Positron Emission Tomography in Human Auditory Cortex in Case of Deafness of Early Onset.* Vol. 623. 1993. https://doi.org/10.1016/0006-8993(93)91439-Y.
66. Kujala T, Huotilainen M, Sinkkonen J, et al. Visual cortex activation in blind humans during sound discrimination. *Neurosci Lett.* 1995;183(1):143–146. https://doi.org/10.1016/0304-3940(94)11135-6.
67. Pons T. Novel sensations in the congenitally blind. *Nature.* 1996;380(6574).
68. Uhl F, Franzen P, Lindinger G, Lang W, Deecke L. On the functionality of the visually deprived occipital cortex in early blind persons. *Neurosci Lett.* 1991;124(2):256–259. https://doi.org/10.1016/0304-3940(91)90107-5.
69. Dormal G, Collignon O. Functional selectivity in sensory-deprived cortices. *J Neurophysiol.* 2011;105(6):2627–2630. https://doi.org/10.1152/jn.00109.2011.
70. Heimler B, Weisz N, Collignon O. Revisiting the adaptive and maladaptive effects of crossmodal plasticity. *Neuroscience.* 2014;283:44–63. https://doi.org/10.1016/j.neuroscience.2014.08.003.
71. Striem-Amit E, Dakwar O, Reich L, Amedi A. The large-scale organization of "visual" streams emerges without visual experience. *Cerebr Cortex.* 2012;22(7):1698–1709. https://doi.org/10.1093/cercor/bhr253.
72. Collignon O, Vandewalle G, Voss P, et al. Functional specialization for auditory-spatial processing in the occipital cortex of congenitally blind humans. *Proc Natl Acad Sci U S A.* 2011;108(11):4435–4440. https://doi.org/10.1073/pnas.1013928108.
73. Renier LA, Anurova I, De Volder AG, Carlson S, VanMeter J, Rauschecker JP. Preserved functional specialization for spatial processing in the middle occipital gyrus of the early blind. *Neuron.* 2010;68(1):138–148. https://doi.org/10.1016/j.neuron.2010.09.021.
74. Matteau I, Kupers R, Ricciardi E, Pietrini P, Ptito M. Beyond visual, aural and haptic movement perception: hMT+ is activated by electrotactile motion stimulation of the tongue in sighted and in congenitally blind individuals. *Brain Res Bull.* 2010;82(5–6):264–270. https://doi.org/10.1016/j.brainresbull.2010.05.001.

75. Amedi A, Stern WM, Camprodon J a, et al. Shape conveyed by visual-to-auditory sensory substitution activates the lateral occipital complex. *Nat Neurosci*. 2007;10(6):687–689. https://doi.org/10.1038/nn1912.
76. Reich L, Szwed M, Cohen L, Amedi A. A ventral visual stream reading center independent of visual experience. *Curr Biol*. 2011;21(5):363–368. https://doi.org/10.1016/j.cub.2011.01.040.
77. Abboud S, Maidenbaum S, Dehaene S, Amedi A. A number-form area in the blind. *Nat Commun*. 2015;6:6026. https://doi.org/10.1038/ncomms7026.
78. Striem-Amit E, Amedi A. *Visual Cortex Extrastriate Body-Selective Area Activation in Congenitally Blind People "Seeing" by Using Sounds*. Vol. 24. 2014. https://doi.org/10.1016/j.cub.2014.02.010.
79. Reich L, Maidenbaum S, Amedi A. The brain as a flexible task machine: implications for visual rehabilitation using noninvasive vs. invasive approaches. *Curr Opin Neurol*. 2012;25(1):86–95. https://doi.org/10.1097/WCO.0b013e32834ed723.
80. Amedi A, Hofstetter S, Maidenbaum S, Heimler B. Task-selectivity as a comprehensive principle for brain organization. *Trends Cogn Sci*. 2017;21(5):307–310.
81. Hannagan T, Amedi A, Cohen L, Dehaene-Lambertz G, Dehaene S. Origins of the specialization for letters and numbers in ventral occipitotemporal cortex. *Trends Cogn Sci*. 2015;19(7):374–382. https://doi.org/10.1016/j.tics.2015.05.006.
82. Heimler B, Striem-Amit E, Amedi A. Origins of task-specific sensory-independent organization in the visual and auditory brain: neuroscience evidence, open questions and clinical implications. *Curr Opin Neurobiol*. 2015;35:169–177. https://doi.org/10.1016/j.conb.2015.09.001.
83. Meredith MA, Kryklywy J, McMillan AJ, Malhotra S, Lum-Tai R, Lomber SG. Crossmodal reorganization in the early deaf switches sensory, but not behavioral roles of auditory cortex. *Proc Natl Acad Sci U S A*. 2011;108(21):8856–8861. https://doi.org/10.1073/pnas.1018519108.
84. Lomber SSG, Meredith MA, Kral A. Cross-modal plasticity in specific auditory cortices underlies visual compensations in the deaf. *Nat Neurosci*. 2010;13(11):1421–1427. https://doi.org/10.1038/nn.2653.
85. Lomber SG, Meredith MA, Kral A. Crossmodal plasticity in deaf auditory cortex mediates enhanced face perception in the congenitally deaf. *J Int Adv Otol*. 2015;11.
86. MacSweeney M, Capek CM, Campbell R, Woll B. The signing brain: the neurobiology of sign language. *Trends Cogn Sci*. 2008;12(11):432–440. https://doi.org/10.1016/j.tics.2008.07.010.
87. Corina DP, Jose-Robertson LS, Guillemin A, High J, Braun AR. Language lateralization in a bimanual language. *J Cogn Neurosci*. 2003;15(5):718–730. https://doi.org/10.1162/jocn.2003.15.5.718.
88. Emmorey K, Grabowski T, McCullough S, et al. Neural systems underlying lexical retrieval for sign language. *Neuropsychologia*. 2003;41(1):85–95. https://doi.org/10.1016/S0028-3932(02)00089-1.
89. Emmorey K, Mehta S, Grabowski TJ. The neural correlates of sign versus word production. *Neuroimage*. 2007;36(1):202–208. https://doi.org/10.1016/j.neuroimage.2007.02.040.
90. McGuire PK, Robertson D, Thacker a, et al. Neural correlates of thinking in sign language. *Neuroreport*. 1997;8(3):695–698. https://doi.org/10.1097/00001756-199702100-00023.
91. Petitto LA, Zatorre RJ, Gauna K, Nikelski EJ, Dostie D, Evans AC. Speech-like cerebral activity in profoundly deaf people processing signed languages: implications for the neural basis of human language. *Proc Natl Acad Sci U S A*. 2000;97(25):13961–13966. https://doi.org/10.1073/pnas.97.25.13961.
92. San José-Robertson L, Corina DP, Ackerman D, Guillemin A, Braun AR. Neural systems for sign language production: mechanisms supporting lexical selection, phonological encoding, and articulation. *Hum Brain Mapp*. 2004;23(3):156–167. https://doi.org/10.1002/hbm.20054.
93. MacSweeney M, Woll B, Campbell R, et al. Neural systems underlying British Sign Language and audio-visual English processing in native users. *Brain*. 2002;125(Pt 7):1583–1593. https://doi.org/10.1093/brain/awf153.
94. Neville HJ, Bavelier D, Corina D, et al. Cerebral organization for language in deaf and hearing subjects: biological constraints and effects of experience. *Proc Natl Acad Sci U S A*. 1998;95(3):922–929. https://doi.org/10.1073/pnas.95.3.922.
95. Sakai KL, Tatsuno Y, Suzuki K, Kimura H, Ichida Y. Sign and speech: amodal commonality in left hemisphere dominance for comprehension of sentences. *Brain*. 2005;128(6):1407–1417. https://doi.org/10.1093/brain/awh465.
96. Bellugi U, Poizner H, Klima ES. Brain organization for language: clues from sign aphasia. *Hum Neurobiol*. 1983;2(3):155–170. http://www.ncbi.nlm.nih.gov/pubmed/6668233.
97. Damasio A, Bellugi U, Damasio H, Poizner H, Gilder J Van. Sign language aphasia during left-hemisphere Amytal injection. *Nature*. 1986;322(6077):363–365. https://doi.org/10.1038/322363a0.

98. Hickok G, Bellugi U, Klima ES. The neural organization of language: evidence from sign language aphasia. *Trends Cogn Sci.* 1998;2(4):129–136. https://doi.org/10.1016/S1364-6613(98)01154-1.
99. Marshall J, Atkinson J, Smulovitch E, Thacker A, Woll B. Aphasia in a user of British sign language: dissociation between sign and gesture. *Cogn Neuropsychol.* 2004;21(5):537–554. https://doi.org/10.1080/02643290342000249.
100. Bottari D, Heimler B, Caclin A, Dalmolin A, Giard MH, Pavani F. Visual change detection recruits auditory cortices in early deafness. *Neuroimage.* 2014;94:172–184. https://doi.org/10.1016/j.neuroimage.2014.02.031.
101. Näätänen R, Gaillard AWK, Mäntysalo S. Early selective-attention effect on evoked potential reinterpreted. *Acta Psychol.* 1978;42(4):313–329. https://doi.org/10.1016/0001-6918(78)90006-9.
102. Näätänen R, Tervaniemi M, Sussman E, Paavilainen P, Winkler I. "Primitive intelligence" in the auditory cortex. *Trends Neurosci.* 2001;24(5):283–288. https://doi.org/10.1016/S0166-2236(00)01790-2.
103. Kimura M, Schröger E, Czigler I. Visual mismatch negativity and its importance in visual cognitive sciences. *Neuroreport.* 2011;22(14):669–673. https://doi.org/10.1097/WNR.0b013e32834973ba.
104. Bola Ł, Zimmermann M, Mostowski P, et al. Task-specific reorganization of the auditory cortex in deaf humans. *Proc Natl Acad Sci U S A.* 2017;114(4):E600–E609. https://doi.org/10.1073/pnas.1609000114.
105. Benetti S, van Ackeren MJ, Rabini G, et al. Functional selectivity for face processing in the temporal voice area of early deaf individuals. *Proc Natl Acad Sci.* 2017;114(31):E6437–E6446.
106. Benetti S, Novello L, Maffei C, Rabini G, Jovicich J, Collignon O. White matter connectivity between occipital and temporal regions involved in face and voice processing in hearing and early deaf individuals. *Neuroimage.* 2018;179:263–274.
107. Striem-Amit E, Ovadia-Caro S, Caramazza A, Margulies DS, Villringer A, Amedi A. Functional connectivity of visual cortex in the blind follows retinotopic organization principles. *Brain.* 2015;138(6):1679–1695. https://doi.org/10.1093/brain/awv083.
108. Striem-Amit E, Almeida J, Belledonne M, et al. Topographical functional connectivity patterns exist in the congenitally, prelingually deaf. *Sci Rep.* 2016;6:29375. https://doi.org/10.1038/srep29375.
109. Röder B, Stock O, Bien S, Neville H, Rösler F. Speech processing activates visual cortex in congenitally blind humans. *Eur J Neurosci.* 2002;16(5):930–936. https://doi.org/10.1046/j.1460-9568.2002.02147.x.
110. Bedny M, Pascual-Leone A, Dodell-Feder D, Fedorenko E, Saxe R. Language processing in the occipital cortex of congenitally blind adults. *Proc Natl Acad Sci U S A.* 2011;108(11):4429–4434. https://doi.org/10.1073/pnas.1014818108.
111. Amedi A, Floel A, Knecht S, Zohary E, Cohen LG. Transcranial magnetic stimulation of the occipital pole interferes with verbal processing in blind subjects. *Nat Neurosci.* 2004;7(11):1266–1270. https://doi.org/10.1038/nn1328.
112. Amedi A, Raz N, Pianka P, Malach R, Zohary E. Early "visual" cortex activation correlates with superior verbal memory performance in the blind. *Nat Neurosci.* 2003;6(7):758–766. https://doi.org/10.1038/nn1072.
113. Watkins KE, Cowey A, Alexander I, et al. Language networks in anophthalmia: maintained hierarchy of processing in "visual" cortex. *Brain.* 2012. https://doi.org/10.1093/brain/aws067.
114. Burton H, Sinclair RJ, Agato A. Recognition memory for Braille or spoken words: an fMRI study in early blind. *Brain Res.* 2012. https://doi.org/10.1016/j.brainres.2011.12.032.
115. Bedny M. Evidence from blindness for a cognitively pluripotent cortex. *Trends Cogn Sci.* 2017. https://doi.org/10.1016/j.tics.2017.06.003.
116. Kanjlia S, Lane C, Feigenson L, Bedny M. Absence of visual experience modifies the neural basis of numerical thinking. *Proc Natl Acad Sci.* 2016. https://doi.org/10.1073/pnas.1524982113.
117. Raz N, Amedi A, Zohary E. V1 activation in congenitally blind humans is associated with episodic retrieval. *Cerebr Cortex.* 2005;15(9):1459–1468. https://doi.org/10.1093/cercor/bhi026.
118. Weaver KE, Stevens A a. Attention and sensory interactions within the occipital cortex in the early blind: an fMRI study. *J Cogn Neurosci.* 2007;19:315–330. https://doi.org/10.1162/jocn.2007.19.2.315.
119. Loiotile RE, Bedny M. "Visual" cortices of congenitally blind adults respond to executive demands. *bioRxiv.* 2018. https://doi.org/10.1101/390450.
120. Vetter P, Smith FW, Muckli L. Decoding sound and imagery content in early visual cortex. *Curr Biol.* 2014. https://doi.org/10.1016/j.cub.2014.04.020.
121. Cheung SH, Fang F, He S, Legge GE. Retinotopically specific reorganization of visual cortex for tactile pattern recognition. *Curr Biol.* 2009. https://doi.org/10.1016/j.cub.2009.02.063.

122. Kim JS, Kanjlia S, Merabet LB, Bedny M. Development of the visual word form area requires visual experience: evidence from blind Braille readers. *J Neurosci*. 2017. https://doi.org/10.1523/JNEUROSCI.0997-17.2017.
123. Kanjlia S, Pant R, Bedny M. Sensitive period for cognitive repurposing of human visual cortex. *bioRxiv*. 2018:402321.
124. Sigalov N, Maidenbaum S, Amedi A. Reading in the dark: neural correlates and cross-modal plasticity for learning to read entire words without visual experience. *Neuropsychologia*. 2016;83:149–160. https://doi.org/10.1016/j.neuropsychologia.2015.11.009.
125. Lewis T, Maurer D. Multiple sensitive periods in human visual development: evidence from visually deprived children. *Dev Psychobiol*; 2005. http://onlinelibrary.wiley.com/doi/10.1002/dev.20055/full.
126. Lyness CR, Woll B, Campbell R, Cardin V. How does visual language affect crossmodal plasticity and cochlear implant success? *Neurosci Biobehav Rev*. 2013;37(10):2621–2630. https://doi.org/10.1016/j.neubiorev.2013.08.011.
127. Gaylor JM, Raman G, Chung M, et al. Cochlear implantation in adults: a systematic review and meta-analysis. *JAMA Otolaryngol Head Neck Surg*. 2013;139(3):265–272. https://doi.org/10.1001/jamaoto.2013.1744.
128. Lee DS, Lee JS, Oh SH, et al. Cross-modal plasticity and cochlear implants. *Nature*. 2001;409(6817):149–150. https://doi.org/10.1038/35051650.
129. Lee HJ, Giraud AL, Kang E, et al. Cortical activity at rest predicts cochlear implantation outcome. *Cerebr Cortex*. 2007. https://doi.org/10.1093/cercor/bhl001.
130. Kral A, Sharma A. Developmental neuroplasticity after cochlear implantation. *Trends Neurosci*. 2012;35(2):111–122. https://doi.org/10.1016/j.tins.2011.09.004.
131. Heming JE, Brown LN. Sensory temporal processing in adults with early hearing loss. *Brain Cogn*. 2005;59(2):173–182. https://doi.org/10.1016/j.bandc.2005.05.012.
132. Strelnikov K, Rouger J, Demonet JF, et al. Visual activity predicts auditory recovery from deafness after adult cochlear implantation. *Brain*. 2013. https://doi.org/10.1093/brain/awt274.
133. Hassanzadeh S. Outcomes of cochlear implantation in deaf children of deaf parents: comparative study. *J Laryngol Otol*. 2012;126(10):989–994. https://doi.org/10.1017/S0022215112001909.
134. Isaiah A, Vongpaisal T, King AJ, Hartley DEH. Multisensory training improves auditory spatial processing following bilateral cochlear implantation. *J Neurosci*. 2014;34(33):11119–11130. https://doi.org/10.1523/JNEUROSCI.4767-13.2014.
135. Gregory RL, Wallace JG. Recovery from early blindness. *Exp Psychol Soc Monogr*. 1963;2:65–129.
136. Fine I, Wade AAR, Brewer AA, et al. Long-term deprivation affects visual perception and cortex. 2003;6(9):915–916. https://doi.org/10.1038/nn1102.
137. Huber E, Webster JM, Brewer Aa, et al. A lack of experience-dependent plasticity after more than a decade of recovered sight. *Psychol Sci*. 2015;26(4):393–401. https://doi.org/10.1177/0956797614563957.
138. Dormal G, Lepore F, Harissi-Dagher M, et al. Tracking the evolution of crossmodal plasticity and visual functions before and after sight restoration. *J Neurophysiol*. 2015. https://doi.org/10.1152/jn.00420.2014.
139. Le Grand R, Mondloch CJ, Maurer D, Brent HP. Impairment in holistic face processing following early visual deprivation. *Psychol Sci*. 2004. https://doi.org/10.1111/j.0956-7976.2004.00753.x.
140. Le Grand R, Mondloch CJ, Maurer D, Brent HP. Neuroperception: early visual experience and face processing. *Nature*. 2001. https://doi.org/10.1038/35073749.
141. Bourne RRA, Flaxman SR, Braithwaite T, et al. Magnitude, temporal trends, and projections of the global prevalence of blindness and distance and near vision impairment: a systematic review and meta-analysis. *Lancet Glob Heal*. 2017. https://doi.org/10.1016/S2214-109X(17)30293-0.
142. Tinga AM, Visser-Meily JMA, van der Smagt MJ, Van der Stigchel S, van Ee R, Nijboer TCW. Multisensory stimulation to improve low-and higher-level sensory deficits after stroke: a systematic review. *Neuropsychol Rev*. 2016;26(1):73–91.
143. Papageorgiou E, Hardiess G, Schaeffel F, et al. Assessment of vision-related quality of life in patients with homonymous visual field defects. *Graefes Arch Clin Exp Ophthalmol*. 2007. https://doi.org/10.1007/s00417-007-0644-z.
144. Frassinetti F, Bolognini N, Bottari D, Bonora A, Làdavas E. Audiovisual integration in patients with visual deficit. *J Cogn Neurosci*. 2005. https://doi.org/10.1162/0898929054985446.
145. Passamonti C, Frissen I, Làdavas E. Visual recalibration of auditory spatial perception: two separate neural circuits for perceptual learning. *Eur J Neurosci*. 2009. https://doi.org/10.1111/j.1460-9568.2009.06910.x.
146. Keller I, Lefin-Rank G. Improvement of visual search after audiovisual exploration training in hemianopic patients. *Neurorehabilitation Neural Repair*. 2010. https://doi.org/10.1177/1545968310372774.

CHAPTER

16

Crossmodal neuroplasticity in deafness: evidence from animal models and clinical populations

Stephen G. Lomber[1], *Blake E. Butler*[2], *Hannah Glick*[3], *Anu Sharma*[3]

[1]Departments of Physiology and Pharmacology and Psychology, National Centre for Audiology, Brain and Mind Institute, University of Western Ontario, London, ON, Canada; [2]Department of Psychology, National Centre for Audiology, Brain and Mind Institute, University of Western Ontario, London, ON, Canada; [3]Department of Speech, Language, & Hearing Science, Institute of Cognitive Science, Center for Neuroscience, University of Colorado at Boulder, Boulder, CO, United States

Introduction

A profound hearing loss acquired during development can have widespread, devastating consequences that impact a child and their family for a lifetime. Perhaps most importantly, hearing loss can prevent a child from acquiring spoken language, which has a number of subsequent developmental and psychosocial consequences (for review[1]). Fortunately, interventions have been developed that bypass damaged peripheral structures and allow for the restoration of auditory input. In fact, if implanted within a sensitive period for normal development, children with cochlear implants (CIs) typically go on to display expressive and receptive language skills similar to those of normal hearing children by the time they are school-aged.[2] However, successful intervention requires that the remaining auditory structures are of sufficient anatomical integrity and functional state. For example, while CIs have been successfully applied in cases of cochlear degeneration, they require intact spiral ganglion neurons to be present to function.

Much of what we know about the changes to auditory structures that result from deafness, and how these changes have informed the design of cochlear prostheses has come from studies

Multisensory Perception
https://doi.org/10.1016/B978-0-12-812492-5.00016-4

in animal models. Fortunately, the auditory system is highly conserved among mammals,[3] such that a number of animal models exist that can inform our understanding of its structure and function. Moreover, a number of deaf animal models exist which closely resemble common morphologies of human disease (e.g., BALB/c mice, *deafness* mice, deaf-white cats[4]). However, it is important to note that changes in the anatomy and function of peripheral and central auditory structures depend highly on a number of factors, including the time of onset of hearing loss and the specific nature of the impairment. This chapter aims to address changes that occur in response to congenital, early-onset, or adult-onset deafness.

Neuroplasticity describes changes that occur in the function and organization of the brain as a consequence of experience. One of the most profound changes in experience is the loss of an entire sensory modality, such as deafness or blindness. Crossmodal plasticity occurs when a sensory brain region is deprived of normal input and is innervated by another sensory modality.

This reorganization can be considered as either adaptive or maladaptive: adaptive crossmodal plasticity refers to enhanced perceptual performance in the intact sensory systems whereby one or more intact modalities compensate for the impaired sensory system. Thus, in the deaf, the remaining sensory modalities adapt and are enhanced through interaction with the environment. However, reorganization can also be considered maladaptive, as crossmodal plasticity can limit an impaired sensory system's ability to adapt to restoration. Either the establishment of novel neural responses and pathways or the unmasking and strengthening of existing connections may accompany functional crossmodal plasticity in the auditory cortex of the deaf. In this chapter, we will discuss crossmodal reorganization that appears to follow hearing impairment with consideration of potential mechanisms.

Crossmodal neuroplasticity in animal models of deafness

Changes in auditory cortex as a consequence of deafness: structure and neural function

Deafness has specific influences on the location, size, function, and connectivity of auditory cortex.[5] In congenitally deaf cats, primary auditory cortex (A1) has a similar laminar structure to that of hearing animals.[6] Electrophysiological studies have suggested that the area occupied by A1 increases slightly following neonatal deafening,[7] while the size of A1 in congenitally deaf animals appears to be no different than in hearing animals.[8] Conversely, anatomical studies have demonstrated a trend toward an overall decrease in the size of auditory cortex in the deaf. This magnitude of this effect appears to be inversely correlated with the age of deafness onset and is driven in large part by significant reductions in the size of A1 following both early- and late-onset deafness.[9] In addition, congenitally deaf animals present with reductions in both the number of primary dendrites and in the span of dendritic trees in primary auditory cortex relative to hearing controls.[10] Thus, while gross-level anatomical similarities may exist between hearing and nonhearing animals, functional connectivity differs greatly between the two. For example, inputs to layers III/IV of A1 are present in congenitally deaf animals, as are subsequent inputs to more superficial, supragranular layers.[11] However, activity in deeper, infragranular layers is significantly decreased,[8,12] and synaptic current latencies are significantly longer (after controlling for brainstem latency shifts[12,13]), suggesting that connections between superficial and deeper layers do not mature. In hearing animals, the infragranular layers of A1 are the source of descending, feedback projections. Thus, inactivity in these layers following auditory deprivation suggests that subcortical feedback loops are not driven by auditory input in a manner similar to that observed in hearing animals.

In hearing animals, supragranular layers of A1 project to higher-order areas of auditory cortex. The presence of supragranular activity in electrically stimulated deaf animals suggests that feed-forward connections persist between A1 and secondary auditory areas in deaf animals, at least early in development. Feedback projections from these higher-order auditory areas primarily target the deep layers of A1.[14] Inactivity in the infragranular layers of A1 suggests that these feedback projections and the associated top-down modulation of activity in A1 do not develop in deaf animals.[15] In support of this idea, in vitro electrophysiological examination of hearing-deprived auditory cortex has demonstrated that layer V neurons are incapable of undergoing the sort of long-term potentiation that normally underlies synaptic plasticity.[16]

Functional changes in the primary auditory cortex of congenitally deaf animals have been explored using in vitro electrophysiological techniques, as well as through the introduction of peripheral electrical stimulation via a CI. Multiunit recordings from A1 in deaf animals show slightly increased spontaneous firing rates when compared with hearing animals, which may reflect upregulated spontaneous activity in thalamic inputs.[17] Additionally, the excitability of A1 neurons has been shown to increase following deprivation of afferent activity, while inhibition is decreased.[7,16,18] Together, these results suggest that cortical neurons favor excitability, likely as a response to reduced cochlear excitation. However, when driven via electrical stimulation, evoked neural activity is decreased in congenitally deaf animals compared with electrically evoked activity in implanted hearing controls.[18]

Apart from these changes in the rate of activity, the rudimentary properties of A1 neuron responses appear to be present in congenitally deaf animals, despite a complete, and in some cases long-term, lack of auditory experience. For example, the rate-intensity and latency-intensity functions of electrically stimulated deaf A1 neurons are similar to those of hearing animals.[17,19] Interestingly, there are no reports of changes in the temporal processing of electrically stimulated A1 neurons, despite changes in downstream structures.

Electrophysiological investigations of crossmodal plasticity in the auditory cortex of deaf cats,[17,20–24] mice,[25] and ferrets[26] have identified various levels of increased responsiveness to visual stimuli. Early studies utilized bipolar (transcortical) electrodes to record visually evoked potentials (VEPs) in deaf cats and compare them to hearing animals. These experiments determined that VEP components originating in A1 were elicited by photic stimuli in congenitally deaf animals as well as cats deafened during the first three postnatal weeks.[24] However, no VEP component from A1 was observed in white cats with incomplete deafness, pigmented cats deafened after 4 weeks of age, or in hearing controls.[23,24] Single-unit studies of crossmodal plasticity provide the most direct measure of function at the neuronal level and provide a mechanistic understanding of the changes that occur subsequent to hearing loss onset. Investigations across species (cats, ferrets, and mice) find that the overwhelming proportion of units in auditory cortex are unimodal, responding only to auditory stimuli.[21,22,25,27–32] However, in the normally developed auditory cortex, there are also unimodal neurons that respond to tactile[22,25] or visual[21,30] stimulation. Furthermore, neurons can also be identified that are multimodal (responding to both acoustic and nonacoustic stimuli).[21,25,30]

Following deafness, the relative number of neurons in auditory cortex that respond to tactile and visual stimuli increases. Single-unit recordings in the core auditory cortex of ferrets (A1 and the anterior auditory field (AAF)) with late-onset, partial deafness (189–240 days postnatal; ~32 dB deficit) have shown an average of 165% increase in the number of action potentials evoked by visual stimulation (flashed or moving light) when compared with controls.[33] Moreover, deaf ferrets show a dramatic decrease in the number of unimodal auditory neurons compared with normal hearing animals (31% vs. 65%, respectively) and an increase in the number of neurons that respond to visual and/or somatosensory in addition to auditory stimuli (68% vs. 34%, respectively[33]). A similar shift has been observed in core auditory areas of congenitally deaf mice. Hunt and colleagues[25] observed that while 91% of neurons in A1 and AAF of intact mice are unimodal auditory units, neurons in A1 and AAF responded predominantly to somatosensory stimuli (50%–68%) or to somatosensory and visual stimuli (18%–40%) in deaf animals. These same deaf animals also showed an increase in the cortical volume occupied by the primary visual cortex (V1) relative to controls. Similarly, the AAF of hearing controls contains 93% unimodal auditory neurons with the remaining 7% being nonresponsive to sensory stimulation; in stark contrast, 83% of the neurons in the AAF of early-deaf cats respond to somatosensory cues and 44% are driven by visual stimulation (these numbers include both unimodal and bimodal neurons).[22] Interestingly, studies of A1[17,34] and of the second auditory cortex[34] (A2) in the deaf cat suggest that units in these regions do not respond to either visual or tactile stimuli, suggesting a lack of crossmodal reorganization. However, subsequent analyses of this work[5] have raised the possibility that visually evoked activity may have been suppressed by the anesthetic regime.

In the cat, crossmodal plasticity has also been assessed in non-primary regions. In the auditory field of the anterior ectosylvian sulcus (FAES) of deaf animals, 68% of neurons are driven by visual stimulation while 34% respond to tactile stimulation (including bimodal neurons that respond to visual plus somatosensory stimulation), and a small percentage are unresponsive.[21] This is in contrast to the FAES of hearing animals, where the majority of neurons (~60%) are unimodal auditory units, while approximately 20% are responsive to a combination of auditory and visual inputs.[21] Similarly, the dorsal zone of the auditory cortex (area DZ) also presents more visually responsive units in deaf cats than in hearing cats.[20] Land and colleagues[20] also demonstrated that CIs evoke similar firing rates and a comparable number of responsive sites in the DZ of deaf and hearing controls, suggesting that acoustically-responsive units are retained following deafness. Therefore, both core and non-primary regions of auditory cortex show similar levels of crossmodal plasticity following deafness.

The auditory cortex of congenitally deaf animals maintains a rudimentary representation of tonotopy (organization based on the preferred frequency of neurons), even after extensive periods of hearing loss,[6,8,35,36] with a similar spread of activity in response to electrical stimulation to that observed in implanted hearing controls.[18] Conversely, neonatally deafened animals show a near-complete loss of tonotopic organization and a rostrocaudal spread of activation in A1.[7,37,38] However, chronic intracochlear electrical stimulation after an extended period of deafness is able to restore tonotopy in some (but not all) cases: this has implications for the performance of patients implanted with a cochlear prosthetic after an extended period of deafness.[37,38] Tonotopic organization of the inferior colliculus (the principal source of ascending inputs to the auditory thalamus) remains following neonatal deafening, and thalamocortical projections to A1 have been shown to be relatively normal in deafened animals[39,40]; this suggests that differences in A1 tonotopy are the result of reorganization at the level of the thalamus or within A1 itself, serving to increase the overlap between adjacent basilar membrane representations.

Crossmodal reorganization in auditory cortex following deafness: behavior and psychophysics

Throughout the auditory system, there appears to be a rudimentary organization that results from a genetic blueprint, which is established before the onset of hearing. In normally developing animals, this organization undergoes experience-dependent refinement, such that adult-like perception is achieved only after hearing onset. As with other sensory systems, congenital deprivation results in an immature system that appears to persist for some time following the normal point of hearing onset. However, if sensory input is not restored before the end of the sensitive period for normal development, many auditory structures may be recruited by another sensory modality. This crossmodal reorganization of cortical structures is thought to underlie behavioral enhancements observed in the remaining sensory modalities of both animal models[41] and of humans.[42]

In hearing animals, the response properties of A1 neurons remain dynamic into adulthood, undergoing rapid changes to optimize auditory perception. For example, animals trained to detect a tone of a particular frequency within a complex soundscape show facilitated processing for that frequency in A1.[43] However, in animals with congenital or neonatal deafness, A1 does not appear to be subject to crossmodal reorganization with the visual or somatosensory systems, even following extended periods of auditory deprivation,[17,23,34,44] and despite the presence of crossmodal projections from somatosensory areas.[26,27] This is in accordance with research in the visual system: congenital blindness leads to the processing of auditory stimuli in areas of cortex which normally process visual information in both cats[31] and humans.[45] However, such crossmodal reorganization is limited to higher-order visual areas, with no change in primary visual cortex.[46,47] This is despite evidence that primary sensory areas are *capable* of processing information from remaining sensory modalities when that information is introduced via surgical manipulation of afferent inputs.[48,49]

Unlike A1, there is evidence that higher-order auditory areas process nonauditory stimuli in deaf animals. For example, it has been demonstrated that recruitment of auditory areas typically involved in sound localization, including the posterior auditory field[41,50] (PAF) and the FAES,[21] underlies enhanced peripheral localization of visual stimuli in deaf animals. Moreover, the DZ of the auditory cortex, which lies adjacent to the visual motion processing regions of the middle suprasylvian sulcus,[51] has been shown to mediate enhanced visual motion sensitivity in deaf animals[41] (Fig. 16.1A), similar to the enhancement of visual motion sensitivity observed in deaf humans[52] (Fig. 16.1B). Lomber and colleagues[51] not only showed that a particular enhanced function could be localized within the auditory cortex of deaf cats, but also that the two different compensatory visual effects, visuospatial localization and visual motion discrimination, could be localized to two distinct regions. These results demonstrate a double dissociation of visual functions in reorganized auditory cortex of the deaf cat (Fig. 16.2), suggesting that two *auditory* cortical regions mediate independent *visual* functions/behaviors in deafness.

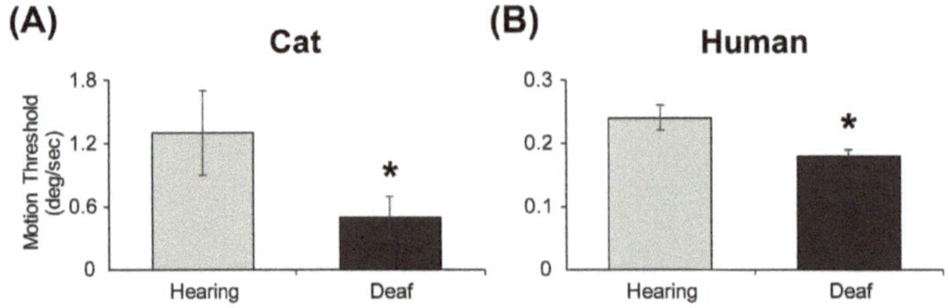

FIGURE 16.1 **Mean motion detection threshold ? s.e.m. for hearing (light gray) and deaf (dark gray) cats (A) and human subjects (B).** Data from cat (Lomber SG, Meredith MA, Kral A Crossmodal plasticity in specific auditory cortices underlies visual compensations in the deaf. *Nat Neurosci* 2010;13:1421–1429) and human (Shiell MM, Champoux F, Zatorre RJ. Enhancement of visual motion detection thresholds in early deaf people. *PLoS One*, 2014;9(2):e90498) studies. Asterisks = significant differences.

What is the anatomical basis for crossmodal reorganization following deafness?

The structural or anatomical basis of the crossmodal reorganization described above remains an issue of debate. Rauschecker[53] described several possible cortical mechanisms,

Task	PAF Deactivation	DZ Deactivation
Visual Localization in the Peripheral Field	Deficit	No Deficit
Movement Detection	No Deficit	Deficit

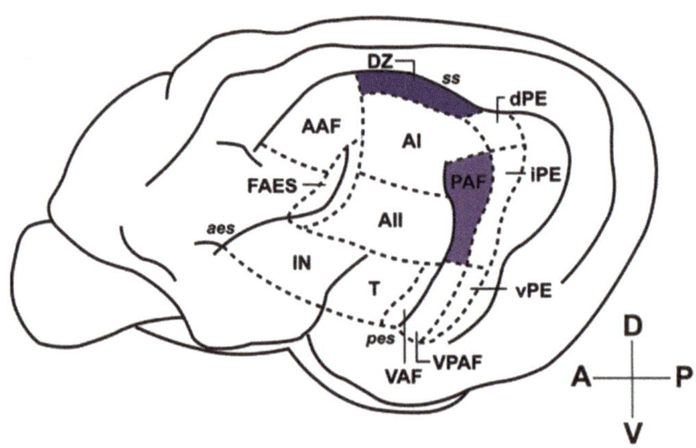

FIGURE 16.2 **Summary diagram illustrating the double-dissociation of visual functions in auditory cortex of the deaf cat.** Bilateral deactivation of posterior auditory cortex (area PAF), but not dorsal auditory cortex (area DZ), results in the loss of enhanced visual localization in the far periphery. On the other hand, bilateral deactivation of DZ, but not PAF, results in higher movement detection thresholds. Lower panel shows a lateral view of the cat cerebrum highlighting the locations of PAF and DZ. *Figure based on data from Lomber SG, Meredith MA, Kral A Crossmodal plasticity in specific auditory cortices underlies visual compensations in the deaf.* Nat Neurosci *2010;13:1421–1429.*

including unmasking of silent inputs, stabilization of normally transient connections, sprouting of new axons, or some combination of these processes. Indeed, anatomical studies in nonrodent species have demonstrated that nonprimary regions of auditory, visual, and somatosensory cortex are connected to one another both directly[26,27,54–56] and via multisensory cortical areas.[57] However, this is not generally thought to be the case with primary sensory cortices.[58] Thus, it is possible that intermodal connections that are normally latent or transient may underlie reorganization. Such reorganization is often examined using tracer injections designed to determine whether the number of neurons connecting sensory areas is increased following deafness. Indeed, the number of thalamic and cortical neurons projecting to A1,[40] A2,[59] DZ,[60–62] and PAF[63,64] has been extensively studied. All of these studies have very similar findings and report little difference between hearing and deaf cats in the pattern or quantity of neurons projecting to these cortical fields. However, this does not mean that the patterns or numbers of neurons projecting to all regions of deaf auditory cortex are unaffected by hearing loss. There are significant changes in the numbers of neurons projecting to the AAF from somatosensory cortex following early hearing loss.[65] These changes may underlie the somatosensory crossmodal plasticity functionally identified in early-deaf AAF at the single-unit level.[22]

In addition to changes in numbers of neurons projecting between areas, intersensory connections might also be strengthened via increases in dendritic branching and synapse number. Thus, anterograde tracing and analysis of changes in the number of terminal boutons would provide a fuller insight into the role of intracortical connections in crossmodal plasticity. This proposal is supported by the recent studies of Clemo and colleagues[66,67] demonstrating increased supragranular dendritic spine density in A1 and FAES of the early-deaf cat. As outlined earlier, while evidence of functional plasticity within A1 is limited, FAES has been demonstrated to undergo significant visual crossmodal plasticity at both the single-unit level and it has been shown to mediate behavioral plasticity as well.[21]

Conversely, it has also been suggested that cortical reorganization may result from changes in subcortical inputs.[27] For example, both the cochlear nucleus[68] and inferior colliculus[69] have been shown to respond to somatosensory inputs in hearing animals, and this response is enhanced following hearing loss.[70,71] In the absence of auditory input, subcortical nuclei may respond to inputs from other sensory modalities, and the reorganization of auditory cortex may simply reflect downstream processing of these changes. Cortical and subcortical mechanisms for reorganization are by no means mutually exclusive; it is likely that crossmodal plasticity involves some combination of mechanisms that depend, at least in part, on the nature of the hearing impairment, the timing of auditory deprivation, and the replacement sensory modality involved.

Crossmodal neuroplasticity in clinical populations with deafness and hearing loss

It is estimated that permanent bilateral hearing loss affects 1–3 per 1000 children.[72–74] Unilateral hearing loss also affects a high portion of the population, with a prevalence of 3%–6% among school-aged children and rising statistics with increased age, with a reported 6.5 million adults reporting hearing loss in one ear.[75] In adults, age-related hearing loss (ARHL) is also remarkably pervasive.[76] The previous section of this chapter described crossmodal plasticity in animal models of deafness, in which auditory deprivation results in the recruitment and repurposing of auditory cortical regions for visual and somatosensory processing. The research in deaf animals has inspired studies of crossmodal plasticity and reorganization in human clinical populations with deafness and hearing loss. As we will show, crossmodal cortical plasticity is evident in both pediatric and adult-onset hearing loss. Furthermore, crossmodal reorganization has been documented across the severity spectrum of hearing loss, including bilateral deafness, single-sided deafness (SSD), and early-stage, mild-moderate, ARHL. In this section, we will describe the relationship between crossmodal plasticity and behavioral outcomes in individuals with impaired hearing, including those fitted with CIs.

The CI is the most advanced neural prosthesis in existence, having restored hearing for hundreds of thousands of deaf adults and children worldwide. A CI consists of an externally worn microphone and speech processor that picks up sound, processes it, and converts it into an electrical signal; a receiver-stimulator implanted under the skin receives and transmits this signal to an electrode array implanted in the inner ear which stimulates the cochlear nerve fibers, bypassing the damaged inner ear. In this section, we will describe how CIs provide a platform to understand deprivation-induced neuroplasticity in the central auditory system. We will describe studies highlighting crossmodal reorganization by vision and somatosensation in adults and children with hearing loss, including those fitted with CIs. Finally, in line with advances in technology and medicine over the past several decades and the growing corpus of neuroscience literature examining neural plasticity in hearing loss, we direct this chapter toward the next frontier: Developing targeted and individualized intervention and rehabilitation for patients with hearing loss; toward slowing, arresting, or reversing negative consequences of auditory deprivation; and toward harnessing the neuroplastic nature of the human brain to promote brain development and maintenance within the context of healthy aging.

Many neuroimaging methods are employed in the study of crossmodal neuroplasticity. While it is limited in spatial resolution, electroencephalography (EEG) offers several benefits over other imaging modalities for study of the auditory system in human populations due to its excellent temporal resolution, non-magnetic compatibility with CIs, minimization of electrical artifact generated by CIs, and its inexpensive, non-invasive, easy application in infants and children. In the following sections, we describe evidence of developmental and crossmodal neuroplasticity in hearing loss in human EEG studies, as well as findings using other neuroimaging modalities that can provide more detailed spatial localization of cortical activity.

Congenital auditory deprivation and neuroplasticity

Pediatric hearing loss may be congenital or acquired. More than 50% of congenital sensorineural hearing loss of the inner ear is of genetic origin (80% recessive, 70% nonsyndromic), including mutation of the GJB2 gene which is estimated to account for 30%–50% of nonsyndromic sensorineural hearing loss.[77] Nongenetic congenital hearing loss may arise from intrauterine infections, meningitis, hyperbilirubinemia, ototoxic drugs, and auditory neuropathy.[77] Auditory deprivation during development fundamentally alters normal extrinsic (experience-driven) development of the central auditory pathways. Audiological intervention in children has been shown to be most effective when delivered within a 3.5-year sensitive period in which there is peak neural plasticity and the central auditory pathways are most amenable to change.[78–83]

Evidence of the sensitive period for central auditory development in children stems from EEG studies using the P1 component of the cortical auditory evoked potential (CAEP). The P1 component of the CAEP is a noninvasively recorded event-related potential that occurs in response to auditory stimulation. Comparing the latency of this component to normative data from an age-matched population[79–81,84] can provide information about maturation of the primary auditory cortex and feed-forward, thalamocortical connections. The latency of this response occurs around 300 ms in children and systematically decreases rapidly over the first few years of life, gradually reaching adult-like latencies by ages 12–16 years.[85–93] This age-related decrease in P1 latency coincides with developmental periods of synaptic pruning, increased neural efficiency, and experience-dependent exposure to sound during early infancy and childhood.

Children who do not receive consistent quality and quantity of input during the sensitive period rarely exhibit normal auditory cortical development. Late-implanted congenitally deaf children (implanted after age 7 years) continue to exhibit delayed P1 latencies even after years of CI use,[79–83] whereas the majority of early-implanted children (implanted before age 3.5 years) exhibit rapid decreases in P1 latency shortly after implantation, developing a normal response within 6–8 months,[82] suggestive of a sensitive period of 3.5 years for central auditory development in children with CIs. Together with behavioral data documenting better listening and spoken language outcomes in early- compared to late-implanted children,[84,94–100] neuroimaging evidence in children documents the importance of early audiological intervention and habilitation within this sensitive period.

Unlike the central auditory pathways which are at least partially formed at birth, data from animal studies suggest that the capacity to integrate multisensory information is not present at birth, but develops only after experience (see chapter by Stein and Rowland, this volume).[101,102] For example, neurons in the cat superior colliculus do not integrate multisensory input until months after birth, after receiving specifically timed multisensory experiences.[103–105] In fact, it appears that multisensory integration is a process that emerges in childhood but does not fully develop until midadolescence, around 12–15 years of age in humans.[103–105] Research indicates that multisensory development of the secondary auditory cortex can be disrupted by childhood hearing loss. For example, late-implanted children (implanted >7 years) rarely develop later CAEP components (e.g., N1/P2 complex),[92,106–108] reflecting abnormalities of the corticocortical and corticothalamic connections required for higher-level auditory and multimodal processing, such as that required for successful acquisition of spoken language and advanced skills such as those involved in literacy.[87,109–112] Findings of an absent N1/P2 complex in late-implanted children[106,108] are consistent with evidence of altered feed-forward and feedback cortical circuitry in animal models of deafness, a neural framework vital for auditory predictive coding and learning.[113]

Compensatory crossmodal plasticity in pediatric deafness

Altered feed-forward and feedback circuitry in the central auditory system as a result of deafness or hearing loss may contribute to compensatory, crossmodal reorganization. In pediatric deafness, auditory deprivation induces compensatory, selective repurposing of auditory cortical regions by intact sensory modalities (e.g., vision, somatosensation).[44] For example, a recent study examined cortical visual evoked potentials (CVEPs) to an apparent motion stimulus in a group of children with CIs (n = 14) and a group of age-matched children with normal hearing (n = 41), using high-density EEG.[114] The children with CIs, on average, received their first device toward the end of the sensitive period (average = 3.12 - years, sd = ± 2.27 years) and their second CI well after the sensitive period (average = 6.20 - years, sd = ± 3.45 years). Whereas children with normal hearing showed the expected pattern of activity elicited by visual motion in visual cortical regions, the children with CIs showed more widespread activity across both visual cortex and regions of the temporal lobe, including the auditory cortex (hereafter the term "temporal regions" will be used to refer to auditory cortex and surrounding areas; Fig. 16.3A).[114] These results are consistent with positron emission tomographic imaging findings evidencing higher levels of resting glucose metabolism over temporal, frontal, and visual cortices in the deaf[115–117] and in pediatric CI recipients with poor speech perception outcomes.[118–120]

Crossmodal plasticity involving the somatosensory modality has also been documented in pediatric deafness.[108] For example, both parietal and temporal regions respond to somatosensory inputs (250 Hz vibrotactile stimuli) in children implanted with a CI toward the end of the sensitive period (average age at first implant = 3.90 years, sd = ±4.03 years; average age at second implant = 7.33 years, sd = ±4.47 years), while age-matched normal hearing children exhibit responses restricted to parietal cortex (n = 35, average age at test = 10.54 years, sd = ±4.03 years).[115] Crossmodal recruitment of temporal regions by somatosensory stimuli has been similarly reported in congenitally deaf adults.[121–123]

There is evidence to suggest that crossmodal plasticity may be related to functional reliance on vision and/or enhanced auditory–visual integration in hearing impaired children. Late-implanted children are better lipreaders than early-implanted children and derive greater benefit from vision when facial cues are added to auditory stimuli for speech perception.[124] Furthermore, children with CIs continue to exhibit improvements in lipreading and auditory–visual speech perception even years after implantation.[125] Children with CIs show greater reliance on vision during auditory–visual integration, as demonstrated by the McGurk effect, compared with normally hearing children.[126] Taken together, these studies suggest a link between compensatory crossmodal neural plasticity in pediatric deafness and functional dependence on vision in some patients, whereby children may rely on other sensory inputs to disambiguate degraded auditory signals.[127] It is possible that auditory–somatosensory plasticity may also hold functional significance. For example, vibrotactile inputs have been shown to play an important role in the differentiation between same-gender talkers[128] and timbre differentiation between musical instruments[129] in deaf listeners. While somatosensory enhancement following hearing loss has not been well studied, there have been some reports of enhanced tactile discrimination in deaf children.[130,131] Thus, consistent with animal models,[22,27,68,132,133] deaf children show behavioral evidence of enhanced somatosensation, which may reflect recruitment and repurposing of auditory brain regions. Relying more on intact sensory inputs such as vision and touch may be a natural mechanism employed to compensate for impaired audition in real-world, multisensory listening situations.

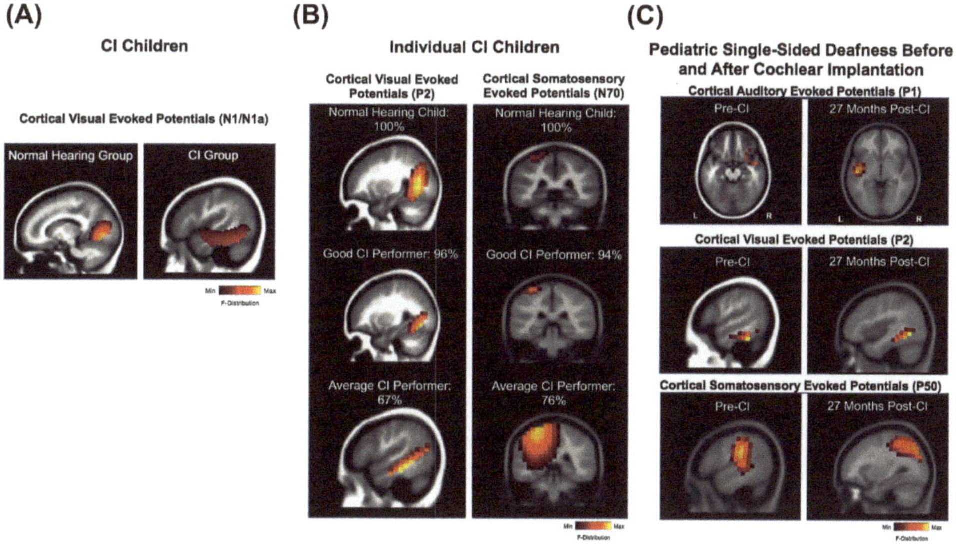

FIGURE 16.3 **Panel A: Visual crossmodal reorganization in CI children.** N1 cortical visual evoked potential (CVEP) current density source reconstructions for a group of normal hearing children (n = 41) and a group of cochlear implant (CI) children (n = 14) in response to a visual motion stimulus. Normally, hearing children show activity restricted to visual cortex, whereas CI children show additional recruitment of temporal regions, including auditory cortex, for the processing visual motion stimuli suggestive of crossmodal recruitment. **Panel B: Visual and somatosensory crossmodal reorganization in individual CI children.** Current density source reconstructions for the P2 CVEP (left panel) and current density source reconstructions for the N70 cortical somatosensory evoked potential (CSSEP) in six individual children (right panel). While the two normally hearing children (upper panels) and the two CI children with good speech perception abilities (96% and 94% accuracy on unimodal auditory tests of speech perception; middle panels) show CVEP activity restricted to visual cortex and CSSEP activity restricted to parietal cortex, the two average performing CI children (67% and 76% accuracy on unimodal auditory tests of speech perception; lower panels) show more widespread activation of temporal regions for both visual and somatosensory processing. **Panel C: Crossmodal reorganization in pediatric single-sided deafness before and after cochlear implantation.** Current density source reconstructions for the P1 auditory evoked potential (CAEP) (top panel), P2 CVEP (middle panel), and P50 somatosensory evoked potential (bottom panel) in a child with progressive single-sided deafness in the right ear before CI (left panels) and 27 months post-CI (right panels). Before CI, stimulation of the deaf ear elicits activation of the ipsilateral temporal (including auditory cortex) and frontal cortex, evidence of atypical auditory dominance patterns that are a result of unilateral auditory deprivation; visual stimulation elicits visual and temporal regions and vibrotactile stimulation elicits parietal and temporal cortical activation. However, within 27 months of CI, more typical contralateral dominance patterns in the temporal (auditory) cortex is observed with auditory stimulation of the deaf ear. Furthermore, a reduction in temporal (auditory) cortical recruitment for visual and somatosensory processing is evident within 27 months of CI use, indicating alterations in crossmodal neuroplasticity with CI treatment. Such changes in neuroplasticity coincide with the child's improvements in auditory-only speech perception and auditory localization after treatment with CI. *Panel A: Adapted from Campbell J, Sharma A. Visual crossmodal Re-organization in children with cochlear implants.* PLoS One *2016;11(1):e0147793 with permission. Panel B: Adapted from Sharma A, Campbell J, Cardon G. Developmental and crossmodal plasticity in deafness: evidence from the P1 and N1 event related potentials in cochlear implanted children.* Int J Psychophysiol *2015;95(2):135–144 with permission. Panel C: Adapted from Sharma A, Glick H, Campbell J, et al. Cortical plasticity and reorganization in pediatric single-sided deafness pre- and post-cochlear implantation: a case study.* Otol Neurotol *2016;37:e26–34 with permission.*

Clinical implications of compensatory crossmodal plasticity in pediatric hearing loss

Using crossmodal reorganization to direct individualized intervention and habilitation for children with hearing loss

Despite a push by the Federal Drug Administration to allow children with hearing loss to receive a CI as early as 12 months of age, and a growing emphasis on population-based newborn hearing screening to detect permanent childhood hearing loss,[134] there still exists a great deal of variability in performance outcomes for children with CIs. For instance, it has been reported that less than 50% of the variability in listening and spoken language outcomes is accounted for by demographic factors such as age of implantation.[135] Crossmodal reorganization is one factor that may well contribute the variability in behavioral outcomes among children with CIs. For example, in the study described above, early CVEP latencies (suggestive of auditory–visual crossmodal plasticity) were related to poorer auditory speech perception in background noise in children with CIs,[114] and similar findings have been observed in adults with hearing loss.[136,137]

Crossmodal reorganization may have important implications regarding habilitation for pediatric patients. Fig. 16.3B highlights how the neurophysiological profile of crossmodal reorganization in individual children may be used to predict behavioral outcomes.[108] Fig. 16.3B (*left panel*) depicts current source density reconstructions (CDRs) for the P2 CVEP component in a normal hearing child (age 10 years), a pediatric CI user with excellent auditory speech perception outcomes (age 8 years; 96% on the Lexical Neighborhood Test), and a pediatric CI user with average performance on an auditory unimodal test of speech perception (age 6 years; 67% on the Multisyllabic Lexical Neighborhood Test), in response to the same visual motion stimulus described above.[114] While the child with normal hearing and the child with a CI and excellent speech perception exhibit activation of visual cortical regions associated with visual motion processing (e.g., occipital gyrus, fusiform gyrus, lingual gyrus), the child with a CI and average speech performance exhibits more widespread activation across visual and temporal regions, suggestive of auditory–visual crossmodal plasticity. Fig. 16.3B (*right panel*) also shows CDRs for the N70 component of the cortical somatosensory evoked potential in three separate children: a child with normal hearing (age 7 years), a child with a CI exhibiting excellent auditory speech perception (age 13 years; 94% on the Consonant Nucleus Consonant (CNC) test), and a child with a CI exhibiting average performance on an auditory unimodal test of speech perception (age 15 years; 76% on the CNC test), in response to a vibrotactile stimulus delivered to the right index finger.[138] Where the child with normal hearing and child with a CI and excellent speech perception showed expected activation in cortical regions associated with somatosensory processing (e.g., postcentral gyrus), the child with a CI and average auditory speech perception exhibited additional recruitment of temporal regions (e.g., superior temporal gyrus, transverse temporal gyrus) suggestive of auditory–somatosensory crossmodal reorganization. The aforementioned results demonstrate that crossmodal reorganization in deafness may be associated with less than optimal performance on current clinical tests of speech perception in noise without multimodal cues. However, these results also highlight that the clinical audiology community needs to develop new, ecologically valid tests which measure multisensory speech perception, as crossmodal plasticity may enhance multisensory speech perception.[139] While single-case study evidence should be interpreted with the appropriate caution, this clinical research highlights the potential for neurophysiological biomarkers of crossmodal plasticity to predict clinical outcomes following audiological intervention.

Knowledge about the extent of crossmodal reorganization in individual patients may help develop targeted, individualized therapeutic plans following audiological intervention. For example, there still exists great debate over decisions on modes of communication and choice of rehabilitation programs for pediatric populations with hearing loss. With further research, objective markers of crossmodal reorganization may help to objectively guide rehabilitation decisions after an intervention has been chosen. For example, it is possible that patients who continue to exhibit widespread crossmodal plasticity related to vision and somatosensation after CI may perform better with a multimodal approach to speech and language acquisition. Clinical recommendations about rehabilitative training programs or paradigms for CI patients currently lack evidence-based scientific rigor.

Crossmodal reorganization as an indicator of efficacy of cochlear implantation

There is some evidence to suggest that crossmodal plasticity is reversible following audiological intervention, at least in the case of SSD. Unilateral hearing loss has been historically underdiagnosed and undertreated among both children and adults alike, despite compelling evidence that unilateral auditory deprivation during childhood is associated with deficits in sound localization,[140–142] educational delays in school,[143–146] and delays in language development[147] that persists into adulthood.[148] Despite deafness in one ear, intervention options are limited and CI is not currently FDA-approved for these patients.[149–155] Fig. 16.3C depicts CDRs to visual and somatosensory stimuli from a child with SSD.[138] The child had a progressive SSD in the right ear and received a CI at age 9.86 years. Before CI, an immature CAEP response was observed in the impaired ear, marked by a borderline delayed P1 latency and absence of the N1/P2 complex. Furthermore, the typical pattern of bilateral activity, wherein activity is greater in the contralateral hemisphere, was not apparent before CI; rather, stimulation of the child's impaired ear elicited activity confined to the ipsilateral temporal cortex (Fig. 16.3C *upper left panel*). These findings are consistent with studies of unilateral hearing arising from long delays between first and second CI (e.g., >1.5 years) in bilaterally deaf children, in which atypical development of the central auditory pathways has been found.[156,157] Before CI, visual motion stimulation elicited activation of both visual and temporal regions in this child with SSD, and vibrotactile stimulation elicited activation of parietal and temporal regions (Fig. 16.3C *middle and lower left panels*), suggesting that auditory cortex had been reorganized to contribute to both visual and somatosensory perception. However, restoration of more typical pattern of activity when stimulating the implanted ear, marked by activation of contralateral cortex, was observed following 27 months of CI use (Fig. 16.3C *upper right panel*). A reduction in temporal recruitment in response to visual and somatosensory stimulation was also evident (Fig. 16.3C *middle and lower right panels*). These changes coincided with the child's improvements in auditory speech perception in background noise and auditory localization.[138]

Similar results have been documented in a recent study of five children with SSD.[158] Here, children with a moderate to severe hearing impairment in one ear were fitted with a CI, and CAEPs were recorded at the time of implantation, 1 month after CI, and 6 months after CI. As in the case study described above, responses at implantation did not reflect the pattern of contralateral dominance shown in children with normal hearing and were suggestive of recruitment of higher-level cortical areas involved in arousal and attention. However, more normal patterns of activity emerged over this relatively short period of bilateral sound experience, suggestive of a remarkable level of plasticity. A subsequent study of 10 children with asymmetric hearing (a single-sided impairment of ~45 dB) confirmed these findings and suggested that the degree to which a normal pattern of contralateral dominance was established following CI was inversely correlated with the duration of asymmetric hearing.[159] Future studies should extend on this work to systematically examine the relationship between crossmodal plasticity before and after intervention as it relates to functional outcomes of unimodal and multisensory auditory speech perception and auditory–visual integration, in clinical populations with hearing loss.

Age-related hearing loss and neuroplasticity

The prevalence of ARHL increases with each decade of life (34%, 53%, and 77% of adults in their 40s, 50s, and 60s, respectively).[76] ARHL, or presbycusis, is the most common form of hearing loss among older adults and typically presents as a bilateral, gradual, progressive sensorineural hearing loss often affecting the high frequencies first, and reflecting the multifactorial accumulation of normal aging, genetic factors, noise exposure, use of ototoxic medications, and presence of comorbid conditions such as diabetes over the lifetime.[158] While traditionally diagnosed using audiometric testing, new research suggests that traditional audiological testing may not be sensitive to early neural changes induced by auditory deprivation. For example, noise-induced hearing loss may result in permanent synaptic and central nervous system changes despite recovery of audiometric thresholds.[159,160] Anatomical changes occurring more centrally, such as the loss of spiral ganglion cells that often follows hair cell loss, may in turn affect auditory processing throughout the central auditory pathways, including at the cortical level.[161] Alongside functional changes, structural brain imaging studies have also demonstrated accelerated atrophy and decreased volume of gray matter in auditory cortex in adults with hearing loss.[162–164] While some of the behavioral consequences of these central changes in ARHL are unknown, recent research suggests that ARHL increases the risk of cognitive decline, including all-cause dementia and Alzheimer's disease, with hazard ratios for all-cause dementia increasing from 1.89 to 4.94 among people with mild and severe ARHL, respectively.[165,166] One theory describing a possible mechanistic link between hearing loss and cognitive decline is that hearing loss results in effortful listening, depleting cognitive spare capacity, and potentially accelerating cognitive decline beyond normal aging.[165,167–169] In light of these concerns, it is important that hearing losses in the elderly are also met with appropriate intervention in a timely manner.

Compensatory crossmodal plasticity in adult-onset deafness

Both pre- and postlingually deaf adults exhibit evidence of crossmodal neuroplasticity between auditory and visual brain regions.[115,116,137,170,171] For instance, individuals with adult-onset hearing loss and subsequent CI exhibit reduced CVEP amplitudes and spread of activation over visual cortex, and earlier latencies and increased activation over temporal cortex in response to visual stimuli relative to listeners with normal hearing. Furthermore, the magnitude of visually evoked activity over temporal cortex was negatively correlated with auditory-only speech perception outcomes in these recipients, whereas the magnitude of visually evoked activity over visual cortex was positively correlated with speech perception outcomes.[115] Such evidence of crossmodal recruitment of temporal regions in deafness is evident in other studies, where prelingually deaf adults with CIs show increased CVEP amplitudes over temporal cortex.[137] Similar findings have been reported in postlingually deaf adults with CIs.[171,172] Similar to children with CIs, more extensive crossmodal reorganization appears to correlate with poorer unimodal auditory-only speech perception performance. Doucet and colleagues,[172] for example, showed that deaf adults with CIs and excellent auditory speech perception exhibited cortical activation patterns restricted to visual cortex in response to visual motion stimuli, whereas deaf adults with CIs and poor auditory speech perception outcomes exhibited more pervasive spread of cortical activation into temporal cortices, suggestive of crossmodal reorganization. An interesting laterality effect exists with respect to crossmodal plasticity in human cortex; right temporal (auditory) cortical areas appear to be more susceptible to crossmodal recruitment by vision, with CVEP amplitudes/latencies over right temporal cortex more strongly correlated with auditory speech perception performance than those recorded over the left hemisphere.[114,171]

Similarly, crossmodal recruitment of temporal regions for tactile processing has been reported in adults with congenital and early-onset deafness,[82,122,173] suggestive of crossmodal plasticity between the auditory and somatosensory systems. Reorganization in deaf adults has been documented using neuroimaging techniques including functional near-infrared spectroscopy,[174,175] EEG,[137,171–173,176] and functional magnetic resonance imaging.[172] Preliminary case reports and group-level evidence also show evidence of crossmodal reorganization of auditory cortex by both visual and somatosensory modalities following adult-onset SSD, suggesting that unilateral deafness may also lead to deprivation-induced plasticity.

Visual crossmodal reorganization may be related to increased behavioral reliance on visual cues in postlingually deafened adults with CIs[177,178] and in deaf adults whose mode of communication is American Sign Language.[179] Additionally, the enhanced auditory–visual integration described above following early-onset hearing loss has also been reported in adult-onset deafness.[177,180–183] Deaf adults exhibited enhanced selective responses to faces when compared with nonface stimuli (e.g., houses) in the right midlateral superior temporal gyrus, a brain region normally involved in voice perception.[184] These findings have also been replicated by Strophal and colleagues[177] in a high-density EEG study, where deaf adults exhibit more anterior, right lateralized responses to faces that were selectivity localized to right temporal regions. Auditory–visual crossmodal reorganization in the deaf has been shown to have functional consequences: adults with more visually evoked activity over temporal cortices have also been shown to integrate auditory and visual signals more readily, as evidenced by paradigms like the McGurk effect, which asks participants to respond to incongruous bimodal stimuli. Future studies should focus on developing clinically relevant tests of multisensory speech perception that can elucidate the extent to which auditory and visual signals are rebalanced following sensory impairment to optimize multisensory communication.

Compensatory crossmodal plasticity in early-stage, mild-moderate, age-related hearing loss

Until now, all evidence of crossmodal plasticity has been discussed in children and adults with profound hearing loss or deafness. However, ARHL typically starts out as mild impairment and gradually progresses to moderate, severe, and then profound hearing loss or deafness over several years or even decades. Indeed, recent studies provide evidence of crossmodal plasticity in early-stage, mild to moderate, ARHL. Fig. 16.4 depicts CDRs in response to a moving visual stimulus in a group of adults with normal hearing and a group of adults with mild to moderate hearing loss,[136] many of whom were unaware that they had a hearing loss upon time of enrollment in the study. While the normally hearing group demonstrated a pattern of activation confined to cortical areas typically associated with visual motion processing for all CVEP components examined (P1, N1, P2) (Fig. 16.4), the mild to moderate hearing loss group showed more widespread activation across temporal regions for the later occurring CVEP components (N1, P2; Fig. 16.4C *left and right panels*). Furthermore, decreased CVEP latencies were negatively correlated with speech-in-noise perception in individuals with hearing loss, suggesting that changes in visual processing negatively impact auditory perception following hearing restoration.[136,137]

While these results were the first to suggest that crossmodal recruitment by vision occurs in early-stage, mild to moderate, ARHL, a recent EEG study[183] replicated this effect, demonstrating that while adults with mild to moderate hearing loss do not show crossmodal recruitment to the same extent as postlingually deafened CI recipients, they do show more extensive crossmodal activation than normal hearing listeners, supporting the idea that even mild auditory deprivation can induce crossmodal recruitment of the auditory system by vision. A recent case study of sudden adult-onset hearing loss suggests that crossmodal recruitment by vision is apparent within 3 months after hearing loss onset, coinciding with enhancement in auditory–visual speech perception.[185] While adult-onset auditory deprivation induces crossmodal recruitment of the auditory cortex for vibrotactile processing in animal models[27,66] and in completely deaf humans,[82,173,186] these studies have not been replicated in lesser degrees of hearing loss. Furthermore, the time course of crossmodal recruitment by somatosensation after adult-onset hearing loss is not well understood, though animal studies suggest repurposing of auditory pathways occurs as soon as 16 days after deafness onset.[27]

Another neuroplastic compensatory change in partial hearing loss is the recruitment of additional brain networks for auditory processing. In the same group of adults with mild to moderate hearing loss described above,[136] CAEPs were recorded in response to auditory speech stimuli.[187] Decreased activation over temporal regions was observed in the group with hearing loss compared with the group with normal hearing in response to speech stimuli, despite loudness balancing (Fig. 16.4A *left and right panels*), consistent with MRI studies showing accelerated atrophy and decreased volume of gray matter over right temporal lobes in the hearing-impaired adults.[162–164] Additionally, the group with hearing loss exhibited increased activity over frontal cortical regions (e.g., inferior and superior frontal gyrus), something that was not evident in the group with normal hearing (Fig. 16.4B *left and right panels*), suggestive of changes in cortical neurodynamics. Increased frontal cortical recruitment has been observed in degraded auditory conditions in both adults with normal hearing and adults with hearing loss.[163,188] Activity within these frontal regions is often accompanied by pupil dilation, a commonly used physiological indicator of increased cognitive load,[189] suggesting that the differential patterns of activity recorded in hearing-impaired adults may reflect more effortful listening. Additionally, changes in resting-state attention and default mode networks have been observed in clinical populations with hearing loss.[190] Thus, recruitment of crossmodal and frontal networks and/or other changes in cortical networks that occur subsequent to hearing loss may allow those with impaired hearing to

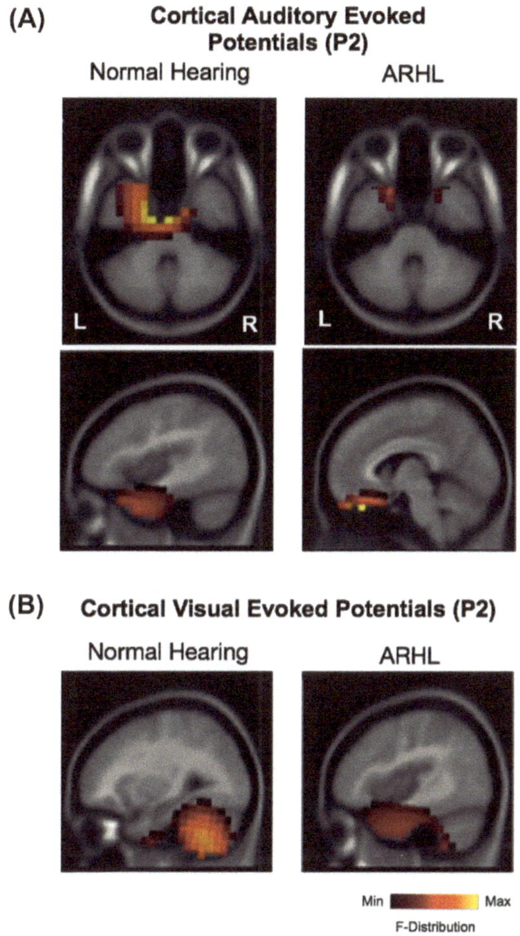

FIGURE 16.4 **Cortical brain changes in early-stage, mild to moderate, age-related hearing loss.** Panel (A): Horizontal and sagittal slices depicting cortical auditory evoked potential (CAEP) source reconstructions for the P2 components in a group of normal hearing adults (n = 8) (left top and bottom panels) and a group of hearing loss adults with bilateral, early-stage, mild to moderate hearing loss (n = 9) (right top and bottom panels). Auditory stimulation elicits decreased activity in temporal regions in the hearing loss group compared with the normal hearing group. Furthermore, auditory stimulation is associated with increased frontal cortical activity in the hearing loss group compared to the normal hearing group, indicating alterations in typical cortical resource allocation. Panel (B): Cortical visual evoked potential (CVEP) source reconstructions for the same group of normal hearing adults (left panel) and hearing loss adults (right panel) depicted in Fig. 16.4 Panel (A). Visual motion stimuli results in more widespread activity across visual and temporal regions in the hearing loss group, evidencing crossmodal recruitment, whereas the normal hearing group shows visual-driven activity restricted to the visual cortex. *Adapted from Campbell J, Sharma A. Compensatory changes in cortical resource allocation in adults with hearing loss.* Front Syst Neurosci. *2013;7:71 and Campbell J, Sharma A. Crossmodal Re-organization in adults with early stage hearing loss.* PLoS One *2014;9(2):e90594, with permission.*

compensate in degraded listening conditions, possibly invoking greater cognitive resources to accomplish difficult perceptual tasks.[167,191]

Even adults with mild to moderate hearing loss have been shown behaviorally to rely more on visual cues than their normal hearing counterparts.[192] Accordingly, enhanced auditory–visual integration in adult-onset hearing loss has also been reported in multisensory conditions.[177,180–183] In a study cited above,[180] auditory–visual integration was measured using the McGurk effect, and source localization of the N170 component elicited by faces and was compared between a group of adults with mild to moderate ARHL, a control group with normal hearing, and a group of postlingually deaf adults with CIs. The group with mild to moderate hearing loss showed improved audio–visual integration compared to normal hearing controls, though this effect was less pronounced than in the completely deaf adults with CIs. Interestingly, behavioral measures of auditory–visual integration and crossmodal recruitment of auditory cortex by vision were uncorrelated in the mild to moderate hearing loss group, perhaps suggesting that while crossmodal changes occur very early following hearing loss onset, behavioral dependence on vision (e.g., lipreading) and enhanced auditory–visual integration may not emerge until the degree of hearing loss begins to impact speech perception in real-world scenarios. While understudied, there is some evidence to suggest that vibrotactile (somatosensory) inputs may enhance speech perception in adult-onset hearing loss and deafness.[182,183] Future research is needed to investigate the possible clinical implications for speech perception of crossmodal reorganization involving the somatosensory system.

Clinical implications of crossmodal plasticity in age-related hearing loss

Using crossmodal reorganization to direct individualized rehabilitation for age-related hearing loss

Given the close relationship between crossmodal reorganization and auditory–visual speech perception, these findings may inform rehabilitation for patients with adult-onset hearing loss who are receiving intervention with hearing aids or CIs. Despite rapid improvement in auditory perceptual abilities in postlingually deaf adults following CI, there exists a large variability in performance outcomes.[181] Even after implantation, postlingually deaf CI users have been shown to continue to rely on visual cues for auditory–visual speech perception, particularly in degraded listening situations like in the presence of background noise.[181,193] Studies show that with ongoing CI use, crossmodal recruitment of auditory cortex by vision gradually decreases in parallel to auditory recovery.[181] Furthermore, greater levels of activity in visual brain regions elicited by speech and lower levels of crossmodal recruitment of temporal regions by vision before CI were predictive of auditory recovery 6 months postimplant,[194] suggesting that remapping between the visual and auditory cortices may impact speech perception outcomes following CI.[193] Given low rates (<10%) of individualized auditory rehabilitation therapy for adults receiving hearing aids and CIs,[195,196] greater resources should be devoted to systematically examining whether rehabilitation strategies that target and scaffold auditory-integration abilities yield better outcomes in adult patients receiving audiological intervention. While speech perception outcomes in clinical populations with hearing loss are almost exclusively measured under auditory-only conditions, emerging studies are examining the clinical utility of measuring auditory–visual integration abilities in CI patients.[139] While the use of tactile aids was introduced before the advent of the CI,[197] the concept of leveraging auditory–somatosensory integration has recently been reintroduced. A recent study by Huang and colleagues (2017) combined electrotactile stimulation with CI use[198] and found that the addition of a tactile stimulator conveying fundamental frequency information to the index finger significantly improved speech perception in noise when compared with CI stimulation alone, indicating a potential benefit of multimodal intervention in CI rehabilitation.[198]

Conclusions

Perhaps the most remarkable ability of the brain is its dynamic capacity for change. During development, neural plasticity serves as the catalyst for the formation and refinement of auditory pathways and the establishment of neural networks. Via sensory input beginning in childhood and lasting through to adulthood, neural plasticity allows for structural and functional changes within the central auditory system and helps promote language development, learning, and memory. Conferring adaptability to intrinsic and extrinsic forces, and the capability to adjust to insult and injury, neural plasticity is a critical component of brain development and maintenance across the life span. Neural plasticity attempts to compensate for abnormal periods of experience but can also give rise to a variety of disorders and disabilities. Indeed, several neurodevelopmental and neurodegenerative conditions reflect the cascading effects of aberrations in neural development over time.

The absence of auditory input that accompanies hearing impairment causes long-term changes to the structure and function of the auditory system. The exact nature of these changes depends on factors such as the etiology and onset time of the impairment and can have significant developmental and psychosocial consequences. Interventions including amplification and CI may mediate these consequences, but each depends critically on the integrity and function of remaining auditory structures. Studies undertaken in deaf animal models have provided much of what is known about the function of the brain in deafness and have informed the development and design of hearing prostheses. Perhaps most interestingly, these studies have informed our understanding of sensitive periods in development and their role in functional recovery following the provision of a hearing aid and/or CI. The animal studies described here have illustrated the importance of early intervention both in terms of minimizing structural and functional damage within auditory structures, as well as recovering temporal cortical areas that might otherwise be recruited by other sensory systems. However, the effects of deafness on higher-order cortical areas and the exact mechanism(s) underlying crossmodal plasticity are not yet fully understood. Thus, research using animal models will continue to inform our understanding of the far-reaching consequences of deafness as the field moves forward.

We have provided evidence of compensatory crossmodal plasticity across the life span, from congenital deafness in children to adult-onset, ARHL—and across the entire hearing loss severity spectrum—from mild to moderate hearing loss to unilateral hearing loss to bilateral profound deafness. In this chapter, we have highlighted how crossmodal plasticity may be reversible following CI in some clinical patients but may chronically persist in others, reflecting continued reliance on intact sensory modalities to maximize function in ecologically valid, real-world listening situations. It is essential that clinical otology and audiology develop new measures of patient performance in real-world multimodal communication situations and of the neuroplastic changes underlying these outcomes. Ultimately, audiological intervention relies on the principles of neuroplasticity and the ability for the brain to adapt to new or restored auditory input. Over the next decade, our knowledge of neural plasticity should lead to tailored and individualized multisensory rehabilitation for patients with hearing loss.

Acknowledgments

Research supported by the US National Institutes of Health, the Canadian Institutes of Health Research, and the Natural Sciences and Engineering Research Council of Canada.

References

1. Möeller MP. Current state of knowledge: psychosocial development in children with hearing impairment. *Ear Hear*. 2007;28:729–739.
2. Svirsky MA, Teoh SW, Neuburger H. Development of language and speech perception in congenitally, profoundly deaf children as a function of age at cochlear implantation. *Audiol Neuro Otol*. 2004;9:224–233.
3. Glendenning KK, Masterton RB. Comparative morphometry of mammalian central auditory systems: variation in nuclei and form of the ascending system. *Brain Behav Evol*. 1998;51:58–89.
4. Kral A, Lomber SG. Deaf white cats. *Curr Biol*. 2015;25:R351–R353.
5. Butler BE, Lomber SG. Functional and structural changes throughout the auditory system following congenital and early-onset deafness: implications for hearing restoration. *Front Syst Neurosci*. 2013;7:92. https://doi.org/10.3389/fnsys.2013.00092, 2013.
6. Hartmann R, Shepherd RK, Heid S, et al. Response of the primary auditory cortex to electrical stimulation of the auditory nerve in the congenitally deaf white cat. *Hear Res*. 1997;112:115–133.
7. Raggio MW, Schreiner CE. Neuronal responses in cat primary auditory cortex to electrical cochlear stimulation. III. Activation patterns in short- and long-term deafness. *J Neurophysiol*. 1999;82:3506–3526.
8. Kral A, Hartmann R, Tillein J, et al. Hearing after congenital deafness: central auditory plasticity and sensory deprivation. *Cerebr Cortex*. 2002;12:797–807, 2002.
9. Wong C, Chabot N, Kok MA, et al. Modified areal cartography in auditory cortex following early and late-onset deafness. *Cerebr Cortex*. 2014;24:1778–1792.
10. Kral A, Hartmann R, Klinke R. Recruitment of the auditory cortex in congenitally deaf cats. In: Lomber SG, Eggermont JJ, eds. *Reprogramming the Cerebral Cortex*. London, England: Oxford University Press; 2006:191–210.
11. Klinke R, Kral A, Heid S, et al. Recruitment of the auditory cortex in congenitally deaf cats by long-term cochlear electrostimulation. *Science*. 1999;285:1729–1733.
12. Kral A, Hartmann R, Tillein J, et al. Congenital auditory deprivation reduces synaptic activity within the auditory cortex in a layer-specific manner. *Cerebr Cortex*. 2000;10:714–726.
13. Klinke R, Hartmann R, Heid S, et al. Plastic changes in the auditory cortex of congenitally deaf cats following cochlear implantation. *Audiol Neuro Otol*. 2001;6:203–206, 2001.
14. Rouiller EM, Simm GM, Villa AE, et al. Auditory corticocortical interconnections in the cat: evidence for parallel and hierarchical arrangement of the auditory cortical areas. *Exp Brain Res*. 1991;86:483–505, 1991.
15. Raizada RDS, Grossberg S. Towards a theory of the laminar architecture of cerebral cortex: computational clues from the visual system. *Cerebr Cortex*. 2003;13:100–113.
16. Kotak VC, Breithaupt AD, Sanes DH. Developmental hearing loss eliminates long-term potentiation in the auditory cortex. *Proc Nat Acad Sci USA*. 2007;104:3550–3555.
17. Kral A, Schroder J-H, Klinke R, et al. Absence of crossmodal reorganization in the primary auditory cortex of congenitally deaf cats. *Exp Brain Res*. 2003;153:605–613.
18. Kral A, Tillein J, Heid S, et al. Postnatal cortical development in congenital auditory deprivation. *Cerebr Cortex*. 2005;15:552–562.
19. Raggio MW, Schreiner CE. Neuronal responses in cat primary auditory cortex to electrical cochlear stimulation. I. Intensity dependence of firing rate and response latency. *J Neurophysiol*. 1994;72:2334–2359.

20. Land R, Baumhoff P, Tillein J, et al. Crossmodal plasticity in higher-order auditory cortex of congenitally deaf cats does not limit auditory responsiveness to cochlear implants. *J Neurosci*. 2016;36:6175–6185.
21. Meredith MA, Kryklywy J, McMillan, et al. Crossmodal reorganization in the early deaf switches sensory, but not behavioural roles of auditory cortex. *Proc Natl Acad Sci USA*. 2011;108:8856–8861.
22. Meredith MA, Lomber SG. Somatosensory and visual crossmodal plasticity in the anterior auditory field of early-deaf cats. *Hear Res*. 2011;280:38–47.
23. Rebillard G, Carlier E, Rebillard M, et al. Enhancement of visual responses on the primary auditory cortex of the cat after an early destruction of cochlear receptors. *Brain Res*. 1977;129:162–164.
24. Rebillard G, Rebillard M, Pujol R. Factors affecting the recording of visual-evoked potentials from the deaf cat primary auditory cortex (AI). *Brain Res*. 1980;188:252–254.
25. Hunt DL, Yamoah EN, Krubitzer L. Multisensory plasticity in congenitally deaf mice: how are cortical areas functionally specified? *Neuroscience*. 2006;139:1507–1524.
26. Meredith MA, Allman BL. Early hearing-impairment results in crossmodal reorganization of ferret core auditory cortex. *Neural Plast*. 2012;2012:601591. https://doi.org/10.1155/2012/601591.
27. Allman BL, Keniston LP, Meredith MA. Adult deafness induces somatosensory conversion of ferret auditory cortex. *Proc Natl Acad Sci USA*. 2009;106:5925–5930.
28. Carriere BN, Royal DW, Perrault TJ, et al. Visual deprivation alters the development of cortical multisensory integration. *J Neurophysiol*. 2007;98:2858–2867.
29. Las L, Shapira AH, Nelken I. Functional gradients of auditory sensitivity along the anterior ectosylvian sulcus of the cat. *J Neurosci*. 2008;28:3657–3667.
30. Meredith MA, Allman BL. Subthreshold multisensory processing in cat auditory cortex. *Neuroreport*. 2009;20:126–131.
31. Rauschecker JP, Korte M. Auditory compensation for early blindness in cat cerebral cortex. *J Neurosci*. 1993;13:4538–4548.
32. Wallace MT, Carriere BN, Perrault Jr TJ, et al. The development of cortical multisensory integration. *J Neurosci*. 2006;26:11844–11849.
33. Meredith MA, Keniston LP, Allman BL. Multisensory dysfunction accompanies crossmodal plasticity following adult hearing impairment. *Neuroscience*. 2012;214:136–148.
34. Stewart DL, Starr A. Absence of visually influenced cells in auditory cortex of normal and congenitally deaf cats. *Exp Neurol*. 1970;28:525–528.
35. Kral A, Hartmann R, Tillein J, et al. Delayed maturation and sensitive periods in the auditory cortex. *Audiol Neuro Otol*. 2001;6:346–362.
36. Tillein J, Hubka P, Syed E, et al. Cortical representation of interaural time difference in congenital deafness. *Cerebr Cortex*. 2010;20:492–506.
37. Fallon JB, Irvine DRF, Shepherd RK. Cochlear implant use following neonatal deafness influences the cochleotopic organization of the primary auditory cortex in cats. *J Comp Neurol*. 2009;512:101–114.
38. Fallon JB, Shepherd RK, Irvine DR. Effects of chronic cochlear electrical stimulation after an extended period of profound deafness on primary auditory cortex organization in cats. *Eur J Neurosci*. 2014;39:811–820.
39. Stanton SG, Harrison RV. Projections from the medial geniculate body to primary auditory cortex in neonatally deafened cats. *J Comp Neurol*. 2000;426:117–129, 2000.
40. Chabot N, Butler BE, Lomber SG. Differential modification of cortical and thalamic projections to cat primary auditory cortex following early- and late-onset deafness. *J Comp Neurol*. 2015;523:2297–2320.
41. Lomber SG, Meredith MA. Kral A Crossmodal plasticity in specific auditory cortices underlies visual compensations in the deaf. *Nat Neurosci*. 2010;13:1421–1429.
42. Bavelier D, Dye MWG, Hauser PC. Do deaf individuals see better? *Trends Cogn Sci*. 2006;10:512–518.
43. Fritz J, Shamma S, Elhilali M, et al. Rapid task-related plasticity of spectrotemporal receptive fields in primary auditory cortex. *Nat Neurosci*. 2003;6:1216–1223.
44. Bavelier D, Neville HJ. Crossmodal plasticity: where and how? *Nat Rev Neurosci*. 2002;3:443–452.
45. Röder B, Teder-Sälejärvi W, Sterr A, et al. Improved auditory spatial tuning in blind humans. *Nature*. 2000;400:162–166.
46. Yaka R, Yinon U, Wollberg Z. Auditory activation of cortical visual areas in cats after early visual deprivation. *Eur J Neurosci*. 1999;11:1301–1312.

47. Weeks R, Horwitz B, Aziz-Sultan A, et al. A positron emission tomographic study of auditory localization in the congenitally blind. *J Neurosci*. 2000;20:2664–2672.
48. Sur M, Garraghty PE, Roe AW. Experimentally induced visual retinal projections into auditory thalamus and cortex. *Science*. 1988;242:1437–1441.
49. Ptito M, Giguère J-F, Boire D, et al. When the auditory cortex turns visual. *Prog Brain Res*. 2002;134:447–458.
50. Lomber SG, Meredith MA, Kral A. Adaptive crossmodal plasticity in deaf auditory cortex: areal and laminar contributions to supranormal vision in the deaf. *Prog Brain Res*. 2011;191:251–270.
51. Lomber SG. Behavioural cartography of visual functions in cat parietal cortex: areal and laminar dissociations. *Prog Brain Res*. 2001;134:265–284.
52. Shiell MM, Champoux F, Zatorre RJ. Enhancement of visual motion detection thresholds in early deaf people. *PLoS One*. 2014;9(2):e90498.
53. Rauschecker JP. Compensatory plasticity and sensory substitution in the cerebral cortex. *Trends Neurosci*. 1995;18:36–43.
54. Falchier A, Clavagnier S, Barone P, et al. Anatomical evidence of multimodal integration in primate striate cortex. *J Neurosci*. 2002;22:5749–5759.
55. Rockland KS, Ojima H. Multisensory convergence in calcarine visual areas in macaque monkey. *Int J Psychophysiol*. 2003;50:19–26.
56. Hall AJ, Lomber SG. Auditory cortex projections target the peripheral field representation of primary visual cortex. *Exp Brain Res*. 2008;190:413–430.
57. Cappe C, Barone P. Heteromodal connections supporting multisensory integration at low levels of cortical processing in the monkey. *Eur J Neurosci*. 2005;22:2886–2902.
58. Meredith MA, Lomber SG. Species-dependent role of crossmodal connectivity among the primary sensory cortices. *Hear Res*. 2017;343:83–91.
59. Butler BE, de la Rua A, Ward-Able T, Lomber SG. Cortical and thalamic connectivity to the second auditory cortex of the cat is resilient to the onset of deafness. *J Comp Neurol*. 2018:819–835.
60. Barone P, Lacassagne L, Kral A. Reorganization of the connectivity of cortical field DZ in congenitally deaf cat. *PLoS One*. 2013;8(4):e60093.
61. Kok MA, Chabot N, Lomber SG. Crossmodal reorganization of cortical afferents to dorsal auditory cortex following early- and late-onset deafness. *J Comp Neurol*. 2014;522:654–675.
62. Kok MA, Lomber SG. Origin of the thalamic projection to dorsal auditory cortex in hearing and deaf cats. *Hear Res*. 2017;343:108–117, 2017.
63. Butler BE, Chabot N, Lomber SG. Quantifying and comparing the pattern of thalamic and cortical projections to the posterior auditory field in hearing and deaf cats. *J Comp Neurol*. 2016;524:3042–3063.
64. Butler BE, Chabot N, Kral A, et al. Origins of thalamic and cortical projections to the posterior auditory field in congenitally-deaf cats. *Hear Res*. 2017;343:118–127.
65. Wong C, Chabot N, Kok M, et al. Amplified somatosensory and visual cortical projections to a core auditory area, the anterior auditory field, following early- and late-onset deafness. *J Comp Neurol*. 2015;523:1925–1947.
66. Clemo HR, Lomber SG, Meredith MA. Synaptic distribution and plasticity in primary auditory cortex (A1) exhibits laminar and cell-specific changes in the deaf. *Hear Res*. 2017:122–134.
67. Clemo HR, Lomber SG, Meredith MA. Synaptic basis for crossmodal plasticity: enhanced supragranular dendritic spine density in anterior ectosylvian auditory cortex of the early-deaf cat. *Cerebr Cortex*. 2016;26:1365–1376.
68. Shore SE, Zhou J. Somatosensory influence on the cochlear nucleus and beyond. *Hear Res*. 2006;216:90–99.
69. Aitkin LM, Kenyon CE, Philpott P. The representation of the auditory and somatosensory systems in the external nucleus of the cat inferior colliculus. *J Comp Neurol*. 1981;196:25–40.
70. Shore SE, Koehler S, Oldakowski M, et al. Dorsal cochlear nucleus responses to somatosensory stimulation are enhanced after noise-induced hearing loss. *Eur J Neurosci*. 2008;27:155–168.
71. Zeng C, Yang Z, Shreve L, et al. Somatosensory projections to cochlear nucleus are upregulated after unilateral deafness. *J Neurosci*. 2012;32:15791–15801, 2012.

72. Fortnum HM, Summerfield AQ, Marshall DH, et al. Prevalence of permanent childhood hearing impairment in the United Kingdom and implications for universal neonatal hearing screening: questionnaire based ascertainment study. *BMJ*. 2001;323(7312):536.
73. Prieve BA, Stevens F. The New York State universal newborn hearing screening demonstration project: introduction and overview. *Ear Hear*. 2000;21(2):85–91.
74. Watkin PM, Baldwin M. Identifying deafness in early childhood: requirements after the newborn hearing screen. *Arch Dis Child*. 2011;96(1):62–66.
75. Ross DS, Visser SN, Holstrum WJ, et al. Highly variable population-based prevalence rates of unilateral hearing loss after the application of common case definitions. *Ear Hear*. 2010;31(1):126–133.
76. Agrawal Y, Platz EA, Niparko JK. Prevalence of hearing loss and differences by demographic characteristics among US adults: data from the National Health and Nutrition Examination Survey, 1999–2004. *Arch Intern Med*. 2008;168(14):1522–1530.
77. Nikolopoulos TP, Vlastarakos PV. Treating options for deaf children. *Early Hum Dev*. 2010;86(11):669–674.
78. Kral A, Sharma A. Developmental neuroplasticity after cochlear implantation. *Trends Neurosci*. 2012;35(2):111–122.
79. Sharma A, Dorman MF, Spahr AJ. A sensitive period for the development of the central auditory system in children with cochlear implants: implications for age of implantation. *Ear Hear*. 2002a;23(6):532–539.
80. Sharma A, Dorman MF, Spahr AJ. Rapid development of cortical auditory evoked potentials after early cochlear implantation. *Neuroreport*. 2002b;13(10):1365–1368.
81. Sharma A, Dorman M, Spahr A, et al. Early cochlear implantation in children allows normal development of central auditory pathways. *Ann Otol Rhinol Laryngol*. 2002c;S189:38–41.
82. Sharma A, Gilley PM, Dorman MF, et al. Deprivation-induced cortical reorganization in children with cochlear implants. *Int J Audiol*. 2007;46(9):494–499.
83. Sharma A, Nash AA, Dorman M. Cortical development, plasticity and re-organization in children with cochlear implants. *J Commun Disord*. 2009;42(4):272–279.
84. Gilley PM, Sharma A, Dorman MF. Cortical reorganization in children with cochlear implants. *Brain Res*. 2008;1239:56–65.
85. Eggermont JJ, Salamy A. Maturational time course for the ABR in preterm and full term infants. *Hear Res*. 1988;33(1):35–47.
86. Eggermont JJ, Ponton CW. The neurophysiology of auditory perception: from single units to evoked potentials. *Audiol Neuro Otol*. 2002;7(2):71–99.
87. Eggermont JJ, Ponton CW. Auditory-evoked potential studies of cortical maturation in normal hearing and implanted children: correlations with changes in structure and speech perception. *Acta Otolaryngol*. 2003;123(2):249–252.
88. Ponton CW, Don M, Eggermont JJ, et al. Maturation of human cortical auditory function: differences between normal-hearing children and children with cochlear implants. *Ear Hear*. 1996;17:430–437.
89. Ponton CW, Eggermont JJ, Kwong B, et al. Maturation of human central auditory system activity: evidence from multi-channel evoked potentials. *Clin Neurophysiol*. 2000;111(2):220–236.
90. Ponton C, Eggermont JJ, Khosla D, et al. Maturation of human central auditory system activity: separating auditory evoked potentials by dipole source modeling. *Clin Neurophysiol*. 2002;113(3):407–420.
91. Ceponiene R, Cheour M, Näätänen R. Interstimulus interval and auditory event-related potentials in children: evidence for multiple generators. *Electroencephalogr Clin Neurophysiol*. 1998;108(4):345–354.
92. Ceponiene R, Rinne T, Näätänen R. Maturation of cortical sound processing as indexed by event-related potentials. *Clin Neurophysiol*. 2002;113(6):870–882.
93. Liégeois-Chauvel C, Musolino A, Badier JM, et al. Evoked potentials recorded from the auditory cortex in man: evaluation and topography of the middle latency components. *Electroencephalogr Clin Neurophysiol*. 1994;92(3):204–214.
94. Dunn CC, Walker EA, Oleson J, et al. Longitudinal speech perception and language performance in pediatric cochlear implant users: the effect of age at implantation. *Ear Hear*. 2014;35(2):148–160.
95. Geers AE. Factors influencing spoken language outcomes in children following early cochlear implantation. *Adv Oto Rhino Laryngol*. 2006;64:50–65.

96. Harrison RV, Gordon KA, Mount RJ. Is there a critical period for cochlear implantation in congenitally deaf children? Analyses of hearing and speech perception performance after implantation. *Dev Psychobiol.* 2005;46(3):252–261.
97. Holt RF, Svirsky MA. An exploratory look at pediatric cochlear implantation: is earliest always best? *Ear Hear.* 2008;29(4):492–511.
98. Niparko JK, Tobey EA, Thal DJ, et al. Spoken language development in children following cochlear implantation. *J Am Med Assoc.* 2010;303(15):1498–1506.
99. Robbins AM, Green JE, Waltzman SB. Bilingual oral language proficiency in children with cochlear implants. *Arch Otolaryngol Head Neck Surg.* 2004;130(5):644–647.
100. Wang NY, Eisenberg LS, Johnson KC. Tracking development of speech recognition: longitudinal data from hierarchical assessments in the childhood development after Cochlear Implantation Study. *Otol Neurotol.* 2008;29(2):240–245.
101. Wallace MT, Stein BE. Sensory and multisensory responses in the newborn monkey superior colliculus. *J Neurosci.* 2001;21(22):8886–8894.
102. Stein BE, Stanford TR, Rowland BA. Development of multisensory integration from the perspective of the individual neuron. *Nat Rev Neurosci.* 2014;15(8):520–535.
103. Wallace MT, Meredith MA, Stein BE. Multisensory integration in the superior colliculus of the alert cat. *J Neurophysiol.* 1998;80:1006–1010.
104. Wallace MT, Stein BE. Onset of crossmodal synthesis in the neonatal superior colliculus is gated by the development of cortical influences. *J Neurophysiol.* 2000;83:3578–3582.
105. Wallace MT, Stein BE. Early experience determines how the senses will interact. *J Neurophysiol.* 2007;97(1):921–926.
106. Gilley PM, Sharma A, Dorman M, et al. Developmental changes in refractoriness of the cortical auditory evoked potential. *Clin Neurophysiol.* 2005;116(3):648–657.
107. King KA, Campbell J, Sharma A, et al. The representation of voice onset time in the cortical auditory evoked potentials of young children. *Clin Neurophysiol.* 2008;119(12):2855–2861.
108. Sharma A, Campbell J, Cardon G. Developmental and crossmodal plasticity in deafness: evidence from the P1 and N1 event related potentials in cochlear implanted children. *Int J Psychophysiol.* 2015;95(2):135–144.
109. Naatanen R, Picton T. The N1 wave of the human electric and magnetic response to sound: a review and an analysis of the component structure. *Psychophysiology.* 1987;24(4):375–425.
110. Gomot M, Bruneau N, Laurent JP, et al. Left temporal impairment of auditory information processing in prematurely born 9-year-old children: an electrophysiological study. *Int J Psychophysiol.* 2007;64(2):123–129.
111. Tong Y, Melara RD, Rao A. P2 enhancement from auditory discrimination training is associated with improved reaction times. *Brain Res.* 2009;1297:80–88.
112. Moore JK, Guan YL. Cytoarchitectural and axonal maturation in human auditory cortex. *J Assoc Res Otolaryngol.* 2001;2(4):297–311.
113. Kral A, Yusuf PA, Land R. Higher-order auditory areas in congenital deafness: top-down interactions and corticocortical decoupling. *Hear Res.* 2017;343:50–63.
114. Campbell J, Sharma A. Visual crossmodal Re-organization in children with cochlear implants. *PLoS One.* 2016;11(1):e0147793.
115. Sandmann P, Dillier N, Eichele T, et al. Visual activation of auditory cortex reflects maladaptive plasticity in cochlear implant users. *Brain.* 2012;135:555–568.
116. Finney EM, Clementz B, Hickok G, et al. Visual stimuli activate auditory cortex in deaf subjects: evidence from MEG. *Neuroreport.* 2003;14:1425–1427.
117. Fine I, Finney EM, Boynton GM, et al. Comparing the effects of auditory deprivation and sign language within the auditory and visual cortex. *J Cogn Neurosci.* 2005;17:1621–1637.
118. Lee H, Kang E, Oh SH, et al. Preoperative differences of cerebral metabolism relate to the outcome of cochlear implants in congenitally deaf children. *Hear Res.* 2005;203:2–9.
119. Lee HJ, Giraud AL, Kang E, et al. Cortical activity at rest predicts cochlear implantation outcome. *Cerebr Cortex.* 2007;17(4):909–917.
120. Giraud A, Lee H. Predicting cochlear implant outcome from brain organisation in the deaf. *Restor Neurol Neurosci.* 2007;25:381–390.

121. Levänen S, Hamdorf D. Feeling vibrations: enhanced tactile sensitivity in congenitally deaf humans. *Neurosci Lett*. 2001;301:75–77.
122. Levänen S, Jousmäki V, Hari R. Vibration-induced auditory-cortex activation in a congenitally deaf adult. *Curr Biol*. 1998;8:869–872.
123. Schürmann M, Caetano G, Hlushchuk Y, et al. Touch activates human auditory cortex. *Neuroimage*. 2006;30:1325–1331.
124. Bergeson TR, Pisoni DB, Davis RAO. Development of audiovisual comprehension skills in prelingually deaf children with cochlear implants. *Ear Hear*. 2005;26:149–164.
125. Tyler R, Fryauf-Bertschy H, Kelsay D, et al. Speech perception by prelingually deaf children using cochlear implants. *Otolaryngol Head Neck Surg*. 1997;117:180–187.
126. Schorr EA, Fox NA, Wassenhove V, et al. Auditory-visual fusion in speech perception in children with cochlear implants. *Proc Natl Acad Sci*. 2007;102:18748–18750.
127. Peelle JE, Sommers MS. Prediction and constraint in audiovisual speech perception. *Cortex*. 2015;68:169–181.
128. Ammirante P, Russo FA, Good A, Fels DI. Feeling voices. *PLoS One*. 2013;8(1):e53585.
129. Russo FA, Ammirante P, Fels DI. Vibrotactile discrimination of musical timbre. *J Exp Psychol Hum Percept Perform*. 2012;38(4):822–826.
130. Chakravarty A. Influence of tactual sensitivity on tactual localization, particularly of deaf children. *J Gen Psychol*. 1968;78:219–221.
131. Schiff W, Dytell RS. Deaf and hearing children's performance on a tactual perception battery. *Percept Mot Skills*. 1972;35(3):683–706.
132. Dehmel S, Cui Y, Shore S. Crossmodal interactions of auditory and somatic inputs in the brainstem and midbrain and their imbalance in tinnitus and deafness. *Am J Audiol*. 2008;17:1–40.
133. Kanold PO, Young ED. Proprioceptive information from the pinna provides somatosensory input to cat dorsal cochlear nucleus. *J Neurosci*. 2001;21:7848–7858.
134. Nelson HD, Bougatsos C, Nygren P. Universal newborn hearing screening: systematic review to update the 2001 US Preventive Services Task Force recommendation. *Pediatrics*. 2008;122:e266–e276.
135. Geers AE, Moog JS, Biedenstein J, et al. Spoken language scores of children using cochlear implants compared to hearing age-mates at school entry. *J Deaf Stud Deaf Educ*. 2009;14(3):371–385.
136. Campbell J, Sharma A. Crossmodal Re-organization in adults with early stage hearing loss. *PLoS One*. 2014;9(2):e90594. s).
137. Buckley KA, Tobey EA. Crossmodal plasticity and speech perception in pre- and postlingually deaf cochlear implant users. *Ear Hear*. 2010;32:2–15.
138. Sharma A, Glick H, Campbell J, et al. Cortical plasticity and reorganization in pediatric single-sided deafness pre- and post-cochlear implantation: a case study. *Otol Neurotol*. 2016;37:e26–34.
139. Dorman MF, Liss J, Wang S, Berisha V, Ludwig C, Natale SC. Experiments on auditory-visual perception of sentences by users of unilateral, bimodal, and bilateral cochlear implants. *J Speech Lang Hear Res*. 2016;59(6):1505–1519.
140. Döge J, Baumann U, Weissgerber T, et al. Single-sided deafness: impact of cochlear implantation on speech perception in complex noise and on auditory localization accuracy. *Otol Neurotol*. 2017;38(10):e563–e569.
141. Zeitler DM, Dorman MF, Natale SJ, et al. Sound source localization and speech understanding in complex listening environments by single-sided deaf listeners after cochlear implantation. *Otol Neurotol*. 2015;36(9):1467–1471.
142. Grossmann W, Brill S, Moeltner A, et al. Cochlear implantation improves spatial release from masking and restores localization abilities in single-sided deaf patients. *Otol Neurotol*. 2016;37(6):658–664.
143. Lieu JE, Tye-Murray N, Karzon RK, et al. Unilateral hearing loss is associated with worse speech-language scores in children. *Pediatrics*. 2010;125(6):e1348–e1355.
144. Bess FH, Tharpe AM. Unilateral hearing impairment in children. *Pediatrics*. 1984;74(2):206–216.
145. Oyler RF, Oyler AL, Matkin ND. Unilateral hearing loss: demographics and educational impact. *Lang Speech Hear Serv Sch*. 1988;19:201–210.
146. Brookhouser PE, Worthington DW, Kelly WJ. Unilateral hearing loss in children. *The Laryngoscope*. 1991;101(12 Pt 1):1264–1272.

147. Borg E, Risberg A, McAllister B, et al. Language development in hearing-impaired children. Establishment of a reference material for a 'Language test for hearing-impaired children', LATHIC. *Int J Pediatr Otorhinolaryngol.* 2002;65(1):15–26.
148. Lieu JE, Tye-Murray N, Fu Q. Longitudinal study of children with unilateral hearing loss. *Laryngoscope.* 2012;122(9):2088–2095.
149. Thomas JP, Neumann K, Dazert S, et al. Cochlear implantation in children with congenital single-sided deafness. *Otol Neurotol.* 2017;38(4):496–503.
150. Arndt S, Aschendorff A, Laszig R, et al. Comparison of pseudobinaural hearing to real binaural hearing rehabilitation after cochlear implantation in patients with unilateral deafness and tinnitus. *Otol Neurotol.* 2011;32(1):39–47.
151. Arndt S, Prosse S, Laszig R, et al. Cochlear implantation in children with single-sided deafness: does aetiology and duration of deafness matter? *Audiol Neuro Otol.* 2015;20(S1):21–30.
152. Cadieux JH, Firszt JN, Reeder RM. Cochlear implantation in non-traditional candidates: preliminary results in adolescents with asymmetric hearing loss. *Otol Neurotol.* 2013;34:408–415.
153. Dorman MF, Zeitler D, Cook SJ, et al. Interaural level difference cues determine sound source localization by single-sided deaf patients fit with a cochlear implant. *Audiol Neurotol.* 2015;20:183–188.
154. Friedmann D, Ahmed O, McMenomey S. Single-sided deafness cochlear implantation: candidacy, evaluation, and outcomes in children and adults. *Otol Neurotol.* 2016;37:e154–e160.
155. Távora-Vieira D, Marino R, Krishnaswamy J, et al. Cochlear implantation for unilateral deafness with and without tinnitus: a case series. *Laryngoscope.* 2013;123:1251–1255.
156. Gordon KA, Wong DD, Papsin BC. Bilateral input protects the cortex from unilaterally-driven reorganization in children who are deaf. *Brain.* 2013;136(5):1609–1625.
157. Gordon KA, Jiwani S, Papsin BC. Benefits and detriments of unilateral cochlear implant use on bilateral auditory development in children who are deaf. *Front Psychol.* 2013;4:719.
158. Polonenko MJ, Gordon KA, Cushing SL, Papsin BC. Cortical organization restored by cochlear implantation in young children with single sided deafness. *Sci Rep.* 2017. Article ID - 16900.
159. Polonenko MJ, Papsin BC, Gordon KA. Cortical plasticity with bimodal hearing in children with asymmetric hearing loss. *Hear Res.* 2019:88–98.
160. Gates GA, Mills JH. Presbycusis. *Lancet.* 2005;366(9491):1111–1120.
161. Bharadwaj HM, Masud S, Mehraei G, et al. Individual differences reveal correlates of hidden hearing deficits. *J Neurosci.* 2015;35(5):2161–2172.
162. Lin FR, Ferrucci L, An Y, et al. Association of hearing impairment with brain volume changes in older adults. *Neuroimage.* 2014;90:84–92.
163. Peelle JE, Troiani V, Grossman M, et al. Hearing loss in older adults affects neural systems supporting speech comprehension. *J Neurosci.* 2011;31:12638.
164. Wingfield A, McCoy SL, Peelle JE, Tun PA, Cox LC. Effects of adult aging and hearing loss on comprehension of rapid speech varying in syntactic complexity. *J Am Acad Audiol.* 2006;17(7):487–497.
165. Lin FR, Metter EJ, O'Brien RJ, et al. Hearing loss and incident dementia. *Arch Neurol.* 2011a;68(2):214–220.
166. Lin FR, Niparko JK, Ferrucci L. Hearing loss prevalence in the United States. *Arch Intern Med.* 2011b;171(20):1851–1852.
167. Cardin V. Effects of aging and adult-onset hearing loss on cortical auditory regions. *Front Neurosci.* 2016;10:199.
168. Pichora-Fuller MK, Levitt H. Speech comprehension training and auditory and cognitive processing in older adults. *Am J Audiol.* 2012;21(2):351–357.
169. Wayne RV, Johnsrude IS. A review of causal mechanisms underlying the link between age-related hearing loss and cognitive decline. *Ageing Res Rev.* 2015;23(B):154–166.
170. Finney EM. Visual stimuli activate auditory cortex in the deaf. *Nat Neurosci.* 2001;4:1171–1173.
171. Kim M, Shim H, Jin SH, et al. Crossmodal and intra-modal characteristics of visual function and speech perception performance in postlingually deafened cochlear implant users. *PLoS One.* 2016;11:1–19.
172. Doucet ME, Bergeron F, Lassonde M, et al. Crossmodal reorganization and speech perception in cochlear implant users. *Brain.* 2006;129:3376–3383.
173. Karns CM, Dow MW, Neville HJ. Altered crossmodal processing in the primary auditory cortex of congenitally deaf adults: a visual-somatosensory fMRI study with a double-flash illusion. *J Neurosci.* 2012;32:9626–9638.

174. Chen L, Sandmann P, Thorne JD, et al. Crossmodal functional reorganization of visual and auditory cortex in adult cochlear implant users identified with fNIRS. *Neural Plast*. 2016. Article ID - 4382656.
175. Dewey RS, Hartley DEH. Cortical crossmodal plasticity following deafness measured using functional near-infrared spectroscopy. *Hear Res*. 2015;325:55–63.
176. Neville HJ, Lawson D. Attention to central and peripheral visual space in a movement detection task. Separate effects of auditory deprivation and acquisition of a visual language. *Brain Res*. 1987;405:284–294.
177. Stropahl M, Plotz K, Schönfeld R, et al. Crossmodal reorganization in cochlear implant users: auditory cortex contributes to visual face processing. *Neuroimage*. 2015;121:159–170.
178. Hay-McCutcheon M, Pisoni D, Kirk K. Audiovisual speech perception in elderly cochlear implant recipients. *The Laryngoscope*. 2005;115:1887–1894.
179. Mitchell T, Letourneau S, Maslin M. Behavioral and neural evidence of increased attention to the bottom half of the face in deaf signers. *Restor Neurol Neurosci*. 2013;31:125–139.
180. Stropahl M, Debener S. Auditory crossmodal reorganization in cochlear implant users indicates audio-visual integration. *Neuroimage*. 2017;16:514–523.
181. Rouger J, Lagleyre S, Fraysse B, et al. Evidence that cochlear-implanted deaf patients are better multisensory integrators. *Proc Natl Acad Sci*. 2007;104:7295–7300.
182. Rouger J, Lagleyre S, Démonet JF, et al. Evolution of crossmodal reorganization of the voice area in cochlear-implanted deaf patients. *Hum Brain Mapp*. 2012;33:1929–1940.
183. Stropahl M, Chen LC, Debener S. Cortical reorganization in postlingually deaf cochlear implant users: intra-modal and crossmodal considerations. *Hear Res*. 2017;343:128–137.
184. Benetti S, Van Ackeren MJ, Rabini G, et al. Functional selectivity for face processing in the temporal voice area of early deaf individuals. *Proc Natl Acad Sci*. 2017;114(31):E6437–E6446.
185. Glick H, Sharma A. Crossmodal plasticity in developmental and age-related hearing loss: clinical implications. *Hear Res*. 2017;343:191–201.
186. Cardon GC. *Somatosensory Crossmodal Reorganization in Children with Cochlear Implants [dissertation]*. Boulder: Unviersity of Colorado; 2015.
187. Campbell J, Sharma A. Compensatory changes in cortical resource allocation in adults with hearing loss. *Front Syst Neurosci*. 2013;7:71.
188. Wingfield A, Peelle JE. The effects of hearing loss on neural processing and plasticity. *Front Syst Neurosci*. 2015;9:35.
189. Zekveld AA, Rudner M, Kramer SE, Lyzenga J, Rönnberg J. Cognitive processing load during listening is reduced more by decreasing voice similarity than by increasing spatial separation between target and masker speech. *Front Neurosci*. 2014;8:88.
190. Husain FT, Carpenter-Thompson JR, Schmidt SA. The effect of mild-to-moderate hearing loss on auditory and emotion processing networks. *Front Syst Neurosci*. 2014;8:10.
191. Rudner M, Lunner T. Cognitive spare capacity and speech communication: a narrative overview. *BioMed Res Int*. 2014:869726.
192. Tye-Murray N, Sommers MS, Spehar B. Audiovisual integration and lipreading abilities of older adults with normal and impaired hearing. *Ear Hear*. 2007;28:656–668.
193. Barone P, Deguine O. Multisensory processing in cochlear implant listeners. In: F. G, Zeng FD, Fay R, Popper A, eds. *Springer Handbook of Auditory Research. Auditory Prostheses: Cochlear Implants and beyond*. New York: Springer-Verlag; 2011:365–382.
194. Strelnikov K, Rouger J, Demonet JF, et al. Visual activity predicts auditory recovery from deafness after adult cochlear implantation. *Brain*. 2013;136:3682–3695.
195. Henshaw H, Ferguson MA. Efficacy of individual computer-based auditory training for people with hearing loss: a systematic review of the evidence. *PLoS One*. 2013;8(5):e62836.
196. Sweetow R, Palmer CV. Efficacy of individual auditory training in adults: a systematic review of the evidence. *J Am Acad Audiol*. 2005;16(7):494–504.
197. Weisenberger JM, Miller JD. The role of tactile aids in providing information about acoustic stimuli. *J Acoust Soc Am*. 1987;82(3):906–916.
198. Huang J, Sheffield B, Lin P, Zeng F. Electro-tactile stimulation enhances cochlear implant speech recognition in noise. *Nature Scientific Rep*. 2017;7:2196.

Further reading

1. Nardini M, Bales J, Mareschal D. Integration of audio-visual information for spatial decisions in children and adults. *Dev Neurosci.* 2015;19:803–816.
2. Brandwein AB, Foxe JJ, Russo NN, et al. The development of audiovisual multisensory integration across childhood and early adolescence: a high-density electrical mapping study. *Cerebr Cortex.* 2011;21(5):1042–1055.
3. Burr D, Gori M. Multisensory integration develops late in human. In: Murray MM, Wallace MT, eds. *The Neural Bases of Multisensory Processes.* Boca Raton: CRC Press/Taylor & Francis; 2012.
4. Kujawa SG, Liberman MC. Adding insult to injury: cochlear nerve degeneration after "temporary" noise-induced hearing loss. *J Neurosci.* 2009;29(45):14077–14085.
5. Kujawa SG, Liberman MC. Synaptopathy in the noise-exposed and aging cochlea: primary neural degeneration in acquired sensorineural hearing loss. *Hear Res.* 2015;330:191–199.
6. Shiell MM, Champoux F, Zatorre RJ. Reorganization of auditory cortex in early-deaf people: functional connectivity and relationship to hearing aid use. *J Cogn Neurosci.* 2015;27:150–163.
7. Gick B, Derrick D. Aero-tactile integration in speech perception. *Nature.* 2009;462:502–504.
8. Ito T, Tiede M, Ostry DJ. Somatosensory function in speech perception. *Proc Natl Acad Sci.* 2009;106:1245–1248.

CHAPTER

17

Neurodevelopmental and neuropsychiatric disorders affecting multisensory processes

Carissa J. Cascio[1], *David M. Simon*[2], *Lauren K. Bryant*[3], *Gabriella DiCarlo*[4], *Mark T. Wallace*[5]

[1]Vanderbilt University Medical Center, Nashville, TN, United States; [2]Axial Healthcare, Nashville, TN, United States; [3]Microsoft, Inc. Seattle, WA, United States; [4]Vanderbilt University School of Medicine, Nashville, TN, United States; [5]Vanderbilt Brain Institute, Vanderbilt University, Nashville, TN, United States

Introduction

Although the vast arrays of neurodevelopmental and neuropsychiatric conditions are diagnosed primarily on the basis of dysfunction in higher-order cognitive abilities, it has long been known that many of these disorders are also accompanied by changes in sensory function. In the neurodevelopmental realm, one of the best examples of this is autism spectrum disorder (ASD), which is diagnosed on the basis of weaknesses in social communication and the presence of restricted interests and repetitive behaviors. However, it is well established that sensory disturbances are a prevalent feature of autism, with nearly 90% of those on the spectrum showing measurable difference in sensory function.[1] Despite the inclusion of sensory features in the first clinical accounts of ASD,[2] they were only recently included in the Diagnostic and Statistical Manual of Mental Disorders, in its latest revision, the DSM-5.[3] In the neuropsychiatric arena, perhaps the best example of sensory issues can be found in schizophrenia (SZ). Although diagnosed based on the presence of symptoms such as delusions, hallucinations, disorganized thought and speech, and flat affect, those with SZ frequently exhibit differences in sensory sensitivities and sensory function.

In the current chapter, we look at two disorders—autism and SZ—through the lens of these differences in sensory function to provide an overview of how such differences may play an important role in the canonical social-cognitive-affective symptoms that characterize these two groups. The choice of autism and SZ as exemplars is motivated simply

Multisensory Perception
https://doi.org/10.1016/B978-0-12-812492-5.00017-6

by the amount of evidence that has been gathered about sensory disturbances in these conditions. However, we argue that these conditions may be representative of the broader arrays of neurodevelopmental and neuropsychiatric conditions, with this view predicated on the core concept that sensory function serves to scaffold higher-order abilities, and that disturbances in sensory processes are likely to have cascading effects on higher-order function.

This chapter addresses these issues not from a sensory-specific perspective (i.e., with a focus on one sense), but rather from a multisensory perspective. Such a multisensory emphasis is motivated by two factors: (1) the viewpoint that our perceptual gestalt of the world is built on the synthesis of information from a number of different sensory systems and (2) the growing body of evidence for multisensory differences in both ASD and SZ.

A great deal of behavioral and physiological research has focused on the issue of multisensory processing, and we have learned a significant amount about the behavioral and perceptual advantages that can be gained under multisensory circumstances, and about how the typical brain is organized to carry out these remarkable integrative feats. From a behavioral and perceptual perspective, the presence of information from two or more senses has been shown to increase the detectability of stimuli, speed our responses to them, improve our ability to localize them, and enable greater discriminability and disambiguation.[4] One of the most tangible of these benefits is in speech perception, where the ability to see the lips of a speaker results in an effective gain in the intelligibility of the auditory signal by up to 15 dB. Along with these perceptual improvements are examples in which the combination of stimuli from multiple senses can result in categorical changes in perception. One of the best examples of this is the so-called McGurk effect, in which the combination of discordant auditory and visual speech tokens often results in the perception of a third token. In the most common instantiation of this illusion, the combination of an auditory /ba/ with the lip movements of /ga/ frequently results in the report of the novel /da/.

From a neurobiological perspective, we have learned much about the neural architecture and principles that support basic multisensory functions. One of the foundational concepts emerging from this work is that multisensory stimulus combinations that effect behavioral and perceptual improvements often also result in neurophysiological changes. For example, when multisensory (e.g., visual and auditory) stimuli are paired in a spatially and temporally concordant manner (a combination that generally results in behavioral/perceptual improvements), there are concomitant changes in measures of information encoding such as neuronal firing, local field and evoked potentials, network synchrony, and hemodynamic responses. These changes with spatiotemporal contingency are adaptive, as environmental stimuli that emanate from the same location and time are highly likely to be a result of the same event. Hence, these statistical features represent critical cues for the brain to decide what information to "bind" together.

In addition to basic science work in animal models and human subjects that has focused on the fundamental properties of how the nervous system processes multisensory input, there has been a growing interest in how these multisensory processes differ from their typical counterparts in neurodevelopmental and neuropsychiatric conditions. As alluded to earlier, perhaps the best-studied of these to date have been autism and SZ, and the remainder of this chapter is devoted to a review of how multisensory processing differs in these two clinical groups from what is seen in neurotypical individuals.

Autism spectrum disorder

Audiovisual integration in autism spectrum disorder

Audiovisual sensory signals are ubiquitous in the everyday environment and make substantial contributions to human perception and behavior. As highlighted earlier, a simple example of an audiovisual signal with substantial ecological importance is human speech, in which mouth movements strongly correlate with the dynamics of the auditory signal.[5] If audiovisual integration is impaired in autism, this might not only affect simple comprehension of human speech, but may also cascade into the higher-order deficits that characterize ASD, such as impaired social communication skills. For this reason, audiovisual integration is the most widely studied multisensory process in ASD.

Differences in integration of simple audiovisual stimuli

One approach to examining audiovisual integration is construction of psychophysical experiments utilizing low-level stimuli such as flashes and beeps and well-constrained tasks such as speeded reaction times or visual search. These highly controlled stimuli, which can be envisioned as impulses (on-off stimuli), have been utilized to interrogate deficits in audiovisual integration in the absence of social cues or other top-down effects associated with more complex or naturalistic stimuli. Experiments utilizing such well-constrained approaches provide evidence that the integration of sensory inputs across the auditory and visual modalities is impaired in ASD. For example, in a simple visual search task referred to as "pip and pop," the presence of an irrelevant auditory input enhances the speed of visual search compared to visual inputs alone.[6] In the visual-alone search version of this task, individuals with ASD outperform their typically developing (TD) peers.[7,8] However, the addition of auditory information provides no facilitation of search speed in the ASD group, indicating a failure to utilize crossmodal cues to benefit behavior. Importantly, behavioral facilitation in this task is believed to be "bottom-up", or primarily stimulus-driven, as spatial congruence of the auditory signal is not required for facilitation. This task is thus a strong indication that low-level, primarily feed-forward, audiovisual integration is impaired in ASD. However, it is unknown whether this is a true multisensory group difference, or an example of inverse effectiveness (i.e., multisensory benefit being greater for weaker stimuli; see chapter by Stein and Rowland, this volume), given the enhanced visual performance in the ASD group.

Audiovisual illusions can also be constructed from these relatively simple stimuli. In these illusion-based experimental approaches, multisensory integration results in a percept that is nonveridical. An example of such an illusion is the sound-induced flash illusion, in which the addition of sounds to a single visual flash results in the illusory percept of multiple flashes.[9] Children with ASD report the illusory flash less frequently than TD individuals, indicating a reduction in low-level integration.[10] Intriguingly, adults with ASD do not appear to differ from controls in their perception of this illusion,[11,12] suggesting a developmental contribution to low-level differences in audiovisual integration in ASD (see developmental section below).

Speeded reaction time tasks using simple audiovisual stimuli also provide evidence for altered audiovisual integration in ASD. In experiments measuring simple reaction times to auditory stimuli alone, visual stimuli alone, and the combination of both auditory and visual stimuli, the presence of both typically results in reaction times faster than to either the auditory or visual stimulus alone. Children with ASD, however, show reduced behavioral facilitation of reaction times (i.e., less speeding) in this task relative to TD children.[13] The ASD group in this study also failed to show facilitation of RTs beyond what could be expected by statistical facilitation[14]—the predicted gain based on the two unisensory signals. This finding illustrates that one of the classical hallmarks of multisensory behavioral facilitation, the so-called race model violation, is diminished in individuals with ASD.

Physiological recordings using this task mirror these results. In TD children, event-related potentials (ERPs) recorded during this task are characterized by superadditivity (i.e., AV response > A + V responses[15]), but this marker of neural integration was diminished in children with ASD at frontal and occipital electrode sites which putatively index auditory and visual processing[13] (Fig. 17.1). This study shows that basic neural processes for integration of simple audiovisual stimuli are different in ASD and that these differences may contribute to the observed lack of behavioral facilitation. Electrophysiological measures in this study also correlate with metrics of symptom severity,[16] serving as evidence that low-level deficits in multisensory neural integration are related to higher-order characteristics of the disorder.

Differences in integration of audiovisual speech signals

Speech signals represent one of the most ecologically important types of audiovisual signals present in the everyday environment due to their fundamental contributions to communication and social interaction. As described earlier, visual speech (i.e., mouth movements) makes substantial contributions to speech perception, and these contributions are largest in noisy auditory conditions.[17] Reductions in this benefit might substantially impair speech perception in ASD. Indeed, some of the best evidence that audiovisual speech processing is impaired in ASD comes from speech-in-noise tasks, in which children with ASD show reduced perceptual gain from the availability of visual speech.[18–20] In some of these experiments, younger children with ASD show remarkably large deficits in their ability to utilize visual speech to facilitate correct word identification[21] (Fig. 17.2). In addition, this last study further suggests age-related differences in the degree of impairment, in that the perceptual deficits for audiovisual speech in noise are largely resolved by the late teen years.[21]

Additional evidence of altered audiovisual speech integration in individuals with ASD comes from experiments utilizing the McGurk illusion (see above).[22] A number of studies have demonstrated differences in McGurk fusion rates (i.e., the proportion of trials on which the illusion is experienced) in both children and adults with ASD.[19,23–28] Somewhat puzzling has been the differences across these studies, in which some report similar or higher McGurk fusion rates for the ASD group (i.e., Woynaroski et al.[28]), while others report lower fusion rates (i.e., Irwin et al. and Stevenson et al.[19,26]). A recent study has made progress in resolving these differences through the development of a noisy encoding model of the McGurk effect, which takes into account differences in the stimulus properties used across these studies.[29] The most important finding from this study is that the model strongly reinforces that greater noise in the encoding of audiovisual sensory information is a core characteristic of individuals with ASD.

Differences in audiovisual speech processing are also supported by physiological investigations, which suggest that differences in multisensory integration might emerge at relatively late speech processing stages associated with lexical or semantic processing.[30,31] These findings

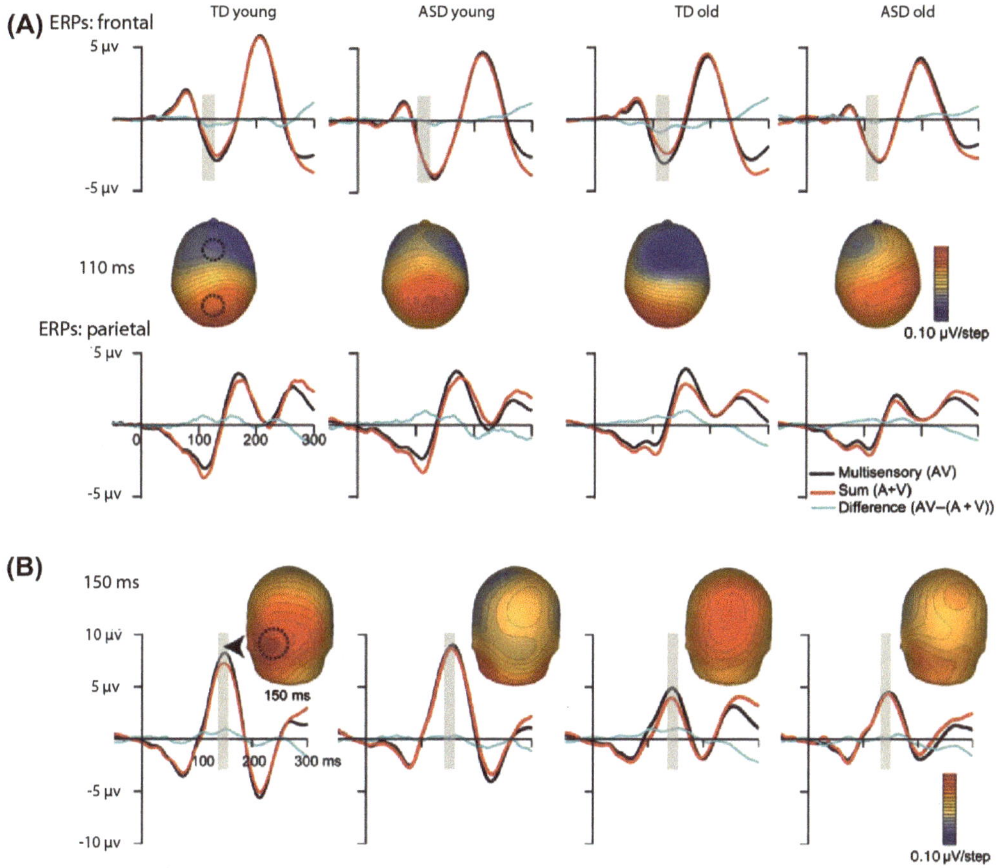

FIGURE 17.1 Event-related potentials (ERPs) for simple audiovisual (AV) stimuli. (A) Midline frontal and parietal ERPs in response to AV multisensory stimuli (black) or the sum of separate auditory and visual ERPs (red). Displayed topographies correspond with 110 ms after stimulus onset. Nonlinear interactions are absent in the ASD group. (B) Left occipital ERPs displayed as in panel (A) Displayed topographies correspond with 150 ms after stimulus onset. *µV*, microvolts; *ASD*, autism spectrum disorder, *Ms*, milliseconds; *TD*, typically developing. *Figure modified with permission from Brandwein AB, Foxe JJ, Butler JS, et al. The development of multisensory integration in high-functioning autism: high-density electrical mapping and psychophysical measures reveal impairments in the processing of audiovisual inputs.* Cereb Cortex. *2013;23(6):1329–1341. https://doi.org/10.1093/cercor/bhs109.*

are congruent with evidence that audiovisual integration at the level of single syllables might be more preserved in ASD than integration at the level of the whole word.[32] Furthermore, studies of audiovisual speech processing also provide complementary evidence that the developmental trajectory of audiovisual integration for speech is altered in ASD. This is embodied by the developmental trajectory of the McGurk fusion effect in individuals with ASD, which either fails to mature completely[26] or matures much later in life.[27] These findings are in agreement with those described above for the integration of both more simplistic stimuli and audiovisual speech stimuli, thus supporting the presence of developmental contributions to differences in audiovisual integration. We discuss the importance of considering developmental trajectories in light of

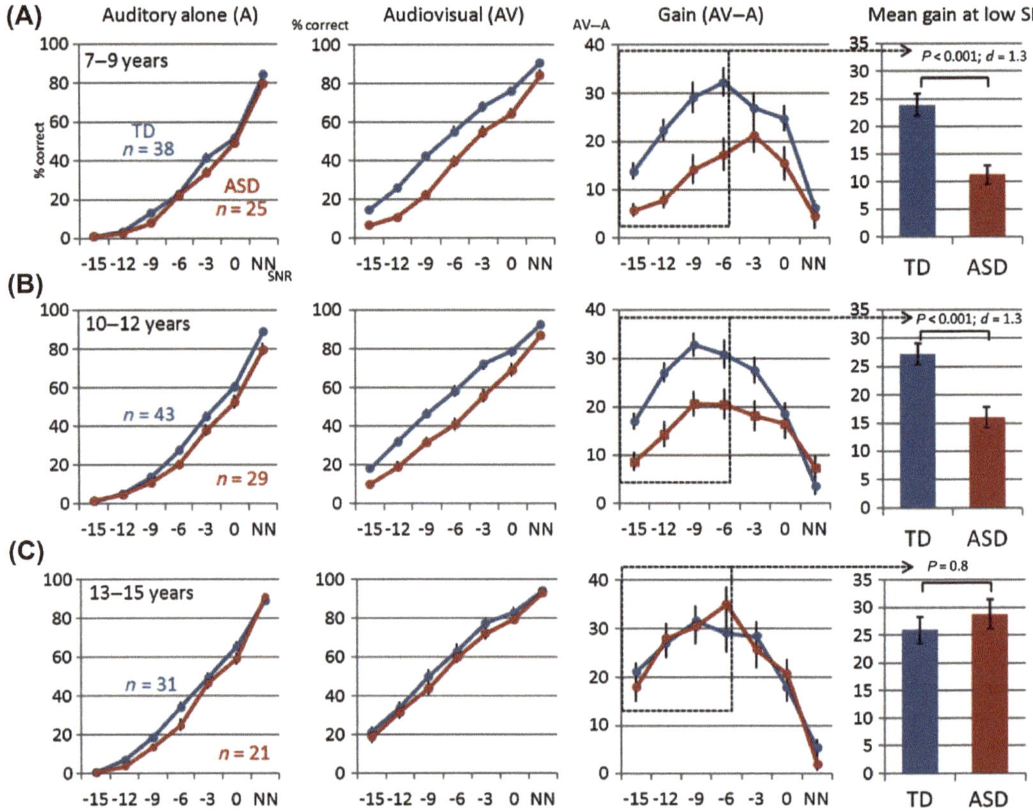

FIGURE 17.2 Speech-in-noise performance for auditory and audiovisual speech. (A) Rates of correct speech identification for auditory only and audiovisual speech for children 7–9 years of age. Performance for typically developing (TD) children is in blue, while for children with autism spectrum disorder (ASD) is in red. Note the substantial difference in audiovisual gain between TD and ASD for stimuli with low signal to noise ratios. (B) Children from 10 to 12 years of age as in panel (A) (C) Children from 13 to 15 years of age as in panel (A) Audiovisual gain does not differ between the TD and ASD groups in this age range. *Figure reproduced with permission from Foxe JJ, Molholm S, Del Bene VA, et al. Severe multisensory speech integration deficits in high-functioning school-aged children with Autism Spectrum Disorder (ASD) and their resolution during early adolescence.* Cereb Cortex. *2015;25(2):298–312. https://doi.org/10.1093/cercor/bht213.*

attentional findings below. Finally, the presence of integrative differences in ASD may not be restricted to audiovisual speech stimuli, as there is evidence that the integration of speech with visual nonspeech signals such as gestures may also be impaired in ASD.[33]

Disrupted audiovisual temporal processing

The temporal concordance between sensory inputs is an important cue for the integration of both simplistic and naturalistic audiovisual signals.[34] That is, audiovisual inputs which are temporally aligned (i.e., occur at around the same time) are generally bound together into a unified percept, while temporally disparate inputs are not. Importantly, this process is characterized by a degree of temporal tolerance to account for the statistics of the natural environment, in which auditory and visual inputs originating from a common source are not always precisely aligned due to differences in energy propagation time (light travels faster than sound). By manipulating the temporal relationship between auditory and visual stimuli, researchers have uncovered a basic difference in the temporal interval over which individuals with ASD are carrying out this binding process, such that the temporal window of integration (also known as the temporal binding window [TBW]) appears to be larger for individuals with ASD (Fig. 17.3). An example of such a finding is evident in a study by Foss-Feig et al.,[35] which studied the sound-induced flash

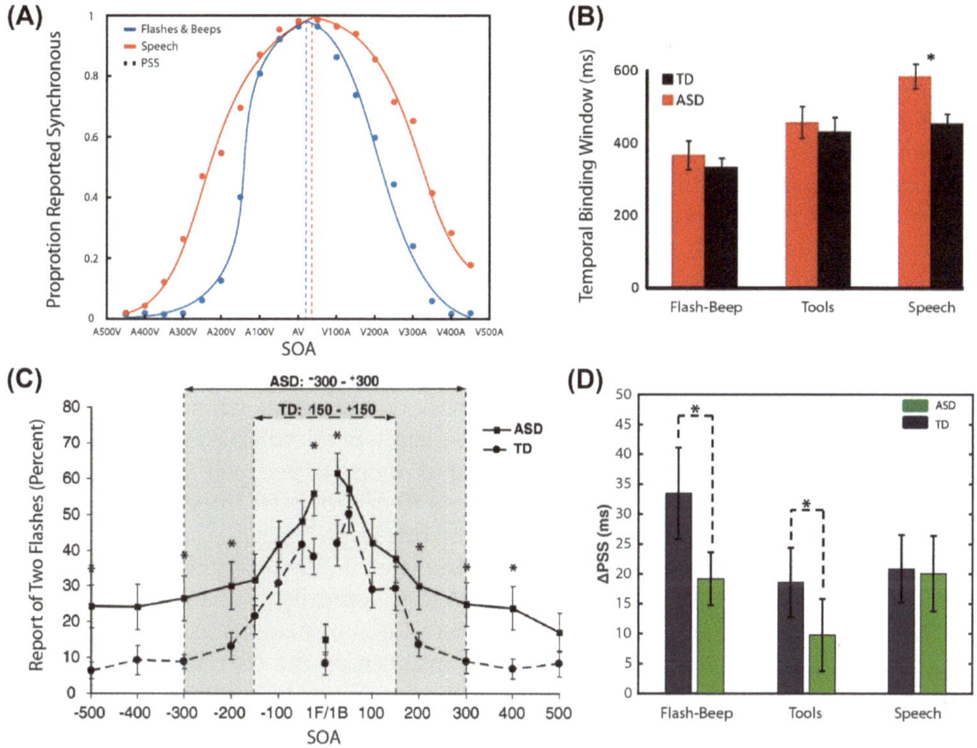

FIGURE 17.3 Altered audiovisual temporal processing in autism spectrum disorder (ASD). (A) Simulated temporal binding window (TBW) for flashes and beeps (orange) and speech (blue) illustrating the differences typically observed between these stimulus types during simultaneity judgment. (B) TBW width for ASD children (red) and typically developing (TD) children (black). The widths are similar for flash-beep stimuli and tool stimuli, but children with ASD have wider TBWs for speech stimuli. (C) Enlarged temporal window for the sound-induced flash illusion. Compared to TD children, children with ASD report illusory flashes in the presence of a second beep at substantially larger temporal offsets (SOAs). (D) Atypical rapid recalibration in ASD. Children with ASD show reduced single trial adaptation to flash-beep and tool stimuli when compared to TD children. Adaptation to speech stimuli, however, is identical between groups. ΔPSS: Change in point of subjective simultaneity. *PSS*, point of subjective simultaneity; *SOA*, stimulus onset asynchrony. *(B) Reproduced with permission from Stevenson RA, Siemann JK, Schneider BC, et al. Multisensory temporal integration in autism spectrum disorders.* J Neurosci. *2014;34(3):691-697. https://doi.org/10.1523/JNEUROSCI.3615-13.2014. (C) Reproduced with permission from Foss-Feig JH, Kwakye LD, Cascio CJ, et al. An extended multisensory temporal binding window in autism spectrum disorders.* Exp Brain Res. *2010;203(2):381-389. https://doi.org/10.1007/s00221-010-2240-4. (D) Reproduced with permission from Noel JP, De Niear MA, Stevenson R, Alais D, Wallace MT. Atypical rapid audio-visual temporal recalibration in autism spectrum disorders. Autism Res. 2016. https://doi.org/10.1002/aur.1633.*

illusion[9] for stimuli with varying temporal offsets and found that children with ASD continued to report the illusion at substantially larger temporal offsets when compared with their TD counterparts. In a second study using a paradigm in which auditory stimuli assist in temporal separation of visual stimuli (also known as temporal ventriloquism[36]), children with ASD were found to receive benefits from auditory stimuli at larger temporal offsets than TD controls.[37] Adults with ASD have also been shown to demonstrate diminished sensitivity to temporal offsets during temporal order judgment (TOJ) of simple flashes and beeps,[38] although this finding has recently been disputed.[39] Together, these studies suggest that the temporal constraints for multisensory integration differ for individuals with ASD.

Differences in multisensory temporal processing in ASD might also be dependent on the nature of the stimulus, which is known to be an important factor affecting temporal acuity.[40] Specifically, when asked to explicitly judge the simultaneity of paired audiovisual stimuli (which can be presented synchronously or asynchronously), individuals with ASD perform similarly to controls for simple impulse stimuli (i.e., short-duration flashes and beeps), but present an enlarged temporal window of integration for more complex audiovisual stimuli such as speech.[41] A similar study utilizing the McGurk illusion,[22] in which children with ASD perceived the illusion over a larger range of temporal offsets,[28] affirms this finding. Further reinforcing the presence of atypical multisensory temporal processing in ASD are the findings of strong correlations between McGurk fusion rates and TBW size (i.e., the larger the window the less fusion—Stevenson et al., 2014b) and evidence that audiovisual synchrony does not bias passive eye gaze as strongly in ASD when compared with TD.[42] This divergence between simple stimuli and more complex speech stimuli has led to the proposal that the complexity of stimuli plays an important role in audiovisual integration deficits in ASD.[41,43] Speech-in-noise experiments further support this idea, as these experiments demonstrate that deficits in audiovisual integration at the level of words are more severe when compared to the phonemic level.[32]

These studies have additionally revealed that trial-to-trial flexibility in audiovisual temporal processing is reduced in individuals with ASD. Typically, perception of audiovisual simultaneity depends not only on the current stimulus, but also on the temporal structure of recent audiovisual information, such that performance on the current trial changes based on whether the previous trial was visual-leading or auditory-leading.[44,45] Individuals with ASD do not perform this trial-by-trial adjustment in temporal perception for simple short-duration (i.e., impulse) stimuli,[46,47] but appear to do so for audiovisual speech stimuli.[46] Interpreted in the light of Bayesian inference,[48] which has been proposed to be disrupted in ASD,[49,50] this finding indicates that individuals with ASD may substantially overweight immediate sensory inputs during these temporal judgments. These effects may be most noticeable for impulse stimuli because human participants lack extensive experience with stimuli such as paired flashes and beeps. In contrast, our extensive experience with audiovisual speech stimuli may be sufficient to overcome these weighting differences.

Developmental considerations

Autism is a developmental disorder that unfolds in the first few years of life, thus particular attention is warranted to the common thread of developmental differences in the studies reviewed above. Evidence for differences in audiovisual processing very early in life suggests that the development of audiovisual integration in ASD involves an early divergence from the trajectory of TD children followed by either an arrest or delay in developmental progression. "High-risk" infants, who have an older sibling with a diagnosis of ASD, do not orient preferentially to congruent audiovisual information at 9 months of age,[51] and converging evidence that synchrony between sounds and biological motion has no effect on orienting in 3-year-old children with ASD suggests the presence of differences in multisensory temporal processing within the first few years of life.[52] Preschool children with ASD also demonstrate a positive association between preference for synchronous speech and language ability (i.e., higher sensitivity to the temporal alignment of speech is associated with better language ability).[53] This suggests that deficits in the ability to temporally integrate auditory and visual speech stimuli contribute to impaired language development early in life.

A possible developmental explanation for the eventual "normalization" of audiovisual integration seen in some studies[21,27] revolves around the maturation of attentional systems. Specifically, enhanced attentional resource allocation might form a compensatory strategy in older children with ASD that is unavailable to younger children with less mature attentional brain networks. This hypothesis is supported by the strong dependency of many forms of audiovisual integration on attention[54] (reviewed by Talsma et al.[55]), and by demonstrations that focused attention can provide multisensory processing benefits in ASD equivalent to those seen in TD individuals.[56] Additional support for this perspective comes from physiological investigations demonstrating that attentional load impacts audiovisual integration in ASD,[57] and that attentional allocation between auditory and visual modalities, as indexed by alpha oscillations, is atypical in ASD.[58] Collectively, these studies highlight the importance of considering attentional resources when interpreting multisensory function. Developmental and attentional findings thus converge to highlight the importance of future comprehensive evaluations of the neural processes contributing to audiovisual integration in ASD.

Summary of audiovisual integration in autism spectrum disorder

Taken together, these works suggest that the integration of audiovisual inputs is impacted in ASD, with a specific finding that this integration occurs over an extended window of time. Furthermore, even when inputs occur in relative temporal proximity, both low-level feed-forward and higher-order integrative processes appear to be disrupted, including the flexibility of temporal perception to adapt to the statistics of the natural environment. Allocation of attentional resources may play an important role in allowing older individuals to compensate for these deficits in some tasks, but such increased attentional allocation has its limitations in naturalistic environments such as a classroom. These deficits likely impair the ability of individuals with autism to utilize complex sensory statistics, such as the temporal correlations between audition and vision found in natural speech,[5] to facilitate perception and behavior. Over the course of development such differences in sensory processing might also make substantial contributions to higher-order cognitive differences such as language ability and social function.[43,59]

Integration of extrapersonal and peripersonal sensory input in ASD

While audiovisual integration is the most thoroughly studied multisensory process in autism and has clear links to impaired social communication that is fundamental to ASD, multisensory interactions taking place across other senses are also important. In addition to vision and audition, in which physical sources of stimulus energy are distal to the body of the perceiver, are the more peripersonal or proximal senses such as touch, proprioception, and vestibular sense. Indeed, the interaction between the extrapersonal and the peripersonal senses is essential to the construction of coherent perceptual self-awareness that forms much of the foundation for social and cognitive development (Fig. 17.4). For example, perspective taking, a core component of social interactions, requires the ability to use multisensory input to distinguish self from others[60–62] using converging input from proximal proprioceptive and somatosensory cues and distal visual information.[63–65] Additionally, postural awareness and control, which rely on integration of somatosensory, vestibular, and visual information, are impacted in autism and linked with social functioning.[66–70] Finally, the integration of visual and proprioceptive input for self–other distinction and comparison has been linked to the development of higher-order social cognitive abilities such as empathy and theory of mind.[71] The role of perceptual integration of both proximal and distal sensory cues in social cognition throughout development has clear relevance for ASD and is commonly experimentally tested in paradigms that combine visual and somatic (somatosensory, proprioceptive) inputs.

Visual–somatic integration in autism spectrum disorder

Similar to the utility of illusions like the sound-induced flash illusion and the McGurk effect described above for the study of atypical audiovisual integration, the rubber hand

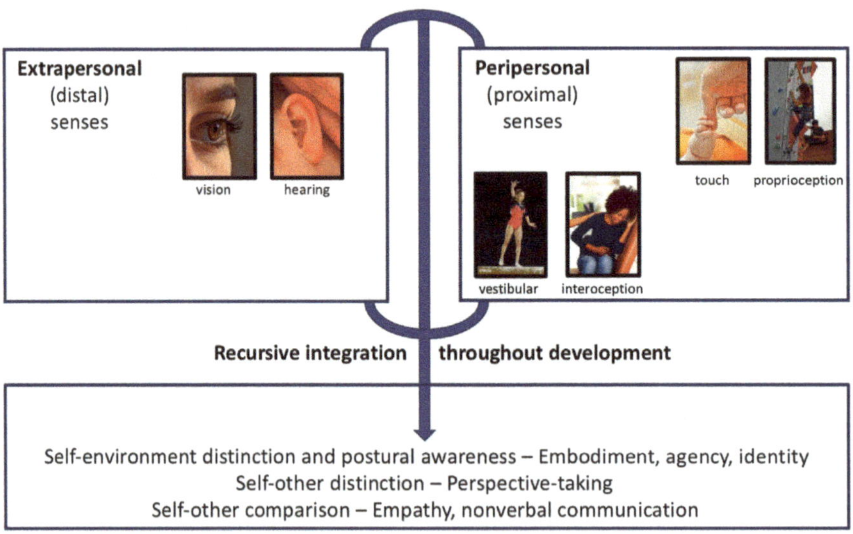

FIGURE 17.4 Model of the developmental sequelae of multisensory integration between extrapersonal and somatic (or peripersonal) sensory sources. Over the course of development, the integration of somatic and environmental sensory cues shapes complex social-cognitive abilities such as agency, perspective-taking, and empathy.

illusion (RHI)[72] involves the interplay between visual, tactile, and proprioceptive input and thus presents a unique opportunity to study multisensory processing in ASD as it applies to the sense of self. In the RHI, synchronous touch applied to an individual's obscured hand and a visible rubber hand tends to induce the experience of a touch located on the rubber hand or a fusion of the real with the rubber hand (Fig. 17.5). The RHI is thought to reflect modulation of perceived body ownership due to multisensory integration such that the rubber hand is incorporated into the body schema. The extent to which an individual experiences the RHI relies on spatiotemporal congruency between visual and tactile stimuli and is quantifiable by a drift in perceived hand location (proprioception) toward the rubber hand.[73,74] Individuals with ASD experience this illusion, but to a lesser extent than their TD counterparts.[75–77] Explanatory frameworks for reduced susceptibility to the RHI in ASD emphasize either sensory input and the contribution of bottom-up sensory processing or the top-down modulation of input based on internal factors like prior knowledge and attention. Both perspectives are considered below.

Bottom-up influences on visual–somatic integration

As described above, the TBW for multisensory stimuli is defined by the onsets and offsets of distinct stimulus inputs and reflects sensory-driven, bottom-up processing. To elicit the RHI, there must be both spatial and temporal congruency such that the visual and tactile input are perceived as perceptually bound. For example, temporal asynchrony of greater than 300 ms fails to induce the illusion.[78] For these reasons, asynchronous or spatially incongruent brushing frequently serves as control conditions in the RHI task. In ASD, these temporal and spatial constraints required for the RHI appear to be altered. Results from Cascio et al. (2012), over two 3-min

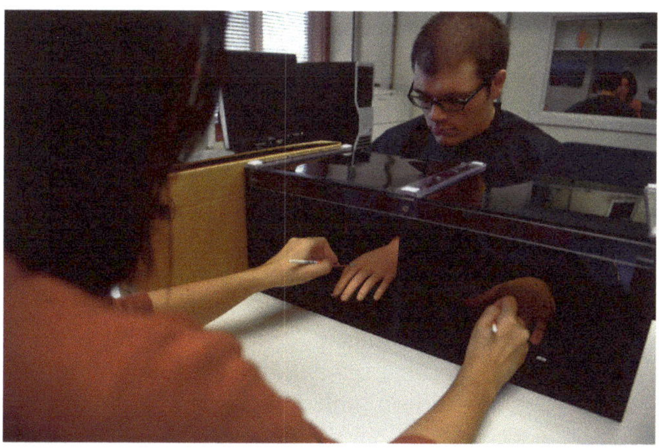

FIGURE 17.5 The rubber hand illusion. The experimenter (foreground) synchronously strokes the rubber hand (left side of figure) and the participant's actual left hand with a brush. The participant is instructed to attend to the visual input of the brushing on his right (rubber hand), while his left hand is visually obscured from view. A period of synchronous brushing tends to produce a perception of embodiment of the rubber hand. *Reproduced with permission from Vanderbilt University.*

exposure periods to visuo-tactile stimulation, demonstrated that ASD participants were initially more likely than TD peers to experience effects of the illusion during asynchronous brushing. This suggests a wider TBW, wherein asynchronous brush strokes were perceived more synchronously at first in the ASD group, but there are no studies that explicitly test for differences in the RHI TBW in ASD. Additionally, research in healthy subjects demonstrating experience-dependent malleability of the audiovisual TBW suggests that spatiotemporal binding windows are plastic.[79,80] The RHI is one opportunity to interrogate TBWs that incorporates peripersonal and extrapersonal inputs in ASD and can be modified with more explicit manipulation of effects of timing and training to shed light on this kind of multisensory integration in ASD. A related study reported enlarged visual-interoceptive TBWs in ASD for judgments of synchrony between visual stimuli and heartbeats,[81] adding to the evidence that wider TBWs in autism span the perceptual space within and outside of the perceiver's body.

Top-down influences on visual-somatic integration

Preferential weighting of sensory cues, a process involving top-down cognitive modulation of stimulus input, has also been proposed to explain ASD performance on tasks requiring the integration of visual, proprioceptive, and tactile stimuli. Cascio et al. (2012) also noted a delayed effect in the ASD group, with controls exhibiting the RHI after 3 minutes of brushing in the synchronous condition, but the ASD group requiring three additional minutes of exposure to produce a comparable effect. The delayed effect of the RHI in children with ASD could reflect a tendency to more greatly weight proprioceptive signals in the presence of competing input from other modalities, reducing susceptibility to bias resulting from discrepant tactile and visual inputs. This interpretation is supported by research in other multisensory and sensorimotor paradigms.[82–84] This kind of veridical sensory processing was also proposed in a study of the crossed-hands illusion in autism. In the crossed-hands illusion, crossing the hands produces errors in order judgments of stimuli delivered in rapid succession to each hand.[85] Individuals with ASD did not exhibit the expected reversal of TOJs, indicating a heavier weighting on somatic versus visual input.[86] These differences may, over the course of development, impact perceptual reference frames, leading to altered representations of the self and its relation to the environment.

Research has yet to provide clear evidence as to whether differential weighting is a result of increased attentional (or other cognitive) resources allocated to one sense over the other in a top-down manner, or if it is the result of more stimulus-directed bottom-up processes. As researchers continue to tease apart the mechanisms involved, a review of both perspectives indicates that differences in visual, tactile, and proprioceptive integration are likely influenced by a combination of both processes. In fact, a modified version of the RHI task aimed at distinguishing evidence of overreliance on proprioceptive information (top-down) and visuo-tactile temporal binding (bottom-up) differences in children with ASD indicates joint contributions.[87]

Stimulus considerations

As reviewed in earlier sections, the complexity of stimuli plays an important role in audiovisual deficits in ASD. Although illusory tasks like the RHI have a degree of ecological validity and are readily mapped to higher-order social cognitive constructs that are impacted by ASD, an investigation of low-level stimulus processing is essential to understanding atypical visuotactile and proprioceptive integration. Such studies, however, are scarce in individuals with ASD. Poole et al.[88] investigated the impact of temporal and spatial modulation on low-level visual-tactile integration in adults with ASD using a visual-tactile variant of the sound-induced flash illusion. Participants were asked to judge whether they felt a single or double tactile vibration while presented with light flashes that were either congruent (e.g., one vibration, one flash) or incongruent (one vibration, two flashes). The position of the lights was also manipulated along the proximal-distal axis from the participants. Results revealed intact temporal modulation of visual–tactile interactions but differences in spatial modulation, where individuals with ASD tended to integrate visual and tactile stimuli over greater spatial distances. In another study using simple, nonsocial stimuli, adults with ASD showed superior performance in a haptic-to-visual delayed shape-matching task compared to adults without ASD.[89] Evidence of intact and or superior performance in ASD individuals during multisensory tasks involving low-level stimuli emphasizes the importance of additional research aimed at understanding fundamental integration patterns necessary for more complex tasks and higher-order behaviors.

Summary of extrapersonal-peripersonal multisensory processing in ASD

Behavioral studies of multisensory integration outside of the audiovisual domain suggest partial disruption that may echo altered TBWs described in audiovisual paradigms, or may be influenced by differential weighting of unisensory inputs. Presently, there is very little neural evidence of altered multisensory processing outside of audiovisual integration to clarify the nature of these behavioral differences. A study of audio-tactile integration using electroencephalography (EEG) demonstrated delayed integration in children with ASD,[90] consistent with wider TBWs. However, additional research on the neural components of nonaudiovisual multisensory integration is necessary if we are to understand whether perceptual differences are universal, modality-specific, or dependent upon unique pairings of the senses. There is also a surprising dearth of information on the development of extrapersonal-peripersonal multisensory integration. Given the developmental trajectory of unisensory processing and well-documented developmental differences in the audiovisual literature, this area is also in need of further exploration. Continued investigation of multisensory integration beyond audiovisual sensations and across the entire distal–proximal sensory continuum will aid in generating a more holistic understanding of atypical multisensory processing in ASD.

Multisensory integration in animal models of autism spectrum disorder

While work in humans provides valuable insight into the neurological and behavioral changes in multisensory integration observed in individuals with neurodevelopmental and neuropsychiatric disorders, to study the mechanistic underpinnings of these changes at the cellular and molecular levels requires the use of animal models. Rodent models offer a host of advantages in this regard, including short maturational timelines, genetic tractability, and ease of causal manipulations to examine neuronal circuits (i.e., optogenetics). These models can provide valuable data necessary for the establishment of causal links between gene variation or drug exposure and the phenotypes associated with these disorders, including altered multisensory processing. It must be stressed that while these models provide insight as to the potential underpinnings of ASD, ASD is a complex polygenic disorder that can be either sporadic or familial and is heterogeneous in its clinical presentation. As such, no single rodent model can fully replicate this disorder, and the concept of an animal "model" of autism is fraught with a litany of conceptual and practical caveats. However, these models are useful for the study of potential causal links between specific gene variants or environmental exposures and the physiological and behavioral characteristics associated with ASD, including altered sensory processing.[91] Currently, mouse models exist for idiopathic and secondary ASD. Secondary ASD can be defined as ASD arising from exposure to a known environmental agent, chromosomal abnormalities, or single-gene disorders. Idiopathic ASD refers to ASD that occurs when no causal factor can be positively identified. It is important to consider models of both idiopathic and secondary ASD when examining the mechanisms that may underlie its clinical presentation. While idiopathic ASD is far more common than secondary ASD, the study of single-gene disorders and environmental exposures associated with ASD can elucidate the mechanisms that may play a role in idiopathic ASDs. The models considered here demonstrate behavioral characteristics associated with ASD, including altered social behavior and repetitive or ritualistic behaviors. We will focus on one model of idiopathic ASD and one single-gene variant model, with an emphasis on either the behavioral or molecular evidence for altered multisensory processing exhibited by these models.

The BTBR T + tf/J mouse is an inbred model that has been studied extensively as a model of *idiopathic* ASD. These animals display reduced social approach, low social interactions, impaired juvenile play, and repetitive behaviors (i.e., high levels of self-grooming).[92] To explore alterations in sensory processing and multisensory integration in this model, Gogolla and colleagues mapped postnatal maturation of sensory representations and integration in the insular cortex. In humans, insular cortex consistently demonstrates hypoactivity in people with ASD compared to typically developed individuals.[93] In rodents, insular cortex contains both auditory and somatosensory maps that are partially overlapping, with the region of spatial overlap exhibiting nonlinear responses indicative of multisensory integration.[94] While wild-type mice exhibit normal audiotactile integration in the insular cortex, such integration is lost in the BTBR mouse.[95] The authors of this last study further demonstrated that BTBR mice have prolonged and incomplete refinement of insular auditory fields (but not somatosensory fields) as compared to wild-type controls, indicating that impaired postnatal maturation may underlie the multisensory abnormalities present in the insular cortex of the BTBR mouse (Fig. 17.6). This study provides evidence in support of functional deficits in cortical multisensory processing in a model exhibiting behavioral characteristics associated with ASD. Given that the BTBR mouse is an animal model arising from inbreeding of strains with relevant traits rather than by manipulating a candidate gene derived from the human population, it is unclear how fully reflective this model is of human populations with ASD.

To directly address this question, it becomes useful to consider animal models expressing gene variants associated with ASD in human clinical populations. One such animal is the SERT G56A mouse, which expresses a serotonin transporter with a glycine-to-alanine substitution at position 56. This mutation has been found in individuals with ASD, and has been associated specifically with rigid-compulsive behavior and sensory aversion,[96] in humans. In mice, this variant has been shown to drive aberrant social behaviors and repetitive behaviors, as well as alterations to the serotonin (5-HT) system, including enhanced clearance rates of 5-HT and hyperserotonemia (which is also found in one-third of humans with ASD).[97,98] More recently, work performed by Siemann and colleagues demonstrated that the SERT G56A mouse exhibits disrupted multisensory processing when examined from a behavioral perspective. In this study, wild-type mice displayed behavioral gains (i.e., increased response accuracy) under multisensory conditions in an audiovisual operant task, while SERT G56A mice failed to demonstrate these behavioral gains.[99] This provides evidence for both the involvement of the serotonin system in multisensory processing and altered multisensory processing as a direct result of an ASD-associated gene variant.

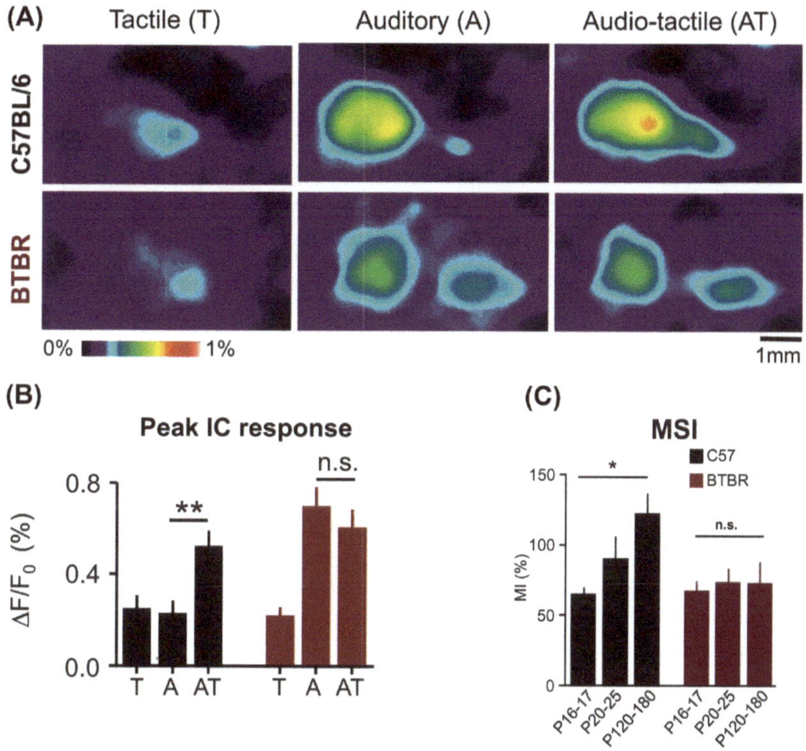

FIGURE 17.6 Multisensory processing in the insular cortex (IC) of BTBR and C57 mouse strains. (A) To assess multisensory processing in the IC, Gogolla and colleagues compared responses to pure tones of varying sound frequencies and intensities, air puffs administered to the front paw, or both audio-tactile stimuli applied simultaneously. Here we see activation patterns upon tactile (T), auditory (A), and audio-tactile (AT) stimulation in adult C57 (top row) and BTBR mice (bottom row). (B) Peak response fluorescence (ΔF/F0) in the IC upon A, T, and AT stimulation. Adult C57 mice exhibited superadditive multisensory responses in the IAF (insular auditory field) with amplitudes that were greater than the arithmetic sum of the two sensory modalities applied alone. This pattern is not observed in adult BTBR mice (t test; $^{**}P < .01$; n.s. = not significant, $P > .05$). (C) To study the development of integration, Gogolla and colleagues compared multisensory index (MI = [AT/(A + T)] × 100) at key developmental time points across groups. MSI increases with age in the IC of C57 but not BTBR mice. One-way ANOVA (C57: $^{*}P < .05$, BTBR: not significant, n.s.). *IAF*, Insular auditory field. *Figure reproduced with permission from Gogolla N, Takesian AE, Feng G, Fagiolini M, Hensch TK. Sensory integration in mouse insular cortex reflects GABA circuit maturation.* Neuron. *2014;83(4):894–905. https://doi.org/10.1016/j.neuron.2014.06.033*

Schizophrenia

Multisensory integration deficits have also been described in individuals on the SZ spectrum. As in autism, a prevailing hypothetical framework is that basic sensory differences cascade into higher-order cognitive, behavioral, and affective symptoms that characterize the disorder. Both SZ and autism are developmental in nature, autism having an age of onset in very early childhood, while the first episode of SZ typically occurs in late adolescence or early adulthood.[3] Both are defined by widespread behavioral symptoms that can be conceptualized as positive (repetitive behaviors in autism; delusions, hallucinations, and disorganized behavior in SZ), negative (reduced eye contact or conversational bids in autism; blunted affect, avolition, or anhedonia in SZ), or cognitive[100]. Both disorders involve significant deficits in social cognition and social interaction,[101] although there are both converging and distinct impairments in specific social skills.[102] The relevance of multisensory processing for both positive and negative symptoms of SZ has driven extensive investigation, much of which has focused on socially relevant stimuli such as audiovisual voice/face combinations, and visual/tactile/proprioceptive integration that informs perception of the bodily self. Simpler low-level stimuli have also been examined, to test the specificity of multisensory processing differences and define the boundaries of the cascading deficits framework. Given the age of onset for SZ, unless otherwise noted, all studies reviewed below focused on adults.

Low-level multisensory integration

Studies employing simple, low-level stimuli have largely suggested that multisensory integration is impaired in SZ, even at the most basic level. Patients with SZ showed reduced facilitation effects of bimodal relative to unimodal targets in an audiovisual speeded reaction time task, an effect that was more pronounced in patients with more severe negative symptoms.[103] Converging evidence for reduced influence of auditory input on visual perception was also seen in a sound-induced flash illusion study that considered both fission and fusion of flashes by added sounds, noting decreased influence of auditory input for both.[104] Several studies point toward an enlarged temporal integration window for audiovisual stimuli using the stream-bounce illusion (in which two identical visual targets moving across each other can be perceived as either streaming through or bouncing off of one another, and for which the probability of the latter can be increased by adding a brief tone at the moment the targets cross),[105] the sound-induced flash illusion,[106] and a simultaneity judgment task,[107,108] though the last two studies noted impaired unimodal temporal processing as well, suggesting a nonspecific temporal sensory deficit. This nonspecific temporal sensory deficit is also suggested by evidence for prolonged processing times of unimodal stimuli leading to deficits in both unimodal and multimodal sensory gating.[109]

Evidence against multimodal deficits is also seen in some studies: de Boer-Schellekens et al.[110] reported decreased sensitivity in a visual-only TOJ but comparable improvement to controls with added sounds in a multisensory condition. A study of multisensory facilitation in a target detection task failed to replicate Williams et al.'s early result,[103] finding similarly enhanced reaction times in both SZ and control groups, along with comparable ERP measures of multisensory integration.[111] Finally, a recent study suggests the multisensory deficits associated with subclinical SZ symptoms may be complex and nonlinear, finding that deviations in either direction from an intermediate-sized TBW and an associated measure of auditory cortical connectivity predicted higher levels of schizotypic symptoms.[112] This more nuanced finding is indicative of the need for improvement on case–control and linear-dependent experimental designs, better accounting for individual variability and the emerging continuum model of psychiatric nosology.[113]

Complex stimulus multisensory integration

Given the social deficits that are associated with SZ, the bulk of multisensory studies in this population have employed more complex, socially relevant stimuli such as faces and voices. For relatively simple audiovisual speech stimuli, the McGurk and ventriloquist effects have shed light on integration of voice and lip movements for spoken syllables in the temporal/spectral, and spatial domains, respectively. Intact spatial integration was reported in an early study of the ventriloquist effect, along with a reduction in the fused percepts during a McGurk-like paradigm, suggesting decreased influence of visual input on auditory perception of the spectral, but not spatial, properties of speech.[114] This was corroborated by a more recent study of the McGurk effect,[115] and another study that included adolescents with SZ,[116] but see Refs. 117 and 118 for nonreplications of reduced fused percepts. In spite of comparable behavioral performance on the McGurk paradigm, both of these studies and others have reported additional evidence of impaired audiovisual integration of speech stimuli in patients with SZ. Both Martin and colleagues[117] and an independent study by Stevenson and colleagues found prolonged windows of perceived simultaneity for asynchronous audiovisual syllables,[108] in keeping with findings from lower-level stimuli reviewed above. Roa Romero et al.[118] found that, despite intact perception, both short-latency ERP and longer-latency alpha-band EEG oscillations in response to the stimuli were abnormal in the SZ group. Abnormal ERP response to synchronous[119] and asynchronous[120] audiovisual speech stimuli has also been noted, particularly in later components.

Beyond single syllables, more complex, ecologically valid, social multisensory paradigms include speech in noise, stimuli that integrate speech and gesture, or stimuli introducing emotion and prosody into the perceptual paradigm. In a speech-in-noise paradigm, patients with SZ showed reduced benefit from watching visual articulation.[121] Patients with SZ show reduced hemodynamic response[122] in the middle temporal gyrus and inferior frontal gyrus, and reduced connectivity[123] between posterior superior temporal sulcus and inferior frontal gyrus when processing metaphoric, but not concrete gestures. Patients also show aberrant congruency effects of mismatched audiovisual words, both for overall blood oxygenation level-dependent (BOLD) responses in the superior temporal gyrus as well as frontal, visual, and limbic regions,[124] and for connectivity of the superior temporal sulcus and Broca's area.[125] The neural connectivity correlates of altered processing of audiovisual mismatch are illustrated in Fig. 17.7.

There may be differential magnitudes and directions of influence between specific sensory modalities for these more complex stimuli. Classification of a visual facial expression was less influenced by an emotional voice, but classification of an emotional voice was more influenced by facial expression in patients with SZ,[126] an effect that may be mediated by modality-specific attention.[127] In support of this interpretation, deficits in attention and other cognitive control

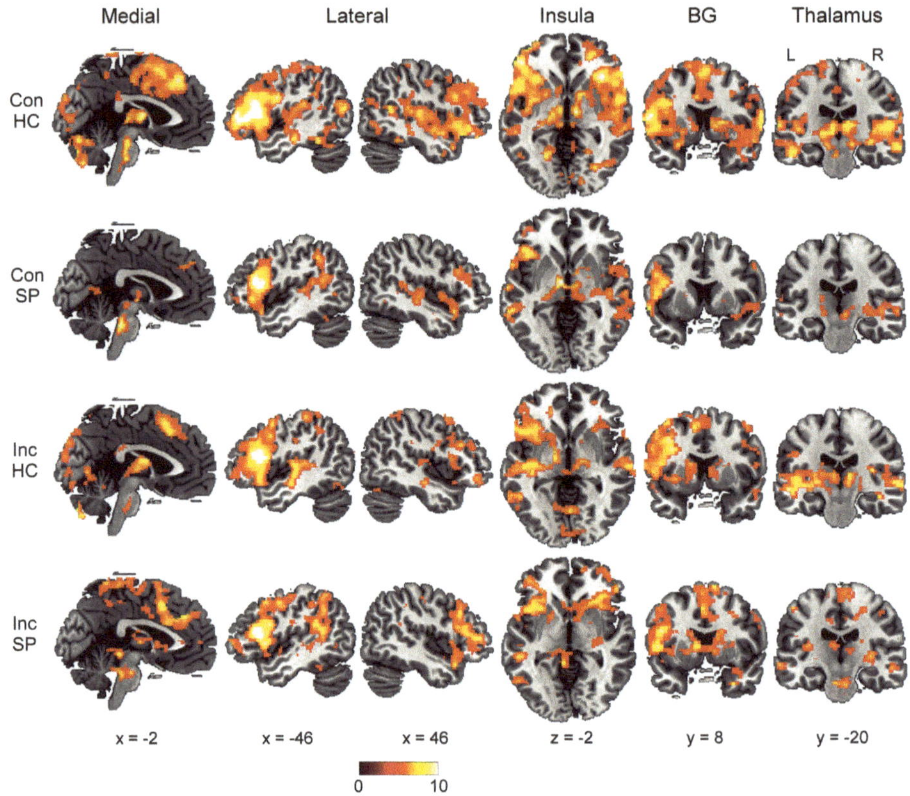

FIGURE 17.7 Altered functional connectivity during audiovisual processing in schizophrenia patients (SPs). Brain connectivity with the left inferior frontal gyrus (IFG, Broca's area) during exposure to congruent (Con) and incongruent (Inc) audiovisual speech stimuli in SPs and healthy controls (HCs) is displayed. The first row illustrates widespread connectivity between the IFG and medial frontal regions, superior temporal cortex, insula, basal ganglia (BG) and thalamus (THL) for the congruent stimuli in the HC group. This connectivity is substantially diminished in the SP group in this condition, which is much less differentiated from the Inc condition in SP than HC. Coordinates are in MNI (Montreal Neurological Institute) space and color map indicates t values ($P < .005$, FDR corrected). *Figure reproduced with permission from Szycik GR, Ye Z, Mohammadi B, et al. Maladaptive connectivity of Broca's area in schizophrenia during audiovisual speech perception: an fMRI study.* Neuroscience. *2013;253:274–282. https://doi.org/10.1016/j.neuroscience.2013.08.041.*

systems in the context of multimodal sensory processing are associated with dysfunction in modality-specific cortices but not in nodes of the cognitive control network.[128] However, for highly complex audiovisual tasks such as prosody discrimination and emotion recognition, global executive function may play a more important role.[129] When recognizing emotions in whole-body stimuli rather than faces alone, the visual dominance of auditory information may switch in favor of auditory dominance, especially when the auditory stimulus is a human voice.[130] Thus, up to a certain level of complexity, dysfunctional multisensory processing of audiovisual socially relevant stimuli may be mediated primarily by dynamics between unimodal neural systems, but beyond this level, top-down cognitive factors also come into play.

As discussed in the sections above, one of the prominent top-down factors is the cultivation of expectation based on prior experience. Of note, a recent study of hallucinations, a positive symptom of SZ that is far less conducive to experimental study than the aspects explored in the studies reviewed above, provides compelling evidence of hallucinations as aberrant top-down influences on perception (Fig. 17.8), mediated by unusually strong perceptual priors in a Bayesian predictive

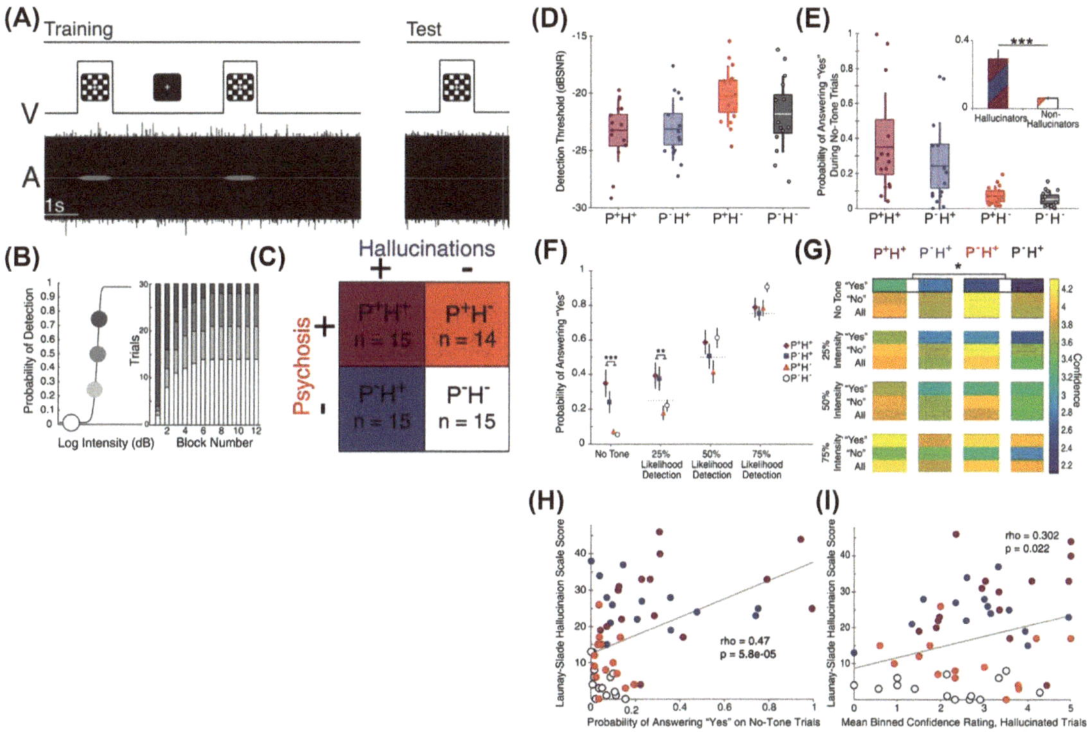

FIGURE 17.8 Multisensory associative learning-induced hallucinations in people with and without schizophrenia. (A) Establishing strong perceptual priors using paired auditory (A) and visual (V) stimuli, then delivering the V stimulus alone to determine whether the exposure induced an A hallucination. (B) Twelve blocks of conditioning trials systematically varied the intensity of auditory stimuli based on individual thresholds (right). Early blocks presented mostly stimuli that had approximately a 75% probability of detection, while later blocks increased the proportion of "absent" (0% probability of detection) stimuli. (C) Participants were in one of four groups, defined by two dimensions: the presence (+) or absence (−) of auditory (verbal) hallucinations (H) in daily life (blue) and the presence (+) or absence (−) of a psychotic (P) spectrum diagnosis (red). (D) Auditory detection thresholds did not differ among the four groups. (E) The probability of reporting hearing a tone that was not actually presented was significantly higher in the hallucinatory (H+) groups, regardless of whether they had psychosis. (F) The H+ groups were only statistically significantly different from the H- groups in the tone-absent or 25% likelihood detection conditions, illustrating that the difference was specific to absent or very faint auditory stimuli, rather than a global tendency to answer yes more than the H+ group. (G) Hallucinators (H+) were more confident than H- groups when responding "yes" to report an auditory stimulus that did not exist. (H)-(I) Both the probability of reporting conditioned hallucinations (H) in the experiment, and the confidence with which they were reported (I) correlated with a measure of hallucination severity in daily life. *Figure reproduced with permission from Powers AR, Mathys C, Corlett PR. Pavlovian conditioning–induced hallucinations result from overweighting of perceptual priors.* Science. *2017;357(6351):596–600. https://doi.org/10.1126/science.aan3458.*

framework.[131,132] This study capitalized on individuals without SZ who do hallucinate (self-described psychics who hear voices from the dead) as a comparison group and found that individuals who hallucinate in daily life (psychics and schizophrenic patients with reported hallucinations) are more prone to an induced auditory illusion brought on by associative learning (repeated pairings of visual and threshold or subthreshold auditory stimuli, at different probability levels, followed by visual stimuli alone). In the visual-only condition after the learning trials, the two groups of hallucinators were more likely to experience the nonexistent auditory stimuli than the two groups of nonhallucinators (typical adults and SZ patients for whom illusions are not part of the symptom profile). While psychics are not a clinical group, the effective use of an isolated behavioral/perceptual phenomenon that (a) is not limited to a single clinical diagnostic category and (b) can be mapped to specific neural circuits, is a good example of the current emphasis on dimensionality in clinical neuroscience.[113]

Multisensory integration relevant for self-perception

Both the positive and negative symptoms of SZ point toward an altered perceptual sense of the self.[133] Multisensory paradigms that manipulate this complex perceptual phenomenon typically focus on the interaction between somatic senses (touch, proprioception) and vision, as reviewed above. Several studies have reported a stronger RHI effect in patients with SZ,[134–136] and the multisensory nature of this phenomenon has been corroborated by the opposite trend (weaker illusion) in a related paradigm that utilizes expected, but not actual, touch.[137]

This information has been integrated into a theoretical proposal suggesting that patients with SZ have weaker or less predictable internal body representations and rely more heavily on external input (e.g., visual) to perceive the self.[138] While consistent with behavioral results reviewed above, this model does not account for neuroimaging reports of *decreased* connectivity between visual and sensorimotor cortices in patients with SZ.[139] While the mechanism remains unclear, there is widespread agreement that dysfunctional multisensory integration is likely to contribute to the altered sense of self, and resulting positive and negative symptoms, observed in SZ.[140] A clearer understanding of this phenomenon represents great therapeutic potential for addressing these symptomatic domains by way of their multisensory underpinnings.

Multisensory integration in animal models of schizophrenia

Having reviewed the literature on multisensory processing in human subjects with SZ, we turn now to animal models of the disorder, as we did previously for autism. As with rodent models of other complex, heterogeneous neuropsychiatric disorders (such as ASD), it is clear that no single model can fully replicate this disorder. Animal models instead provide insight as to neurophysiological mechanisms that may be at play in contributing to the phenotype of this disorder. Rodent models of SZ have been shown to demonstrate aberrant behaviors in paradigms designed to test the behaviors corresponding to the positive, negative, and cognitive symptoms of SZ. Locomotor activity paradigms assay psychomotor agitation to represent the positive symptoms of SZ. Aggressive behaviors and social withdrawal in rodents represent the negative symptoms associated with SZ, broadly framed as social deficits. Finally, spatial learning and memory tasks (i.e., Morris Water Maze or Barnes Maze), working memory tasks (i.e., passive and active avoidance tasks), and sensorimotor gating tasks (i.e., prepulse inhibition) assay the cognitive symptoms associated with SZ.

One commonly used rodent model of SZ involves subchronic administration of ketamine, a noncompetitive NMDA receptor antagonist. Ketamine has been shown to induce psychosis in humans that closely resembles SZ and exacerbates symptoms in patients with SZ, supporting its use in rodents to induce the symptoms of SZ and to study possible mechanisms contributing toward the behaviors associated with SZ. In rodents, subchronic application of ketamine drives aberrant behaviors that can be associated with all the symptom domains associated with the SZ phenotype. Specifically, these animals demonstrate increases in locomotion (positive symptom domain), increases in aggressive behaviors (interpreted in the negative symptom domain), and disruption of latent inhibition (cognitive symptom domain).[141,142]

Work by Cloke and colleagues in ketamine-treated rats supports findings related to altered multisensory processing in humans with SZ and points toward a possible underlying mechanism.[143] In this study, ketamine-treated rats displayed selective impairments in a multisensory object oddity task (specifically in detecting novel tactile-visual and olfactory-visual combinations). The authors demonstrate that these behavioral changes are accompanied by a reduction in GABAergic currents in the orbitofrontal cortex (OFC) of ketamine-treated rats, with reductions in parvalbumin-positive GABAergic interneuron populations in lateral OFC. The authors then show that administration of nicotinic acetylcholine receptor agonists ameliorates these deficits in multisensory task performance and normalizes GABAergic currents in these animals. These findings support the role of both prefrontal cortical and GABAergic function in multisensory task performance, particularly in the context of SZ.

Basic and clinical neuroscience links for multisensory integration

Recent work in rodents has looked to further dissect the mechanisms and locations of multisensory integration at the cortical and subcortical levels. This foundational work provides new evidence for understanding altered multisensory integration in clinical populations. In mouse sensory cortex, multisensory enhancement is found primarily in supragranular (layer 2/3) pyramidal cells and is uncommon in parvalbumin-expressing interneurons.[144] Further, the authors of this study show that lack of integration in these inhibitory interneurons is required to enable multisensory enhancement in neighboring pyramidal neurons. As an imbalance between excitation and inhibition has been proposed to contribute to both the schizophrenic and the ASD phenotype,[145–148] this work provides important clues as to how such an imbalance may contribute specifically to the multisensory behavioral and neurological changes observed in these clinical groups.

Subcortically, neurons capable of visual and tactile sensory integration have been identified in the dorsomedial striatum of mice.[149] The authors of this study demonstrated through anatomical tracing that the dorsal striatum receives projections from primary visual and somatosensory cortices. Extracellular recordings in the dorsal striatum demonstrated that maximal response amplitudes to visual-tactile stimuli were smaller and earlier than those predicted by the linear sum of visual and tactile stimuli alone in the dorsomedial striatum, indicating multisensory integration. Striatal abnormalities are implicated in both ASD and schizophrenic[150–153] symptoms including cognitive dysfunction and repetitive behaviors. The multisensory capabilities of striatal neurons present an important lead to follow when considering the mechanism of multisensory processing changes observed in both SZ and ASD.

Conclusions

In conclusion, the current literature points to global deficits in multisensory processing in both autism and SZ, as the most well-researched exemplars of neurodevelopmental and neuropsychiatric disorders, respectively. Prevailing theories of the etiology of these disorders emphasize sensory processing differences as putative drivers of the complex behavioral phenotypes that characterize them. Given the ontological primacy of sensory systems, if they do not develop properly, the ramifications for later-developing, more complex behaviors that they scaffold are severe. This is particularly relevant for SZ and autism, which both unfold with predictable developmental courses. An important theme that recurred in our review is that, while unisensory processing may be intact (and in ASD sometimes even enhanced, as in superior visual search[154]), multisensory processing, even of very simple, low-level stimuli, is consistently affected. This is reflected in faulty "binding" or "separating" of stimuli that occur close together in space and/or time, diminished perceptual benefits of multisensory input as in noisy speech, or from altered "weighting" of individual sensory modalities when resolving conflicting perceptual cues. Another key theme we touched on was the integration of input from extrapersonal/distal (auditory and visual) with peripersonal/proximal/somatic (touch, proprioception, vestibular sense, interoception) sources, and the importance of this kind of multisensory processing for the development of key aspects of social cognition such as self-other distinction, perspective-taking, and empathy that are impacted by these disorders. Finally, the perceptual phenomena discussed in the preceding paragraphs have strong parallels in their corresponding neural pathways, at both cortical and subcortical levels, and can be assessed in humans through approaches such as EEG and functional magnetic resonance imaging (fMRI), and more directly in animal models of these conditions. The integration of basic science work in animals using sophisticated approaches such as optogenetics, pharmacology, and intracranial electrophysiology, with the rigorous assessment of human perception using psychophysics and neuroimaging, promises to advance our understanding of the role multisensory processing plays in these and other conditions affecting behavior, and ultimately to inform more effective treatment and intervention.

References

1. Leekam SR, Nieto C, Libby SJ, Wing L, Gould J. Describing the sensory abnormalities of children and adults with autism. *J Autism Dev Disord*. 2007;37(5):894–910. https://doi.org/10.1007/s10803-006-0218-7.
2. Kanner L, others. *Autistic Disturbances of Affective Contact*. publisher not identified; 1943. http://neurodiversity.com/library_kanner_1943.pdf.
3. American Psychiatric Association. *Diagnostic and Statistical Manual of Mental Disorders*. Washington, DC: DSM-5; 2013.
4. Calvert G, Stein B, Spence C. *Handbook of Multisensory Processes*. Cambridge, MA: MIT Press; 2004.
5. Chandrasekaran C, Trubanova A, Stillittano S, Caplier A, Ghazanfar AA. The natural statistics of audiovisual speech. *PLoS Comput Biol*. 2009;5(7):e1000436. https://doi.org/10.1371/journal.pcbi.1000436.

6. Van der Burg E, Olivers CN, Bronkhorst AW, Theeuwes J. Pip and pop: nonspatial auditory signals improve spatial visual search. *J Exp Psychol Hum Percept Perform*. 2008;34(5):1053–1065. https://doi.org/10.1037/0096-1523.34.5.1053.
7. Collignon O, Charbonneau G, Peters F, et al. Reduced multisensory facilitation in persons with autism. *Cortex*. 2013;49(6):1704–1710. https://doi.org/10.1016/j.cortex.2012.06.001.
8. Joseph RM, Keehn B, Connolly C, Wolfe JM, Horowitz TS. Why is visual search superior in autism spectrum disorder? *Dev Sci*. 2009;12(6):1083–1096. https://doi.org/10.1111/j.1467-7687.2009.00855.x.
9. Shams L, Kamitani Y, Shimojo S. Illusions. What you see is what you hear. *Nature*. 2000;408(6814):788. https://doi.org/10.1038/35048669.
10. Stevenson RA, Siemann JK, Woynaroski TG, et al. Evidence for diminished multisensory integration in autism spectrum disorders. *J Autism Dev Disord*. 2014;44(12):3161–3167. https://doi.org/10.1007/s10803-014-2179-6.
11. Keane BP, Rosenthal O, Chun NH, Shams L. Audiovisual integration in high functioning adults with autism. *Res Autism Spectr Disord*. 2010;4(2):276–289. https://doi.org/10.1016/j.rasd.2009.09.015.
12. van der Smagt MJ, van Engeland H, Kemner C. Brief report: can you see what is not there? Low-level auditory-visual integration in autism spectrum disorder. *J Autism Dev Disord*. 2007;37(10):2014–2019. https://doi.org/10.1007/s10803-006-0346-0.
13. Brandwein AB, Foxe JJ, Butler JS, et al. The development of multisensory integration in high-functioning autism: high-density electrical mapping and psychophysical measures reveal impairments in the processing of audiovisual inputs. *Cerebr Cortex*. 2013;23(6):1329–1341. https://doi.org/10.1093/cercor/bhs109.
14. Miller J. Divided attention: evidence for coactivation with redundant signals. *Cogn Psychol*. 1982;14(2):247–279.
15. Stevenson RA, Ghose D, Fister JK, et al. Identifying and quantifying multisensory integration: a tutorial review. *Brain Topogr*. 2014;27(6):707–730. https://doi.org/10.1007/s10548-014-0365-7.
16. Brandwein AB, Foxe JJ, Butler JS, et al. Neurophysiological indices of atypical auditory processing and multisensory integration are associated with symptom severity in autism. *J Autism Dev Disord*. 2015;45(1):230–244. https://doi.org/10.1007/s10803-014-2212-9.
17. Ross LA, Saint-Amour D, Leavitt VM, Javitt DC, Foxe JJ. Do you see what I am saying? Exploring visual enhancement of speech comprehension in noisy environment. *Cerebr Cortex*. 2007;17(5):1147–1153. https://doi.org/10.1093/cercor/bhl024.
18. Alcantara JI, Weisblatt EJ, Moore BC, Bolton PF. Speech-in-noise perception in high-functioning individuals with autism or Asperger's syndrome. *J Child Psychol Psychiatry*. 2004;45(6):1107–1114. https://doi.org/10.1111/j.1469-7610.2004.t01-1-00303.x.
19. Irwin JR, Tornatore LA, Brancazio L, Whalen DH. Can children with autism spectrum disorders "hear" a speaking face? *Child Dev*. 2011;82(5):1397–1403. https://doi.org/10.1111/j.1467-8624.2011.01619.x.
20. Smith EG, Bennetto L. Audiovisual speech integration and lipreading in autism. *J Child Psychol Psychiatry*. 2007;48(8):813–821. https://doi.org/10.1111/j.1469-7610.2007.01766.x.
21. Foxe JJ, Molholm S, Del Bene VA, et al. Severe multisensory speech integration deficits in high-functioning school-aged children with Autism Spectrum Disorder (ASD) and their resolution during early adolescence. *Cerebr Cortex*. 2015;25(2):298–312. https://doi.org/10.1093/cercor/bht213.
22. McGurk H, MacDonald J. Hearing lips and seeing voices. *Nature*. 1976;264(5588):746–748.
23. Bebko JM, Schroeder JH, Weiss JA. The McGurk effect in children with autism and Asperger syndrome. *Autism Res*. 2014;7(1):50–59. https://doi.org/10.1002/aur.1343.
24. Mongillo EA, Irwin JR, Whalen DH, Klaiman C, Carter AS, Schultz RT. Audiovisual processing in children with and without autism spectrum disorders. *J Autism Dev Disord*. 2008;38(7):1349–1358. https://doi.org/10.1007/s10803-007-0521-y.
25. Saalasti S, Katsyri J, Tiippana K, Laine-Hernandez M, von Wendt L, Sams M. Audiovisual speech perception and eye gaze behavior of adults with asperger syndrome. *J Autism Dev Disord*. 2012;42(8):1606–1615. https://doi.org/10.1007/s10803-011-1400-0.
26. Stevenson RA, Siemann JK, Woynaroski TG, et al. Brief report: arrested development of audiovisual speech perception in autism spectrum disorders. *J Autism Dev Disord*. 2014;44(6):1470–1477. https://doi.org/10.1007/s10803-013-1992-7.
27. Taylor N, Isaac C, Milne E. A comparison of the development of audiovisual integration in children with autism spectrum disorders and typically developing children. *J Autism Dev Disord*. 2010;40(11):1403–1411. https://doi.org/10.1007/s10803-010-1000-4.

28. Woynaroski TG, Kwakye LD, Foss-Feig JH, Stevenson RA, Stone WL, Wallace MT. Multisensory speech perception in children with autism spectrum disorders. *J Autism Dev Disord*. 2013;43(12):2891–2902. https://doi.org/10.1007/s10803-013-1836-5.
29. Magnotti JF, Beauchamp MS. The noisy encoding of disparity model of the McGurk effect. *Psychon Bull Rev*. 2015;22(3):701–709. https://doi.org/10.3758/s13423-014-0722-2.
30. Magnee MJCM, de Gelder B, van Engeland H, Kemner C. Audiovisual speech integration in pervasive developmental disorder: evidence from event-related potentials. *J Child Psychol Psychiatry*. 2008;49(9):995–1000. https://doi.org/10.1111/j.1469-7610.2008.01902.x.
31. Megnin O, Flitton A, Jones CR, de Haan M, Baldeweg T, Charman T. Audiovisual speech integration in autism spectrum disorders: ERP evidence for atypicalities in lexical-semantic processing. *Autism Res*. 2012;5(1):39–48. https://doi.org/10.1002/aur.231.
32. Stevenson RA, Baum SH, Segers M, Ferber S, Barense MD, Wallace MT. Multisensory speech perception in autism spectrum disorder: from phoneme to whole-word perception. *Autism Res*. 2017;10(7):1280–1290. https://doi.org/10.1002/aur.1776.
33. Silverman LB, Bennetto L, Campana E, Tanenhaus MK. Speech-and-gesture integration in high functioning autism. *Cognition*. 2010;115(3):380–393. https://doi.org/10.1016/j.cognition.2010.01.002.
34. Vroomen J, Keetels M. Perception of intersensory synchrony: a tutorial review. *Atten Percept Psychophys*. 2010;72(4):871–884. https://doi.org/10.3758/APP.72.4.871.
35. Foss-Feig JH, Kwakye LD, Cascio CJ, et al. An extended multisensory temporal binding window in autism spectrum disorders. *Exp Brain Res*. 2010;203(2):381–389. https://doi.org/10.1007/s00221-010-2240-4.
36. Morein-Zamir S, Soto-Faraco S, Kingstone A. Auditory capture of vision: examining temporal ventriloquism. *Cogn Brain Res*. 2003;17(1):154–163. https://doi.org/10.1016/S0926-6410(03)00089-2.
37. Kwakye LD, Foss-Feig JH, Cascio CJ, Stone WL, Wallace MT. Altered auditory and multisensory temporal processing in autism spectrum disorders. *Front Integr Neurosci*. 2011;4:129. https://doi.org/10.3389/fnint.2010.00129.
38. de Boer-Schellekens L, Eussen M, Vroomen J. Diminished sensitivity of audiovisual temporal order in autism spectrum disorder. *Front Integr Neurosci*. 2013;7:8. https://doi.org/10.3389/fnint.2013.00008.
39. Poole D, Gowen E, Warren PA, Poliakoff E. Brief report: which came first? Exploring crossmodal temporal order judgements and their relationship with sensory reactivity in autism and neurotypicals. *J Autism Dev Disord*. 2017;47(1):215–223. https://doi.org/10.1007/s10803-016-2925-z.
40. Stevenson RA, Wallace MT. Multisensory temporal integration: task and stimulus dependencies. *Exp Brain Res*. 2013;227(2):249–261. https://doi.org/10.1007/s00221-013-3507-3.
41. Stevenson RA, Siemann JK, Schneider BC, et al. Multisensory temporal integration in autism spectrum disorders. *J Neurosci*. 2014;34(3):691–697. https://doi.org/10.1523/JNEUROSCI.3615-13.2014.
42. Grossman RB, Steinhart E, Mitchell T, McIlvane W. "Look who's talking!" gaze patterns for implicit and explicit audio-visual speech synchrony detection in children with high-functioning autism. *Autism Res*. 2015;8(3):307–316. https://doi.org/10.1002/aur.1447.
43. Wallace MT, Stevenson RA. The construct of the multisensory temporal binding window and its dysregulation in developmental disabilities. *Neuropsychologia*. 2014;64C:105–123. https://doi.org/10.1016/j.neuropsychologia.2014.08.005.
44. Simon DM, Noel JP, Wallace MT. Event related potentials index rapid recalibration to audiovisual temporal asynchrony. *Front Integr Neurosci*. 2017;11:8. https://doi.org/10.3389/fnint.2017.00008.
45. Van der Burg E, Alais D, Cass J. Rapid recalibration to audiovisual asynchrony. *J Neurosci*. 2013;33(37):14633–14637. https://doi.org/10.1523/JNEUROSCI.1182-13.2013.
46. Noel JP, De Niear MA, Stevenson R, Alais D, Wallace MT. Atypical rapid audio-visual temporal recalibration in autism spectrum disorders. *Autism Res*. 2016. https://doi.org/10.1002/aur.1633.
47. Turi M, Karaminis T, Pellicano E, Burr D. No rapid audiovisual recalibration in adults on the autism spectrum. *Sci Rep*. 2016;6. https://doi.org/10.1038/Srep21756.
48. Stocker AA Simoncelli EP. *Sensory Adaptation Within a Bayesian Framework for Perception*. 2006.
49. Pellicano E, Burr D. When the world becomes "too real": a Bayesian explanation of autistic perception. *Trends Cogn Sci*. 2012;16(10):504–510. https://doi.org/10.1016/j.tics.2012.08.009.
50. Rosenberg A, Patterson JS, Angelaki DE. A computational perspective on autism. *Proc Natl Acad Sci U A*. 2015;112(30):9158–9165. https://doi.org/10.1073/pnas.1510583112.

51. Guiraud JA, Tomalski P, Kushnerenko E, et al. Atypical audiovisual speech integration in infants at risk for autism. *PLoS One*. 2012;7(5). https://doi.org/10.1371/journal.pone.0036428.
52. Falck-Ytter T, Rehnberg E, Bolte S. Lack of visual orienting to biological motion and audiovisual synchrony in 3-year-olds with autism. *PLoS One*. 2013;8(7):e68816. https://doi.org/10.1371/journal.pone.0068816.
53. Patten E, Watson LR, Baranek GT. Temporal synchrony detection and associations with language in young children with ASD. *Autism Res Treat*. 2014;2014:678346. https://doi.org/10.1155/2014/678346.
54. Alsius A, Navarra J, Campbell R, Soto-Faraco S. Audiovisual integration of speech falters under high attention demands. *Curr Biol*. 2005;15(9):839–843. https://doi.org/10.1016/j.cub.2005.03.046.
55. Talsma D, Senkowski D, Soto-Faraco S, Woldorff MG. The multifaceted interplay between attention and multisensory integration. *Trends Cogn Sci*. 2010;14(9):400–410. https://doi.org/10.1016/j.tics.2010.06.008.
56. Dunn MA, Gomes H, Gravel J. Mismatch negativity in children with autism and typical development. *J Autism Dev Disord*. 2008;38(1):52–71. https://doi.org/10.1007/s10803-007-0359-3.
57. Magnee MJCM, de Gelder B, van Engeland H, Kemner C. Multisensory integration and attention in autism spectrum disorder: evidence from event-related potentials. *PLoS One*. 2011;6(8). https://doi.org/10.1371/journal.pone.0024196.
58. Murphy JW, Foxe JJ, Peters JB, Molholm S. Susceptibility to distraction in autism spectrum disorder: probing the integrity of oscillatory alpha-band suppression mechanisms. *Autism Res*. 2014;7(4):442–458. https://doi.org/10.1002/aur.1374.
59. Thye MD, Bednarz HM, Herringshaw AJ, Sartin EB, Kana RK. The impact of atypical sensory processing on social impairments in autism spectrum disorder. *Dev Cogn Neurosci*. 2017. https://doi.org/10.1016/j.dcn.2017.04.010.
60. Brass M, Ruby P, Spengler S. Inhibition of imitative behaviour and social cognition. *Philos Trans R Soc Lond B Biol Sci*. 2009;364(1528):2359–2367. https://doi.org/10.1098/rstb.2009.0066.
61. Paladino M-P, Mazzurega M, Pavani F, Schubert TW. Synchronous multisensory stimulation blurs self-other boundaries. *Psychol Sci*. 2010;21(9):1202–1207. https://doi.org/10.1177/0956797610379234.
62. Spengler S, von Cramon DY, Brass M. Resisting motor mimicry: control of imitation involves processes central to social cognition in patients with frontal and temporo-parietal lesions. *Soc Neurosci*. 2010;5(4):401–416. https://doi.org/10.1080/17470911003687905.
63. Jeannerod M. Visual and action cues contribute to the self-other distinction. *Nat Neurosci*. 2004;7(5):422–423. https://doi.org/10.1038/nn0504-422.
64. Maravita A, Spence C, Driver J. Multisensory integration and the body schema: close to hand and within reach. *Curr Biol CB*. 2003;13(13):R531–R539.
65. Schütz-Bosbach S, Musil JJ, Haggard P. Touchant-touché: the role of self-touch in the representation of body structure. *Conscious Cognit*. 2009;18(1):2–11. https://doi.org/10.1016/j.concog.2008.08.003.
66. Casartelli L, Molteni M, Ronconi L. So close yet so far: motor anomalies impacting on social functioning in autism spectrum disorder. *Neurosci Biobehav Rev*. 2016;63:98–105. https://doi.org/10.1016/j.neubiorev.2016.02.001.
67. Memari AH, Ghanouni P, Shayestehfar M, Ghaheri B. Postural control impairments in individuals with autism spectrum disorder: a critical review of current literature. *Asian J Sports Med*. 2014;5(3):e22963. https://doi.org/10.5812/asjsm.22963.
68. Molloy CA, Dietrich KN, Bhattacharya A. Postural stability in children with autism spectrum disorder. *J Autism Dev Disord*. 2003;33(6):643–652.
69. Peterka RJ. Sensorimotor integration in human postural control. *J Neurophysiol*. 2002;88(3):1097–1118.
70. Weimer AK, Schatz AM, Lincoln A, Ballantyne AO, Trauner DA. "Motor" impairment in Asperger syndrome: evidence for a deficit in proprioception. *J Dev Behav Pediatr JDBP*. 2001;22(2):92–101.
71. Williams JHG. Self-other relations in social development and autism: multiple roles for mirror neurons and other brain bases. *Autism Res Off J Int Soc Autism Res*. 2008;1(2):73–90. https://doi.org/10.1002/aur.15.
72. Botvinick M, Cohen J. Rubber hands "feel" touch that eyes see. *Nature*. 1998;391(6669):756. https://doi.org/10.1038/35784.
73. Kammers MPM, Kootker JA, Hogendoorn H, Dijkerman HC. How many motoric body representations can we grasp? *Exp Brain Res*. 2010;202(1):203–212. https://doi.org/10.1007/s00221-009-2124-7.
74. Tsakiris M, Haggard P. The rubber hand illusion revisited: visuotactile integration and self-attribution. *J Exp Psychol Hum Percept Perform*. 2005;31(1):80–91. https://doi.org/10.1037/0096-1523.31.1.80.

75. Cascio CJ, Foss-Feig JH, Burnette CP, Heacock JL, Cosby AA. The rubber hand illusion in children with autism spectrum disorders: delayed influence of combined tactile and visual input on proprioception. *Autism Int J Res Pract.* 2012;16(4):406–419. https://doi.org/10.1177/1362361311430404.
76. Palmer CJ, Paton B, Kirkovski M, Enticott PG, Hohwy J. Context sensitivity in action decreases along the autism spectrum: a predictive processing perspective. *Proc Biol Sci.* 2015;282(1802). https://doi.org/10.1098/rspb.2014.1557.
77. Paton B, Hohwy J, Enticott PG. The rubber hand illusion reveals proprioceptive and sensorimotor differences in autism spectrum disorders. *J Autism Dev Disord.* 2012;42(9):1870–1883. https://doi.org/10.1007/s10803-011-1430-7.
78. Shimada S, Fukuda K, Hiraki K. Rubber hand illusion under delayed visual feedback. *PLoS One.* 2009;4(7):e6185. https://doi.org/10.1371/journal.pone.0006185.
79. Mégevand P, Molholm S, Nayak A, Foxe JJ. Recalibration of the multisensory temporal window of integration results from changing task demands. *PLoS One.* 2013;8(8):e71608. https://doi.org/10.1371/journal.pone.0071608.
80. Powers AR, Hillock AR, Wallace MT. Perceptual training narrows the temporal window of multisensory binding. *J Neurosci Off J Soc Neurosci.* 2009;29(39):12265–12274. https://doi.org/10.1523/JNEUROSCI.3501-09.2009.
81. Noel J-P, Lytle M, Cascio C, Wallace MT. Disrupted integration of exteroceptive and interoceptive signaling in autism spectrum disorder. *Autism Res Off J Int Soc Autism Res.* 2017. https://doi.org/10.1002/aur.1880.
82. Glazebrook C, Gonzalez D, Hansen S, Elliott D. The role of vision for online control of manual aiming movements in persons with autism spectrum disorders. *Autism Int J Res Pract.* 2009;13(4):411–433. https://doi.org/10.1177/1362361309105659.
83. Haswell CC, Izawa J, Dowell LR, Mostofsky SH, Shadmehr R. Representation of internal models of action in the autistic brain. *Nat Neurosci.* 2009;12(8):970–972. https://doi.org/10.1038/nn.2356.
84. Masterton BA, Biederman GB. Proprioceptive versus visual control in autistic children. *J Autism Dev Disord.* 1983;13(2):141–152.
85. Yamamoto S, Kitazawa S. Reversal of subjective temporal order due to arm crossing. *Nat Neurosci.* 2001;4(7):759–765. https://doi.org/10.1038/89559.
86. Wada M, Suzuki M, Takaki A, Miyao M, Spence C, Kansaku K. Spatio-temporal processing of tactile stimuli in autistic children. *Sci Rep.* 2014;4:5985. https://doi.org/10.1038/srep05985.
87. Greenfield K, Ropar D, Smith AD, Carey M, Newport R. Visuo-tactile integration in autism: atypical temporal binding may underlie greater reliance on proprioceptive information. *Mol Autism.* 2015;6:51. https://doi.org/10.1186/s13229-015-0045-9.
88. Poole D, Gowen E, Warren PA, Poliakoff E. Investigating visual-tactile interactions over time and space in adults with autism. *J Autism Dev Disord.* 2015;45(10):3316–3326. https://doi.org/10.1007/s10803-015-2492-8.
89. Nakano T, Kato N, Kitazawa S. Superior haptic-to-visual shape matching in autism spectrum disorders. *Neuropsychologia.* 2012;50(5):696–703. https://doi.org/10.1016/j.neuropsychologia.2011.12.024.
90. Russo N, Foxe JJ, Brandwein AB, Altschuler T, Gomes H, Molholm S. Multisensory processing in children with autism: high-density electrical mapping of auditory-somatosensory integration. *Autism Res Off J Int Soc Autism Res.* 2010;3(5):253–267. https://doi.org/10.1002/aur.152.
91. Crawley JN. Mouse behavioral assays relevant to the symptoms of autism*. *Brain Pathol.* 2007;17(4):448–459. https://doi.org/10.1111/j.1750-3639.2007.00096.x.
92. McFarlane HG, Kusek GK, Yang M, Phoenix JL, Bolivar VJ, Crawley JN. Autism-like behavioral phenotypes in BTBR T+tf/J mice. *Genes Brain Behav.* 2008;7(2):152–163. https://doi.org/10.1111/j.1601-183X.2007.00330.x.
93. Di Martino A, Ross K, Uddin LQ, Sklar AB, Castellanos FX, Milham MP. Functional brain correlates of social and nonsocial processes in autism spectrum disorders: an activation likelihood estimation meta-analysis. *Biol Psychiatry.* 2009;65(1):63–74. https://doi.org/10.1016/j.biopsych.2008.09.022.
94. Rodgers KM, Benison AM, Klein A, Barth DS. Auditory, somatosensory, and multisensory insular cortex in the rat. *Cereb Cortex N Y NY.* 2008;18(12):2941–2951. https://doi.org/10.1093/cercor/bhn054.
95. Gogolla N, Takesian AE, Feng G, Fagiolini M, Hensch TK. Sensory integration in mouse insular cortex reflects GABA circuit maturation. *Neuron.* 2014;83(4):894–905. https://doi.org/10.1016/j.neuron.2014.06.033.
96. Sutcliffe JS, Delahanty RJ, Prasad HC, et al. Allelic heterogeneity at the serotonin transporter locus (SLC6A4) confers susceptibility to autism and rigid-compulsive behaviors. *Am J Hum Genet.* 2005;77(2):265–279. https://doi.org/10.1086/432648.

97. Anderson GM, Freedman DX, Cohen DJ, et al. Whole blood serotonin in autistic and normal subjects. *J Child Psychol Psychiatry*. 1987;28(6):885–900. https://doi.org/10.1111/j.1469-7610.1987.tb00677.x.
98. Veenstra-VanderWeele J, Muller CL, Iwamoto H, et al. Autism gene variant causes hyperserotonemia, serotonin receptor hypersensitivity, social impairment and repetitive behavior. *Proc Natl Acad Sci U S A*. 2012;109(14):5469–5474. https://doi.org/10.1073/pnas.1112345109.
99. Siemann JK, Muller CL, Forsberg CG, Blakely RD, Veenstra-VanderWeele J, Wallace MT. An autism-associated serotonin transporter variant disrupts multisensory processing. *Transl Psychiatry*. 2017;7(3):e1067. https://doi.org/10.1038/tp.2017.17.
100. Foss-Feig JH, McPartland JC, Anticevic A, Wolf J. Re-conceptualizing ASD within a dimensional framework: positive, negative, and cognitive feature clusters. *J Autism Dev Disord*. 2016;46(1):342–351. https://doi.org/10.1007/s10803-015-2539-x.
101. Sasson NJ, Pinkham AE, Carpenter KLH, Belger A. The benefit of directly comparing autism and schizophrenia for revealing mechanisms of social cognitive impairment. *J Neurodev Disord*. 2011;3(2):87–100. https://doi.org/10.1007/s11689-010-9068-x.
102. Morrison KE, Pinkham AE, Penn DL, Kelsven S, Ludwig K, Sasson NJ. Distinct profiles of social skill in adults with autism spectrum disorder and schizophrenia. *Autism Res Off J Int Soc Autism Res*. 2017;10(5):878–887. https://doi.org/10.1002/aur.1734.
103. Williams LE, Light GA, Braff DL, Ramachandran VS. Reduced multisensory integration in patients with schizophrenia on a target detection task. *Neuropsychologia*. 2010;48(10):3128–3136. https://doi.org/10.1016/j.neuropsychologia.2010.06.028.
104. Vanes LD, White TP, Wigton RL, Joyce D, Collier T, Shergill SS. Reduced susceptibility to the sound-induced flash fusion illusion in schizophrenia. *Psychiatry Res*. 2016;245:58–65. https://doi.org/10.1016/j.psychres.2016.08.016.
105. Zvyagintsev M, Parisi C, Mathiak K. Temporal processing deficit leads to impaired multisensory binding in schizophrenia. *Cogn Neuropsychiatry*. 2017:1–12. https://doi.org/10.1080/13546805.2017.1331160.
106. Haß K, Sinke C, Reese T, et al. Enlarged temporal integration window in schizophrenia indicated by the double-flash illusion. *Cogn Neuropsychiatry*. 2017;22(2):145–158. https://doi.org/10.1080/13546805.2017.1287693.
107. Foucher JR, Lacambre M, Pham B-T, Giersch A, Elliott MA. Low time resolution in schizophrenia: lengthened windows of simultaneity for visual, auditory and bimodal stimuli. *Schizophr Res*. 2007;97(1):118–127. https://doi.org/10.1016/j.schres.2007.08.013.
108. Stevenson RA, Park S, Cochran C, et al. The associations between multisensory temporal processing and symptoms of schizophrenia. *Schizophr Res*. 2017;179:97–103. https://doi.org/10.1016/j.schres.2016.09.035.
109. Haß K, Bak N, Szycik GR, Glenthøj BY, Oranje B. Deficient prepulse inhibition of the startle reflex in schizophrenia using a cross-modal paradigm. *Biol Psychol*. 2017;128:112–116. https://doi.org/10.1016/j.biopsycho.2017.07.016.
110. de Boer-Schellekens L, Stekelenburg JJ, Maes JP, Van Gool AR, Vroomen J. Sound improves diminished visual temporal sensitivity in schizophrenia. *Acta Psychol*. 2014;147:136–142. https://doi.org/10.1016/j.actpsy.2013.06.013.
111. Wynn JK, Jahshan C, Green MF. Multisensory integration in schizophrenia: a behavioural and event-related potential study. *Cogn Neuropsychiatry*. 2014;19(4):319–336. https://doi.org/10.1080/13546805.2013.866892.
112. Ferri F, Nikolova YS, Perrucci MG, et al. A neural "tuning curve" for multisensory experience and cognitive-perceptual schizotypy. *Schizophr Bull*. 2017. https://doi.org/10.1093/schbul/sbw174.
113. Cuthbert BN, Insel TR. Toward the future of psychiatric diagnosis: the seven pillars of RDoC. *BMC Med*. 2013;11:126. https://doi.org/10.1186/1741-7015-11-126.
114. de Gelder B, Vroomen J, Annen L, Masthof E, Hodiamont P. Audio-visual integration in schizophrenia. *Schizophr Res*. 2003;59(2):211–218. https://doi.org/10.1016/S0920-9964(01)00344-9.
115. White TP, Wigton RL, Joyce DW, et al. Eluding the illusion? Schizophrenia, dopamine and the McGurk effect. *Front Hum Neurosci*. 2014;8:565. https://doi.org/10.3389/fnhum.2014.00565.
116. Pearl D, Yodashkin-Porat D, Katz N, et al. Differences in audiovisual integration, as measured by McGurk phenomenon, among adult and adolescent patients with schizophrenia and age-matched healthy control groups. *Compr Psychiatr*. 2009;50(2):186–192. https://doi.org/10.1016/j.comppsych.2008.06.004.

117. Martin B, Giersch A, Huron C, van Wassenhove V. Temporal event structure and timing in schizophrenia: preserved binding in a longer "now. *Neuropsychologia*. 2013;51(2):358–371. https://doi.org/10.1016/j.neuropsychologia.2012.07.002.
118. Roa Romero Y, Keil J, Balz J, Niedeggen M, Gallinat J, Senkowski D. Alpha-band oscillations reflect altered multisensory processing of the McGurk illusion in schizophrenia. *Front Hum Neurosci*. 2016;10:41. https://doi.org/10.3389/fnhum.2016.00041.
119. Liu T, Pinheiro AP, Zhao Z, Nestor PG, McCarley RW, Niznikiewicz M. Simultaneous face and voice processing in schizophrenia. *Behav Brain Res*. 2016;305:76–86. https://doi.org/10.1016/j.bbr.2016.01.039.
120. Stekelenburg JJ, Maes JP, Van Gool AR, Sitskoorn M, Vroomen J. Deficient multisensory integration in schizophrenia: an event-related potential study. *Schizophr Res*. 2013;147(2–3):253–261. https://doi.org/10.1016/j.schres.2013.04.038.
121. Ross LA, Saint-Amour D, Leavitt VM, Molholm S, Javitt DC, Foxe JJ. Impaired multisensory processing in schizophrenia: deficits in the visual enhancement of speech comprehension under noisy environmental conditions. *Schizophr Res*. 2007;97(1–3):173–183. https://doi.org/10.1016/j.schres.2007.08.008.
122. Straube B, Green A, Sass K, Kirner-Veselinovic A, Kircher T. Neural integration of speech and gesture in schizophrenia: evidence for differential processing of metaphoric gestures. *Hum Brain Mapp*. 2013;34(7):1696–1712. https://doi.org/10.1002/hbm.22015.
123. Straube B, Green A, Sass K, Kircher T. Superior temporal sulcus disconnectivity during processing of metaphoric gestures in schizophrenia. *Schizophr Bull*. 2014;40(4):936–944. https://doi.org/10.1093/schbul/sbt110.
124. Szycik GR, Münte TF, Dillo W, et al. Audiovisual integration of speech is disturbed in schizophrenia: an fMRI study. *Schizophr Res*. 2009;110(1–3):111–118. https://doi.org/10.1016/j.schres.2009.03.003.
125. Szycik GR, Ye Z, Mohammadi B, et al. Maladaptive connectivity of Broca's area in schizophrenia during audiovisual speech perception: an fMRI study. *Neuroscience*. 2013;253:274–282. https://doi.org/10.1016/j.neuroscience.2013.08.041.
126. de Gelder B, Vroomen J, de Jong SJ, Masthoff ED, Trompenaars FJ, Hodiamont P. Multisensory integration of emotional faces and voices in schizophrenics. *Schizophr Res*. 2005;72(2):195–203. https://doi.org/10.1016/j.schres.2004.02.013.
127. de Jong JJ, Hodiamont PPG, de Gelder B. Modality-specific attention and multisensory integration of emotions in schizophrenia: reduced regulatory effects. *Schizophr Res*. 2010;122(1–3):136–143. https://doi.org/10.1016/j.schres.2010.04.010.
128. Mayer AR, Hanlon FM, Teshiba TM, et al. An fMRI study of multimodal selective attention in schizophrenia. *Br J Psychiatry J Ment Sci*. 2015;207(5):420–428. https://doi.org/10.1192/bjp.bp.114.155499.
129. Castagna F, Montemagni C, Maria Milani A, et al. Prosody recognition and audiovisual emotion matching in schizophrenia: the contribution of cognition and psychopathology. *Psychiatry Res*. 2013;205(3):192–198. https://doi.org/10.1016/j.psychres.2012.08.038.
130. Van den Stock J, de Jong SJ, Hodiamont PPG, de Gelder B. Perceiving emotions from bodily expressions and multisensory integration of emotion cues in schizophrenia. *Soc Neurosci*. 2011;6(5–6):537–547. https://doi.org/10.1080/17470919.2011.568790.
131. Powers AR, Mathys C, Corlett PR. Pavlovian conditioning–induced hallucinations result from overweighting of perceptual priors. *Science*. 2017;357(6351):596–600. https://doi.org/10.1126/science.aan3458.
132. Powers AR, Kelley M, Corlett PR. Hallucinations as top-down effects on perception. *Biol Psychiatry Cogn Neurosci Neuroimaging*. 2016;1(5):393–400. https://doi.org/10.1016/j.bpsc.2016.04.003.
133. Noel J-P, Cascio CJ, Wallace MT, Park S. The spatial self in schizophrenia and autism spectrum disorder. *Schizophr Res*. 2017;179:8–12. https://doi.org/10.1016/j.schres.2016.09.021.
134. Germine L, Benson TL, Cohen F, Hooker CI. Psychosis-proneness and the rubber hand illusion of body ownership. *Psychiatry Res*. 2013;207(1–2):45–52. https://doi.org/10.1016/j.psychres.2012.11.022.
135. Peled A, Ritsner M, Hirschmann S, Geva AB, Modai I. Touch feel illusion in schizophrenic patients. *Biol Psychiatry*. 2000;48(11):1105–1108.
136. Thakkar KN, Nichols HS, McIntosh LG, Park S. Disturbances in body ownership in schizophrenia: evidence from the rubber hand illusion and case study of a spontaneous out-of-body experience. *PLoS One*. 2011;6(10):e27089. https://doi.org/10.1371/journal.pone.0027089.
137. Ferri F, Costantini M, Salone A, et al. Upcoming tactile events and body ownership in schizophrenia. *Schizophr Res*. 2014;152(1):51–57. https://doi.org/10.1016/j.schres.2013.06.026.

138. Klaver M, Dijkerman HC. Bodily experience in schizophrenia: factors underlying a disturbed sense of body ownership. *Front Hum Neurosci*. 2016;10:305. https://doi.org/10.3389/fnhum.2016.00305.
139. Chen X, Duan M, Xie Q, et al. Functional disconnection between the visual cortex and the sensorimotor cortex suggests a potential mechanism for self-disorder in schizophrenia. *Schizophr Res*. 2015;166(1–3):151–157. https://doi.org/10.1016/j.schres.2015.06.014.
140. Postmes L, Sno HN, Goedhart S, van der Stel J, Heering HD, de Haan L. Schizophrenia as a self-disorder due to perceptual incoherence. *Schizophr Res*. 2014;152(1):41–50. https://doi.org/10.1016/j.schres.2013.07.027.
141. Becker A, Peters B, Schroeder H, Mann T, Huether G, Grecksch G. Ketamine-induced changes in rat behaviour: a possible animal model of schizophrenia. *Prog Neuropsychopharmacol Biol Psychiatry*. 2003;27(4):687–700. https://doi.org/10.1016/S0278-5846(03)00080-0.
142. Irifune M, Shimizu T, Nomoto M. Ketamine-induced hyperlocomotion associated with alteration of presynaptic components of dopamine neurons in the nucleus accumbens of mice. *Pharmacol Biochem Behav*. 1991;40(2):399–407. https://doi.org/10.1016/0091-3057(91)90571-I.
143. Cloke JM, Nguyen R, Chung BYT, et al. A novel multisensory integration task reveals robust deficits in rodent models of schizophrenia: converging evidence for remediation via nicotinic receptor stimulation of inhibitory transmission in the prefrontal cortex. *J Neurosci*. 2016;36(50):12570–12585. https://doi.org/10.1523/JNEUROSCI.1628-16.2016.
144. Olcese U, Iurilli G, Medini P. Cellular and synaptic architecture of multisensory integration in the mouse neocortex. *Neuron*. 2013;79(3):579–593. https://doi.org/10.1016/j.neuron.2013.06.010.
145. Foss-Feig JH, Adkinson BD, Ji JL, et al. Searching for cross-diagnostic convergence: neural mechanisms governing excitation and inhibition balance in schizophrenia and autism spectrum disorders. *Biol Psychiatry*. 2017;81(10):848–861. https://doi.org/10.1016/j.biopsych.2017.03.005.
146. Hoftman GD, Datta D, Lewis DA. Layer 3 excitatory and inhibitory circuitry in the prefrontal cortex: developmental trajectories and alterations in schizophrenia. *Biol Psychiatry*. 2017;81(10):862–873. https://doi.org/10.1016/j.biopsych.2016.05.022.
147. Nelson SB, Valakh V. Excitatory/inhibitory balance and circuit homeostasis in autism spectrum disorders. *Neuron*. 2015;87(4):684–698. https://doi.org/10.1016/j.neuron.2015.07.033.
148. Tabuchi K, Blundell J, Etherton MR, et al. A neuroligin-3 mutation implicated in autism increases inhibitory synaptic transmission in mice. *Science*. 2007;318(5847):71–76. https://doi.org/10.1126/science.1146221.
149. Reig R, Silberberg G. Multisensory integration in the mouse striatum. *Neuron*. 2014;83(5):1200–1212. https://doi.org/10.1016/j.neuron.2014.07.033.
150. Haznedar MM, Buchsbaum MS, Hazlett EA, LiCalzi EM, Cartwright C, Hollander E. Volumetric analysis and three-dimensional glucose metabolic mapping of the striatum and thalamus in patients with autism spectrum disorders. *Am J Psychiatry*. 2006;163(7):1252–1263. https://doi.org/10.1176/appi.ajp.163.7.1252.
151. Langen M, Bos D, Noordermeer SDS, Nederveen H, van Engeland H, Durston S. Changes in the development of striatum are involved in repetitive behavior in autism. *Biol Psychiatry*. 2014;76(5):405–411. https://doi.org/10.1016/j.biopsych.2013.08.013.
152. Levitt JJ, Nestor PG, Levin L, et al. Reduced structural connectivity in frontostriatal white matter tracts in the associative loop in schizophrenia. *Am J Psychiatry*. 2017;174(11):1102–1111. https://doi.org/10.1176/appi.ajp.2017.16091046.
153. Simpson EH, Kellendonk C, Kandel E. A possible role for the striatum in the pathogenesis of the cognitive symptoms of schizophrenia. *Neuron*. 2010;65(5):585–596. https://doi.org/10.1016/j.neuron.2010.02.014.
154. Kaldy Z, Giserman I, Carter AS, Blaser E. The mechanisms underlying the ASD advantage in visual search. *J Autism Dev Disord*. 2016;46(5):1513–1527. https://doi.org/10.1007/s10803-013-1957-x.

CHAPTER

18

Disorders of body representation

Laura K. Case[1], *Marco Solcà*[2,3], *Olaf Blanke*[2,4], *Nathan Faivre*[2,5]

[1]Pain and Integrative Neuroscience Branch, National Center for Complementary and Integrative Health, Bethesda, MD, United States; [2]Center for Neuroprosthetics, Laboratory of Cognitive Neuroscience, Brain-Mind Institute, École Polytechnique Fédérale de Lausanne-EPFL, Lausanne, Switzerland; [3]Department of Psychiatry, University Hospital Geneva, Geneva, Switzerland; [4]Department of Neurology, University Hospital Geneva, Geneva, Switzerland; [5]Laboratoire de Psychologie et Neurocognition, LPNC CNRS 5105, Université Grenoble Alpes, France

Introduction

Our brain constantly receives and sends a flow of multisensory information including proprioceptive, tactile, visual, vestibular, auditory, olfactory, visceral, and motor-related signals. The integration of these signals into multisensory representations is responsible not only for the way our body is represented but also for the way it is consciously experienced. Conscious body experience includes the experience that a "real me" "resides" in "my" body and perceives the world from the perspective of that body, a phenomenon called bodily self-consciousness or corporeal awareness.[1,2] Recent studies in cognitive neuroscience have shown that it is possible to modulate bodily self-consciousness by experimentally manipulating these multisensory bodily signals (see Chapter 8). During such manipulations, healthy individuals may transiently experience (1) ownership of another body or body part (i.e., self-identification), (2) changes in where they feel their body located in space (i.e., self-location), or (3) modulation of the perspective from where they perceive the world (i.e., first-person perspective). These different experimental protocols have provided valuable insights into the neural mechanisms that generate and modulate bodily self-perception in the normal brain.

Another strategy consists in studying the phenomenology of individuals presenting altered perceptions of their bodies. Disorders of body representation take various forms and have a rich phenomenology, with impacts on distinct body parts and different levels of disorder awareness. Disorders may be limited to the upper limb, relate to internal organs, or involve the whole body. Moreover, patients may report an "absence" of a body part,

Multisensory Perception
https://doi.org/10.1016/B978-0-12-812492-5.00018-8

describe supernumerary phantom limbs, or perceive a double of themselves in extrapersonal space. Finally, some patients notice distorted or unusual body representations, sometimes being critical and able to rationally describe how they perceive their body as abnormal.[3] Others are rather indifferent (i.e., anosognosia) or hold false beliefs regarding the very existence of an alteration (i.e., delusion).

In this chapter, we will review the main clinical alterations of body representation. First, we will describe instances of altered body representations in neurological conditions, either constrained to a specific part or impacting the whole body. In the second part, we will present body representation disorders associated with other diseases, namely chronic pain and psychiatric conditions.

Neurological disorders of body representation

Unilateral disorder of body representation

One of the most common cases of altered body representation in neurology is the perceived absence of a body part, as if it was not part of the body, or at least not completely. This entails inattention toward a given body part (i.e., personal neglect), the vivid sensation that a body part has disappeared (i.e., the feeling of amputation), or the misattribution of a limb to someone else (i.e., somatoparaphrenia). In contrast to these cases in which body perception is diminished, some disorders of body representation imply abnormally increased bodily percepts, like sensations in a nonexisting limb (i.e., phantom limbs and supernumerary phantom limbs) or overestimation of perceived body size (macrosomatognosia). In the following section, we describe the main unilateral disorders of body representation.

Personal neglect

The term personal neglect was coined by Zingerle[4] in reference to a neuropsychological disorder characterized by the inattention toward one part or an entire half of the body[5,6] (see also Chapter 19). Personal neglect typically concerns the left body side and is associated with right hemispheric brain lesions. The clinical manifestations indicative of personal neglect are indifference, forgetfulness, or unawareness for the left hemi-body. Classically, patients forget to comb, shave, or make up the left side of their face or leave their left foot out of the wheelchair rest. Although inattention is striking at the behavioral level, patients are not aware of their deficit and do not report the unattended body part as missing from their body representation. In contrast to somatoparaphrenic patients (see next section), patients with personal neglect do not manifest disownership for the affected hemi-body, and acknowledge under request that the disregarded body part belongs to them, even if they behave as if it did not exist.

Lesion analysis in patients with personal neglect revealed the role of the right inferior parietal cortex including the supramarginal and postcentral gyri.[6] Lesions were also found in the underlying white matter, suggesting that neglect may result from a disconnection between the postcentral gyrus coding for proprioceptive and somatosensory signals, and areas linked to more abstract body representations. Subsequent lesion analyses have confirmed the importance of parietal regions and underlying white matter, extending to temporal areas.[7,8]

Feeling of amputation, hemi-depersonalization

Neurological patients may experience the sensation from a body part as numbed or completely absent. As opposed to personal neglect, patients fully appreciate the illusory nature of their sensation. This disorder is considered as the reverse of the well-known phantom limb sensation experienced by most amputees (see below). Other related phenomena include the feeling that a limb is no longer attached to the rest of the body, as if it were floating at some distance, or the feeling that the whole body is split into two halves.[9,10] These symptoms are usually of short duration and appear mostly during epileptic seizures, migraine events, or vascular stroke affecting premotor, primary motor, or parietal cortex, as well as subcortical structures of either hemisphere.

Somatoparaphrenia

The term somatoparaphrenia was introduced by the neurologist Joseph Gerstmann[11] in reference to patients presenting an abnormal sense of disownership for their contralesional hemi-body. Somatoparaphrenic patients claim that their own limb does not belong to them, and more explicitly that it belongs to someone else like the doctor, a nurse, a roommate, or some undetermined person.[12] Somatoparaphrenia is characterized by a distal-to-proximal gradient, with greater prevalence for the hands, followed by an entire limb (arm/leg) and only rarely the whole hemi-body. Patients can display strong emotional reactions and develop feelings of hostility against the affected body part, manifested as verbal or physical aggressive behaviors (i.e., misoplegia). Most of the reported cases of somatoparaphrenia involve extensive frontotemporoparietal lesions, with a prominent role of the temporoparietal junction (TPJ) in the genesis of the delusion.[12] More sporadically, deep cortical regions such as the insular cortex or subcortical regions including the basal ganglia have also been involved.[13,14]

Phantom limbs and supernumerary phantom limbs

The majority of amputees experience persistent and vivid sensations in their physically absent limb, referred to as a "phantom limb".[15–17] The phantom limb is usually clearly perceived and is similar in shape, size, and posture to the physical limb before amputation, although distorted perception of the phantom limb can also occur (see Ref.18 for review). In rare cases, "supernumerary" phantom limbs are experienced by nonamputated patients, described as an additional body part, felt as an entity, and sharing the properties of the real body.[19–21] Supernumerary limbs are mostly perceived on the same side as a paralyzed limb and typically remain immobile although movements have been occasionally described.[22] Supernumerary phantom limbs have been reported following lesions of the basal ganglia,[23] capsulolenticular region,[22,24] thalamus,[19] supplementary motor area,[25] bilateral parietal lobe,[26] spinal cord,[27] and following motor cortex stimulation.[28] In all cases, the reduplicated physical body part is always injured, deafferented, or paretic. Some authors have proposed that supernumerary phantom limbs are due to a mismatch between the perceived paretic or deafferented limb and its brain representation.[29]

Macro- and microsomatognosia

In rare occasions, some patients misperceive the size and weight of specific body parts. Microsomatognosia refers to the subjective experience of perceiving a body part as smaller than usual, whereas macrosomatognosia is used for patients describing a limb that is increased in size and often in weight.[30] Frederiks[30] proposed that such misperception is typically paroxysmal, occurs in both halves of the body, and occurs in an unclouded mind. Similarly to what is observed for supernumerary phantom limbs, patients with macro- or microsomatognosia are usually fully aware of the illusory nature of their percepts. Typical causes include migraines and epileptic seizures.[31] Rare cases have been reported following toxoplasmosis or typhoid infections, mesencephalic lesions, and damage to sensorimotor structures in either hemisphere.[30]

Global body representation disorder

Most of the body representation disorders described so far can conceptually be extended to the full body. For instance, macrosomatognosia can concern the entire body in patients with Alice in Wonderland syndrome, who have an erroneous perception of their whole body size with respect to the external environment.[32] Similarly, extreme forms of depersonalization in which patients claim to be nonexistent or dead (i.e., Cotard's syndrome)[33] can be considered as an equivalent of the feeling of amputation described earlier, but concerning the whole body.

In the next section, we will focus on a particular form of full-body hallucination in which patients experience illusory duplications of their own body. Most duplications are predominantly visual, commonly referred to as "autoscopic phenomena" (i.e., autoscopy, heautoscopy, and out-of-body experience (OBE)), but can also be nonvisual, like in the feeling of presence (sensorimotor phenomenon; i.e., Ref. 34).

Autoscopic hallucinations

During an autoscopic hallucination, people experience seeing an image of themselves in the extrapersonal space (i.e., the space that is far away from the subject and that cannot be directly acted on by the body), as if they were looking into a mirror, without the experience of leaving their body (i.e., no disembodiment). Patients with autoscopic hallucinations see the world from their habitual perspective and their "self" is perceived as located inside their physical body. Therefore, autoscopic hallucinations are mostly visual phenomena (with multisensory components), with no change in the bodily self. They usually last a few seconds or minutes and may occur repeatedly. A few case studies have reported persistent autoscopic hallucinations over time with a visual double present over months and even years.[35,36]

Visual field deficits and visual hallucinations are frequently associated with autoscopic hallucinations.[37,38] Based on this observation, it has been proposed that autoscopic hallucinations relate to visual deficits including abnormal visual imagery or defective plasticity following a lesion in the visual cortex.[36] Others have proposed that this phenomenon is linked to defective multisensory integration of signals from vision, proprioception, and touch.[37,39,40]

Autoscopic hallucinations can occur after neurologic disorders such as migraine and epilepsy as well as brain damage in the occipital and/or parietal lobe.[36,40,41] A recent quantitative lesion analysis study investigated the brain correlates of autoscopic hallucinations in a group of seven patients. The authors found that damage affecting the superior occipital gyrus and the cuneus in the visual cortex of the right hemisphere was involved.[42]

Heautoscopy

Unlike autoscopic hallucinations, people experiencing heautoscopy self-identify with a body seen in the extrapersonal space. It is usually difficult for the observer to determine whether he/she is disembodied or not and whether the center of conscious experience (i.e., the self) is localized within the physical body or in the autoscopic body.[41] During heautoscopy, patients may even experience so-called bilocation (i.e., the feeling of existing at two places at the same time), often associated with the experience of seeing from different visuospatial perspectives (i.e., from the physical and autoscopic bodies), in an alternating or even simultaneous fashion.[41–45] This phenomenon can be considered as an intermediate form between autoscopic hallucination (where the self is located in the physical body) and OBE (see next section) (where the physical body is completely abandoned by the self).

Heautoscopy has been reported in patients suffering from parietal or temporal lobe epilepsies, neoplastic lesions of the insular cortex, and migraine in association with a psychiatric disorder.[44,46–49] A recent lesion study found that heautoscopy was associated with damage to the left posterior insula.[42] Patients with heautoscopy often present altered perception of visceral information such as palpitation,[45] which is in line with the involvement of the insular cortex and its role in interoceptive processing and the encoding of emotionally relevant information for self and other.[50–52] Further corroborating the link between insular cortex, interoception, and heautoscopy, a recent report described a patient with a selective right insular tumor in whom a mild form of heautoscopy, including bilocation and body reduplication, could be experimentally induced based on cardiovisual stimulation (i.e., participants observe a virtual body illuminated in synchrony with their heartbeat).[53]

Out-of-body experience

An OBE can be defined as a waking experience combining disembodiment (i.e., the feeling of being outside one's physical body), elevated perspective (i.e., the perceived location of the self at a distanced and elevated visuospatial perspective), and autoscopy (i.e., the experience of seeing of one's own body from this elevated perspective). Subjects experiencing OBEs always localize the self outside their physical body, usually as if located in an elevated position looking at the physical body on the bed or the ground.[41,54]

OBEs have been reported predominantly in patients with epilepsy and migraine.[54] Although many brain regions have been linked to OBEs (e.g., frontotemporal cortex,[48] parietal lobe,[55] temporal lobe[56]), the TPJ seems to play a crucial role[40,41,41a], with a right hemispheric predominance.[57] Notably, electrical stimulation of the right TPJ induced an OBE in a patient presenting with intractable epilepsy.[58] Other cases of OBEs induced by brain stimulation of the TPJ have been reported.[59] Importantly, OBEs are not only found in clinical populations but also appear in approximately 5%-10% of the healthy population across the majority of the world's cultures.[60] A variant of OBE called the full-body illusion can be experimentally induced in healthy volunteers by providing conflicting multisensory signals (Ref. 61; see Chapter 8, for further details). Brain imaging during this illusion confirmed the role of the TPJ for OBEs.[62]

With respect to other autoscopic phenomena, OBEs are characterized by specific vestibular sensations.[41,63] These are feelings of elevation, floating, and a 180 degrees inversion of the body and its visuospatial perspective in extrapersonal space. Otolithic dysfunctions are therefore likely to contribute to OBEs.[1] In addition to these vestibular disturbances, OBEs are sometimes accompanied by paroxysmal visual body–part illusions such as supernumerary phantom limbs and illusory limb transformations.[41,48,55,58] These observations suggest that visual illusions of body parts and autoscopic phenomena may share similar neural origins.[45] Based on the association of OBEs with visuo-somatosensory illusions, abnormal vestibular sensations,[63] and the well-known role of the TPJ in multisensory integration,[63a,63b] it has been proposed that OBEs are caused by disturbed multisensory integration of bodily signals.[1,41]

Feeling of presence

Initially described by the psychiatrist Karl Theodor Jaspers,[64] the "feeling of a presence" (FoP) refers to the distinct feeling of the physical presence of another person or "being" in the near extracorporeal space although nobody is actually around.[45] Importantly and in contrast to autoscopic phenomena, this illusion is not experienced visually as the person is "sensed" but usually not seen. This "presence" can be felt behind, sideways, or in front of one's physical body and may even involve multiple entities.[47] Authors have named this illusion of a sensorimotor double "hallucination du compagnon,"[65] idea of a presence[66] or presence hallucination[66a]. The FoP has been described in several psychiatric conditions,[34,64,66–68] neurological patients suffering from epilepsy, stroke, or Parkinson's disease[67,69,70] and healthy individuals mostly during periods of physical exhaustion.[64,66,67]

The mechanisms underlying the FoP are the topic of recent research, highlighting the role of multisensory integration and body representation. Electrical stimulation of the TPJ induced FoP in a single case study during presurgical investigations.[71] This finding was recently confirmed by a lesion analysis study in 12 FoP patients: focal brain lesions overlapped in the temporoparietal, frontoparietal, and insular cortex (of either hemisphere).[72] Additional analysis in control patients revealed that from the three lesion-overlap zones only the frontoparietal site was specifically associated with the FoP. Interestingly, the temporoparietal cortex,[62] insula,[73] and frontoparietal cortex[74] are known to integrate multisensory bodily signals and are considered as neural loci of bodily self-consciousness. As for OBEs, mild forms of FoP can now be induced noninvasively in healthy volunteers. Using a robotic system generating specific sensorimotor conflicts, Blanke and collaborators were recently able to experimentally induce the FoP and related illusory own-body perceptions.[72] In this experiment, blindfolded participants moved a master robotic device in front of them while receiving delayed tactile stimuli on their back. During such spatiotemporal mismatch between motor–proprioceptive signals (participant's movements in front of them) and their sensory consequences (tactile feedback on their back) subjects reported being in the presence of another person behind them and being touched by this invisible presence. A prominent model for motor control and bodily experience posits that in self-generated movement, efference copy signals from the sensorimotor system are used to make predictions about the sensory consequences of movement and that such integration is fundamental for normal self-generated experience.[75,76] Collectively, these suggest that the FoP might be the consequence of a misperception of the source and identity of signals of one's own body.[72]

Body representation disturbance in chronic pain

More than any other sensation, pain is inextricably linked to the body, which constitutes the reference and the object of any painful sensation.[77] Neuroimaging studies have shown that the link between body and pain underlies a partial overlap and mutual connections between central pain representations (the so-called pain matrix; i.e., a network of brain areas activated by nociceptive inputs including brainstem and thalamic nuclei, primary and secondary somatosensory areas, insular, and anterior cingulate cortices) and central body representations (i.e., the body matrix, a network of multisensory regions processing bodily related inputs, such as the posterior parietal cortex, the somatosensory cortices, and the insula).[78] This link at the neural level is supported by behavioral evidence in patients experiencing pain over a prolonged period and beyond the expected time for healing (i.e., chronic pain), who also demonstrate abnormalities in their body representation. Indeed, patients with chronic pain often misperceive their affected body part in size or shape, reporting feelings of foreignness, strangeness, or even hostility toward the painful limb. In the following section, we present and discuss the main changes in body representation occurring in three different chronic pain states, namely complex regional pain syndrome (CRPS), phantom limb pain (PLP), and spinal cord injury (SCI).

Complex regional pain syndrome

CRPS is a chronic pain condition usually affecting one limb, characterized by pain in combination with sensory, autonomic, trophic, and motor abnormalities.[79] Body perception disturbances are frequent in such patients who, for instance, report their affected limb to be larger than it really is.[80] In addition to size distortions, some patients also demonstrate disturbances in how they perceive the shape of their affected body part, for instance, describing a missing segment in the affected limb or having difficulties in determining its position.[81] Moreover, patients with CRPS have reduced abilities to determine the laterality of pictured hands, implying the existence of underlying altered spatial representations; the degree of this disturbance is directly influenced by the intensity of pain.[82,83]

Patients with CRPS show an important cortical reorganization with reduced representation of their affected limb in the primary sensory and motor cortices. Studies in the CRPS population revealed that the amount of cortical reorganization directly correlates with pain intensity and that these cortical changes are normalized during recovery.[84–86] However, the directionality of this link is unclear, and whether these cortical changes cause or are caused by chronic pain remain to be tested.

An interesting clinical feature of CRPS is that patients tend to neglect their affected limb and report finding their hand "foreign," "strange," or "as if someone had sewed a foreign hand on it."[87–89] This clinical manifestation observed in more than half the patients is called "neglect-like syndrome" and shares similarities with symptoms observed following right parietal damage. However, (1) this feeling of foreignness is observed independently of the affected side, (2) is not associated with hemispatial deficit (e.g., patients show no bias in the bisection task, see Chapter 19), and (3) patients are typically fully aware of their deficit and realize the irrational nature of their feeling.[87] Together, these three points make the neglect-like syndrome distinct from the conditions of personal neglect and somatoparaphrenia described in Unilateral disorder of body representation section.

Phantom limb pain

A striking example of body misperception in a clinical population is phantom limb sensation, defined as the sensation that a missing body part is still present. PLP occurs in up to 80% of amputees.[15,90] Over the past decades, several studies reported multiple cortical changes in PLP. Since seminal animal studies, it is indeed well established that cortical reorganization occurs following amputation, with an invasion of adjacent body part representations into the cortical representation of the deafferented body part.[91] For instance, upper limb amputees show a shift of their facial representations in somatosensory and motor cortex into the digit and hand area.[92,93] More controversial is the relation between such cortical reorganization and chronic pain. Some authors reported that amputees with PLP have a greater shift of their mouth cortical representation into the hand area in motor and somatosensory cortex than amputees without pain.[94] Moreover, this cortical reorganization appeared to be correlated with the level of pain.[95] Based on the relation between the degree of cortical reorganization and the level of pain, Flor and colleagues have proposed maladaptive changes as the neural basis of PLP.[96] More recently, it has been claimed by Makin and colleagues that PLP maintains local cortical representations but disrupts interregional connectivity.[97–99] For instance, it was reported that functional connectivity between the representation of the missing limb and the rest of the sensorimotor network is decreased.[98] A recent study furthermore showed that somatosensory regions are functionally disconnected from the posterior parietal cortex in amputees, the latter being a key region for the integration of multisensory bodily signals.[100] Collectively, these results underline the role of cortical body representations (unimodal or multimodal) in PLP.

As for the previous disorders, there is a clear link between PLP and altered body representation. For instance, amputees typically report their missing limb as heavy, swollen, stuck in a given position, or shortened.[18,101] The feeling of telescoping is a commonly reported symptom with significant association with PLP, where patients experience their phantom has shrunken with just the more distal portion floating near, attached to, or "within" the stump.[16,96,102] It is estimated that about 50% of amputees perceive their phantom limb to be telescoped; the telescoping process generally begins within the first few weeks postamputation.[15,103] Some authors have proposed that telescoping originates from the disparity in brain representation of the different limb segments, with an overrepresentation of distal (i.e., the hand) compared with proximal parts.[15] Neuroimaging data showed that telescoping is associated with cortical reorganization in which distal representations invade brain regions representing proximal body parts. For instance, imaginary movement of a completely telescoped phantom arm induces activity in the shoulder area.[96]

Based on the hypothesis that cortical reorganization and phantom pain are related, a range of novel therapies have been developed to diminish PLP by targeting maladaptive cortical reorganization. These include sensory[104] and motor training,[105] peripheral[106] or cortical stimulation,[107] or combined visuomotor stimulation using a mirror box setup[92] (see Chapter 20).

Spinal cord injury

Spinal cord damage can cause permanent loss of sensorimotor function and, in about 65% of patients, chronic neuropathic pain.[108,109] Similarly to amputees, SCI patients may experience vivid phantom sensations in the deafferented body part.[17,110,111] However, they commonly describe their phantom occupying anatomically unrealistic and unnatural postures: for example, patients may feel that their legs are "twisted" or "blown up." They may perceive their "toes turned down under the bottom of the foot" or their digit somehow twisted so that "each toe pointed in a different way."[110] Moreover, patients often report their phantom to be larger than the actual in size or in movement.[110–112] This is in contrast with phantom limb sensations in amputees, which occur in a plausible body space and are reduced in size (i.e., telescoping).[18,111]

Several studies have demonstrated that functional and structural cortical reorganizations occur following SCI.[113–115] These changes are in line with the modifications described in amputees, that is an invasion of the adjacent cortical representation into that of the deafferented body part. Neuroimaging studies showed shifts of functional motor and sensory cortical representations that relate to the severity of SCI.[116] Moreover, these cortical changes also appear significantly correlated with ongoing pain intensity levels in SCI.[117] Recently, Scandola and colleagues meticulously examined bodily misperceptions in a group of 49 patients with SCI.[118] They reported various corporeal illusions involving body form (sensations of body loss and body-part misperceptions), body motion (illusory motion), and body ownership (disownership-like feelings and somatoparaphrenia-like feelings) that were related to neuropathic pain. The authors hypothesized that these body misrepresentations reflect uncontrolled neuroplastic changes.

Based on the observation that multisensory processing and body representation are impaired in SCI patients,[119,120] a recent study investigated how body ownership and neuropathic pain can be modulated by multisensory stimulation. Using immersive virtual reality (VR), Pozeg and colleagues manipulated the sense of leg ownership and global body ownership in SCI patients applying synchronous visuotactile stimulation (i.e., creating a virtual leg illusion[121] or full-body illusion[61]). Compared with healthy subjects, SCI patients showed reduced sensitivity to multisensory stimulation inducing illusory leg ownership but preserved ability in global ownership manipulation. In addition, leg ownership decreased with time since SCI. This study, among others, suggests that manipulations of bodily self-consciousness are likely to be of high relevance to alleviate pain, given that these effects were achieved after even short periods of multisensory VR exposure.

Body representation disturbance in psychiatric disorders

Representations of the body are altered in a number of psychiatric conditions. Here, we review studies on anorexia nervosa (AN) and schizophrenia (SZ), two conditions for which a great deal of research has been conducted on body representation. We will also briefly discuss alterations in body representation seen in gender dysphoria (GD).

Anorexia

Patients with AN show extreme dissatisfaction with their body size, despite being underweight. There is a long-standing debate about whether this dissatisfaction is purely cognitive-affective or whether there is also a perceptual distortion of body size.[122] This distinction fits with current conceptualizations of body representation in the brain as shaped both by bottom-up sensory input and by top-down cognitive, semantic, and affective representations[123]. An abundance of data confirm differences in the former, "attitudinal," component of body image in AN.[124] Across studies of AN, attitudinal body dissatisfaction shows a larger effect size than visual distortion[125] and is observed in more studies and more patients.[126,127] Yet there is evidence of perceptual body distortion in AN as well. We will focus on this perceptual component of body representation, while acknowledging that the affective component plays a prominent role in AN.

The majority of research on body representation in AN has probed visual body representation. Numerous studies have examined visual estimations of body size by asking patients to draw the width of their body, select a body outline matching their shape, or adjust a photograph, mirror, or video image until it is perceived to be the patient's size. A number of studies report visual body size differences in AN[125]. Some observe these measures to be positively correlated with attitudinal measures of body dissatisfaction,[125,128,129] suggesting a causal relationship of some kind between these components of the body image. Quite a few other studies, however, do not observe distortions in the visual body image or observe it only in a subset of patients.[127,130] The presence of visual body distortion in AN is thus controversial and certainly not universal, suggesting that it is not the primary cause of body image dissatisfaction. AN patients have been found to show selective deficits in visually processing upright—but not upside-down—bodies, suggesting difficulties with configural processing that may be related to a more detail-oriented approach to viewing bodies.[131]

Brain imaging has also been used to investigate visual body representation in AN. An occipitotemporal pathway including the extrastriate body area (EBA) and fusiform body area is key to detecting body-related information, while a parietofrontal pathway is closely linked to body identification and self-other discrimination.[124] Differences in processing visual images of bodies have been found in individuals with AN in the body-shape processing network,[132–134] as well as in the insula, for self-images.[135,136] Mohr and colleagues suggest that difficulty retrieving multimodal body image representations from the precuneus and posterior parietal cortex may underlie deficits in body size estimation.[135] In addition, visual body shape comparison tasks show more activation of right hemisphere sensorimotor regions in AN, including hyperactivation of the insula, but hypoactivation of the anterior cingulate cortex. This finding may relate to altered interoceptive or motivational processes in AN.[124] Finally, alterations in the structure of the EBA, located in the lateral occipital cortex, have been observed.[137]

While researchers have traditionally focused on visual body distortion in AN, more recent research efforts have turned to somatosensory body representation. Most studies of primary tactile perception do not find deficits in AN, although slight deficits in more difficult versions of a finger identification task have been documented in AN patients before treatment.[138] However, several studies have reported differences in secondary tactile perception, which involves perceptual scaling of a tactile stimulus to compute and represent its size (and other characteristics).[139] Gaudio and colleagues review evidence from 13 studies examining nonvisual multisensory alteration of body perception in AN and conclude that there are tactile and proprioceptive differences that may be associated with alterations in parietal cortex functioning in AN patients.[140]

Patients with AN overestimate the distances between points applied both on the arm and on the abdomen, suggesting an enlarged tactile body representation. Tactile overestimation is correlated with body dissatisfaction, suggesting a connection between tactile body maps and attitudinal aspects of body image.[141,142] To see whether deficits in primary tactile perception might underlie this effect, Keizer and colleagues studied touch detection and found a higher threshold for two-point discrimination on the arm and abdomen in AN, as well as a lower pressure detection threshold on the abdomen in patients with AN.[142] These inconsistent findings suggest alterations in primary tactile perception that may impact tactile body distortion. More recently it was found that tactile overestimation occurs only in the horizontal direction of the body, suggesting a warping of tactile body image by specific cultural body fears.[143]

Sensory information also comes from the inside of the body, through interoception. Several studies document difficulties with interoceptive awareness in AN. These difficulties include reduced sensitivities to sensations of hunger and satiety,[144–146] difficulty recognizing signs of physiological stress such as an increased heart rate[147,148] and altered processing of taste and pain.[149,150]

Patients with AN also show differences in integrating visual and proprioceptive information. The size-weight illusion (SWI) arises from visual and haptic comparison of two objects of equal weight but different physical size. Typically, the smaller object feels heavier due to an implicit expectation that weight is proportional to size. AN patients show a reduced SWI despite normal discrimination of mass, suggesting decreased integration of visual and proprioceptive information in AN.[151] This result could imply that individuals with AN have more difficulty taking their appearance (visual feedback) into account when judging their body size and might rely to a greater extent on internal sensory cues.

Haptic perception involves active sensorimotor exploration of the surface of an object. Deficits in integrating visual and haptic information are reported by Grunwald and colleagues, who found that patients with AN had difficulty drawing objects that they explore through touch[152] and reproducing angles through haptic perception.[153] Patients in this study also showed reduced parietal activation during this task.[152] In contrast, no deficits have been observed in haptic recognition of simple shapes.[154,155]

The rubber hand illusion (RHI) involves integrating visual and tactile input (see Chapter 8). Patients with AN show a stronger RHI than controls. Greater proprioceptive drift and greater embodiment of the hand both correlated with symptoms of AN.[156] The authors suggest these results indicate that the bodily self is more plastic in individuals with an eating disorder. Indeed, heightened malleability of the body persisted beyond recovery, suggesting a trait phenomenon.[157]

There is also evidence of altered sensorimotor and spatial orientation representations of the body in AN. Individuals with AN judged they would be unable to fit their body through an aperture that was easily wide enough,[158] showing distortions in body schema. Nico and colleagues found that AN patients showed selective distortions of their left body boundary when judging whether an approaching visual stimulus would contact their body. This performance paralleled that of right—but not left—parietal patients, suggesting alterations in right hemisphere processing of the body schema.[159] Other investigators[155 160] demonstrated an effect of body tilt on the visual and tactile sense of verticality in AN patients, showing deficits in integrating visual, tactile, and gravitational information and using the body as a frame of reference. In contrast, another study[138] found no differences on cognitive and body-related spatial tasks in AN patients after treatment, and during acute illness found differences only when tasks required an executive function load in additional body schema–related processes. Body schema dysfunction may thus reflect broader cognitive dysfunction during acute states of AN. Stimulation of the vestibular system alters representation of body parts. Noting the high comorbidity of vestibular dysfunctions and psychiatric symptoms, Mast et al. postulate that the vestibular system plays an integral role in multisensory coordination of body representation and may also play a role in AN.[161]

In sum, individuals with AN show significant affective bodily dissatisfaction but also evidence of perceptual distortions in body representation. There is evidence of distorted bodily perception in visual, tactile, and motor domains as well as altered multisensory body representations. The causality of these distortions for affective body dissatisfaction and progression of AN is unclear.

Schizophrenia

SZ is a severe psychological disorder characterized by abnormal social behavior and unusual or confused thoughts. Common symptoms include "positive symptoms" such as hallucinations and delusions as well as "negative symptoms" such as reduced movement and emotional responsiveness. Cognitive neuroscience approaches to SZ have amassed evidence that core features of SZ may arise from cognitive dysfunction.[162] Cognitive and perceptual declines are found in most individuals with SZ; indeed, cognitive impairment is more common in SZ than psychotic symptoms.[163] Accordingly, disruptions in multisensory body perception may underlie certain symptoms of SZ (see also Chapter 17 by Cascio et al., this volume).

SZ is strongly associated with anomalous self-perception. Patients with SZ often experience problems with self-recognition and self-attribution of thoughts and actions.[164] A theme of blurred boundaries between self and other ties together many symptoms of SZ including auditory hallucinations, thought insertion, thought broadcasting, and the influence of others on the patient's thoughts, actions, or emotions. With regard to body perception, there is evidence of altered body structural description in SZ.[165,166] In addition, patients more frequently report feelings of strangeness toward their faces than healthy controls.[167] Bodily delusions and hallucinations are also not uncommon in SZ.

Perception of bodily touch in patients with SZ reveals altered multisensory representations and impairment of self-other distinction. Patients with SZ show reduced ventral premotor cortex response to observed touch of the body and abnormal responses to bodily touch and observed touch in the posterior insula.[168] The RHI has been found to be affected in SZ, with studies differing with regard to being stronger[169] or weaker[170] in SZ patients than in healthy controls, suggesting, at the very least, altered mechanisms of body representation that require further study. Multisensory perception of bodily movement is also disrupted in SZ. Results from a number of studies suggest that patients experiencing hallucinations or delusions of control frequently misattribute their own actions to others.[171] In healthy controls, tracking of self- versus other-generated hand movements activates the angular gyrus and insula. SZ patients do not show this pattern, suggesting abnormal tracking of self-generating movement.[171] This may relate to the frequent experience in SZ of personal actions not feeling under one's control. Indeed, SZ patients experiencing feelings of alien control of self-generated movements show hyperactivity in the right inferior parietal lobule.[172]

Multisensory integration is a foundational capacity for a normal experience of self. A bottom-up account of SZ postulates that perceptual deficits impact higher-level cognitive processes whose disruption leads to symptoms of SZ.[173] For examples, Postmes and colleagues suggest that failures of multisensory integration may underpin disrupted experiences of self commonly seen in SZ such as depersonalization, diminished feelings of agency, and loose associations.[174] Many examples of deficits in multisensory integration have been found in SZ (see chapter by Cascio et al., this volume). Patients with SZ show reduced audiovisual binding and deficits in the network subserving audiovisual integration.[175,176] They also exhibit reduced facilitation of reaction time for detecting bimodal targets relative to unimodal targets, and those with more negative symptoms show the least degree of benefit from bimodal cues.[177] Relatedly, patients with SZ show impairments in recognizing whole-body expressions and impairments in integrating affective visual and vocal cues from the same source (such as a face or body along with a human vocalization).[178,179] The bottom-up account of SZ is also supported by functional brain imaging data showing disrupted resting state networks that particularly affected visual, auditory, and crossmodal binding networks. These disruptions were correlated with negative symptoms, positive symptoms, and hallucinations in individuals with SZ.[173]

In sum, patients with SZ show altered bodily perception and difficulties relating to distinguishing self from other. Differences in visual, tactile, and sensorimotor representation of the body have been observed. Multisensory integration is also altered. Problems with sensory binding correlate with many clinical symptoms of SZ and may play a causal role in these symptoms.

Gender dysphoria

People who are transgender experience a marked discrepancy between their experienced or expressed gender, and the gender assigned to them at birth. When this discrepancy causes significant distress or problems in functioning, it may be diagnosed in the DSM-5 as GD.[180] The biological mechanisms of GD are not known, and research in this area is nascent. Most studies have explored differences in brain structure in FtM (female to male) and MtF (male to female) individuals. Overall, these studies show a mixed pattern of masculine and feminine cortical thickness and white matter tracts, different from both cisgender men and women.[181] The incongruence between the perceived and physical body frequently leads to body dysphoria and body-related avoidance, such as avoidance of looking in the mirror.[180,182,183] For FtM individuals, breasts and genitals cause the greatest dissatisfaction.[184–186] Problematic areas for MtF individuals include genitals, face, and hair.[187] In contrast,[184] identify socially visible characteristics such as voice, hair, and muscularity as most predictive of overall body satisfaction. Most transgender individuals feel more like "themselves" and experience a more positive body image after physically transitioning their body to better align with their gender.[188] Numerous studies document improved quality of life for transgender individuals following hormone therapy and gender-confirming surgery.[189]

Initial work on GD by Ramachandran and colleagues has found evidence of altered body representation aligned with gender identity. Some presurgical FtM individuals reported the feeling of having a penis, despite being clearly aware it is not physically present.[190] Ramachandran and McGeoch note a parallel to the experience of a phantom limb after amputation, suggesting that just as the neural representation of a body part lingers after it is removed, body maps in the brain might be altered to align with gender identity in individuals with GD. Indeed, FtM and MtF individuals may have lower rates of phantom breasts and penises after they are removed during a gender-confirming surgery than do cisgender individuals who have these body parts removed for other medical reasons, suggesting altered neural representation of these body parts before their removal.[190,191]

To test whether somatosensory processing is altered for incongruent-feeling body parts, Case and colleagues compared processing of tactile input to the breast in presurgical FtM individuals compared with cisgender female participants.[192] Breasts were rated as highly incongruent for all FtM men and genderqueer individuals in the study, but not for the cisgender women. Magnetoencephalography recordings of brain responses to tactile stimulation of the breast showed reduced response to the tactile input in the supramarginal gyrus and secondary somatosensory cortex, but increased activation at the temporal pole, near the amygdala, in the FtM group. No such differences were seen following tactile stimulation of the hand. These results suggest reduced sensory integration and more anxiety or alarm for sensation from this body part. Furthermore, altered white matter connectivity (measured by diffusion tensor imaging) was found in these same brain areas, suggesting that altered sensory processing could be related to underlying structural differences in these brain regions. These results suggest that the experience of bodily incongruence may include altered integration of tactile sensation.

Several groups have now examined differences in resting state connectivity in transgender individuals, as related to body representation. Lin and colleagues found that transgender participants showed higher centrality of the primary somatosensory cortex and superior parietal lobule, as well as greater recruitment of visual and auditory regions in the body network.[193] These results suggest greater multisensory influences on body representation in transgender individuals. Manzouri and colleagues found evidence that FtM individuals may have weaker connections between body perception networks and body self-ownership networks as well as reduced functional connectivity between regions involved in body perception and emotion.[194] A similar attempt to characterize functional connectivity in adolescents with GD identified sex-atypical connectivity patterns within the visual network, the sensorimotor network, and the posterior default mode network (DMN). Interestingly, these networks, which are sexually dimorphic between cisgender male and female adolescents, did not differ between prepubertal children with and without GD.[195] Feusner and colleagues also attempt to identify neurobiological correlates of the subjective incongruence between body and self in FtM individuals.[196] They report decreased connectivity within the DMN in FtM individuals as well as decreased connectivity in occipital and temporal regions. Furthermore, they report correlations between higher ratings of "self" for gendered body images and greater connectivity within the anterior cingulate cortex in FtM individuals. Similar to an earlier report,[192] this study[196] suggests that individuals with GD may not incorporate physical traits of their assigned birth into their neural self-representation.

In sum, individuals with GD show high levels of body dissatisfaction, related particularly to sexually dimorphic body features. Individuals with GD show evidence that multisensory neural body representation is altered in the brain and is less connected with areas related to emotions and representations of "self." Further work is needed to investigate the neural representation of the desired body form and its impact on body image and body schema.

References

1. Blanke O. Multisensory brain mechanisms of bodily self-consciousness. *Nat Rev Neurosci*. 2012;13:556–571.
2. Blanke O, Slater M, Serino A. Behavioral, neural, and computational principles of bodily self-consciousness. *Neuron*. 2015;88:145–166.
3. Dieguez S, Staub F, Bogousslavsky J. Asomatognosia. In: Godefroy O, Bogousslavsky J, eds. *The Behavioral and Cognitive Neurology of Stroke*. 2007:215–253.
4. Zingerle H. Ueber Störungen der Wahrnehmung des eigenen Körpers bei organischen Gehirnerkrankungen. (Part 1 of 2). *Eur Neurol*. 1913;34:13–24.
5. Bisiach E, Perani D, Vallar G, Berti A. Unilateral neglect: personal and extra-personal. *Neuropsychologia*. 1986;24:759–767.
6. Committeri G, et al. Neural bases of personal and extrapersonal neglect in humans. *Brain J Neurol*. 2007;130:431–441.
7. Baas U, et al. Personal neglect-a disorder of body representation? *Neuropsychologia*. 2011;49:898–905.
8. Rousseaux M, Allart E, Bernati T, Saj A. Anatomical and psychometric relationships of behavioral neglect in daily living. *Neuropsychologia*. 2015;70:64–70.
9. Heydrich L, Dieguez S, Grunwald T, Seeck M, Blanke O. Illusory own body perceptions: case reports and relevance for bodily self-consciousness. *Conscious Cognit*. 2010;19:702–710.
10. Podoll K, Robinson D. Splitting of the body image as somesthetic aura symptom in migraine. *Cephalalgia Int J Headache*. 2002;22:62–65.
11. Gerstmann J. PROBLEM OF IMPERCEPTION OF DISEASE AND OF IMPAIRED BODY TERRITORIES WITH ORGANIC LESIONS: RELATION TO BODY SCHEME AND ITS DISORDERS. *Arch Neurol Psychiatr*. 1942;48:890–913.
12. Vallar G, Ronchi R. Somatoparaphrenia: a body delusion. A review of the neuropsychological literature. *Exp Brain Res*. 2009;192:533–551.
13. Bottini G, Bisiach E, Sterzi R, Vallar G. Feeling touches in someone else's hand. *Neuroreport*. 2002;13:249–252.
14. Cereda C, Ghika J, Maeder P, Bogousslavsky J. Strokes restricted to the insular cortex. *Neurology*. 2002;59:1950–1955.
15. Ramachandran VS, Hirstein W. The perception of phantom limbs. The D. O. Hebb lecture. *Brain J Neurol*. 1998;121(Pt 9):1603–1630.
16. Flor H. Phantom-limb pain: characteristics, causes, and treatment. *Lancet Neurol*. 2002;1:182–189.
17. Melzack R. Phantom limbs and the concept of a neuromatrix. *Trends Neurosci*. 1990;13:88–92.
18. Giummarra MJ, Gibson SJ, Georgiou-Karistianis N, Bradshaw JL. Central mechanisms in phantom limb perception: the past, present and future. *Brain Res Rev*. 2007;54:219–232.
19. Bakheit AMO, Roundhill S. Supernumerary phantom limb after stroke. *Postgrad Med J*. 2005;81:e2.
20. Miyazawa N, Hayashi M, Komiya K, Akiyama I. Supernumerary phantom limbs associated with left hemispheric stroke: case report and review of the literature. *Neurosurgery*. 2004;54:228–231.
21. Sakagami Y, Murai T, Sugiyama H. A third arm on the chest: implications for the cortical reorganization theory of phantom limbs. *J Neuropsychiatry Clin Neurosci*. 2002;14:90–91.
22. Khateb A, et al. Seeing the phantom: a functional magnetic resonance imaging study of a supernumerary phantom limb. *Ann Neurol*. 2009;65:698–705.
23. Halligan PW, Marshall JC, Wade DT. Three arms: a case study of supernumerary phantom limb after right hemisphere stroke. *J Neurol Neurosurg Psychiatry*. 1993;56:159–166.

24. Staub F, et al. Intentional motor phantom limb syndrome. *Neurology*. 2006;67:2140–2146.
25. McGonigle DJ, et al. Whose arm is it anyway? An fMRI case study of supernumerary phantom limb. *Brain J Neurol*. 2002;125:1265–1274.
26. Vuilleumier P, Reverdin A, Landis T. Four legs: illusory reduplication of the lower limbs after bilateral parietal lobe damage. *Arch Neurol*. 1997;54:1543–1547.
27. Curt A, Yengue CN, Hilti LM, Brugger P. Supernumerary phantom limbs in spinal cord injury. *Spinal Cord*. 2011;49:588–595.
28. Canavero S, Bonicalzi V, Castellano G, Perozzo P, Massa-Micon B. Painful supernumerary phantom arm following motor cortex stimulation for central poststroke pain. Case report. *J Neurosurg*. 1999;91:121–123.
29. Giummarra MJ, Gibson SJ, Georgiou-Karistianis N, Bradshaw JL. Mechanisms underlying embodiment, disembodiment and loss of embodiment. *Neurosci Biobehav Rev*. 2008;32:143–160.
30. Frederiks JA. Macrosomatognosia and microsomatognosia. *Psychiatr Neurol Neurochir*. 1963;66:531–536.
31. Robinson D, Podoll K. Macrosomatognosia and microsomatognosia in migraine art. *Acta Neurol Scand*. 2000;101:413–416.
32. Mastria G, Mancini V, Viganò A, Di Piero V. Alice in wonderland syndrome: a clinical and pathophysiological review. *BioMed Res Int*. 2016;2016:8243145.
33. Debruyne H, Portzky M, Van den Eynde F, Audenaert K. Cotard's syndrome: a review. *Curr Psychiatr Rep*. 2009;11:197–202.
34. Blanke O, Arzy S, Landis T. Illusory reduplications of the human body and self. *Handb Clin Neurol*. 2008;88:429–458.
35. Conrad C. Un cas singulier de 'fantôme spéculaire'. *Encephale*. 1953;42:338–352.
36. Zamboni G, Budriesi C, Nichelli P. 'Seeing oneself': a case of autoscopy. *Neurocase*. 2005;11:212–215.
37. Blanke O, Mohr C. Out-of-body experience, heautoscopy, and autoscopic hallucination of neurological origin Implications for neurocognitive mechanisms of corporeal awareness and self-consciousness. *Brain Res Brain Res Rev*. 2005;50:184–199.
38. Kölmel HW. Complex visual hallucinations in the hemianopic field. - ProQuest. *J Neurol Neurosurg Psychiatry*. 1985;29.
39. Bolognini N, Làdavas E, Farnè A. Spatial perspective and coordinate systems in autoscopy: a case report of a 'fantome de profil' in occipital brain damage. *J Cogn Neurosci*. 2011;23:1741–1751.
40. Maillard L, Vignal JP, Anxionnat R, Taillandier L, Vespignani H. Semiologic value of ictal autoscopy. *Epilepsia*. 2004;45:391–394.
41. Blanke O, Landis T, Spinelli L, Seeck M. Out-of-body experience and autoscopy of neurological origin. *Brain J Neurol*. 2004;127:243–258.
41a. Brandt C, Brechtelsbauer D, Bien CG, Reiners Nervenarzt K. 2005;76(1259):1261–1262.
42. Heydrich L, Blanke O. Distinct illusory own-body perceptions caused by damage to posterior insula and extrastriate cortex. *Brain J Neurol*. 2013;136:790–803.
43. Brugger P. Reflective mirrors: perspective-taking in autoscopic phenomena. *Cogn Neuropsychiatry*. 2002;7:179–194.
44. Brugger P, Agosti R, Regard M, Wieser HG, Landis T. Heautoscopy, epilepsy, and suicide. *J Neurol Neurosurg Psychiatry*. 1994;57:838–839.
45. Brugger P, Regard M, Landis T. Illusory reduplication of one's own body: phenomenology and classification of autoscopic phenomena. *Cogn Neuropsychiatry*. 1997;2:19–38.
46. Anzellotti F, et al. Autoscopic phenomena: case report and review of literature. *Behav Brain Funct BBF*. 2011;7:2.
47. Brugger P, Blanke O, Regard M, Bradford DT, Landis T. Polyopic heautoscopy: case report and review of the literature. *Cortex*. 2006;42:666–674.
48. Devinsky O, Feldmann E, Burrowes K, Bromfield E. Autoscopic phenomena with seizures. *Arch Neurol*. 1989;46:1080–1088.
49. Tadokoro Y, Oshima T, Kanemoto K. Postictal autoscopy in a patient with partial epilepsy. *Epilepsy Behav*. 2006;9:535–540.
50. Craig AD. How do you feel? Interoception: the sense of the physiological condition of the body. *Nat Rev Neurosci*. 2002;3:655–666.
51. Damasio A. Feelings of emotion and the self. *Ann N Y Acad Sci*. 2003;1001:253–261.
52. Damasio A, Carvalho GB. The nature of feelings: evolutionary and neurobiological origins. *Nat Rev Neurosci*. 2013;14:143–152.

53. Ronchi R, et al. Right insular damage decreases heartbeat awareness and alters cardio-visual effects on bodily self-consciousness. *Neuropsychologia*. 2015;70:11–20.
54. Bünning S, Blanke O. The out-of body experience: precipitating factors and neural correlates. *Prog Brain Res*. 2005;150:331–350.
55. Lunn V. Autoscopic phenomena. *Acta Psychiatr Scand*. 1970;46(Suppl. 219):118–125.
56. Daly DD. Ictal affect. *Am J Psychiatry*. 1958;115:171–181.
57. Ionta S, Gassert R, Blanke O. Multi-sensory and sensorimotor foundation of bodily self-consciousness - an interdisciplinary approach. *Front Psychol*. 2011;2:383.
58. Blanke O, Ortigue S, Landis T, Seeck M. Stimulating illusory own-body perceptions. *Nature*. 2002;419:269–270.
59. Nakul E, Lopez C. Commentary: out-of-body experience during awake craniotomy. *Front Hum Neurosci*. 2017;11.
60. Sheils D. A cross-cultural study of beliefs in out-of-the-body experiences, waking and sleeping. *J Soc Psych Res*. 1978.
61. Lenggenhager B, Tadi T, Metzinger T, Blanke O. Video ergo sum: manipulating bodily self-consciousness. *Science*. 2007;317:1096–1099.
62. Ionta S, et al. Multisensory mechanisms in temporo-parietal cortex support self-location and first-person perspective. *Neuron*. 2011;70:363–374.
63. Lopez C, Halje P, Blanke O. Body ownership and embodiment: vestibular and multisensory mechanisms. *Neurophysiol Clin Clin Neurophysiol*. 2008;38:149–161.
63a. Bremmer F, Schlack A, Shah NJ, Zafiris O, Kubischik M, Hoffmann KP, Fink GR. Polymodal motion processing in posterior parietal and premotor cortex: a human fMRI study strongly implies equivalencies between humans and monkeys. *Neuron*. 2001;29(1):287–296.
63b. Calvert GA, Campbell R, Brammer MJ. Evidence from functional magnetic resonance imaging of crossmodal binding in the human heteromodal cortex. *Curr Biol*. 2000;10:649–657.
64. Jaspers K. Über leibhaftige Bewusstheiten (Bewusstheitstäuschungen), ein psychopathologisches Elementarsymptom. *Z Pathopsychol*. 1913;2:150–161.
65. Lhermitte J. Les visions d'un poète. *Rev Oto-Neuro-Ophtalmol (Paris)*. 1939;17:81–96.
66. Critchley M. The idea of a presence. *Acta Psychiatr Neurol Scand*. 1955;30:155–168.
66a. James W. *The Variety of Religious Experience*. New York: Coller McMillan; 1961.
67. Brugger P, Regard M, Landis T. Unilaterally felt 'presences': the neuropsychiatry of one's invisible Doppelganger. *Cogn Behav Neurol*. 1996;9:114.
68. Lhermitte J. Visual hallucinations of the self. *Br Med J*. 1951;1:431–434.
69. Fénelon G, Soulas T, De Langavant LC, Trinkler I, Bachoud-Lévi A-C. Feeling of presence in Parkinson's disease. *J Neurol Neurosurg Psychiatry*. 2011;82:1219–1224.
70. Williams D. The structure of emotions reflected in epileptic experiences. *Brain J Neurol*. 1956;79:29–67.
71. Arzy S, Seeck M, Ortigue S, Spinelli L, Blanke O. Induction of an illusory shadow person. *Nature*. 2006;443:287.
72. Blanke O, et al. Neurological and robot-controlled induction of an apparition. *Curr Biol*. 2014;24:2681–2686.
73. Tsakiris M, Hesse MD, Boy C, Haggard P, Fink GR. Neural signatures of body ownership: a sensory network for bodily self-consciousness. *Cereb Cortex N Y N*. 1991;17:2235–2244 (2007).
74. Ehrsson HH, Holmes NP, Passingham RE. Touching a rubber hand: feeling of body ownership is associated with activity in multisensory brain areas. *J Neurosci*. 2005;25:10564–10573.
75. Blakemore SJ, Wolpert DM, Frith CD. Central cancellation of self-produced tickle sensation. *Nat Neurosci*. 1998;1:635–640.
76. Blakemore SJ, Wolpert DM, Frith CD. Abnormalities in the awareness of action. *Trends Cogn Sci*. 2002;6:237–242.
77. Haggard P, Iannetti GD, Longo MR. Spatial sensory organization and body representation in pain perception. *Curr Biol*. 2013;23:R164–R176.
78. Longo MR, Iannetti GD, Mancini F, Driver J, Haggard P. Linking pain and the body: neural correlates of visually induced analgesia. *J Neurosci*. 2012;32:2601–2607.
79. Marinus J, et al. Clinical features and pathophysiology of complex regional pain syndrome. *Lancet Neurol*. 2011;10:637–648.
80. Moseley GL. Distorted body image in complex regional pain syndrome. *Neurology*. 2005;65:773.

81. Lewis JS, et al. Wherever is my arm? Impaired upper limb position accuracy in complex regional pain syndrome. *Pain*. 2010;149:463–469.
82. Schwoebel J, Friedman R, Duda N, Coslett HB. Pain and the body schema: evidence for peripheral effects on mental representations of movement. *Brain J Neurol*. 2001;124:2098–2104.
83. Schwoebel J, Coslett HB, Bradt J, Friedman R, Dileo C. Pain and the body schema: effects of pain severity on mental representations of movement. *Neurology*. 2002;59:775–777.
84. Juottonen K, et al. Altered central sensorimotor processing in patients with complex regional pain syndrome. *Pain*. 2002;98:315–323.
85. Maihöfner C, Handwerker HO, Neundörfer B, Birklein F. Patterns of cortical reorganization in complex regional pain syndrome. *Neurology*. 2003;61:1707–1715.
86. Maihöfner C, Handwerker HO, Neundörfer B, Birklein F. Cortical reorganization during recovery from complex regional pain syndrome. *Neurology*. 2004;63:693–701.
87. Förderreuther S, Sailer U, Straube A. Impaired self-perception of the hand in complex regional pain syndrome (CRPS). *Pain*. 2004;110:756–761.
88. Galer BS, Butler S, Jensen MP. Case reports and hypothesis: a neglect-like syndrome may be responsible for the motor disturbance in reflex sympathetic dystrophy (Complex Regional Pain Syndrome-1). *J Pain Symptom Manag*. 1995;10:385–391.
89. Galer BS, Jensen M. Neglect-like symptoms in complex regional pain syndrome: results of a self-administered survey. *J Pain Symptom Manag*. 1999;18:213–217.
90. Ephraim PL, Wegener ST, MacKenzie EJ, Dillingham TR, Pezzin LE. Phantom pain, residual limb pain, and back pain in amputees: results of a national survey. *Arch Phys Med Rehabil*. 2005;86:1910–1919.
91. Merzenich MM, et al. Somatosensory cortical map changes following digit amputation in adult monkeys. *J Comp Neurol*. 1984;224:591–605.
92. Ramachandran VS, Rogers-Ramachandran D, Cobb S. Touching the phantom limb. *Nature*. 1995;377:489–490.
93. Yang TT, et al. Noninvasive detection of cerebral plasticity in adult human somatosensory cortex. *Neuroreport*. 1994;5:701–704.
94. Flor H, et al. Phantom-limb pain as a perceptual correlate of cortical reorganization following arm amputation. *Nature*. 1995;375:482–484.
95. Karl A, Birbaumer N, Lutzenberger W, Cohen LG, Flor H. Reorganization of motor and somatosensory cortex in upper extremity amputees with phantom limb pain. *J Neurosci*. 2001;21:3609–3618.
96. Flor H, Nikolajsen L, Staehelin Jensen T. Phantom limb pain: a case of maladaptive CNS plasticity? *Nat Rev Neurosci*. 2006;7:873–881.
97. Makin TR, et al. Phantom pain is associated with preserved structure and function in the former hand area. *Nat Commun*. 2013;4:1570.
98. Makin TR, et al. Network-level reorganisation of functional connectivity following arm amputation. *Neuroimage*. 2015;114:217–225.
99. Makin TR, Scholz J, Henderson Slater D, Johansen-Berg H, Tracey I. Reassessing cortical reorganization in the primary sensorimotor cortex following arm amputation. *Brain J Neurol*. 2015;138:2140–2146.
100. Serino A, et al. Upper limb cortical maps in amputees with targeted muscle and sensory reinnervation. *Brain*. 2017;140:2993–3011.
101. Fraser C. Fact and fiction: a clarification of phantom limb phenomena. *Br J Occup Ther*. 2002;65:256–260.
102. Grüsser SM, et al. The relationship of perceptual phenomena and cortical reorganization in upper extremity amputees. *Neuroscience*. 2001;102:263–272.
103. Carlen PL, Wall PD, Nadvorna H, Steinbach T. Phantom limbs and related phenomena in recent traumatic amputations. *Neurology*. 1978;28:211–217.
104. Flor H, Diers M. Sensorimotor training and cortical reorganization. *NeuroRehabilitation*. 2009;25:19–27.
105. Giraux P, Sirigu A. Illusory movements of the paralyzed limb restore motor cortex activity. *Neuroimage*. 2003;20(Suppl 1):S107–S111.
106. Mulvey MR, et al. Transcutaneous electrical nerve stimulation for phantom pain and stump pain in adult amputees. *Pain Pract*. 2013;13:289–296.
107. Töpper R, Foltys H, Meister IG, Sparing R, Boroojerdi B. Repetitive transcranial magnetic stimulation of the parietal cortex transiently ameliorates phantom limb pain-like syndrome. *Clin Neurophysiol Off J Int Fed Clin Neurophysiol*. 2003;114:1521–1530.

108. Siddall PJ, Taylor DA, McClelland JM, Rutkowski SB, Cousins MJ. Pain report and the relationship of pain to physical factors in the first 6 months following spinal cord injury. *Pain*. 1999;81:187–197.
109. Siddall P, Yezierski R, Loeser J. Taxonomy and epidemiology of spinal cord injury pain. In: *Progress in Pain Research and Management*. Vol. 23. 2002:9–23. Yezierski BKJR.
110. Conomy JP. Disorders of body image after spinal cord injury. *Neurology*. 1973;23:842–850.
111. Fuentes CT, Pazzaglia M, Longo MR, Scivoletto G, Haggard P. Body image distortions following spinal cord injury. *J Neurol Neurosurg Psychiatry*. 2013;84:201–207.
112. Burke D, Woodward J. Pain and phantom sensation in spinal paralysis. In: Vinken PJ, ed. *Injuries of the Spine and Spinal Cord Part II*. 1976:489–499. Bruyn GW.
113. Ding Y, Kastin AJ, Pan W. Neural plasticity after spinal cord injury. *Curr Pharmaceut Des*. 2005;11:1441–1450.
114. Lotze M, Laubis-Herrmann U, Topka H. Combination of TMS and fMRI reveals a specific pattern of reorganization in M1 in patients after complete spinal cord injury. *Restor Neurol Neurosci*. 2006;24:97–107.
115. Moore CI, et al. Referred phantom sensations and cortical reorganization after spinal cord injury in humans. *Proc Natl Acad Sci USA*. 2000;97:14703–14708.
116. Nardone R, et al. Functional brain reorganization after spinal cord injury: systematic review of animal and human studies. *Brain Res*. 2013;1504:58–73.
117. Wrigley PJ, et al. Anatomical changes in human motor cortex and motor pathways following complete thoracic spinal cord injury. *Cerebr Cortex*. 2009;19:224–232.
118. Scandola M, et al. Corporeal illusions in chronic spinal cord injuries. *Conscious Cognit*. 2017;49:278–290.
119. Lenggenhager B, Pazzaglia M, Scivoletto G, Molinari M, Aglioti SM. The sense of the body in individuals with spinal cord injury. *PLoS One*. 2012;7:e50757.
120. Scandola M, Aglioti SM, Pozeg P, Avesani R, Moro V. Motor imagery in spinal cord injured people is modulated by somatotopic coding, perspective taking, and post-lesional chronic pain. *J Neuropsychol*. 2017;11:305–326.
121. Pozeg P, Galli G, Blanke O. Those are your legs: the effect of visuo-spatial viewpoint on visuo-tactile integration and body ownership. *Front Psychol*. 2015;6:1749.
122. Longo MR, Azañón E, Haggard P. More than skin deep: body representation beyond primary somatosensory cortex. *Neuropsychologia*. 2010;48:655–668.
123. Dijkerman HC, De Haan EH. Somatosensory processing subserving perception and action: dissociations, interactions, and integration. *Behav Brain Sci*. 2007;30:224–230.
124. Friederich H-C, et al. Neural correlates of body dissatisfaction in anorexia nervosa. *Neuropsychologia*. 2010;48:2878–2885.
125. Cash TF, Deagle EA. The nature and extent of body-image disturbances in anorexia nervosa and bulimia nervosa: a meta-analysis. *Int J Eat Disord*. 1997;22:107–126.
126. Ben-Tovim DI, Walker MK, Murray H, Chin G. Body size estimates: body image or body attitude measures? *Int J Eat Disord*. 1990;9:57–67.
127. Skrzypek S, Wehmeier P, Remschmidt H. Body image assessment using body size estimation in recent studies on anorexia nervosa. A brief review. *Eur Child Adolesc Psychiatry*. 2001;10:215–221.
128. Benninghoven D, Raykowski L, Solzbacher S, Kunzendorf S, Jantschek G. Body images of patients with anorexia nervosa, bulimia nervosa and female control subjects: a comparison with male ideals of female attractiveness. *Body Image*. 2007;4:51–59.
129. Sunday SR, Halmi KA, Werdann L, Levey C. Comparison of body size estimation and eating disorder inventory scores in anorexia and bulimia patients with obese, and restrained and unrestrained controls. *Int J Eat Disord*. 1992;11:133–149.
130. Probst M, Vandereycken W, Van Coppenolle H, Pieters G. Body size estimation in anorexia nervosa patients: the significance of overestimation. *J Psychosom Res*. 1998;44:451–456.
131. Urgesi C, et al. Impaired configural body processing in anorexia nervosa: evidence from the body inversion effect. *Br J Psychol*. 2014;105:486–508.
132. Beato-Fernández L, et al. Changes in regional cerebral blood flow after body image exposure in eating disorders. *Psychiatry Res Neuroimaging*. 2009;171:129–137.
133. Uher R, et al. Functional neuroanatomy of body shape perception in healthy and eating-disordered women. *Biol Psychiatry*. 2005;58:990–997.

134. Wagner A, Ruf M, Braus DF, Schmidt MH. Neuronal activity changes and body image distortion in anorexia nervosa. *Neuroreport*. 2003;14:2193–2197.
135. Mohr HM, et al. Separating two components of body image in anorexia nervosa using fMRI. *Psychol Med*. 2010;40:1519–1529.
136. Sachdev P, Mondraty N, Wen W, Gulliford K. Brains of anorexia nervosa patients process self-images differently from non-self-images: an fMRI study. *Neuropsychologia*. 2008;46:2161–2168.
137. Suchan B, et al. Reduction of gray matter density in the extrastriate body area in women with anorexia nervosa. *Behav Brain Res*. 2010;206:63–67.
138. Epstein J, et al. Neurocognitive evidence favors "top down" over "bottom up" mechanisms in the pathogenesis of body size distortions in anorexia nervosa. *Eat Weight Disord*. 2001;6:140–147.
139. Spitoni GF, Galati G, Antonucci G, Haggard P, Pizzamiglio L. Two forms of touch perception in the human brain. *Exp Brain Res*. 2010;207:185–195.
140. Gaudio S, Brooks SJ, Riva G. Nonvisual multisensory impairment of body perception in anorexia nervosa: a systematic review of neuropsychological studies. *PLoS One*. 2014;9:e110087.
141. Keizer A, et al. Tactile body image disturbance in anorexia nervosa. *Psychiatry Res*. 2011;190:115–120.
142. Keizer A, Smeets MAM, Dijkerman HC, van Elburg A, Postma A. Aberrant somatosensory perception in anorexia nervosa. *Psychiatry Research*. 2012;200:530–537.
143. Spitoni GF, et al. The two dimensions of the body representation in women suffering from Anorexia Nervosa. *Psychiatry Research*. 2015;230:181–188.
144. Fassino S, Pierò A, Gramaglia C, Abbate-Daga G. Clinical, psychopathological and personality correlates of interoceptive awareness in anorexia nervosa, bulimia nervosa and obesity. *Psychopathology*. 2004;37:168–174.
145. Matsumoto R, et al. Regional cerebral blood flow changes associated with interoceptive awareness in the recovery process of anorexia nervosa. *Prog Neuro Psychopharmacol Biol Psychiatry*. 2006;30:1265–1270.
146. Pollatos O, et al. Reduced perception of bodily signals in anorexia nervosa. *Eat Behav*. 2008;9:381–388.
147. Miller SP, Redlich AD, Steiner H. The stress response in anorexia nervosa. *Child Psychiatr Hum Dev*. 2003;33:295–306.
148. Zonnevylle-Bender MJ, et al. Adolescent anorexia nervosa patients have a discrepancy between neurophysiological responses and self-reported emotional arousal to psychosocial stress. *Psychiatry Res*. 2005;135:45–52.
149. Papežová H, Yamamotova A, Uher R. Elevated pain threshold in eating disorders: physiological and psychological factors. *J Psychiatr Res*. 2005;39:431–438.
150. Wagner A, et al. Altered insula response to taste stimuli in individuals recovered from restricting-type anorexia nervosa. *Neuropsychopharmacology*. 2008;33:513.
151. Case LK, Wilson RC, Ramachandran VS. Diminished size-weight illusion in anorexia nervosa: evidence for visuo-proprioceptive integration deficit. *Exp Brain Res*. 2012;217:79–87. https://doi.org/10.1007/s00221-011-2974-7.
152. Grunwald M, et al. Deficits in haptic perception and right parietal theta power changes in patients with anorexia nervosa before and after weight gain. *Int J Eat Disord*. 2001;29:417–428.
153. Grunwald M, et al. Angle paradigm: a new method to measure right parietal dysfunctions in anorexia nervosa. *Arch Clin Neuropsychol*. 2002;17:485–496.
154. Goldzak-Kunik G, Friedman R, Spitz M, Sandler L, Leshem M. Intact sensory function in anorexia nervosa. *Am J Clin Nutr*. 2012;95:272–282.
155. Guardia D, Carey A, Cottencin O, Thomas P, Luyat M. Disruption of spatial task performance in anorexia nervosa. *PLoS One*. 2013;8:e54928.
156. Eshkevari E, Rieger E, Longo MR, Haggard P, Treasure J. Increased plasticity of the bodily self in eating disorders. *Psychol Med*. 2012;42:819–828.
157. Eshkevari E, Rieger E, Longo MR, Haggard P, Treasure J. Persistent body image disturbance following recovery from eating disorders. *Int J Eat Disord*. 2014;47:400–409.
158. Guardia D, et al. Anticipation of body-scaled action is modified in anorexia nervosa. *Neuropsychologia*. 2010;48:3961–3966.
159. Nico D, et al. The role of the right parietal lobe in anorexia nervosa. *Psychol Med*. 2010;40:1531–1539.
160. Guardia D, Cottencin O, Thomas P, Dodin V, Luyat M. Spatial orientation constancy is impaired in anorexia nervosa. *Psychiatry Research*. 2012;195:56–59.

161. Mast FW, Preuss N, Hartmann M, Grabherr L. Spatial cognition, body representation and affective processes: the role of vestibular information beyond ocular reflexes and control of posture. *Front Integr Neurosci.* 2014;8.
162. Silverstein S, Wang Y, Keane BP. Cognitive and neuroplasticity mechanisms by which congenital or early blindness may confer a protective effect against schizophrenia. *Front Psychol.* 2013;3:624.
163. Palmer BW, Dawes SE, Heaton RK. What do we know about neuropsychological aspects of schizophrenia? *Neuropsychol Rev.* 2009;19:365–384.
164. Schneider K. *Die Psychopatischen Personlichkeiten.* 9th ed. Deuticke; 1950.
165. NEL250404R01_Esch-Stefano_p_.pdf.
166. Graham-Schmidt KT, Martin-Iverson MT, Holmes NP, Waters F. Body representations in schizophrenia: an alteration of body structural description is common to people with schizophrenia while alterations of body image worsen with passivity symptoms. *Cogn Neuropsychiatry.* 2016;21:354–368.
167. Bortolon C, et al. Mirror self-face perception in individuals with schizophrenia: feelings of strangeness associated with one's own image. *Psychiatry Res.* 2017;253:205–210.
168. Ebisch SJ, et al. Out of touch with reality? Social perception in first-episode schizophrenia. *Soc Cogn Affect Neurosci.* 2012;8:394–403.
169. Thakkar KN, Nichols HS, McIntosh LG, Park S. Disturbances in body ownership in schizophrenia: evidence from the rubber hand illusion and case study of a spontaneous out-of-body experience. *PLoS One.* 2011;6:e27089.
170. Ferri F, et al. Upcoming tactile events and body ownership in schizophrenia. *Schizophr Res.* 2014;152:51–57.
171. Farrer C, et al. Neural correlates of action attribution in schizophrenia. *Psychiatry Res Neuroimaging.* 2004;131:31–44.
172. Spence SA, et al. A PET study of voluntary movement in schizophrenic patients experiencing passivity phenomena (delusions of alien control). *Brain: A Journal of Neurology.* 1997;120:1997–2011.
173. Orliac F, et al. Network modeling of resting state connectivity points towards the bottom up theories of schizophrenia. *Psychiatry Res Neuroimaging.* 2017;266:19–26.
174. Postmes L, et al. Schizophrenia as a self-disorder due to perceptual incoherence. *Schizophr Res.* 2014;152:41–50.
175. Stekelenburg JJ, Maes JP, Van Gool AR, Sitskoorn M, Vroomen J. Deficient multisensory integration in schizophrenia: an event-related potential study. *Schizophr Res.* 2013;147:253–261.
176. Tschacher W, Bergomi C. Cognitive binding in schizophrenia: weakened integration of temporal intersensory information. *Schizophr Bull.* 2011;37:S13–S22.
177. Williams LE, Light GA, Braff DL, Ramachandran VS. Reduced multisensory integration in patients with schizophrenia on a target detection task. *Neuropsychologia.* 2010;48:3128–3136.
178. De Jong J, Hodiamont P, Van den Stock J, de Gelder B. Audiovisual emotion recognition in schizophrenia: reduced integration of facial and vocal affect. *Schizophr Res.* 2009;107:286–293.
179. Van den Stock J, de Jong SJ, Hodiamont PP, de Gelder B. Perceiving emotions from bodily expressions and multisensory integration of emotion cues in schizophrenia. *Soc Neurosci.* 2011;6:537–547.
180. Association AP. *Diagnostic and Statistical Manual of Mental Disorders (DSM-5®).* American Psychiatric Pub; 2013.
181. Guillamon A, Junque C, Gómez-Gil E. A review of the status of brain structure research in transsexualism. *Arch Sex Behav.* 2016;45:1615–1648.
182. Cohen-Kettenis PT, Pfäfflin F. The DSM diagnostic criteria for gender identity disorder in adolescents and adults. *Arch Sex Behav.* 2010;39:499–513.
183. Coleman E, et al. Standards of care for the health of transsexual, transgender, and gender-nonconforming people, version 7. *Int J Transgenderism.* 2012;13:165–232.
184. Becker I, et al. Body image in young gender dysphoric adults: a European multi-center study. *Arch Sex Behav.* 2016;45:559–574.
185. Davis SA, Colton Meier S. Effects of testosterone treatment and chest reconstruction surgery on mental health and sexuality in female-to-male transgender people. *Int J Sex Health.* 2014;26:113–128.
186. Dutton L, Koenig K, Fennie K. Gynecologic care of the female-to-male transgender man. *J Midwifery Women's Health.* 2008;53:331–337.
187. van de Grift TC, et al. Body satisfaction and physical appearance in gender dysphoria. *Arch Sex Behav.* 2016;45:575–585.
188. Kraemer B, Delsignore A, Schnyder U, Hepp U. Body image and transsexualism. *Psychopathology.* 2008;41:96–100.

189. Wylie K, et al. Serving transgender people: clinical care considerations and service delivery models in transgender health. *The Lancet*. 2016;388:401–411.
190. Ramachandran V, McGeoch PD. Phantom penises in transsexuals. *J Conscious Stud*. 2008;15:5–16.
191. Ramachandran V, McGeoch PD. Occurrence of phantom genitalia after gender reassignment surgery. *Med Hypotheses*. 2007;69:1001–1003.
192. Case LK, Brang D, Landazuri R, Viswanathan P, Ramachandran VS. Altered white matter and sensory response to bodily sensation in female-to-male transgender individuals. *Arch Sex Behav*. 2017;46:1223–1237.
193. Lin C-S, et al. Neural network of body representation differs between transsexuals and cissexuals. *PLoS One*. 2014;9:e85914.
194. Manzouri A, Kosidou K, Savic I. Anatomical and functional findings in female-to-male transsexuals: testing a new hypothesis. *Cerebr Cortex*. 2017;27:998–1010.
195. Nota NM, et al. Brain functional connectivity patterns in children and adolescents with gender dysphoria: sex-atypical or not? *Psychoneuroendocrinology*. 2017.
196. Feusner JD, et al. Intrinsic network connectivity and own body perception in gender dysphoria. *Brain Imag Behav*. 2017;11:964–976.

CHAPTER

19

Hemianopia, spatial neglect, and their multisensory rehabilitation

Nadia Bolognini[1,2], *Giuseppe Vallar*[1,2]

[1]Department of Psychology & NeuroMi, University of Milan – Bicocca, Milano, Italy; [2]Istituto Auxologico Italiano, IRCCS, Laboratory of Neuropsychology, Milano, Italy

Introduction

In this chapter, we present evidence showing the existence of spared multisensory abilities in adult stroke patients with postchiasmatic visual field defects (VFDs) and unilateral spatial neglect (USN), and discuss the therapeutic potential of multisensory integration for the development of novel rehabilitation procedures for these neuropsychological disorders. VFDs are discussed first, given the more advanced status of multisensory research and rehabilitation in this area. Subsequently, current knowledge on the multisensory aspects of the syndrome of USN is considered, for which less evidence is available, both as to its assessment and to the development of specific, multisensory-based, treatments. However, USN is of particular interest for multisensory research, because the cerebral areas typically damaged in this condition are crucially involved in the multisensory representation of space. Indeed, USN can virtually affect all sensory modalities, separately or jointly; however, there is also evidence that the multisensory binding of spatial and temporal information from the different senses may be largely spared in patients with USN, opening new perspectives for their rehabilitation.

Multisensory rehabilitation for central visual field defects

Visual field defects: clinical features and anatomy

VFDs consist in the loss of vision in sectors of the visual field. The optic chiasm is used as the anatomical landmark to differentiate the "prechiasmatic" and the "postchiasmatic" visual pathways, and, in turn, the damage leading to "prechiasmatic" and "postchiasmatic" VFDs.[1]

Multisensory Perception
https://doi.org/10.1016/B978-0-12-812492-5.00019-X

Unilateral lesions of the prechiasmatic pathway affect only the visual hemifield ipsilateral to the side of the lesion (ipsilesional), leading to a partial or complete visual field loss that is monocular. Instead, unilateral lesions of the postchiasmatic pathway cause the loss of conscious vision in the contralateral visual hemifield. This loss is unilateral and homonymous: this implies that the visual deficit affects the same region of the visual field in both eyes; this is due to the fact that fibers from the nasal hemi-retinas (representing the lateral or temporal visual field) from each eye cross in the optic chiasm, while fibers from the temporal hemi-retina (representing the medial or nasal visual field) remain ipsilateral (Fig. 19.1).[1,2] Hence, the right visual hemifield is represented in the left visual cortex and the left visual hemifield in the right visual cortex. The next sections will focus on postchiasmatic VFDs.

Based on the extent of the lesion of the postchiasmatic pathways, VFDs can vary from complete hemianopia, the loss of the entire half of the visual field, to quadrantanopia,

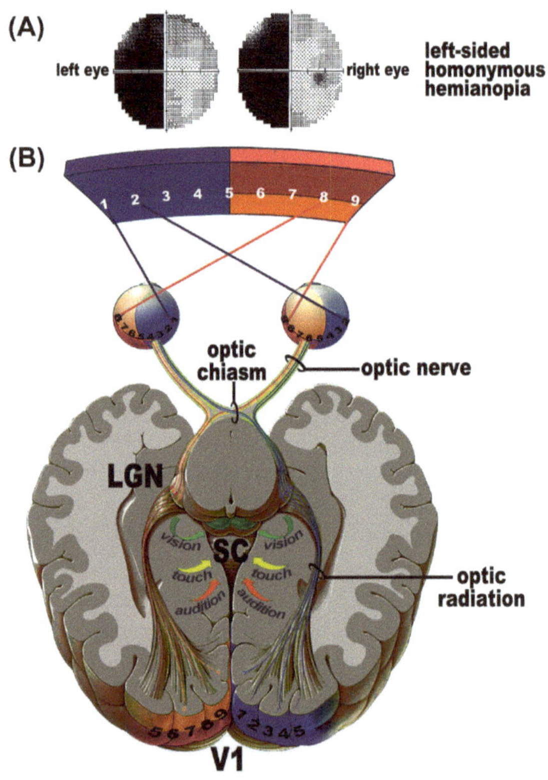

FIGURE 19.1 (A) Example of left homonymous hemianopia following a right occipital lobe stroke evaluated with Humphrey's automated perimetry. (B) In the undamaged brain, the main retinofugal pathway is the retino-geniculo-striate pathway which comprises the majority (>90%) of retinofugal fibers. It supplies the primary visual cortex (V1). A smaller amount of retinofugal fibers (probably <10%) reaches the superior colliculus (SC), where inputs from different senses converge and are integrated by multisensory neurons. The SC is the first station in a subcortical relay of retinal information to the extrastriate visual cortex (i.e., the retino-collicular-extrastriate pathway), bypassing V1.[16] *LGN*, Lateral geniculate nucleus.

encompassing the upper or the lower visual quadrant, to a paracentral scotoma (not exceeding 10 degrees).[1,2] In patients with unilateral postchiasmatic damage, foveal sparing (0.5–1 degree) is always present. However, due to the overrepresented central visual field in the visual cortex and the dual blood supply by the posterior cerebral artery and the deep branch of the middle cerebral artery to the occipital lobe, macular sparing (1–5 degrees) may be found with lesions of the occipital lobe, and also following damage to the optic radiations or tracts.[2–4] Even without macular sparing, a VFD itself does not usually reduce visual acuity. If this is the case, an additional lesion involving the prechiasmatic visual pathway should be suspected.[1]

In adults, the most common cause of postchiasmatic VFDs is stroke (about 70% of patients). Approximately 8%–10% of stroke patients have permanent VFDs. The most frequent locations of lesions resulting in hemianopia are in the occipital lobe (45%) and in the optic radiation (32%), followed by damage to the optic tract (10%), and to the lateral geniculate nucleus (LGN, 1.3%); in about 11% of patients, the damage involves a combination of several areas.[2,5] Damage to the primary visual cortex (V1) occurs in more than 40% of patients following a stroke in the vascular territory of the posterior cerebral artery. Other common causes of acquired VFDs in adulthood include traumatic brain injury (14%), tumors (11%), brain surgery (2.5%), demyelination (1.5%), and other rare causes (1.5%).[2,5]

VFDs from vascular disease seem to have a poor prognosis for a spontaneous visual field recovery, while a more remarkable recovery is observed after a traumatic damage.[6] Spontaneous recovery typically occurs within the first month after stroke, decreasing as time from the injury increases; after 6 months spontaneous visual field restitution is improbable. Overall, complete resolution is rare (<5% of cases).[2,6,7]

Visual field testing can be performed by various methods, starting from visual field assessment by confrontation testing, as a part of the neurological examination,[8] but the gold standard is the automated (computerized) perimetry.[1] Perimetry refers to the systematic measurement of the visual field. Automated perimetry allows assessing differential light sensitivities (i.e., stimulus luminance threshold values) in the visual field through the detection of visual targets presented on a defined background. Although perimetry can be also performed manually, the automated version has the advantage of the standardization of both stimulus presentation and response recording, leading to more reproducible results. Microperimetry[9] is used for assessing macular function, while electrophysiological techniques, such as visual evoked potentials (VEPs), are useful for the differential diagnosis of VFDs and USN (see below).

Patients with postchiasmatic VFDs are usually aware of their blind field, which causes many difficulties in everyday activities. They complain about seeing people or things "too late," describing a sort of "slow" vision; usually they report difficulties in reading and navigating in complex visual environments, getting lost in crowded or open places, especially if unfamiliar, and problems with avoiding obstacles and searching for objects.[10] Indeed, VFDs are frequently associated with other visually related complaints and dysfunctions. Among them, hemianopic dyslexia and disorders of oculomotor visual exploration are of major importance for their debilitating consequences on quality of life.

Hemianopic dyslexia is an acquired lower-level reading disorder, which involves impairments of prelexical (visual) processes.[11] In languages where reading occurs from left-to-right, patients with right-sided hemianopia are more impaired than patients with left-sided hemianopia. This finding is in line with evidence for a major role of left hemispheric areas (striate and extrastriate cortex; posterior parietal cortex, PPC) for controlling the oculomotor exploration of a line of text. This asymmetry may be functional and culturally determined, as suggested by investigations in healthy individual and patients, across reading systems. The degree of macular sparing also influences reading: when the central visual field sparing is > 10 degrees, reading is rarely compromised.[11]

Patients with VFDs also present with impaired visual scanning behavior, which impacts the development of oculomotor compensatory strategies.[10,12] Oculomotor compensation represents an adaptive visual scanning behavior, allowing patients to explore the blind visual area through eye movements. However, oculomotor behavior in the presence of homonymous hemianopia is characterized by longer scan paths and search times, along with smaller, less regular, and less accurate saccades directed toward the blind field, and a higher number of fixations and refixations. As a result, the scanning of the visual field by hemianopic patients is usually unsystematic, irregular, and very time-consuming. This largely explains these patients' complaints about having a limited overview and a "slow" vision. Although adaptive eye-movement strategies develop spontaneously over time, only about 40% of patients with chronic hemianopia show a normal visual search.[12,13]

Such impaired visual search is occasionally related to the location/side of the brain lesion: occipitoparietal and posterior thalamic injuries may underlie inefficient spatial orienting and oculomotor responses.[10] However, defective visual scanning may also represent a direct knock-on effect of the visual field loss,[14] as it may manifest even without higher-order dysfunctions.[10]

It is worth mentioning the phenomenon of *blindsight*, given its possible role in visual rehabilitation and in multisensory processing in the area of vision loss. Blindsight refers to various forms of "unconscious vision" that may be encountered in patients with postchiasmatic VFDs.[15] Despite cortical blindness, certain levels of perceptual processing are indeed maintained inside the blind visual sector. Behaviorally, blindsight manifests itself in the patients' preserved ability of detecting, localizing, and even discriminating visual stimuli in the blind field, in the face of having no awareness of them.[15] This condition is thought to be sustained by different neural substrates: either by alternative visual pathways bypassing the damaged V1 (i.e., extrageniculate activity via subcortical pathways, or geniculoextrastriate involvement), or by partial sparing of V1, with preservation of cortical processing sufficient for stimuli to reach a "subjective threshold," so that patients are not aware of the stimulus, as indexed by their verbal report, but their behavior shows responses at better than chance level.[15,16] Blindsight is typically examined by forced-choice guessing tasks, measures of implicit processing (e.g., pupillary reflex), or assessing the effect of the visual stimulus in the blind visual hemifield on the conscious response to the seen stimulus in the intact hemifield.[15]

Multisensory perception in hemianopia

So far, multisensory perception in stroke patients with postchiasmatic VDFs has been investigated only at a behavioral level.[17] All in all, substantial multisensory processing may take place in the blind sector of the visual field. The facilitatory effects of nonvisual stimuli on impaired visual functions are relevant from a rehabilitative perspective. This line of investigation was originally prompted by the discovery of multisensory neurons in the superior colliculus (SC). The SC is a subcortical structure of the extrastriate visual pathways mediating some unconscious visual functions, such as the abovementioned blindsight. The SC is also involved in both initiation and execution of saccades, as well as in target selection and spatial orienting.[18,19]

The deep layers of the SC contain multisensory neurons which integrate multisensory cues, so that a spatially and temporally congruent nonvisual stimulus (e.g., auditory or tactile) enhances their visual responsiveness, and facilitates orientation behavior. An important and fascinating property of such neurons is known as the "principle of inverse effectiveness," namely: multisensory stimuli enhance the detection and localization of suboptimal stimuli in one or another modality, and do so most strongly when the stimuli involved are least effective (see chapter by Stein and Rowland, this volume).[20] This neural property is of great relevance with respect to the brain's capability to compensate for unisensory impairments through multisensory integration.

The first original study in patients with chronic poststroke VFDs showed that visual detection in the blind hemifield is improved by auditory stimuli.[21] The auditory-driven enhancement emerges only when the sound is spatially and temporally aligned to the unseen target. Moreover, the multisensory facilitation shares an inverse relationship with the amount of visual impairment, as well as with the extent of the brain injury. First, the lower the probability of visual detection in the blind field, the greater the multisensory benefit, as predicted by the principle of inverse effectiveness. Second, the sound improves visual detection in the blind hemifield only if the lesion is confined to the occipital cortex; instead, when the lesion extends to frontotemporoparietal areas, there is no multisensory facilitation.[21] Because primary cortical sensory regions are interconnected in complex multisensory networks, the occipital cortex deprived of its sensory input (as a consequence of the acquired focal lesion) may rely on spared connections, or it may even establish novel functional connections, to access sensory information from the intact senses. Such alternative pathways represent an attempt by the multisensory network to reconnect the various senses bypassing the injured area. The concurrent damage to higher-level, intrinsically supramodal, frontoparietal areas may prevent this rearrangement of multisensory interactions outside the injured occipital cortex; under these conditions, crossmodal stimuli are no longer able to influence visual processing.[17]

Even the mere exposure to passive unimodal auditory stimulation can be beneficial. In one study,[22] after 1 hour of exposure to repetitive trains of sound pulses presented on the side of the blind field, patients with chronic hemianopia showed an improvement in visual detection by almost 100% within 30 minutes after the passive auditory stimulation. Such visual enhancement disappeared about 90 minutes later. A trend for a shift of the visual field border toward the blind hemifield was also observed.[22] Passive auditory stimulation may induce some activation of residual visual networks with multisensory proprieties, rendering them sensitive to nonvisual (in this case auditory) stimulation.

Another study[23] showed the opposite crossmodal effect, namely the influence of unseen visual stimuli on auditory processing. Patients with chronic VFDs due to occipital lobe damage (in some cases, extending to the parietal and frontal cortices) were asked to localize auditory targets, which could be presented alone or paired with a visual stimulus. Under this latter condition, the visual cue from the blind hemifield improves auditory localization; again, the spatial proximity of the auditory and of the visual stimuli drives the multisensory facilitation. A similar effect emerges after a brief (4 minutes) period of exposure to congruent multisensory stimuli: the adaptation to spatially coincident audiovisual stimuli presented in the hemianopic field improves auditory localization in that hemifield.[24] In sum, an unseen visual stimulus, processed unconsciously due to the injury to V1, is still effective in improving auditory function in the blind field.

The neural underpinnings of the improvement of visual perception by auditory stimuli in VFDs, as well as of the improvement of auditory processing by visual stimuli, still need to be uncovered. Likely, the observed multisensory enhancement effects rely on spared extrastriate pathways, bypassing the damaged striate cortex; this circuit mediates blindsight[15] and involves multisensory neurons of the SC.[20,25]

The link between unconscious vision and multisensory integration is further supported by evidence from the study of those perceptual illusions known to involve crossmodal interactions in primary visual areas. The "spatial ventriloquism" effect represents the best-known example of how the perceptual system deals with a multisensory spatial conflict: the presentation of synchronous visual and auditory stimuli in slightly separate (incongruent) locations leads to an immediate crossmodal bias, with the auditory stimulus being "captured" by the visual event (i.e., the sound is colocalized with the visual stimulus). This phenomenon is absent in stroke patients with hemianopia: a visual stimulus in the blind field does not cause any perceptual bias in auditory localization in incongruent audiovisual trials, although it can improve auditory localization whenever the visual stimulus is spatially aligned to the auditory target.[23,24]

In the same way, the "sound-induced flash illusion" is perceived less reliably by patients with postchiasmatic VFDs.[26] Two (or more) rapid tones (beeps) accompanying a single brief visual flash induce the illusory perception of seeing a double flash ("fission"); conversely, two (or more) flashes are perceived as one when presented with a single beep ("fusion"). The "fission" and the "fusion" illusions are weakened in patients with VFDs, and their disruption is proportional to the extent of the occipital damage.[26] Together, findings from the ventriloquism and the sound-induced flash illusions suggest that the integrity of occipital cortical activity is a crucial factor for some multisensory perceptual functions, requiring sensory (visual) awareness, to emerge.

Not only audition but also proprioception and touch may be helpful. A patient with left-sided homonymous hemianopia caused by a posterior cerebral artery infarct (involving the right occipital lobe, including V1) showed an improved ability to consciously detect left-sided visual stimuli when he extended the left arm into the blind hemifield, placing the hand nearby unseen visual targets.[27] This proprioceptive-tactile enhancement has been related to the activity of visuotactile neurons, found in dorsal regions of the visual cortex in the monkey; these neurons are involved in the multisensory, body-centered, representation of space.[28] Yet, such an intriguing finding was not replicated in a larger sample of patients with homonymous hemianopia (four patients with ischemic infarcts, one with left thalamic and intraventricular hemorrhage).[29]

Another study has documented the effect of hand proximity on unconscious residual vision in two patients with dense postchiasmatic VFDs.[30] One patient presented with upper-left quadrantanopia following surgical treatment that severed Meyer's loop (i.e., the optic radiations that course through the temporal lobe); the second patient presented with a left hemianopia following childhood meningitis, which had caused lesions extending from the right occipital cortex into the temporal cortex and the PPC, in turn isolating much of V1. Objects of different sizes were presented in the blind hemifield. Patients were asked to estimate the size of each object, using their right hand, by adjusting the thumb-finger distance or by reaching and grasping the object. Target size was processed more accurately in the blind hemifield only when the patients' left hand, contralateral to the side of the lesion (contralesional), was placed near the object. Such a benefit emerged in both visuomotor and size judgment tasks. However, hand proximity did not change the patients' conscious experience of stimulus size. Therefore, the mere presence of a motor effector in the blind hemifield, and the potential for hand–target interaction, influence how features of nearby visual targets, such as size, are unconsciously processed. This implies that visuotactile neurons are recruited by extrageniculate projections (e.g., from the SC to the PPC, via the thalamic pulvinar nucleus), their activation requiring neither intact optic radiations nor an intact striate cortex (damaged in the studied patients). Overall, there is evidence for crossmodal interactions between tactile-proprioceptive and visual systems that may modulate at least some residual visual functions in the blind hemifield.[27,30]

Finally, an injured occipital lobe may also promote the emergence of abnormal multisensory processes in the blind field. This is the condition of the "autoscopic" hallucination, a complex perceptual experience involving the illusory reduplication of one's own face or body in the extrapersonal space (see chapter in this volume by Case et al.). This condition, associated with lesions to occipital areas, seems to be caused by an abnormal integration of multisensory body-related inputs, that follows the visual field loss. Higher-order cortical visual areas, deprived of their visual input due to the brain injury, seem to establish (or strengthen) abnormal crossmodal connections with cortical areas involved in body representation, giving rise to a multisensory illusory reduplication of one's own body in extrapersonal space.[31]

In summary, various multisensory phenomena arise in patients with postchiasmatic VFDs. The spared ability of combining inputs from the damaged visual modality with input from intact sensory systems may offer a powerful mechanism to compensate for the loss of vision. Importantly, multisensory function can be trained to reinforce impaired visual functions, as exemplified below.

Multisensory rehabilitation of VFDs

Overview of standard rehabilitation approaches

The visual system has been for a long time believed to have limited chances for recovery after injury; the generally accepted notion is that patients are permanently left with irreparable blindness. Despite this lingering conviction, a growing body of evidence indicates that targeted treatments may effectively improve both the pathophysiological underpinnings and the behavioral manifestations of VFDs.[10,16]

Treatment options for VFDs can be categorized into three main types: "substitutive," "restitutive," and "compensatory." The first category includes optical devices, such as prisms, used to reallocate, or expand, the field of view. Optical correction represents the best option for those patients with little or no potential for rehabilitation.[2] Restitutive training approaches encompass various behavioral methods sharing the goal of promoting vision recovery, also termed "vision restoration." Vision restoration does not assume a complete return to normal function. Rather, this term is used to emphasize the residual potential of the damaged central visual system to improve its function.[16] Essentially, restitutive therapies take advantage of various psychophysical paradigms to expand the visual field by the reactivation of surviving, functionally intact, neural tissue bordering the lesion (the so-called "transition zone"). These treatments typically involve a massed stimulation of the "transition zone," which is carried out for up to 60 minutes daily for weeks and months. This strategy may lead to enlargements of the visually responsive zone at the boundary with the blind field, averaging around 5 degrees of visual angle.[16,32] However, patients with homonymous hemianopia may develop adaptive eye movement strategies that involve frequent involuntary saccades toward their blind hemifield. Saccadic eye movements may therefore explain, or at least contribute to, the visual field enlargements on conventional perimetry detected after the restitution therapy.[33] Improvement in activities of daily living, mostly based on patients' subjective testimonials, has been reported, but they are also controversial, as they may likely reflect a placebo effect.[33]

Other restitutive options include intensive training, involving stimuli deep in the blind field (rather than stimulating the "transition zone"), with the goal of activating blindsight-like responses mediated by multiple cortical and subcortical regions.[34,35] Overall, controversy has arisen regarding the efficacy of restitutive therapies, in particular with respect to the reliability of the reported expansion of the visual field, and its impact on daily living.[2]

Finally, compensatory approaches train intact visual functions to promote compensatory behaviors, with the aim of overcoming the visual field loss. Regardless of the specific paradigm adopted, the goal is the improvement of oculomotor visual field scanning, by increasing the number, amplitude, and speed of saccades into the blind hemifield, and reinforcing visual attention and spatial orienting skills toward it. Compensatory therapies can enlarge the visual search field, namely the part of the visual field actively scanned via eye movements, up to 30 degrees.[10] Notably, compensatory therapies also ameliorate patients' mobility, navigation, and reading; all in all, they may lead to better performance in activities of daily living, as assessed by measurement scales and specific tasks. In one study, 91% of participants were able to return to part-time work following a compensatory training method.[10,32] Training approaches to correct specific impairments, such as hemianopic dyslexia, have been also employed.[10]

Thus far, there is not enough evidence to support the superiority of one of the three rehabilitation approaches (substitutive, restitutive, and compensatory) over the others.[36]

Multisensory compensatory training

Combing sensory information from multiple modalities enhances the likelihood of detecting and localizing an event or object, increases perceptual saliency, and speeds up reaction times. Furthermore, sensory processing is further facilitated in the event of sensory impairment or loss,[17] as in the case of VFDs. With these premises, multisensory stimulation has been used to enhance the efficacy of compensatory therapies. The choice of using multisensory stimuli in compensatory training has been driven by the idea of putative shared pathways, recruited by multisensory interactions guiding eye movement and spatial orienting, and of intact visual functions, activated by compensatory therapies, with emphasis on the role of the SC.

The first original study developed an audiovisual apparatus to deliver visual and auditory stimuli during oculomotor training.[37] Patients are seated centrally at the concave face of an ellipse-shaped apparatus (30 cm height X 200 cm length), over which visual and acoustic stimuli are positioned. Sounds (white-noise bursts, duration 100 ms) are delivered by eight loudspeakers (covered to prevent visual cues about their position), located horizontally at the patient's ear level, at eccentricities of 8 degrees, 24 degrees, 40 degrees, and 56 degrees, in both the hemianopic and the intact hemifields. Eight lights delivering the visual stimulus (duration 100 ms) are placed at the same locations as the loudspeakers. During training, patients are instructed to look at the fixation point at the center of the apparatus (which can be moved by ± 30 degrees of vertical eccentricity, for inferior and superior quadrantanopia) and to explore the blind hemifield by shifting the gaze toward the visual target, without head movements. Patients are also instructed to ignore the sound because it is not (always) predictive of the presence of the visual target. Indeed, during training, the visual target is presented alone or coupled with the sound; sounds without visual targets ("catch" trials) are also presented. To boost oculomotor exploration of the hemianopic field, a greater proportion of stimuli are presented in that area. Training lasts 2 weeks, at a rate of ~4 hours/working day. In the original study, eight patients with chronic poststroke VFDs (average duration of disease of 12 months) showed, after training, increased visual detection in the hemianopic field when they could use eye movements to search for the visual target, even in the absence of the adjuvant auditory cue. Moreover, the treatment improves visual search and reading and reduces self-perceived disability in many daily activities affected by the visual field loss (e.g., crossing the street, finding objects). The improvements remain stable for at least 1 month.[37] The efficacy of the multisensory compensatory training has been confirmed by a number of studies, in both chronic and acute stroke patients,[38–41] as well in children and adolescents with acquired VFDs.[42] Overall, the areas of improvement brought about by the training include oculomotor scanning (featured, after the therapy, by fewer fixations and refixations, faster and larger saccades, reduced scan-path length), reading (with different effects, depending on the damaged hemisphere: increased saccadic amplitude in patients with a left-sided hemianopia, reduced number of saccades during the return sweep in the case of right-sided hemianopia), and functional abilities in daily living. All these benefits persist up to a year.[41] Crucially, the audiovisual training proved to be superior to an analogous, but purely visual, training.[40]

At the electrophysiological level, multisensory compensatory training affects event-related potentials (ERPs) elicited by seen visual stimuli. In particular, the therapy reduces the amplitude of the component P3 in response to visual stimuli in the intact hemifield; P3 is a wave of the visual ERP related to activity in the PPC, indexing the allocation of attentional resources.[39] Patients with hemianopia typically tend to direct the focus of their attention to the intact hemifield[16]; the intensive audiovisual stimulation is able to reverse this attentional bias (reduced P3 amplitude), facilitating exogenous, stimulus-driven, shifts of attention toward the blind hemifield, and further consolidating oculomotor scanning of the blind area.[39]

It is noteworthy that visual improvements emerge only in tasks requiring the use of eye movements to compensate for the visual field loss. Whenever eye movements are prevented, there is no enhancement of visual detection.[37] It follows that the multisensory training does not induce any visual restitution, rather its mechanism of action more likely relies on the activation of the visual responsiveness of the oculomotor system.

This proposal is supported by evidence in animals. A multisensory (audiovisual) training was used to reinstate visuomotor abilities in cats rendered hemianopic by complete unilateral ablation of the visual cortex.[43] This cortical removal is accompanied by the loss of visual responsiveness of neurons in the deep layers of the ipsilesional SC; this condition is reversed by the multisensory training. In particular, hemianopic cats were trained in a task requiring repeated, food-rewarded, orienting movements to an auditory cue, presented together with a spatially and temporally coincident (but task-irrelevant) visual cue, delivered within the blind hemifield. This training reinstated basic visuomotor competencies, by facilitating the reemergence of the lost (due to the cortical lesion) visual activity in the deep layers of the SC. In fact, the training-induced behavioral improvements were linked to changes in the descending influences from the association cortex (i.e., the anterior ectosylvian sulcus); such changes rendered their target SC neurons once more capable of transforming visual cues into appropriate orientation behaviors. These findings underscore the inherent plasticity of visuomotor circuits that can express residual compensatory visual capabilities overcoming visual field loss.[43]

In humans, similar mechanisms may be activated by multisensory compensatory training. Converging evidence reveals the pivotal role of the human SC in integrating audiovisual spatiotemporal coincident stimuli[25], and the relevance of the temporoparietal and posterior parietal areas in mediating covert and overt orienting toward audiovisual stimuli.[44] Because the retino-colliculo-extrastriate pathway is usually functionally and anatomically spared in patients with VFDs, intensive multisensory stimulation may facilitate the activity of this intact circuit, allowing the development of more proficient compensatory oculomotor strategies, which are often compromised, and hence ineffective, without a rehabilitative intervention (Fig. 19.2).

Recent developments of multisensory compensatory training include the use of more ecological settings, for instance, larger training boards (2 × 2 m), mimicking more natural exploration conditions (Fig. 19.3).[45] This strategy may aid the transfer of practice effects to

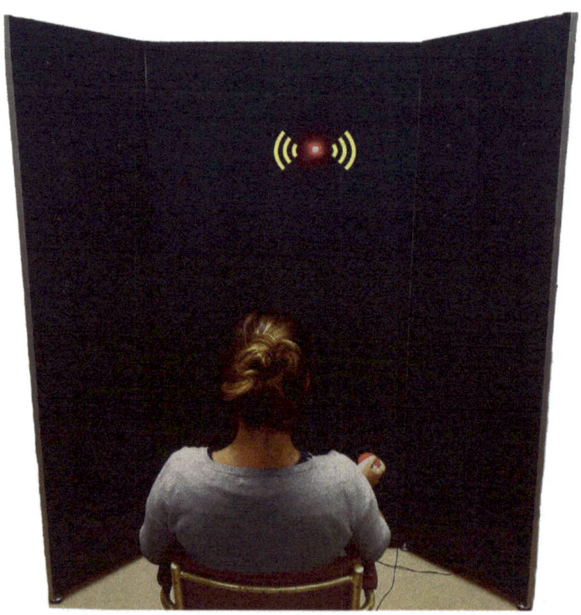

FIGURE 19.2 Overview of the audiovisual apparatus used for multisensory compensatory training. An audiovisual, spatially aligned, stimulus is portrayed.

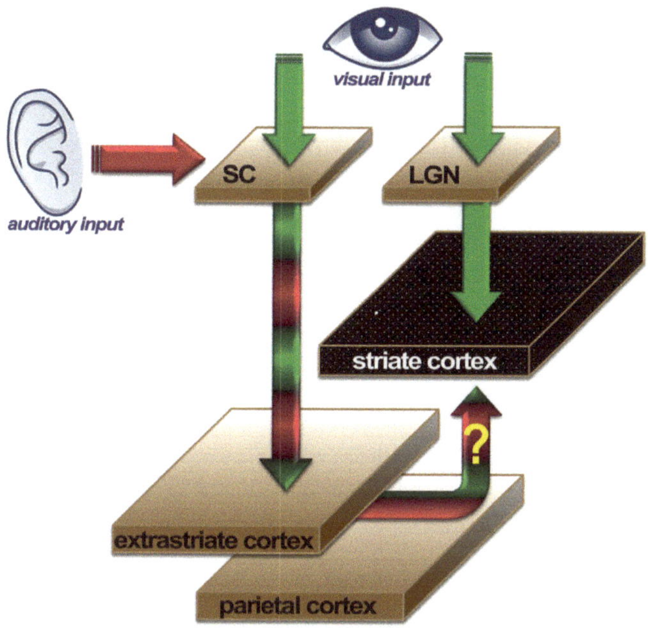

FIGURE 19.3 Simplified schema showing the putative substrate activated by multisensory compensatory training. Audiovisual stimuli are integrated by superior colliculus (SC) neurons, enhancing their neural activity. Projections from the SC reach extrastriate visual areas and the posterior parietal cortex, outflanking the damaged (black) striate cortex. This pathway mediates different kinds of residual visual functions (i.e., blindsight), and is crucial for eye movements and spatial orienting. This route could also include feedback projections from higher-order cortical areas to V1 (yellow question mark), supporting an indirect modulation of V1 responsiveness to multisensory stimuli. This proposal still requires empirical support. *LGN*, lateral geniculate nucleus.

functional skills of daily living. Moreover, from our clinical experience, shorter (~1 hour) daily training sessions are also effective, and more practical, especially for patients who are not hospitalized. The adjuvant use of noninvasive brain stimulation during training is also under investigation.[45]

Unilateral spatial neglect

USN is a deficit of exploration of the contralesional side of space, and of awareness of sensory events taking place in that portion of space. USN is more frequent and severe after lesions of the right cerebral hemisphere and is not due to sensory or motor deficits, although these disorders (hemianopia, hemianesthesia, hemiparesis, or hemiplegia) are often associated with USN.[46,47]

Clinical presentation

The abovementioned definition does not make specific reference to the sensory features of USN, other than distinguishing between "input," with a defective ability to orient and report sensory events, and "output," with an impairment in setting up motor plans and actions, in the neglected side of space. USN has been regarded as a deficit with a "polysensorial nature,"[48] meaning that it can show up in different sensory domains.

The multisensory features of USN are revealed by observing the patient's behavior during a neurological examination or spontaneous activities. Patients may show a deviation of the head and eyes toward the ipsilesional side of space, rightward in right brain-damaged patients. Addressed from the left side, patients may fail to respond or look for the speaker in the right side of the room, turning the head and eyes rightward. Patients do not pick up food from the left half of the plate. Given a crossword puzzle, they may complete only the squares to the right. If walking is not prevented by left hemiparesis, patients may lose their bearings, making no use of left-sided cues, and systematically turning to the right, instead of to the left when appropriate.[47] USN is mainly assessed by paper-pencil tasks requiring the exploration of space, such as target cancellation, and search tasks. Patients show USN not only in the visual modality (visuospatial neglect) but also when they are blindfolded during the execution of the task.[47]

Fig. 19.4 shows clinical examples of left USN.

Unisensory and multisensory perception in USN

VISUAL deficits

USN may mimic VFDs, bringing about a "pseudohemianopia" as assessed both behaviorally and electrophysiologically, which is distinguished from genuine sensory hemianopia due

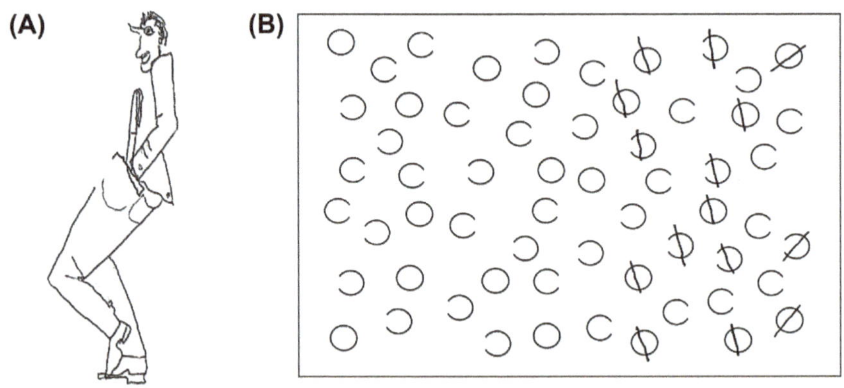

FIGURE 19.4 Visual unilateral spatial neglect (USN) as revealed by a patient's performance on two clinical tests. (A) A drawing from memory of a human figure[115,116] by a 64-year-old, male patient (TI), an engineer, with a right hemorrhagic stroke due to the rupture of a middle cerebral artery aneurysm; left USN is shown by omission of parts of the left-hand side of the figure (the trunk, the arm, and part of the upper left trouser leg). (B) Cancellation performance on a task,[117] in which patient TI was required to mark all complete circles. The center of the sheet was aligned with the midsagittal plane of the patient's body. The stimuli included 20 complete circles, 20 with a left-sided, and 20 with a right-sided missing portion. The patient omitted all circles on the left-hand side of the sheet, indicating USN in "egocentric" coordinates, i.e., with reference to the patient's trunk; on the right hand-side the patient crossed out nine complete circles and six circles with a left-sided missing portion, indicating a left-sided, "object-based" USN, with reference to the circle stimuli; no circle with a right-sided missing portion was crossed out. *Source: Vallar G, Calzolari E. Unilateral spatial neglect after posterior parietal damage.* Handb Clin Neurol. *2018;151:287–312, with permission of Elsevier.*

to damage to the retrogeniculate visual pathways up to the primary visual cortex (V1). Pseudohemianopia diminishes when patients look rightward, because, with this gaze orientation, left-sided stimuli in the retinotopic reference frame are presented on the right (intact) side in the spatial frame.[49,50] Patients with a pseudohemianopia show largely preserved VEPs, reflecting activity from the striate and extrastriate areas.[51,52] In these patients, electrical activity is altered at later stages of processing, mainly in the PPC.[53–55]

USN is overall more severe when vision is available.[56] However, in search and exploratory tasks, patients may be selectively impaired either when vision is available or when they operate without visual control,[57–59] suggesting that the multisensory signs of USN are an association of modality-specific deficits that may occur in isolation.

Auditory deficits

In the auditory modality,[47,58] two main clinical signs are *alloacusis* (left-sided stimuli are reported, but localized in the right side of space) and *auditory extinction*, which is the defective report of a stimulus delivered to the contralesional ear (the "extinguished" stimulus), when another stimulus is simultaneously presented to the other ear. The report of unilateral stimuli to either ear is preserved, ruling out a sensory deficit. Auditory extinction can be brought about by lesions to either hemisphere but has been found to be more long-lasting after damage to the right hemisphere.[60] In right brain-damaged patients with left USN, auditory extinction mainly concerns the spatial aspects of the contralesional, "extinguished" stimulus, rather than the processing of its content[61]; however, stimulus content modulates the severity of USN.[62] Sensory extinction, which manifests in different modalities as a deficit in the processing of multiple (typically two), briefly presented targets, with a failure to report the contralesional one, is distinct from impairment of the active exploration of space, and the serial detection of targets, such as in cancellation tasks, usually referred to as USN.[63]

Early observations suggested a role of damage to the temporal lobe in bringing about mislocalization errors of free-field acoustic stimuli in the contralateral side of space, with no definite hemispheric differences; later studies indicate, however, a prominent role of right brain damage.[64,65] A mislocalization, with an ipsilesional rightward displacement, has been found in tasks requiring to set the position of a binaural auditory stimulus in the coronal plane, with the lateralization of the sound being achieved based on interaural differences of intensity[66] (see Fig. 19.5), and timing[67] of the sounds. Such a displacement occurs both in the front and in the back halves of space, with respect to the body of the participant.[68] Right brain-damaged patients miss interruptions of sounds delivered to the contralesional left ear[69] and, with binaural stimulation, point to the right of the actual source of a sound from a loudspeaker in the left side of space.[58] Right brain-damaged patients with visual USN fail to judge accurately whether the lateral position of two auditory stimuli, presented in sequence on the contralesional side of space, is the same or not.[70]

Tactile deficits

The failure to report tactile stimuli delivered to the left side of the body may be due to USN, rather than to a primary somatosensory deficit. This, similarly to the case of "pseudohemianopia," has been demonstrated by showing, in right brain-damaged patients, preserved somatosensory evoked potentials to unreported, "neglected," stimuli[51] and evidence of implicit processing from an autonomic index.[71]

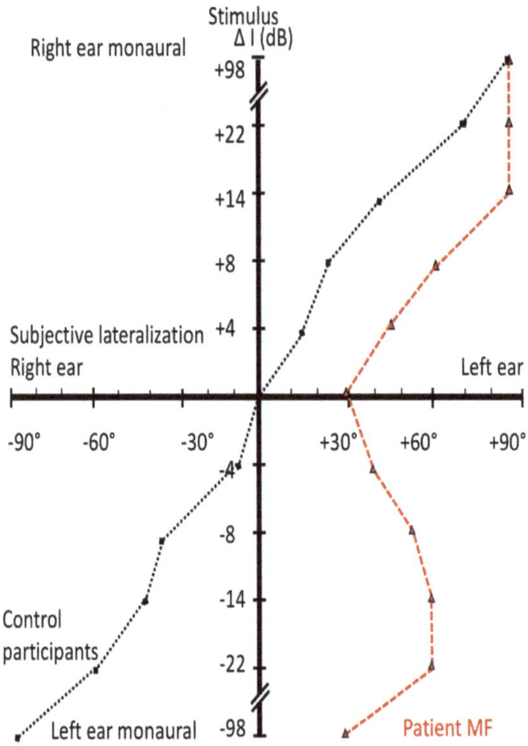

FIGURE 19.5 Auditory neglect. Ordinate: ΔI, interaural intensity difference [positive values (+98 dB to + 4 dB): right ear more intense; negative values (−4 dB to −98 dB): left ear more intense. Abscissa: perceived lateralization [positive values (up to +90 degrees): lateralization to right; negative values (up to −90 degrees): lateralization to left; origin of the axes: ΔI = 0 dB, midline localization]. Average lateralization curves of control participants (N = 30, black dotted line) and of right brain-damaged patient MF, with left visual unilateral spatial neglect (red dashed line). Patient MF disproportionately lateralizes all binaural, and even left monaural (−98 ΔI dB), stimuli rightward, toward the nonneglected right side of space. *Data from Bisiach E, Cornacchia L, Sterzi R, Vallar G. Disorders of perceived auditory lateralization after lesions of the right hemisphere.* Brain. *1984;107 (Pt 1):37–52.*

The spatial, USN-related, component of this "pseudohemianesthesia" has also been revealed by experiments where spatial and somatotopic reference frames are dissociated by the crossing of the patient's arms, with the left hand being placed in the right side of space, in right brain-damaged patients with left USN. This maneuver increases the patients' ability to report touches to the left hand, in response to both single and double simultaneous stimulation, with a decrease in extinction of left-sided (in the somatotopic reference frame) stimuli.[72,73]

The main manifestations of USN, related to sensory input modality, are summarized in Table 19.1.

TABLE 19.1 Manifestations of unilateral spatial neglect (USN) specific to single sensory modalities.

	Extrapersonal space	Personal/bodily, Near-peripersonal space
Modality-specific signs of USN	Visuospatial neglect (Assessed by exploratory, search, and estimation of extent—line bisection—tasks)	Proprioceptive-somatosensory, haptic spatial neglect (Idem, in blindfolded condition)
Modality-specific, neglect-related impairments mimicking primary sensory disorders	Neglect • Visual (*pseudohemianopia*) • Auditory • Olfactory§ • Gustatory¥ (Unawareness of contralateral visual, auditory, olfactory and gustatory stimuli, not due to a primary sensory deficit)	Tactile/somatosensory neglect (*pseudohemianesthesia*) (Unawareness of tactile and proprioceptive stimuli to the contralateral side of the body, not due to a primary sensory deficit)
Domain-specific forms of visual neglect	• Neglect dyslexia (written word, nonword letter strings; sentences; passages of prose) ◦ Omission and substitution errors • Facial neglect (faces)	

§,¥, The existence of these forms of sensory neglect has been questioned (see Ref. 47 for discussion).

Perceptual sensory awareness

USN is not only independent of primary sensory deficits, as it may manifest in patients who do not show such disorders, but it can also mimic them, with "pseudohemianopia" and "pseudohemianesthesia." In such patients, primary sensory processing is, at least in part, preserved, as indexed by electrophysiological parameters.[51,52,71,74] The ERP abnormalities (reduced amplitudes of potentials evoked by visual and auditory left-sided stimuli) diminish along with the recovery from left USN.[75] These signals from the left side of space do not appear to gain access to conscious perceptual awareness that supports the report of stimulus detection and overt recognition, with the processing being regarded as "implicit," and its abnormalities as electrophysiological indexes of defective awareness.[76] In the visual domain, the largely preserved activity in primary striate and extrastriate occipital areas may be contrasted with electrophysiological abnormalities involving activity in the dorsal PPC.[53] In line with the spatial key feature of left USN, a reduced mismatch negativity (an electrophysiological slow potential which automatically and preattentively indexes deviations from regularity in the acoustic environment) to left-sided auditory stimuli has been found, indicating an early (preattentive) space-specific information processing deficit. The asymmetry is specific for spatial location and is not present for deviations of other perceptual dimensions, such as stimulus pitch and duration.[77]

The neural bases of the neglect syndrome

The neuropathological correlates of USN are summarized in Fig. 19.6. Relevant regions include the PPC,[78,79] particularly the inferior parietal lobule at the temporoparietal junction, as indicated by early studies.[47,80–82] Current evidence indicates that other cortical regions, including the superior temporal gyrus, the frontal premotor cortex, their white matter connections, and subcortical gray nuclei, are also involved. Damage to the primary sensory and motor cortices is neither necessary nor sufficient to bring about USN.[47,83,84]

Of interest, many areas damaged in USN are also involved in multisensory processing,[85,86] among which are the inferior parietal lobule, the intraparietal sulcus, the superior temporal sulcus, the premotor cortex, the ventrolateral prefrontal cortex, and dorsal and ventral frontoparietal networks.[47,83] The PPC especially plays a major role in the integration of multisensory inputs from the environment and the body, relevant for the building up of representations of near-peripersonal space and the body, and supporting spatial attention[87–89] and action.[90]

The extensive network involved in USN may account for the different clinical manifestations of the syndrome, which are rarely purely modality-specific.

Multisensory perception and its potential for neglect rehabilitation

Multisensory perception seems to be preserved in patients with USN, as shown by the few studies assessing the susceptibility of these patients to perceptual illusions. For instance, the

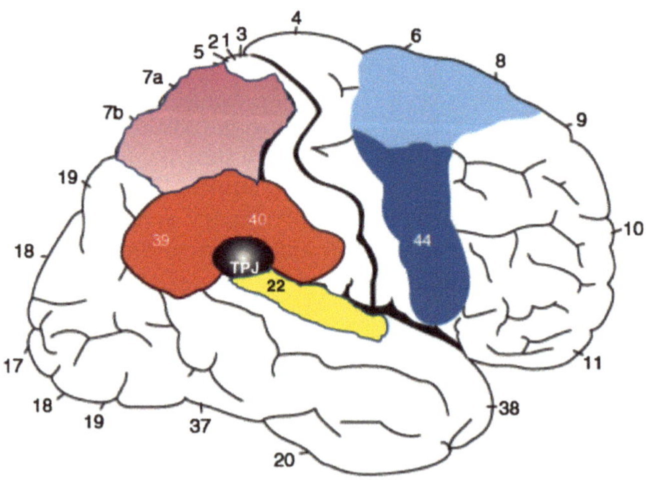

FIGURE 19.6 Lateral view of the right hemisphere, showing cortical regions (colored), whose damage is associated with specific patterns of impairment. Numbers refer to Brodmann areas (BAs). Inferior parietal lobule (IPL, red): BAs 39 (angular gyrus, AG), and 40 (supramarginal gyrus, SMG). Superior parietal lobule (SPL, pink): BAs 7 and 5. Premotor cortex: ventral BA 44 (blue), dorsal BAs 6 and 8 (azure). Left extrapersonal unilateral spatial neglect (USN) in near space, within arm and hand reach: estimation of lateral extent, as assessed by line bisection (IPL, particularly posteriorly, the AG, at the border with the lateral occipital lobe); spatial exploration, as assessed by target cancellation, and use of a spatial working space, as assessed by drawing tasks (IPL, lateral premotor cortex, dorsolateral prefrontal cortex). Allocentric/object-based USN: more ventral regions (deep and middle temporal) than those associated with egocentric USN. Personal USN: IPL (particularly SMG, BA 40) and the TPJ[78].

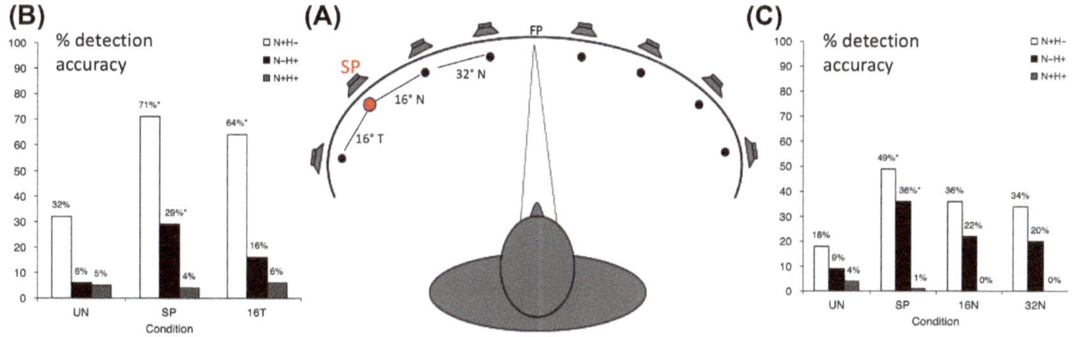

FIGURE 19.7 (A) Bird's eye schematic view of the position of light displays (black circles) and loudspeakers (trapezoids). A visual illustrative left-sided target is marked red; a single auditory stimulus may be simultaneously presented in the same position (SP red) or in different positions [16 degrees temporal (T), 16 degrees or 32 degrees nasal (N)]. (B) Percent detection accuracy of the visual stimulus alone (unimodal condition, UN) or associated with an SP or 16°T auditory stimulus, by patients with and without spatial neglect (N+/-), and with and without hemianopia (H+/-). (C) Same as (B), with associated 16°N or 32°N auditory stimuli. Compared with the UN condition, detection performance of N+ and H+ patients significantly improves (*), both when an auditory stimulus is simultaneously presented in SP, with a major increase, and, although at a lesser extent, in a close position (16°T in N+ patients only). *Modified from Frassinetti F, Bolognini N, Bottari D, Bonora A, Ladavas E. Audiovisual integration in patients with visual deficit.* J Cogn Neurosci. *2005;17(9):1442–1452.*

perceptual consequences of audiovisual interactions, the "McGurk illusion,"[91] and the "ventriloquist effect,"[92] both consisting in the binding of conflicting audiovisual stimuli, are found in patients with USN,[93–95] with visual stimuli delivered in the right, preserved, side of space improving the report of previously neglected left-sided stimuli.

Another instance of preserved audiovisual integration is provided by the "sound-induced flash illusion."[96] Right brain-damaged patients with left USN exhibit both the "fission" and the "fusion" aspects of the illusion; these crossmodal effects may be even disproportionately enhanced in right brain-damaged patients with left USN, in particular the "fusion" effects.[26]

There is also evidence, more directly relevant for rehabilitation, that audiovisual integration may temporarily reduce visuospatial USN, as also observed in hemianopia. Visual detection improves when visual and auditory stimuli originate from the same, or spatially close, positions in space, with a larger improvement for the more peripheral positions in the left (neglected) half-field.[21,97] The effect is absent when left USN is associated with left hemianopia (see Fig. 19.6).[21] In line with these findings, left hemianopia may exacerbate some manifestations of left USN (Fig. 19.7).[98]

Multisensory effects are also preserved when the tactile modality is involved.[99] Right brain-damaged patients with left USN show illusory effects with the Judd variant of the Müller-Lyer illusion, not only with visual and haptic stimuli, but also in a visuohaptic condition, in which the illusion is visually induced on a line explored haptically; the subjective midpoint of the line is symmetrically illusorily displaced leftward and rightward, in both neglect patients and healthy controls. As for left USN, this is present in the visual modality and absent in both the haptic and the visuohaptic (bimodal) conditions; the haptic input appears then to be able to counteract the visual impairment[99] (see Fig. 19.8).

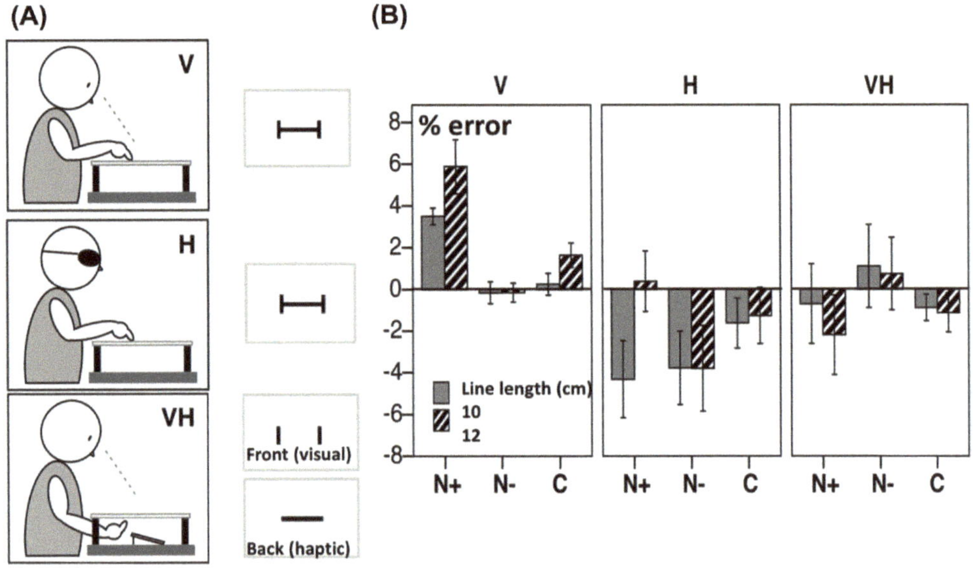

FIGURE 19.8 Differences related to sensory modality in line bisection tasks. With visual presentation, right brain-damaged patients with left unilateral spatial neglect (USN) (N+) make a disproportionate rightward error, indicating underestimation of the left side of the line. With haptic and visuohaptic presentations, N+ patients' performance is within the normal range. In the multisensory condition the haptic signal may prevent USN occurring, even when a visual input is also provided. (A) Stimuli and apparatus for setting the subjective midpoint of horizontal line stimuli with vertical ends, under visual (V), haptic (H), and visuohaptic (VH) presentations. In the visuohaptic condition, the ends were glued on the front of the board, and the horizontal shaft to be bisected on the back, in the corresponding positions. (B) Mean percent error (±SEM) in line bisection, for two different line lengths (10, 12 cm), in three groups (N+, USN patients; N-, patients without USN; C, control neurologically unimpaired participants), and three presentation modalities (visual, V; haptic, H, visuohaptic, VH). Negative/positive score: leftward/rightward error. *From Mancini F, Bricolo E, Mattioli FC, Vallar G. Visuo-haptic interactions in unilateral spatial neglect: the cross modal Judd illusion. Front Psychol. 2011;2:341.*

Another visuotactile effect investigated in USN is the rubber hand illusion (RHI)[100]: synchronous touches, applied to a visible rubber hand and to the participant's real hand, positioned under the rubber hand, and hidden from view, produce the sensation that the felt touches originate from the rubber hand, and a feeling of ownership of the artificial hand develops (see chapter by Ehrsson, this volume). Two right brain-damaged patients with left sensorimotor deficits and left visuospatial and personal USN showed an RHI in the left hand, as control participants did, although this hand was unattended, due to left USN.[101]

Specific multisensory-based procedures for USN rehabilitation have been not developed so far. This lack is likely due to the scarce research conducted to precisely assess the multisensory abilities of USN patients in clinical practice, beyond the established unisensory impairments that may affect different sensory modalities separately.

However, many studies have shown beneficial crossmodal effects on modality-specific unisensory symptoms of USN. Here, we use the term "crossmodal" to refer to situations in which the presentation of a stimulus in one sensory modality influences the perception of, or the ability to respond to, stimuli presented in another sensory modality[91]; this is different from the effect of multisensory integration, which involves quantitative (enhancement), and sometimes qualitative (illusions), modifications of unisensory processing. A paradigmatic example is prism adaptation (PA), an effective and widely used technique for the rehabilitation of USN. PA is based on the use of goggles fitted with prismatic lenses, which induce a lateral displacement of the visual field.[102] PA to a lateral displacement of the visual field is known to directionally bias sensory-motor correspondences. In a standard PA paradigm for left USN,[103] patients wear goggles with lenses consisting in optical prisms that displace the visual field rightward (ipsilesionally). While wearing the goggles, patients are required to make repeated visuomotor ballistic pointing movements toward a visual target: due to the prism-induced displacement of the visual scene, patients make initial rightward pointing errors, missing the target. In the early pointing trials, both patients and healthy participants performing this task detect the visuomotor error and quickly correct it strategically, in a conscious manner, with a rapid reduction in the pointing errors made (recalibration phase or error correction). In the subsequent pointing trials, an automatic (unconscious) remapping of the visuoproprioceptive coordinates occurs (realignment phase or true adaptation). These two components of PA rely on PPC-cerebellar networks, with a more relevant role of the PPC for recalibration, and of the cerebellum for realignment. After removal of the rightward-displacing prisms, participants make leftward pointing errors (after-effects), in the direction opposite to the prism-induced visual displacement to which they had adapted. These after-effects are characteristic of longer-term plastic changes in visuomotor mapping. Thanks to such directional effects, PA has been successfully used in the treatment of the spatial lateralized deficits of USN.[103] A number of clinical symptoms of USN can be temporarily reduced after a session of PA, while repeated sessions of PA (typically, 10 daily sessions) may induce long-lasting improvements,[104,105] even with treatment at the patients' home.[106] Of interest, although vision is the sense primarily being manipulated through prism exposure, PA not only improves the visuospatial manifestations of USN but also crossmodally ameliorates nonvisual disorders, such as unisensory auditory[107] and tactile[108] extinction. This evidence suggests that the therapeutic effects of PA expand to unexposed sensory systems.[109] How the recalibration of visuomotor transformations induced by prism exposure generates such crossmodal effects remains an open issue, but it likely involves bottom-up multisensory representations driving a spatial remapping.

An option to potentiate the clinical effects of PA treatment on USN symptoms is the use of different types of sensory stimulation for the adaptation phase, besides the time-honored PA to visual targets. For instance, using auditory targets could be more effective for improving visual deficits in patients with no sign of auditory neglect.[110] The rationale here is to facilitate PA crossmodally, through an intact sensory system, which may be used as a compensatory route in patients with visual USN. Alternatively, multisensory, audiovisual, targets could be used with the aim of boosting multisensory enhancements of spatial localization and sensorimotor pointing activity, a strategy that may likely augment the clinical benefits of PA. Preliminary data show that healthy individuals adapt more quickly and are more precise in pointing to audiovisual targets, as compared with unisensory ones.[110] This evidence also suggests that the use of multisensory stimuli during PA could allow to reduce the duration of the treatment sessions, resulting in a more sustainable and less tedious training that may increase the patient's compliance to it.

It is worth mentioning that crossmodal improvements in the manifold manifestations of USN can be induced also by other sensory side-specific stimuli. Visual optokinetic stimulation, with movement of the visual stimuli toward the contralesional, neglected, side of space, reduces proprioceptive deficits.[111] Proprioceptive stimulation, evoked by active and passive movements of the contralesional upper limb, improves visual USN.[112,113] Vestibular stimulation ameliorates visual[114] and somatosensory impairments.[115] These crossmodal effects, while short-lasting in the abovementioned studies, support the view that a supramodal (in the sense of modality-independent) modulation of defective spatial processing of inputs from individual sensory modalities takes place. To further support the existence of crossmodal effects, vestibular caloric[116] and optokinetic stimulation temporarily improve even contralesional left motor deficits of brain-damaged patients with left USN.[117]

Novel perspectives for the rehabilitation of the USN syndrome are offered by paradigms based on illusions.[118] Patients with left USN are susceptible to perceptual crossmodal illusions arising from intersensory conflicts, as healthy participants are, in the face of the unisensory disturbances that characterize the syndrome.[47,118] In this scenario, "fooling" the brain with multisensory illusions may offer a strategy to boost and wake up spared multisensory capabilities that might strengthen visuospatial attention and awareness. Some preliminary evidence supports this view. In a right brain-damaged patient with left USN, the RHI (with the rubber hand located in the contralesional left hemispace) was shown to induce an immediate, temporary, amelioration of visuospatial USN, as assessed with target cancellation and midline pointing tasks.[119] The improvement of USN could have been caused by RHI-induced changes in the patient's egocentric reference frames, by a cueing of spatial attention driven by the RHI, or by a combination of both mechanisms. Another promising tool is the Mirror Box Illusion (MBI, see chapter by Altschuler et al. in this volume). In a recent clinical trial,[120] a group of stroke patients, with USN and with thalamic and parietal lobe lesions, received an MBI-based treatment for 1−2 hours a day, for 4 weeks (5 days/week). During the treatment, patients had to perform flexion and extension movements of the unaffected hand while looking at its reflection in the mirror. As compared to a control treatment (comprising similar motor exercises for the same time period, but using the nonreflecting side of the mirror), the MBI treatment ameliorates visuospatial deficits (assessed by the star cancellation and the line bisection tests, and a picture identification task), with clinical benefits persisting in the long term up to 6 months. Moreover, the MBI treatment was shown to facilitate the recovery of functional independence. From a multisensory perspective, the MBI may augment attentional demands for the integration of conflicting visual and proprioceptive signals from the contralesional space, in turn facilitating leftward visuospatial orienting and awareness of the affected limb.[118] In line with this proposal, neuroimaging evidence indicates that the MBI treatment increases neural activity in multisensory areas (among which is the precuneus) associated with self-awareness and spatial attention.[121]

In conclusion, based on interpretations of the USN syndrome as a supramodal attentional and representational imbalance,[47] asymmetrical multisensory stimulation (consisting of increasing multisensory inputs from the affected side, perhaps along with a decrease from the intact side) may promote an automatic, multisensory-driven recalibration of sensorimotor mechanisms controlling space and body representation and awareness, which may be more effective than the most common unisensory (crossmodal) strategies applied to date.

Perspectives for multisensory rehabilitation research

The study of multisensory integration offers novel and valuable therapeutic options for brain-damaged patients with VFDs and USN. Growing evidence shows the efficacy of the multisensory compensatory therapy for homonymous hemianopia, although randomized clinical trials are needed to further confirm its clinical validity. The investigation of the neural underpinnings of multisensory-induced benefits on VFDs is now mandatory, not only for a mechanistic understanding but also for its clinical relevance. An in-depth knowledge of the mechanisms at the basis of this therapy will allow us to identify clinical and anatomical predictors of its usefulness in stroke patients, in turn optimizing its therapeutic success. In addition, new multisensory therapeutic strategies should be explored, in light of the considerable potential for vision restoration and recovery in adulthood. The increasing focus on the use/experience-dependent plastic capability of repair of the human visual system,[122] along with the fact that substantial multisensory convergence and integration occurs even within the striate cortex,[123,124] may guide the development of paradigms aimed at a sort of "multisensory boosting" of plasticity in the injured occipital cortex for vision restitution.

In USN, the research has focused on the unisensory impairments affecting different sensory modalities, rather than on multisensory integration. However, some evidence suggests that multisensory perception may be preserved or even disproportionately increased, in USN. Further research is needed to explore how to take advantage of these phenomena in rehabilitation settings, as done so far so for VFDs.

References

1. Walsh T, ed. *Visual Fields: Examination and Interpretation*. 3rd ed. Oxford University Press; 2010.
2. Goodwin D. Homonymous hemianopia: challenges and solutions. *Clin Ophthalmol*. 2014;8:1919–1927.
3. Leff A. A historical review of the representation of the visual field in primary visual cortex with special reference to the neural mechanisms underlying macular sparing. *Brain Lang*. 2004;88(3):268–278.
4. Rowe FJ, Wright D, Brand D, et al. A prospective profile of visual field loss following stroke: prevalence, type, rehabilitation, and outcome. *BioMed Res Int*. 2013;2013:719096.
5. Zhang X, Kedar S, Lynn MJ, Newman NJ, Biousse V. Homonymous hemianopias: clinical-anatomic correlations in 904 cases. *Neurology*. 2006;66(6):906–910.
6. Zhang X, Kedar S, Lynn MJ, Newman NJ, Biousse V. Natural history of homonymous hemianopia. *Neurology*. 2006;66(6):901–905.
7. Zihl J, von Cramon D. Visual field recovery from scotoma in patients with postgeniculate damage. A review of 55 cases. *Brain*. 1985;108(Pt 2):335–365.
8. DeMyer W. *Technique of the Neurological Examination*. 5th ed. New York: McGraw-Hill Medical; 2013.
9. Laishram M, Srikanth K, Rajalakshmi AR, Nagarajan S, Ezhumalai G. Microperimetry – a new tool for assessing retinal sensitivity in macular diseases. *J Clin Diagn Res*. 2017;11(7):NC08–NC11.
10. Zihl J, ed. *Rehabilitation of Visual Disorders after Brain Injury*. 2nd ed. Psychology Press; 2010.
11. Schuett S, Heywood CA, Kentridge RW, Zihl J. The significance of visual information processing in reading: insights from hemianopic dyslexia. *Neuropsychologia*. 2008;46(10):2445–2462.
12. Zihl J. Oculomotor scanning performance in subjects with homonymous visual field disorders. *Vis Impair Res*. 1999;1:23–31.
13. Machner B, Sprenger A, Sander T, et al. Visual search disorders in acute and chronic homonymous hemianopia: lesion effects and adaptive strategies. *Ann N Y Acad Sci*. 2009;1164:419–426.
14. Tant ML, Cornelissen FW, Kooijman AC, Brouwer WH. Hemianopic visual field defects elicit hemianopic scanning. *Vision Res*. 2002;42(10):1339–1348.

15. Cowey A, Stoerig P. The neurobiology of blindsight. *Trends Neurosci.* 1991;14(4):140–145.
16. Sabel BA, Henrich-Noack P, Fedorov A, Gall C. Vision restoration after brain and retina damage: the residual vision activation theory. *Prog Brain Res.* 2011;192:199–262.
17. Bolognini N, Convento S, Rossetti A, Merabet LB. Multisensory processing after a brain damage: clues on post-injury crossmodal plasticity from neuropsychology. *Neurosci Biobehav Rev.* 2013;37(3):269–278.
18. Krauzlis RJ, Liston D, Carello CD. Target selection and the superior colliculus: goals, choices and hypotheses. *Vision Res.* 2004;44(12):1445–1451.
19. Stein BE, Meredith MA. *The Merging of the Senses.* Cambridge, MA, USA: MIT Press; 1993.
20. Stein BE. Neural mechanisms for synthesizing sensory information and producing adaptive behaviors. *Exp Brain Res.* 1998;123(1–2):124–135.
21. Frassinetti F, Bolognini N, Bottari D, Bonora A, Ladavas E. Audiovisual integration in patients with visual deficit. *J Cogn Neurosci.* 2005;17(9):1442–1452.
22. Lewald J, Tegenthoff M, Peters S, Hausmann M. Passive auditory stimulation improves vision in hemianopia. *PLoS One.* 2012;7(5):e31603.
23. Leo F, Bolognini N, Passamonti C, Stein BE, Ladavas E. Cross-modal localization in hemianopia: new insights on multisensory integration. *Brain.* 2008;131(Pt 3):855–865.
24. Passamonti C, Frissen I, Ladavas E. Visual recalibration of auditory spatial perception: two separate neural circuits for perceptual learning. *Eur J Neurosci.* 2009;30(6):1141–1150.
25. Maravita A, Bolognini N, Bricolo E, Marzi CA, Savazzi S. Is audiovisual integration subserved by the superior colliculus in humans? *Neuroreport.* 2008;19(3):271–275.
26. Bolognini N, Convento S, Casati C, Mancini F, Brighina F, Vallar G. Multisensory integration in hemianopia and unilateral spatial neglect: evidence from the sound induced flash illusion. *Neuropsychologia.* 2016;87:134–143.
27. Schendel K, Robertson LC. Reaching out to see: arm position can attenuate human visual loss. *J Cogn Neurosci.* 2004;16(6):935–943.
28. Graziano MS, Gross CG. A bimodal map of space: somatosensory receptive fields in the macaque putamen with corresponding visual receptive fields. *Exp Brain Res.* 1993;97(1):96–109.
29. Smith DT, Lane AR, Schenk T. Arm position does not attenuate visual loss in patients with homonymous field deficits. *Neuropsychologia.* 2008;46(9):2320–2325.
30. Brown LE, Kroliczak G, Demonet JF, Goodale MA. A hand in blindsight: hand placement near target improves size perception in the blind visual field. *Neuropsychologia.* 2008;46(3):786–802.
31. Bolognini N, Ladavas E, Farne A. Spatial perspective and coordinate systems in autoscopy: a case report of a "fantome de profil" in occipital brain damage. *J Cogn Neurosci.* 2011;23(7):1741–1751.
32. Kerkhoff G, Munssinger U, Meier EK. Neurovisual rehabilitation in cerebral blindness. *Arch Neurol.* 1994;51(5):474–481.
33. Reinhard J, Schreiber A, Schiefer U. Does visual restitution training change absolute homonymous visual field defects? A fundus controlled study. *Br J Ophthalmol.* 2005;89(1):30–35.
34. Huxlin KR, Martin T, Kelly K, et al. Perceptual relearning of complex visual motion after V1 damage in humans. *J Neurosci.* 2009;29(13):3981–3991.
35. Sahraie A, Trevethan CT, MacLeod MJ, Murray AD, Olson JA, Weiskrantz L. Increased sensitivity after repeated stimulation of residual spatial channels in blindsight. *Proc Natl Acad Sci USA.* 2006;103(40):14971–14976.
36. Pollock A, Hazelton C, Henderson CA, et al. Interventions for visual field defects in patients with stroke. *Cochrane Database Syst Rev.* 2011;10:CD008388.
37. Bolognini N, Rasi F, Coccia M, Ladavas E. Visual search improvement in hemianopic patients after audio-visual stimulation. *Brain.* 2005;128(Pt 12):2830–2842.
38. Dundon NM, Ladavas E, Maier ME, Bertini C. Multisensory stimulation in hemianopic patients boosts orienting responses to the hemianopic field and reduces attentional resources to the intact field. *Restor Neurol Neurosci.* 2015;33(4):405–419.
39. Grasso PA, Ladavas E, Bertini C. Compensatory recovery after multisensory stimulation in hemianopic patients: behavioral and neurophysiological components. *Front Syst Neurosci.* 2016;10:45.
40. Keller I, Lefin-Rank G. Improvement of visual search after audiovisual exploration training in hemianopic patients. *Neurorehabilitation Neural Repair.* 2010;24(7):666–673.

41. Passamonti C, Bertini C, Ladavas E. Audio-visual stimulation improves oculomotor patterns in patients with hemianopia. *Neuropsychologia*. 2009;47(2):546–555.
42. Tinelli F, Purpura G, Cioni G. Audio-visual stimulation improves visual search abilities in hemianopia due to childhood acquired brain lesions. *Multisensory Res*. 2015;28(1–2):153–171.
43. Jiang H, Stein BE, McHaffie JG. Multisensory training reverses midbrain lesion-induced changes and ameliorates haemianopia. *Nat Commun*. 2015;6:7263.
44. Driver J, Noesselt T. Multisensory interplay reveals crossmodal influences on 'sensory-specific' brain regions, neural responses, and judgments. *Neuron*. 2008;57(1):11–23.
45. Bolognini N, Fregni F, Casati C, Olgiati E, Vallar G. Brain polarization of parietal cortex augments training-induced improvement of visual exploratory and attentional skills. *Brain Res*. 2010;1349:76–89.
46. Heilman KM, Watson RT, Valenstein E. Neglect and related disorders. In: Heilman KM, Valenstein E, eds. *Clinical Neuropsychology*. 4th ed. New York: Oxford University Press; 2003:296–346.
47. Vallar G, Bolognini N. Unilateral spatial neglect. In: Nobre AC, Kastner S, eds. *The Oxford Handbook of Attention*. Oxford: Oxford University Press; 2014:972–1027.
48. Hécaen H, Albert ML. *Human Neuropsychology*. New York: John Wiley; 1978.
49. Kooistra CA, Heilman KM. Hemispatial visual inattention masquerading as hemianopia. *Neurology*. 1989;39(8):1125–1127.
50. Nadeau SE, Heilman KM. Gaze-dependent hemianopia without hemispatial neglect. *Neurology*. 1991;41(8):1244–1250.
51. Vallar G, Sandroni P, Rusconi ML, Barbieri S. Hemianopia, hemianesthesia, and spatial neglect: a study with evoked potentials. *Neurology*. 1991;41(12):1918–1922.
52. Viggiano MP, Spinelli D, Mecacci L. Pattern reversal visual evoked potentials in patients with hemineglect syndrome. *Brain Cogn*. 1995;27(1):17–35.
53. Di Russo F, Aprile T, Spitoni G, Spinelli D. Impaired visual processing of contralesional stimuli in neglect patients: a visual-evoked potential study. *Brain*. 2008;131(Pt 3):842–854.
54. Lhermitte F, Turell E, LeBrigand D, Chain F. Unilateral visual neglect and wave P 300. A study of nine cases with unilateral lesions of the parietal lobes. *Arch Neurol*. 1985;42(6):567–573.
55. Verleger R, Heide W, Butt C, Wascher E, Kompf D. On-line brain potential correlates of right parietal patients' attentional deficit. *Electroencephalogr Clin Neurophysiol*. 1996;99(5):444–457.
56. Gainotti G. The role of automatic orienting of attention towards ipsilesional stimuli in non-visual (tactile and auditory) neglect: a critical review. *Cortex*. 2010;46(2):150–160.
57. Cubelli R, Nichelli P, Bonito V, De Tanti A, Inzaghi MG. Different patterns of dissociation in unilateral spatial neglect. *Brain Cogn*. 1991;15(2):139–159.
58. Pavani F, Husain M, Ladavas E, Driver J. Auditory deficits in visuospatial neglect patients. *Cortex*. 2004;40(2):347–365.
59. Vallar G, Rusconi ML, Geminiani G, Berti A, Cappa SF. Visual and nonvisual neglect after unilateral brain lesions: modulation by visual input. *Int J Neurosci*. 1991;61(3–4):229–239.
60. De Renzi E, Gentilini M, Pattacini F. Auditory extinction following hemisphere damage. *Neuropsychologia*. 1984;22(6):733–744.
61. Deouell LY, Soroker N. What is extinguished in auditory extinction? *Neuroreport*. 2000;11(13):3059–3062.
62. Grandjean D, Sander D, Lucas N, Scherer KR, Vuilleumier P. Effects of emotional prosody on auditory extinction for voices in patients with spatial neglect. *Neuropsychologia*. 2008;46(2):487–496.
63. de Haan B, Karnath HO, Driver J. Mechanisms and anatomy of unilateral extinction after brain injury. *Neuropsychologia*. 2012;50(6):1045–1053.
64. Altman JA, Balonov LJ, Deglin VL. Effects of unilateral disorder of the brain hemisphere function in man on directional hearing. *Neuropsychologia*. 1979;17(3–4):295–301.
65. Tanaka H, Hachisuka K, Ogata H. Sound lateralisation in patients with left or right cerebral hemispheric lesions: relation with unilateral visuospatial neglect. *J Neurol Neurosurg Psychiatry*. 1999;67(4):481–486.
66. Bisiach E, Cornacchia L, Sterzi R, Vallar G. Disorders of perceived auditory lateralization after lesions of the right hemisphere. *Brain*. 1984;107(Pt 1):37–52.
67. Bellmann A, Meuli R, Clarke S. Two types of auditory neglect. *Brain*. 2001;124(Pt 4):676–687.
68. Vallar G, Guariglia C, Nico D, Bisiach E. Spatial hemineglect in back space. *Brain*. 1995;118(Pt 2):467–472.
69. De Renzi E, Gentilini M, Barbieri C. Auditory neglect. *J Neurol Neurosurg Psychiatry*. 1989;52(5):613–617.

70. Pavani F, Meneghello F, Ladavas E. Deficit of auditory space perception in patients with visuospatial neglect. *Neuropsychologia*. 2001;39(13):1401–1409.
71. Vallar G, Bottini G, Sterzi R, Passerini D, Rusconi ML. Hemianesthesia, sensory neglect, and defective access to conscious experience. *Neurology*. 1991;41(5):650–652.
72. Aglioti S, Smania N, Peru A. Frames of reference for mapping tactile stimuli in brain-damaged patients. *J Cogn Neurosci*. 1999;11(1):67–79.
73. Smania N, Aglioti S. Sensory and spatial components of somaesthetic deficits following right brain damage. *Neurology*. 1995;45(9):1725–1730.
74. Angelelli P, De Luca M, Spinelli D. Early visual processing in neglect patients: a study with steady-state VEPs. *Neuropsychologia*. 1996;34(12):1151–1157.
75. Tarkka IM, Luukkainen-Markkula R, Pitkanen K, Hamalainen H. Alterations in visual and auditory processing in hemispatial neglect: an evoked potential follow-up study. *Int J Psychophysiol*. 2011;79(2):272–279.
76. Deouell LY. Pre-requisites for conscious awareness: clues from electrophysiological and behavioral studies of unilateral neglect patients. *Conscious Cognit*. 2002;11(4):546–567.
77. Deouell LY, Bentin S, Soroker N. Electrophysiological evidence for an early(pre-attentive) information processing deficit in patients with right hemisphere damage and unilateral neglect. *Brain*. 2000;123(Pt 2):353–365.
78. Vallar G, Calzolari E. Unilateral spatial neglect after posterior parietal damage. *Handb Clin Neurol*. 2018;151:287–312.
79. Vallar G, Coslett HB. The parietal lobe. In: *Handbook of Clinical Neurology*. Amsterdam: Elsevier; 2018.
80. Hecaen H, Penfield W, Bertrand C, Malmo R. The syndrome of apractognosia due to lesions of the minor cerebral hemisphere. *AMA Arch Neurol Psychiatry*. 1956;75(4):400–434.
81. Heilman KM, Valenstein E. Auditory neglect in man. *Arch Neurol*. 1972;26(1):32–35.
82. Vallar G, Perani D. The anatomy of unilateral neglect after right-hemisphere stroke lesions. A clinical/CT-scan correlation study in man. *Neuropsychologia*. 1986;24(5):609–622.
83. Karnath HO, Rorden C. The anatomy of spatial neglect. *Neuropsychologia*. 2012;50(6):1010–1017.
84. Vuilleumier P. Mapping the functional neuroanatomy of spatial neglect and human parietal lobe functions: progress and challenges. *Ann N Y Acad Sci*. 2013;1296:50–74.
85. Jacobs S, Brozzoli C, Farne A. Neglect: a multisensory deficit? *Neuropsychologia*. 2012;50(6):1029–1044.
86. Tang X, Wu J, Shen Y. The interactions of multisensory integration with endogenous and exogenous attention. *Neurosci Biobehav Rev*. 2016;61:208–224.
87. Macaluso E. Orienting of spatial attention and the interplay between the senses. *Cortex*. 2010;46(3):282–297.
88. Macaluso E, Maravita A. The representation of space near the body through touch and vision. *Neuropsychologia*. 2010;48(3):782–795.
89. Talsma D, Senkowski D, Soto-Faraco S, Woldorff MG. The multifaceted interplay between attention and multisensory integration. *Trends Cogn Sci*. 2010;14(9):400–410.
90. Huang RS, Sereno MI. Multisensory and sensorimotor maps. *Handb Clin Neurol*. 2018;151:141–161.
91. McGurk H, MacDonald J. Hearing lips and seeing voices. *Nature*. 1976;264(5588):746–748.
92. Howard IP, Templeton WB. *Human Spatial Orientation*. London: Wiley; 1966.
93. Calamaro N, Soroker N, Myslobodsky MS. False recovery from auditory hemineglect produced by source misattribution of auditory stimuli (the ventriloquist effect). *Restor Neurol Neurosci*. 1995;7(3):151–156.
94. Soroker N, Calamaro N, Myslobodsky M. "McGurk illusion" to bilateral administration of sensory stimuli in patients with hemispatial neglect. *Neuropsychologia*. 1995;33(4):461–470.
95. Soroker N, Calamaro N, Myslobodsky MS. Ventriloquist effect reinstates responsiveness to auditory stimuli in the 'ignored' space in patients with hemispatial neglect. *J Clin Exp Neuropsychol*. 1995;17(2):243–255.
96. Shams L, Kamitani Y. Illusions. What you see is what you hear. *Nature*. 2000;408:788.
97. Frassinetti F, Pavani F, Ladavas E. Acoustical vision of neglected stimuli: interaction among spatially converging audiovisual inputs in neglect patients. *J Cogn Neurosci*. 2002;14(1):62–69.
98. Doricchi F, Galati G, DeLuca L, Nico D, D'Olimpio F. Horizontal space misrepresentation in unilateral brain damage. I. Visual and proprioceptive-motor influences in left unilateral neglect. *Neuropsychologia*. 2002;40(8):1107–1117.
99. Mancini F, Bricolo E, Mattioli FC, Vallar G. Visuo-haptic interactions in unilateral spatial neglect: the cross modal Judd illusion. *Front Psychol*. 2011;2:341.
100. Botvinick M, Cohen J. Rubber hands 'feel' touch that eyes see. *Nature*. 1998;391(6669):756.

101. Bolognini N, Ronchi R, Casati C, Fortis P, Vallar G. Multisensory remission of somatoparaphrenic delusion: my hand is back!. *Neurol Clin Pract*. 2014;4(3):216–225.
102. Redding GM, Rossetti Y, Wallace B. Applications of prism adaptation: a tutorial in theory and method. *Neurosci Biobehav Rev*. 2005;29(3):431–444.
103. Redding GM, Wallace B. Prism adaptation and unilateral neglect: review and analysis. *Neuropsychologia*. 2006;44(1):1–20.
104. Fortis P, Maravita A, Gallucci M, et al. Rehabilitating patients with left spatial neglect by prism exposure during a visuomotor activity. *Neuropsychology*. 2010;24(6):681–697.
105. Frassinetti F, Angeli V, Meneghello F, Avanzi S, Ladavas E. Long-lasting amelioration of visuospatial neglect by prism adaptation. *Brain*. 2002;125(Pt 3):608–623.
106. Fortis P, Ronchi R, Velardo V, et al. A home-based prism adaptation training for neglect patients. *Cortex*. 2018.
107. Jacquin-Courtois S, Rode G, Pavani F, et al. Effect of prism adaptation on left dichotic listening deficit in neglect patients: glasses to hear better? *Brain*. 2010;133(Pt 3):895–908.
108. Maravita A, McNeil J, Malhotra P, Greenwood R, Husain M, Driver J. Prism adaptation can improve contralesional tactile perception in neglect. *Neurology*. 2003;60(11):1829–1831.
109. Jacquin-Courtois S, O'Shea J, Luaute J, et al. Rehabilitation of spatial neglect by prism adaptation: a peculiar expansion of sensorimotor after-effects to spatial cognition. *Neurosci Biobehav Rev*. 2013;37(4):594–609.
110. Calzolari E, Albini F, Bolognini N, Vallar G. Multisensory and modality-specific influences on adaptation to optical prisms. *Front Hum Neurosci*. 2017;11:568.
111. Vallar G, Antonucci G, Guariglia C, Pizzamiglio L. Deficits of position sense, unilateral neglect and optokinetic stimulation. *Neuropsychologia*. 1993;31(11):1191–1200.
112. Frassinetti F, Rossi M, Ladavas E. Passive limb movements improve visual neglect. *Neuropsychologia*. 2001;39(7):725–733.
113. Robertson IH, North N. Active and passive activation of left limbs: influence on visual and sensory neglect. *Neuropsychologia*. 1993;31(3):293–300.
114. Rubens AB. Caloric stimulation and unilateral visual neglect. *Neurology*. 1985;35(7):1019–1024.
115. Vallar G, Bottini G, Rusconi ML, Sterzi R. Exploring somatosensory hemineglect by vestibular stimulation. *Brain*. 1993;116(Pt 1):71–86.
116. Rode G, Charles N, Perenin MT, Vighetto A, Trillet M, Aimard G. Partial remission of hemiplegia and somatoparaphrenia through vestibular stimulation in a case of unilateral neglect. *Cortex*. 1992;28(2):203–208.
117. Vallar G, Guariglia C, Nico D, Pizzamiglio L. Motor deficits and optokinetic stimulation in patients with left hemineglect. *Neurology*. 1997;49(5):1364–1370.
118. Bolognini N, Russo C, Vallar G. Crossmodal illusions in neurorehabilitation. *Front Behav Neurosci*. 2015;9:212.
119. Kitadono K, Humphreys GW. Short-term effects of the 'rubber hand' illusion on aspects of visual neglect. *Neurocase*. 2007;13(4):260–271.
120. Pandian JD, Arora R, Kaur P, Sharma D, Vishwambaran DK, Arima H. Mirror therapy in unilateral neglect after stroke (MUST trial): a randomized controlled trial. *Neurology*. 2014;83(11):1012–1017.
121. Michielsen ME, Smits M, Ribbers GM, et al. The neuronal correlates of mirror therapy: an fMRI study on mirror induced visual illusions in patients with stroke. *J Neurol Neurosurg Psychiatry*. 2011;82(4):393–398.
122. Li W. Perceptual learning: use-dependent cortical plasticity. *Annu Rev Vis Sci*. 2016;2:109–130.
123. Convento S, Vallar G, Galantini C, Bolognini N. Neuromodulation of early multisensory interactions in the visual cortex. *J Cogn Neurosci*. 2013;25(5):685–696.
124. Murray MM, Thelen A, Thut G, Romei V, Martuzzi R, Matusz PJ. The multisensory function of the human primary visual cortex. *Neuropsychologia*. 2016;83:161–169.

CHAPTER

20

Mirror therapy

Christian Dohle[1,2], *Eric Altschuler*[3], *Vilayanur S. Ramachandran*[4]

[1]MEDIAN Klinik Berlin-Kladow, Berlin, Germany; [2]Center for Stroke Research Berlin, Charité – University Medicine Berlin, Berlin, Germany; [3]Department of Physical Medicine and Rehabilitation, Metropolitan Hospital, New York, NY, United States; [4]Department of Psychology and Center for Brain and Cognition, University of California, San Diego, CA, United States

Introduction

During mirror therapy (MT), a mirror is placed in the patient's midsagittal plane in such a way that the mirror image of the unaffected limb looks as if it were optically superimposed on the affected one (Fig. 20.1). MT was first observed to relieve painful spasms and phantom pain in arm amputees.[1] Soon afterward, its potential benefits for treatment of stroke-induced hemiparesis was postulated and demonstrated as well.[2,3] Over the last few years, a constantly rising number of high-quality, double-blind placebo trials have been performed and published, providing good evidence for beneficial effects of MT on several pathological conditions, primarily involving stroke-induced hemiparesis and chronic pain syndromes.[4,5] However, in spite of this substantial evidence for the effectiveness of MT, the underlying neurophysiological mechanisms are still not entirely clear. In this article, we present a brief summary of the actual state of current findings, along with a spectrum of opinions on how MT works.

Apart from its clinical significance, these findings demand a radical revision of the way in which we theoretically conceptualize brain function and plasticity. Instead of being composed of hardwired, autonomous, specialized modules, the different modules are in a state of dynamic equilibrium with each other and the environment, perpetually changing their connections in response to environmental challenges. Brain disease often results not from "losing a module = losing a function" but rather from temporary shifts in this equilibrium and can sometimes be ameliorated by relatively simple procedures.

Multisensory Perception
https://doi.org/10.1016/B978-0-12-812492-5.00020-6

FIGURE 20.1 Principle of mirror therapy.

Neural mechanisms of mirror visual feedback

Ramachandran first introduced the basic concept of using mirror visual feedback (MVF)—or indeed, any kind of visual feedback (e.g., virtual reality), in the treatment of neurological ailments—particularly pain and paralysis. He and his colleagues went on to demonstrate its efficacy in a small sample of patients.[1,2] In addition to independent confirmation of clinical utility by several groups, there followed a number of high-quality imaging studies which helped us to understand the effect of movement mirroring on brain activation patterns. It has long been established that movement observation and movement imagery share a significant proportion of neural circuits with actual movement execution.[6] The question of lateralization, however, i.e., the differences in brain activation between right-hand and left-hand movements, has historically received considerably less attention. Fortunately, this has changed since the discovery of the powerful effects of MVF.

Deconinck and coworkers undertook a systematic review of brain imaging studies of MVF.[7] They identified a total of 33 studies, employing a variety of different techniques, including functional magnetic resonance imaging, positron emission tomography, functional near-infrared spectroscopy, and transcranial magnetic stimulation (TMS). At first glance, these studies provided no conclusive results regarding the brain structures involved. However, after comparing different experimental setups carefully, the picture becomes clearer. The primary motor cortices (M1) and the precunei (PC) of both hemispheres (area V6/V6A) seem to play the decisive role. Although both M1 and PC showed bilateral activation, there were still distinct patterns of lateralization. M1 activation was shown to be predominantly contralateral to the moving limb, irrespective of the limb being observed, while activation of the PC was predominantly contralateral to the observed limb, irrespective of which limb was moving.[8–10] This activation pattern only occurs during self-induced movements, not during movement observation.[11] Consequently, if the visual feedback is inverted as during MT, this leads to an additional activation of the precuneus contralateral to the visually perceived limb.

Some authors also reported direct activation of the primary motor cortex induced by the MVF which is important and challenges the dogma of clear anatomical segregation of "motor versus sensory"—one of the axioms of neuroscience. However, in these studies, the MVF was generally implemented using a real mirror, actually comparing the image of two versus one moving limb(s).[12] In addition, some studies deliberately restricted their analysis to the primary motor and sensory cortices of either hemisphere.[13–15] However, there is some evidence that MVF works differently on the motor and sensory cortices.[16] Thus, an analysis which combines activity from both areas has to be interpreted with caution. Increased activation levels of the primary motor cortex as measured with TMS occurred only when compared to conditions without equivalent visual feedback.[17–19]

When examining the mirror transformation process, however, it has to be considered that the mirrored image of the arm's movement actually combines two transformations simultaneously. First, the body laterality is changed (e.g., the right arm appears as a left arm and vice versa). Second, an additional coordinate transformation occurs: movements to the right appear as movements to the left and vice versa. Functional imaging studies hint that these transformations are accomplished by different neural structures. Whereas the PCs of either hemisphere convey the limb laterality (right arm/left arm), an additional activation in the insula only occurs when movement mirroring is demanded.[20] This is reflected in the behavioral data: the visually perceived limb laterality determines the shape of the movement performance—independent of its implementation. Contrary, the need for the additional mirror transformation process reduces movement accuracy.[20] This might be caused by the fact that execution of goal-directed movement under mirrored visual feedback requires the attribution of formerly coupled coordinate parameters (movement direction and distance) to different modalities (visual and proprioceptive).[21]

Regarding clinical application, there are four subgroups of stroke patients for whom MT provides a benefit: patients with chronic pain syndromes, patients with hemiparesis after stroke, patients with somatosensory deficits, and patients with visual hemineglect. Behavioral data confirm the effects on reaching tasks and hemiakinesia. These effects are detailed below.

Effects of mirror therapy on pain syndromes

The first and still one of the most popular applications of MT is the treatment of different pain syndromes, especially phantom limb pain (PLP) and complex regional pain syndrome (CRPS).[22] Even as there is a mounting body of evidence supporting the use of MT for pain syndromes,[23] it is not as extensive as one might like. A recent systematic review of MT and other movement representation techniques such as movement imagery for different pain syndromes[4] only identified a total of six randomized controlled studies on MT by itself for treatment of PLP,[24–26] CRPS,[27] or CRPS after stroke.[28,29] Two of these six studies are only available as published doctoral theses in German.[26,27]In addition to these studies, MT was evaluated as part of the Graded Motor Imagery program, consisting of 2-week periods of laterality judgment, movement imagery, and MT each.[30–32] Still, there is the need to explore the effects of MT in greater detail through further large-scale trials with sufficient numbers of patients.

The neural mechanisms underlying the benefits of MT for pain syndromes are unclear. For PLP, it has been shown that the magnitude of pain is associated with the amount of cortical reorganization in the primary sensory cortex.[33] This might involve accidental cross-innervation of high-threshold pain neurons by low-threshold touch neurons, leading a form of "touch-pain synesthesia" and allodynia. One hypothesis is that the MVF of a "normal" movement can lead to a reduction of this cortical reorganization. However, further evidence is required to establish this claim.

In summary, in addition to striking anecdotal clinical results, we also need double-blind, placebo-controlled trials to further establish the effectiveness of these treatments on different clinical groups. The only exception is CRPS after stroke, whose efficacy is far more clear-cut, as discussed in more detail below. New studies should also include potential predictors to identify those patients that are likely to benefit in comparison with those who are not.

Effects of mirror therapy on hemiparesis after stroke

The evidence for the application of MT to improve motor deficits after stroke has virtually exploded over the last few years. A systematic review in 2012 (based on a literature search in June 2011) identified 14 studies with 567 participants.[5] A recent update based on a search of the literature in August 2017 found 62 studies with a total number of 1982 participants.[34] Right now, MT is the technique in neurorehabilitation with the highest number of separate studies and largest total number of participants for rehabilitation of upper extremity function, even larger than for constraint-induced movement therapy,[35] which used to be the "gold standard."

Reviews by Thieme and coworkers[34] provide a substantial body of evidence that MT can lead to persistent improvements of motor function and of motor impairment after stroke when administered as adjunct therapy. However, for clinical purposes, we would also like to know if MT on its own is more effective than other forms of therapy. Apparently not: for patients with moderate arm paresis, the clinical effect of MT is comparable to that of a therapy with direct view of the affected limb.[34,36] What is critical is the visual feedback, whether delivered through a mirror, direct viewing, or virtual reality. Clear effects of MT are recorded when compared with a sham therapy without any visual feedback of the affected limb (covered mirror).[34]

However, there is one subgroup of paretic stroke patients for whom MT provides a substantial benefit: patients with severe arm paresis when active movements are not possible at all.[37] In these patients, the MVF provides a real advantage compared with other forms of therapy, presumably because it allows active movement performance which the patient attributes to himself. Further clinical studies are necessary to provide more evidence for the effect of MT in this patient group specifically. Furthermore, MT seems to have a synergistic effect with guided imagery; they seem especially effective when combined.[32]

This, however, highlights a further issue: even though we have some evidence for the benefit of MT in certain patient groups, we still are unable to make predictions regarding the recovery of individual patients. Especially in patients with severe arm paresis, the variability of "spontaneous recovery" is high, because some patients remain hemiplegic even after intensive rehabilitation, whereas others show substantial recovery.[38] We know that MT can provide recovery for a larger proportion of patients overall and result in better final performance values, as measured by conventional tests.[37] However, we still do not know the specific characteristics of the patients who show good recovery (Fig. 20.2).

Imaging techniques might help in this differentiation process. In a prospective pilot study, it was found that those patients whose brain activation patterns in response to MVF most closely resembled those of healthy subjects also showed the most striking recovery.[39] It is also possible that those who benefit most from MT are the ones with the largest either "dormant" or inhibited ipsilateral pathways,[40] which are then activated by MVF. If this is true, one could use activation of ipsilateral pathways using TMS as a predictor of good outcome. Alternatively, it has been proposed that the observation of the reflection of the healthy/intact limb moving synchronously with motor commands provides "correct visual feedback," harmonizing input and motor output of the affected limb, thus improving its motion.[41] Further studies are needed to distinguish between these hypotheses, e.g., by

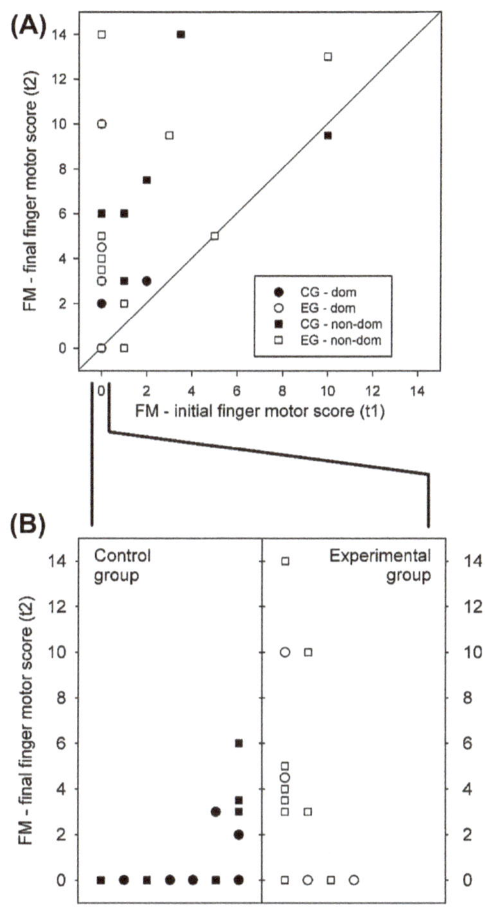

FIGURE 20.2 Results on motor recovery of individual patients (Fugl-Meyer finger subscore) (0–4 points). Upper panel (A): initial versus final value, each data point representing one patient. Circle: lesion of the dominant hemisphere, square: lesion of the nondominant hemisphere. Filled symbols: control therapy; empty symbols: mirror therapy. Note that some data pairs occur in more than one patient, thus the number of data points in the figure is less than the number of patients examined. Lower panel (B): final Fugl-Meyer score (expanded data points) for those 27 patients who were initially distally plegic. *Data from Dohle C, Püllen J, Nakaten A, Küst J, Rietz C, Karbe H. Mirror therapy promotes recovery from severe hemiparesis: a randomized controlled trial.* Neurorehabil Neural Repair. *2009;23(3):209-217. https://doi.org/10.1177/1545968308324786.*

measuring ipsilateral TMS responses or magnetoencephalographic activation before and after different stages of MT in individual patients. Ultimately, this type of investigations will have a major impact on practice in neurorehabilitation.

Effects of mirror therapy on other symptoms after stroke

The vast majority of the abovementioned studies focused on rehabilitation of motor deficits. However, as we know, symptoms after stroke are not restricted to motor deficits. Other symptoms include sensory deficits, spatial hemineglect, and central pain.

There are a number of studies suggesting that somatosensory impairments,[37,42–44] or motor incoordination resulting from impaired somatosensory processing[45] can also improve after application of MT. This shows us that the real nervous system does not respect clear distinction between motor and sensory modalities, as textbooks proclaim. Neurophysiological data and data from other pathological conditions give[45,46] hints on the mechanisms for recovery of these deficits. Apparently, improvements in the sensory domain can occur on much shorter time scales,[46] possibly due to direct stimulation of bilateral representations.[16] These effects certainly warrant further investigation, which is in part complicated by the variety of sensory deficits and the lack of generally accepted scales for their measurement.

Recovery from spatial hemineglect is another clinical enigma. The common occurrence of such recovery, even in the absence of therapy, provides grounds for optimism in neurology and deserves further exploration. The basic therapeutic intervention—inspired more by common sense than deep insight—involves visual feedback from the affected hemifield. There are some data from clinical studies that this process can be accelerated by MT.[37,47–50] As noted above, this effect is produced by placing the mirror in the affected hemifield, so as to view the reflection of the unaffected, moving limb, creating illusory movements in the affected limb. However, there are also sound theoretical reasons for placing the mirror in the unaffected visual hemifield (usually the right) to create awareness for the affected hemifield via the mirror.[51] This approach also needs investigation using formal clinical trials.

Effects of mirror therapy on pain after stroke

Stroke patients quite frequently suffer from pain. A closer analysis of the data suggests that such pain may arise from a variety of sources. For instance, lesions can be due to different reasons. Lesions of certain structures, e.g., the thalamus, can lead to central poststroke pain (PSP). Severe arm paresis can also cause shoulder-hand syndromes (SHSs), including (sub) luxation of the glenohumeral joint. Finally, stroke patients can also develop the full clinical picture of excruciating pain, paralysis, and inflammation that characterizes CRPS. However, pain is a highly subjective experience. The established definition of CRPS provided by the International Associations of the Study of Pain requires the presence of disproportionately high pain in response to seemingly trivial injury in combination with sensory, vasomotor, sudomotor, and motor deficits.[22] However, all these symptoms can also occur as a direct consequence of stroke, blurring the distinctions between PSP, SHS, and CRPS.

In spite of these terminological ambiguities, clinical studies have shown MT to be effective in treating stroke patients suffering from chronic pain.[28,29] However, if there is no pain before the intervention, MT does not seem to be effective in preventing its emergence.[37,42,49,52]

Effects of mirror therapy on other conditions

Most of the abovementioned studies have been performed on adult patients. There is good evidence, however, that MT is also beneficial in children, e.g., those suffering from cerebral palsy.[53] More recently, case reports and RCTs of MT for orthopedic injuries have been reported as well, confirming the potential of MT not only for improvement of painful conditions but also for recovery of impaired function.[54–56]

How should mirror therapy be performed?

Even though the underlying principle of MVF is well-defined, its implementation may differ considerably from one practitioner to another, according to personal taste, fads, and fashions. Specifically, the following aspects may vary:

(1) Unimanual versus bimanual visual feedback

In the original version of the "mirror box" invented by Ramachandran and his group, the patient's affected (or amputated) limb is placed behind the mirror, hidden from view, while the unaffected limb is placed facing the mirror, providing the illusion of two symmetrical working limbs. However, later studies have employed large-scale mirrors that cover nearly the entire field of view of the affected side. Apart from the size of the mirror, these two approaches differ in terms of their theoretical justification: the mirror box providing the visual image of two simultaneously moving hands was sometimes interpreted as visual bilateral training. Contrarily, large-scale mirrors usually require attention to the mirror image only, i.e., unilateral feedback. For stroke patients, there is recent clinical evidence that unilateral feedback outperforms bilateral feedback.[36].

(2) Voluntary movement performed by the unaffected limb

Subjects can be required to perform either intransitive or transitive movements with their unaffected limb. Generally, transitive approaches tend to be preferred in neurorehabilitation.[57] However, for tasks under mirrored visual control, manipulating objects viewed in a mirror introduces a second difficulty, namely the attribution of different movement coordinates, such as distance or azimuth to different modalities (visual and proprioceptive).[21] As detailed above, this transformation process seems to be accomplished by neural structures different from those activated by limb mirroring.[20] There is clinical evidence that some stroke patients are just not able to perform this coordination procedure, in a closely related syndrome called "mirror agnosia."[58,59] Thus, the introduction of object-oriented actions should be carefully considered.

(3) Using the unaffected limb to coax synchronized movements of the affected limb (or phantom)

In amputees, the mirror can optically "resurrect" the missing limb.[1] In other conditions, such as hemiparesis, however, the affected limb is still present and could be integrated into the task. But should that be done? Common sense suggests that simultaneous visual and kinesthetic feedback should enhance MVF. However, clinical studies propose the opposite.[36] If the visual and the kinesthetic information are not exactly the same (sensory incongruence), this can lead to unpleasant sensations and even pain.[60] There is evidence from imaging studies that the right dorsolateral prefrontal cortex plays a decisive role for this matching process.[61] Thus, there ought to be synergistic clinical improvement only if precise synchrony of vision and proprioception is ensured.

(4) Instructions to the patient

What instructions should be given to patients? Should they be asked to simply move the unaffected limb and watch its reflection to create the illusion that the affected limb is moving? Or should they be asked to send symmetrical commands to both limbs to convey the illusion that the affected limb is obeying the commands sent by the contralateral hemisphere? Or should they be asked to actively imagine that the "virtual" limb is moving as visible in the mirror? There is some evidence that when you move your limb while watching it, versus imagining that you are seeing it move, distinct but partially overlapping populations of neurons are activated.[62,63] In particular, as Shakespeare said, "You cannot cloy the hungry edge of appetite by bare imagination of a feast." Likewise, while imagining the affected limb moving can partially help relieve pain and paralysis, it works especially well when actual visual feedback is provided using a mirror. Obviously, the main focus of MT is attention to the images viewed in the mirror. Movement imagery can be added, although additional difficulty might be introduced by this. A combination of the two procedures, called "guided visual imagery," might prove even more valuable in the clinic.[32]

Even as there is considerable evidence that different protocols might have different clinical effects, there is a lack of standardization of procedures used.[64] For patients with pain symptoms, Grünert-Plüss and coworkers published the "St. Gallen protocol."[65] For stroke patients, Dohle's group presented the "Bonn protocol,"[66] later modified to the "Berlin protocol."[67] Unfortunately, both protocols are only available in German right now. Based on a survey among therapists, Rothgangel and Braun proposed a protocol that included additional variations, depending on the major presenting symptoms.[68]

Conclusion

The discovery of MVF and its subsequent implementations in a variety of clinical contexts has two important implications for neurology—(1) clinical and (2) theoretical.

(1) Clinical

MT is easy to implement. There is sound evidence for its beneficial effects in a variety of conditions. Furthermore, there are a now a number of different approaches that might help us to unravel the neural mechanisms underlying the effects of MVF. MVF began as a tentative hypothesis and a small-scale case study that was published in NATURE nearly two decades ago,[1] in which the focus was on PLP. It subsequently occurred to us that the treatment might also help patients recover from other neurological disorders and we suggested its use in accelerating recovery from hemiparesis,[2,3] neglect,[51] and CRPS.[41] Nearly all of 90-odd papers published since then have shown statistically significant improvement with MVF compared with placebos.[4,34] Most intriguing of these results is that MVF can not only help eliminate phantom pain but also pain in a real limb—i.e., CRPS/RSD, a notoriously intractable, cruel disorder,[4,69] for which most drugs are only partially effective, if at all, and the one surgical procedure that works—sympathetic ganglion injection—is invasive and has not been evaluated against placebos. On the other hand, MVF is simple, noninvasive, and inexpensive and can be administered by a spouse or caregiver at home. In addition, as noted above, several high-quality placebo-controlled studies have vindicated its use.[23] Despite the inertia of new discoveries making it into the clinic, MVF is now the first line of treatment for CRPS recommended in mainstream clinical practice.[70]

(2) Theoretical

The dogma in clinical neurology, and indeed in AI, until two decades ago, involves the following assumption. The sensory brain consists of a large number of autonomous, hardwired modules that are organized in a serial, hierarchical manner. Each module performs its computation, makes explicit some aspect of sensory information, and sends it to the next stage in the hierarchy. This process repeats itself, culminating in the act of perception and cognition.

Our studies with MVF in both normal and abnormal populations, together with the research done by many other groups (as discussed in elsewhere in this book), is that, far from being autonomous, the so-called modules interact to a substantial extent. For instance, the visual appearance of movement can profoundly modulate, and in some instances even cure pain originating in the somatosensory system. Additionally, far from being a hierarchy, messages from every stage in processing, after partial computation, get fed back to the earlier stages to bias processing, thereby iteratively reducing the problem space, as the organism interacts with and perceives its environment (e.g., imagining a limb seems to involve activation via back projections from higher cortical areas down to primary visual cortex (V1)).

Thus, the brain, far from resembling a digital computer, made up of hardwired modules that process sensory information, is more like an enormous, complex ecosystem that is in a state of perpetual dynamic equilibrium, constantly modifying its connections in responses to new environmental challenges. Disease, in this view, results not from "punch out a module—lose a function," but from shifts in equilibria, and if so, a relatively minor procedure can be used to restore equilibrium.

Indeed, even as basic a distinction as that which exists between sensory and motor, dissolves under scrutiny—as when visual feedback accelerates recovery from hemiparesis. Taking it even further, one could argue that the modules not only interact with each other but also with the skin and bones, as evidenced by the effect of MVF on the inflammation, pain, and paralysis of an intact swollen limb (RSD/CRPS). Of course, the pathways that mediate these effects remain obscure and demand further investigation.

Finally and most remarkably, if an amputee with excruciating cramping pain in his phantom watches his spouse massaging her own intact hand, he experiences a phantom massage which alleviates his phantom cramp. Thus, in addition to all the aforementioned interactions within an organism, these brain modules also interact with other brains through the mirror neuron system. Taken collectively, these observations have both practical utility as well as extraordinary theoretical implications.

References

1. Ramachandran VS, Rogers-Ramachandran D, Cobb S. Touching the phantom limb. *Nature*. 1995;377:489–490.
2. Altschuler EL, Wisdom SB, Stone L, et al. Rehabilitation of hemiparesis after stroke with a mirror. *Lancet*. 1999;353(9169):2035–2036.
3. Ramachandran VS. Phantom limbs, neglect syndromes, repressed memories, and Freudian psychology. *Int Rev Neurobiol*. 1994;37:291–333.
4. Thieme H, Morkisch N, Rietz C, Dohle C, Borgetto B. The efficacy of movement representation techniques for treatment of limb pain—a systematic review and meta-analysis. *J Pain*. 2016;17(2):167–180. https://doi.org/10.1016/j.jpain.2015.10.015.
5. Thieme H, Mehrholz J, Pohl M, Behrens J, Dohle C. Mirror therapy for improving motor function after stroke. In: The Cochrane Collaboration, ed. *Cochrane Database of Systematic Reviews*. Chichester, UK: John Wiley & Sons, Ltd; 2012. http://doi.wiley.com/10.1002/14651858.CD008449.pub2.
6. Grèzes J, Decety J. Functional anatomy of execution, mental simulation, observation and verb generation of actions: a meta-analysis. *Hum Brain Mapp*. 2001;12:1–19.
7. Deconinck FJA, Smorenburg ARP, Benham A, Ledebt A, Feltham MG, Savelsbergh GJP. Reflections on mirror therapy: a systematic review of the effect of mirror visual feedback on the brain. *Neurorehabilitation Neural Repair*. 2015;29(4):349–361. https://doi.org/10.1177/1545968314546134.
8. Dohle C, Kleiser R, Seitz RJ, Freund H-J. Body scheme gates visual processing. *J Neurophysiol*. 2004;91(5):2376–2379. https://doi.org/10.1152/jn.00929.2003.
9. Matthys K, Smits M, Van der Geest JN, et al. Mirror-induced visual illusion of hand movements: a functional magnetic resonance imaging study. *Arch Phys Med Rehabil*. 2009;90(4):675–681. https://doi.org/10.1016/j.apmr.2008.09.571.
10. Mehnert J, Brunetti M, Steinbrink J, Niedeggen M, Dohle C. Effect of a mirror-like illusion on activation in the precuneus assessed with functional near-infrared spectroscopy. *J Biomed Opt*. 2013;18(6):066001. https://doi.org/10.1117/1.JBO.18.6.066001.
11. Wang J, Fritzsch C, Bernarding J, et al. A comparison of neural mechanisms in mirror therapy and movement observation therapy. *J Rehabil Med*. 2013;45(4):410–413. https://doi.org/10.2340/16501977-1127.
12. Shinoura N, Suzuki Y, Watanabe Y, et al. Mirror therapy activates outside of cerebellum and ipsilateral M1. *NeuroRehabilitation*. 2008;23(3):245–252.
13. Diers M, Christmann C, Koeppe C, Ruf M, Flor H. Mirrored, imagined and executed movements differentially activate sensorimotor cortex in amputees with and without phantom limb pain. *Pain*. 2010;149(2):296–304. https://doi.org/10.1016/j.pain.2010.02.020.
14. Imai I, Takeda K, Shiomi T, Taniguchi T, Kato H. Sensorimotor cortex activation during mirror therapy in healthy right-handed subjects: a study with near-infrared spectroscopy. *J Phys Ther Sci*. 2008;20(2):141–145. https://doi.org/10.1589/jpts.20.141.
15. Merians AS, Tunik E, Fluet GG, Qiu Q, Adamovich SV. Innovative approaches to the rehabilitation of upper extremity hemiparesis using virtual environments. *Eur J Phys Rehabil Med*. 2009;45(1):123–133.
16. Fritzsch C, Wang J, Dos Santos LF, Mauritz K-H, Brunetti M, Dohle C. Different effects of the mirror illusion on motor and somatosensory processing. *Restor Neurol Neurosci*. 2014;32(2):269–280. https://doi.org/10.3233/RNN-130343.
17. Carson RG, Ruddy KL. Vision modulates corticospinal suppression in a functionally specific manner during movement of the opposite limb. *J Neurosci*. 2012;32(2):646–652. https://doi.org/10.1523/JNEUROSCI.4435-11.2012.
18. Fukumura K, Sugawara K, Tanabe S, Ushiba J, Tomita Y. Influence of mirror therapy on human motor cortex. *Int J Neurosci*. 2007;117(7):1039–1048. https://doi.org/10.1080/00207450600936841.
19. Funase K, Tabira T, Higashi T, Liang N, Kasai T. Increased corticospinal excitability during direct observation of self-movement and indirect observation with a mirror box. *Neurosci Lett*. 2007;419(2):108–112. https://doi.org/10.1016/j.neulet.2007.04.025.

20. Dohle C, Stephan KM, Valvoda JT, et al. Representation of virtual arm movements in precuneus. *Exp Brain Res.* 2011;208(4):543–555. https://doi.org/10.1007/s00221-010-2503-0.
21. Snijders HJ, Holmes NP, Spence C. Direction-dependent integration of vision and proprioception in reaching under the influence of the mirror illusion. *Neuropsychologia.* 2007;45(3):496–505. https://doi.org/10.1016/j.neuropsychologia.2006.01.003.
22. Bruehl S, Harden RN, Galer BS, et al. External validation of IASP diagnostic criteria for complex regional pain syndrome and proposed research diagnostic criteria. International association for the study of pain. *Pain.* 1999;81(1–2):147–154.
23. Sayegh SA, Filén T, Johansson M, Sandström S, Stiewe G, Butler S. Mirror therapy for Complex Regional Pain Syndrome (CRPS)-A literature review and an illustrative case report. *Scand J Pain.* 2013;4(4):200–207. https://doi.org/10.1016/j.sjpain.2013.06.002.
24. Chan BL, Witt R, Charrow AP, et al. Mirror therapy for phantom limb pain. *N Engl J Med.* 2007;357(21):2206–2207.
25. Finn SB, Perry BN, Clasing JE, et al. A randomized, controlled trial of mirror therapy for upper extremity phantom limb pain in male amputees. *Front Neurol.* 2017;8. https://doi.org/10.3389/fneur.2017.00267.
26. Stein R. *Einfluss nichtmedikamentöser Verfahren (repetitive transkranielle Magnetstimulation bzw. Spiegeltherapie) auf die Schmerzintensität und die schmerzassoziierte zentralnervöse Plastizität bei Patienten mit neuropathischen Schmerzen;* October 2013. http://www-brs.ub.ruhr-uni-bochum.de/netahtml/HSS/Diss/SteinRobert/diss.pdf.
27. Michenthaler MC. *Spiegeltherapeutische Interventionen bei Komplexem Regionalem Schmerzsyndrom (CRPS) Eine randomisierte kontrollierte Längsschnittstudie zur motorischen Funktionalität;* 2015. https://online.medunigraz.at/mug_online/wbAbs.showThesis?pThesisNr=49113&pOrgNr=1&pPersNr=51205.
28. Cacchio A, De Blasis E, De Blasis V, Santilli V, Spacca G. Mirror therapy in complex regional pain syndrome type 1 of the upper limb in stroke patients. *Neurorehabilitation Neural Repair.* 2009;23:792–799.
29. Cacchio A, De Blasis E, Necozione S, di Orio F, Santilli V. Mirror therapy for chronic complex regional pain syndrome type 1 and stroke. *N Engl J Med.* 2009;361:634–636.
30. Moseley GL. Graded motor imagery for pathological pain. A randomized controlled trial. *Neurology.* 2006.
31. Moseley GL. Is successful rehabilitation of complex regional pain syndrome due to sustained attention to the affected limb? A randomised clinical trial. *Pain.* 2005;114:54–61.
32. Moseley GL. Graded motor imagery is effective for long-standing complex regional pain syndrome: a randomised controlled trial. *Pain.* 2004;108:192–198.
33. Flor H, Elbert T, Knecht S, et al. Phantom-limb pain as a perceptual correlate of cortical reorganization following arm amputation. *Nature.* 1995;375(6531):482–484. https://doi.org/10.1038/375482a0.
34. Thieme H, Morkisch N, Mehrholz J, et al. Mirror therapy for improving motor function after stroke. *Cochrane Database Syst Rev.* 2018;7:CD008449. https://doi.org/10.1002/14651858.CD008449.pub3.
35. Corbetta D, Sirtori V, Castellini G, Moja L, Gatti R. Constraint-induced movement therapy for upper extremities in people with stroke. *Cochrane Database Syst Rev.* 2015;10:CD004433. https://doi.org/10.1002/14651858.CD004433.pub3.
36. Selles RW, Michielsen ME, Bussmann JBJ, et al. Effects of a mirror-induced visual illusion on a reaching task in stroke patients: implications for mirror therapy training. *Neurorehabilitation Neural Repair.* 2014;28(7):652–659. https://doi.org/10.1177/1545968314521005.
37. Dohle C, Püllen J, Nakaten A, Küst J, Rietz C, Karbe H. Mirror therapy promotes recovery from severe hemiparesis: a randomized controlled trial. *Neurorehabilitation Neural Repair.* 2009;23(3):209–217. https://doi.org/10.1177/1545968308324786.
38. Heller A, Wade DT, Wood VA, Sunderland A, Hewer RL, Ward E. Arm function after stroke: measurement and recovery over the first three months. *J Neurol Neurosurg Psychiatry.* 1987;50(6):714–719. https://doi.org/10.1136/jnnp.50.6.714.
39. Brunetti M, Morkisch N, Fritzsch C, et al. Potential determinants of efficacy of mirror therapy in stroke patients - a pilot study. *Restor Neurol Neurosci.* 2015;33(4):421–434. https://doi.org/10.3233/RNN-140421.
40. Bawa P, Hamm JD, Dhillon P, Gross PA. Bilateral responses of upper limb muscles to transcranial magnetic stimulation in human subjects. *Exp Brain Res.* 2004;158:385–390.
41. Ramachandran VS, Altschuler EL. The use of visual feedback, in particular mirror visual feedback, in restoring brain function. *Brain.* 2009;132(7):1693–1710. https://doi.org/10.1093/brain/awp135.
42. Acerra NE. *Sensorimotor Dysfunction in CRPS1 and Stroke: Characterisation, Prediction and Intervention.* July 2007.

43. Colomer C, NOé E, Llorens R. Mirror therapy in chronic stroke survivors with severely impaired upper limb function: a randomized controlled trial. *Eur J Phys Rehabil Med*. 2016;52(3):271–278.
44. Wu C-Y, Huang P-C, Chen Y-T, Lin K-C, Yang H-W. Effects of mirror therapy on motor and sensory recovery in chronic stroke: a randomized controlled trial. *Arch Phys Med Rehabil*. 2013;94(6):1023–1030. https://doi.org/10.1016/j.apmr.2013.02.007.
45. Sathian K, Greenspan AI, Wolf SL. Doing it with mirrors: a case study of a novel approach to neurorehabilitation. *Neurorehabilitation Neural Repair*. 2000;14(1):73–76.
46. Moseley GL, Wiech K. The effect of tactile discrimination training is enhanced when patients watch the reflected image of their unaffected limb during training. *Pain*. 2009;144(3):314–319. https://doi.org/10.1016/j.pain.2009.04.030.
47. Moustapha A, Rousseaux M. Immediate Effects of Mirror Therapy on Spatial Neglect. /data/revues/18770657/v55sS1/S1877065712005969/. 2012. http://www.em-consulte.com/en/article/749751.
48. Pandian JD, Arora R, Kaur P, Sharma D, Vishwambaran DK, Arima H. Mirror therapy in unilateral neglect after stroke (MUST trial) a randomized controlled trial. *Neurology*. 2014;83(11):1012–1017.
49. Thieme H, Bayn M, Wurg M, Zange C, Pohl M, Behrens J. Mirror therapy for patients with severe arm paresis after stroke—a randomized controlled trial. *Clin Rehabil*. 2013;27(4):314–324.
50. Tyson S, Thomas N, Wilkinson J, et al. The effects of patient-directed mirror therapy on upper limb impairments and activity. *Int J Stroke*. 2014;9:16–17.
51. Ramachandran VS, Altschuler EL, Stone L, Al-Aboudi M, Schwartz E, Siva N. Can mirrors alleviate visual hemineglect? *Med Hypotheses*. 1999;52:303–305.
52. Michielsen ME, Selles RW, van der Geest JN, et al. Motor recovery and cortical reorganization after mirror therapy in chronic stroke patients: a phase II randomized controlled trial. *Neurorehabilitation Neural Repair*. 2011;25(3):223–233. https://doi.org/10.1177/1545968310385127.
53. Park E, Baek S, Park S. Systematic review of the effects of mirror therapy in children with cerebral palsy. *J Phys Ther Sci*. 2016;28(11):3227–3231. https://doi.org/10.1589/jpts.28.3227.
54. Altschuler EL, Hu J. Mirror therapy in a patient with a fractured wrist and no active wrist extension. *Scand J Plast Reconstr Surg Hand Surg*. 2008;42(2):110–111. https://doi.org/10.1080/02844310701510355.
55. Rosén B, Lundborg G. Training with a mirror in rehabilitation of the hand. *Scand J Plast Reconstr Surg Hand Surg*. 2005;39(2):104–108. https://doi.org/10.1080/02844310510006187.
56. Rostami HR, Arefi A, Tabatabaei S. Effect of mirror therapy on hand function in patients with hand orthopaedic injuries: a randomized controlled trial. *Disabil Rehabil*. 2013;35(19):1647–1651. https://doi.org/10.3109/09638288.2012.751132.
57. Veerbeek JM, van Wegen E, van Peppen R, et al. What is the evidence for physical therapy poststroke? A systematic review and meta-analysis. *PLoS One*. 2014;9(2). https://doi.org/10.1371/journal.pone.0087987.
58. Binkofski F, Buccino G, Dohle C, Seitz RJ, Freund HJ. Mirror agnosia and mirror ataxia constitute different parietal lobe disorders. *Ann Neurol*. 1999;46(1):51–61.
59. Ramachandran VS, Altschuler EL, Hillyer S. Mirror agnosia. *Proc Biol Sci*. 1997;264:645–647.
60. McCabe CS, Haigh RC, Halligan PW, Blake DR. Simulating sensory-motor incongruence in healthy volunteers: implications for a cortical model of pain. *Rheumatol Oxf*. 2005;44:509–516.
61. Fink GR, Marshall JC, Halligan PW, et al. The neural consequences of conflict between intention and the senses. *Brain J Neurol*. 1999;122(Pt 3):497–512.
62. Pearson J, Naselaris T, Holmes EA, Kosslyn SM. Mental imagery: functional mechanisms and clinical applications. *Trends Cogn Sci*. 2015;19(10):590–602. https://doi.org/10.1016/j.tics.2015.08.003.
63. Farah MJ, Heberlein AS. Personhood and neuroscience: naturalizing or nihilating? *Am J Bioeth*. 2007;7(1):37–48. https://doi.org/10.1080/15265160601064199.
64. Rothgangel AS, Braun SM, Beurskens AJ, Seitz RJ, Wade DT. The clinical aspects of mirror therapy in rehabilitation: a systematic review of the literature. *Int J Rehabil Res*. 2011;34(1):1–13. https://doi.org/10.1097/MRR.0b013e3283441e98.
65. Grünert-Plüss N, Hufschmid U, Santschi L, Grünert J. Mirror therapy in hand rehabilitation: a review of the literature, the St Gallen protocol for mirror therapy and evaluation of a case series of 52 patients. *Br J Hand Ther*. 2008;13(1):4–11. https://doi.org/10.1177/175899830801300101.
66. Nakaten A, Govers J, Dohle C. *Spiegeltherapie in der Neurorehabilitation*. 1. Aufl. Idstein. Schulz-Kirchner; 2009.

67. Morkisch N, Dohle C. *BeST - Berliner Spiegeltherapieprotokoll Ein wissenschaftlich evaluiertes Manual zur Durchführung der Spiegeltherapie*. Bad Honnef: Hippocampus Verlag; 2015.
68. Rothgangel A, Braun S. Mirror therapy. In: *Practical Protocol for Stroke Rehabilitation*. 2013. https://doi.org/10.12855/ar.sb.mirrortherapy.e2013.
69. McCabe CS, Haigh RC, Ring EFJ, Halligan PW, Wall PD, Blake DR. A controlled pilot study of the utility of mirror visual feedback in the treatment of complex regional pain syndrome (type 1). *Rheumatol Oxf Engl*. 2003;42(1):97–101.
70. Bruehl S. Complex regional pain syndrome. *BMJ*. 2015;351. https://www.jstor.org/stable/26522525.

Index

'*Note:* Page numbers followed by "f" indicate figures and "t" indicate table.'

T

U

V

W

Kolibri 360 GmbH
Ferdinand-Jühlke-Straße 7
99095 Erfurt, Deutschland
produktsicherheit@kolibri360.de